Room Temperature Organic Synthesis

Room Temperature Organic Synthesis

Goutam Brahmachari

ELSEVIER

AMSTERDAM • BOSTON • HEIDELBERG • LONDON • NEW YORK • OXFORD
PARIS • SAN DIEGO • SAN FRANCISCO • SINGAPORE • SYDNEY • TOKYO

Elsevier
Radarweg 29, PO Box 211, 1000 AE Amsterdam, Netherlands
The Boulevard, Langford Lane, Kidlington, Oxford OX5 1GB, UK
225 Wyman Street, Waltham, MA 02451, USA

Library of Congress Cataloging-in-Publication Data
A catalog record for this book is available from the Library of Congress

British Library Cataloguing in Publication Data
A catalogue record for this book is available from the British Library

For information on all Elsevier publications
visit our website at http://store.elsevier.com/

ISBN: 978-0-12-801025-9

Dedication

To my wife *PIYASI* who cares for me a lot!

Contents

Foreword

The topic of this book is both important and significant because it allows us to isolate an aspect of synthetic elegance that is all too often overlooked. The transformations of molecules that have been mastered over the past two centuries by synthetic chemists border on the miraculous. We are at a stage where nearly any chemical that can be drawn or represented computationally can be synthesized. And for that reason, the field of synthetic chemistry should be tremendously proud.

However, we also are well aware that many of the methods that have been invented over the years have required extraordinary measures in order to achieve these astonishing results. In some cases, the synthesis would be an extremely large number of steps while in other examples it may be that the results were achieved with an extremely poor yield. But one factor that is often employed is that many of the reactions that have allowed a significant number of the realized achievements over the years is the use of high temperatures.

The use of high temperature in synthesis when necessary us is understandable in order to obtain the desired molecule. However, the brilliance of the chemistry represented in this book, is that it represents the unwillingness of chemists to compromise in the never-ending quest for the ideal. By not accepting that synthetic methods have to be run at high temperatures and can be done at ambient conditions, is an important step toward along the path of continuous improvement.

It is true that running reactions at unnecessarily high temperatures is a waste of energy and is in opposition to the *Principles of Green Chemistry*. But even beyond that, it ignores the fact that so much of the most elegant chemistry, the most complex chemistry, the most important chemistry, is carried out by *Nature* at ambient temperatures. To ignore *Nature* as our most important chemistry teacher, mentor, and inspiration is unwise at best, and harmful and futile at worst.

My compliments and congratulations to Prof. Brahmachari for the important contribution that this book will be for our collective scientific knowledge and may it serve as a catalyst for further improvements in our collective mission in synthetic chemistry.

Paul T. Anastas
Yale University
New Haven
CT, USA

How to Read

As stated in the Preface, the book *Room Temperature Organic Synthesis* is designed with a motto to offer recent cutting-edge advances in developing organic synthetic protocols under ambient conditions, thereby satisfying the *sixth principle of green chemistry*, and features an in-depth and thorough coverage of a huge number of organic synthetic protocols at room temperature. This definite volume incorporates more than 200 sincerely screened organic synthetic protocols at room temperature for generation of carbon-carbon and carbon-heteroatom bonds resulting a wide spectrum of chemical compounds — aliphatic, aromatic, alicyclic and heterocycles. The reactions are classified in eight distinct chapters (Chapter 1 to 8) based on their bond forming strategies such as C – C, C – N, C – O, C – S, C – P, C – halogen, C – B, and miscellaneous bond forming reactions (*viz.* C – H, N – H, O – Si, S – N, S – O and S – S). Clearly structured for easy access to the information, each selected reaction is discussed in a very compact manner through point-wise discussion such as: Reaction type; Reaction conditions; Reaction strategy; Catalyst; Keywords; General reaction scheme; Mechanism; Representative examples; Experimental procedure; Characterization data of representative entries; Reference. Reaction scheme, plausible mechanism (if any) and illustrative examples relating to a particular reaction are presented under that reaction and are self-explanatory in nature. Each organic synthesis is supplemented with all its details including experimental procedure, representative examples and their physical and spectral properties so that one can reproduce the same with ease.

Preface

Organic synthesis is a special branch of chemical synthesis involved with the construction of organic compounds following certain distinct methodologies! It is truly a hardcore laboratory based set-up to carry out the desired organic reactions for the purpose. Over the past two centuries, synthetic organic chemistry has seen a tremendous all-round development, and the credits obviously go to the synthetic chemists at large!

Organic molecules often find their uses as key starting materials for a great number of major chemical industries. The production of organic chemicals as raw materials or reagents for other applications is a major sector of manufacturing polymers, pharmaceuticals, pesticides, paints, artificial fibres, food additives, *etc.* The overall outcome of an organic synthesis, *i.e.* the productivity, cost, safety, hazards, energy-efficiency, environmental-concerns *etc.*, is largely dependent on the generality and effectiveness of its synthetic method. Organic synthesis involves breakage and formation of chemical bonds! Formations of new bonds are broadly categorized as carbon-carbon and carbon-heteroatom bond forming reactions resulting a wide spectrum of chemical compounds — aliphatic, aromatic, alicyclic, heterocycles and so on. Hence, proper information on the recent development of novel and useful methods for carbon-carbon and carbon-heteroatom bond formation is of prime importance to the synthetic chemists.

With the advent of the twenty-first century, Anastas and Warner through their "*The Twelve Principles of Green Chemistry*" have developed a concept of "*Green Chemistry*" or "*Environmentally Benign Chemistry*". This serves as guidelines for practicing chemists in developing and assessing how green a synthesis, compound, process, or technology is! The first principle espouses the basic tenet of green chemistry, namely, pollution prevention. Other principles deal with topics such as atom economy, toxicity, the use of solvents and auxiliary agents, energy usage, renewable *versus* non-renewable feedstocks, and decomposition of compounds into non-toxic, environmentally benign substances. The last decade has seen a huge interest in green chemistry, particularly as organic chemists look to "green" the methodologies/techniques for the synthesis of organic molecules of interest! As a result several methodologies/concepts such as solvent-free synthesis, organic reactions under aqueous conditions, organic reactions in non-aqueous media, greener organic reagents, development of heterogeneous catalysts and ionic liquids, waste management/recycling strategies, application of ultrasound, and microwave technology as a greener heating tool have emerged and are in extensive use at present-time.

The *sixth principle* is dedicated to "*Design of Energy Efficiency*". That is to develop synthetic strategies that require less/minimum amount of energy to carry out a specific reaction with optimum productivity. During large-scale industrial production, energy usage is also a vital concern. For this purpose, use of microwave and application of ultrasound are in use as substitutes of conventional heating. *However, the most effective way-out to save energy is to develop strategies/protocols that are capable enough to carry out the transformations at ambient temperature!* In addition, room temperature condition is a mild reaction strategy, essentially required for many temperature-sensitive organic substrates as a key step in multi-step sequence reactions. That is why, as part of on-going developments on green synthetic strategies, designing for room temperature conditions coupled with other green aspects is also an area of current choice. The concept of developing room temperature reaction strategy is now an emerging field of research in organic chemistry and is progressing largely.

The book *Room Temperature Organic Synthesis* is a first-time attempt to offer recent cutting-edge advances in developing organic synthetic protocols under ambient conditions, thereby satisfying the *sixth principle of green chemistry*, and features an in-depth and thorough coverage of a huge number of organic synthetic protocols at ambient temperature. This book is unique in its authoritative, thorough, and comprehensive inclusion of a wide variety of more than 200 sincerely screened organic synthetic protocols at room temperature for generation of carbon-carbon and carbon-heteroatom bonds resulting a wide spectrum of chemical compounds — aliphatic, aromatic, alicyclic and heterocycles. Clearly structured for easy access to the information, each selected reaction is discussed in a very compact manner through point-wise discussion such as: Reaction type; Reaction conditions; Reaction strategy; Catalyst; Keywords; General reaction scheme; Mechanism; Representative examples; Experimental procedure; Characterization data of representative entries; Reference.

The book successfully integrates research advances in designing energy-efficient reaction procedures for useful organic transformations to satisfy one of the notable green chemistry principles (*i.e.* reactions at ambient conditions) and their feasibility into industrial applications and process developments. A wide spectrum of important synthetic methodologies involving carbon-carbon and carbon-heteroatom bond forming reactions under mild conditions dealt in this book would surely make the work much interesting to the scientists deeply engaged in organic synthesis and related fields. This timely volume serves the purpose of also an outstanding source of information with regard to the industrial applications. It will serve not only as a valuable resource for researchers in their own fields, but also motivates young scientists to the dynamic field of organic synthesis and practice of green chemistry.

I would like to express my sincere thanks and deep sense of gratitude to Professor Paul T. Anastas, Director, Center for Green Chemistry and Green Engineering, Yale University, New Haven, CT, USA for his keen interest in the manuscript and for writing foreword to the book.

I would also like to express my deep sense of appreciation to all of the editorial and publishing staff members associated with Elsevier Inc., United States of America, for their keen interest in publishing the work as well as their all-round help so as to ensure that the highest standards of publication have been maintained in bringing out this book. My effort will be successful only when it is found helpful to the readers at large. Every step has been taken to make the manuscript error-free; in spite of that, some errors might have crept in. Any remaining error is, of course, of my own. Constructive comments and approach of the book from the readers will be highly appreciated.

Finally, I should thank my wife and my son for their well understanding and allowing me enough time throughout the entire period of writing; without their support, this work would not have been successful.

Goutam Brahmachari
Chemistry Department, Visva-Bharati University
Santiniketan, West Bengal, India
February 2015

Chapter 1

Carbon – Carbon Bond Forming Reactions at Room Temperature

C-C bond_1

Type of reaction: C-C bond formation
Reaction conditions: Solvent-free, room temperature
Synthetic strategy: One-step condensation
Catalyst: Niobium pentachloride ($NbCl_5$)
Keywords: 1,4-Dicarbonyls, aldehydes, niobium pentachloride ($NbCl_5$), homogeneous catalysis, solvent-free, room temperature, condensation, Knoevenagel reaction, trisubstituted alkenes

General reaction scheme

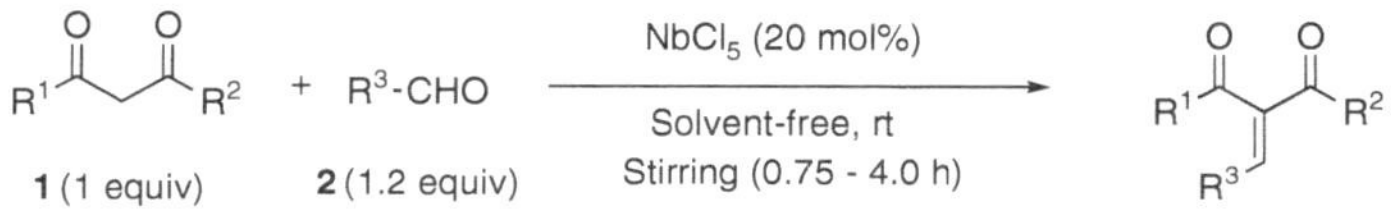

$R^1 = C_6H_5$, Me, OEt
$R^2 = C_6H_5$, Me, OMe, OEt
$R^3 = C_6H_5$, 4-$CH_3C_6H_4$, 4-ClC_6H_4, 2-$NO_2C_6H_4$, 4-$OCH_3C_6H_4$, *n*-heptyl, cyclohexyl

Representative entries

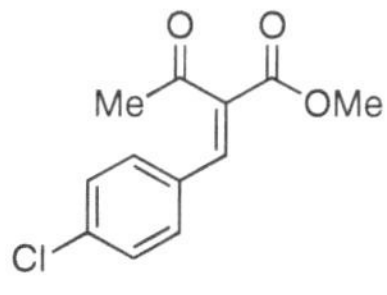

3a

(*E*)-Methyl 2-(4-chlorobenzylidene)-3-oxobutanoate (**3a**): Reddish solid; mp 80 °C; reaction time: 1.5 h; yield: 80%

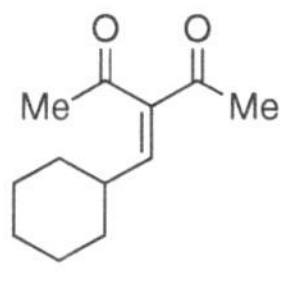

3b

3-(Cyclohexylmethylene)pentane-2,4-dione (**3b**): Reaction time: 4.0 h; yield: 62%

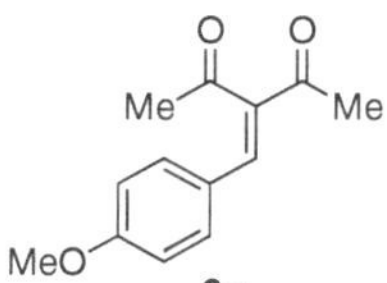

3c

3-(4-Methoxybenzylidene)pentane-2,4-dione (**3c**): Yellow liquid; reaction time: 1.5 h; yield: 80%

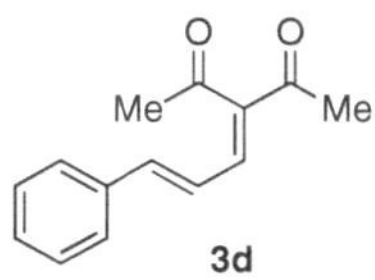

3d

(*E*)-3-(3-Phenylallylidene)pentane-2,4-dione (**3d**): Reaction time: 3.0 h; yield: 65%

Experimental procedure

A round-bottomed flask was charged with a magnetic stir bar, 1,3-carbonyl compound (**1**; 1 mmol), and aldehyde (**2**; 1.2 mmol), followed by 20 mol% of $NbCl_5$. The reaction mixture

Room Temperature Organic Synthesis. http://dx.doi.org/10.1016/B978-0-12-801025-9.00001-2

was then stirred at room temperature for required time-frame (0.75-4.0 h). After completion of the reaction, it was quenched by the addition of saturated aqueous $NaHCO_3$ solution and the product (**3**) was extracted with ethyl acetate. The product was purified by column chromatography using ethyl acetate/hexane as eluent. Structure of each of the products was verified from IR, ^1H NMR, ^{13}C NMR and HRMS studies.

Characterization data of two representative compounds

(*E*)-Methyl 2-(4-chlorobenzylidene)-3-oxobutanoate (**3a**): Reddish solid; mp 80 °C; IR (KBr): 821, 1088, 1250, 1618, 1654, 1733, 2362, 2855, 2932 cm^{-1}; ^1H NMR (300 MHz, $CDCl_3$): δ 1.16 (s, 3H), 2.58 (s, 3H), 6.11–6.12 (m, 4H), 6.27 (s, 1H); ^{13}C NMR (75 MHz, $CDCl_3$): δ 26.5, 52.6, 129.1, 130.6, 136.8, 134.5, 140.0, 169.7, 194.3; ESIMS: *m*/*z* 261 (M^+ + Na); HRMS: Calcd for $C_{12}H_{11}O_3NaCl$, 261.0294; Found 261.0289.

3-(4-Methoxybenzylidene)pentane-2,4-dione (**3c**): Yellow liquid; IR (neat): 831, 1108, 1174, 1258, 1258, 1513, 1599, 1653, 1706, 2844, 2926, 3004 cm^{-1}; ^1H NMR (300 MHz, $CDCl_3$): δ 2.28 (s, 3H), 2.26 (s, 3H), 3.83 (s, 3H), 6.87 (d, J=9.0 Hz, 2H), 7.31 (d, J=9.0 Hz, 2H), 7.37 (s, 1H); ^{13}C NMR (75 MHz, $CDCl_3$): δ 26.2, 31.6, 55.3, 114.5, 125.2, 131.6, 139.7, 140.6, 161.6, 196.4, 206.1; ESIMS: *m*/*z* 241 (M^+ + Na); HRMS: Calcd for $C_{13}H_{14}O_3Na$, 241.0840; Found 241.0833.

Reference

Yadav, J. S., Bhunia, D. C., Singh, V. K., and Srihari, P. (2009). Solvent-free $NbCl_5$ catalyzed condensation of 1,3-dicarbonyl compounds and aldehydes: A facile synthesis of trisubstituted alkenes. *Tetrahedron Lett.*, **50**, 2470–2473.

C-C bond_2

Type of reaction: C-C bond formation
Reaction conditions: Water, neutral conditions, room temperature
Synthetic strategy: One-step condensation

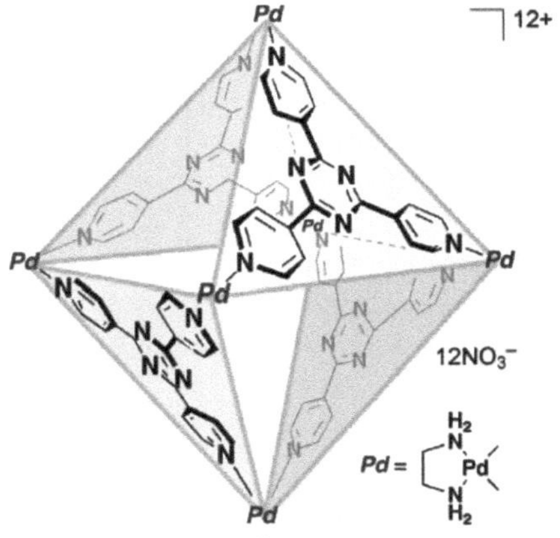

1

[Cage **1**: adapted from Murase *et al.* (2012), *J. Am. Chem. Soc.*, **134**, 162]

Catalyst: Heterogeneous, a cationic coordination cage (**1**) (12+ charged M_6L_4 cage) [Cage components: (ethylene diamine)Pd(NO_3)$_2$ and 2,4,6-tripyridyl-1,3,5-triazine]

Keywords: Aromatic aldehydes, Meldrum's acid, 12+ charged cage **1** (catalyst), water, room temperature, condensation, Knoevenagel reaction, trisubstituted alkenes

General reaction scheme

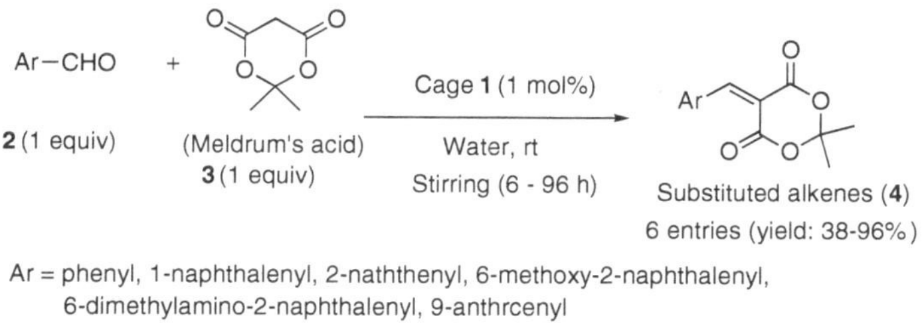

Representative entries

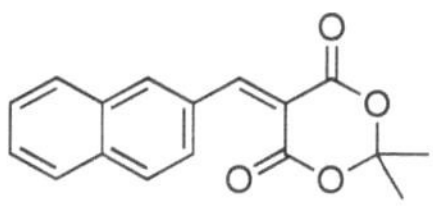

4a

2,2-Dimethyl-5-(naphthalene-2-ylmethylene)-1,3-dioxane-4,6-dione (**4a**): Pale yellow solid; mp 150.8-151.5 °C; reaction time: 6 h; yield: 96% (NMR), 86%(isolated)

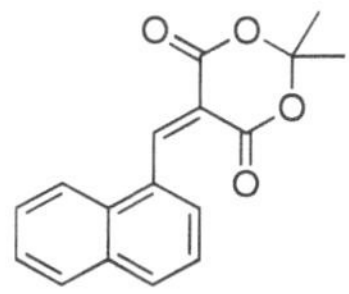

4b

2,2-Dimethyl-5-(naphthalene-1-ylmethylene)-1,3-dioxane-4,6-dione (**4b**): Solid, mp 144.9-145.4 °C; reaction time: 24 h; yield: 67% (NMR)

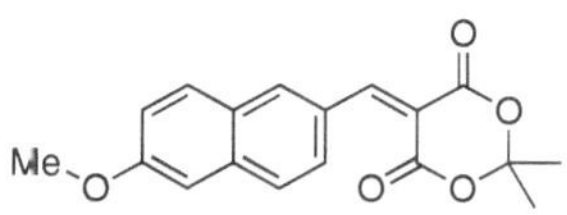

4c

5-((6-Methoxynaphthalen-2-yl)methylene)-2,2-dimethyl-1,3-dioxane-4,6-dione (**4c**): Solid, mp 191.2-192.2 °C; reaction time: 6 h; yield: 96% (NMR, 88% (isolated)

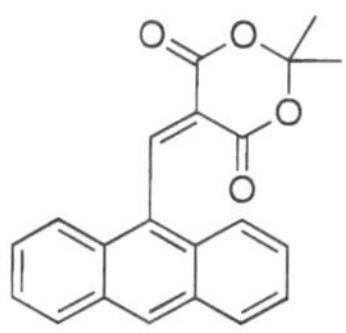

4d

5-(Anthracen-9-ylmethylene)-2,2-dimethyl-1,3-dioxane-4,6-dione (**4d**): Solid, mp 193.6-194.3 °C; reaction time: 96 h; yield: 63% (NMR)

Experimental procedure

Aldehyde (**2**; 0.5 mmol) and Meldrum's acid (**3**; 0.5 mmol) were added to an aqueous solution (5 mL) of cage (**1**; 15.0 mg, 5.00×10^{-3} mmol; 1 mol%), and the reaction mixture was stirred at room temperature for stipulated time (6-96 h). The product was extracted with $CHCl_3$ (2×5 mL), and the organic layer was evaporated *in vacuo* to furnish the condensation product **4** (38-96% yield). The obtained product **4** was purified by recrystallization from refluxing ethanol. Each of the products was fully characterized from its detailed spectral studies including IR, ^{1}H NMR, ^{13}C NMR, MS, and also from analytical analyses.

Cage 1 – a synthetic mimic of enzymes: This technique demonstrated a unique dehydration condensation under neutral conditions in water catalyzed by the water-soluble synthetic cationic host **1**. An aromatic aldehyde substrate (**2**; an electron-rich guest) first became efficiently encapsulated into the host's (**1**) hydrophobic cavity, which after then attacked by the enolate of Meldrum's acid (**3**) to generate oxyanion intermediate. The condensation reaction seems to be facilitated by the anionic intermediate in the cationic environment of the cage. The eventual loss of water molecule occurs smoothly within the hydrophobic cavity to form the dehydrated product (**4**) which is too large for the cavity and is spontaneously released from the cage, and a new incoming substrate molecule (**2**) occupies the position. The overall phenomenon follows the tricks of enzyme-like catalysis.

Characterization data of two representative compounds

5-((6-Methoxynaphthalen-2-yl)methylene)-2,2-dimethyl-1,3-dioxane-4,6-dione (**4c**): Solid, mp 191.2-192.2 °C; IR (ATR): 2992, 2946, 1756, 1725, 1587, 1402, 1341, 1291, 1222, 1158, 1023, 1010 cm^{-1}; ^{1}H NMR (500 MHz, $CDCl_3$): δ 8.56 (br s,1H), 8.55 (br s,1H), 8.18 (dd, $J=8.8$, 1.7 Hz, 1H), 7.85 (d, $J=9.0$ Hz, 1H), 7.76 (d, $J=8.8$ Hz, 1H), 7.21 (dd, $J=9.0$, 2.4 Hz, 1H,), 7.16 (d, $J=2.4$ Hz, 1H), 3.97 (s, 3H), 1.83 (s, 6H); ^{13}C NMR (125 MHz, $CDCl_3$): δ 163.8 (C), 160.9 (C), 160.3 (C), 158.5 (CH), 137.8 (C), 137.7 (CH), 131.7 (CH), 129.4 (CH), 128.1 (C), 127.3 (C), 127.1 (CH), 120.0 (CH), 112.7 (C), 106.0 (CH), 104.4 (C), 55.6 (CH_3), 27.6 (CH_3); GC-MS (EI): $m/z=312$ $[M]^+$. Anal. Calcd. for $C_{18}H_{16}O_5$: C, 69.22; H, 5.16. Found: C, 69.11; H, 5.32.

5-(Anthracen-9-ylmethylene)-2,2-dimethyl-1,3-dioxane-4,6-dione (**4d**): Solid, mp 193.6-194.3 °C; IR (ATR): 3049, 3031, 3013, 2925, 2854, 1764, 1737, 1631, 1446, 1396, 1385, 1365, 1334, 1292, 1218, 1202, 1123, 1067, 1028; ^{1}H NMR (500 MHz, $CDCl_3$): δ 9.47 (s, 1H), 8.55 (s, 1H), 8.06 (d, $J=8.8$ Hz, 2H), 7.84 (d, $J=9.1$ Hz, 2H), 7.52 (t, $J=9.1$ Hz, 1H), 7.50 (t, $J=8.8$ Hz, 1H), 1.90 (s, 6H); ^{13}C NMR (125 MHz, $CDCl_3$): δ 161.91 (C), 157.89 (C), 157.75 (CH), 130.94 (C), 130.29 (CH), 129.34 (CH), 128.63 (C), 127.09 (CH), 126.80 (C), 125.58 (CH), 124.52 (CH), 121.02 (C), 104.97 (C), 28.17 (CH_3); GC-MS (EI): $m/z=332$ $[M]^+$. Anal. Calcd. for $C_{21}H_{16}O_4$: C, 75.89; H, 4.85; Found: C, 75.66; H, 5.03.

Reference

Murase, T., Nishijima, Y., and Fujita, M. (2012). Cage-catalyzed knoevenagel condensation under neutral conditions in water. *J. Am. Chem. Soc.*, **134**, 162–164.

C-C bond_3

Type of reaction: C-C bond formation
Reaction conditions: Ethanol / dichloromethane, room temperature
Synthetic strategy: *E* or *Z*-Selective Knoevenagel condensation
Catalyst: Piperidine
Keywords: Aldehydes, *O*-acetoacetylTEMPO (2,2,6,6-tetramethylpiperidin-1-yl 3-oxobutanoate), *N*-methoxy-*N*-methyl-3-oxobutanamide, piperidine, ethanol, room temperature, *E*- and *Z*-selective Knoevenagel condensation, trisubstituted *E* or *Z*-2-alkenes

General reaction scheme

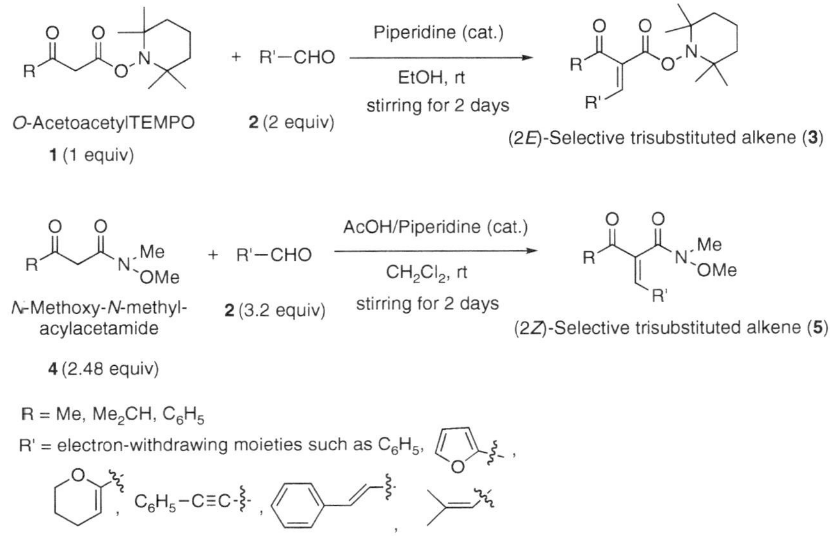

Representative entries

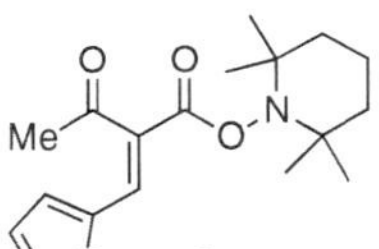

3a

(*E*)-2,2,6,6-Tetramethylpiperidin-1-yl 2-(furan-2-ylmethylene)-3-oxobutanoate (**3a**): Solid, mp 131-132 °C; reaction time: 2 days; yield: 76%

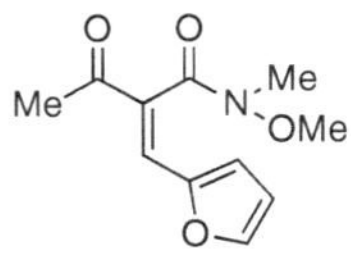

5a

(*Z*)-2-(Furan-2-ylmethylene)-*N*-methoxy-*N*-methyl-3-oxobutanamide (**5a**): Solid, mp 101-103 °C; reaction time: 2 days; yield: 69%

Experimental procedure

A typical procedure for Knoevenagel condensaton of *O*-acetoacetylTEMPO (1). To a cooled (0-4 °C) mixture of *O*-acetoacetylTEMPO (**1**; 243 mg, 1.0 mmol), 2-furaldehyde (**2**; 192 mg, 2.0 mmol), and ethanol (4 drops, 20 mg) was added piperidine (two drops, 10 mg). The mixture was allowed to warm gradually to room temperature and was stirred for 2 days. The reaction was quenched with cold aqueous ammonium chloride, and the products were extracted with ethyl acetate. The extracts were washed with brine, dried ($MgSO_4$), and concentrated. The crude products were purified by column chromatography (SiO_2, hexane-ethyl acetate, increasing the gradient from 10:1 to 3:1 v/v) to give 243 mg (76%) of (*E*)-2,2,6,6-tetramethylpiperidin-1-yl 2-(furan-2-ylmethylene)-3-oxobutanoate (**3**; R_f=0.34, hexane-ethyl acetate 5:1) and trace of *Z*-isomer (R_f=0.13) (*E*:*Z* ratio 38:1).

A typical procedure for Knoevenagel condensaton of *N*-methoxy-*N*-methylacetoacetamide (5). To a mixture of *N*-methoxy-*N*-methylacetoacetamide (**4**; 360 mg,

2.48 mmol) and 2-furaldehyde (**2**; 309 mg, 3.2 mmol) in CH_2Cl_2 (0.2 mL) were added consecutively AcOH (two drops) and piperidine (three drops) at 0-4 °C. The mixture was stirred at room temperature (28-30 °C) for 2 days and was quenched with aqueous NH_4Cl. Products were extracted with ethyl acetate and were worked up. The crude products were purified by column chromatography (SiO_2, hexane-ethyl acetate, increasing the gradient from 10:1 to 1:3 v/v) to give 382 mg (69%) of (*Z*)-2-(furan-2-ylmethylene)-*N*-methoxy-*N*-methyl-3-oxobutanamide (**5**; R_f=0.21, hexane-ethyl acetate 1:1) almost exclusively (*E*:*Z* ratio 1:>50).

All the products were characterized from their detailed spectral studies including IR, ^{1}H NMR, ^{13}C NMR and HRMS.

Characterization data of the representative compounds

(*E*)-2,2,6,6-tetramethylpiperidin-1-yl 2-(furan-2-ylmethylene)-3-oxobutanoate (**3a**): Solid, mp 131-132 °C; IR (KBr): 3141, 3100, 2976, 1739, 1625, 1551, 1473, 1381, 1365, 1233, 1159, 1071, 1043, 1014, 957, 905, 873, 778, 753 cm^{-1}; ^{1}H NMR (300 MHz): δ 1.11, 1.14 (s, 12H), 1.38-1.79 (m, 6H), 2.54 (s, 3H), 6.49 (d,d, J=3.3, 1.6 Hz, 1H), 6.74 (d, J=3.3 Hz, 1H), 7.40 (s, 1H, CH), 7.53 (m, 1H); ^{13}C NMR (75.5 MHz): δ 16.7, 20.6 (2C), 31.0, 31.7, 38.8 (2C), 60.3 (2C), 112.5, 117.6, 126.0, 128.6, 146.0, 149.1, 164.7 (^3J(COO, H)= 7.8 Hz), 201.2. HRMS (ESI): Calcd for $C_{18}H_{25}NO_4$ $(M+H)^+$ 320.1862; Found 320.1838 $(M+H)^+$.

(*Z*)-2-(Furan-2-ylmethylene)-*N*-methoxy-*N*-methyl-3-oxobutanamide (**5a**): Solid, mp 101-103 °C (partially decompose); IR (KBr): 3155, 3119, 3049, 2975, 1665, 1625, 1547, 1482, 1429, 1393, 1371, 1335, 1283, 1256, 1209, 1185, 1154, 1081, 1022, 993, 951, 932, 915, 883, 827, 761, 703 cm^{-1}; ^{1}H NMR (300 MHz) (a ca. 5.5:1 rotatory mixture on amide C-N bond): δ 2.34 (major) and 2.38 (minor) (s, 3H), 3.12 (minor) and 3.36 (major) (s, 3H), 3.47 (major) and 3.91 (minor) (s, 3H), 6.59 (major) and 6.52 (minor) (m, 1H), 6.78 (major) and 6.85 (minor) (d, J=3.6Hz, 1H), 7.31 (s, 1H), 7.53 (major) and 7.57 (minor) (d, J=1.6 Hz); ^{13}C NMR (75.5 MHz) (major isomer): δ 26.4, 32.5, 61.4, 112.7, 117.7, 125.3, 131.7, 146.1, 149.3, 168.7 (^3J(COO, H)=10.4 Hz), 194.2. HRMS (ESI): Calcd for $C_{11}H_{14}NO_4$ $(M+H)^+$ 224.0923; Found 224.0928 $(M+H)^+$.

Reference

Inokuchi, T., and Kawafuchi, H. (2006). *E*- or *Z*-selective Knoevenagel condensation of acetoacetic derivatives: Effect of acylated substituent, that is, TEMPO and amines, as an auxiliary, and new accesses to trisubstituted *E*- and *Z*-2-alkenals and furans. *J. Org. Chem.*, **71**, 947–953.

C-C bond_4

Type of reaction: C-C bond formation
Reaction conditions: Toluene /DMF, room temperature
Synthetic strategy: One-step condensation
Catalyst: Heterogeneous, modified Mg-Al hydrotalcite (MHT)
Keywords: Aromatic carbonyl compounds, malononitrile, ethyl cyanoacetate, modified and activated Mg-Al hydrotalcite (MHT), toluene or DMF, room temperature, condensation, Knoevenagel reaction, substituted alkenes

General reaction scheme

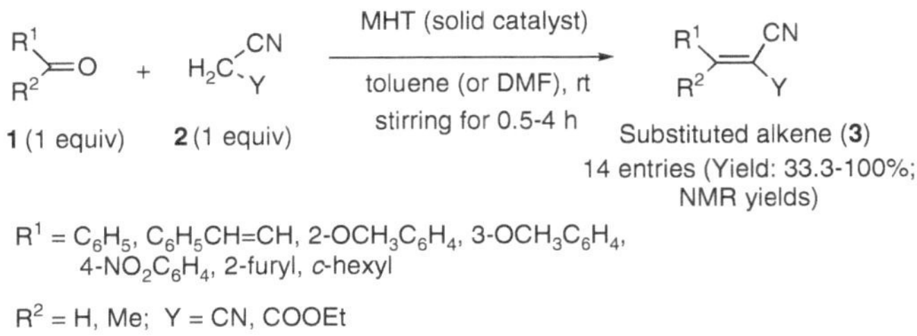

Representative entries

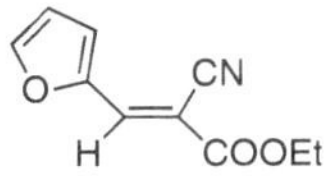

3a

(*E*)-Ethyl 2-cyano-3-(furan-2-yl)acrylate (**3a**): Reaction time: 2 h in toluene; NMR yield: 100%

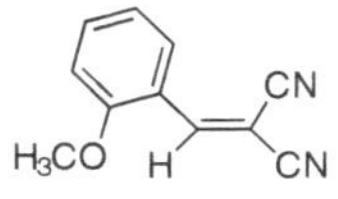

3b

2-(2-Methoxybenzylidene)malononitrile (**3b**): Reaction time: 1 h in toluene; isolated yield: 96%

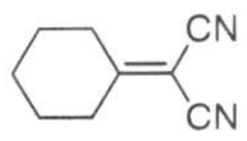

3c

2-Cyclohexylidenemalononitrile (**3c**): Reaction time: 0.5 h in toluene; NMR yield: 100%

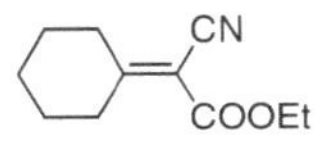

3d

2-Ethyl 2-cyano-2-cyclohexylideneacetate (**3d**): Reaction time: 1 h in DMF; NMR yield: 33.3%

Experimental procedure

To a mixture of an aromatic carbonyl compound (**1**; 2.0 mmol) and an active methylene compound (**2**; 2.0 mmol) in 10 mL of dry toluene (in case of most of the entries) was added 0.05 g of solid activated catalyst MHT at room temperature. The reaction mixture was then stirred for stipulated time (0.5 to 4 h) to complete the reaction. The Knoevenagel product was obtained in excellent yield after usual work-up and proper purification. Physical and spectral properties of all the products were found to be identical when compared with those of authentic samples. After completion of each reaction, the solid catalyst was filtered off followed by its activation before further use.

Reference

Lakshmi Kantam, M., Choudary, B. M., Venkat Reddy, Ch., Koteswara Rao, K., and Figueras, F. (1998). Aldol and Knoevenagel condensations catalysed by modified Mg-Al hydrotalcite: A solid base as catalyst useful in synthetic organic chemistry. *Chem. Commun.*, 1033–1034.

Further information

Rare-earth exchanged NaY zeolite [1], and $CeCl_3.7H_2O$-NaI [2] catalyzed Knoevenagel condensation at room temperature were also reported.

[1] Indrasena Reddy, T., and Varma, R. S. (1997). Rare-earth (RE) exchanged NaY zeolite promoted Knoevenagel condensation. *Tetrahedron Lett.*, **38**, 1721–1724.

[2] Bartoli, G., Beleggia, R., Giuli, S., Giuliani, A., Marcantoni, E., Massaccesi, M., and Paoletti, M. (2006). The $CeCl_3.7H_2O$–NaI system as promoter in the synthesis of functionalized trisubstituted alkenes via Knoevenagel condensation. *Tetrahedron Lett.*, **47**, 6501–6504.

C-C bond_5

Type of reaction: C-C bond formation
Reaction conditions: Dichloromethane, room temperature
Synthetic strategy: One-pot dimerization
Catalyst: Tungsten(II) carbonyl complex [$(CO)_4$ W(μ-Cl_3)W($SnCl_3$)]
Keywords: Norbornene, Tungsten(II) carbonyl complex catalyst, dichlorometane, room temperature, dimerization, C-H bond activation

Reaction scheme

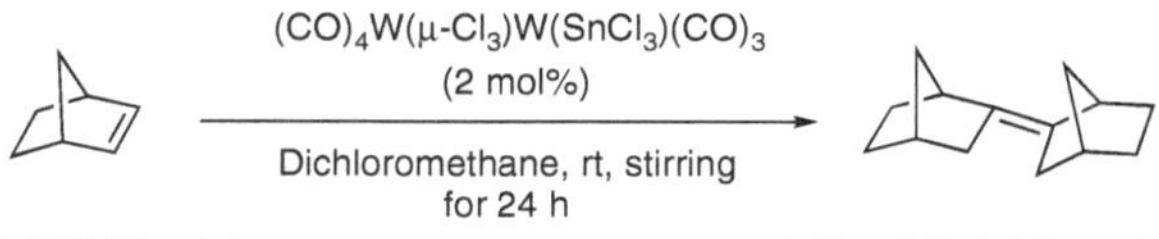

Bicyclo[2.2.1]hept-2-ene (norbornene, **1**)

2-Bicyclo[2.2.1.]hept-2-ylidene-bicyclo[2.2.1]heptane (**2**; 68%)
[colorless oil, a mixture of four stereoisomers (*syn-trans*, *anti-trans*, *syn-cis*, and *anti-cis*)]

Experimental procedure

A dichloromethane solution (10 mL) of norbornene (**1**; 0.24 g, 2.5 mmol) and *n*-heptane (0.3 mL as internal chromatographic standard) was added under an atmosphere of nitrogen to a 50 mL Schlenk tube containing the catalyst, tungsten(IV) carobonyl complex (0.045 g, 0.05 mmol). The reaction mixture was stirred at room temperature until completion (ca. 24 h, GC-FID monitoring). The solvent was then evaporated *in vacuum* and the crude reaction product extracted with *n*-heptane. Fractional distillation of this crude product at 65 °C under reduced pressure afforded pure **2** (colorless oil) as a mixture of four stereoisomers identified by GC-MS analysis: GC-MS for the of **2** ($C_{14}H_{20}$, $M_r = 188.32$) appeared as three signals at retention time 11.48 min (43.2%), 11.53 min (23.3%), and 11.63 (33.5%), *m/z* (%relative intensity): 41(14), 67(22), 79(39), 91(33), 105(11), 131(18), 159(100), 188(35).

NMR characterization data of the four stereoisomers of 2-bicyclo[2.2.1]hept-2-ylidenebicyclo [2.2.1]-heptane (**2**):

^{1}H NMR ($CDCl_3$): (1) *syn-trans*: δ 2.54 (s, 2H-1), 2.32 (s, 2H-4), 2.12 (*d*, $J_{H\text{-}H} = 15$ Hz, 2H-3), 1.71 (*d*, $J_{H\text{-}H} = 15$ Hz, 2H-3′), 1.54 (m, 4H-5,6), 1.27 (m, 2H-7), 1.22 (m, 6H-5,6,7); (2) *anti-trans*: δ 2.58 (s, 2H, H-1), 2.32 (s, 2H-4, 2.03 (*d*, $J_{H\text{-}H} = 15$ Hz, 2H-3), 1.80 (*d*, $J_{H\text{-}H} = 15$ Hz, 2H3′), 1.54 (m, 4H-5,6), 1.27 (m, 2H-7), 1.22 (m, 6H-5,6,7); (3) *syn-cis*: δ 2.76 (s, 2H-1), 2.32 (s, 2H-4), 1.94 (*d*, $J_{H\text{-}H} = 15$ Hz, 2H-3), ca. 1.6, 2H-3′, 1.54 (m, 4H-5,6), 1.27 (m, 2H-7), 1.22 (m, 6H-5,6,7); (4) *anti-cis*: δ 2.79 (s,2H-1), 2.32 (s,2H-4), 1.94 (*d*, $J_{H\text{-}H} = 15$ Hz, 2H-3), ca. 1.6 (2H-3′), 1.54 (m, 4H-5,6),1.27 (m, 2H-7), 1.22 (m, 6H-5,6,7); ^{13}C NMR ($CDCl_3$): (1) *syn-trans*: δ 131.53 (C-2), 41.01 (C-1), 39.03 (C-7), 37.16 (C-3), 36.78 (C-4), 29.07 (C-5), 28.59 (C-6); (2) *anti-trans*: δ 131.43 (C-2), 41.37 (C-1), 39.26 (C-7), 37.08 (C-3), 36.70 (C-4), 29.77 (C-5), 28.55 (C-6); (3) *syn-cis*: δ 131.95 (C-2), 41.82 (C-1), 39.89 (C-7), 37.04 (C-3), 36.64 (C-4), 28.93 (C-5), 28.67 (C-6); (4) *anti-cis*: δ 131.84 (C-2), 41.87 (C-1), 39.69 (C-7), 37.01 (C-3), 36.59 (C-4), 29.73 (C-5), 28.72 (C-6).

Reference

Malinowska, A., Czeluśniak, I., Górski, M., and Szymańska-Buzar, T. (2005). A novel catalytic route to 2-bicyclo[2.2.1]hept-2-ylidenebicyclo[2.2.1]-heptane involving C-H bond activation of bicyclo[2.2.1] hept-2-ene. *J. Mol. Catal. A: Chem.*, **226**, 259–262.

C-C bond_6

Type of reaction: C-C bond formation
Reaction conditions: Solvent-free, room temperature
Synthetic strategy: One-pot condensation
Catalyst: Cetyltrimethylammonium hydroxide (CTAOH)
Keywords: Nitroalkanes, α,β-unsaturated ketones, aldehydes, CTAOH, solvent-free, room temperature, Michael reaction, Henry reaction, β-nitroalcohols

Reaction scheme

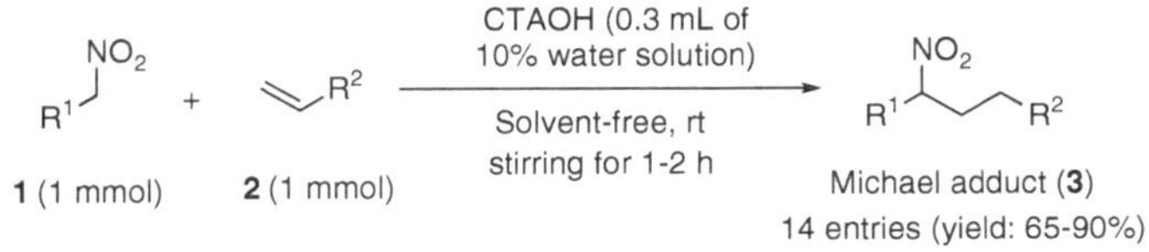

R^1 = CH_3, C_2H_5, *n*-Pr, *n*-Bu, *n*-Pentyl, $CH_3OCO(CH_2)_3CH_2$, $CH_3COCH_2CH_2$, $CH_3CH(OH)CH_2CH_2$
R^2 = $COOCH_3$, $COOC_2H_5$, COC_2H_5, CN, SO_2Ph, $COCH_3$
(2-cyclohexenone was also used in two entries)

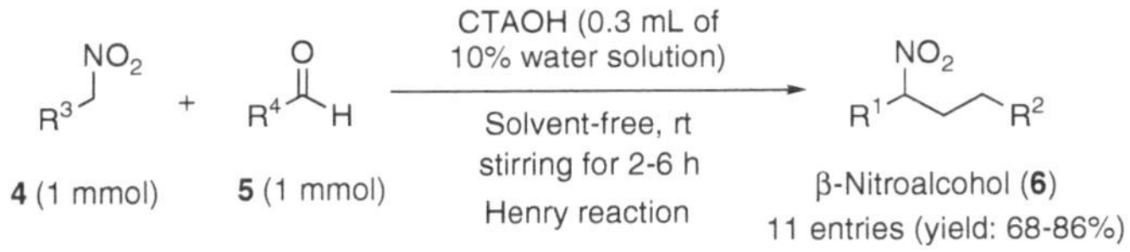

R^3 = C_2H_5, *n*-Pr, *n*-Bu, *n*-Pentyl, $CH_3OCO(CH_2)_3CH_2$, C_2H_5OCO, $PhCH_2$; nitrocyclopentane was also used in one entry)
R^4 = H, *n*-Pr, $CH_3(CH_2)_8$, $CH_3(CH_2)_4$, Ph, $PhCH_2CH_2$, 4-$NO_2C_6H_4$,

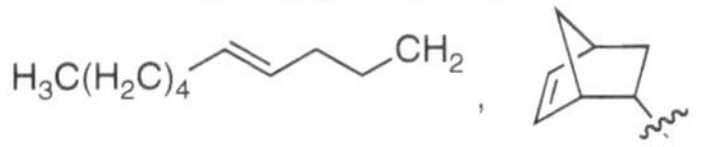

Representative entries

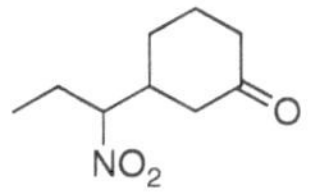

3a

3-(1-Nitropropyl)-1-cyclohexanone (**3a**): Diastereomeric mixture, yellow oil; reaction time: 1 h; yield: 80%

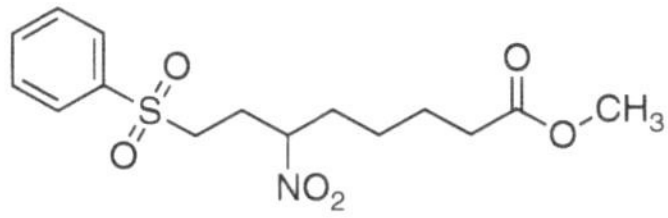

3b

Methyl-6-nitro-8-(phenylsulfonyl)octanoate (**3b**): Yellow oil; reaction time: 2 h; yield: 73%

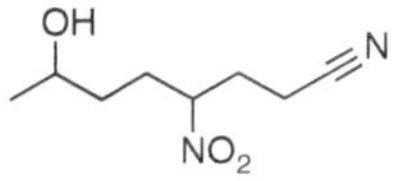

3c

7-Hydroxy-4-nitrooctanenitrile (**3c**): Diastereomeric mixture, yellow oil; reaction time: 1 h; yield: 72%

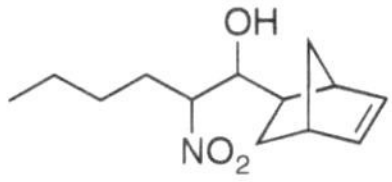

6a

1-Bicyclo[2.2.1]hept-5-en-2-yl)-2-nitrohexan-1-ol (**6a**): White solid, mp 43-45 °C; reaction time: 3 h; yield: 85%

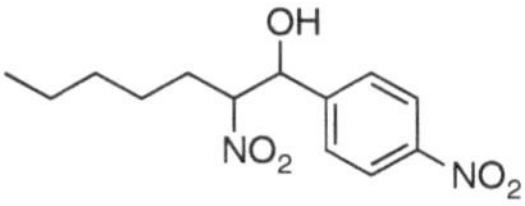

6b

2-Nitro-1-(4-nitrophenyl)-1-heptanol (**6b**): Yellow oil; reaction time: 4 h; yield: 76%

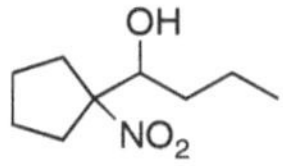

6c

1-(1-Nitrocyclopentyl)-1-butanol (**6c**): Colorless oil; reaction time: 4 h; yield: 83%

Experimental procedure

Michael adduct: α,β-Unsaturated ketone, the Michael acceptor (**2**; 1 mmol), was added to a stirred mixture of nitrocompound (**1**; 1 mmol) in a 10% water solution of hexadecyltrimethyl ammoniumhydroxide (CTAOH, 0.3 mL) at room-temperature under solvent-free conditions. The reaction progress was monitored by both GC and TLC. After completion of the reaction (within 1-2 h), the resulting solution was treated with brine (10 mL) and extracted by dichloromethane (3 × 25 mL). The organic layer was dried over anhydrous sodium sulfate, filtered, concentrated under vacuum to afford the Michael adduct **3**, which was purified on flash chromatography using cyclohexane-ethyl acetate (yield 65-90%).

Henry adduct: The aldehyde (**5**; 1 mmol) was added to a stirred mixture of nitrocompound (**4**; 1 mmol) in a 10% water solution of hexadecyltrimethyl ammonium hydroxide (0.300 mL) at room-temperature under solvent-free conditions. The reaction progress was monitored by TLC. After completion of the reaction (within 2-6 h), the resulting solution was treated with brine (10 mL) and extracted by dichloromethane (3 × 25 mL). The organic layer was dried over anhydrous sodium sulfate, filtered, concentrated under vacuum to afford β-nitroalcohol **6**, the Henry adduct, which was purified on flash chromatography using cyclohexane-ethyl acetate (yield 68-86%).

Characterization data of representative entries

Methyl-6-nitro-8-(phenylsulfonyl)octanoate (**3b**): Yellow oil; IR (neat): 1308, 1556, 1732 cm^{-1}; ^1H NMR ($CDCl_3$, 300 MHz): δ 1.22-1.41 (m, 2H), 1.58-1.82 (m, 3H), 1.90-2.09 (m, 1H), 2.21-2.39 (m, 4H), 3.11 (t, $J = 7.4$ Hz, 2H), 3.67 (s, 3H), 4.58-4.67 (m, 1H), 7.58-7.93 (m, 5H); ^{13}C NMR ($CDCl_3$, 75 MHz): δ 24.2, 25.2, 26.6, 33.5, 33.6, 51.8, 52.5, 86.5, 128.2, 129.8, 134.4, 138.8, 173.7; EI-MS: *m/z* 265, 171, 143, 123, 95, 77 (100), 67, 55, 41. Anal. calcd for $C_{15}H_{21}NO_6S$: C, 52.47; H, 6.16; N, 4.08; Found: C, 52.59; H, 6.30; N, 3.99.

7-Hydroxy-4-nitrooctanenitrile (**3c**): Diastereomeric mixture, yellow oil; IR (neat): 1376, 1552, 2250, 3422 cm^{-1}; ^1H NMR ($CDCl_3$, 300 MHz): δ 1.21 (d, $J = 6.2$ Hz, 3H), 1.41-1.55 (m, 2H), 1.71-1.82 (bs, 1H), 1.95-2.49 (m, 6H), 3.71-3.92 (m, 1H), 4.58-4.76 (m, 1H); ^{13}C NMR ($CDCl_3$, 75 MHz): δ 14.5, 24.0, 24.1, 29.2, 29.4, 29.8, 30.3, 34.5, 34.9, 66.8, 67.4, 86.6, 87.2, 118.1; EI-MS: *m/z* 171, 122, 96, 82, 67, 55, 45 (100), 39. Anal. calcd for $C_8H_{14}N_2O_3$: C, 51.60; H, 7.58; N, 15.04; Found: C, 51.77; H, 7.66; N, 14.95.

1-Bicyclo[2.2.1]hept-5-en-2-yl)-2-nitrohexan-1-ol (**6a**): White solid, mp 43-45 °C; IR (neat): 1365, 1543, 1625, 3040, 3420 cm^{-1}; ^1H NMR ($CDCl_3$, 300 MHz): δ 0.83-1.05 (m, 3H), 1.22-1.57 (m, 7H), 1.70-2.38 (m, 5H), 2.82-2.94 (m, 1H), 3.02-3.12 (m, 1H), 4.38-4.76 (m, 2H), 5.82-6.27 (m, 2H); ^{13}C NMR ($CDCl_3$, 75 MHz): δ 13.9, 22.3, 28.0, 28.1, 28.4, 29.6, 29.8, 30.7, 42.6, 44.4, 49.1, 49.2, 50.1, 50.2, 75.8, 76.0, 91.3, 91.4, 132.5, 132.6, 138.3, 138.5. Anal. calcd for $C_{13}H_{21}NO_3$: C, 65.25; H, 8.84; N, 5.85; Found: C, 65.39; H, 8.99; N, 5.77.

1-(1-Nitrocyclopentyl)-1-butanol (**6c**): Colorless oil; IR (neat): 1357, 1537, 3445 cm^{-1}; ^1H NMR ($CDCl_3$, 300 MHz): δ 0.92 (t, $J = 7.0$ Hz, 3H), 1.22-1.91 (m, 9H), 2.03-2.20 (m, 1H),

2.37-2.62 (m, 3H), 3.79-3.91 (m, 1H); ^{13}C NMR ($CDCl_3$, 75 MHz): δ 14.0, 19.7, 24.8, 25.0, 33.8, 34.7, 35.6, 75.7, 103.8. Anal. calcd for $C_9H_{17}NO_3$: C, 57.73; H, 9.15; N, 7.48; Found: C, 57.89; H, 9.27; N, 7.33.

Reference

Ballini, R., Fiorini, D., Gil, M. V., and Palmieri, A. (2004). Cetyltrimethylammonium hydroxide (CTAOH) as a general, ecofriendly catalyst for the formation of carbon–carbon bond through nitroalkanes. *Tetrahedron*, **60**, 2799–2804.

C-C bond_7

Type of reaction: C-C bond formation
Reaction conditions: Dichloromethane, room temperature
Synthetic strategy: One-pot C-H activation
Catalyst: Cobalt acetylacetonate [Co(acac)$_2$]
Keywords: 3-Oxo-1,5-pentanedioic acid dimethylester, isocyanates, dichloromethane, Co (acac)$_2$, room temperature, C-H activation, β,β′-tricarbonyl derivatives

Reaction scheme

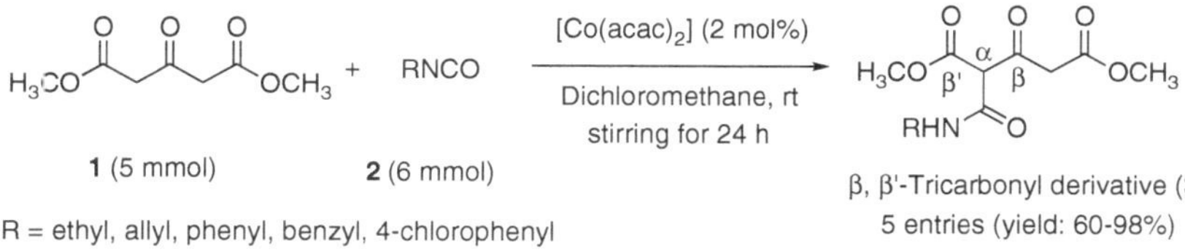

Representative entries

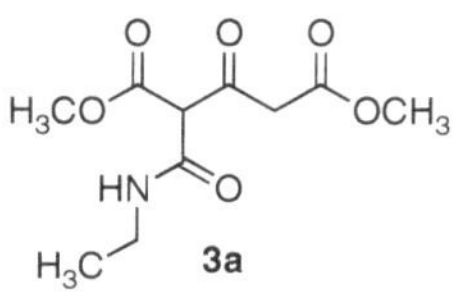

Dimethyl 2-(ethylcarbamoyl)-3-oxopentanedioate (**3a**): Colorless crystals; mp 50-52 °C; reaction time: 24 h; yield: 98%

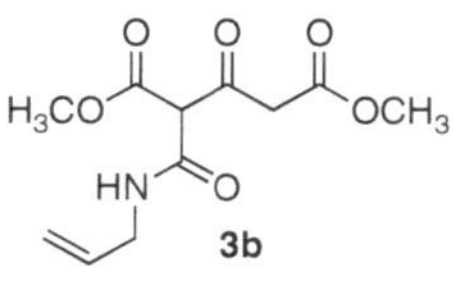

Dimethyl 2-(allylcarbamoyl)-3-oxopentanedioate (**3b**): Yellow oil; reaction time: 24 h; yield: 92%

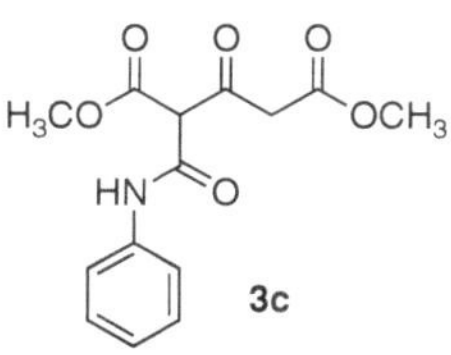

Dimethyl 3-oxo-2-(phenylcarbamoyl) pentanedioate (**3c**): Colorless crystals; mp 58-61 °C; reaction time: 24 h; yield: 60%

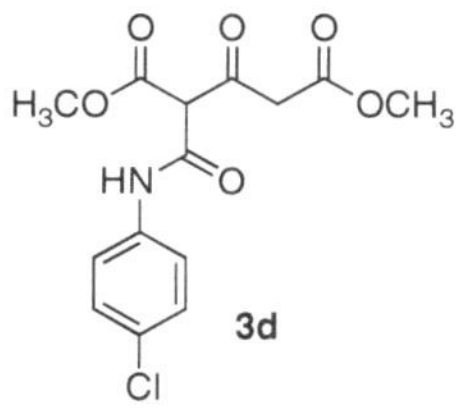

Dimethyl 3-oxo-2-((4-chlorophenyl)carbamoyl) pentanedioate (**3d**): Colorless crystals; mp 118-119 °C; reaction time: 24 h; yield: 65%

Experimental procedure

To a solution of 3-oxo-1,5-pentanedioic acid dimethylester (**1**; 0.87 g, 5.0 mmol) in anhydrous dichloroethane (2 mL), isocyanate (**2**; 6.0 mmol) and [Co(acac)$_2$] (2 mol%) were added. The reaction mixture was stirred at room temperature for 24 h, and then diluted with diethyl ether. The resulting suspension was filtered on Celite™ and the filtrate on concentrated under reduced pressure afforded β,β′-tricarbonyl derivative **3** in good yields (60-98%). All the products were characterized by analytical and spectral (IR, ^{1}H- and ^{13}C-NMR) studies.

Characterization data of representative entries

Dimethyl 2-(allylcarbamoyl)-3-oxopentanedioate (**3b**): Yellow oil; IR (oil): 3450 (br),1750 (br),1570 (br) cm^{-1}; ^{1}H NMR (300.1 MHz, CDCl$_3$): δ 3.69 (s, 3H, OMe), 3.72 (s, 3H, OMe), 3.74 (s, 2H, CH$_2$), 3.96 (m, 2H, CH$_2$N), 5.1-5.3 (m, 2H, CH$_2$=), 5.7-5.9 (m, 1H, CH), 9.2 (br, 1H, NH),18.68 (s, 1H, OH); ^{13}C NMR (75.5 MHz, CDCl$_3$): δ 41.1, 44.8, 50.7, 51.6, 92.7, 115.9, 132.6, 167.5, 168.1, 171.6, 186.3. Anal. Calcd for $C_{11}H_{15}NO_6$: C, 51.36; H, 5.88; N, 5.45%; Found: C, 50.78; H, 6.14; N, 5.17%.

Dimethyl 3-oxo-2-((4-chlorophenyl)carbamoyl)pentanedioate (**3d**): Colorless crystals, mp 118-119 °C; IR (KBr): 3100 (br), 1730, 1670, 1540 cm^{-1}; ^{1}H NMR (300.1 MHz, DMSO-*d$_6$*): δ 3.62 (s, 3H, OMe), 3.71 (s, 5H, OMe+CH$_2$), 7.40-7.60 (m, 4H, Ph), 10.54 (br, 1H, NH), 16.81 (br, 1H, OH); ^{13}C NMR (75.5 MHz, CDCl$_3$): δ 45.3, 51.8, 52.4, 95.6, 121.3, 122.7, 129.1, 135.1, 168.3, 168.5, 170.5, 186.7. Anal. Calcd for $C_{14}H_{14}ClNO_6$: C, 51.31; H, 4.31; N, 4.27%; Found: C, 51.40; H, 3.77; N, 4.91%.

Reference

Veronese, A. C., Durini, E., Bertolasi, V., Basato, M., and Tubaro, C. (2010). Metal catalysed reactions of β, β′-tricarbonyl derivatives with isocyanates. *Tetrahedron*, **66**, 8313–8316.

C-C bond_8

Type of reaction: C-C bond formation
Reaction conditions: Dioxane, room temperature
Synthetic strategy: Regio- and chemoselective one-step Suzuki-Miyaura cross-coupling between two sp^3-carbons
Catalyst: Pd-catalyst [Pd-PEPPSI-IPr] (**1**)/Pd[P(*t*-Bu)$_3$]$_2$ (**2**) in the presence of aqueous KOH

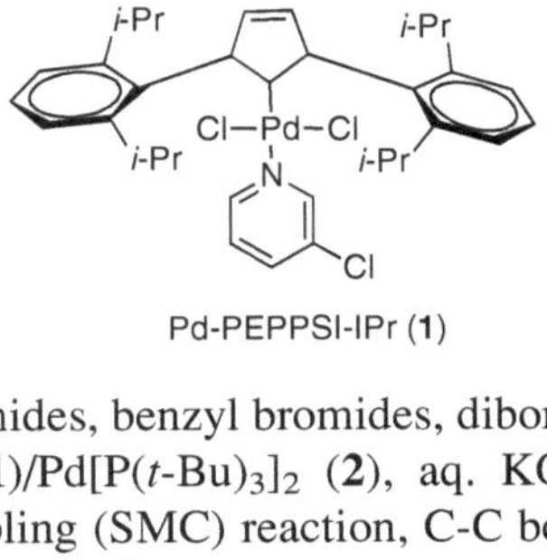

Pd-PEPPSI-IPr (**1**)

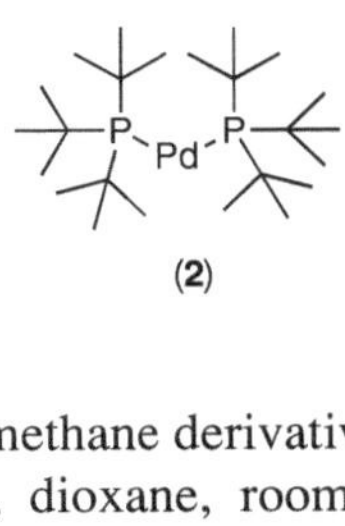

(**2**)

Keywords: Allyl bromides, benzyl bromides, diborylmethane derivative, Pd-catalysts such as [Pd-PEPPSI-IPr] (**1**)/Pd[P(*t*-Bu)$_3$]$_2$ (**2**), aq. KOH, dioxane, room temperature, Suzuki-Miyaura cross-coupling (SMC) reaction, C-C bond formation, regio- and chemoseletivity, homoallylboronates, alkylboronates

Reaction scheme

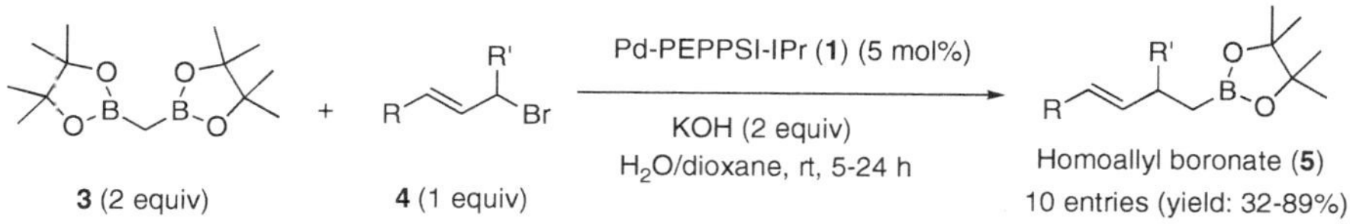

R, R' = C_6H_5, H; 4-MeC_6H_4, H; 4-*t*-BuC_6H_4, H; 4-OMeC_6H_4,, H; 4-ClC_6H_4, H; 4-$CF_3C_6H_4$, H; 2-thienyl, H; 2-naphthyl, H; C_6H_5, Me; $C_6H_5CH_2CH_2$, H

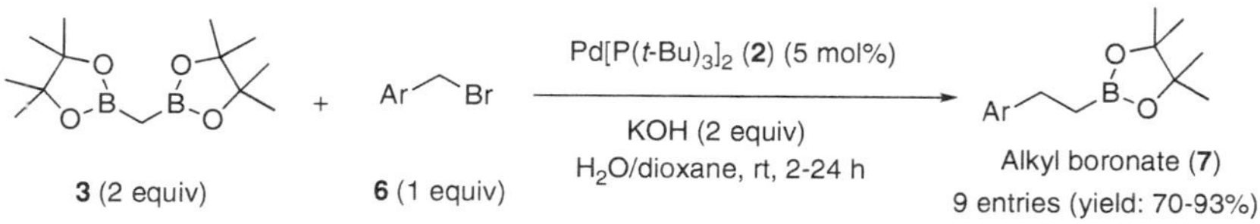

Ar = C_6H_5; 4-MeC_6H_4; 4-*t*-BuC_6H_4; 4-ClC_6H_4; 4-FC_6H_4; 4-$NO_2C_6H_4$; 4-CNC_6H_4;4-COOMeC_6H_4; 2-naphthyl

Representative entries

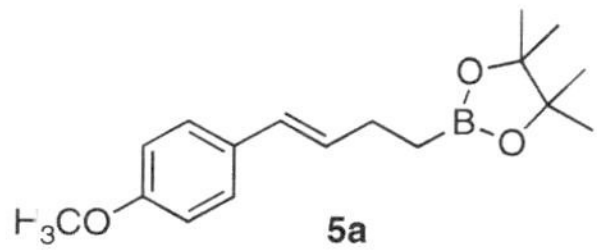

5a

(*E*)-2-(4-(4-Methoxyphenyl)but-3-en-1-yl)-4,4,5,5-tetramethyl-1,3,2-dioxaborolane (**5a**): Yellow oil; reaction time: 6 h; yield: 92%

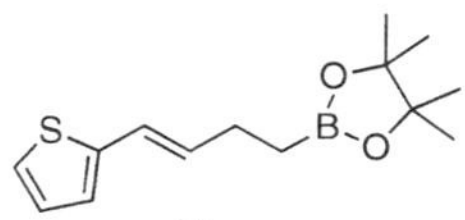

5b

(*E*)-4,4,5,5-Tetramethyl-2-(4-(thiophene-2-yl)-but-3-en-1-yl)-1,3,2-dioxaborolane (**5b**): Yellow oil; reaction time: 24 h; yield: 39%

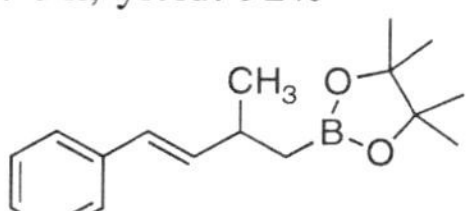

5c

4-(2-(4,4,5,5-tetramethyl-1,3,2-dioxaborolan-2-yl) ethyl)benzonitrile (**5c**): Yellow oil; reaction time: 24 h; yield: 75%

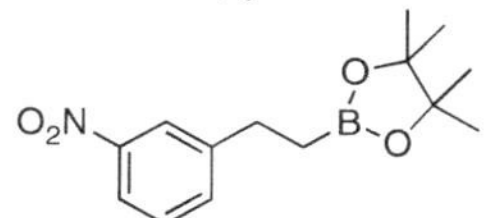

7a

4,4,5,5-Tetramethyl-2-(3-nitrophenethyl)-1,3,2-dioxaborolane (**7a**): Reaction time: 5 h; yield: 70%

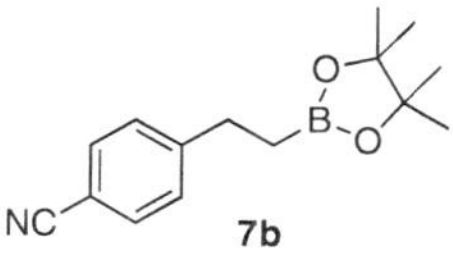

7b

4-(2-(4,4,5,5-Tetramethyl-1,3,2-dioxaborolan-2-yl) ethyl)benzonitrile (**7b**): Colorless oil, reaction time: 5 h; yield: 85%

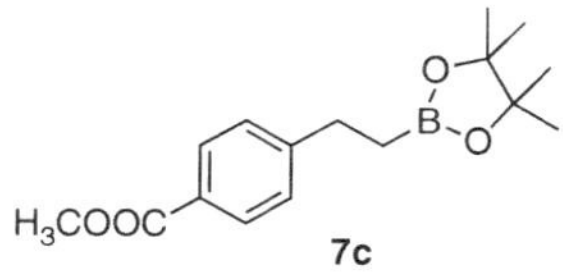

7c

Methyl 4-(2-(4,4,5,5-tetramethyl-1,3,2-dioxaborolan-2-yl)ethyl)benzoate (**7c**): Colorless oil, reaction time: 24 h; yield: 77%

Experimental procedures

Synthesis of homoallylboronate (**5**): To a solution of diborylmethane (**3**; 2 equiv), allyl bromide (**4**; 1 equiv) and Pd-PEPPSI-IPr (**1**; 2.5 mg, 3.5 μmol, 5 mol %) in dioxane (0.75 mL) was added 8N aqueous KOH (18 μL, 0.14 mmol, 2 equiv) at room temperature. The mixture was stirred for stipulated periods (5-24 h) and on completion of reaction (monitored by NMR) it was filtered through a pad of silica gel with ether. The crude product, obtained on removal of solvents under reduced pressure, was then purified using silica gel column chromatography (hexane/ether = 10/1) to afford homoallylboronate **5** in 32-89% yield. Each of the products was characterized by detailed spectral studies including IR, ^{1}H NMR, ^{13}C NMR, ^{11}B NMR and HRMS.

Synthesis of alkylboronate (**7**): To a solution of diborylmethane (**3**; 2 equiv), benzyl bromide (**5**; 1 equiv) and Pd[P(t-Bu)$_3$]$_2$ (**2**; 2.5 mg, 3.5 μmol, 5 mol %) in dioxane (1 mL) was added 8N aqueous KOH (24 μL, 0.2 mmol, 2 equiv) at room temperature. The mixture was stirred for stipulated periods (2-24 h) and on completion of reaction (monitored by NMR) it was filtered through a pad of silica gel with ether. The crude product, obtained on removal of solvents under reduced pressure, was then purified using silica gel column chromatography (hexane/ether = 10/1) to afford alkylboronate **7** in 32-89% yield. Each of the products was characterized by detailed spectral studies including IR, ^{1}H NMR, ^{13}C NMR, ^{11}B NMR and HRMS.

Characterization data of representative entries

(E)-4,4,5,5-Tetramethyl-2-(4-(thiophene-2-yl)-but-3-en-1-yl)-1,3,2-dioxaborolane (**5b**): Yellow oil; IR (neat): 2978, 2934, 1730, 1378, 1275, 1145, 965, 848, 749 cm^{-1}; ^{1}H NMR (400 MHz, $CDCl_3$): δ 7.07 (d, J = 5.2 Hz, 1H), 6.94-6.90 (m, 1H), 6.86-6.83 (m, 1H), 6.50 (d, J = 15.6 Hz, 1H), 6.13 (dt, J = 6.4 Hz, 15.6 Hz, 1H), 2.31-2.29 (m, 2H), 1.25 (s, 12H), 0.97 (t, J = 7.6 Hz, 2H); ^{13}C NMR (100 MHz, $CDCl_3$): δ 132.7, 128.8, 127.1, 124.0, 122.9, 122.1, 83.1, 27.0, 24.8; ^{11}B NMR (128 MHz, $CDCl_3$): δ 34.3; HRMS (FAB, positive) at m/z calcd for $C_{14}H_{21}^{10}BO_2S$: 263.1392; found 263.1392.

Methyl 4-(2-(4,4,5,5-tetramethyl-1,3,2-dioxaborolan-2-yl)ethyl)benzoate (**7c**): Colorless oil; IR (neat): 2979, 2950, 1739, 1373, 1145, 1110, 968,704 cm^{-1}; ^{1}H NMR (400 MHz, $CDCl_3$): δ 7.94-7.92 (m, 2H), 7.29-7.26 (m, 2H), 3.89 (s, 3H), 2.80 (t, J = 8.4 Hz, 2H), 1.21 (s, 12H), 1.18-1.13 (m, 2H); ^{13}C NMR (100 MHz, $CDCl_3$): δ 167.2, 149.9, 129.6, 128.0, 127.5, 83.2, 51.9, 30.0, 24.8; ^{11}B NMR (128 MHz, $CDCl_3$): δ 34.4; HRMS (FAB, positive) at m/z calcd for $C_{16}H_{24}^{10}BO_4$ $[M+H]^+$ 290.1804; found 290.1805.

Reference

Endo, K., Ohkubo, T., Ishioka, T., and Shibata, T. (2012). Cross coupling between sp^3-carbon and sp^3-carbon using a diborylmethane derivative at room temperature. *J. Org. Chem.*, **77**, 4826–4831.

C-C bond_9

Type of reaction: C-C bond formation
Reaction conditions: Tetrahydrofuran (THF), room temperature
Synthetic strategy: "Copper-free Sonogashira" — aryl bromide – alkyne cross-coupling
Catalyst: Pd(I)-dimer catalyst (**1**) promoted by $ZnCl_2$
Keywords: Aryl bromides, alkynes, Pd(I)-catalyst, $ZnCl_2$, HN(i-Pr)$_2$, THF, room temperature, Sonogashira cross-coupling, substituted alkynes

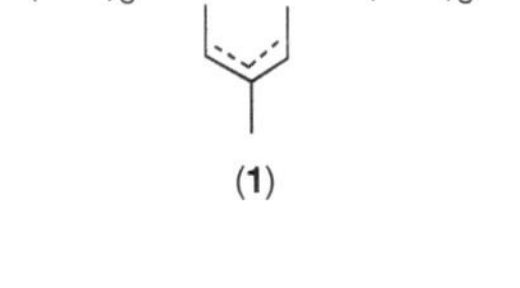

(**1**)

Reaction scheme

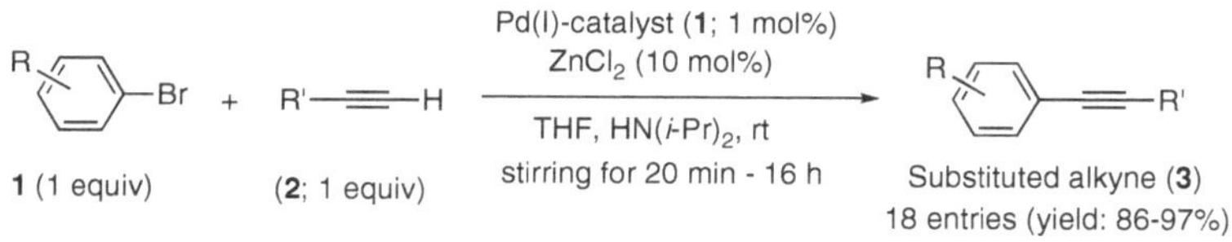

R = H, 2-OMe, 4-OMe, 3-OTf, 4-OTf, 2-Me, 2,4,6-tri-Me, 4-CHO, 4-COMe, 4-COOMe, 4-NO_2, 2-Ph
In addition to: 1-naphthyl bromide, 9-anthracyl bromide, and 2-thienyl bromide
R' = Ph, TMS, *n*-Bu

Representative entries

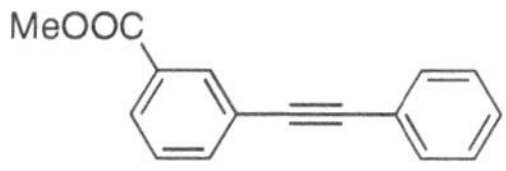

3a

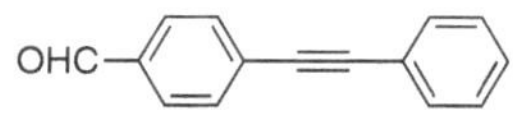

3b

Methyl 3-(phenylethynyl)benzoate (**3a**): Yellow solid; mp 76-78 °C; reaction time: 30 min; yield: 97%

4-(Phenylethynyl)benzaldehyde (**3b**): Yellow solid; mp 98-100 °C; reaction time: 20 min; yield: 93%

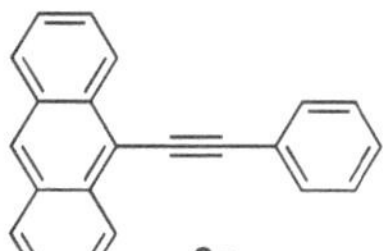

3c

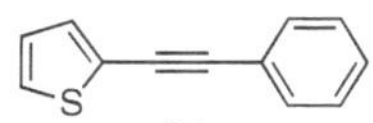

3d

9-(Phenylethynyl)anthracene (**3c**): Highly fluorescent yellow solid; mp 110-112 °C; reaction time: 5 h; yield: 86%

9-(Phenylethynyl)thiophene (**3d**): White solid; mp 50-51 °C; reaction time: 1 h; yield: 91%

Experimental procedure

In a glove box, a 20 mL vial with magnetic stir bar and PTFE/silicone septum-lined cap was charged successively with the aryl bromide (**1**; 3 mmol, 1 equiv), $ZnCl_2$ (41 mg, 0.3 mmol, 0.1 equiv), THF (7 mL), HN(*i*-Pr)$_2$ (3 mL), and Pd(I) dimer **1** (21 mg, 0.03 mmol, 0.01 equiv). The solution was stirred to dissolve the solids, and then the alkyne (**2**; 3 mmol, 1 equiv) was added. The alkyne **3** was added lastly so as to avoid catalyst decomposition and to avoid excess alkyne addition yielding also byproduct(s). The vial was then capped, removed from the glove box, and stirred at room temperature for specified time period (20 min to 16 h for varying entries). The solution was diluted with EtOAc, and passed through a plug of silica gel (10 g) and eluted with 100 mL EtOAc to remove H_2N(*i*-Pr$_2$)Br and Zn salt. The solution was evaporated, and the residue was purified by column chromatography on silica gel. Each of the products was characterized from its physical properties as well as its ^{1}H NMR and ^{13}C NMR spectral studies.

Characterization data of representative entries

Methyl 3-(phenylethynyl)benzoate (**3a**): Yellow solid; mp 76-78 °C; ^{1}H NMR (500 MHz, $CDCl_3$): δ 8.21 (s, 1H), 8.00 (d, J = 7.9 Hz, 1H), 7.71 (d, J = 7.7 Hz, 1H), 7.56-7.53 (m, 2H), 7.44 (t, J = 7.8 Hz, 1H), 7.38-7.36 (m, 3H), 3.94 (s, 3H); ^{13}C NMR (126 MHz, $CDCl_3$): δ 166.6, 135.8, 132.9, 131.8, 130.6, 129.3, 128.70, 128.64, 128.55, 123.9, 123.0, 90.4, 88.4, 52.4.

9-(Phenylethynyl)anthracene (**3c**): Highly fluorescent yellow solid; mp 110-112 °C; ^{1}H NMR (500 MHz, $CDCl_3$): δ 7.55-7.52 (m, 2H), 7.39-7.35 (m, 3H), 7.30 (d, J = 4.1 Hz, 2H),

7.04-7.02 (m, 1H);^{13}C NMR (126 MHz, $CDCl_3$): δ 132.0, 131.5, 128.54, 128.50, 127.38, 127.22, 123.4, 123.0, 93.2, 82.7.

Reference

Finke, A. D., Elleby, E. C., Boyd, M. J., Weissman, H., and Moore, J. S. (2009). Zinc chloride-promoted aryl bromide-alkyne cross-coupling reactions at room temperature. *J. Org. Chem.*, **74**, 8897–8900.

Further information

Copper-free Sonogashira coupling of aryl bromide with alkynes using $(AllylPdCl)_2/P(t\text{-}Bu)_3$ catalyst in the presence of piperidine or DABCO in acetonitrile at room temperature was also reported by Soheiji *et al* [1]. Another mild protocol for the copper-free Sonogashira coupling of aryl iodides with terminal acetylenes in water under aerobic conditions was accomplished using 1 mol % $PdCl_2$ in the presence of pyrrolidine at room temperature [2].

References

[1] Soheili, A., Albaneze-Walker, J., Murry, J. A., Dormer, P. G., and Hughes, D. L. (2003). Efficient and general protocol for the copper-free Sonogashira coupling of aryl bromides at room temperature. *Org. Lett.*, **5**, 4191–4194.

[2] Liang, B., Dai, M., Chen, J., and Yang, Z. (2005). Copper-free Sonogashira coupling reaction with $PdCl_2$ in water under aerobic conditions. *J. Org. Chem.*, **70**, 391–393.

C-C bond_10

Type of reaction: C-C bond formation
Reaction conditions: Dioxane, room temperature
Synthetic strategy: Aryl bromide – alkyne Sonogashira cross-coupling
Catalyst: $Pd(PhCN)_2Cl_2$ / $P(t\text{-}Bu)_3$
Keywords: Aryl bromides, alkynes, $Pd(PhCN)_2Cl_2$ / $P(t\text{-}Bu)_3$, CuI, $HN(i\text{-}Pr)_2$, dioxane, room temperature, Sonogashira cross-coupling, substituted alkynes

Reaction scheme

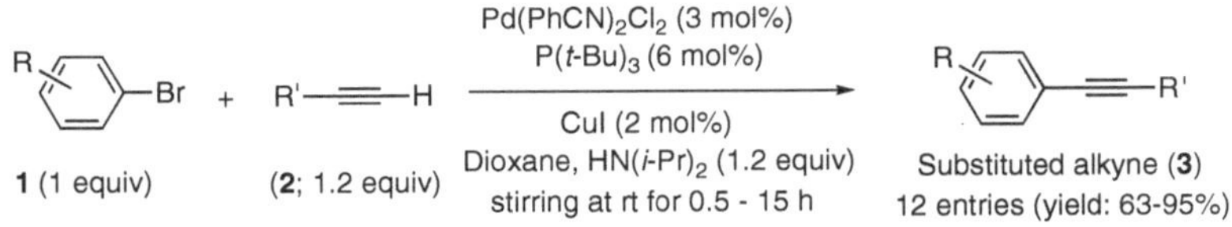

R = H, 4-OMe, 2-Me, 2,6-di-Me, 4-COMe, 4-NMe_2
R' = *n*-hexyl, $Me_2C(OH)$, Ph

Representative entries

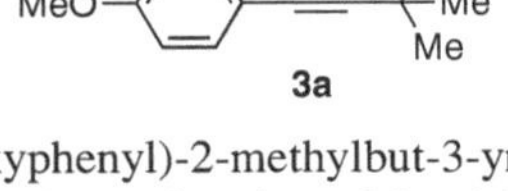

3a

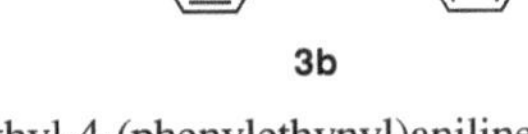

3b

4-(4-Methoxyphenyl)-2-methylbut-3-yn-2-ol (**3a**): Colorless solid; reaction time: 8 h; yield: 95%

N,N-Dimethyl-4-(phenylethynyl)aniline (**3b**): Slightly-yellow solid; reaction time: 2 h; yield: 94%

Characterization data of two representative compounds

(E)-1-Phenylhexa-1,5-dien-3-ol (**3c**): ^{1}H NMR (400 MHz, $CDCl_3$): δ 2.09 (br s, 1H), 2.35-2.44 (m, 2H), 4.34 (br s, 1H), 5.14-5.19 (m, 2H), 5.79-5.90 (m, 1H), 6.23 (dd, *J* = 7.9, 15.9 Hz, 1H), 6.59 (d, J = 15.9 Hz, 1H), 7.21-7.37 (m, 5H).

tert-Butyl ((2*S*)-3-hydroxy-1-phenylhex-5-en-2-yl)carbamate (**3d**): $[\alpha]_D^{24}$ −21.2° ($CHCl_3$, *c* 1.02); ^{1}H NMR (500 MHz, $CDCl_3$): δ 1.35 (s, 9H), 2.20-2.30 (m, 2H), 2.83-2.94 (m, 2H), 3.59 (td, *J* = 1.5, 7.0 Hz, 1H), 3.73(br q, *J* = 8.0 Hz, 1H), 4.88 (d, *J* = 9.5 Hz, 1H), 5.10-5.15 (m, 2H), 5.70-5.80 (m, 1H), 7.18-7.32 (m, 5H).

Reference

Zhang, X. (2008). Mild and efficient allylation of aldehydes with allyltributylstannane promoted by MgI_2. $(OEt)_n$ etharate. *Synlett*, 65–68.

C-C bond_18

Type of reaction: C-C bond formation
Reaction conditions: Solvent-free, room temperature
Synthetic strategy: One-step allylation of ketones
Catalyst: Zinc-copper couple
Keywords: Ketones, allyl bromide, Zn-Cu couple, solvent-free, room temperature, allylation, Barbier-type reaction, homoallylic alcohols

General reaction scheme

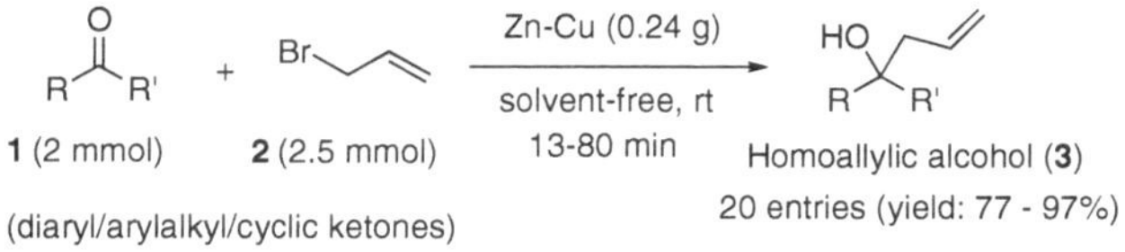

Representative entries

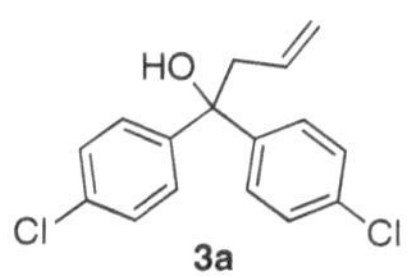

3a

1,1-bis(4-Chlorophenyl)but-3-en-1-ol (**3a**):
Reaction time: 15 min; yield: 96%

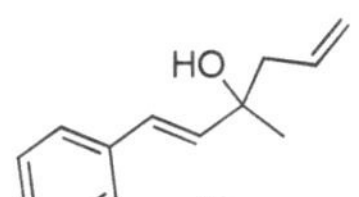

3b

(*E*)-3-Methyl-1-phenylhexa-1,5-dien-3-ol (**3b**):
Reaction time: 17 min; yield: 91%

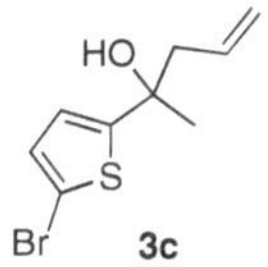

3c

2-(5-Bromothiophen-2-yl)pent-4-en-2-ol (**3c**):
Reaction time: 15 min; yield: 86%

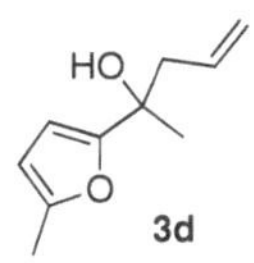

3d

2-(5-Methylfuran-2-yl)pent-4-en-2-ol (**3d**):
Reaction time: 14 min; yield: 83%

Experimental procedure

A mixture of zinc–copper couple (0.24 g), allyl bromide (**2**; 2.5 mmol) and ketone (**1**; 2 mmol) was stirred in a dried round-bottom flask at room temperature for stipulated time (13-80 min). On completion of the reaction (as monitored by TLC), saturated brine (10 mL) was poured into the mixture, and it was extracted with Et_2O (3×10 mL); the organic layer was separated, dried over anhydrous $MgSO_4$, filtered, and evaporated. The pure product of homoallylic alcohol **3** was obtained by flash chromatography over silica gel, and characterized by analytical analyses as well as by spectral studies including IR, ^{1}H NMR, ^{13}C NMR and MS.

Characterization data of two representative compounds

2-(5-Bromothiophen-2-yl)pent-4-en-2-ol (**3c**): IR: 3406, 3077, 2977, 1644, 1439, 1372, 1114, 923, 796 cm^{-1}; ^{1}H NMR (400 MHz, $CDCl_3$): δ 6.88 (d, J=3.6 Hz, 1H), 6.63 (d, J=3.6 Hz, 1H), 5.67-5.78 (m, 1H), 5.12-5.19 (m, 2H), 2.57-2.67 (m, 1H), 2.48-2.54 (m, 1H), 2.34 (s, 1H), 1.57 (s, 3H); ^{13}C NMR (100 MHz, $CDCl_3$): δ 154.6, 132.8, 129.5, 122.5, 120.2, 110.5, 73.0, 48.7, 30.1; EI-MS: *m/z* (rel.%) 248 (0.3) [M^+ + 2], 246 (0.2) [M^+], 231 (0.3) [M^+ – Me], 229 (0.3) [M^+ – OH], 207 (33.7), 43 (100), 39 (21.7). Anal. Calcd for $C_9H_{11}BrOS$: C, 43.74; H, 4.49; Found: C, 43.29; H, 3.93.

2-(5-Methylfuran-2-yl)pent-4-en-2-ol (**3d**): IR: 3411, 3077, 2981, 1684, 1373, 1097, 921, 786 cm^{-1}. ^{1}H NMR (400 MHz, $CDCl_3$): δ 6.05 (s, 1H), 5.87 (s, 1H), 5.68-5.72 (m, 1H), 5.11-5.15 (m, 2H), 2.64-2.69 (m, 1H), 2.51-2.56 (m, 1H), 2.28 (s, 3H), 2.14 (s, 1H), 1.51 (s, 3H); ^{13}C NMR (100 MHz, $CDCl_3$): δ 157.3, 151.2, 133.5, 119.0, 105.8, 105.3, 70.5, 46.0, 26.3, 13.5; EI-MS: *m/z* (rel.%) 166 (0.7) [M^+], 149 (1.7) [M^+ – OH], 135 (0.7), 125 (91.5), 109 (10.3), 95 (2.1), 43 (100), 39 (20.6). Anal. Calcd for $C_{10}H_{14}O_2$: C, 72.26; H, 8.49; Found: C, 72.28; H, 8.64.

Reference

Zhou, W., Yan, W., Wang, J.-X., and Wang, K. (2008). Efficient solvent-free synthesis of homoallylic alcohols mediated by zinc-copper couple. *Synlett*, 137–141.

C-C bond_19

Type of reaction: C-C bond formation
Reaction conditions: Dichloromethane, room temperature
Synthetic strategy: Lewis acid-mediated ring expansion
Catalyst: Titanium chloride ($TiCl_4$)
Keywords: Vinylidenecyclopropanes, $TiCl_4$, dichloromethane, room temperature, ring expansion, naphthalenacarbocycles

General reaction scheme

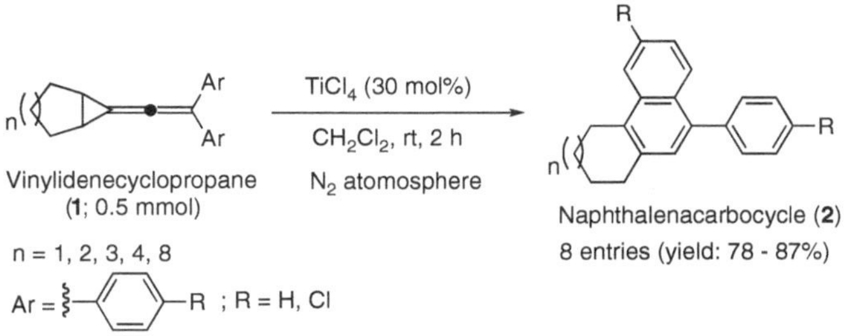

Representative entries

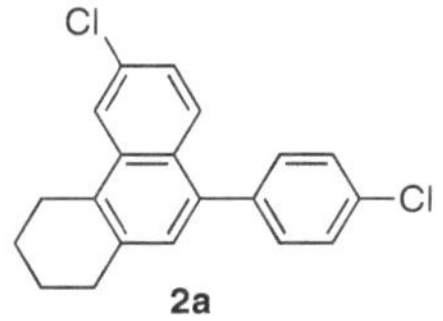

2a

6-Chloro-9-(4-chlorophenyl)-1,2,3,4-tetrahydrcphenanthrene (**2a**): Yield: 87%

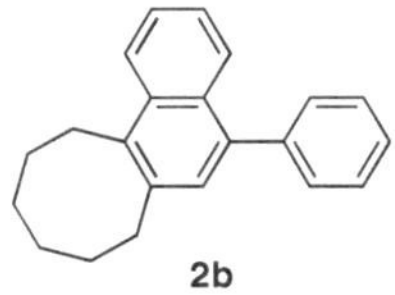

2b

5-Phenyl-7,8,9,10,11,12-hexahydrocycloocta[*a*]naphthalene (**2b**): Yield: 78%

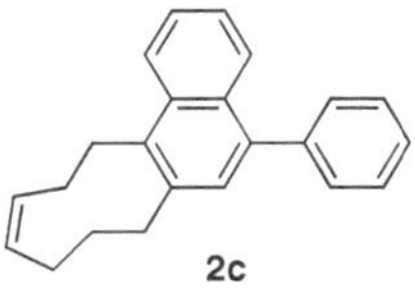

2c

(*Z*)-5-Phenyl-8,9,12,13-tetrahydro-7*H*-cyclonona[*a*]naphthalene (**2c**): Yield: 85%

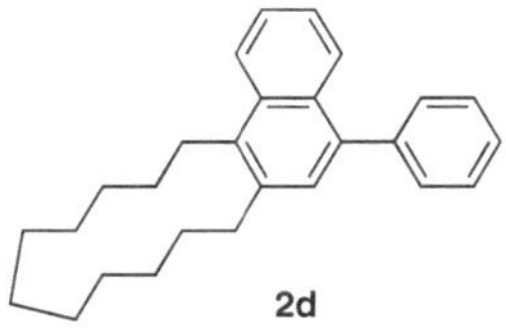

2d

5-Phenyl-8,9,10,11,12,13,14,15,16,17-decahydro-7*H*-cyclotrideca[*a*]naphthalene (**2d**): Yield: 82%

Experimental procedure

To a solution of bicyclic vinylidenecyclopropane (**1**; 0.5 mmol) in anhydrous dichloromethane (3 mL) under nitrogen atmosphere was added titanium chloride ($TiCl_4$) (28.4 mg, 0.15 mmol) under a nitrogen atmosphere. The mixture was then stirred at room temperature for 2 h to complete the reaction. On completion of the reaction, the resulting mixture was concentrated and the residue was purified by flash chromatography to afford medium- and large-size naphthalenacarbocycle **2** with 78-87% yield. The products were characterized on the basis of detailed spectral studies.

Characterization data of one representative compound

5-Phenyl-8,9,10,11,12,13,14,15,16,17-decahydro-7H-cyclotrideca[a]naphthalene (**2d**): IR (KBr): 2929, 2856, 1461, 1383, 762, 707 cm^{-1}; ^1H NMR (400 MHz, $CDCl_3$): δ 8.07 (d, $J = 8.0$ Hz, 1H), 7.87 (d, $J = 8.4$ Hz, 1H), 7.42-7.53 (m, 5H), 7.36-7.42 (m, 1H), 7.30-7.36 (m, 1H), 7.25 (s, 1H), 3.00-3.09 (m, 2H), 2.70-2.80 (m, 2H), 1.74-1.89 (m, 4H), 1.64-1.73 (m, 2H), 1.48-1.64 (m, 6H), 1.34-1.48 (m, 6H); ^{13}C NMR (100 MHz, $CDCl_3$): δ 23.2, 23.4, 24.5, 24.6, 26.4, 26.6, 26.9, 27.0, 27.1, 27.8, 32.3, 123.9, 124.5, 125.6, 126.6, 126.9, 128.1, 129.8, 130.2, 130.6, 132.8, 135.1, 137.4, 138.2, 141.1; MS (70 eV, EI): *m*/*z* 356 [M^+], 356 (100); HRMS (EI): *m*/*z* [M^+] Calcd for $C_{27}H_{32}$: 356.2504; Found: 356.2512.

Reference

Huang, X., Su, C., Liu, Q., and Song, Y. (2008). A facile access to medium- and large-size naphthalenacarbocycles via Lewis acid mediated ring-expansion reaction of bicyclic vinylidenecyclopropanes. *Synlett*, 229–232.

C-C bond_20

Type of reaction: C-C bond formation
Reaction conditions: THF-H_2O (9:1), room temperature
Synthetic strategy: One-step hydroxymethylation reaction

Catalyst: Scandium(III) fluoride (ScF_3)
Keywords: Dimethylsilyl enolates, formaldehyde, scandium(III) fluoride, THF-H_2O (9:1), room temperature, hydroxymethylation, β-hydroxy ketones

General reaction scheme

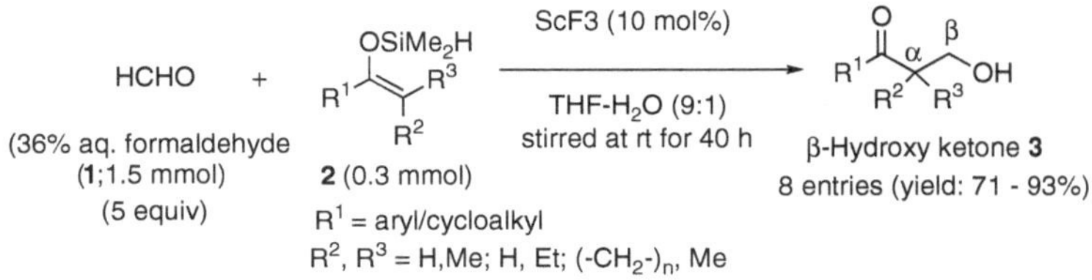

Representative entries

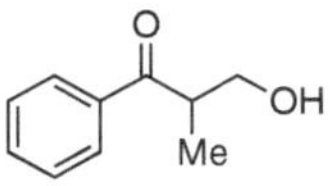

3a

3-Hydroxy-2-methyl-1-phenylpropane-1-one (**3a**): Yield: 89%

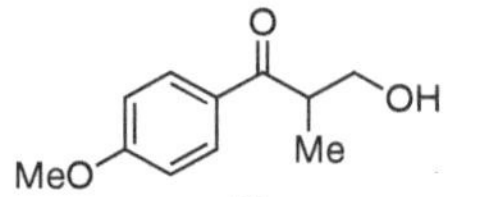

3b

3-Hydroxy-1-(4-methoxyphenyl)-2-methylpropan-1-one (**3b**): Yield: 78%

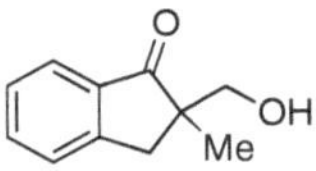

3c

2-(Hydroxymethyl)-2-methyl-2,3-dihydro-1H-inden-1-one (**3c**): Yield: 81%

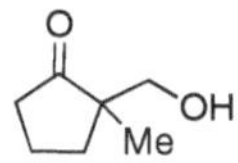

3d

2-(Hydroxymethyl)-2-methylcyclopentanone (**3d**): Yield: 89%

Experimental procedure

To a solution of scandium fluoride (0.03 mmol; 10 mol%) in THF-H_2O (0.22 mL) was added 36% aqueous solution of formaldehyde (**1**; 1.5 mmol; 5 equiv.) and dimethylsilyl enolate (**2**; 0.3 mmol). The mixture was stirred for 40 h at room temperature to complete the reaction. The mixture was then diluted with water (10 mL), and the aqueous layer was extracted with dichloromethane (3 × 20 mL). The combined organic extracts were washed with brine, and dried over anhyd. sodium sulfate. After filtration, the solvent was removed in *vacuo*. The crude residue was purified by preparative TLC (elution with *n*-hexane-EtOAc) to furnish β-hydroxy ketone **3** with good yield (71-93%). All the compounds were characterized from comparison of their physical and spectral properties with those reported for authentic samples in literature.

Reference

Kokubo, M., and Kobayashi, S. (2008). Scandium(III) fluoride as a novel catalyst for hydroxymethylation of dimethylsilyl enolates in aqueous media. *Synlett*, 1562–1564.

C-C bond_21

Type of reaction: C-C bond formation
Reaction conditions: *N,N*-Dimethylformamide (DMF), room temperature

Synthetic strategy: Mukaiyama-type Mannich reaction
Catalyst: Lithium chloride (LiCl)
Keywords: Arylaldemines, trimethylsilyl ketene acetal, lithium chloride, DMF, room temperature, Mukaiyama-type Mannich reaction, β-amino esters

General reaction scheme

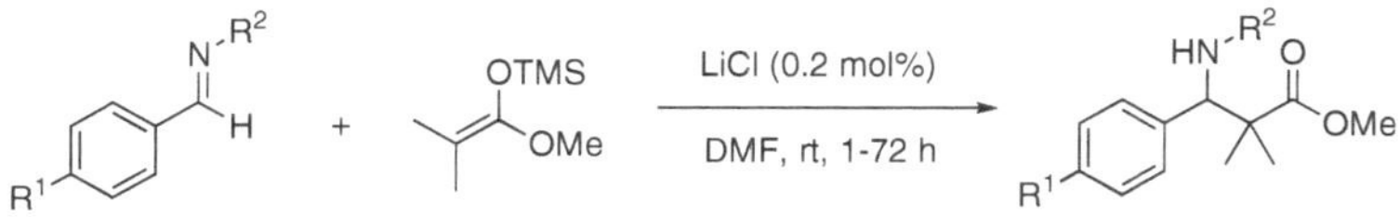

(**1**; 1 equiv) **2** (1.5 equiv) β-Amino ester **3**

11 entries (yield: 58 - 95%)

R^1 = H, Cl, MeO, NO_2, AcO, TBDMSO, MeS
R^2 = Ts, SO_2Ph, Bz, Boc

Representative entries

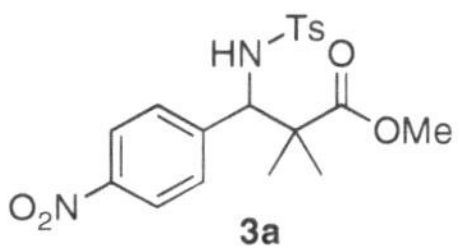

3a

Methyl 2,2-dimethyl-3-(4-methylphenylsulfonamido)-3-(4-nitrophenyl) propanoate (**3a**): Reaction time: 1 h; yield: 95%

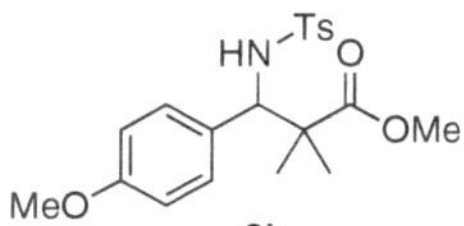

3b

Methyl 3-(4-methoxyphenyl)-2,2-dimethyl-3-(4-methylphenylsulfonamido)propanoate (**3b**): Reaction time: 72 h; yield: 76%

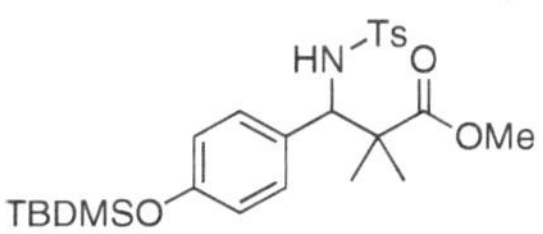

3c

Methyl 3-(4-((*tert*-butyldimethylsilyl)oxy)phenyl)-2,2-dimethyl-3-(4-methylphenylsulfonamido) propanoate (**3c**): Reaction time: 24 h; yield: 70%

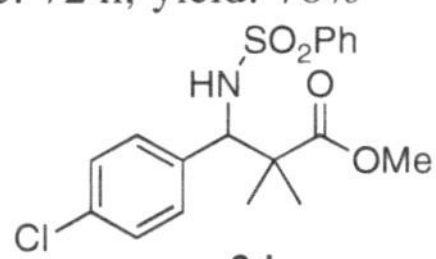

3d

Methyl 3-(4-chlorophenyl)-2,2-dimethyl-3-(phenylsulfonamido)propanoate (**3d**): Reaction time: 12 h; yield: 93%

Experimental procedure

To a stirred solution of arylaldimine (**1**; 1 equiv) and lithium chloride (0.2 equiv) in DMF was added trimethylsilyl ketene acetal (**2**; 1.5 equiv) at room temperature under nitrogen atmosphere, and stirring was continued for requisite time (1-72 h). On completion of the reaction, it was quenched by adding water. The product was extracted with ethyl acetate and the combined organic layer was washed with water and brine, and then evaporated to dryness. The crude product on column chromatographic resolution (EtOAc–*n*-hexane) afforded pure β-amino ester **3** with 58-95% yield. All the compounds were characterized from comparison of their physical and spectral properties with those reported for authentic samples in literature.

Reference

Hagiwara, H., Iijima, D., Awen, B. Z. S., Hoshi, T., and Suzuki, T. (2008). Expedient Mukaiyama-type Mannich reaction catalyzed by lithium chloride. *Synlett*, 1520–1522.

C-C bond_22

Type of reaction: C-C bond formation
Reaction conditions: Tetrhydrofuran (THF), room temperature
Synthetic strategy: Nucleophilic addition reaction
Catalyst: Tetrabutylammonium difluorotriphenylsilicate (TBAT)
Keywords: *N*-Phosphinoyl benzaldimines, trimethyl(trichloromethyl)silane, tetrhydrofuran, room temperature, nucleophilic addition reaction, *N*-phosphinoyl-α-(trichloromethyl) benzylamines

General reaction scheme

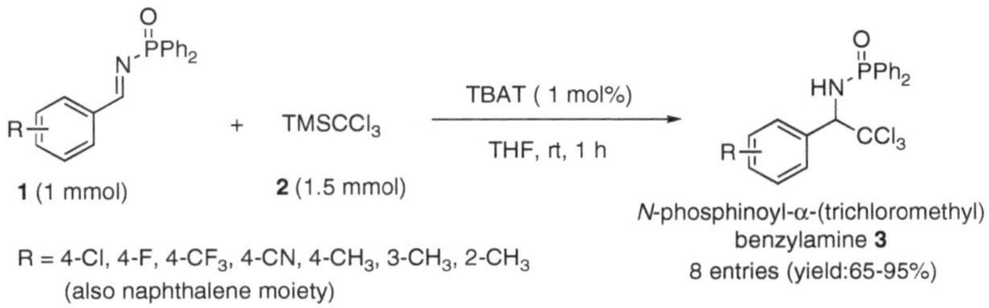

Representative entries

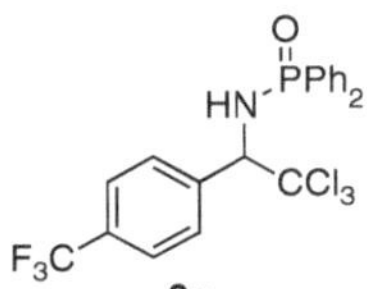

3a

(*P*,*P*-Diphenyl-*N*-(2,2,2-trichloro-1-(4-(trifluoromethyl)phenyl)ethyl)phosphinic amide (**3a**): White solid, mp 201-206 °C; yield: 68%

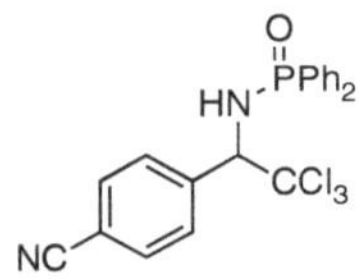

3b

(*P*,*P*-Diphenyl-*N*-(2,2,2-trichloro-1-(4-cyanophenyl)ethyl)phosphinic amide (**3b**): White solid, mp 220-223 °C; yield: 72%

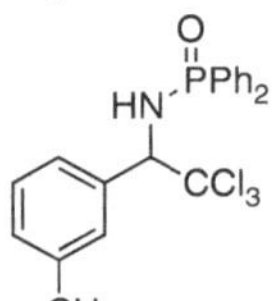

3c

P,*P*-Diphenyl-*N*-(2,2,2-trichloro-1-(*m*-tolyl)ethyl) phosphinic amide (**3c**): White solid, mp 210-213 °C; yield: 72%

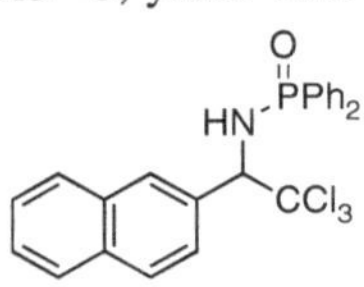

3d

P,*P*-Diphenyl-*N*-(2,2,2-trichloro-1-(naphthalen-2-yl)ethyl)phosphinic amide (**3d**): White solid, mp 259-263 °C; yield: 65%

Experimental procedure

To a mixture of *N*-Phosphinoyl benzaldimines (**1**; 1 mmol) and $TMSCCl_3$ (**2**; 1.5 mmol) in THF (3 mL) was added TBAT (1 mol%, 5.4 mg) in THF (1 mL), and the resulting mixture was stirred at room temperature for 1 h. Then, a saturated solution of ammonium chloride (1 mL) was added and the quenched reaction mixture was extracted with ethyl acetate. The combined organic layers were dried over anhydrous magnesium sulfate. Evaporation of the solvent afforded the crude product which was purified by flash chromatography (95:5 CH_2Cl_2:methanol) to furnish pure *N*-phosphinoyl-α-(trichloromethyl)benzylamine **3** in good yield. Each of the products was characterized by means of analytical and spectral methods.

Characterization data of two representative entries

(*P*,*P*-Diphenyl-*N*-(2,2,2-trichloro-1-(4-(trifluoromethyl)phenyl)ethyl)phosphinic amide (**3a**): White solid, m.p. 201-206 °C; IR ($CHCl_3$): 3020, 1216, 761, 670 cm^{-1}; ^1H NMR (270 MHz, $CDCl_3$): δ 7.87-7.79 (m, 2H, *Ar*), 7.66-7.43 (m, 10H, *Ar*), 7.34-7.26 (m, 2H, *Ar*), 5.03 (dd, $J = 11.1$, 9.6 Hz, 1H, CH), 4.17 (dd, $J = 11.1$, 8.1 Hz, 1H, NH); ^{13}C NMR (101 MHz, $CDCl_3$): δ 140.6, 132.6, 132.3, 132.1, 130.7, 129.7, 128.6, 128.5, 125.0 (q, CF_3), 101.9, 69.4; ^{31}P NMR (162 MHz, $CDCl_3$): δ 23.5; ^{19}F NMR (377 MHz, $CDCl_3$): δ -62.8; HRMS (EI^+): Calcd. for $C_{21}H_{16}Cl_3F_3NOP$ $[M+H]^+$ *m*/*z* 492.0060; Found 492.0054. Anal. Calc. for $C_{21}H_{17}Cl_3F_3NOP$: C, 51.19; H, 3.27; N, 2.84; Found: C, 51.78; H, 3.29; N, 2.63.

P,*P*-Diphenyl-*N*-(2,2,2-trichloro-1-(naphthalen-2-yl)ethyl)phosphinic amide (**3d**): White solid, m.p. 259-263 °C; IR ($CHCl_3$): 3020, 1216, 775, 746, 670 cm^{-1}; ^1H NMR (270 MHz, $CDCl_3$): δ 7.90 -7.75 (m, 6H, *Ar*), 7.68 -7.58 (m, 2H, *Ar*), 7.58-7.42 (m, 6H, *Ar*), 7.41-7.34 (m, 1H, *Ar*), 7.26-7.18 (m, 2H, *Ar*), 5.13 (dd, $J = 10.9$, 10.2 Hz, 1H, CH), 4.17 (dd, $J = 10.9$, 8.5 Hz, 1H, NH); ^{13}C NMR (101 MHz, $CDCl_3$): δ 134.1, 133.4, 132.6, 132.4, 132.3, 132.1, 129.3, 128.6, 128.3, 127.8, 126.8, 126.4, 125.8, 102.8, 70.2; ^{31}P NMR (162 MHz, $CDCl_3$): δ 23.3; HRMS (EI^+): calcd. for $C_{24}H_{20}Cl_3NOP$ $[M+H]^+$ *m*/*z* 474.0343; Found 474.0348. Anal. Calc. for $C_{24}H_{20}Cl_3NOP$: C, 60.72; H, 4.03; N, 2.95; Found: C, 60.08; H, 3.92; N, 2.82.

Reference

Wahl, B., Cabré, A., Woodward, S., and Lewis, W. (2014). Nucleophilic addition of $TMSCCl_3$ to *N*-phosphinoyl benzaldimines: a route to *N*-phosphinoyl-α-(trichloromethyl)benzylamines. *Tetrahedron Lett.*, **55**, 5829–5831.

C-C bond_23

Type of reaction: C-C bond formation
Reaction conditions: *N*,*N*-Dimethyformamide, room temperature
Synthetic strategy: Organocatalytic Michael reaction
Catalyst: 2-Dicyclohexylphosphino-2,4,6-triisopropylbiphenyl (XPhos, a phosphine derivative, **I**)

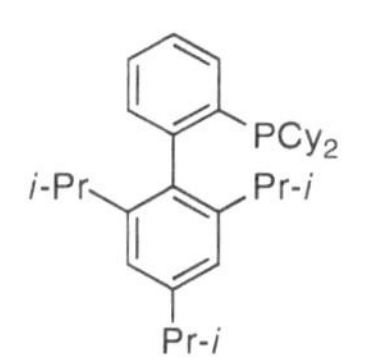

(XPhos, **I**)

Keywords: α-Trifluoromethylated esters, vinyl ketones, XPhos (**I**), organocatalyst, *N*,*N*-dimethyformamide, room temperature, Michael reaction, Michael adducts

General reaction scheme

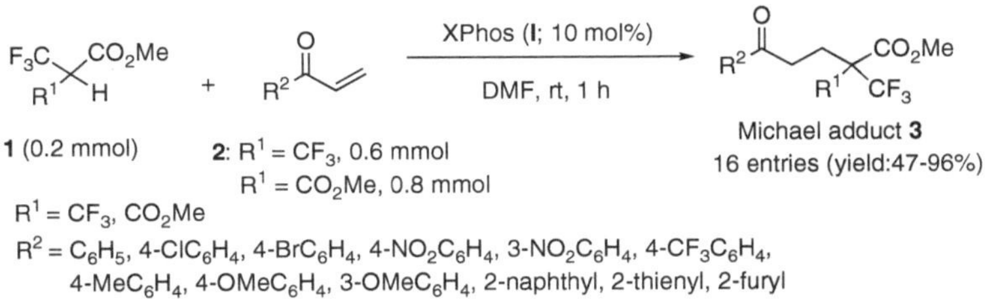

Representative entries

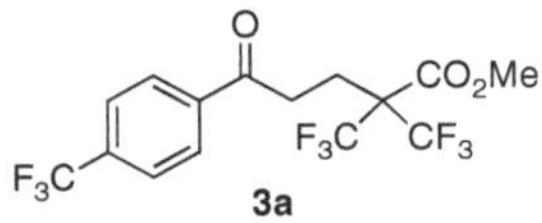

3a

Methyl 5-oxo-2,2-bis(trifluoromethyl)-5-(4-(trifluoromethyl)phenyl)pentanoate (**3a**): White solid, mp 71.8 °C; yield: 72%

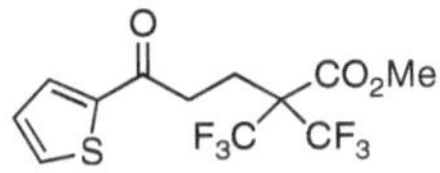

3b

Methyl 5-oxo-5-(thiophen-2-yl)-2,2-bis(trifluoromethyl)pentanoate (**3b**): Colorless oil; yield: 72%

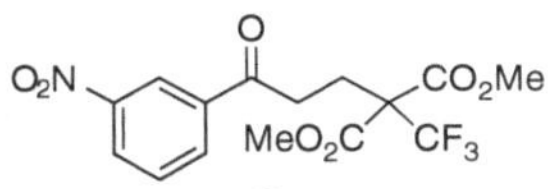

3c

Dimethyl 2-(3-(3-nitrophenyl)-3-oxopropyl)-2-(trifluoromethyl)malonate (**3c**): White solid, yield: 88%

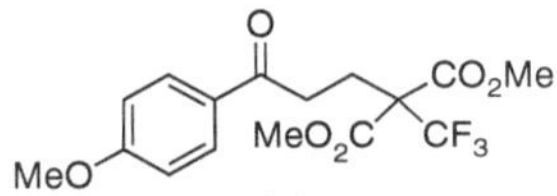

3d

Dimethyl 2-(3-(4-methoxyphenyl)-3-oxopropyl)-2-(trifluoromethyl)malonate (**3d**): White solid; yield: 47%

Experimental procedure

To a solution of XPhos (**I**; 0.01 mmol) in DMF (0.4 mL) were added α-trifluoromethylated ester (**1**; 0.2 mmol) and α,β-unsaturated ketone (**2**; 0.6 or 0.8 mmol), and the mixture was then stirred at room temperature for 1 h (monitored by ^{19}F NMR spectroscopy). Upon completion of the reaction, it was diluted with water and extracted with ethyl acetate. The combined organic extract was washed with brine, dried over sodium sulfate, and then concentrated in *vacuo*. The residue was purified by flash column chromatography on silica gel (petroleum ether/EtOAc) to yield the desired product **3**. All the products were characterized with the help of detailed spectral measurements.

Characterization data of two representative entries

Methyl 5-oxo-2,2-bis(trifluoromethyl)-5-(4-(trifluoromethyl)phenyl)pentanoate (**3a**): White solid, mp 71.8 °C; IR (KBr): 2923.9, 2853.5, 1755.4, 1698.7, 1490.3, 1195.3, 1081.8, 856.7, 775.6, 542.9 cm^{-1}; ^{1}H NMR (300 MHz, $CDCl_3$): δ 8.06 (d, J=8.1 Hz, 2H), 7.75 (d, J=8.1 Hz, 2H), 3.92 (s, 3H), 3.17 (t, J=7.8 Hz, 2H), 2.66 (t, J=7.8 Hz, 2H); ^{19}F NMR (282 MHz, $CDCl_3$): δ −63.2 (s, 3 F), −66.2 (s, 6 F); ^{13}C NMR (100 MHz, $CDCl_3$): δ 196.1, 162.6, 145.4, 138.8, 128.4, 125.8 (q, J=3.7 Hz), 122.6 (q, J=277.3 Hz), 122.1 (q, J=280.2 Hz), 60.2 (m), 54.0, 33.2, 22.5; MS (EI) *m/z* 410 $[M]^+$; HRMS (EI-TOF-MS) calcd for $C_{15}H_{11}F_9O_3$ [M] 410.0564; Found 410.0567.

Dimethyl 2-(3-(4-methoxyphenyl)-3-oxopropyl)-2-(trifluoromethyl)malonate (**3d**): White solid; IR (KBr): 2959.3, 2952.8, 1748.4, 1680.1, 1602.0, 1511.8, 1421.1, 1260.3, 1173.5, 1112.3, 1027.5 cm^{-1}; ^{1}H NMR (300 MHz, $CDCl_3$): δ 7.93 (d, $J=6.6$ Hz, 2H), 6.93 (d, $J=6.6$ Hz, 2H), 3.86 (s, 3H), 3.83 (s, 6H), 3.12 (t, $J=6$ Hz, 2H), 2.61 (t, $J=6$ Hz, 2H); ^{19}F NMR (282 MHz, $CDCl_3$): δ −67.1 (s); ^{13}C NMR (100 MHz, $CDCl_3$): δ 196.2, 165.2, 163.6, 130.2, 129.4, 123.4 (q, $J=287.0$ Hz), 113.7, 62.6 (q, $J=26.0$ Hz), 55.4, 53.5, 33.2, 25.2; MS (ESI) *m/z* 363.0 $(M+1)^+$; HRMS (ESI-FTICR-MS) calcd for $C_{16}H_{17}F_3O_6Na$ $[M+Na]^+$ 385.0869; Found 385.0880.

Reference

Wang, Q., Huan, F., Shen, H., Xiao, J.-C., Gao, M., Yang, X., Murahashi, S.-I., Chen, Q.-Y., and Guo, Y. (2013). Organocatalytic reactions of α-trifluoromethylated esters with terminal alkenes at room temperature. *J. Org. Chem.*, **78**, 12525–12531.

C-C bond_24

Type of reaction: C-C bond formation
Reaction conditions: Dichloromethane, room temperature
Synthetic strategy: Oxidative direct *ortho*-C-H ethoxycarboxylation of anilides
Catalyst: $Pd(TFA)_2$/ $(NH_4)_2S_2O_8$/ *p*-TsOH.H_2O/ $Cu(TFA)_2.H_2O$
Keywords: Anilides, diethyl azodicarboxylate (DEAD), $Pd(TFA)_2$/ $(NH_4)_2S_2O_8$/ *p*-TsOH. H_2O/ $Cu(TFA)_2.H_2O$, dichloromethane, room temperature, oxidative direct *ortho*-C-H ethoxycarboxylation, substituted anthranilic acid derivatives

General reaction scheme

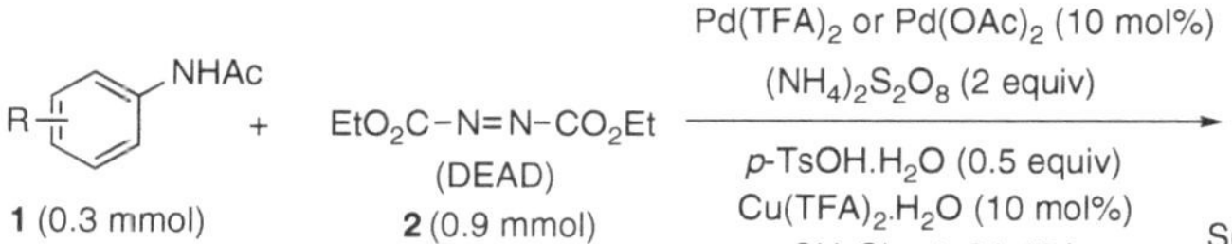

1 (0.3 mmol) **2** (0.9 mmol)

Substituted anthranilic acid **3**
15 entries (yield: 42-83%)

R = H, 3-Me, 4-Me, 4-^{t}Bu, 4-OMe, 4-OBu^n, 4-OBn, 4-C_6H_5, 3-F, 4-F, 3-Cl, 3,4-di-Me, (3-Cl,4-OMe), (3-Br,4-Me), 4-CH_2CO_2Et

Representative entries

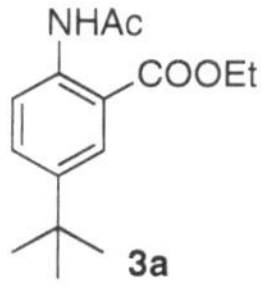

3a

Ethyl 2-acetamido-5-(*tert*-butyl)benzoate (**3a**): White solid, mp. 88-89 °C; yield: 73%

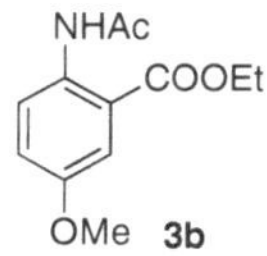

3b

Ethyl 2-acetamido-5-methoxybenzoate (**3b**): White solid, mp. 88-91 °C; yield: 62%

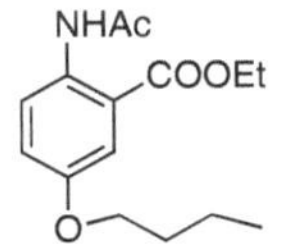

3c

Ethyl 2-acetamido-5-butoxybenzoate (**3c**): White solid, mp. 68-70 °C; yield: 60%

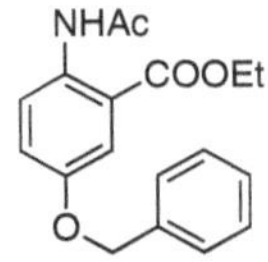

3d

Ethyl 2-acetamido-5-(benzyloxy)benzoate (**3d**): White solid, mp. 88-90 °C; yield: 52%

Experimental procedure

A Schlenk tube with a magnetic stir bar was charged with $Pd(TFA)_2$ or $Pd(OAc)_2$, ammonium persulfate (136.9 mg, 0.6 mmol) or Selectfluor (212.6 mg, 0.6 mmol), acetanilide (**1**; 0.3 mmol), $Cu(TFA)_2 \cdot H_2O$ (8.7 mg, 0.03 mmol), *p*-TsOH$\cdot H_2O$ (28.5 mg, 0.15 mmol) and dichloromethane (1 mL). After stirring the reaction mixture for 1 min at room temperature, DEAD (**2**; 156.7 mg, 0.9 mmol) was added, and The resulting mixture was stirred at room temperature for stipulated time (24-48 h). The solvent was evaporated and the residue was diluted with 30 mL dichloromethane, filtered through a Celite pad, and washed with dichloromethane (10-20 mL). The combined organic phases were concentrated and the resulting residue was purified by column chromatography on silica gel to provide the desired product **3**. All the products were characterized by means of spectral studies.

Characterization data of two representative entries

Ethyl 2-acetamido-5-methoxybenzoate (**3b**): White solid, mp. 88-91 °C; ^{1}H NMR (400 MHz, $CDCl_3$): δ 1.41 (t, $J=7.2$ Hz, 3H), 2.21 (s, 3H), 3.81 (s, 3H), 4.38 (q, $J=7.2$ Hz, 2H), 7.11 (dd, $J=2.8$ Hz, 9.2 Hz, 1H), 7.52 (d, $J=2.8$ Hz, 1H), 8.61 (d, $J=9.2$ Hz, 1H), 10.80 (s, 1H); ^{13}C NMR (100 MHz, $CDCl_3$): δ 14.3, 25.4, 55.7, 61.6, 114.9, 116.2, 120.6, 121.9, 135.3, 154.4, 168.0, 168.7; HRMS (ESI$^+$): Calcd for $C_{12}H_{15}NO_4Na$ $[M+Na]^+$ 260.0899; Found 260.0897.

Ethyl 2-acetamido-5-(benzyloxy)benzoate (**3d**): White solid, mp. 88-90 °C; ^{1}H NMR (400 MHz, $CDCl_3$): δ 1.41 (t, $J=7.2$ Hz, 3H), 2.21 (s, 3H), 4.37 (q, $J=7.2$ Hz, 2H), 5.06 (s, 2H), 7.17 (dd, $J=2.8$ Hz, 8.8 Hz, 1H), 7.33-7.44 (m, 5H), 7.62 (d, $J=2.8$ Hz, 1H), 8.61 (d, $J=9.2$ Hz, 1H), 10.82 (s, 1H); ^{13}C NMR (100 MHz, $CDCl_3$): δ 14.3, 25.5, 61.6, 70.6, 116.3, 116.4, 121.4, 122.0, 127.7, 128.3, 128.8, 135.6, 136.7, 153.6, 168.0, 168.8; HRMS (ESI$^+$): Calcd for $C_{18}H_{19}NO_4Na$ $[M+Na]^+$ 336.1212; Found 336.1208.

Reference

Huang, Y., Li, G., Huang, J., and You, J. (2014). Palladium-catalyzed direct *ortho*-C–H ethoxycarboxylation of anilides at room temperature. *Org. Chem. Front.*, **1**, 347–350.

C-C bond_25

Type of reaction: C-C bond formation
Reaction conditions: Solvent-free, room temperature
Synthetic strategy: One-pot multicomponent
Catalyst: Mg-Al hydrotalcite (Mg-Al HT)

Keywords: Aromatic aldehydes, malononitrile, nitromethane, solvent-free, room temperature, Mg-A HT, heterogeneous catalysis, condensation, Michael addition, cyclohexenes

General reaction scheme

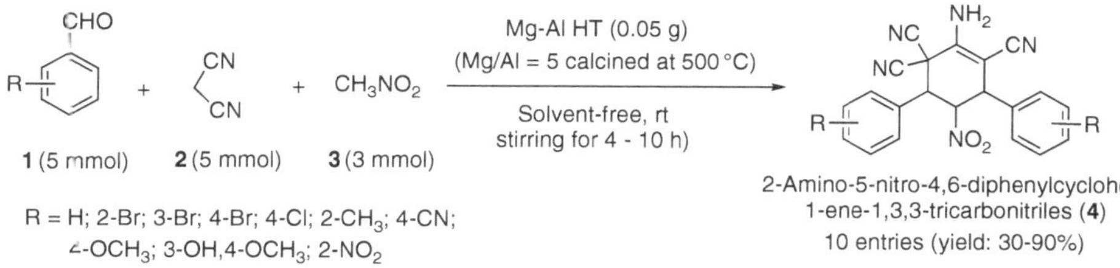

Representative entries

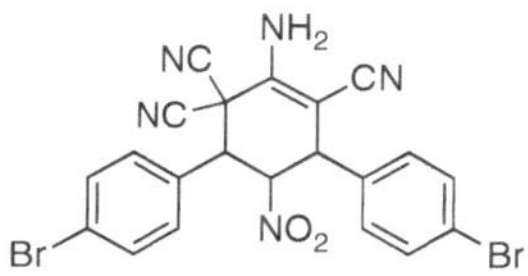

4a

5′-Amino-4,4″-dibromo-2′-nitro-2′,3′-dihydro-[1,1′:3′,1″-terphenyl]-4′,4′,6′(1′*H*)-tricarbonitrile (**4a**): Solid, mp 225-227 °C; reaction time: 4 h; yield: 90%

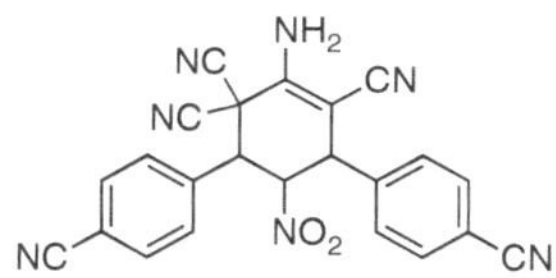

4b

5′-Amino-2′-nitro-2′,3′-dihydro-[1,1′:3′,1′-terphenyl]-4,4′,4′,4″,6′(1′*H*)-pentacarbonitrile (**4b**): Solid, mp 230-234 °C; reaction time: 5 h; yield: 81%

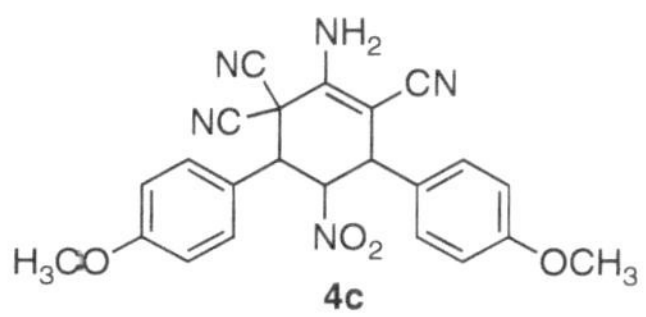

4c

5′-Amino-4,4″-dimethoxy-2′-nitro-2′,3′-dihydro-[1,1′:3′,1″-terphenyl]-4′,4′,6′(1′*H*)-tricarbonitrile (**4c**): Solid, mp 243-244 °C; reaction time: 7 h; yield: 72%)

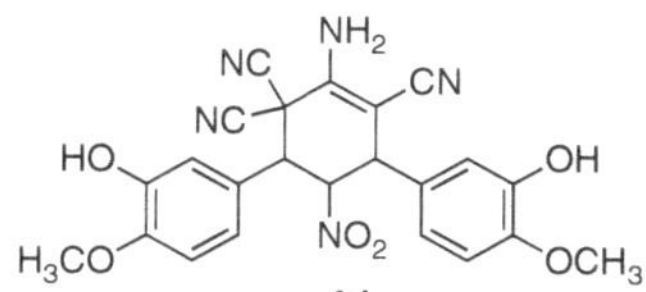

4d

5′-Amino-3,3″-dihydroxy-4,4″-dimethoxy-2′-nitro-2′,3′-dihydro-[1,1′:3′,1″-terphenyl]-4′,4′,6′(1′*H*)-tricarbonitrile (**4d**): Solid, mp 225-226 °C; reaction time: 10 h; yield: 30%

Experimental procedure

In a 50 mL round-bottomed flask a thorough mixture of aldehyde (**1**; 5 mmol), malononitrile (**2**, 5 mmol), and the catalyst Mg-Al HT [(Mg/Al = 5 calcined at 500 °C] (0.05 g) was stirred at room temperature under solvent-free conditions till the consumption of aldehyde (as monitored by TLC). After then nitromethane (**3**, 3 mmol) was added and the reaction mixture was stirred again at room temperature (29 °C) to complete the reaction (TLC). To the resulting mixture chloroform (20 mL) was added to dissolve the organic entities, and the catalyst was filtered off and washed with chloroform (10 mL). The filtrate was concentrated under reduced pressure and the crude product was then subjected to column chromatography on silica gel (60-120 mesh) using petroleum ether/ethyl acetate (10:9) as eluent to obtain the pure product **4**. All the products were characterized from their spectral studies including FT-IR and ^{1}H NMR.

Characterization data of two representative compounds

5′-Amino-4,4″-dibromo-2′-nitro-2′,3′-dihydro-[1,1′:3′,1″-terphenyl]-4′,4′,6′(1′*H*)-tricarbonitrile (**4a**): Solid, mp 225-227 °C; FT-IR (KBr): 3453, 3313, 3212, 2225, 1653, 1562, 1362 cm^{-1}; ^1H NMR (DMSO-d_6, 400 MHz): δ 4.34 (d, $J=10$ Hz, 1H), 4.97 (d, $J=12$ Hz, 1H), 5.83 (dd, $J=10$ & 12 Hz, 1H), 7.31-7.72 (m, 10H).

5′-Amino-2′-nitro-2′,3′-dihydro-[1,1′:3′,1″-terphenyl]-4,4′,4′,4″,6′(1′*H*)-pentacarbonitrile (**4b**): Solid, mp 230-234 °C; FT-IR (KBr): 3449, 3345, 3215, 2208, 1645, 1557, 1362 cm^{-1}; ^1H NMR (DMSO-d_6, 400 MHz): δ 4.33 (d, $J=10$ Hz, 1H), 4.95 (d, $J=12$ Hz, 1H), 6.04 (br s, 1H), 7.31-7.69 (m, 10H).

Reference

Kshirsagar, S. W., Patil, N. R., and Samant, S. D. (2010). One-pot synthesis of 2-amino-5-nitro-4,6-diarylcyclohex-1-ene-1,3,3-tricarbonitriles by condensation of aldehyde, malononitrile, and nitromethane in the presence of Mg-Al HT under solvent-free condition. *Tetrahedron Lett.*, **51**, 2924–2927.

C-C bond_26

Type of reaction: C-C bond formation
Reaction conditions: Nitromethane, room temperature
Synthetic strategy: One-pot cyclisation
Catalyst: Tin(IV) triflimidate [$Sn(NTf_3)_4$]
Keywords: *gem*-(dialkoxymethyl)-1,6-dienes, nitromethane, tin(IV) triflimidate, room temperature, 6-*enoxo*-*endo*-trig cycloisomerisation, six-membered carbocyclic isomers

Reaction scheme

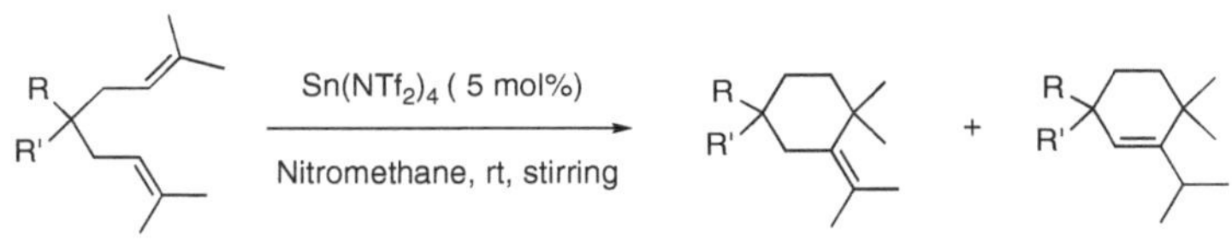

gem-(dialkoxymethyl)-1,6-diene (**1**) carbocyclic isomer **2** carbocyclic isomer **3**

Entry	R	R′	Time	Yield (GC-FID)	Isomeric ratio (**2**:**3**)
1	MeO	MeO	24 h	86%	1:1.7
2	EtO	EtO	24 h	74%	1:2.1
3	AcO	AcO	72 h	82%	10:1
4	MeO	H	6 h	80%	5:3

Experimental procedure

$Sn(NTf_2)_4$.6DMSO (0.015 mmol, 26 mg) was placed in a Schlenk tube closed with a septum in a nitrogen atmosphere. Nitromethane (0.5 mL), previously dried over $CaCl_2$, was added with a syringe and the mixture stirred at room temperature until dissolution of the catalyst. Diene (**1**; 0.3 mmol) in nitromethane (0.1 mL) was added to the solution with a syringe. The reaction progress was monitored by GC-FID analysis. After completion, the reaction mixture was filtrated over a pad of silica gel and eluted with diethyl ether (10 mL). After solvent removal, the crude mixture was subjected to column chromatography over silica gel with a gradient Et_2O/pentane 1:9 to 1:1 as eluent to afford the mixture **2** and **3**, followed by characterization by spectral means.

Characterization data of one representative entry

Mixture of isomers, 4,4-dimethoxy-1,1-dimethyl-2-(propan-2-ylidene)cyclohexane (**2a**) and 1-isopropyl-3,3-dimethoxy-6,6-dimethylcyclohex-1-ene (**3a**), obtained from 5,5-dimethoxy-2,8-dimethylnona-2,7-diene (**1a**):

4,4-Dimethoxy-1,1-dimethyl-2-(propan-2-ylidene)cyclohexane (**2a**): ^{1}H NMR ($CDCl_3$, 200 MHz, 20 °C): δ 4.76 (s, 1H), 4.55 (s, 1H), 3.27 (s, 2H), 3.26 (s, 6H), 3.08 (s, 2H), 1.90 (m, 1H), 1.65 (s, 3H), 1.55-1.00 (m, 6H), 0.82 (s, 3H), 0.78 (s, 3H); ^{13}C NMR ($CDCl_3$, 50 MHz, 20 °C): δ 146.3, 111.6, 79.1, 71.7, 58.4, 47.4, 36.7, 32.7, 38.1, 30.6, 24.4, 30.2, 22.9, 19.2; MS (EI, 70 eV, %rel. intensity): *m/z* 240(0) [M^+], 208(74), 193(5), 176(5), 163(100), 133(16), 121 (30), 107(95), 93(76), 79(58), 71(65); HRMS calcd for $C_{15}H_{28}O_2$: 240.2089; Found: 240.2091.

1-Isopropyl-3,3-dimethoxy-6,6-dimethylcyclohex-1-ene (**3a**): ^{1}H NMR ($CDCl_3$, 200 MHz, 20 °C): δ 5.20 (s, 1H), 3.27 (s, 6H), 3.25-3.00 (m, 4H), 2.24 (hept, $J=6.8$ Hz, 1H), 1.75-1.05 (m, 4H), 0.93 (d, $J=6.8$ Hz, 6H), 0.93 (s, 6H); ^{13}C NMR ($CDCl_3$, 50 MHz, 20 °C): δ 153.5, 119.9, 75.4, 58.4, 39.6, 34.4, 22.8, 34.3, 26.9, 26.8, 24.2; MS (EI, 70 eV, %rel. intensity): *m/z* 240(1) [M^+], 208(1), 195(100), 163(24), 133(20), 121(92), 111(23), 107(87), 93(87), 79 (32), 69(22); HRMS calcd for $C_{15}H_{28}O_2$: 240.2089; Found: 240.2091.

Reference

Vece, V., Hadj Hassen, K. B., Antoniotti, S., and Duñach, E. (2012). C–O and C–C bond formation in the cyclisation of *gem*-(dialkoxymethyl)-1,6-dienes catalysed by tin(IV) triflimidate at room temperature. *Tetrahedron Lett.*, **53**, 5102–5105.

C-C bond_27

Type of reaction: C-C bond formation
Reaction conditions: Acetonitrile, room temperature
Synthetic strategy: One-pot C-H activation
Catalyst: Ceric ammonium nitrate (CAN)
Keywords: Ferrocenyl alcohol, nucleophiles, acetonitrile, room temperature, C-H activation, substitution reaction, ferrocene functionalities

Reaction scheme

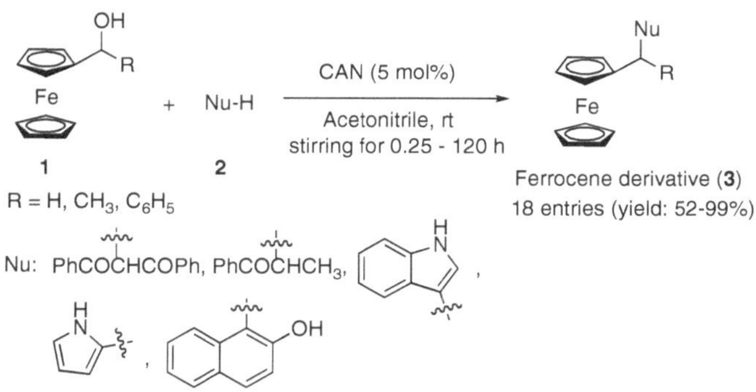

Representative entries

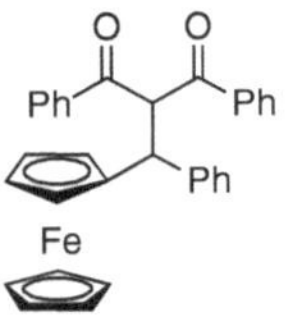

3a

1,3-Diphenyl-2-[phenyl (ferrocenyl) methyl] propane-1,3-dione (**3a**): Orange solid; mp 215-216.2°C; reaction time: 1 h; yield: 87%

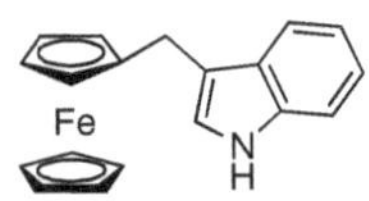

3b

(1*H*-Indol-3-yl)ferrocenylmethane (**3b**): Orange solid; mp 120.8-121.8°C; reaction time: 0.5 h; yield: 66%

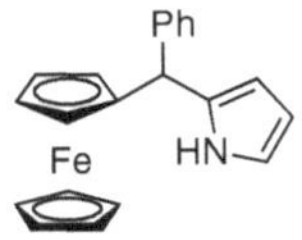

3c

2-[Phenyl(1-ferrocenyl)methyl]-1*H*-pyrrole (**3c**): Orange oil; reaction time: 0.25 h; yield: 88%)

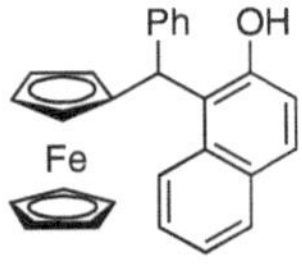

3d

1-[Phenyl(ferrocenyl)methyl]-naphthalen-2-ol (**3d**): Orange solid; mp 159.2-163.1°C; reaction time: 2 h; yield: 56%

Experimental procedure

A mixture of ferrocenyl alcohol (**1**; 0.5 mmol), nucleophile with activated hydrogen atom (**2**; 0.5 mmol) and ceric ammonium nitrate (CAN; 5 mol %) in acetonitrile (2 mL) was stirred at room temperature for an appropriate time (0.25-120 h for varying substrates). Upon completion, monitored by TLC, the solvent was evaporated under the reduced pressure. The crude product was purified by flash column chromatography with ethyl acetate and petroleum ether as eluents to afford ferrocene derivative **3** with good yield (52-99%). Each of the products was characterized by means of spectral studies including IR, ^{1}H NMR, ^{13}C NMR and HRMS.

Characterization data of representative entries

(1*H*-Indol-3-yl)ferrocenylmethane (**3b**): Orange solid, mp 120.8-121.8°C; IR (KBr): 3397, 3057, 1454, 1420, 1104 cm^{-1}; ^{1}H NMR (400 MHz, $CDCl_3$): δ 3.93 (s, 2H, CH_2), 4.06-4.17 (m, 9H, Fc–H), 6.57-7.42 (m, 4H, In–H), 7.64 (d, J=8.4 Hz, 1H, In–H), 7.89 (s, 1H, NH); ^{13}C NMR (100 MHz, DMSO-d_6): δ 141.6, 132.4, 127.9, 126.2, 123.9, 123.6, 120.4, 116.7, 94.3, 73.8, 72.3, 30.6; HRMS (*m/z*): $[M]^+$, calcd for $C_{19}H_{17}NFe$: 315.0710; Found: 315.0703.

1-[Phenyl(ferrocenyl)methyl]-naphthalen-2-ol (**3d**): Orange solid, mp 159.2-163.1°C; IR (KBr): 3443, 3031, 1620,1602, 1468, 1200 cm^{-1}; ^{1}H NMR (400 MHz, $CDCl_3$): δ 3.98-4.30 (m, 9H, Fc–H), 5.48 (s, 1H, OH), 6.18 (s, 1H, CH_3), 7.06 (d, J=8.4 Hz, 1H, Ar–H), 7.22 (m, 1H, Ar–H), 7.29-7.34 (m, 3H, Ar–H), 7.39-7.41 (m, 2H, Ar–H), 7.43-7.48 (m, 1H, Ar–H), 7.70 (d, J=8.8 Hz, 1H, Ar–H), 7.77 (d, J=8.0 Hz, 1H, Ar–H), 8.08 (d, J=8.0 Hz, 1H, Ar–H); ^{13}C NMR (100 MHz, $CDCl_3$): δ 164.7, 164.5, 157.0, 155.2, 154.0, 152.4, 144.3, 132.2, 132.1, 131.0, 130.7, 129.0, 128.8, 128.5, 126.5, 91.5, 69.2, 68.8, 68.4, 68.0, 61.3; HRMS (*m/z*): $[M]^+$, calcd for $C_{27}H_{22}OFe$: 418.1020; Found: 418.1026.

Reference

Xu, X., Jiang, R., Zhou, X., Liu, Y., Ji, S., and Zhang, Y. (2009). Cerium ammonium nitrate: an efficient catalyst for carbon-carbon bond formation from ferrocenyl alcohol substrate. *Tetrahedron*, **65**, 877–882.

C-C bond_28

Type of reaction: C-C bond formation
Reaction conditions: Toluene, room temperature
Synthetic strategy: One-step direct alkylnylation
Catalyst: LiO-*t*-Bu/CuI/1,10-phenanthroline (phen)
Keywords: 1,3,4-Oxadiazoles, alkynyl bromides, LiO-*t*-Bu/CuI/1,10-phenanthroline (phen), toluene, room temperature, C-H activation, direct alkylnylation, heteroaryl-alkynyl linkage

Reaction scheme

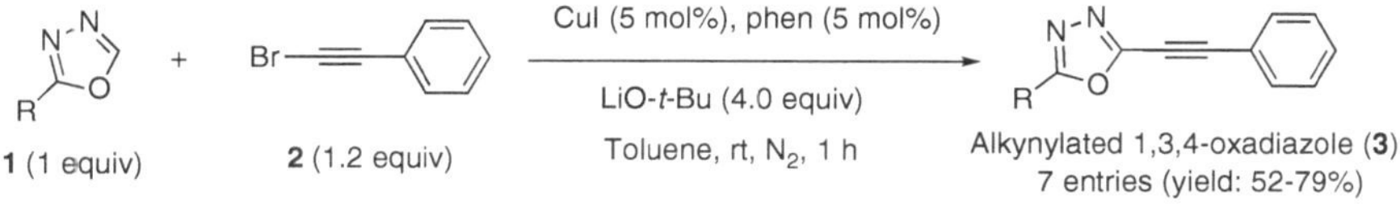

R = C_6H_5, 4-MeC_6H_4, 4-OMeC_6H_4, 4-$CF_3C_6H_4$, 4-ClC_6H_4, $C_6H_4CH_2CH_2$, 1-naphthyl

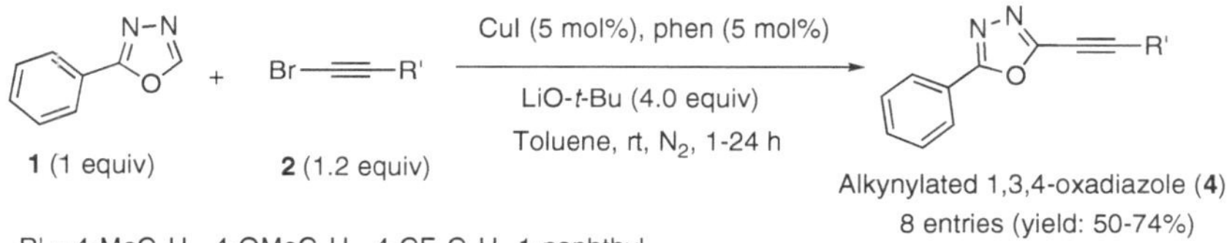

R' = 4-MeC_6H_4, 4-OMeC_6H_4, 4-$CF_3C_6H_4$, 1-naphthyl, *i*-Pr_3Si, *t*-BuMe_2Si, 1-cyclohexenyl, *n*-hexyl

Representative entries

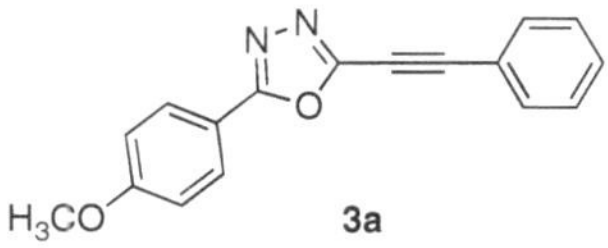
3a

2-(4-Methoxyphenyl)-5-(phenylethynyl)-1,3,4-oxadiazole (**3a**): Solid; mp 140.0-140.5°C; reaction time: 1 h; yield: 64%

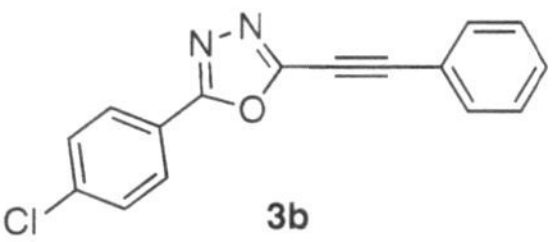
3b

2-(4-Chlorophenyl)-5-(phenylethynyl)-1,3,4-oxadiazole (**3b**): Solid; mp 134.0-134.5 °C; reaction time: 1 h; yield: 79%

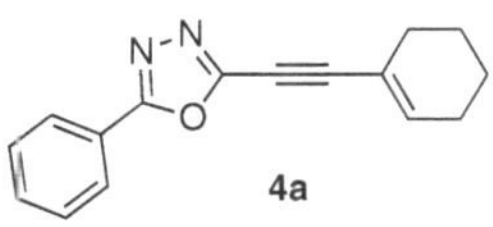
4a

2-(Cyclohex-1-en-1-ylethynyl)-5-phenyl-1,3,4-oxadiazole (**4a**): Solid; mp 85.0-85.5 °C; reaction time: 6 h; yield: 69%

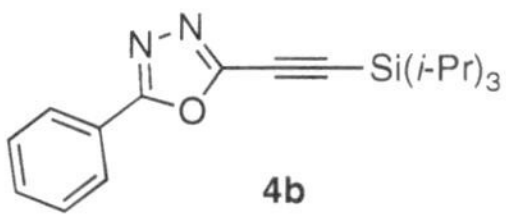
4b

2-Phenyl-5-((triisopropylsilyl)ethynyl)-1,3,4-oxadiazole (**4b**): Reaction time: 24 h; yield: 50%

Experimental procedure

A mixture of 1,3,4-oxadiazole derivative (**1**; 1 equiv.; 0.5 mmol), alkynyl bromide (**2**; 1.2 equiv.; 0.60 mmol), CuI (4.8 mg, 0.025 mmol), 1,10-phenanthroline (4.5 mg, 0.025 mmol), LiO-*t*-Bu (4.0 equiv.; 160 mg, 2.0 mmol) in 2.5 mL of toluene in the presence of 1-methylnaphthalene (ca. 50 mg, internal standard) was placed in a 20 mL two-necked reaction flask. After the mixture was stirred at room temperature for stipulated time frame, the consumption of **1** was checked by

GC analysis. The resulting mixture was then poured into water. The mixture was extracted with ethyl acetate, and the organic layer was dried over sodium sulfate. Concentration *in vacuo* followed by silica gel column purification (hexane:ethyl acetate = 95:5) afforded the desired alkynylated 1,3,4-oxadiazole (**3** or **4**) in moderate to good yields. Each of the products was characterized by means of spectral studies including ^{1}H NMR, ^{13}C NMR and HRMS.

Characterization data of representative entries

2-(4-Methoxyphenyl)-5-(phenylethynyl)-1,3,4-oxadiazole (**3a**): Solid; mp 140.0-140.5°C; ^{1}H NMR (400 MHz, $CDCl_3$): δ 3.88 (s, 3H), 7.00-7.04 (m, 2H), 7.39-7.43 (m, 2H), 7.45-7.49 (m, 1H), 7.63-7.66 (m, 2H), 8.02-8.05 (m, 2H); ^{13}C NMR (100 MHz, $CDCl_3$): δ 55.5, 73.3, 96.9, 114.6, 115.8, 119.9, 128.7, 128.9, 130.5, 132.3, 150.2, 162.7, 164.8; HRMS *m/z* (M^+) calcd for $C_{17}H_{12}N_2O_2$: 276.0899; Found: 276.0901.

2-(Cyclohex-1-en-1-ylethynyl)-5-phenyl-1,3,4-oxadiazole (**4a**): Solid; mp 85.0-85.5 °C; ^{1}H NMR (400 MHz, $CDCl_3$): δ 1.60-1.74 (m, 4H), 2.18-2.28 (m, 4H), 6.51 (dt, $J = 3.0$, 1.8 Hz, 1H), 7.45-7.57 (m, 3H), 8.05-8.08 (m, 2H); ^{13}C NMR (100 MHz, $CDCl_3$): δ 21.1, 21.9, 26.0, 28.0, 70.8, 99.2, 118.7, 123.5, 127.0, 129.1, 132.0, 141.4, 150.9, 164.5; HRMS *m/z* (M^+) calcd for C16H14N2O: 250.1106; Found: 250.1103.

Reference

Kawano, T., Matsuyama, N., Hirano, K., Satoh, T., and Miura, M. (2010). Room temperature direct alkynylation of 1,3,4-oxadiazoles with alkynyl bromides under copper catalysis. *J. Org. Chem.*, **75**, 1764–1766.

C-C bond_29

Type of reaction: C-C bond formation
Reaction conditions: Solvent-free, room temperature
Synthetic strategy: Lewis acid-catalyzed one-step condensation
Catalyst: Molecular iodine (I_2)
Keywords: Indoles, 1,3-dicarbonyl compounds, iodine, solvent-free, room temperature, condensation, 3-alkenylated indoles, bis(indolyl)carbonyl compounds

General reaction scheme

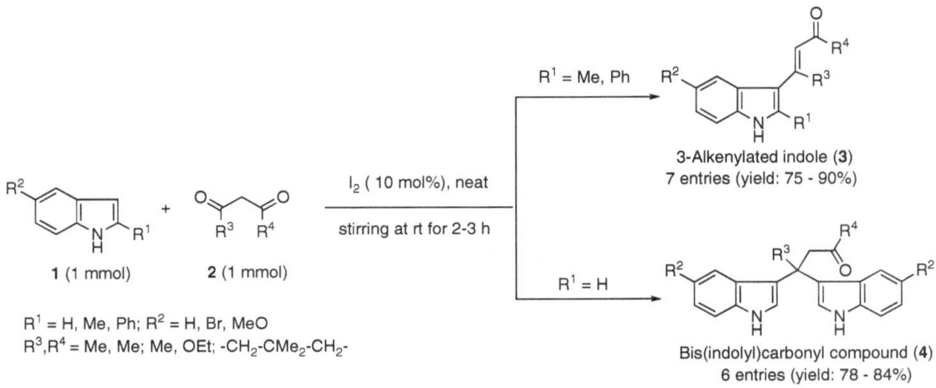

Representative entries

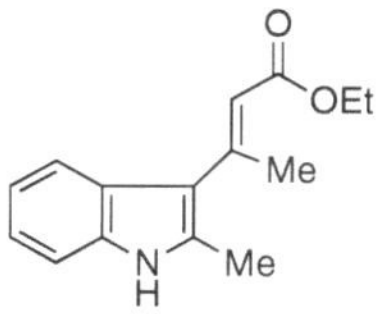

3a

(*E*)-Ethyl 3-(2-methyl-1*H*-indol-3-yl)but-2-enoate (**3a**): Reaction time: 3 h; yield: 76%

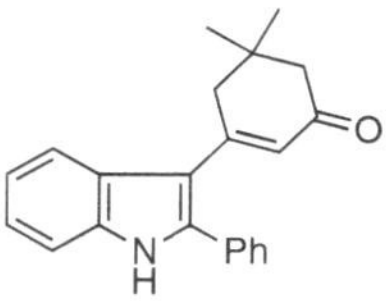

3b

5,5-Dimethyl-3-(2-phenyl-1*H*-indol-3-yl)cyclohex-2-enone (**3b**): Reaction time: 2.5 h; yield: 81%

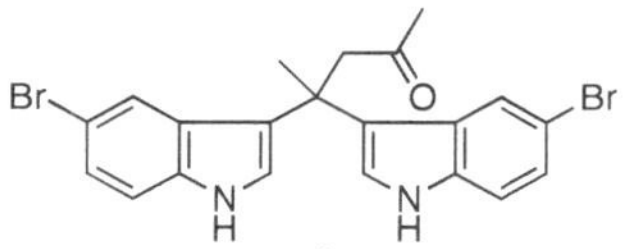

4a

4,4-*Bis*(5-bromo-1*H*-indol-3-yl)pentan-2-one (**4a**): Reaction time: 3 h; yield: 80%)

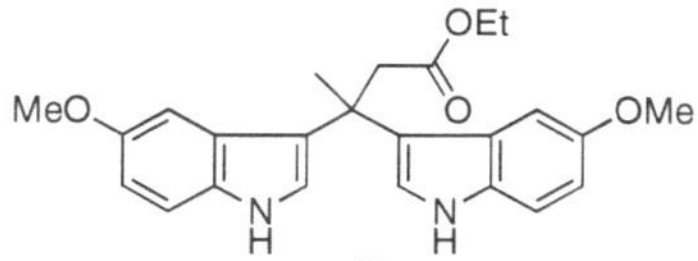

4b

Ethyl 3,3-*bis*(5-methoxy-1*H*-indol-3-yl)butanoate (**4b**): Reaction time: 3 h; yield: 84%)

Experimental procedure

A mixture of indole derivative (**1**; 1 mmol), 1,3-dicarbonyl compound (**2**; 1 mmol) and molecular iodine (10 mol%) was stirred at room temperature in a round-bottom flask for 2-3 h. After completion of the reaction (as monitored by TLC) a saturated aqueous solution of $Na_2S_2O_3$ (2 × 5 mL) was added, and the mixture was extracted with EtOAc (3 × 5 mL). The combined organic phases were dried over anhydrous sodium sulfate, filtered, and evaporated under reduced pressure to obtain the crude product of 3-alkenylated indole (**3**)/3-alkylated indole (**4**) which was then purified using column chromatography. Each of the products (**3** and **4**) was characterized on the basis of spectral and elemental analyses.

Characterization data of representative compounds

(*E*)-Ethyl 3-(2-methyl-1*H*-indol-3-yl)but-2-enoate (**3a**): IR (KBr): 3411, 2933, 1718 cm^{-1}; ^1H NMR (300 MHz, $CDCl_3$): δ 7.99 (br s, 1H), 7.65 (d, J = 7.5 Hz, 1H), 7.29-7.25 (m, 1 H), 7.17-7.09 (m, 2 H), 5.92 (s, 1H), 4.25 (q, J = 7.2 Hz, 2H), 2.66 (s, 3H), 2.51 (s, 3H), 1.35 (t, J = 7.2 Hz, 3H); ^{13}C NMR (75 MHz, $CDCl_3$): δ 167.3, 152.0, 135.0, 132.7, 127.0, 121.7, 120.3, 119.5, 117.1, 116.8, 110.4, 59.5, 20.4, 14.4,13.4. Anal. Calcd for $C_{15}H_{17}NO_2$: C 74.05, H 7.04, N 5.76; Found: C 73.99, H 7.06, N 5.68.

Ethyl 3,3-*bis*(5-methoxy-1*H*-indol-3-yl)butanoate (**4b**): IR (KBr): 3409, 3365, 2978, 1719 cm^{-1}; ^1H NMR (300 MHz, $CDCl_3$): δ 7.87(br s, 2H), 7.19-7.10 (m, 4H), 6.74-6.71 (m, 4H), 3.89 (q, J = 7.2 Hz, 2H), 3.61 (s, 6H), 3.33 (s, 2H), 2.04 (s, 3H), 0.90 (t, J = 7.2 Hz, 3H). Anal. Calcd for $C_{24}H_{26}N_2O_4$: C 70.92, H 6.45, N 6.89; Found: C 70.83, H 6.28, N 6.78.

Reference

Singh, N., and Singh, K. N. (2012). Iodine-catalyzed highly efficient synthesis of 3-alkylated/3-alkenylated indoles from 1,3-dicarbonyl compounds. *Synlett*, **23**, 2116–2120.

C-C bond_30

Type of reaction: C-C bond formation
Reaction conditions: Ethanol, room temperature
Synthetic strategy: One-pot domino synthesis
Catalyst: Piperidine
Keywords: Isatylidene malononitriles, α-azido ketones, ethanol, piperidine, room temperature, one-pot domino process, 3-(aminomethylene)oxindoles

General reaction scheme

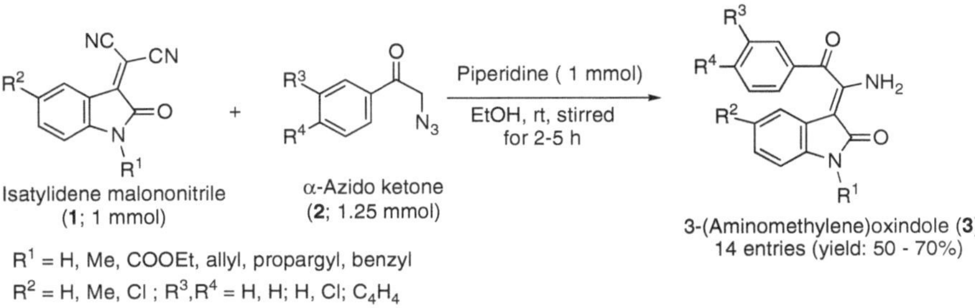

Representative entries

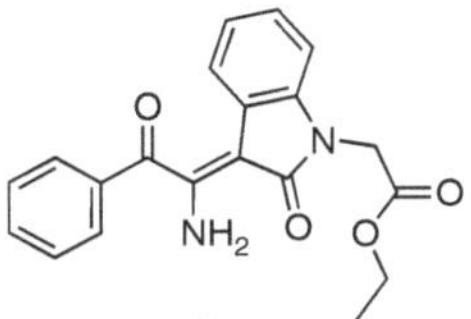

3a

(*Z*)-Ethyl 2-[3-(1-amino-2-oxo-2-phenylethylidene)-2-oxoindolin-1-yl]acetate (**3a**): Pale orange solid; mp 230-232 °C, reaction time: 2.4 h; yield: 70%

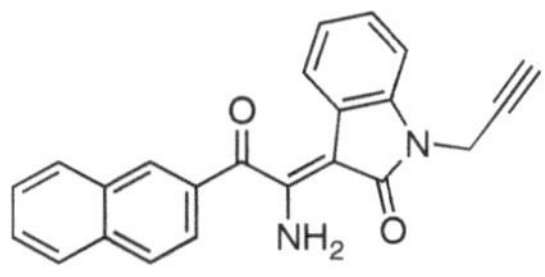

3b

(*Z*)-3-[1-Amino-2-(naphthalen-2-yl)-2-oxoethylidene]-1-(prop-2-ynyl)indolin-2-one (**3b**): Bright orange solid; mp 246-248 °C, reaction time: 5 h; yield: 56%

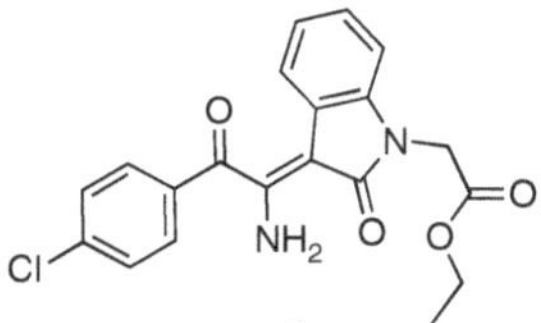

3c

(*Z*)-Ethyl 2-(3-[1-amino-2-(4-chlorophenyl)-2-oxoethylidene]-2-oxoindolin-1-yl)acetate (**3c**): Bright orange solid; mp 221-223 °C, reaction time: 3.2 h; yield: 55%

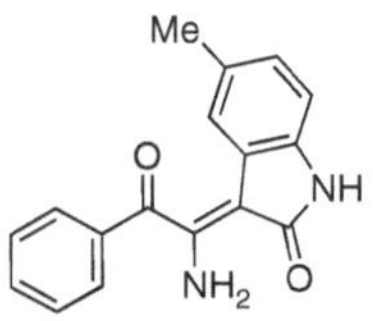

3d

(*Z*)-3-(1-Amino-2-oxo-2-phenylethylidene)-5-methylindolin-2-one (**3d**): Orange solid; mp 210-212 °C, reaction time: 4.5 h; yield: 61%

Experimental procedure

To a stirred solution of isatylidene malononitrile (**1**; 1 mmol) and α-azido ketone (**2**; 1.25 mmol) in EtOH (5 mL) was added piperidine (1.0 mmol), and stirring was continued at room temperature for stipulated time (2-5 h). On completion of the reaction as indicated by TLC, solvent was removed under vacuum and the crude product was purified by column chromatography on silica gel (100–200 mesh) using ethyl acetate –petrol ether (25:75) as eluent to afford pure product of 3-(aminomethylene)oxindole **3** with moderate to good yield (50-70%). Each of the products was characterized on the basis of detailed spectral and elemental analyses.

Characterization data of representative compounds

(*Z*)-3-[1-Amino-2-(naphthalen-2-yl)-2-oxoethylidene]-1-(prop-2-ynyl)indolin-2-one (3b): Bright orange solid; mp 246-248 °C; IR (KBr): 3410, 3297, 3054, 2931, 1669, 1619, 1468, 1357, 1278, 1183, 669, 582 cm^{-1}; ^{1}H NMR (500 MHz, $CDCl_3$): δ 2.23 (t, *J*=2.3 Hz, 1H, ≡C*H*), 4.63 (d, *J*=2.3 Hz, 2H, CH_2), 5.61 (br s, 1H, D_2O exchangeable, N*H*), 6.69-6.74 (m, 2H, Ar-*H*), 7.00-7.05 (m, 2H, Ar-*H*), 7.53 (t, *J*=8.05 Hz, 1H, Ar-*H*), 7.63 (t, *J*=8.05 Hz, 1H, Ar-*H*). 7.86-7.94 (m, 3H, Ar-*H*), 8.14 (d, *J*=8.55 Hz, 1H, Ar-*H*), 8.57 (s, 1H, Ar-*H*), 9.1 (br s, 1H, D_2O exchangeable, NH); ^{13}C NMR (125 MHz, $CDCl_3$): δ 28.7, 71.9, 77.7, 95.6, 108.6, 119.0, 121.8, 121.9, 123.8, 124.5, 127.4, 128.0, 129.6, 129.9, 130.2, 132.5, 134.0, 136.6, 137.1, 153.4, 168.5, 191.7; HRMS (ESI): *m/z* calcd for $C_{23}H_{16}N_2O_2$: 353.1285 $[M+H]^+$; found: 353.1290. Anal. Calcd for $C_{23}H_{16}N_2O_2$: C 78.39, H 4.58, N 7.95; Found: C 78.43, H 4.63, N 7.91.

(*Z*)-Ethyl 2-(3-[1-amino-2-(4-chlorophenyl)-2-oxoethylidene]-2-oxoindolin-1-yl)acetate (**3c**): Bright orange solid; mp 221-223 °C; IR (KBr): 3347, 3194, 2925, 1743, 1663, 1613, 1469, 1368, 1275, 1174, 1092, 1016, 754, 593, 539 cm^{-1}; ^{1}H NMR (500 MHz, DMSO-d_6): δ 1.18 (t, *J*=7.3 Hz, 3H, CH_3), 4.12 (q, *J*=6.85 Hz, 2H, CH_2), 4.63 (s, 2H, CH2), 6.45 (d, *J*=7.65 Hz, 1H, Ar-*H*), 6.68 (t, *J*=7.65 Hz, 1H, Ar-*H*), 6.89-6.93 (m, 2H, Ar-*H*), 7.67 (d, *J*=8.4 Hz, 2H, Ar-*H*), 8.0 (d, *J*=7.6 Hz, 2H, Ar-*H*), 8.59 (br s, 1H, D_2O exchangeable, N*H*), 8.94 (br s, 1H, D_2O exchangeable, N*H*); ^{13}C NMR (125 MHz, DMSO-d_6): δ 14.6, 41.0, 61.5, 92.0, 108.7, 117.4, 121.4, 122.1, 123.9, 130.3, 131.9, 132.7, 137.5, 140.9, 155.0, 169.0, 191.0; HRMS (ESI): *m/z* calcd for $C_{20}H_{17}ClN_2O_4$: 385.0950 $[M+H]^+$; found: 385.0955. Anal. Calcd for $C_{20}H_{17}ClN_2O_4$: C 62.42, H 4.45, N 7.28; Found: C 62.48, H 4.51, N 7.21.

Reference

Kamalraja, J., Hari Babu, T., Muralidharan, D., and Perumal, P. T. (2012). A facile method for the synthesis of 3-(aminomethylene)oxindoles from isatylidene malononitriles and α-azido ketones. *Synlett*, **23**, 1950–1954.

C-C bond_31

Type of reaction: C-C bond formation
Reaction conditions: Ethanol, room temperature
Synthetic strategy: One-pot multicomponent reaction
Catalyst: Catalyst-free
Keywords: Isatin, hydrazine hydrate, 1,1-bis(methylthio)-2-nitroethylene, ethanol, room temperature, one-pot multicomponent reaction, spirooxindole-pyrazolines

General reaction scheme

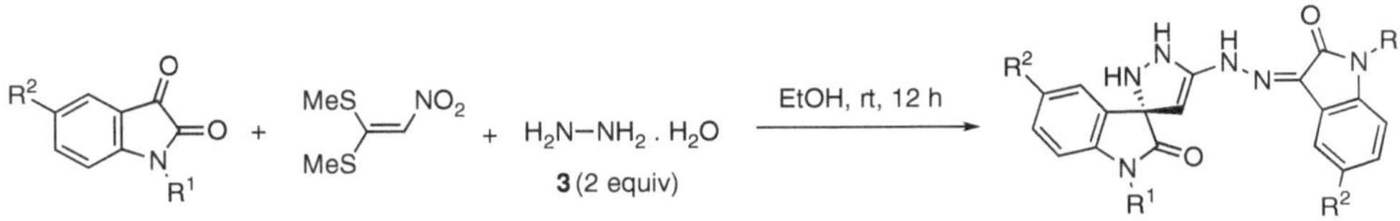

1 (2 equiv) **2** (1 equiv) **3** (2 equiv)

Spirooxindole-pyrazoline (**4**)
8 entries (yield: 60-76%)

R^1 = H, Me, Bn, PMB (*p*-methoxybenzyl)
R^2 = H, Br, NO_2

Representative entries

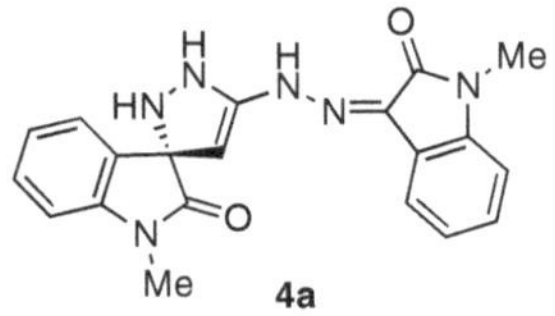

4a

(*S*,*Z*)-1-Methyl-5′-(2-(1-methyl-2-oxoindolin-3-ylidene)hydrazinyl)-1′,2′-dihydrospiro [indoline-3,3′-pyrazol]-2-one (**4a**): Orange powder; mp: >300 °C (dec.); yield: 76%

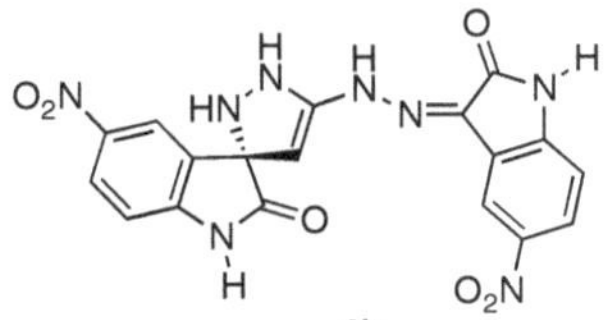

4b

(*S*,*Z*)-5-Nitro-5′-(2-(5-nitro-2-oxoindolin-3-ylidene)hydrazinyl)-1′,2′-dihydrospiro[indoline-3,3′-pyrazol]-2-one (**4b**): Mustard powder; mp: up 300 °C; yield: 64%

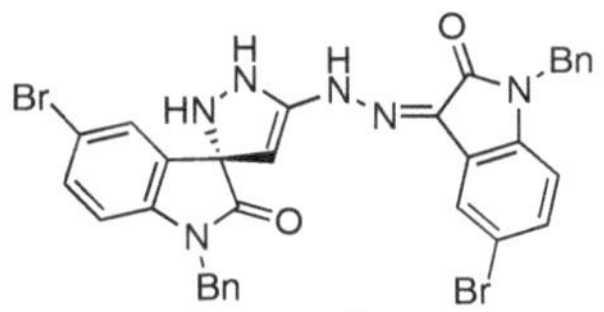

4c

(*S*,*Z*)-1-Benzyl-5′-(2-(1-benzyl-5-bromo-2-oxoindolin-3-ylidene)hydrazinyl)-5-bromo-1′,2′-dihydrospiro[indoline-3,3′-pyrazol]-2-one (**4c**): Orange powder; mp: 220 °C; yield: 60%

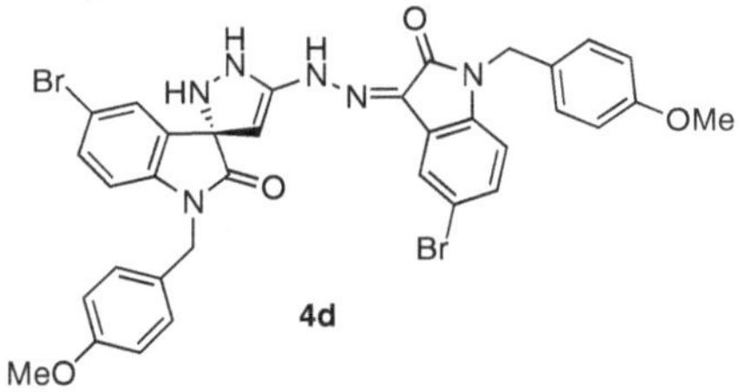

4d

(*S*,*Z*)-5-Bromo-5′-(2-(5-bromo-1-(4-methoxybenzyl)-2-oxoindolin-3-ylidene) hydrazinyl)-1-(4-methoxybenzyl)-1′,2′-dihydrospiro[indoline-3,3′-pyrazol]-2-one (**4d**): Dark brown powder; mp: 163 °C; yield: 63%

Experimental procedure

To a solution of 1,1-bis(thiomethyl)-2-nitroethylene (**2**; 1 mmol) in ethanol (4 mL) in a 10 mL flask was added $NH_2NH_2 \cdot H_2O$ (**3**; 80% aq., 2 mmol) and stirred for 5 h at room temperature. After then, isatin (**1**; 2 mmol) was added to the reaction mixture, and stirring was continued for another 3-7 h. On completion of the reaction, the crude precipitated was filtered and washed with cold ethanol to obtain pure product of spirooxindole-pyrazoline **4**. All the products were characterized by spectral studies.

Characterization data of representative compounds

(*S*,*Z*)-1-Benzyl-5′-(2-(1-benzyl-5-bromo-2-oxoindolin-3-ylidene)hydrazinyl)-5-bromo-1′,2′-dihydrospiro[indoline-3,3′-pyrazol]-2-one (**4c**): Orange powder; mp: 220 °C; IR (KBr): 3437, 1721, 1600, 1562, 1342 cm^{-1}; ^1H NMR (DMSO-d_6, 500 MHz): δ 4.97 (d, $^3J_{HH}$ = 14.6 Hz,

2 × CH_2, 4H), 6.27 (s, 2 × NH, 2H), 6.93-6.96 (m, 2 × CH of Ar, 2H), 7.27-7.34 (m, 10 × CH of Ar, 10H), 7.53 (br s, 3 × CH of Ar, 3H), 8.35 (br, CH of Ar, 1H), 13.41 (s, NH, 1H); ^{13}C NMR (DMSO-d_5, 125.7 MHz,): δ 42.7 (CH_2), 42.8 (CH_2), 74.1 (C_{spiro}), 111.9 (C), 112.6 (C), 114.5 (CH), 115.0 (CH), 117.6 (C), 121.1 (C), 122.7 (C), 127.2 (CH), 127.3 (CH), 127.5 (CH), 127.6 (CH), 128.6 (C), 128.7 (2 CH), 130.9 (CH), 133.7 (C), 133.9 (C), 135.3 (CH), 135.7 (CH), 135.9 (CH), 141.8 (C), 144.2 (C), 148.6 (C), 160.5 (C=O), 162.6 (C=O); MS: *m/z* (% rel) 331 (7), 329 (7), 317 (7), 315 (7), 258 (5), 256 (5), 224 (13), 170 (12), 168 (12), 91 (100), 65 (33).

(*S*,*Z*)-5-Bromo-5′-(2-(5-bromo-1-(4-methoxybenzyl)-2-oxoindolin-3-ylidene) hydrazinyl)-1-(4-methoxybenzyl)-1′,2′-dihydrospiro[indoline-3,3′-pyrazol]-2-one (**4d**): Dark brown powder; mp: 163 °C; IR (KBr): 3308, 3203, 1719, 1604, 1557, 1348 cm^{-1}; ^1H NMR (DMSO-d_6, 500 MHz): δ 2.21 (s, 2 × Me, 6 H), 4.88 (br s, 2 × CH_2, 4H), 6.23-7.47 (m, 14 × CH of Ar, 14H), 8.31 (s, NH, 1H), 10.17 (s, NH, 1H); ^{13}C NMR (DMSO-d_6, 125.7 MHz,): because of insolubility, only some of the resonances are distinguishable including: δ 20.6 (2 Me), 74.1 (C_{spiro}), 127.2 (2 CH of Ar), 127.3 (2 CH of Ar), 129.2 (4 CH of Ar). MS: *m/z* (%rel) 410 (5), 368 (7), 341 (8), 279 (7), 236 (11), 213 (6), 173 (18), 149 (42), 127 (38), 105 (25), 97 (70), 83 (74), 69 (100), 57 (97).

Reference

Alizadeh, A., and Zohreh, N. (2012). A unique approach to catalyst-free, one-pot synthesis of spirooxindole-pyrazolines. *Synlett*, **23**, 428–432.

C-C bond_32

Type of reaction: C-C bond formation
Reaction conditions: Methanol, room temperature
Synthetic strategy: Direct conjugate addition *via* C-H activation
Catalyst: Palladium chloride ($PdCl_2$)
Keywords: Heteroarenes, α,β-unsaturated ketones, methanol, $PdCl_2$, room temperature, Direct conjugate addition, C-H activation, substituted heteroarenes

General reaction scheme

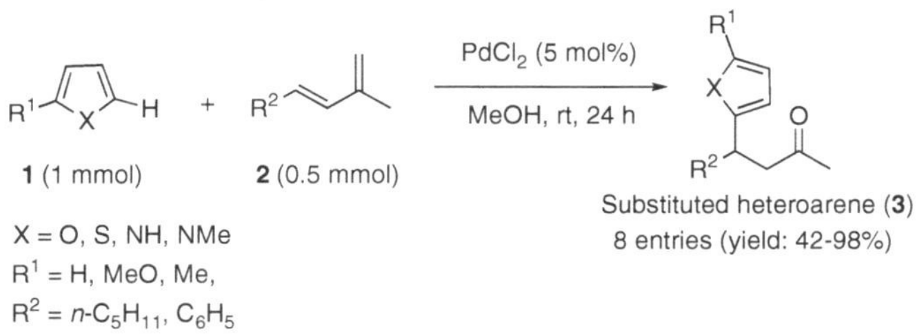

Representative entries

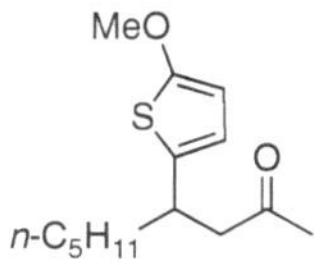

3a

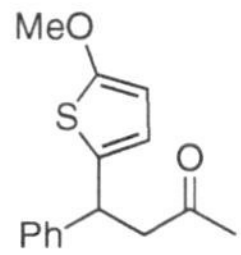

3b

4-(5-Methoxythiophen-2-yl)nonan-2-one (**3a**): Colorless oil; yield: 97%

4-(5-Methoxythiophen-2-yl)-4-phenylbutan-2-one (**3b**): Colorless oil; yield: 76%

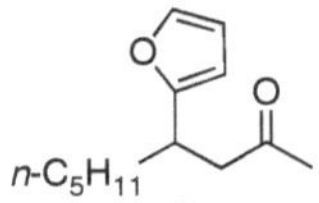

3c

4-(Furan-2-yl)nonan-2-one (**3c**): Colorless oil; yield: 67%

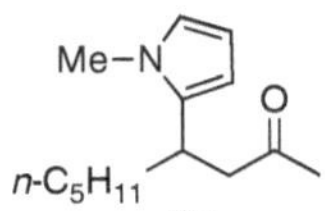

3d

4-(1-Methyl-1*H*-pyrrol-2-yl)nonan-2-one (**3d**): Colorless oil; yield: 70%

Experimental procedure

To a mixture of $PdCl_2$ (0.025 mol, 4.4 mg) and α,β-unsaturated ketone (**2**; 0.5 mmol) in methanol (2.0 ml), heteroarene (**1**; 1 mmol) was added. The resultant mixture was stirred at room temperature for 24 h; the solvent was then removed under vacuum. The residue was purified by flash chromatography on silica gel (eluent: petroleum ether/ethyl acetate = 10:1 to 20:1, v/v) to afford the substituted heteroarene **3**. All the products were characterized by spectral studies.

Characterization data of two representative compounds

4-(5-Methoxythiophen-2-yl)-4-phenylbutan-2-one (**3b**): Colorless oil; IR (neat) 3413, 3026, 2925, 1715, 1559, 1505, 1451, 1430, 1355, 1203, 1153, 990, 762, 698 cm^{-1}; ^1H NMR (400 MHz, $CDCl_3$): δ 7.19-7.31 (m, 5H), 6.37 (d, $J = 3.8$ Hz, 1H), 5.94 (d, $J = 3.8$ Hz, 1H), 4.62 (t, $J = 7.4$ Hz, 1H), 3.80 (s, 3H), 3.18 (dd, $J = 16.6$, 7.5 Hz, 1H), 3.07 (dd, $J = 16.6$, 7.3 Hz, 1H), 2.10 (s, 3H); ^{13}C NMR (100 MHz, $CDCl_3$): δ 206.4, 165.1, 143.4, 134.0, 128.7, 127.6, 126.9, 121.2, 102.8, 60.2, 50.6, 42.0, 30.7; HRMS (EI): Calcd for $C_{15}H_{16}O_2S$ at m/z $[M]^+$ 260.0871; Found 260.0876.

4-(1-Methyl-1*H*-pyrrol-2-yl)nonan-2-one (**3d**): Colorless oil; IR (neat) 2954, 2927, 2856, 1716, 1489, 1359, 1299, 1162, 702 cm^{-1}; ^1H NMR (400 MHz, $CDCl_3$): δ 6.45 (t, $J = 1.8$, 1.9 Hz, 1H), 6.04 (t, $J = 3.1$, 3.0 Hz, 1H), 5.84 (dd, $J = 3.2$, 1.8 Hz, 1H), 3.57 (s, 3H), 3.19-3.26 (m, 1H), 2.74 (dd, $J = 16.7$, 7.5 Hz, 1H), 2.65 (dd, $J = 16.6$, 6.5 Hz, 1H), 2.04 (s, 3H), 1.50-1.56 (m, 2H), 1.19-1.26 (m, 6H), 0.84 (t, $J = 6.5$ Hz, 3H); ^{13}C NMR (100 MHz, $CDCl_3$): δ 208.3, 136.7, 120.9, 106.8, 104.1, 50.2, 36.6, 33.9, 32.0, 31.8, 31.0, 27.1, 22.7, 14.2; HRMS (ESI): Calcd for $C_{14}H_{23}NO$ $[M+Na]^+$ 244.1677; Found 244.1684.

Reference

Zhang, X., Yu, X., Feng, X., and Bao, M. (2012a). Palladium-catalyzed direct 1,4-addition of heteroarenes to α,β-unsaturated ketones via C-H activation. *Synlett*, **23**, 1605–1608.

C-C bond_33

Type of reaction: C-C bond formation
Reaction conditions: Acetic acid, room temperature
Synthetic strategy: Friedel-Crafts alkylation
Catalyst: Acetic acid (AcOH)
Keywords: Indoles, *tert*-enamides, acetic acid, room temperature, Friedel-Crafts alkylation, 2-oxo-1-pyrrolidines

General reaction scheme

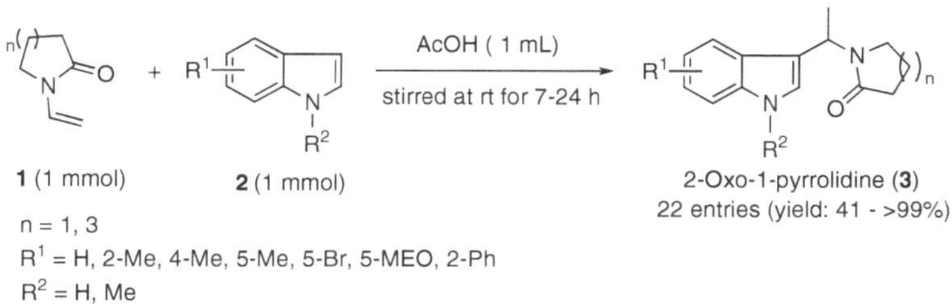

Representative entries

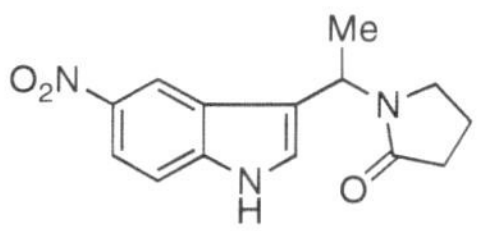

3a

1-(1-(5-Nitro-1*H*-indol-3-yl)ethyl)pyrrolidin-2-one (**3a**): Yellow solid, mp 241-242 °C; reaction time: 12 h; yield: 81%

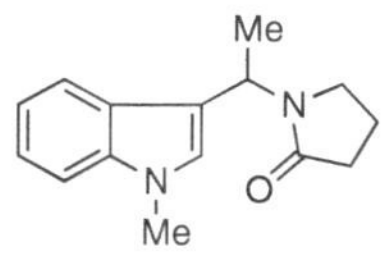

3b

1-(1-(1-Methyl-1*H*-indol-3-yl)ethyl)pyrrolidin-2-one (**3b**): Yellow oil; reaction time: 24 h; yield: >99%

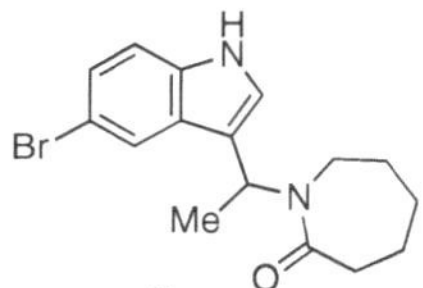

3c

1-(1-(5-Bromo-1*H*-indol-3-yl)ethyl)azepan-2-one (**3c**): White solid, mp 73-74 °C; reaction time: 18 h; yield: 75%

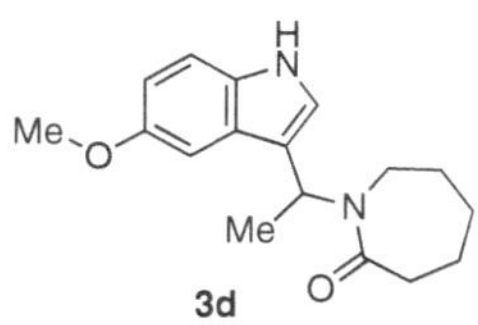

3d

1-(1-(5-Methoxy-1*H*-indol-3-yl)ethyl)azepan-2-one (**3d**): Brown solid, mp 143-144 °C; reaction time: 9 h; yield: 86%

Experimental procedure

A mixture of *tert*-enamide (**1**; 1.0 mmol) and indole derivative (**2**; 1.0 mmol) in acetic acid (1.0 mL) was stirred at room temperature for stipulated time (7-24 h). On completion of the reaction (as monitored by TLC), the reaction mixture was poured into water, and $NaHCO_3$ was added to neutralize the acetic acid. The crude product was extracted with CH_2Cl_2 (15 mL × 2); the organic layer was dried with anhydrous Na_2SO_4. On removal of the solvent under reduced pressure, desired product of 2-oxo-1-pyrrolidine **3** was obtained, which was then purified by flash column chromatography using ethyl acetate and petroleum ether as eluents to afford pure product. All the products were characterized by spectral studies.

Characterization data of two representative compounds

1-(1-(5-Nitro-1*H*-indol-3-yl)ethyl)pyrrolidin-2-one (**3a**): Yellow solid, mp 241-242 °C; IR (KBr): 3379, 2968, 1660, 1522, 1295 cm^{-1}; ^1H NMR (400 MHz, DMSO-d_6): δ 1.54 (d, $J = 6.8$ Hz, 3H), 1.70-1.74 (m, 1H), 1.86-1.90 (m, 1H), 2.20-2.36 (m, 2H), 2.70-2.76 (m, 1H), 3.27-3.34 (m, 1H), 5.56-5.61 (m, 1H), 7.54 (d, $J = 9.2$ Hz, 1H), 7.64 (s, 1H), 8.00 (d, $J = 8.4$ Hz, 1H), 8.45 (s, 1H); ^{13}C NMR (100 MHz, DMSO-d_6): δ 173.1, 140.5, 139.6, 127.2, 125.4, 117.6, 116.8, 115.7, 112.0, 41.4, 41.3, 31.0, 17.3, 16.7; HRMS (EI): *m/z* calcd for $C_{14}H_{15}N_3O_3$: 273.1113; Found: 273.1112.

1-(1-(5-Methoxy-1*H*-indol-3-yl)ethyl)azepan-2-one (**3d**): Brown solid, mp 143-144 °C; IR (KBr): 3232, 3052, 2935, 2855, 2832, 1607, 1490, 1444, 1290, 1216, 1030, 922, 800, 640 cm^{-1}; ^1H NMR (300 MHz, DMSO-d_6): δ 1.34-1.69 (m, 8H), 3.06 (dt, $J=15.1$, 22.2 Hz, 2H), 3.35 (d, $J=6.8$ Hz, 3H), 3.69 (s, 3H), 6.01 (q, $J=6.6$ Hz, 1H), 6.72 (d, $J=8.4$ Hz, 1H), 6.88 (s, 1H), 7.23 (d, J =8.6 Hz, 1H), 7.29 (s, 1H); ^{13}C NMR (75 MHz, DMSO-d_6): δ 174.0, 153.1, 131.5, 126.8, 124.2, 115.0, 112.1, 111.3, 100.7, 55.1, 44.0, 41.9, 37.0, 29.3, 28.6, 23.2, 17.0; HRMS (EI): *m/z* calcd for $C_{17}H_{22}N_2O_2$: 286.1681; Found: 286.1678.

Reference

Zhang, Y., Jiang, J., Chu, X.-Q., Jiang, R., Xu, X.-P., Li, D.-H., and Ji, S.-J. (2012b). Friedel-Crafts alkylation of indoles by *tert*-enamides in acetic acid. *Synlett*, **23**, 751–754.

C-C bond_34

Type of reaction: C-C and C-N bond formation
Reaction conditions: Solvent-free, room temperature
Synthetic strategy: One-pot multicomponent reaction
Catalyst: L-Proline
Keywords: Indoles, aldehydes, secondary amines, L-Proline, solvent-free, room temperature, one-pot multicomponent reaction, Mannich-type reaction, 3-amino alkylated indoles

General reaction scheme

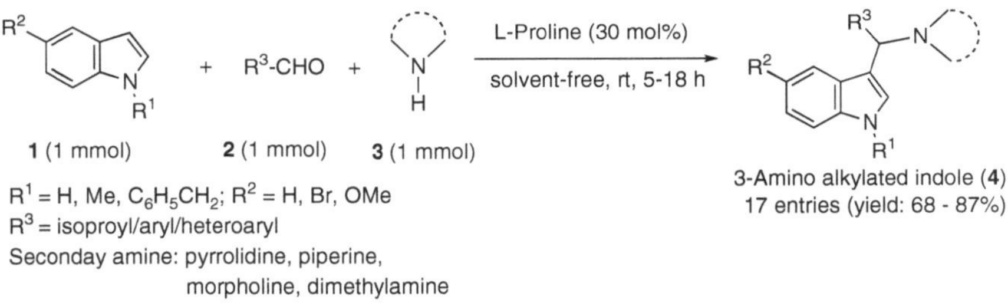

Representative entries

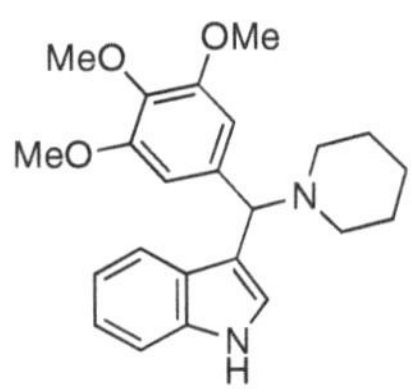

4a

3-(Piperidin-1-yl(3,4,5-trimethoxyphenyl) methyl)-1*H*-indole (**4a**): White solid, mp 112-115 °C; reaction time: 7.5 h; yield: 81%

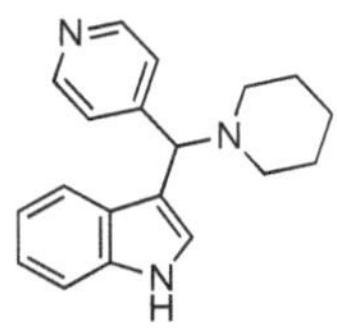

4b

3-(Piperidin-1-yl(pyridin-4-yl)methyl)-1*H*-indole (**4b**): White solid, mp 135-137 °C; reaction time: 6.5 h; yield: 79%

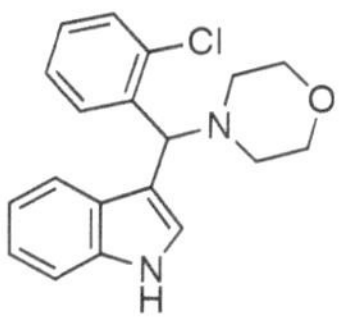

4c

4-((2-Chlorophenyl)(1*H*-indol-3-yl)methyl) morpholine (**4c**): White solid, mp 134-139 °C; reaction time: 5 h; yield: 81%

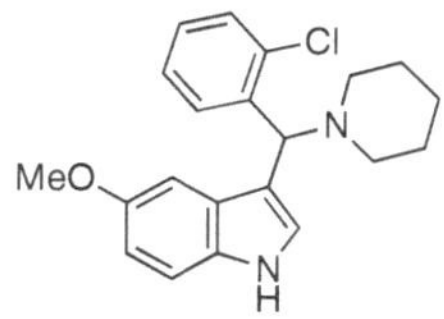

4d

3-((2-Chlorophenyl)(piperidin-1-yl) methyl)-5-methoxy-1*H*-indole (**4d**): White solid, mp 109-113 °C; reaction time: 5 h; yield: 81%

Experimental procedure

To a mixture of aldehyde (**2**; 1 mmol), secondary amine (**3**; 1 mmol) and indole (**1**; 1 mmol) was added with L-proline (30 mol %) in a 25 mL round-bottom flask, and the reaction mixture was stirred at room temperature till the completion of the reaction (5-18 h; monitored by TLC). After completion, the reaction mixture was diluted with water and extracted with ethyl acetate, dried over sodium sulphate and evaporated in *vacuo* to give crude product, which was purified by silica gel (230-400 mesh) column chromatography to afford the corresponding 3-amnio alkylated indole derivative **4** with good yield (68-87%). The products were characterized on the basis of spectral and elemental analyses.

Characterization data of one representative compound

4-((2-Chlorophenyl)(1*H*-indol-3-yl)methyl) morpholine (**4c**): White solid, mp 134-139 °C; IR (KBr): 3415, 3061, 2822, 1492, 1389, 1214, 1164, 749 cm^{-1}; ^{1}H NMR (300 MHz, $CDCl_3$): δ 2.44 (s, 4H), 3.66 (s, 4H), 5.11 (s, 1H), 7.01-7.15 (m, 7H), 7.94 (s, 2H), 8.12 (br, s, 1H); ^{13}C NMR (50 MHz, $CDCl_3$): δ 139.74, 136.14, 133.69, 129.57, 128.87, 127.53, 126.79, 126.23, 123.53, 121.87, 120.09, 119.47, 115.15, 111.16, 67.17, 63.73,52.32; ESIMS: *m/z* 327 (M+H); HRMS (ES): calculated 326.1186; found:326.1172. Analysis calcd. for $C_{19}H_{19}ClN_2O$: C 69.83, H 5.86; N 8.57; Found: C 69.77, H 5.75, N 8.51%.

3-((2-Chlorophenyl)(piperidin-1-yl) methyl)-5-methoxy-1*H*-indole (**4d**): White solid, mp 109-113 °C; IR (KBr): 3419, 3061, 2829, 1489, 1392, 1212, 1160, 749 cm^{-1}; ^{1}H NMR (200 MHz, $CDCl_3$): δ 0.93-1.14 (m, 6H),1.71-1.82 (m, 2H), 1.89-2.07 (m, 2H), 4.58 (s, 1H), 6.21 (s, 1H), 6.41-6.55 (m, 5H), 6.58-6.87 (m, 2H), 7.12 (d, 1H, J=3.3Hz); ^{13}C NMR (50 MHz, $CDCl_3$): δ 153.94,134.05, 132.49, 132.45, 130.46, 130.02, 127.34, 127.28, 125.97, 114.52, 113.36, 113.29, 103.83, 71.28, 56.03, 51.03, 24.68, 23.42; ESIMS: *m/z* 355 (M+H); HRMS (ES): calculated 354.1499;found:354.1483. Analysis calcd. for $C_{21}H_{23}ClN_2O$: C 71.07, H 6.53, N 7.89%; Found: C 71.01, H 6.48, N 7.78%.

Reference

Kumar, A., Gupta, M. K., and Kumar, M. (2012). L-Proline catalysed multicomponent synthesis of 3-amino alkylated indoles via a Mannich-type reaction under solvent-free conditions. *Green Chem*, **24**, 290–295.

C-C bond_35

Type of reaction: C-C bond formation
Reaction conditions: Ionic liquid (IL), room temperature
Synthetic strategy: Friedel-Crafts alkylation of *N*-heterocycles
Catalyst: [Bmim]OTf (IL)
Keywords: *N*-Heterocycles, indoles, pyrroles, epoxides, [Bmim]OTf, room temperature, ring opening, alkylation

General reaction scheme

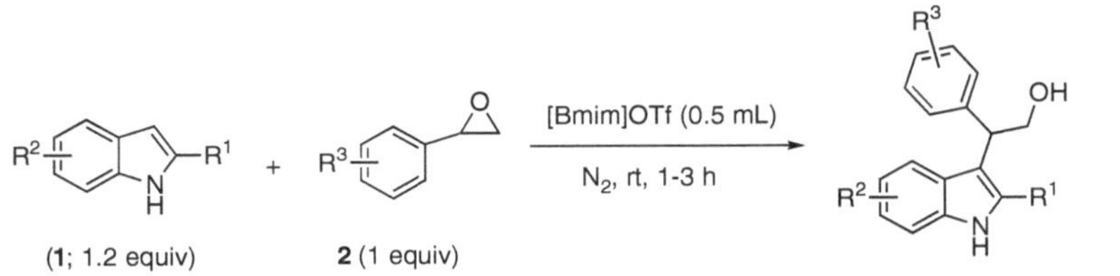

(**1**; 1.2 equiv) **2** (1 equiv)

R^1 = H, Me; R^2 = H, 5-Me, 5-OMe, 4-Br, 5-Br;
R^3= H, 4-F, 4-Cl (besides, naphthyl epoxide)

Alkylated indole **3**
11 entries (yield: 75 - 90%)

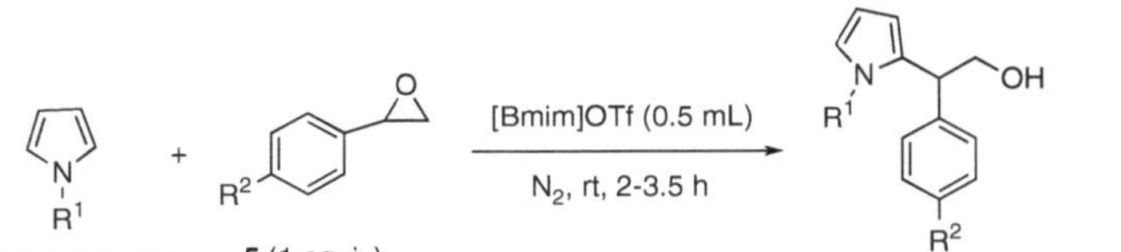

(**4**; 1.2 equiv) **5** (1 equiv)

R^1= H, Me; R^2= H, 4-F, 4-Cl

Alkylated pyrrole **6**
5 entries (yield: 78 - 82%)

Representative entries

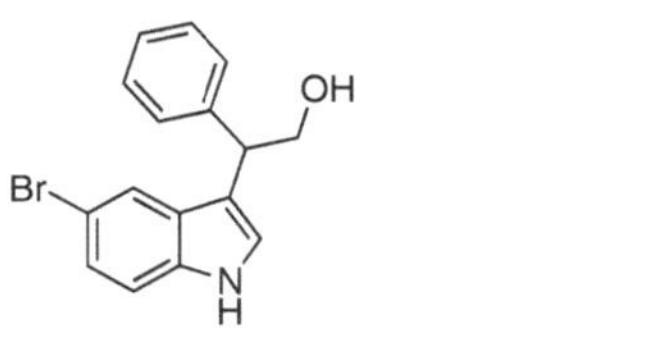

3a

2-(5-Bromo-1*H*-indol-3-yl)-2-phenylethanol (**3a**): Reaction time: 2.5 h; yield: 78%

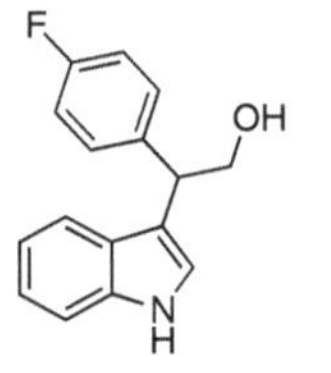

3b

2-(4-Fluorophenyl)-2-(1*H*-indol-3-yl)ethanol (**3b**): Reaction time: 1 h; yield: 82%

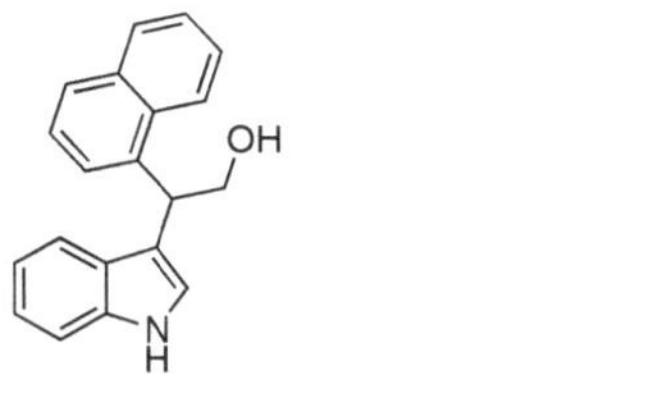

3c

2-(1*H*-Indol-3-yl)-2-(naphthalen-1-yl)ethanol (**3c**): Reaction time: 2.5 h; yield: 78%

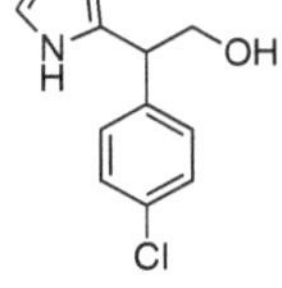

6a

2-(4-Chlorophenyl)-2-(1*H*-pyrrol-2-yl)ethanol (**6a**): Reaction time: 2 h; yield: 80%

Experimental procedure

To a stirred solution of *N*-heterocycle (indoles/pyrroles; **1/4**; 1.2 mmol) in [Bmim]OTf ionic liquid (1.2 mmol, 0.5 mL), epoxide (**2/5**; 1.0 mmol) was added under nitrogen atmosphere and stirred for required time at room temperature. After completion of the reaction, as monitored by TLC, the crude product was extracted with Et_2O (3 × 10 mL), and the combined ether extracts were concentrated in *vacuo*. The crude product was purified by column chromatography on silica gel (100-200 mesh) with EtOAc-*n*-hexane mixture as eluent to afford desired alkylated *N*-hetrocyclic derivative **3/6** in good to moderate yield. The ionic liquid was dried under vacuum and preserved for the reuse. All the compounds were characterized on the basis of analytical and spectral analyses.

Characterization data of two representative entries

2-(4-Fluorophenyl)-2-(1*H*-indol-3-yl)ethanol (**3b**): IR (neat): 3578, 3419, 3069, 2885, 1623, 1556, 1501, 1462, 1412, 1351, 1250, 1106, 1070, 1017, 754 cm^{-1}; ^{1}H NMR (300 MHz, $CDCl_3$): δ 1.81 (br, 1H), 4.06-4.26 (m, 2H), 4.44 (t, *J* = 6.7 Hz, 1H), 6.90-7.37 (m, 9H), 8.05 (br, 1H); ^{13}C NMR (75 MHz, $CDCl_3$): δ 44.83, 66.36, 111.28, 115.23, 115.53, 115.84, 119.30, 119.63, 121.90, 122.42, 126.84, 129.69 (2), 136.53, 137.43, 163.33; MS (EI): *m/z* (%rel. intensity) at 255 (13), 224 (100), 222 (14), 196 (10), 177 (7), 77 (12), 63 (13), 41 (14). Anal. Calcd. for $C_{16}H_{14}FNO$: C70.72, H 5.19, N 5.15; Found: C 70.76, H 5.24, N 5.11.

2-(4-Chlorophenyl)-2-(1*H*-pyrrol-2-yl)ethanol (**6a**): IR (neat): 3346, 2925, 1692, 1490, 1406, 1090, 1058, 1015, 826, 761 cm^{-1}; ^{1}H NMR (300 MHz, $CDCl_3$): δ 1.52 (br,1H), 3.90-4.04 (m 2H,), 4.05-4.12 (m, 1H), 5.91-5.94 (m, 1H), 6.08 (dd, *J* = 6.0, 3.0 Hz, 1H,), 6.64 (m,1H), 7.25 (d, J = 8.8 Hz, 2H), 7.29 (d, *J* = 8.8 Hz, 2H), 8.13 (br, 1H); ^{13}C NMR (75 MHz, $CDCl_3$): δ 56.28, 67.14, 107.0, 109.05, 115.16, 119.08, 130.62, 133.17, 133.24, 156.02; MS (EI): *m/z* (%rel. intensity) at 221 (13), 190 (100), 154 (36), 141 (10), 127 (19), 97 (7), 73 (28), 57 (22), 43 (36). Anal. Calcd. for $C_{12}H_{12}ClNO$: C 65.02, H 5.46, N 6.32; Found: C 65.06, H 5.49, N 6.29.

Reference

Lakshmi kantam, M., Chakravarti, R., Sreedhar, B., and Bhargava, S. (2008). Friedel-Crafts alkylation of nitrogen heterocycles using [Bmim][OTf] as a catalyst and reaction medium. *Synlett*, 1449–1454.

C-C bond_36

Type of reaction: C-C bond formation
Reaction conditions: Water, room temperature
Synthetic strategy: Michael addition of indoles
Catalyst: Glucosyl tolylsulfonyl hydrazine (**1**)

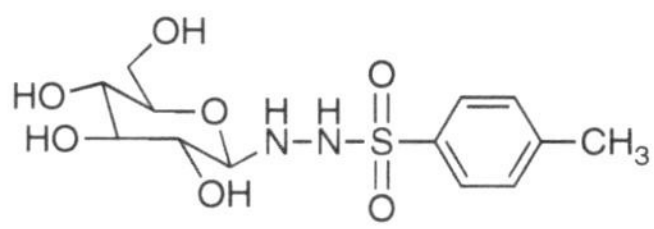

1

Keywords: Carbohydrate-based tolylsulfonyl hydrazine, indoles, electron-deficient olefins, water, room temperature, Michael addition, 3-substituted indoles

General reaction scheme

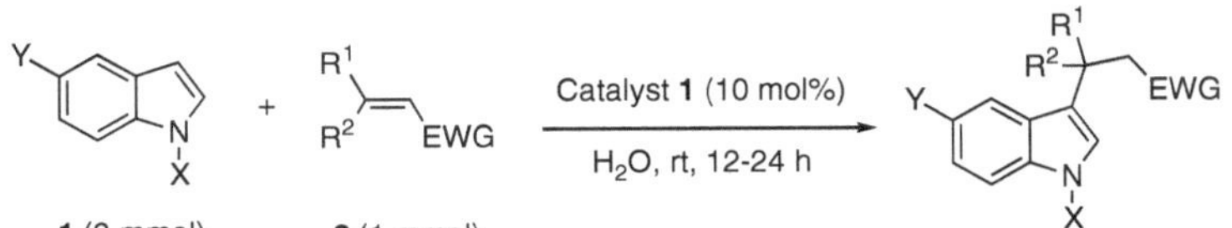

1 (2 mmol) **2** (1 mmol)

3-Substituted indole **3**
19 entries (yield: 27 - 95%)

X,Y = H, H; Me, H; H, Br
Olefins: A variety of nitrostyrenes and cyclic enones
EWG: Electron Withdraing Group

Representative entries

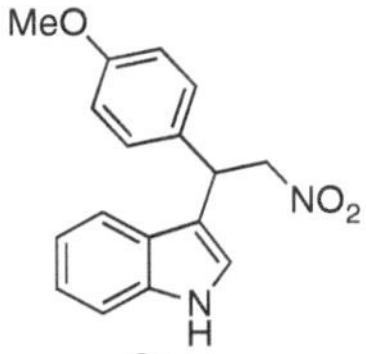

3a

3-(1-(4-Methoxyphenyl)-2-nitroethyl)-1*H*-indole (**3a**): Reaction time: 14 h; yield: 86%

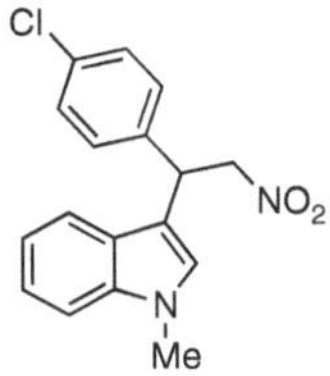

3b

3-(1-(4-Chlorophenyl)-2-nitroethyl)-1-methyl-1*H*-indole (**3b**): Reaction time: 19 h; yield: 90%

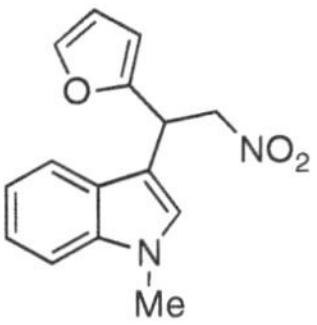

3c

3-(1-(Furan-2-yl)-2-nitroethyl)-1-methyl-1*H*-indole (**3c**): Reaction time: 15 h; yield: 95%

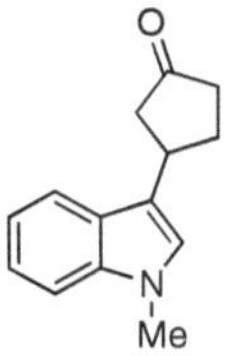

3d

3-(1-Methyl-1*H*-indol-3-yl)cyclopentanone (**3d**): Reaction time: 12 h; yield: 89%

Experimental procedure

A mixture of the olefin (**2**; 1 mmol) and catalyst **1** (10 mol%) in water (1 mL) was added with indole (**1**; 2 mmol), and the resulting suspension was stirred at room temperature until the starting material was consumed (12-24 h). On completion of reaction (as monitored by TLC), the mixture was quenched with a saturated aqueous NaCl solution followed by extraction with EtOAc (3 × 20 mL). The combined organic layers was washed with brine, and dried over anhydrous $MgSO_4$. On removal of the solvent under reduced pressure, a crude product was obtained; this was purified by column chromatographic resolution (silica gel 100–200 mesh ZCX II; hexane-EtOAc mixture as eluent) to afford pure 3-substituted indole **3** with yield up to 95%. All the compounds were characterized from comparison of their physical and spectral properties with those reported for authentic samples in literature.

Reference

Wu, P., Wan, Y., and Cai, J. (2008). Carbohydrate-based tolylsulfonyl hydrazines: Efficient catalysts for Michael addition of indoles to electron-deficient olefins in water. *Synlett*, 1193–1198.

C-C bond_37

Type of reaction: C-C bond formation
Reaction conditions: Acetonitrile, room temperature
Synthetic strategy: Functionalization of indoles
Catalyst: *p*-Tolenesulfonic acid (PTSA)
Keywords: Indoles, tertiary propargylic and benzylic alcohols, PTSA, acetonitrile, room temperature, C-3 selective propargylation and benzylation

General reaction scheme

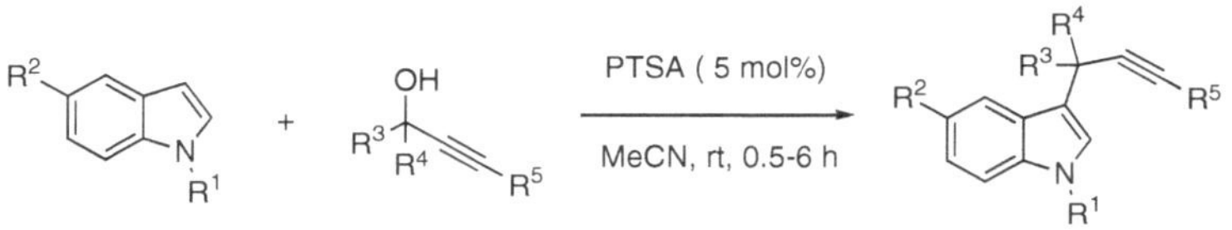

1 (2 mmol) **2** (2.4 mmol)

3-Propargylated indole **3**
18 entries (yield: 46-85%)

R^1 = H, Me; R^2 = H, Br, NO_2, CO_2Me
R^3 = C_6H_5, 4-ClC_6H_4, 2-thienyl
R^4 = Me, Et, *n*-Pr, *i*-Pr
R^5 = C_6H_5, *n*-Bu

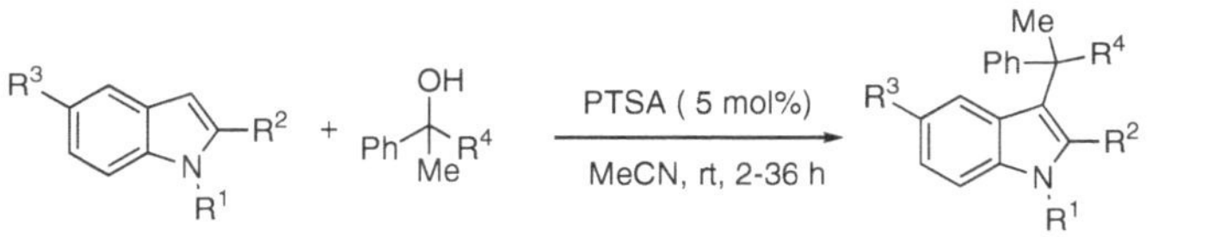

4 (2 mmol) **5** (2.4 mmol)

3-Benzylated indole **6**
7 entries (yield: 60-89%)

R^1 = H, Me; R^2 = H, Me, C_6H_5
R^3 = H, CO_2Me; R^4 = C_6H_5, *n*-Bu

Representative entries

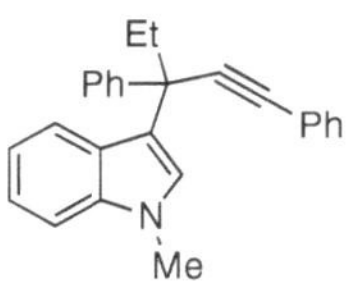

3a

3-(1,3-Diphenylpent-1-yn-3-yl)-1-methyl-1*H*-indole (**3a**): Reaction time: 2 h; yield: 78%

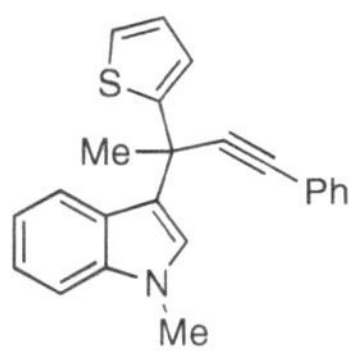

3b

1-Methyl-3-(4-phenyl-2-(thiophen-2-yl)but-3-yn-2-yl)-1*H*-indole (**3b**): Reaction time: 1 h; yield: 85%

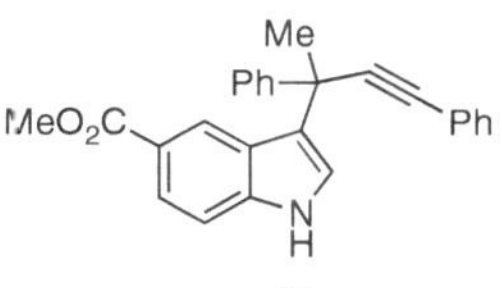

3c

Methyl 3-(2,4-diphenylbut-3-yn-2-yl)-1*H*-indole-5-carboxylate (**3c**): Reaction time: 1 h; yield: 81%

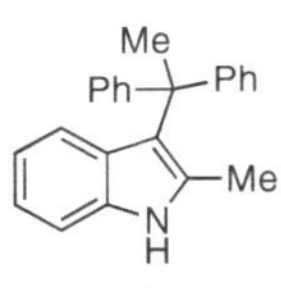

6a

3-(1,1-Diphenylethyl)-2-methyl-1H-indole (**6a**): Reaction time: 24 h; yield: 89%

Experimental procedure

To a mixture of tertiary alcohol (**2/5**; 2.4 mmol) and indole (**1/4**; 2 mmol) in acetonitrile (2 mL), PTSA (0.019 g, 0.1 mmol) was added. The reaction was then stirred at room temperature for stipulated time. On completion of the reaction, the solvent was evapotrated under reduced pressure and the residue was purified by silica gel column chromatography using hexane-Et_2O mixture as eluent to afford substituted indole **3/6**. All the products were characterized by means of spectral analyses.

Characterization data of one representative entry

3-(1,3-Diphenylpent-1-yn-3-yl)-1-methyl-1*H*-indole (**3a**): White solid, mp 124-126 °C. IR (KBr): 2962, 2930, 1488, 1463, 1326, 758, 741, 701 cm^{-1}; 1H NMR (400 MHz, $CDCl_3$): δ 1.32 (t, $J=7.3$ Hz, 3H), 2.56 (dq, $J=7.2$, 14.3 Hz, 1H), 2.83 (dq, $J=7.2$, 14.3 Hz, 1H), 3.82 (s, 3H), 7.16-7.24 (m, 2H), 7.35-7.54 (m, 2H), 7.66-7.74 (m, 2H), 7.81 (d, $J=8.0$ Hz, 1H), 7.86 (d, $J=7.2$ Hz, 2H); ^{13}C NMR (100.6 MHz, $CDCl_3$): δ 10.1 (Me), 32.7 (Me), 34.8 (CH2), 45.4 (C), 84.9 (C), 93.7 (C), 109.3 (CH), 118.8 (CH), 119.5 (C), 121.3 (CH), 121.6 (CH), 124.0 (C), 126.3 (C), 126.4 (CH), 126.5 (CH), 127.3 (2 × CH), 127.8 (CH), 128.1 (2 × CH), 128.3 (2 × CH), 131.7 (2 × CH), 137.7 (C), 144.6 (C); LRMS (EI): *m/z* 349 (9) [M^+], 320 (100); HRMS: *m/z* calcd for $C_{26}H_{23}N$: 349.1830; Found: 349.1836.

Reference

Sanz, R., Miguel, D., Alvarez-Gutierrez, J. M., and Rodriguez, F. (2008). Bronsted acid catalyzed C3-selective propargylation and benzylation of indoles with tertiary alcohols. *Synlett*, 975–978.

C-C bond_38

Type of reaction: C-C bond formation
Reaction conditions: Dichloromethane, room temperature
Synthetic strategy: Cycloaddition reaction
Catalyst: 2,3-Dichloro-5,6-dicyanobenzoquinone (DDQ)/$B(OAc)_3$
Keywords: Juglone, styrenes, DDQ, $B(OAc)_3$, dichloromethane, room temperature, cycloaddition, Diels-Alder reaction, angucyclinones, 11-hydroxybenz[*a*]anthracene-7,12-diones

General reaction scheme

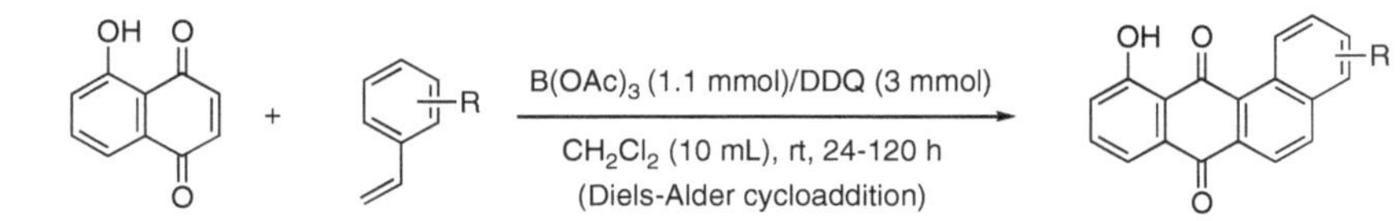

1 (Juglone; 1 mmol) **2** (4 mmol)

R = H; 4-Br; 3-Me; 4-Me; 3-OMe; 4-OMe; 3,4-di-OMe; 3,5-di-OMe; 2-Me,5-OMe; 2-Br,5-OMe; 2-Br,3-Me,5-OMe

11-Hydroxybenz[a]anthracene-7,12-one **3**
11 entries (yield: 32-82%)

Representative entries

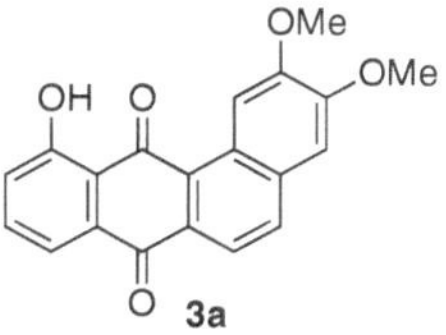

3a

11-Hydroxy-2,3-dimethoxytetraphene-7,12-dione (**3a**): Red solid, mp 268-270 °C; reaction time: 24 h; yield: 61%

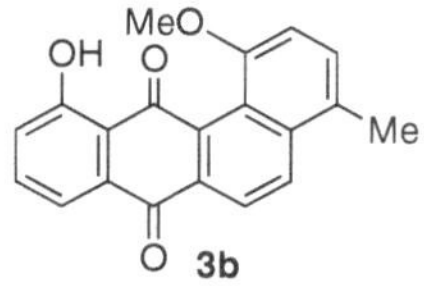

3b

11-Hydroxy-1-methoxy-4-methyltetraphene-7,12-dione (**3b**): Red solid, mp 180-182 °C; reaction time: 24 h; yield: 44%

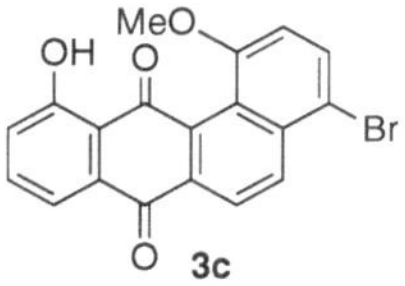

3c

4-Bromo-11-hydroxy-1-methoxytetraphene-7,12-dione (**3c**): Red solid, mp 243-245 °C; reaction time: 48 h; yield: 59%

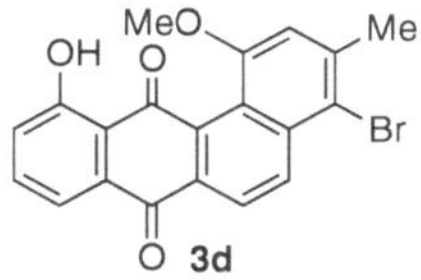

3d

4-Bromo-11-hydroxy-1-methoxy-3-methyltetraphene-7,12-dione (**3d**): Red solid, mp 268-270 °C; reaction time: 24 h; yield: 32%

Experimental procedure

A mixture of juglone (**1**; 1 mmol, 187 mg) and $B(OAc)_3$ (1.1 mmol, 206 mg) in 10 mL of dichloromethane (10 mL) was stirred at room temperature for 10 min and then styrene derivative (**2**; 4 mmol) and DDQ (680 mg, 3 mmol) were added to the reaction mixture. Additional $B(OAc)_3$ (1.1 mmol), styrene derivatives (4 mmol), and DDQ (3 mmol) were added to the reaction mixture on every 24 h stirring until juglone was completely consumed. On completion, the reaction was quenched with saturated aqueous $NaHCO_3$, and extracted with dichloromethane. The combined organic extracts were washed successively with water, brine, dried over anhydrous magnesium sulfate, filtered and concentrated. The crude products of 11-hydroxybenz[*a*]anthracene-7,12-diones (**3**) were recrystallizd from acetonitrile. The filtrate was further purified by flash silica-gel chromatography (hexanes/CH_2Cl_2) to give additional **3**. Analytically pure **3** was obtained by crystallization from CH_2Cl_2/hexane. All the products were characterized by means of detailed spectral studies and elemental analyses.

Characterization data of two representative entries

11-Hydroxy-2,3-dimethoxytetraphene-7,12-dione (**3a**): Red solid, mp 268-270 °C; IR (neat): 1659, 1627, 1585, 1485, 1430, 1368, 1303, 1262, 1215, 1075, 1040, 870, 741 cm^{-1}; ^1H NMR (500 MHz, $CDCl_3$): δ 12.97 (s, 1H), 9.31 (s, 1H), 8.26 (d, $J=8.5$ Hz, 1H), 8.05 (d, $J=8.5$ Hz, 1H), 7.79 (d, $J=7.5$ Hz, 1H), 7.64 (dd, $J=8.5$, 7.5 Hz, 1H), 7.31 (d, $J=8.5$ Hz, 1H), 7.16 (s, 1H), 4.15 (s, 3H), 4.07 (s, 3H); ^{13}C NMR (125 MHz, $CDCl_3$): δ 192.5, 183.3, 162.3, 153.1, 151.3, 136.0, 134.2, 134.1, 132.9, 132.7, 127.1, 126.6, 124.6, 121.7, 118.8, 117.3, 107.0, 106.9, 56.2, 56.0; MS (EI): *m/z* (rel%) 334 (M^+,100), 291 (23), 263 (9), 248 (4), 220 (14), 205 (2), 192 (6), 163 (11), 138 (3), 88 (3); HRMS (EI): *m/z* calcd for $C_{20}H_{14}O_5$ [M^+]: 334.0841. Found 334.0847. Anal. Calcd for $C_{20}H_{14}O_5$: C, 71.85; H, 4.22; Found: C, 71.60; H, 4.30.

4-Bromo-11-hydroxy-1-methoxy-3-methyltetraphene-7,12-dione (**3d**): Red solid, mp 268-270 °C; IR (neat): 1665, 1643, 1602, 1578, 1443, 1360, 1292, 1206, 1078, 846, 740 cm^{-1}; ^1H NMR (500 MHz, $CDCl_3$): δ 11.55 (s, 1H), 8.58 (d, $J=8.5$ Hz, 1H), 8.30 (d, $J=8.5$ Hz, 1H), 7.77 (d, $J=7.0$ Hz, 1H), 7.63 (dd, $J=8.5$, 7.0 Hz, 1H), 7.31 (d, $J=8.5$ Hz, 1H), 7.00 (s, 1H), 3.98 (s, 3H), 2.66 (s, 3H); ^{13}C NMR (125 MHz, $CDCl_3$): δ 188.6, 182.6, 160.1,

156.4, 140.8, 136.4, 135.4, 134.2, 133.6, 133.3, 132.4, 124.0, 123.7, 121.2, 118.9, 117.9, 115.4, 112.3, 56.1, 24.7; MS (EI): *m/z* (rel%) 398 (77), 396 (M^+, 78), 380 (100), 378 (98), 367 (15), 274 (16), 218 (27), 189 (63), 163 (16), 101 (13), 94 (16); HRMS (EI): *m/z* calcd for $C_{20}H_{13}^{79}BrO_4$ [M^+]: 395.9997; Found 396.0006.

Reference

Hsu, D.-S., and Huang, J.-Y. (2012). Room-temperature $B(OAc)_3$-promoted Diels – Alder reaction of juglone with styrenes: Total syntheses of tetrangulol and anhydrolandomycinone. *J. Org. Chem.*, **77**, 2659–2666.

C-C bond_39

Type of reaction: C-C bond formation
Reaction conditions: Dichloromethane, room temperature
Synthetic strategy: One-pot multicomponent [4+3]-cycloaddition reaction
Catalyst: Gallium(III) salts [$GaBr_3$/$Ga(OTf)_3$]
Keywords: Indoles, aldehydes, dienes, Ga(III) salts, dichloromethane, room temperature, one-pot three-component [4+3]-cycloaddition, cyclohepta[*b*]indoles

General reaction scheme

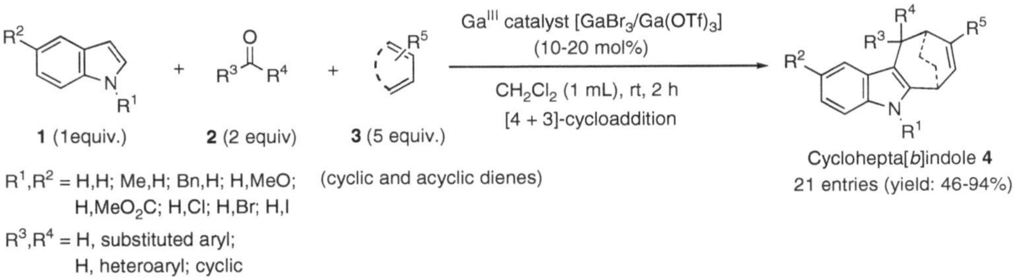

Representative entries

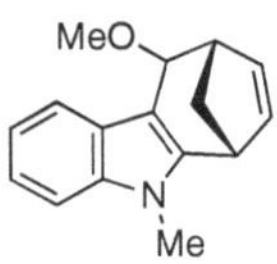

4a

(6*R*,9*S*)-10-Methoxy-5-methyl-5,6,9,10-tetrahydro-6,9-methanocyclohepta[*b*]indole (**4a**): Colorless oil; yield: 75% (10:1 d.r.)

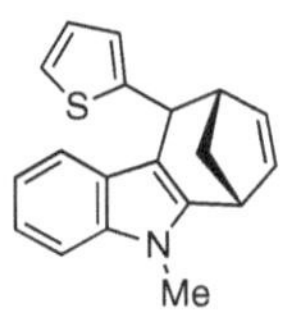

4b

(6*R*,9*S*)-5-Methyl-10-(thiophen-2-yl)-5,6,9,10-tetrahydro-6,9-methanocyclohepta[*b*] indole (**4b**): Light yellow oil; yield: 83% (10:1 d.r.)

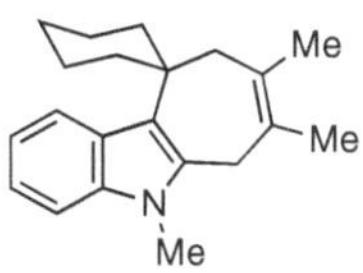

4c

5,7,8-Trimethyl-6,9-dihydro-5*H*-spiro[cyclohepta [*b*]indole-10,1′-cyclohexane] (**4c**): White solid, mp 107 °C; yield: 46%

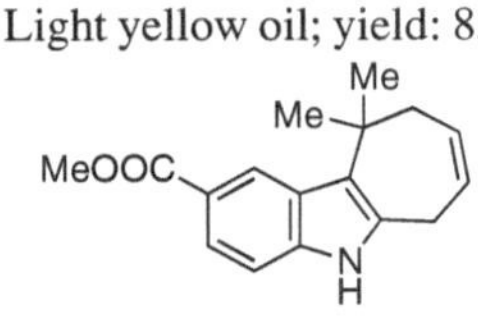

4d

Methyl 10,10-dimethyl-5,6,9,10-tetrahydrocyclohepta[*b*]indole-2-carboxylate (**4d**): White solid, mp 205-206 °C; yield: 76%

Experimental procedure

A mixture of indole (**1**; 1 equiv.), aldehyde (**2**; 2 equiv.) and diene (3; 5 equiv.) in 1 mL of dichloromethane placed in a round-bottom flask was added with Ga(III) salt [$GaBr_3$/Ga $(OTf)_3$] (10-20 mol% as per the diene-type). The mixture was then stirred at room temperature until completion of the reaction (2 h) as monitored by TLC. The crude product of cyclohepta[*b*] indole **4** was purified by silica-gel chromatography (hexanes/EtOAc). All the products were characterized by means of detailed spectral studies.

Characterization data of two representative entries

(6*R*,9*S*)-10-Methoxy-5-methyl-5,6,9,10-tetrahydro-6,9-methanocyclohepta[*b*]indole (**4a**): Colorless oil; IR (film): 2954, 2929, 2856, 1722, 1710, 1611, 1510, 1467, 1248, 1174, 1035, 833, 740 cm^{-1}; ^{1}H NMR (500 MHz, CD_2Cl_2): δ 7.29-7.27 (d, *J*=8.5, Hz, 1H), 7.19-7.16 (dd, *J*=6.8, 1.7 Hz, 2H), 7.08-7.04 (m, 1H), 6.87-6.82 (m, 4H), 6.45-6.43 (q, *J*=2.3 Hz, 1H), 6.07-6.05 (q, *J*=2.7 Hz, 1H), 3.97 (s, 1H), 3.78 (s, 3H), 3.76 (s, 3H), 3.68-3.66 (dd, *J*=4.3, 2.8 Hz, 1H), 2.88-2.86 (dt, *J*=4.7, 1.6 Hz, 1H), 2.18-2.16 (d, *J*=10.0 Hz, 1H), 2.08-2.05 (m, 1H); ^{13}C NMR (125 MHz, $CDCl_3$): δ 158.2, 142.5, 139.7, 138.0, 136.6, 132.4, 129.4, 128.1, 120.0, 118.9, 118.3, 113.7, 109.0, 105.0, 55.3, 49.2, 41.6, 37.6, 37.2, 29.3; HRMS (ES+): *m/z* calcd for [MH^+]: 316.1701. Found 316.1707.

Methyl 10,10-dimethyl-5,6,9,10-tetrahydrocyclohepta[*b*]indole-2-carboxylate (**4d**): White solid, mp 205-206 °C; IR (film): 3326, 2958, 2861, 1710, 1689, 1612, 1451, 1434, 1295, 1262, 1203, 1095, 742 cm^{-1}; ^{1}H NMR (500 MHz, CD_2Cl_2): δ 8.27-8.26 (t, *J*=0.8 Hz, 1H), 8.16 (s, br, 1H), 7.75-7.73 (dd, *J*=8.5, 1.5 Hz, 1H), 7.33-7.31 (dd, *J*=8.5, 0.5 Hz, 1H), 6.45-6.43 (q, *J*=2.8 Hz, 1H), 5.96-5.94 (q, *J*=2.8 Hz,1H), 3.90 (s, 3H), 3.36-3.34 (m, 1H), 2.66-2.64 (t, *J*=3.8 Hz, 1H), 2.29-2.21 (m, 2H), 1.56 (s, 3H), 1.34 (s, 3H); ^{13}C NMR (125 MHz, $CDCl_3$): δ 168.3, 140.4, 139.5, 137.6, 133.0, 127.2, 121.5, 121.4, 114.4, 110.6, 52.7, 51.8, 41.2, 39.1, 35.9, 30.0, 24.4; HRMS (ES+): *m/z* calcd for [MH^+]: 282.1494. Found 282.1499.

Reference

Han, X., Li, H., Hughes, R. P., and Wu, J. (2012). Gallium(III)-catalyzed three-component (4+3)-cycloaddition reactions. *Angew. Chem. Int. Ed.*, **51**, 10390–10393.

C-C bond_40

Type of reaction: C-C and C-N bond formation
Reaction conditions: Neat condition, room temperature
Synthetic strategy: One-pot multicomponent reaction
Catalyst: Ethylammonium nitrate (EAN)
Keywords: Aldehydes, amides/carbamtaes/urea, naphthols/phenols, ethylammonium nitrate (EAN), recyclable ionic liquid catalyst, neat condition, room temperature, one-pot three-component condensation, 1-amino- and 1-carbamato-alkyl naphthols/phenols

General reaction scheme

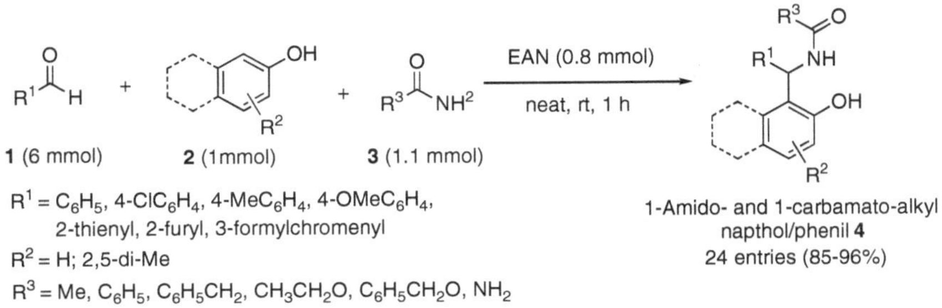

Representative entries

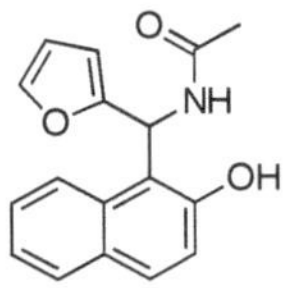

4a

N-(Furan-2-yl(2-hydroxynaphthalen-1-yl)methyl) acetamide (**4a**): Brown solid, mp 226-228 °C; yield: 87%

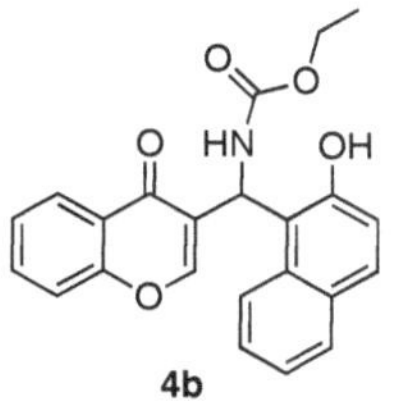

4b

Ethyl ((2-hydroxynaphthalen-1-yl)(4-oxo-4*H*-chromen-3-yl)methyl)carbamate (**4b**): Yellow solid, mp 181-183 °C; yield: 85%

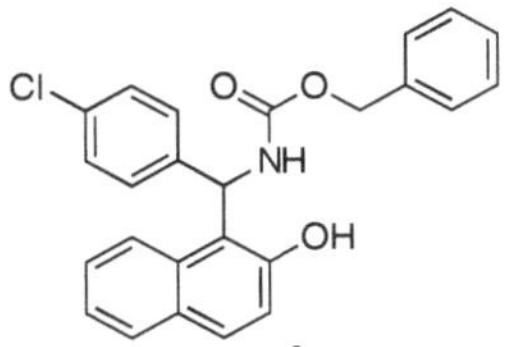

4c

Benzyl ((4-chlorophenyl)(2-hydroxynaphthalen-1-yl)methyl)carbamate (**4c**): White solid, mp 154-156 °C; yield: 90%

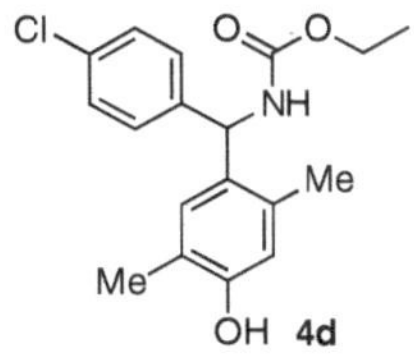

4d

Ethyl ((4-chlorophenyl)(4-hydroxy-2,5-dimethylphenyl)methyl)carbamate (**4d**): Yellow solid, mp 166-168 °C; yield: 95%

Experimental procedure

A mixture of aldehyde (**1**; 6 mmol), beta-naphthol/phenol (**2**; 1 mmol) and amide/carbamate/urea (**3**; 1.1 mmol) and ethylammonium nitrate (EAN; 0.8 mmol) was stirred at room temperature for 1 h (the completion of reaction was monitored by TLC). On completion of reaction, the reaction mixture was extracted thrice with 10 mL ethyl acetate. The extract was dried over anhydrous sodium sulfate, evaporated under vacuum to obtain the crude product which was then purified by chromatographic column on silica gel (hexane/ethyl acetate, 70:30) to yield pure 1-amido- and 1-carbamato-alkyl naphthol/phenol **4**. The recovered EAN was subjected to high vaccum at 80 °C to remove the water and then reused. All the isolated reaction products were characterized and confirmed by NMR.

Characterization data of two representative entries

Ethyl ((2-hydroxynaphthalen-1-yl)(4-oxo-4*H*-chromen-3-yl)methyl)carbamate (**4b**): Yellow solid, mp 181-183 °C; ^{1}H NMR (200 MHz, DMSO-d_6): δ 10.14 (s, 1H), 8.54 (s, 1H), 8.43 (d, 2H), 8.07-8.03 (d, J=9 Hz, 1H),7.82-7.71(m, 3H), 7.69-7.50 (m, 4H), 7.43-7.17

(m, 2H), 7.01 (br s, 1H), 4.03 (q, 2H), 1.21 (t, 3H); ^{13}C NMR (50 MHz, DMSO-d_6): δ 168.39, 152.24, 138.27, 134.60, 131.62, 128.18, 127.65, 127.53, 125.35, 121.70, 117.98, 47.75, 22.24, 19.92.

Ethyl ((4-chlorophenyl)(4-hydroxy-2,5-dimethylphenyl)methyl)carbamate (**4d**): Yellow solid, mp 166-168 °C; ^{1}H NMR (200 MHz, DMSO-d_6): δ 9.17 (s, 1H, OH exchanged with D_2O), 8.01 (d, $J=9$ Hz, 1H, NH exchanged with D_2O), 7.4 (d, $J=8$ Hz, 2H), 7.24 (d, $J=8$ Hz, 2H), 6.85 (s, 1H), 6.60 (s, 1H), 5.91 (d, $J=9$ Hz, 1H), 4.09 (q, 2H), 2.16 (s, 3H), 2.03 (s, 3H), 1.16(t, 3H); ^{13}C NMR (50 MHz, DMSO-d_6): δ 155.92, 154.40, 142.02, 133.65, 131.44, 130.65, 129.77, 129.22, 128.36, 121.20, 116.46, 60.09, 53.76, 18.95, 16.02, 14.86.

Reference

Mulla, S. A. R., Salama, T. A., Pathan, M. Y., Inamdar, S. M., and Chavan, S. S. (2013). Solvent-free, highly efficient one-pot multi-component synthesis of 1-amido- and 1-carbamato-alkyl naphthols/phenols catalyzed by ethyl ammonium nitrate as reusable ionic liquid under neat reaction condition at ambient temperature. *Tetrahedron Lett.*, **54**, 672–675.

C-C bond_41

Type of reaction: C-C bond formation
Reaction conditions: Solvent-free, room temperature
Synthetic strategy: One-pot multicomponent reaction
Catalyst: L-Proline
Keywords: Aldehydes, indoles, C-H activated acids (4-hydroxycoumarin, 4-hydroxy-6-methyl-2*H*-pyran-2-one, dimedone, *N,N*-dimethylbarbituric acid, Meldrum's acid), L-proline, solvent-free, room temperature, grinding, one-pot three-component condensation, *gem*-(β-dicarbonyl)arylmethanes

General reaction scheme

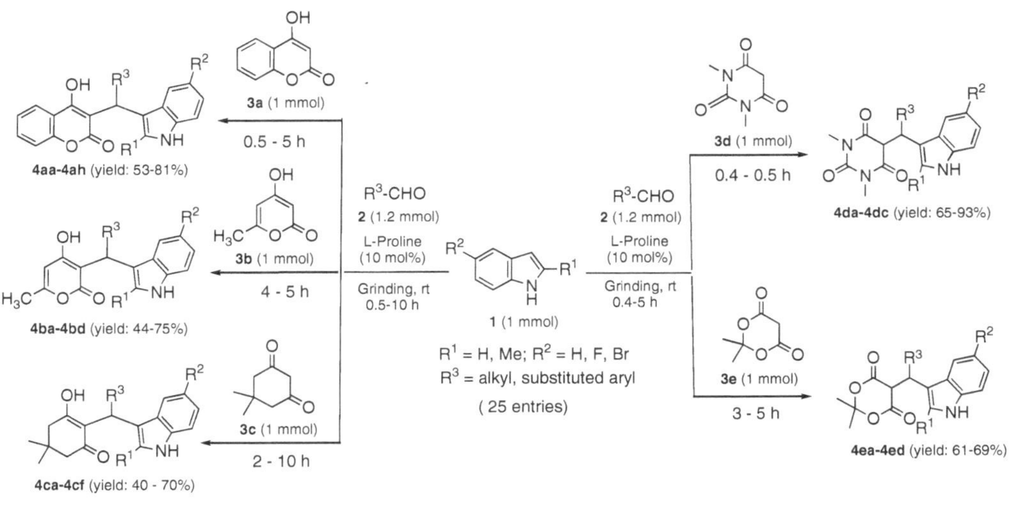

Proposed mechanism

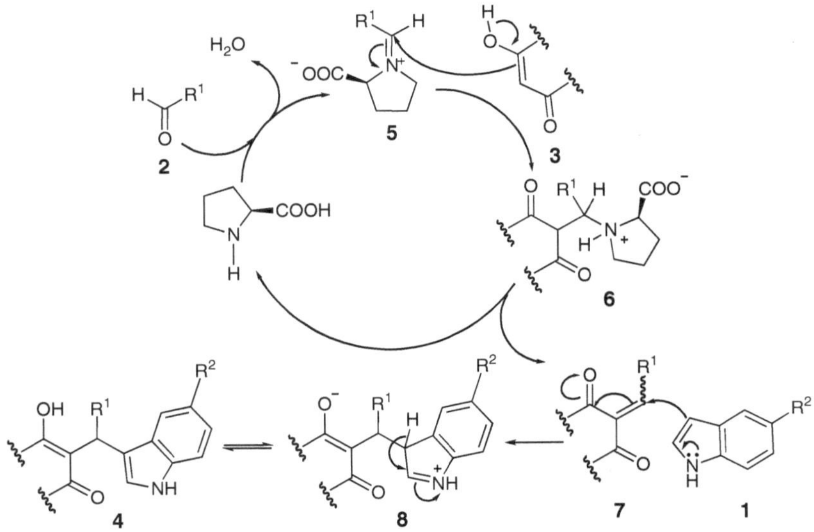

Representative entries

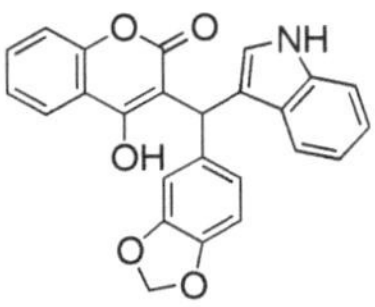

4ae

3-(Benzo[*d*][1,3]dioxol-5-yl(1*H*-indol-3-yl)methyl)-4-hydroxy-2*H*-chromen-2-one (**4ae**): Brownish white solid, mp 242-244 °C; yield: 76%

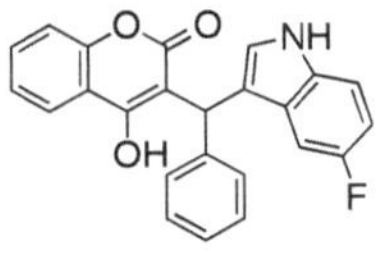

4ah

3-((5-Fluoro-1*H*-indol-3-yl)(phenyl)methyl)-4-hydroxy-2*H*-chromen-2-one (**4ah**): Reddish white solid, mp 178-179 °C; yield: 61%

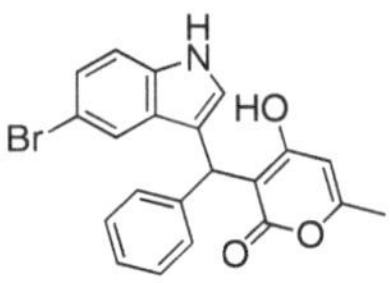

4bc

3-((5-Bromo-1*H*-indol-3-yl)(phenyl)methyl)-4-hydroxy-6-methyl-2*H*-pyran-2-one (**4bc**): Deep brown solid, mp 110-112 °C; yield: 60%

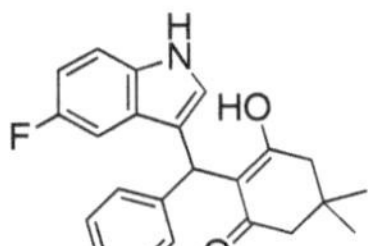

4cf

2-((5-Fluoro-1*H*-indol-3-yl)(phenyl)methyl)-3-hydroxy-5,5-dimethylcyclohex-2-enone (**4cf**): Brownish white solid, mp 169-171 °C; yield: 65%

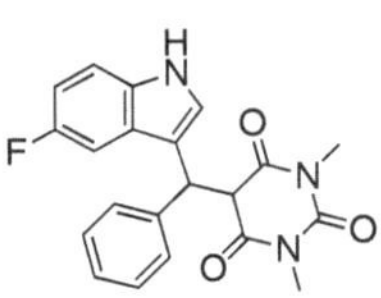

4dc

5-((5-Fluoro-1*H*-indol-3-yl)(phenyl)methyl)-1,3-dimethyl pyrimidine-2,4,6(1*H*,3H,5*H*)-trione (**4 dc**): White solid, mp 174-176 °C; yield: 93%

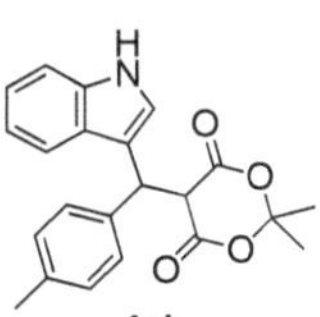

4eb

5-((1*H*-Indol-3-yl)(*p*-tolyl)methyl)-2,2-dimethyl-1,3-dioxane-4,6-dione (**4eb**): White solid, mp 128-130 °C; yield: 63%

Experimental procedure

In an oven-dried watch-glass, C-H activated acid (4-hydroxycoumarine **3a** or 4-hydroxy-6-methyl-2*H*-pyran-2-one **3b** or dimedone **3c** or *N*,*N*-dimethylbarbituric acid **3d** or Meldrum's acid **3e**; 1 mmol), aldehyde (**2**; 1.2 mmol), indole (**1**; 1 mmol), and L-proline (10 mol%) were taken together, and instantly started to grind the mixture by a spatula. After few minutes of scratching/grinding, when the solid reaction mixture liquefied followed by semi-solidification, it was left for stipulated time (0.5 - 10 h) with occasional grinding till it solidified again. After completion of the reaction (as monitored by TLC), the solid reaction mixture was subjected to column chromatography using silica gel (60-120 mesh) and petrol ether-ethyl acetate mixture to furnish the pure products (**4aa-4ed**), characterized by conventional spectroscopic methods and elemental analyses. It was also possible to isolate pure products of **4aa-4af** and **4da-4 dc** without using column chromatographic resolution just by washing the crude products in ethanol — however, with ~5-7% lower yields in comparison to the generalized method, possibly due to partial solubility of these compounds in ethanol. All the products were characterized by means of analytical and spectral studies.

Characterization data of representative entries

3-(Benzo[*d*][1,3]dioxol-5-yl(1*H*-indol-3-yl)methyl)-4-hydroxy-2*H*-chromen-2-one (**4ae**): Brownish white solid, mp 242-244 °C; IR (KBr): 3409, 3295, 3119, 3084, 3049, 2893, 1655, 1636, 1607, 1560, 1491, 1232, 1184, 1040, 933, 764 cm^{-1}; ^1H NMR (400 MHz, DMSO-d_6): δ 5.95 (s, 2H, -CH_2), 6.26 (s, 1H, -CH), 6.62 (d, $J=8$ Hz, 1H, ArH), 6.71 (s, 1H, ArH), 6.76 (d, $J=8.4$, 1H, ArH), 7.31-7.37 (m, 5H, ArH), 7.59 (t, $J=7.2$, 8 Hz, 2H, ArH), 7.91 (d, $J=7.6$ Hz, 2H, ArH), 10.81 (s {weak}, 1H, -NH), 10.89 (s {weak}, 1H, -OH); ^{13}C NMR (100 MHz, DMSO-d_6): δ 35.79, 100.68, 104.33, 107.50, 107.72, 115.93 (2C), 118.00, 119.49 (2C), 123.68 (2C), 123.89 (2C), 128.66, 131.57, 131.83 (2C), 133.91, 145.16, 147.27, 152.23 (2C), 164.70, 165.29; TOF-MS: Calcd for $C_{25}H_{17}NO_5Na$ $[M+Na]^+$: 434.1004; Found: 434.1001. Anal. Calcd. for $C_{25}H_{17}NO_5$: C, 72.99; H, 4.16; N, 3.40; Found: C, 72.95; H, 4.16; N, 3.39.

3-((5-Bromo-1*H*-indol-3-yl)(phenyl)methyl)-4-hydroxy-6-methyl-2*H*-pyran-2-one (**4bc**): Deep brown solid, mp 110-112 °C; IR (KBr): 3398, 3362, 3317, 3076, 3006, 2922, 2851, 1668, 1570, 1556, 1447, 1275, 1099, 1033, 987, 878, 702 cm^{-1}; ^1H NMR (400 MHz, DMSO-d_6): δ 2.15 (s, 3H, -CH_3), 5.76 (s, 1H, -CH), 6.03 (s, 1H, -CH), 7.09 (d, $J=2$ Hz, 1H, ArH), 7.12-7.15 (m, 2H, ArH), 7.23-7.24 (m, 4H, ArH), 7.32 (s, 1H, ArH), 7.34-7.37 (m, 1H, ArH), 11.07 (s, 1H, -NH), 11.46 (s, 1H, -OH); ^{13}C NMR (100 MHz, DMSO-d_6): δ 19.28, 35.91, 101.55, 103.65, 110.84, 113.46, 114.64, 120.56, 123.10, 125.65, 125.90, 126.53, 127.70, 127.96, 128.06, 129.04, 134.62, 142.63, 163.75, 165.38, 166.32; TOF-MS: Calcd for $C_{21}H_{16}BrNO_3Na$ $[M+Na]^+$: 432.0211; Found: 432.0218. Anal. Calcd. for $C_{21}H_{16}BrNO_3$: C, 61.48; H, 3.93; N, 3.41; Found: C, 61.46; H, 3.95; N, 3.43.

2-((5-Fluoro-1*H*-indol-3-yl)(phenyl)methyl)-3-hydroxy-5,5-dimethylcyclohex-2-enone (**4cf**): Brownish white solid, mp 169-171 °C; IR (KBr): 3487, 3366, 2949, 1682, 1548, 1473, 1253, 1163, 1084, 1024, 842, 783 cm^{-1}; ^1H NMR (400 MHz, DMSO-d_6): δ 0.92 (s, 6H, 2 x CH_3), 2.20 (br s, 4H, 2 x CH_2), 5.68 (s, 1H, -CH), 6.77-6.80 (m, 2H, ArH), 6.92 (s, 1H, ArH), 7.01-7.03 (m, 1H, ArH), 7.09-7.13 (m, 4H, ArH), 7.25 (q, $J=4.4$, 4.8, 9.4 Hz, 1H, ArH), 10.49 (br s, 1H, -NH), 10.79 (s, 1H, -OH); ^{13}C NMR (100 MHz, DMSO-d_6): δ 28.37 (2C), 32.08 (2C), 35.38, 44.61, 103.45, 103.68, 108.86, 109.12, 112.57, 112.67, 116.70, 125.45, 126.71, 127.87, 127.99, 128.53, 135.06, 144.63, 155.68, 157.97, 195.23; TOF-MS:

Calcd for $C_{23}H_{22}FNO_2Na$ $[M+Na]^+$: 386.1532; Found: 386.1529. Anal. Calcd. for $C_{23}H_{22}FNO_2$: C, 76.01; H, 6.10; N, 3.85; Found: C, 76.03; H, 6.08; N, 3.87.

5-((5-Fluoro-1H-indol-3-yl)(phenyl)methyl)-1,3-dimethyl pyrimidine-2,4,6(1*H*,3*H*,5*H*)-trione (**4 dc**): White solid, mp 174-176 °C; IR (KBr): 3388, 3122, 2951, 1747, 1668, 1467, 1288, 1170, 1105, 925, 806, 761 cm^{-1}; ^1H NMR (400 MHz, DMSO-d_6): δ 2.84 (s, 3H, $-CH_3$), 2.96 (s, 3H, $-CH_3$), 4.38 (d, $J=3.2$ Hz, 1H,-CH), 5.00 (d, $J=2.8$ Hz, 1H, -CH), 6.79-6.92 (m, 2H, ArH), 7.20-7.28 (m, 5H, ArH), 7.36 (q, $J=4.4$ Hz, 1H, ArH), 7.58 (s, 1H, ArH), 11.18 (s, 1H, -NH); ^{13}C NMR (100 MHz, DMSO-d_6): δ 28.20, 28.31, 46.91, 54.93, 103.45, 103.68, 109.63, 109.89, 112.97, 113.07, 113.45, 127.00, 127.64, 128.54, 133.12, 139.73, 151.51, 155.83, 158.13, 168.32, 169.04; TOF-MS: Calcd for $C_{21}H_{18}FN_3O_3Na$ $[M+Na]^+$: 402.1230; Found: 402.1233. Anal. Calcd. for $C_{21}H_{18}FN_3O_3$: C, 66.48; H, 4.78; N, 11.08; Found: C, 66.46; H, 4.75; N, 11.03.

5-((1*H*-Indol-3-yl)(*p*-tolyl)methyl)-2,2-dimethyl-1,3-dioxane-4,6-dione (**4eb**): White solid, mp 128-130 °C. For *syn* isomer: IR (KBr): 3425, 3181, 2997, 2850, 1776, 1734, 1577, 1504, 1298, 1193, 1063, 902, 850, 740 cm^{-1}; ^1H NMR (400 MHz, DMSO-d_6): δ 1.45 (s, 3H, $-CH_3$), 1.73 (s, 3H, $-CH_3$), 2.21 (s, 3H, $-CH_3$), 5.12 (s, 1H, -CH), 5.31 (s, 1H, -CH), 6.92-6.97 (m, 3H, ArH), 7.09-7.14 (m, 2H, ArH), 6.82 (q, $J=8.4$, 7.6, 7.2, 15.4 Hz, 2H, ArH), 7.19-7.24 (m, 1H, ArH), 7.28-7.39 (m, 1H, ArH), 10.93 (s, 1H, -NH); ^{13}C NMR (100 MHz, DMSO-d_6): δ 27.20, 28.14, 31.14, 21.02, 52.13, 105.13, 111.87, 113.88, 117.07, 118.92, 119.03, 121.57, 124.36, 127.45, 128.81, 128.98, 129.24, 135.79, 136.26, 138.55, 165.48, 166.10. For *anti* isomer: ^1H NMR (400 MHz, DMSO-d_6): δ 2.00 (s, 6H, 2 x CH_3), 2.10 (s, 3H, $-CH_3$), 4.22 (d, $J=12$ Hz, 1H, -CH), 4.69 (d, $J=12.4$ Hz, 1H, -CH), 6.92- 6.97 (m, 3H, ArH), 7.09-7.14 (m, 3H, ArH), 7.19-7.24 (m, 1H, ArH), 7.28-7.33 (m, 1H, ArH), 7.43 (d, $J=7.6$ Hz, 1H, ArH), 10.79 (s, 1H, -NH). TOF-MS: Calcd for $C_{22}H_{21}NO_4Na$ $[M+Na]^+$: 386.1368; Found: 386.1372. Anal. Calcd. for $C_{22}H_{21}NO_4$: C, 72.71; H, 5.82; N, 3.85; Found: C, 72.68; H, 5.80; N, 3.89.

Reference

Brahmachari, G., and Das, S. (2014). L-Proline catalyzed multicomponent one-pot synthesis of *gem*-diheteroarylmethane derivatives using facile grinding operation under solvent-free conditions at room temperature. *RSC Advances*, **4**, 7380–7388.

C-C bond_42

Type of reaction: C-C bond formation
Reaction conditions: Ethanol-water (1:1 v/v), room temperature
Synthetic strategy: One-pot pseudo-multicomponent reaction
Catalyst: Sulfamic acid (NH_2SO_3H)
Keywords: Indoles, istains/acenaphthaquinone, sulfamic acid, ethanol-water (1:1), room temperature, one-pot pseudo-multicomponent reaction, functionalized 3,3-*bis*(indol-3-yl) indolin-2-one scaffolds / 2,2-*bis*(indol-3-yl)acenaphthylen-1(2*H*)-one derivatives

General reaction scheme

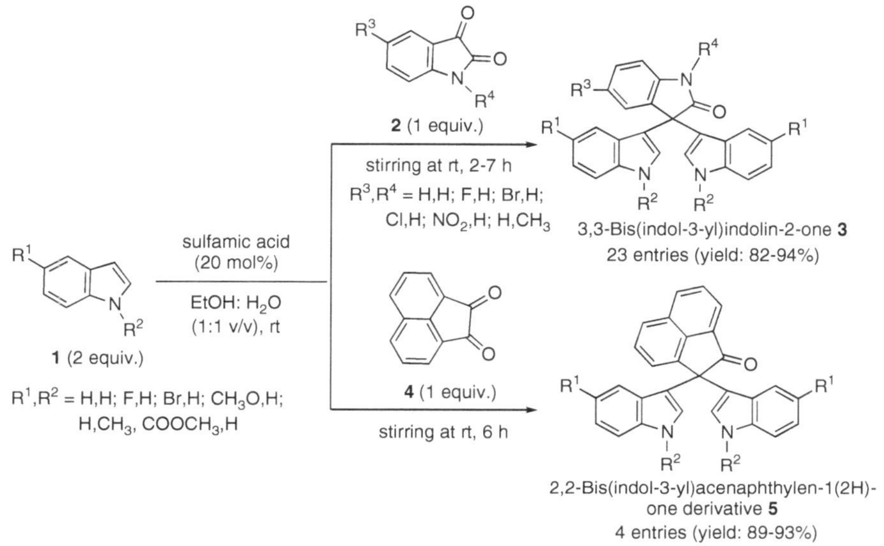

Proposed mechanism

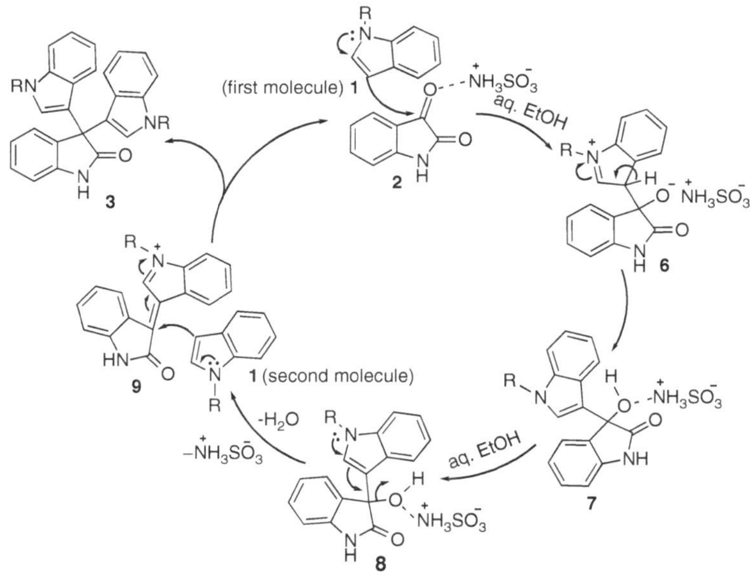

Representative entries

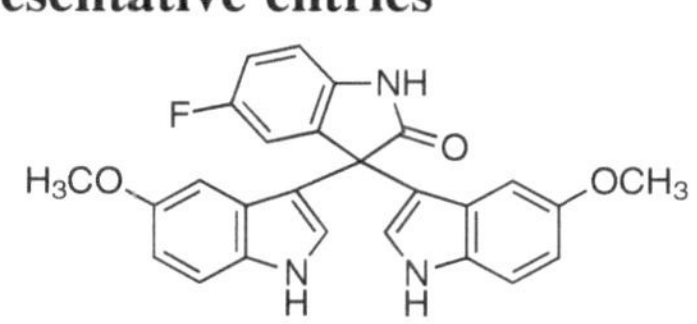

3a

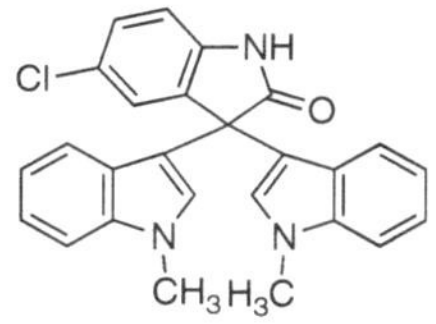

3b

5′-Fluoro-5,5″-dimethoxy-1*H*,1″*H*-[3,3′:3′,3″-terindol]-2′(1′*H*)-one (**3a**): White solid, mp 270-272 °C; reaction time: 2 h; yield: 91%

5′-Chloro-1,1″-dimethyl-1*H*,1″*H*-[3,3′:3′,3″-terindol]-2′(1′*H*)-one (**3b**): White solid, mp >300 °C; reaction time: 3 h; yield: 94%

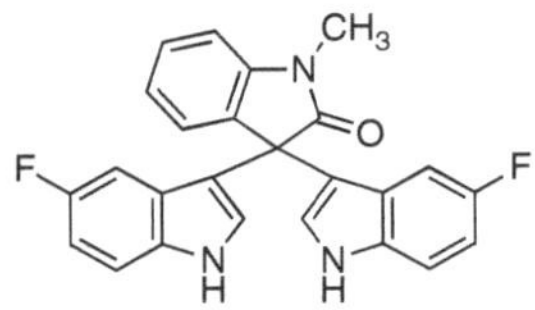

3c

5′-Chloro-1,1″-dimethyl-1*H*,1″*H*-[3,3′:3′,3″-terindol]-2′(1′*H*)-one (**3b**): White solid, mp >300 °C; reaction time: 2 h; yield: 94%

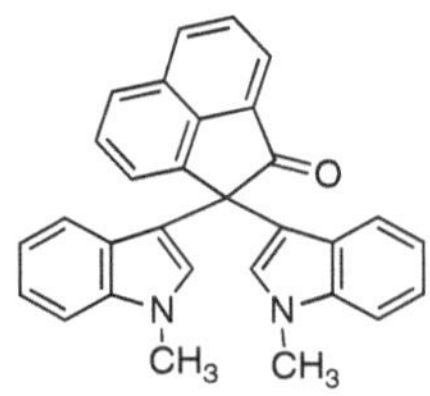

5a

2,2-*Bis*(1-methyl-1*H*-indol-3-yl)acenaphthylen-1 (2*H*)-one (**5a**): Yellow solid, mp 286-288 °C; reaction time: 6 h; yield: 90%

Experimental procedure

An oven-dried screw cap test tube was charged with a magnetic stir bar, indole (**1**, 1 mmol), isatin (**2**, 0.5 mmol)/acenaphthylene-1,2-dione (**4**, 0.5 mmol), sulfamic acid (20 mol% as organo-catalyst), EtOH: H_2O (1:1 v/v; 4 mL) in a sequential manner; the reaction mixture was then stirred vigorously at room temperature for appropriate range of time as indicated in Table 2. The progress of the reaction was monitored by TLC. On completion of reaction, a solid mass precipitated out that was filtered off followed by washing with aqueous ethanol to obtain crude product (**3**/**5**) purified just by re-crystallization from ethanol without carrying out column chromatography. The filtrate containing residual solvent, catalyst and substrates obtained upon filtration of the reaction mixture after completion of reaction could be successfully reused for a particular entry up to three times without appreciable loss of catalytic activity. The structure of each purified products (**3**/**5**) was confirmed by analytical as well as spectral studies including FT-IR, ^{1}H NMR, ^{13}C NMR and TOF-MS.

Characterization data of two representative entries

5,5″-Difluoro-1′-methyl-1*H*,1′*H*-[3,3′:3′,3″-terindol]-2′(1′*H*)-one (**3c**): White solid, mp >300 °C; IR (KBr): 3346, 3115, 2931, 1664, 1629, 1611, 1581, 1539, 1486, 1471, 1455, 1419, 1373, 1353, 1298, 1243, 1223, 1181, 1169, 1121, 1103, 1088, 1069, 1021, 947, 913, 837, 795, 747 cm^{-1}; ^{1}H NMR (400 MHz, DMSO-d_6): δ 11.15 (2H, s, -N*H*-), 7.39-7.35 (3H, m, aromatic H), 7.29 (1H, d, J=7.2 Hz, aromatic H), 7.19 (1H, d, J=8.0 Hz, aromatic H), 7.06 (1H, t, J=7.6 & 7.2 Hz, aromatic H), 6.95 (2H, d, J=2.4 Hz, aromatic H), 6.88 (2H, td, J=8.8, 2.4 & 2.0 Hz, aromatic H), 6.79 (2H, dd, J=10.4 & 2.0 Hz, aromatic H), 3.27 (3H, s, -NCH_3); ^{13}C NMR (100 MHz, DMSO-d_6): δ 177.02, 157.75, 155.46, 143.05, 134.03 (2C), 133.37, 128.66, 126.71 (2C), 126.14, 126.04, 124.95, 122.85, 114.30 (2C), 113.19, 113.09, 109.91, 109.65, 109.31, 105.52, 105.29, 52.17, 26.69; HRMS (ESI-TOF) *m*/*z*: $[M+Na]^+$ Calcd for $C_{25}H_{17}F_2N_3ONa$ 436.1237; Found 436.1235. Elemental analysis: Calcd. (%) for $C_{25}H_{17}F_2N_3O$: C, 72.63; H, 4.14; N, 10.16; Found: C, 72.61; H, 4.11; N, 10.18.

*2,2-*Bis*(1-methyl-1H-indol-3-yl)acenaphthylen-1(2H)-one* **(5a):** Yellow solid, mp 286-288 ° C; IR (KBr): 3053, 2934, 1718, 1609, 1539, 1475, 1421, 1367, 1329, 1250, 1207, 1159, 1128, 1067, 1013, 793, 743 cm^{-1}; ^{1}H NMR (400 MHz, DMSO-d_6): δ 8.36 (1H, d, J=7.6 Hz, aromatic H), 8.01 (2H, d, J=6.4 Hz, aromatic H), 7.89 (1H, t, J=6.8 Hz, aromatic H), 7.69 (1H, t, J=8.0 & 7.2 Hz, aromatic H), 7.56 (1H, d, J=6.0 Hz, aromatic H), 7.37 (2H, d, J=7.6 Hz, aromatic H), 7.05 (4H, t, J=8.4 Hz, aromatic H), 6.88 (2H, br s), 6.79 (2H, t, J=7.2 & 6.0 Hz, aromatic H), 3.67 (6H, s, 2×-NCH_3); ^{13}CNMR (100 MHz, DMSO-d_6): δ 202.87, 144.09, 139.93, 137.77 (2C), 132.55, 130.91, 129.50, 129.30, 129.22 (2C), 126.51 (2C), 124.58 (2C), 122.77, 122.13, 121.57 (2C), 121.17 (2C), 118.98 (2C), 114.34 (2C),

110.27 (2C), 57.86, 32.74 (2C); HRMS (ESI-TOF) *m/z*: $[M+Na]^+$ Calcd for $C_{30}H_{22}N_2ONa$ 449.1630; Found: 449.1628. Elemental analysis: Calcd. (%) for $C_{30}H_{22}N_2O$: C, 84.48; H, 5.20; N, 6.57; Found: C, 84.46; H, 5.17; N, 6.59.

Reference

Brahmachari, G., and Banerjee, B. (2014). Facile and one-pot access of 3,3-bis(indol-3-yl)indolin-2-ones and 2,2-bis(indol-3-yl)acenaphthylen-1(2*H*)-one derivatives via an eco-friendly pseudo-multicomponent reaction at room temperature using sulfamic acid as an organo-catalyst. *ACS Sustainable Chem. Eng.*, **2**, 2802–2812.

C-C bond_43

Type of reaction: C-C bond formation
Reaction conditions: Dichloromethane-water (3:5 v/v), room temperature
Synthetic strategy: Open-flask C-H arylation (functionalization) of electron-deficient heteroarenes
Catalyst: $AgNO_3/K_2S_2O_8$/TFA
Keywords: Electron-deficient heteroarenes, triazenes, silver nitrate, potassium persulfate, trifluoroacetic acid, dichloromethane-water (3:5), room temperature, open-flask C-H functionalization, *C*-arylation, substituted heteroarenes

General reaction scheme

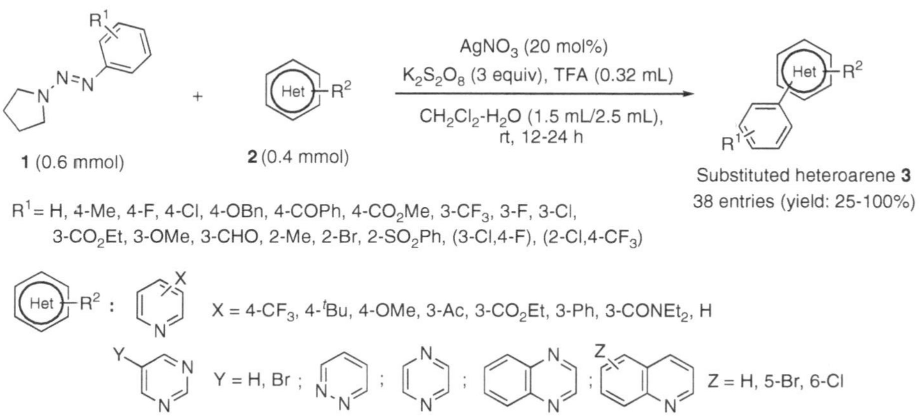

Representative entries

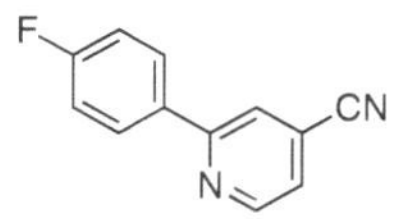

3a

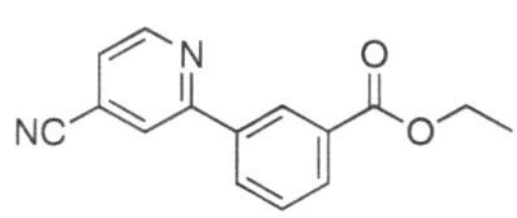

3b

2-(4-Fluorophenyl)isonicotinonitrile (**3a**): Yellow solid; reaction time: 12 h; yield: 100%

Ethyl 3-(4-cyanopyridin-2-yl)benzoate (**3b**): Light yellow oil; reaction time: 12 h; yield: 80%

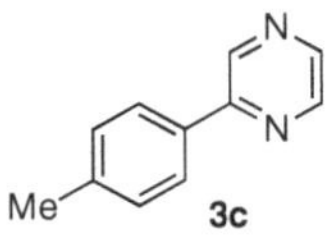

3c

2-(*p*-Tolyl)pyrazine (**3c**): Yellow solid; reaction time: 24 h; yield: 46%

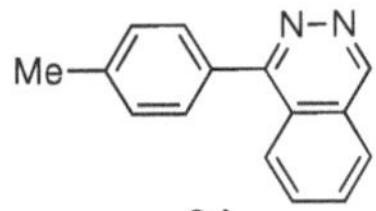

3d

1-(*p*-Tolyl)phthalazine (**3d**): Colorless oil; reaction time: 24 h; yield: 52%

Experimental procedure

To a stirred solution of heteroarene (**2**: 0.4 mmol) in dichloromethane (1.5 mL) at ambient temperature under open-flask was added TFA (0.32 mL) and triazene (**1**; 0.6 mmol), the mixture was then stirred vigorously for 5 min. Then water (1.5 mL) and silver nitrate solution (20 mol % in 1.0 mL water) and potassium persulfate (0.32 g) was added successively, and the stirring was continued up to 24 h (TLC monitored). Upon completion of the reaction, 2N NaOH was added to quench excess TFA, extracted with dichlorometahne (5 x 20 mL), and dried over anhydrous sodium sulfate. Solvent was evaporated under reduced pressure and the crude residue was purified by silica gel flash chromatography (10% MeOH in CH_2Cl_2) to furnish pure substituted heteroaren **3**, characterized by means of spectral studies.

Characterization data of two representative entries

Ethyl 3-(4-cyanopyridin-2-yl)benzoate (**3b**): Light yellow oil; ^{1}H NMR (400 MHz, $CDCl_3$): δ 1.44 (t, J=8.0 Hz, 3H), 4.44 (q, J=8.0 Hz, 2H), 7.49 (d, J=8.0 Hz, 1H), 7.62 (t, J=4.0 Hz, 1H), 8.01 (s, 1H), 8.16 (d, J=4.0 Hz, 1H), 8.22 (d, J=4.0 Hz, 1H), 8.64 (d, J=4.0 Hz, 1H), 8.89 (d, J=4.0 Hz, 1H); ^{13}C NMR (100 MHz, $CDCl_3$): δ 14.4, 61.5, 116.6, 121.4, 122.1, 123.6, 128.0, 129.2, 131.1, 131.2, 131.4, 137.6, 150.8, 157.7, 166.1; ESI-HRMS *m/z* Calcd for $C_{15}H_{12}N_2O_2$ $[M+H]^+$: 253.0972; Found: 253.0968.

1-(*p*-Tolyl)phthalazine (**3d**): Colorless oil; ^{1}H NMR (400 MHz, $CDCl_3$): δ 2.47 (s, 3H), 7.37 (d, J=8.0 Hz, 2H), 7.66 (d, J=8.0 Hz, 2H), 7.92-7.83 (m, 2H), 8.01 (d, J=8.0 Hz, 1H), 8.10 (d, J=8.0 Hz, 1H), 9.50 (s, 1H); ^{13}C NMR (100 MHz, $CDCl_3$): δ 21.4, 125.5, 126.3, 126.6, 127.1, 129.3, 130.3, 132.1, 132.5, 133.2, 139.5, 150.4, 159.9; ESI-HRMS *m/z* Calcd for $C_{15}H_{12}N_2$ $[M+H]^+$: 221.1073; Found: 221.1064.

Reference

Wang, R., and Falck, J. R. (2014). Room temperature, open-flask C–H arylation of electron-deficient heteroarenes with triazenes: rapid synthesis of heterobiaryls. *Org. Chem. Front.*, **1**, 1029–1034.

C-C bond_44

Type of reaction: C-C bond formation
Reaction conditions: Nitromethane, room temperature
Synthetic strategy: Regioselective tandem intramolecular hydroarylation of 1,3-dienes
Catalyst: Bimuth triflate [Bi(OTf)$_3$]
Keywords: Electron-deficient arenes, unactivated 1,3-dienes, bismuth triflate, nitromethane, room temperature, intramolecular double hydroarylation of 1,3-dienes, regioselectivity, substituted indanes

General reaction scheme

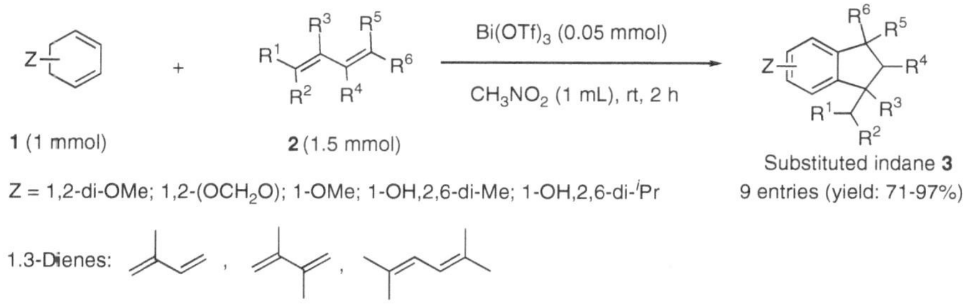

Representative entries

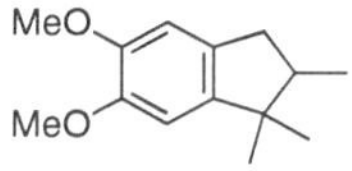

3a

5,6-Dimethoxy-1,1,2-trimethyl-2,3-dihydro-1*H*-indene (**3a**): Colorless oil; yield: 95%

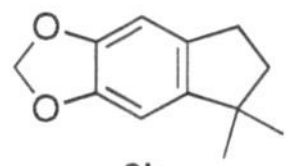

3b

5,5-Dimethyl-6,7-dihydro-5*H*-indeno[5,6-*d*][1,3] dioxole (**3b**): Colorless oil; yield: 91%

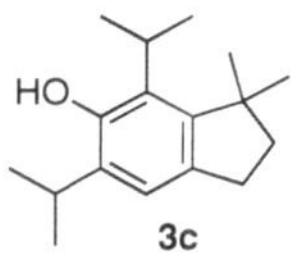

3c

4,6-Diisopropyl-3,3-dimethyl-2,3-dihydro-1*H*-inden-5-ol (**3c**): Colorless oil; yield: 82%

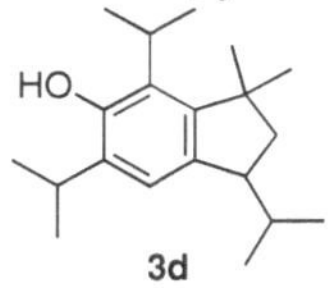

3d

1,4,6-Triisopropyl-3,3-dimethyl-2,3-dihydro-1*H*-inden-5-ol (**3d**): Colorless oil; yield: 94%

Experimental procedure

To a solution of arene (**1**; 1.0 mmol) and bismuth triflate (33 mg, 0.05 mmol) in nitromethane (1 mL) was slowly added 1,3-diene (**2**; 1.5 mmol) in nitromethane (1 mL) over 1 hour at room temperature using a syringe pump. After the addition, the solution was stirred for another 1 h. After complete consumption of the aromatic compound (GC monitoring), the reaction mixture was quenched with a saturated aqueous solution of sodium bicarbonate, extracted thrice with diethyl ether, washed with a saturated aqueous solution of sodium chloride, dried over anhydrous magnesium sulfate and concentrated under reduced pressure. The residue was purified by column chromatography (pentane/Et$_2$O: 100/0 to 90/10) to afford the indane derivative **3**. All the products were characterized by spectral methods.

Characterization data of two representative entries

5,5-Dimethyl-6,7-dihydro-5*H*-indeno[5,6-*d*][1,3]dioxole (**3b**): Colorless oil; ^{1}H NMR (200 MHz, CDCl$_3$): δ 6.65 (s, 1H), 6.61 (s, 1H), 5.90 (s, 2H), 2.78 (t, J=7 Hz, 2H), 1.92 (t, J=7 Hz, 2H), 1.21(s, 3H); ^{13}C NMR (50 MHz, CDCl$_3$): δ 146.4, 146.2, 145.6, 135.2, 105.0, 102.8, 100.8, 43.7, 41.8, 30.0, 28.7 (2C); MS: *m*/*z* (%rel) 190 (26) [M$^+$], 176 (10), 175 (100), 146 (7), 145 (50), 117 (30), 115 (23), 91 (12), 77 (8), 51 (12).

1,4,6-Triisopropyl-3,3-dimethyl-2,3-dihydro-1*H*-inden-5-ol (**3d**): Colorless oil; ^{1}H NMR (200 MHz, CDCl$_3$): δ 6.78 (s, 1H), 4.60 (s, 1H), 3.42 (dt, J=14.3, 7.2 Hz, 1H), 3.21-2.86 (m, 2H), 2.33-2.11 (m, 1H), 1.78 (dd, J=12.4, 7.9 Hz, 2H), 1.48-1.31 (m, 10H), 1.25 (dd, J=6.8, 4.0 Hz, 10H), 1.02 (d, J=6.8 Hz, 4H), 0.76 (d, J=6.8 Hz, 3H); ^{13}C NMR (50 MHz,

CDCl₃): δ 151.0, 147.3, 137.4, 132.3, 129.7, 118.3, 46.6, 44.5, 43.7, 30.3, 28.8, 28.1, 27.8, 27.0, 23.0, 22.8, 21.5, 20.8, 20.8, 16.8; MS: *m/z* (%rel) 288 (20) [M^+], 245 (17), 231 (100), 192 (8), 147 (53), 137 (17), 128 (4), 100 (12), 43 (20), 41 (10).

Reference

Cacciuttolo, B., Ondet, P., Poulain-Martini, S., Lemière, G., and Duñach, E. (2014). $Bi(OTf)_3$-catalysed synthesis of substituted indanes by a double hydroarylation of unactivated 1,3-dienes. *Org. Chem. Front.*, **1**, 765–769.

C-C bond_45

Type of reaction: C-C bond formation
Reaction conditions: Chloroform-acetonitrile (1:1), room temperature
Synthetic strategy: One-pot cascade annulation *via* Diels-Alder reaction
Catalyst: Trifluoroacetic acid (TFA)
Keywords: Dienaminodioate, cinnamaldehydes, allyl amine, trifluoroacetic acid, chloroform-acetonitrile (1:1), room temperature, one-pot cascade annulation, Diels-Alder reaction, air oxidation, polyfunctional biaryl-2-carbaldehydes

General reaction scheme

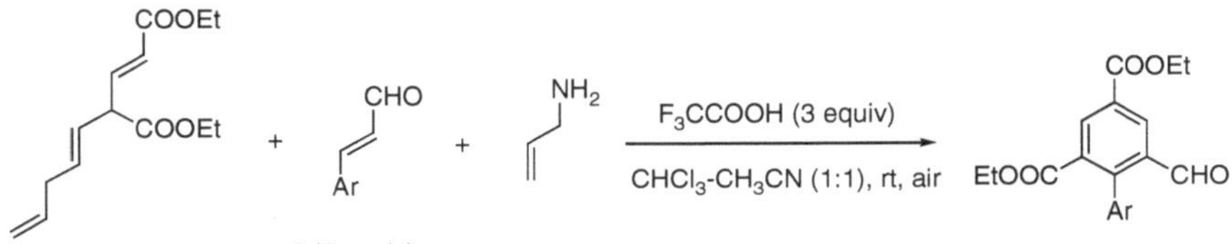

1 (1 equiv) **2** (3 equiv) **3** (3 equiv)
Ar = substituted phenyl

Polyfunctional biaryl-2-carbaldehyde **4**
18 entries (yield: 38-86%)

Representative entries

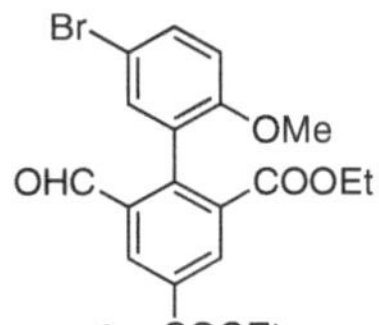

4a

Diethyl 5′-bromo-6-formyl-2′-methoxy-[1,1′-biphenyl]-2,4-dicarboxylate (**4a**): Yield: 85%

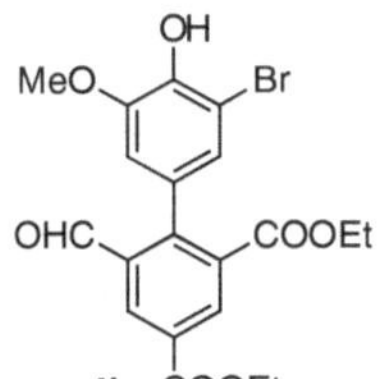

4b

Diethyl 3′-bromo-6-formyl-4′-hydroxy-5′-methoxy-[1,1′-biphenyl]-2,4-dicarboxylate (**4b**): Yield: 38%

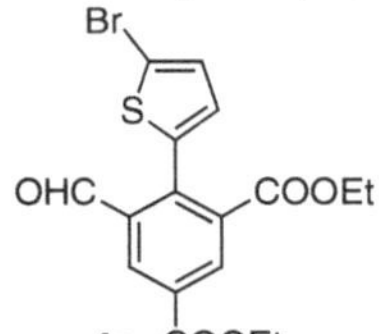

4c

Diethyl 4-(5-bromothiophen-2-yl)-5-formylisophthalate (**4c**): Yield: 49%

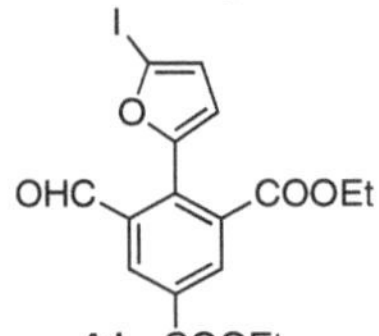

4d

Diethyl 5-formyl-4-(5-iodofuran-2-yl)isophthalate (**4d**): Yield: 55%

Experimental procedure

To a solution of dienaminodiester (**1**; 31 mg, 1 equiv) in chloroform-acetonitrile (1:1) were added cinnamaldehyde (**2**; 3 equiv), allyl amine (3; 27.5 μL, 3 equiv) and TFA (28.0 μL, 3 equiv) in a sequential manner at room temperature. After immediate addition of TFA, the reaction mixture appears intense red in color indicating the formation of trienamine. Upon stirring for appropriate time at room temperature (TLC monitored), the reaction mixture was quenched with saturated aqueous sodium bicarbonate (10 mL) and extracted with dichloromethane (1 x 10 mL). The organic layer was dried over anhydrous magnesium sulfate, concentrated and the crude mixture was subjected to flash column chromatography to afford the desired biaryl **4**, followed by characterization by spectral methods.

Characterization data of two representative entries

Diethyl 3′-bromo-6-formyl-4′-hydroxy-5′-methoxy-[1,1′-biphenyl]-2,4-dicarboxylate (**4b**): ^{1}H NMR (500 MHz, $CDCl_3$): δ 1.11 (t, $J=7.03$, H), 1.44 (t, $J=7.0$, 3H), 3.89 (s, 3H), 4.16 (q, $J=7.0$, 2H), 4.46 (q, $J=7.0$, 2H), 6.12 (bs, 1H), 6.73 (s, 1H), 7.04 (s, 1H), 7.26 (s, 1H), 8.64 (s, 1H), 8.71 (s, 1H), 9.83 (s, 1H); ^{13}C NMR (125 MHz, $CDCl_3$): δ 13.8, 14.2, 56.5, 61.7, 108.2, 111.3, 125.6, 127.5, 130.6, 130.9, 134.0, 134.8, 135.2, 143.5, 146.7, 164.6, 166.4, 190.5; ESI-HRMS: Calcd for $C_{21}H_{23}BrO_8Na$ *m/z* 505.0474 $[M+MeOH+Na]^+$; Found 505.0470.

Diethyl 5-formyl-4-(5-iodofuran-2-yl)isophthalate (**4d**): ^{1}H NMR (500 MHz, $CDCl_3$): δ 1.26 (t, $J=7.0$, 3H), 1.43 (t, $J=7.0$, 3H), 4.28 (q, $J=7.0$, 2H), 4.44 (q, $J=7.0$, 2H), 6.52 (d, $J=3.5$, 1H), 6.76 (d, $J=3.0$, 1H), 8.63 (d, $J=1.5$, 1H), 8.71 (d, $J=1.5$, 1H), 10.06 (s, 1H); ^{13}C NMR (125 MHz, $CDCl_3$): δ 14.1, 14.2, 24.6, 36.6, 61.9, 62.1, 90.5, 116.9, 122.2, 131.2, 131.4, 134.0, 134.6, 134.9, 135.0, 151.4, 162,4, 166.5, 189.9; HRMS: Calcd for $C_{18}H_{19}IO_7Na$ *m/z* 497.0073 $[M+MeOH+Na]^+$; Found 497.0072.

Reference

Challa, C., Vellekkatt, J., Ravindran, J., and Lankalapalli, R. S. (2014). A metal-free one-pot cascade synthesis of highly functionalized biaryl-2-carbaldehydes. *Org. Biomol. Chem.*, **12**, 8588–8592.

C-C bond_46

Type of reaction: C-C bond formation
Reaction conditions: Solvent-free, room temperature
Synthetic strategy: One-pot three-component reaction by grinding operation
Catalyst: Sodium bicarbonate
Keywords: Isatins, malononitrile, 2-pyrazolin-5-ones, sodium bicarbonate, grinding, room temperature, solvent-free, one-pot multicomponent reaction, heterocycles, spiropyranyloxindoles, spiro[indoline-3,4′(1*H*′)-pyrano-[2,3-*c*]pyrazol]-2-ones

General reaction scheme

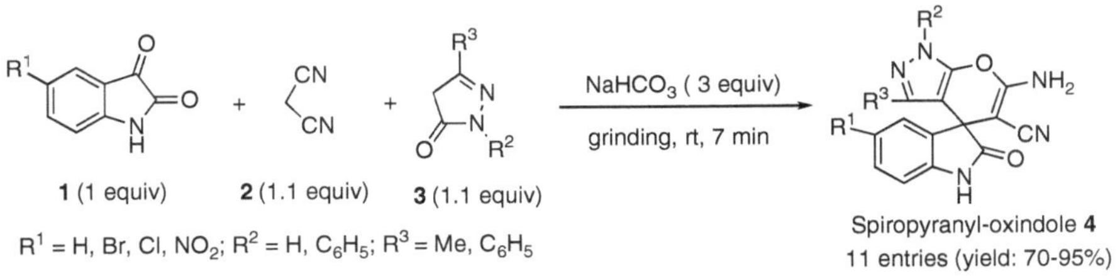

Representative entries

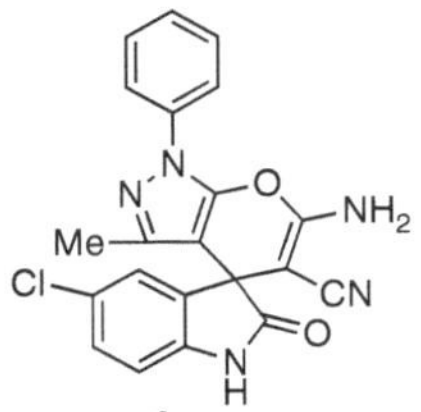

4a

6′-Amino-5-chloro-3′-methyl-2-oxo-1′-phenyl-1′*H*-spiro[indoline-3,4′-pyrano[2,3-*c*]pyrazole]-5′-carbonitrile (**4a**): Solid, mp 223-224 °C; yield: 70%

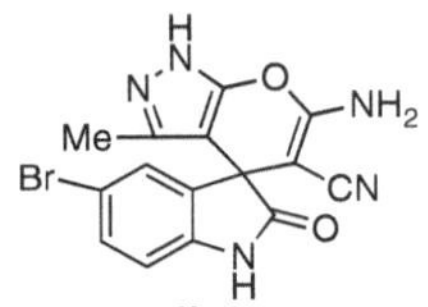

4b

6′-Amino-5-bromo-3′-methyl-2-oxo-1′*H*-spiro[indoline-3,4′-pyrano[2,3-*c*]pyrazole]-5′-carbonitrile (**4b**): Solid, mp 282-283 °C; yield: 95%

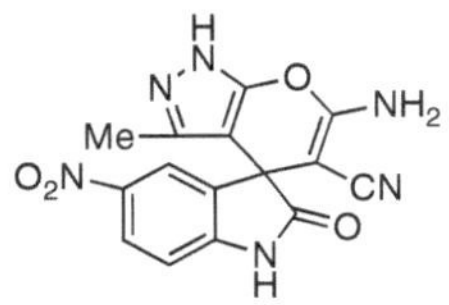

4c

6′-Amino-3′-methyl-5-nitro-2-oxo-1′*H*-spiro[indoline-3,4′-pyrano[2,3-*c*]pyrazole]-5′-carbonitrile (**4c**):): Solid, mp 270-271 °C; yield: 82%

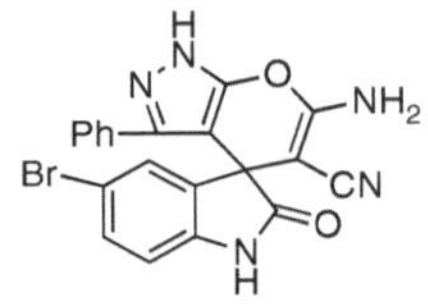

4d

6′-Amino-1′-(4-methoxyphenyl)-3′-methyl-2-oxo-1′*H*-spiro[indoline-3,4′-pyrano[2,3-*c*]pyrazole]-5′-carbonitrile (**4b**):): Solid, mp 258-259 °C; yield: 95%

Experimental procedure

A mixture of isatin (**1**; 2 mmol), malononitrile (**2**;145 mg, 2.2 mmol), 2-pyrazolin-5-one (**3**;2.2 mmol), and sodium bicarbonate (504 mg, 6 mmol) was ground at room temperature with a glass mortar and pestle; the reaction was found to become completed within 7 min (monitored by TLC). Upon completion of the reaction, the precipitate is collected by suction filtration and washed with water. The crude product was recrystallized from ethanol and dried under room temperature to obtain pure spiropyranyl-oxindole **4**. All the products were characterized by spectral studies.

Characterization data of two representative compounds

6′-Amino-5-chloro-3′-methyl-2-oxo-1′-phenyl-1′*H*-spiro[indoline-3,4′-pyrano[2,3-*c*]pyrazole]-5′-carbonitrile (**4a**): Solid, mp 223-224 °C; IR (KBr): 3367, 3186, 2204, 1706, 1658 cm^{-1}; ^1H NMR (500 MHz, DMSO-d_6): δ 1.59 (s, 3H, CH_3), 6.95-6.97 (m, 1H, Ar-H), 7.33-7.37 (m, 3H, Ar-H), 7.51-7.54 (m, 2H, Ar-H), 7.65 (s, 2H, NH_2), 7.77-7.79 (m, 2H, Ar-H), 10.90 (s, 1H, NH).

6′-Amino-3′-methyl-5-nitro-2-oxo-1′*H*-spiro[indoline-3,4′-pyrano[2,3-*c*]pyrazole]-5′-carbonitrile (**4c**):): Solid, mp 270-271 °C; IR (KBr): 3346, 3136, 2182, 1714, 1644 cm^{-1}; ^{1}H NMR (500 MHz, DMSO-d_6): δ 1.58 (s, 3H, CH_3), 6.92-6.93 (m, 1H, Ar-H), 7.13-7.14 (m, 1H, Ar-H), 7.29-7.31 (m, 3H, Ar-H, NH_2), 10.76 (s, 1H, NH), 12.35 (s, 1H, NH).

Reference

Liu, Y., Ren, Z., Cao, W., Chen, J., Deng, H., and Shao, M. (2011). Solvent-free one-pot synthesis of spiro[indoline-3,4′(1*H*′)-pyrano-[2,3-*c*]pyrazol]-2-one derivatives by grinding. *Synth. Commun.*, **41**, 3620–3626.

Chapter 2

Carbon – Nitrogen Bond Forming Reactions at Room Temperature

C-N bond_1

Type of reaction: C-N and C-C bond formation
Reaction conditions: Solvent-free, room temperature
Synthetic strategy: One-pot multicomponent
Catalyst: Zirconyl oxychloride octahydrate ($ZrOCl_2.8H_2O$)
Keywords: Strecker reaction, α-aminonitriles, carbonyl compounds, TMSCN, $ZrOCl_2.8H_2O$, homogeneous catalyst, one-pot synthesis, multicomponent reaction (MCR), solvent-free, room temperature

General reaction scheme

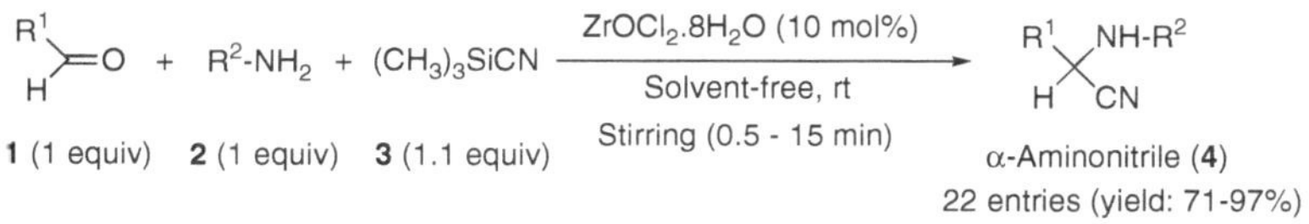

R^1: C_6H_5, $4\text{-}CH_3C_6H_4$, $4\text{-tri-}OCH_3C_6H_2$, $2\text{-}ClC_6H_4$, $4\text{-}ClC_6H_4$, $4\text{-}FC_6H_4$, $4\text{-}NO_2C_6H_4$, $3\text{-}OCH_3\text{-}4\text{-}OHC_6H_3$, $4\text{-}CHOC_6H_4$, $C_6H_5CH{=}CH$, *n*-Pr,

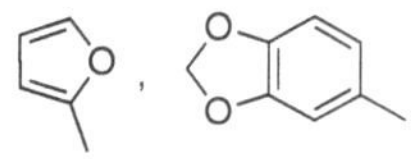

R^2: C_6H_5, $4\text{-}BrC_6H_4$, $4\text{-}FC_6H_4$, $4\text{-}OCH_3C_6H_4$, $(C_2H_4)_2O$ (morpholine), $C_6H_5CH_2$

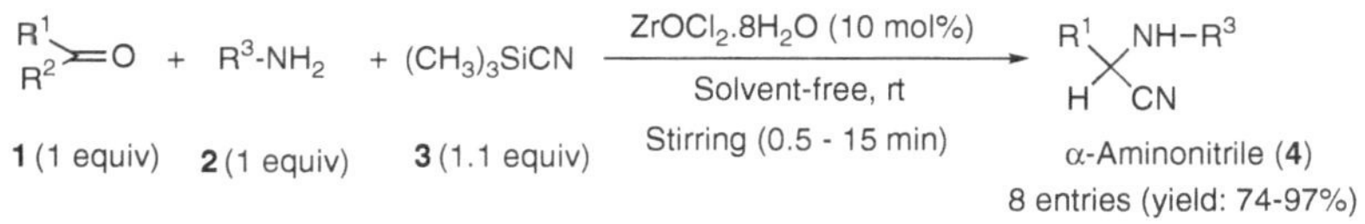

R^1,R^2 (ketone): acetone, cyclohexanone, cyclopentanone, acetophenone
R^3: C_6H_5, $4\text{-}FC_6H_4$, $4\text{-}OCH_3C_6H_4$

Representative new entries

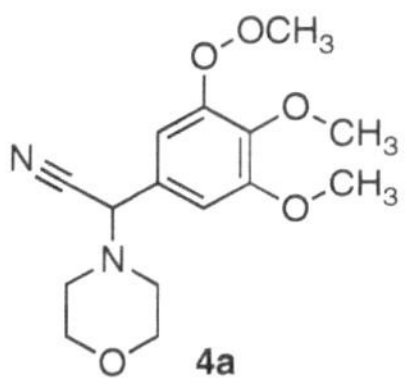

4a

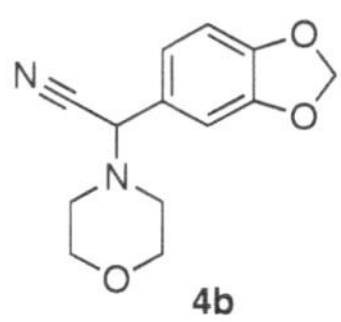

4b

2-Morpholino-2-(3,4,5-trimethoxyphenyl) acetonitrile (**4a**): White solid, mp 138-140 °C; reaction time: 15 min; yield: 92%

2-(3,4-Methylenedioxyphenyl)-2-morpholinoacetonitrile (**4b**): White solid, mp 118-120 °C; reaction time: 10 min; yield: 97%

Room Temperature Organic Synthesis. http://dx.doi.org/10.1016/B978-0-12-801025-9.00002-4

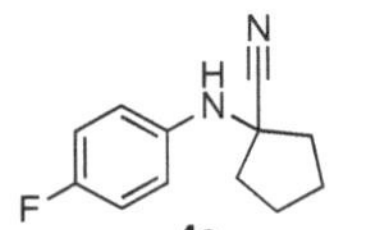

4c

1-(4-Fluorophenylamino)-1-cyclopentanecarbonitrile (**4c**): White solid, mp 72-74 °C; reaction time: 5 min; yield: 97%

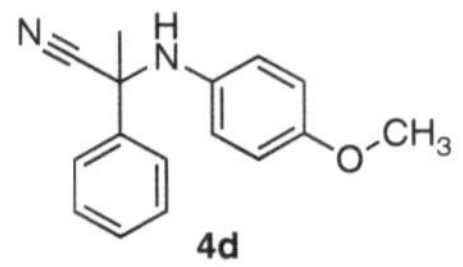

4d

2-(4-Methoxyphenylamino)-2-methyl-2-phenylacetonitrile (**4d**): White solid, mp 88-90 °C; reaction time: 20 min; yield: 74%

Experimental procedure

An oven-dried screw-cap test tube was charged with a magnetic stir bar, the carbonyl compound (**1**; 1 mmol), $ZrOCl_2 \cdot 8H_2O$ (10 mol%), amine (**2**; 1 mmol), and trimethylsilyl cyanide (**3**; 1.1 mmol) sequentially. The reaction mixture was then stirred vigorously at room temperature for required time-frame (0.5-20 min). The progress of the reaction was monitored by TLC, and on completion 20 mL of distilled water was added to the reaction mixture. The mixture was constantly stirred for another 10 min. The solid residue obtained on simple filtration was then recrystallized from aqueous ethanol to afford pure product in most of the cases. Chromatographic purification using silica gel column (60–120 mesh) with ethyl acetate/petroleum ether mixture as eluent was required for few entries; the resulting reaction mixture in each case of these entries was directly extracted with ethyl acetate instead of water, filtered, and the residue obtained on removal of the solvent was then purified by the chromatography. The structure of each purified α-aminonitrile was confirmed by analytical as well as spectral studies.

Characterization data of two representative compounds

2-Morpholino-2-(3,4,5-trimethoxyphenyl)acetonitrile (**4a**): White solid; mp 138-140 °C; IR (KBr): 3080, 2955, 2832, 2230, 1589, 1514, 1458, 1420, 1242, 1126, 995, 926, 874, 851, 716, 581, 527, 449 cm^{-1}; ^1H NMR (500 MHz, $CDCl_3$): δ 6.76 (s, 2H), 4.75 (s, 1 H), 3.89 (s, 6H), 3.84 (s, 3H), 3.79-3.70 (m, 4H), 2.64-2.55 (m, 4H); ^{13}C NMR (125 MHz, $CDCl_3$): δ 153.51, 138.41, 127.96, 115.22, 104.94 (3 × C-OMe), 66.69 (2C), 62.53 (2C), 60.88 (OMe), 56.28 (2 × OMe), 50.02; TOF-MS: 315.15 $[M+Na]^+$. Anal. Calcd (%) for $C_{15}H_{20}N_2O_4$: C, 61.63; H, 6.90; N, 9.58; Found: C, 61.65; H, 6.87; N, 9.59.

1-(4-Fluorophenylamino)-1-cyclopentanecarbonitrile (**4c**): White solid, mp 72-74 °C; IR (KBr): 3368, 3063, 2963, 2228, 1616, 1514, 1443, 1219, 1109, 1045, 949, 824, 777, 716, 517, 461 cm^{-1}; ^1H NMR (500 MHz, $CDCl_3$): δ 6.98-6.93 (m, 2H), 6.81-6.77 (m, 2H), 3.75 (s, 1H), 2.36–2.30 (m, 2 H), 2.12–2.06 (m, 2H), 1.90-1.85 ppm (m, 4H); ^{13}C NMR (125 MHz, $CDCl_3$): δ 158.42, 156.52, 140.28, 140.26, 122.27, 117.75, 117.69, 115.95; 115.77, 58.22, 40.06 (2C), 23.71 (2C); TOF-MS: 227.10 $[M+Na]^+$. Anal. Calcd (%) for $C_{12}H_{13}FN_2$: C 70.57, H 6.42, N 13.72; Found: C 70.56, H 6.43, N 13.71.

Reference

Brahmachari, G., and Banerjee, B. (2012). A comparison between catalyst-free and $ZrOCl_2.8H_2O$-catalyzed Strecker reactions for the rapid and solvent-free one-pot synthesis of racemic α-aminonitrile derivatives. *Asian J. Org. Chem.*, **1**, 521–528.

Further information

Besides MCM-41-SO_3H and $ZrOCl_2.8H_2O$, a variety of homogeneous and heterogeneous catalysts such as $BiCl_3$ [1], montmorillonite KSF clay [2], iodine [3], silica-supported heteropoly acids [4], Indium(III) iodide [5], superparamagnetic iron oxide [6], K_2PdCl_4 [7], $Zr(HSO_4)_4$ [8], nano TiO_2 P 25 [9], sulfuric acid-modified PEG-6000 (PEG-OSO_3H) [10] etc. have also been reported to accomplish the synthesis of α-aminonitrile in the presence of absence of solvents at room temperature.

[1] De, S. K., and Gibbs, R. A. (2004). Bismuth trichloride catalyzed synthesis of α-aminonitriles. *Tetrahedron Lett.*, **45**, 7407–7408.

[2] Yadav, J. S., Subba Reddy, B. V., Eeshwaraiah, B., and Srinivas, M. (2004). Montmorillonite KSF clay catalyzed one-pot synthesis of α-aminonitriles. *Tetrahedron*, **60**, 1767–1771.

[3] Royer, L., De, S. K., and Gibbs, R. A. (2005). Iodine as a novel and efficient reagent for the synthesis of α-aminonitriles by a three-component condensation of carbonyl compounds, amines, and trimethylsilyl cyanide. *Tetrahedron Lett.*, **46**, 4595–4597.

[4] Rafiee, E., Rashidzadeh, S., and Azad, A. (2007). Silica-supported heteropoly acids: Highly efficient catalysts for synthesis of α-aminonitriles, using trimethylsilyl cyanide or potassium cyanide. *J. Mol. Catal. A: Chem.*, **261**, 49–52.

[5] Shen, Z.-L., Ji, S.-J., and Loh, T.-P. (2008). Indium(III) iodide-mediated Strecker reaction in water: an efficient and environmentally friendly approach for the synthesis of α-aminonitrile via a three-component condensation. *Tetrahedron*, **64**, 8159–8163.

[6] Mojtahedi, M. M., Abaee, M. S., and Alishiri, T. (2009). Superparamagnetic iron oxide as an efficient catalyst for the one-pot, solvent-free synthesis of α-aminonitriles. *Tetrahedron Lett.*, **50**, 2322–2325.

[7] Karmakar, B., and Banerji, J. (2010). K_2PdCl_4 catalyzed efficient multicomponent synthesis of α-aminonitriles in aqueous media. *Tetrahedron Lett.*, **51**, 2748–2750.

[8] Hajipour, A. R., Ghayeb, Y., and Sheikhan, N. (2010). $Zr(HSO_4)_4$ catalyzed one-pot Strecker synthesis of α-amino nitriles from aldehydes and ketones under solvent-free conditions. *J. Iran. Chem. Soc.*, **7**, 447–454.

[9] Baghbanian, S. M., Farhang, M., and Baharfar, R. (2011). One-pot three-component synthesis of α-amino nitriles catalyzed by nano powder TiO_2 P 25. *Chin. Chem. Lett.*, **22**, 555–558.

[10] Shekouhy, M. (2012). Sulfuric acid-modified PEG-6000 (PEG-OSO3H): an efficient Bronsted acid-surfactant combined catalyst for the one-pot three component synthesis of α-aminonitriles in water. *Catal. Sci. Technol.*, **2**, 1010–1020.

C-N bond_2

Type of reaction: C-N bond formation
Reaction conditions: Solvent-free, room temperature
Synthetic strategy: One-step conjugate addition
Catalyst: $ZrOCl_2.8H_2O$/Montmorillonite K10
Keywords: Substituted amines, amines, α,β-unsaturated alkenes, $ZrOCl_2.8H_2O$/montmorillonite K10, heterogeneous catalysis, solvent-free, room temperature, Mannich reaction, Michael addition

General reaction scheme

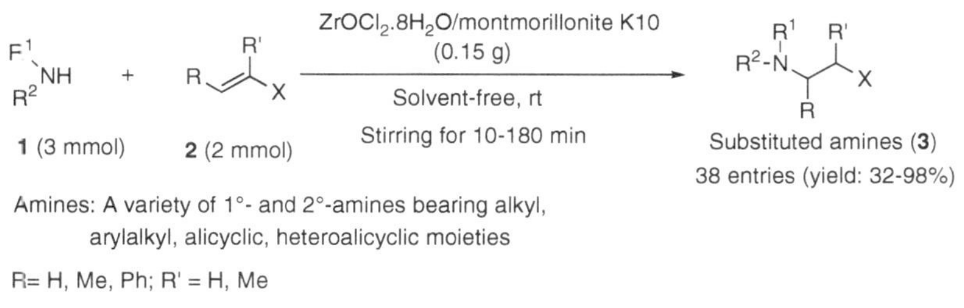

Representative entries

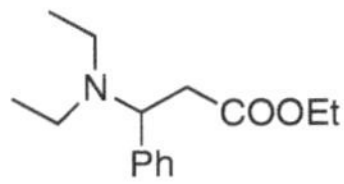

3a

Ethyl 3-(diethylamino)-3-phenylpropanoate (**3a**): Reaction time: 30 min; yield: 74%

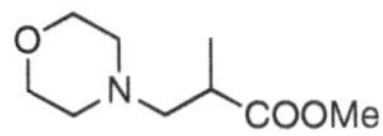

3b

Methyl 2-methyl-3-morpholinopropanoate (**3b**): Reaction time: 30 min; yield: 84%

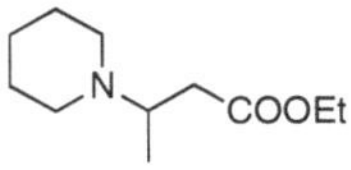

3c

Ethyl 3-(piperidin-1-yl)butanoate (**3c**): Reaction time: 25 min; yield: 84%

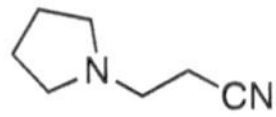

3d

3-(Pyrrolidin-1-yl)propanenitrile (**3d**): Reaction time: 30 min; yield: 98%

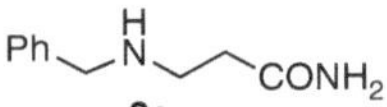

3e

3-(Benzylamino)propanamide (**3e**): Reaction time: 35 min; yield: 83%

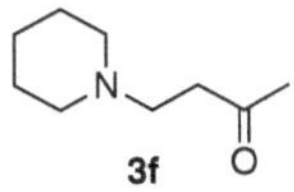

3f

4-(Piperidin-1-yl)butan-2-one (**3f**): Reaction time: 25 min; yield: 72%

Experimental procedure

To a mixture of α,β-unsaturated alkene (**2**; 2 mmol) and catalyst (0.15 g) was added amine (**1**; 3 mmol) and stirred at room temperature under solvent-free conditions for specified period (10-180 min for varying entries). After completion of the reaction, 10 mL of dichlomethane was added to the resulting mixture, and catalyst was removed by filtration. The solvent was evaporated and the product **3** was isolated in almost pure form. Further purification was carried out by short column chromatography on silica gel using ethyl acetate/petroleum ether mixture as eluent. All compounds were characterized by their spectroscopic data (IR, NMR) by comparison with those reported in the literature [1-5].

Reference

Hashemi, M. M., Eftekhari-Sis, B., Abdollahifar, A., and Khalili, B. (2006). $ZrOCl_2.8H_2O$ on montmorillonite K10 accelerated conjugate addition of amines to α,β-unsaturated alkenes under solvent-free conditions. *Tetrahedron*, **62**, 672–677.

References for spectroscopic data

[1] Fadini, L., and Togni, A. (2003). *Chem. Commun.*, 30–31.
[2] Liu, M., and Sibi, M. P. (2002). *Tetrahedron*, **58**, 7991–9035.
[3] Bartoli, G., Bosco, M., Marcantoni, E., Petrini, M., Sambri, L., and Torregiani, E. (2001). *J. Org. Chem.*, **66**, 9052–9055.
[4] Matloubi Moghaddam, F., Mohammadi, M., and Hosseinnia, A. (2000). *Synth. Commun.*, **30**, 643–650.
[5] Cole, D. E. (1994). *Tetrahedron*, **50**, 9517–9582.

C-N bond_3

Type of reaction: C-N bond formation
Reaction conditions: Solvent-free, room temperature
Synthetic strategy: One-pot condensation
Catalyst: Scandium triflate [$Sc(OTf)_3$, reusable catalyst]
Keywords: 1,3-Dicarbonyls, amines, $Sc(OTf)_3$, reusable catalyst, solvent-free, room temperature, condensation reaction, β-enamines

General reaction scheme

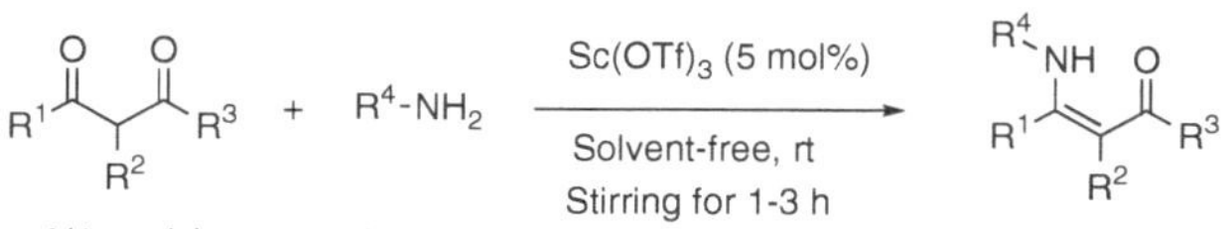

1 (1 equiv) **2** (1 equiv)

β-Enamino derivatives (**3**)
18 entries (yield: 84-95%)

R^1= CH_3, CF_3, C_6H_5, cyclopropyl
R^2= H, CH_3; R3= CH_3, CF_3, C_6H_5, OCH_3, OC_2H_5
R_4= C_6H_5, 2-Cl-C_6H_4, 2-CF_3-C_6H_4, 2,3-di-Cl-C_6H_3, 4-$OCH_3C_6H_4$, 2,5-di-OCH_3-C_6H_3, *n*-butyl, $C_6H_5CH_2$
(piperidine was also used as amine in one entry)

Representative entries

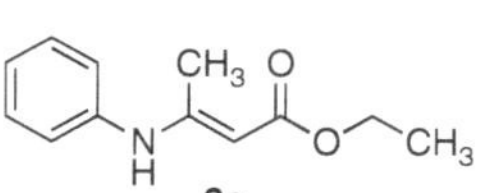

3a

3-Phenylamino-but-2-enoic acid ethyl ester (**3a**): Pale-yellow oil; reaction time: 1 h; yield: 95%

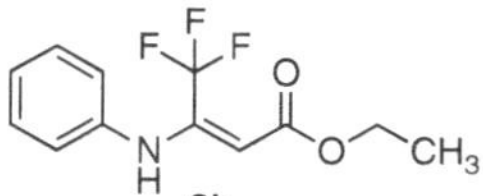

3b

4,4,4-Trifluoro-3-phenylamino-but-2-enoic acid ethyl ester (**3b**): Pale-yellow oil; reaction time: 2.5 h; yield: 92%

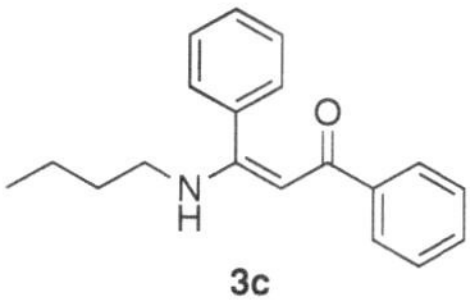

3c

3-(Butylamino)-1,3-diphenylprop-2-en-1-one (**3c**): Reaction time: 3 h; yield: 83%

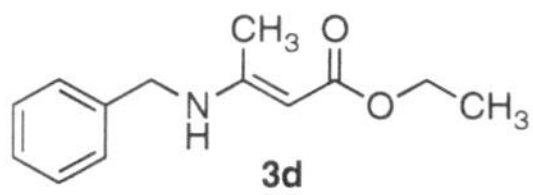

3d

Ethyl 3-(benzylamino)but-2-enoate (**3d**): Semi-solid; reaction time: 3 h; yield: 85%

Experimental procedure

A thorough mixture of 1,3-dicarbonyl compound (**1**; 1.0 mmol) and amine (**2**; 1.0 mmol) was stirred at room temperature in the presence of $Sc(OTf)_3$ (5 mol.%) under solvent-free conditions for the appropriate time-range. After completion of the reaction, as indicated by TLC, the reaction mixture was extracted with diethyl ether (2 × 10 ml). The catalyst recovered was dried in the oven at 80 °C and reused for three to four times without appreciable loss of the activity. The crude products were separated by flash column chromatography on Et_3N pre-treated silica gel using *n*-hexane/EtOAc (10:1) as eluent to afford pure β-enamino compound **3** in excellent yield (84-95%). All the products were characterized from their spectral studies including IR, ^1H NMR, and mass.

Characterization data of representative compounds

3-Phenylamino-but-2-enoic acid ethyl ester (**3a**): Pale yellow oil; IR (neat): 2978, 1648, 1620, 1495, 1441, 1270, 1163, 1062, 751, 698 cm^{-1}; ^{1}HNMR(300 MHz, $CDCl_3$): δ 1.24 (t, J=7.0 Hz, 3H, CH_3), 2.00 (s, 3H, CH_3), 4.12 (q, J=7.0 Hz, 2H, OCH_2), 4.60 (s, 1H, H-C=C), 7.00-7.10 (m, 3H, ArH), 7.28 (m, 2H, ArH), 10.40 (brs, 1H, NH); EIMS: *m/z* (%): 205 (87) M^+, 160 (72), 132 (79), 118 (100), 77 (86), 57 (44).

4,4,4-Trifluoro-3-phenylamino-but-2-enoic acid ethyl ester (**3b**): Pale yellow oil; IR (neat): 3468, 2967, 1657, 1606, 1493, 1445, 1363, 1278, 1173, 1032, 777, 697 cm^{-1}; ^{1}H NMR (400 MHz, $CDCl_3$): δ 1.28 (t, J=7.0 Hz, 3H, CH_3), 4.20 (q, J=7.0 Hz, 2H, OCH_2), 5.32 (s, 1H, H-C=C), 7.20-7.30 (m, 3H, ArH), 7.34-7.40 (m, 2H, ArH), 9.90 (brs, 1H, NH); EIMS: *m/z* (%): 260 (90) M^+, 213 (55), 187 (9), 144 (100), 117 (10), 77 (29), 57 (20).

Reference

Yadav, J. S., Kumar, V. N., Rao, R. S., Priyadarshini, A. D., Rao, P. P., Reddy, B. V. S., and Nagaiah, K. (2006). $Sc(OTf)_3$ catalyzed highly rapid and efficient synthesis of β-enamino compounds under solvent-free conditions. *J. Mol. Catal. A: Chem.*, **256**, 234–237.

C-N bond_4

Type of reaction: C-N bond formation
Reaction conditions: Toluene, room temperature
Synthetic strategy: One-step amidation
Catalyst: $Pd_2(dba)_3.CHCl_3$, ligand **1**

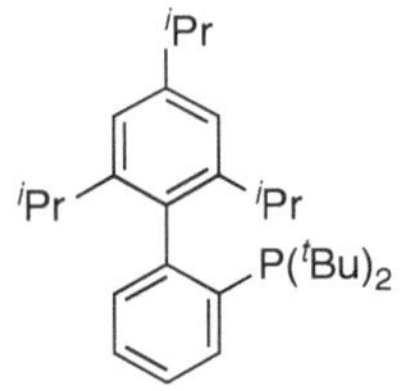

1

Keywords: Aryl bromides, *tert*-butyl carbamate, $Pd_2(dba)_3.CHCl_3$, ligand **1**, ^{t}BuONa, toluene, room temperature, amidation, *N*-Boc-protected anilines

General reaction scheme

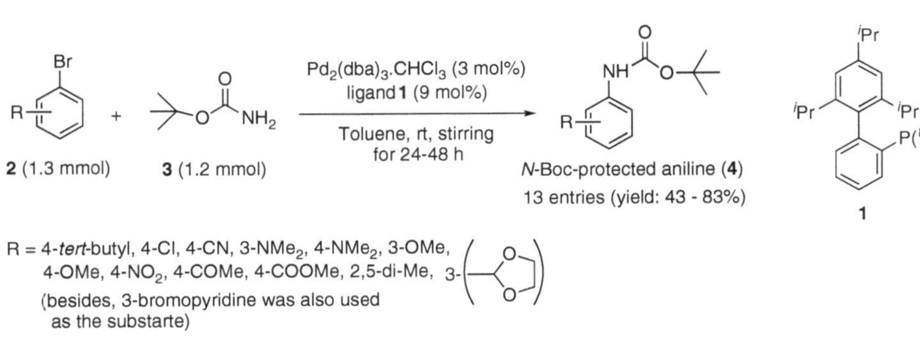

Representative entries

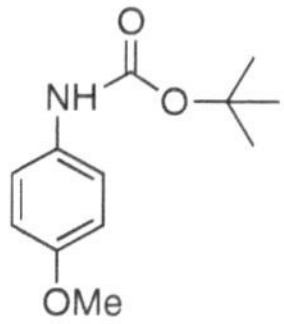

4a

tert-Butyl (4-methoxyphenyl)carbamate (**4a**): Crystalline solid, reaction time: 30 h; yield: 74%

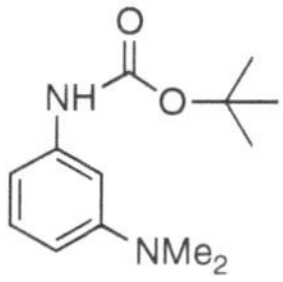

4b

tert-Butyl (3-(dimethylamino) phenyl) carbamate (**4b**): Reaction time: 24 h; yield: 70%

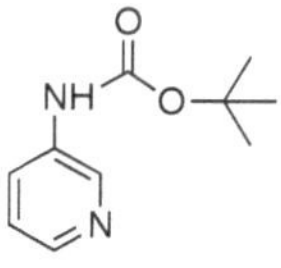

4c

tert-Butyl pyridin-3-ylcarbamate (**4c**): Reaction time: 24 h; yield: 47%

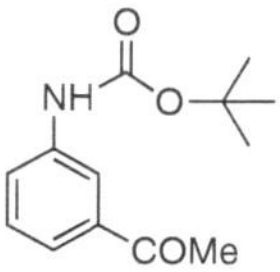

4d

tert-Butyl (3-acetylphenyl)carbamate (**4d**): Reaction time: 48 h; yield: 43%

Experimental procedure

An air-dried glass reaction vessel equipped with a magnetic stir bar was charged successively with $Pd_2(dba)_3.CHCl_3$ (40 mg, 0.039 mmol, 3 mol%), ligand **1** (51 mg, 0.117 mmol, 9 mol%), NaO^tBu (175 mg, 1.82 mmol, 1.4 equiv.), *tert*-butyl carbamate (**3**; 183 mg,1.56 mmol, 1.2 equiv.) and the substituted bromoaniline (**2**; 1.3 mmol). Anhydrous toluene (5.2 mL) was added, and the resultant solution was degassed using one cycle of vacuum and nitrogen purge. The reaction mixture was then stirred at room temperature under nitrogen for 24-48 h. On completion of the reaction as monitored by LC/MS and TLC, the reaction mixture was diluted with 20 mL of diethyl ether and filtered through a pad of Celite. The filtrate was concentrated in *vacuo*, and the residue was purified by flash chromatography to yield *N*-Boc-protected aniline **4**. The products were characterized by from analytical and spectral studies.

Characterization data of representative compounds

tert-Butyl (4-methoxyphenyl)carbamate (**4a**): Crystalline solid, ^1H NMR (400 MHz, $CDCl_3$): δ 7.28-7.26 (m, 2H), 6.85-6.83 (m, 2H), 6.38 (br s, 1H), 3.78 (s, 3H), 1.52 (s, 9H); ^{13}C NMR (100 MHz, $CDCl_3$): δ 155.67, 153.14, 131.40, 120.54, 114.16, 80.20, 55.49, 28.35.

tert-Butyl (3-(dimethylamino) phenyl) carbamate (**4b**): ^1H NMR (400 MHz, $CDCl_3$): δ 7.14 (t, $J=8.1$ Hz, 1 H), 6.89 (br s, 1 H), 6.63 (d, $J=7.9$ Hz, 1 H), 6.43 (dd, $J=8.1$, 2.2 Hz, 2 H), 2.95 (s, 6 H), 1.53 (s, 9 H); ^{13}C NMR (100 MHz, $CDCl_3$): δ 152.7, 151.3, 139.2, 129.4, 107.6, 106.9, 102.7, 80.1, 40.6, 28.4; HRMS (ESI) *m/z* Calcd for $C_{13}H_{21}N_2O_2$ $[M+H]^+$ 237.1603; Found 237.1598.

Reference

Bhagwanth, S., Waterson, A. G., Adjabeng, G. M., and Hornberger, K. R. (2009). Room-temperature Pd-catalyzed amidation of aryl bromides using *tert*-butyl carbamate. *J. Org. Chem.*, **74**, 4634–4637.

C-N bond_5

Type of reaction: C-N bond formation
Reaction conditions: *N,N*-Dimethyl formamide, room temperature
Synthetic strategy: One-step amination *via* coupling reaction
Catalyst: CuBr, *rac*-BINOL (**I**) as ligand

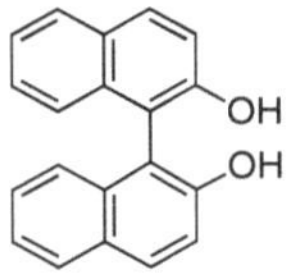

I

Keywords: Aryl iodides, aliphatic amines, CuBr, racemic-1,1′-binaphthyl-2,2′-diol (*rac*-BINOL), K_3PO_4, *N,N*-Dimethyl formamide (DMF), room temperature, coupling reaction, amination, *N*-arylation of aliphatic amines

General reaction scheme

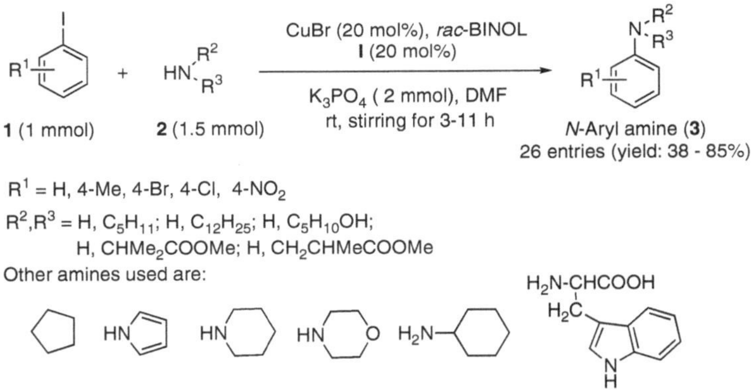

Representative entries

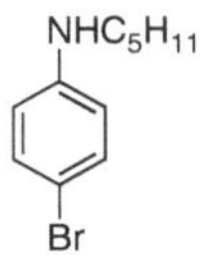

3a

4-Bromo-*N*-pentylaniline (**3a**): Yellow oil, reaction time: 5 h; yield: 80%

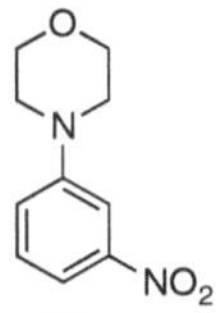

3b

4-(3-Nitrophenyl)morpholine (**3b**): Yellow solid; mp 108-110 °C; reaction time: 11 h; yield: 66%

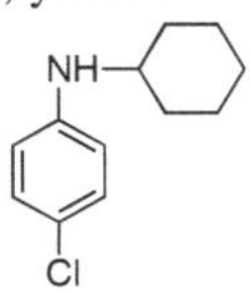

3c

4-Chloro-*N*-cyclohexylaniline (**3c**): White solid; mp 41-43 °C; reaction time: 4 h; yield: 72%

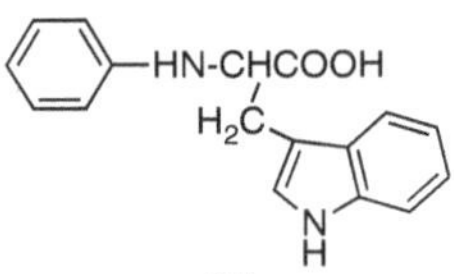

3d

3-(1*H*-Indol-3-yl)-2-(phenylamino) propanoic acid (**3d**): Yellow oil; reaction time: 10 h; yield: 38%

Experimental procedure

A flask was charged with CuBr (28 mg, 0.2 mmol), *rac*-1.1′-binaphthyl-2.2′-diol (**1**; 57 mg, 0.2 mmol), K_3PO_4 (424 mg, 2 mmol), aryl iodide (**2**; 1 mmol), amine (**3**; 1.5 mmol), and DMF (1 mL) under nitrogen atmosphere. The flask was sealed and the mixture was allowed to stir under the nitrogen atmosphere at room temperature for stipulated time-period of 3-11 h. After completion of the reaction, the mixture was diluted with ethyl acetate, the solution was filtered, and the inorganic salts were removed. The residue obtained on removal of the was purified by column chromatography on silica gel using petroleum ether/ethyl acetate (60:1 to 4:1) as eluent to furnish pure *N*-aryl amine **4** with yields of 38-85%. The products were characterized from spectral studies.

Characterization data of representative compounds

4-(3-Nitrophenyl)morpholine (**4b**): Yellow solid; mp 108-110 °C; ^{1}H NMR (300 MHz, $CDCl_3$): δ 7.68 (dd, J=7.92, 1.71 Hz), 7.69 (s, 1H), 7.38 (t, J=7.92, 1H), 7.17 (dd, J=7.92, 1.71 Hz, 1H), 3.87 (t, 4.89Hz, 4H), 3.24 (t, 4.8 Hz, 4H); ^{13}C NMR (75 MHz, $CDCl_3$): δ 151.7, 149.2, 129.7, 120.8, 114.0, 109.3, 66.5, 48.4; HR-EI-MS: M^+ at *m/z* Calcd. for $C_{10}H_{12}N_2O_3$: 208.0848; Found: 208.0848.

3-(1*H*-Indol-3-yl)-2-(phenylamino) propanoic acid (**4d**): Yellow oil; ^{1}H NMR (300 MHz, DMSO-d_6): δ 7.52 (d, 2H, J=7.81 Hz), 7.31 (d, 1H, J=7.82 Hz), 7.16 (s, 1H), 6.92-7.05 (m, 4H), 6.48-6.55 (m, 3H), 4.11 (t, J=6.87 Hz, 1H), 3.51 (br s, 2H), 3.04-3.23 (m, 2H); ^{13}C NMR (75 MHz, DMSO-d_6): δ 175.6, 148.2, 136.5, 129.4, 127.7, 124.2, 121.5, 118.9, 118.7, 116.9, 112.9, 111.9, 110.5, 57.4, 28.5; HR-EI-MS: $[M+H]^+$ *m/z* Calcd. for $C_{17}H_{17}N_2O_2$ 281.1290; Found 281.1300.

Reference

Jiang, D., Fu, H., Jiang, Y., and Zhao, Y. (2007). CuBr/*rac*-BINOL-catalyzed *N*-arylations of aliphatic amines at room temperature. *J. Org. Chem.*, **72**, 672–674.

C-N bond_6

Type of reaction: C-N and C-C bond formation
Reaction conditions: Solvent-free, room temperature
Synthetic strategy: One-pot multi-component nucleophilic addition
Catalyst: No catalyst
Keywords: 1-Substituted imidazoles, isothiocyanates, cyanophenylacetylene, catalyst-free, solvent-free, room temperature, multi-component reaction, *N*-(*Z*-alkenyl)imidazole-2-carbothioamides, stereoselectivity

General reaction scheme

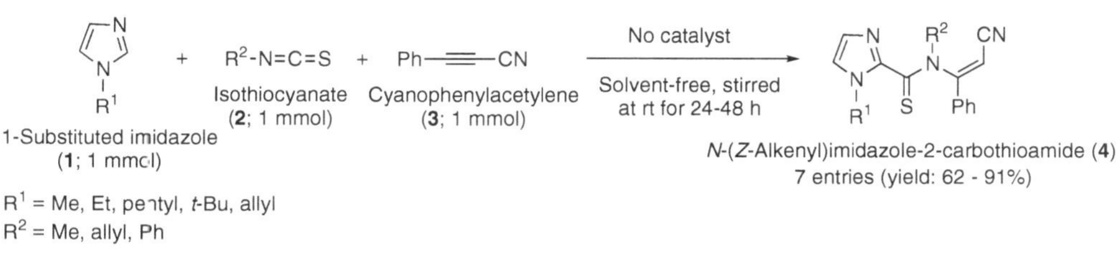

Representative entries

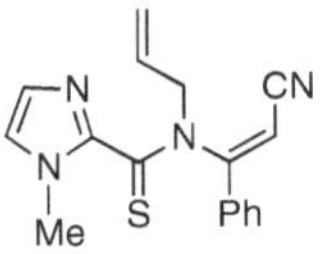

4a

(*Z*)-*N*-Allyl-*N*-(2-cyano-1-phenylvinyl)-1-methyl-1*H*-imidazole-2-carbothioamide (**4a**): Dark yellow oil; reaction time: 24 h; yield: 81%

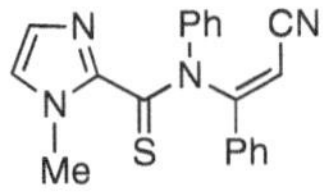

4b

(*Z*)-*N*-(2-Cyano-1-phenylvinyl)-1-methyl-*N*-phenyl-1*H*-imidazole-2-carbothioamide (**4b**): Yellow powder, mp 167-168 °C; reaction time: 48 h; yield: 79%

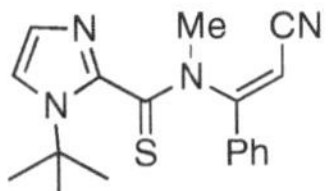

4c

(*Z*)-1-(*tert*-Butyl)-*N*-(2-cyano-1-phenylvinyl)-*N*-methyl-1*H*-imidazole-2-carbothioamide (**4c**): Orange oil; reaction time: 24 h; yield: 74%

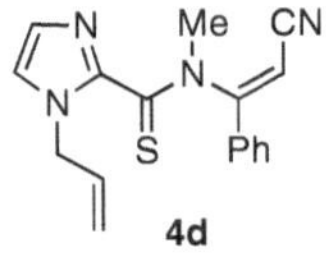

4d

(*Z*)-1-Allyl-*N*-(2-cyano-1-phenylvinyl)-*N*-methyl-1*H*-imidazole-2-carbothioamide (**3d**): Orange oil; reaction time: 24 h; yield: 67%

Experimental procedure

To a mixture of isothiocyanate (**2**; 1 mmol) and cyanophenylacetylene (**3**; 1 mmol) was added 1-substituted imidazole (**1**; 1 mmol). The mixture was stirred at room temperature for 24 h (stirring for 48 h is required when arylisothiocyanates are used). After then the reaction mixture was washed with Et_2O (1×3 mL) to afford *N*-(Z-alkenyl)imidazole-2-carbothioamide **4** in moderate to good yields (62-91%). All the products were characterized by means of elemental analyses and spectral studies including IR, ^{1}H-NMR, and ^{13}C-NMR.

Characterization data of two representative compounds

(*Z*)-*N*-Allyl-*N*-(2-cyano-1-phenylvinyl)-1-methyl-1*H*-imidazole-2-carbothioamide (**4a**): Dark yellow oil; IR (microlayer): 2216 (CN), 1607 (C=C), 1382 (C=S) cm^{-1}; ^{1}H NMR (400.13 MHz, $CDCl_3$): δ 7.60-7.40 (m, 5H, C_8-Ph), 6.83 (s, 2H, 4-H, 5-H), 6.06 (m, 1H, H_X, N_7-allyl]), 5.35 (s, 1H, 9-H), 5.21 (d, $^3J_{HA,HX}$=9.2 Hz, 1H, H_A, N_7-allyl), 5.09 (d, $^3J_{HB,HX}$=16.8 Hz, 1H, H_B, N_7-allyl], 4.05 (m, 2H, -CH_2-, N_7-allyl), 3.88 (s, 3H, N_1-CH_3); ^{13}C NMR (100.62 MHz, $CDCl_3$): δ 189.3 (C-6), 161.9 (C-8), 144.6 (C-2), 137.6 (CH_X, allyl), 133.5 (C_i, C_8-Ph), 131.4 (C_p, C_8-Ph), 129.0 (C_m, C_8-Ph), 127.6 (C-4, C_o, C_8-Ph), 125.1 (C-5), 120.7 ($CH_{A,B}$, allyl), 115.8 (C-10, *C*N), 92.7 (C-9), 55.3 (N_7-CH_2, allyl), 35.9 (N_1-CH_3). Anal. Calcd. for $C_{17}H_{16}N_4S$ (308.40): C 66.21, H 5.23, N 18.17, S 10.40; Found C 66.59, H 5.09, N 18.55, S 10.01.

(*Z*)-1-Allyl-*N*-(2-cyano-1-phenylvinyl)-*N*-methyl-1*H*-imidazole-2-carbothioamide (**3d**): Orange oil; IR (microlayer): 2217 (CN), 1609 (C=C), 1358 (C=S) cm^{-1}; ^{1}H NMR (400.13 MHz, $CDCl_3$): δ 7.50-7.30 (m, 5H, C_8-Ph), 6.85 (s, 2H, 4-H, 5-H), 5.88 (m, 1H, H_X, N_1-allyl), 5.38 (s, 1 H, 9-H), 5.22 (m, 2H, H_A, H_B, N_1-allyl), 4.80 (m, 2H, CH_2, N_1-allyl), 3.64 (s, 3H, N_7-CH_3); ^{13}C NMR (100.62 MHz, $CDCl_3$): δ 189.2 (C-6), 163.8 (C-8), 144.4 (C-2), 136.0 (C_i, C_8-Ph), 132.8 [CH_X, N_1-allyl], 131.3 (C_p, C_8-Ph), 128.9 (C_m, C_8-Ph), 127.6 (C_o, C_8-Ph), 127.4 (C-4), 122.4 (C-5), 119.1 ($CH_{A,B}$, N_1-allyl), 115.6 (C-10, *C*N), 92.0 (C-9), 50.7 (CH_2, N_1-allyl], 42.8 (N_7-CH_3). Anal. Calcd. for $C_{17}H_{16}N_4S$ (308.40): C 66.21, H 5.23, N 18.17, S 10.40; Found C 66.59, H 5.49, N 17.79, S 10.01.

Reference

Trofimov, B. A., Andriyankova, L. V., Nikitina, L. P., Belyaeva, K. V., Mal'kina, A. G., Afonin, A. V., and Ushakov, I. A. (2012). Stereoselective metal-free reaction of imidazoles with isothiocyanates involving cyanophenylacetylene: A shortcut to *N*-(*Z*-alkenyl)imidazole-2-carbothioamides. *Synlett*, **23**, 2069–2072.

C-N bond_7

Type of reaction: C-N bond formation
Reaction conditions: Ionic liquid medium, room temperature
Synthetic strategy: Michael addition
Catalyst: 1-Methyl-3-butylimidazolium hydroxide ([bmIm]OH)
Keywords: Imidazoles, vinyl esters, [bmIm]OH ionic liquid, room temperature, nucleophilic Michael addition, *N*-substituted imidazoles

General reaction scheme

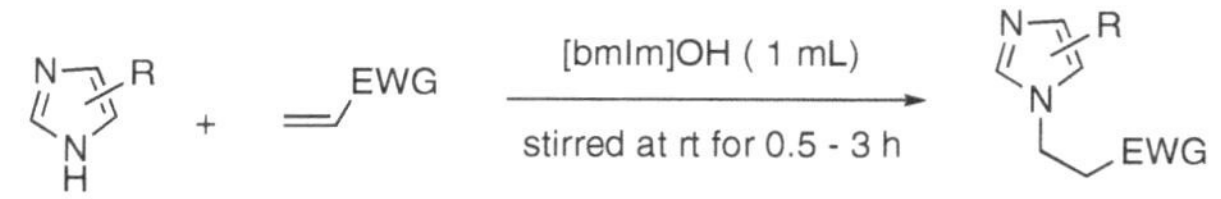

1 (1 mmol) **2** (1.5 mmol) *N*-Substituted imidazole (**3**)
11 entries (yield: 80 - 96%)

R = H; 3-NO_2; 2-CH_3; 5-CH_3,3-NO_2
EWG = CN; $COCH_3$; CO_2CH_3, COC_2H_5; $COOCH{=}CH_2$; $CO_2C_4H_9$

Representative entries

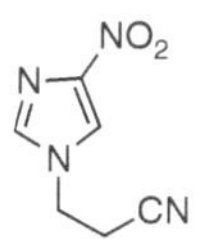

3a

3-(4-Nitro-1*H*-imidazol-1-yl)propanenitrile (**3a**): White solid, mp 105-106 °C; reaction time: 1 h; yield: 90%

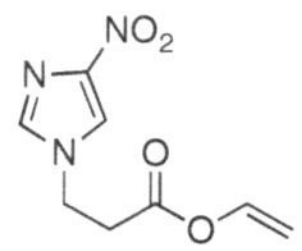

3b

Vinyl 3-(4-nitro-1*H*-imidazol-1-yl)propanoate (**3b**): White solid, mp 94-96 °C; reaction time: 0.5 h; yield: 96%

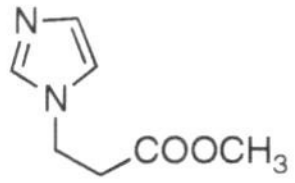

3c

Methyl 3-(1*H*-imidazol-1-yl)propanoate (**3c**): Yellow oil; reaction time: 1 h; yield: 92%

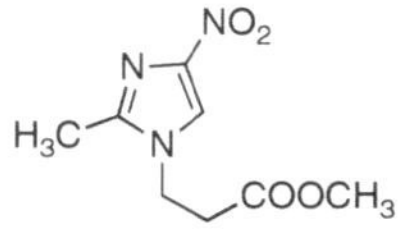

3d

Methyl 3-(2-methyl-4-nitro-1*H*-imidazol-1-yl) propanoate (**3d**): Yellow liquid; reaction time: 3 h; yield: 94%

Experimental procedure

Imidazole derivative (**1**; 1 mmol) and α,β-unsaturated carbonyl compound (**2**; 1.5 mmol) were placed in a 10 mL conical flask containing 1 mL of [bmIm]OH; the mixture was shaken at ambient temperature for 0.5-3 h to complete the reaction. On completion, the reaction mixture was extracted from the ionic liquid phase with ethyl acetate followed by dried over anhydrous sodium sulfate and subsequent evaporation of organic solvent under reduced pressure. The residue was purified by flash column chromatography to afford the corresponding Michael adduct **3** with good to excellent yield. The ionic liquid left in the conical flask was further washed with ether, dried under vacuum at 90 °C for 2 h and reused for subsequent reactions up to fifth run. Each of the products was characterized on the basis of IR, NMR and MS.

Characterization data of two representative compounds

Vinyl 3-(4-nitro-1*H*-imidazol-1-yl)propanoate (**3b**): White solid, mp 94-96 °C; IR (KBr): 1743, 1649, 1526, 1486 cm^{-1}; ^1H NMR ($CDCl_3$): δ 7.88 (s, 1H), 7.54 (s, 1H), 7.26-7.22 (m, 1H), 4.95-4.91, 4.67 (m, 2H), 4.41 (t, $J=6.1$ Hz, 2H), 2.98 (t, $J=6.1$ Hz, 2H); ^{13}C NMR ($CDCl_3$): δ 167.6, 148.9, 140.8, 136.7, 119.8, 99.4, 43.4, 35.1; ESI-MS (*m*/*z*): 212 (M+1).

Methyl 3-(2-methyl-4-nitro-1*H*-imidazol-1-yl)propanoate (**3d**): Yellow liquid; IR (neat): 1729, 1508 cm^{-1}; ^1H NMR ($CDCl_3$): δ 7.76 (s, 1H), 4.25 (t, 2H, J=6.4 Hz, 2H), 3.72 (s, 3H), 2.82 (t, 2H, $J=6.4$ Hz), 2.48 (s, 3H); ^{13}C NMR ($CDCl_3$): δ 173.2, 148.9, 147.8, 123.1, 54.7, 45.0, 36.9, 15.5; ESI-MS (*m*/*z*): 214 (M+1).

Reference

Xu, J.-M., Qian, C., Liu, B.-K., Wu, Q., and Lin, X.-F. (2007). A fast and highly efficient protocol for Michael addition of *N*-heterocycles to α,β-unsaturated compound using basic ionic liquid [bmIm] OH as catalyst and green solvent. *Tetrahedron*, **63**, 986–990.

C-N bond_8

Type of reaction: C-N bond formation
Reaction conditions: Dichloromethane, room temperature
Synthetic strategy: One-step aza-Michael addition
Catalyst: No catalyst
Keywords: Nitroalkenes, pyrazole, catalyst-free, dichloromethane, room temperature, aza-Michael addition, amination

General reaction scheme

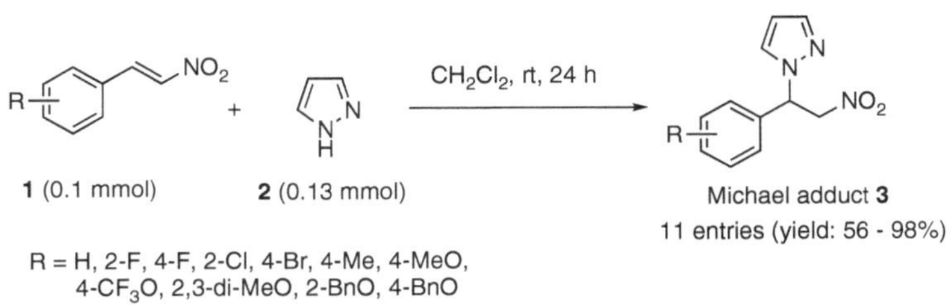

Representative entries

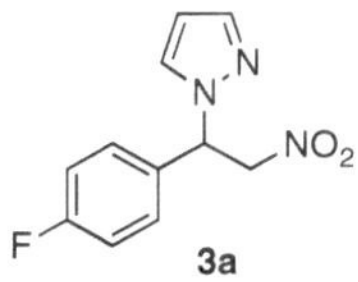

3a

1-(1-(4-Fluorophenyl)-2-nitroethyl)-1*H*-pyrazole (**3a**): Yield: 93%

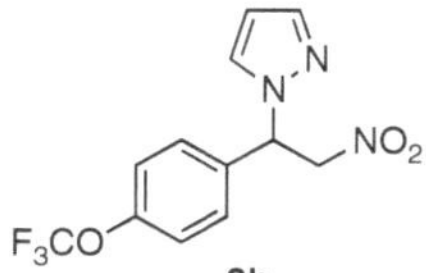

3b

1-(2-Nitro-1-(4-(trifluoromethoxy) phenyl)ethyl)-1*H*-pyrazole (**3b**): Yield: 97%

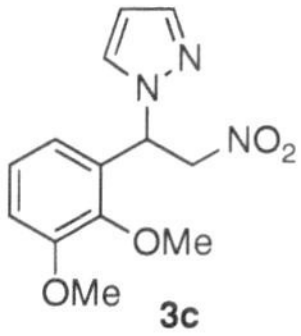

3c

1-(1-(2,3-Dimethoxyphenyl)-2-nitroethyl)-1*H*-pyrazole (**3c**): Yield: 81%

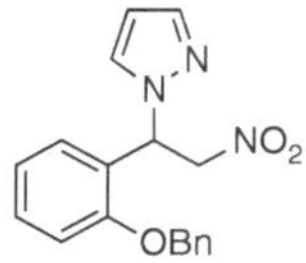

3d

1-(1-(2-(benzyloxy)phenyl)-2-nitroethyl)-1*H*-pyrazole (**3d**): Yield: 56%

Experimental procedure

A mixture of nitroalkene (**1**; 0.1 mmol) and azole (**2**; 0.13 mmol) in dichloromethane (0.1 mL) was stirred at room temperature for 24 h and then the solvent was removed under vacuum. The residue was purified by column chromatography on silica gel to yield the Michael adducts **3**. Each of the products was characterized on the basis of detailed spectral measurements including IR, ^{1}H NMR, ^{13}C NMR and HRMS.

Characterization data of two representative compounds

1-(1-(2,3-Dimethoxyphenyl)-2-nitroethyl)-1*H*-pyrazole (**3c**): IR (KBr): 3120, 2940, 2838, 1556, 1483, 1377, 1273, 1090, 750 cm^{-1}; ^{1}H NMR (400 MHz, $CDCl_3$): δ 7.57 (d, J=2.0 Hz, 1H), 7.46 (d, J=2.4 Hz, 1H), 7.01 (t, J=8.0 Hz, 1H), 6.91 (dd, J=1.6, 8.4 Hz, 1H), 6.69 (dd, J=1.2, 8.0 Hz, 1H), 6.51 (dd, J=4.4, 6.0 Hz, 1H), 6.28 (t, J=2.2 Hz, 1H), 5.57 (dd, J=10.0, 14.0 Hz, 1H), 4.90 (dd, J=4.2, 14.2 Hz, 1H), 3.91 (s, 3H), 3.87 (s, 3H); ^{13}C NMR (100 MHz, $CDCl_3$): δ 152.7, 146.0, 140.1, 130.0, 129.4, 124.7, 119.2, 113.4, 106.4, 76.3, 61.1, 56.9, 56.0; HRMS: $[M+Na]^+$ calcd. for $C_{13}H_{15}N_3O_4Na$ at m/z 300.0960; Found at m/z 300.0963.

1-(1-(2-(benzyloxy)phenyl)-2-nitroethyl)-1*H*-pyrazole (**3d**): IR (KBr): 3066, 3034, 2917, 2873, 1556, 1377, 1243, 1022, 752 cm^{-1}; ^{1}H NMR (400 MHz, $CDCl_3$): δ 7.61 (s, 1H), 7.42-7.29 (m, 7H), 7.02-6.91 (m, 3H), 6.59 (dd, J=3.2, 10.0 Hz, 1H), 6.29 (s, 1H), 5.56 (dd, J=10.2, 13.8 Hz, 1H), 5.18 (s, 2H), 4.92 (dd, J=3.2, 14.0 Hz, 1H); ^{13}C NMR (100 MHz, $CDCl_3$): δ 154.8, 140.1, 136.3, 130.4, 130.3, 129.0, 128.5, 127.9, 127.4, 124.9, 121.7, 112.3, 106.4, 75.8, 70.7, 57.1; IHRMS: $[M+H]^+$ calcd. for $C_{18}H_{18}N_3O_3$ at m/z 324.1348; Found at m/z 324.1349.

Reference

Wu, Y., Wang, J., Li, P., and Kwong, F. Y. (2012). Catalyst-free efficient aza-Michael addition of azoles to nitroalkenes. *Synlett*, **23**, 788–790.

C-N bond_9

Type of reaction: C-N bond formation
Reaction conditions: Ethanol/methanol, room temperature
Synthetic strategy: Sequential multicomponent one-pot synthesis
Catalyst: Catalyst-free
Keywords: Carbonyl compounds, amines, thioacids, alcohol, isonitriles, room-temperature, one-pot multicompinent reaction, Ugi reaction, endothiopeptides

General reaction scheme

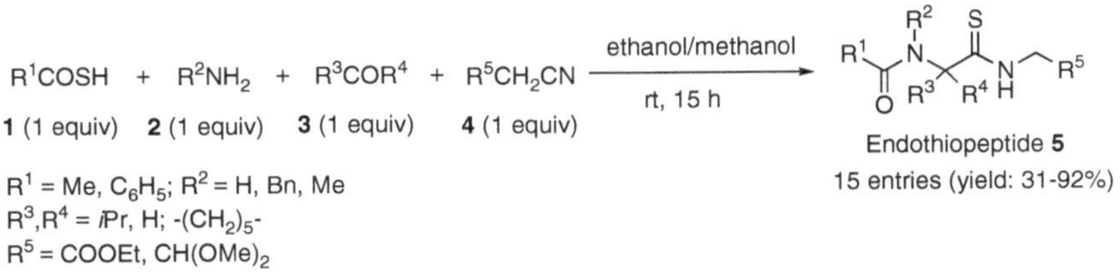

Endothiopeptide **5**
15 entries (yield: 31-92%)

Representative entries

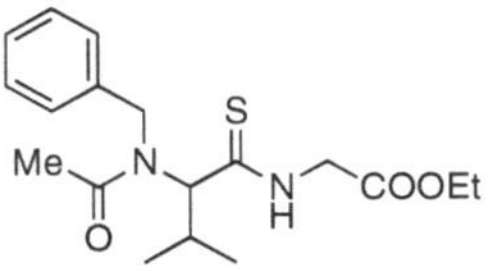

5a

Ethyl 2-(2-(*N*-benzylacetamido)-3-methylbutanethioamido)acetate (**5a**): Yellow oil; yield: 68%

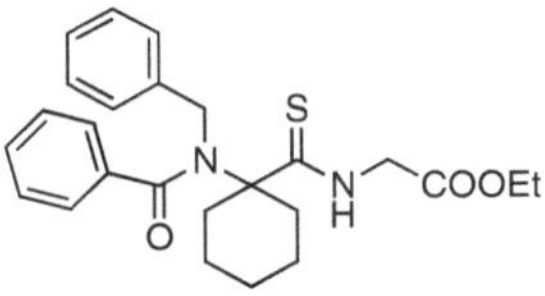

5b

Ethyl 2-(1-(*N*-benzylbenzamido) cyclohexanecarbothioamido)acetate (**5b**): Yield: 55%

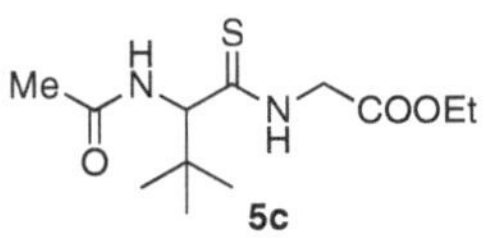

5c

Ethyl 2-(2-acetamido-3,3-dimethylbutanethioamido) acetate (**5c**): Yield: 31%

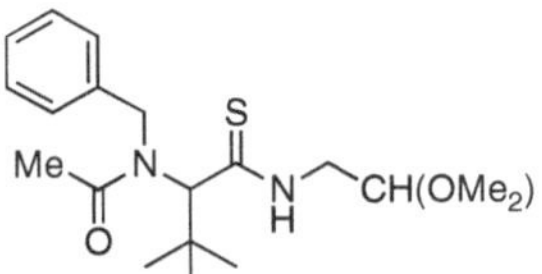

5d

N-Benzyl-*N*-(1-((2,2-dimethoxyethyl)amino)-3,3-dimethyl-1-thioxobutan-2-yl)acetamide (**5d**): White rhombic crystals, mp 110-112 °C; yield: 81%

Experimental procedure

Amine (**2**; 1 equiv.) and carbonyl compound (**3**: 1 equiv.) were dissolved in alcohol (ethanol or methanol) and stirred for 15 min at room temperature. Then the reaction mixture was cooled to 0 °C and the thio acid (**1**; 1 equiv.) was added followed by the isonitrile (**4**; 1 equiv.). The ice-bath was removed and the mixture was stirred overnight at room temperature. Then dichloromethane was added and the organic layer was washed twice with saturated sodium carbonate solution and 1 M $KHSO_4$ solution. After drying the organic layer over sodium sulfate, the solvent was removed in *vacuo* and the crude product was purified by column chromatography and/or recrystallisation. All the products were characterized on the basis of elemental analyses and detailed spectral studies.

Characterization data of two representative compounds

Ethyl 2-(2-(*N*-benzylacetamido)-3-methylbutanethioamido)acetate (**5a**): Yellow oil; ^{1}H NMR (500 MHz, $CDCl_3$): δ 0.67 (d, *J* =6.1Hz, 3H), 0.87 (d, *J*=6.7 Hz, 3H), 1.25 (t, *J*=7.0 Hz, 3H), 2.10 (s, 3H), 2,81 (bs, 1H), 4.18 (q, *J*=7.0 Hz, 2H), 4.29 (dd, *J*=13.1, 5.2 Hz, 2H), 4.35 (d, $J_{4,5}$=4.9 Hz, 1H), 4.57 (s, 2H), 7.16-7.26(m, 5H), 7.97 (bs, 1H); ^{13}C NMR (125 MHz, $CDCl_3$): δ 14.1, 20.3, 28.3, 47.2, 61.6, 125.9, 127.5, 128.7, 138.7, 168.6, 173.8, 202.6; HMRS (CI): Calcd. for $C_{18}H_{26}N_2O_3S$ ($[M]^+$), 350.1664; Found: 350.1653.

N-Benzyl-N-(1-((2,2-dimethoxyethyl)amino)-3,3-dimethyl-1-thioxobutan-2-yl)acetamide (**5d**): White rhombic crystals, mp 110-112 °C; ^{1}H NMR(500 MHz, $CDCl_3$): δ =1.06 (s, 9H), 2.13 (s, 3H), 3.33 (s, 6H), 3.70 (m, 2H), 4.23-4.54 (m, 4H), 7.19-7.26 (m, 5H), 10.82 (bs, 1H); ^{13}C NMR (125 MHz, CD_3OD): δ 23.4, 28.5, 38.2, 54.3, 54.7, 102.2, 129.6, 176.5, 201.5. HRMS (CI): Calcd. for $C_{19}H_{30}N_2O_3S$ ($[M]^+$), 366.1978; found, 366.1978. Elemental analysis: $C_{19}H_{30}N_2O_3S$ (364.51) calcd.: C, 62.61; H, 7.74; N, 7.69; Found: C, 62.10; H, 7.49; N, 7.57.

Reference

Kazmaier, U., and Ackermann, S. (2005). A straightforward approach towards thiazoles and endothiopeptides via Ugi reaction. *Org. Biomol. Chem.*, **3**, 3184–3187.

C-N bond_10

Type of reaction: C-N bond formation
Reaction conditions: Water, room temperature
Synthetic strategy: *N*-arylation of *N*-arylsulfonamides
Catalyst: $CuCl/K_3PO_4$
Keywords: *N*-arylsulfonamides, diaryliodonium triflates, copper(I) chloride, potassium phosphate, water, room temperature, *N,N*-diarylsulfonamides

General reaction scheme

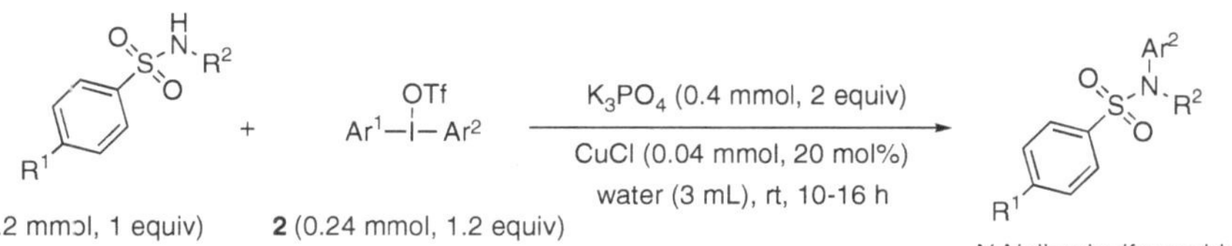

1 (0.2 mmol, 1 equiv) **2** (0.24 mmol, 1.2 equiv)

N,N-diarylsulfonamide **3**
17 entries (yield: 60->99%)

R^1 = H, Me

R^2 = C_6H_5, 4-BrC_6H_4, 4-ClC_6H_4, 4-FC_6H_4, 4-NMe$_2C_6H_4$, 4-OMeC_6H_4, 4-MeC_6H_4

Ar^1, Ar^2 = C_6H_5, C_6H_5; 4-ClC_6H_4, 4-ClC_6H_4; 4-BrC_6H_4, 4-BrC_6H_4; 4-MeC_6H_4, 4-MeC_6H_4; 4-OMeC_6H_4, 4-OMeC_6H_4; 4-OMeC_6H_4, 4-NO$_2C_6H_4$; mesityl, C_6H_5; mesityl, mesityl

Representative entries

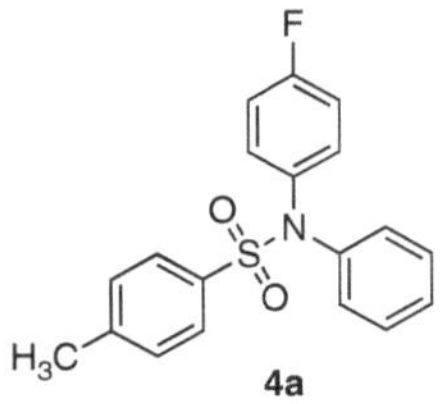

4a

N-(4-Fluorophenyl)-4-methyl-*N*-phenylbenzenesulfonamide (**4a**): White solid, reaction time: 10 h; yield: 97%

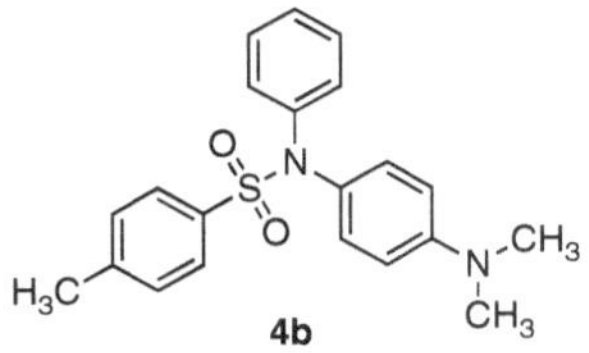

4b

N-(4-(Dimethylamino)phenyl)-4-methyl-*N*-phenylbenzenesulfonamide (**4b**): White solid, reaction time: 10 h; yield: 64%

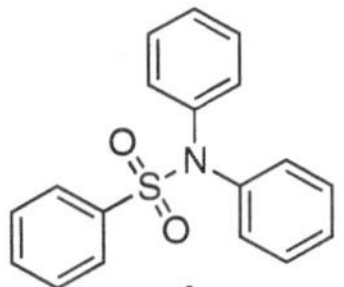

4c

N,N-Diphenylbenzenesulfonamide (**4c**): White solid, reaction time: 10 h; yield: 97%

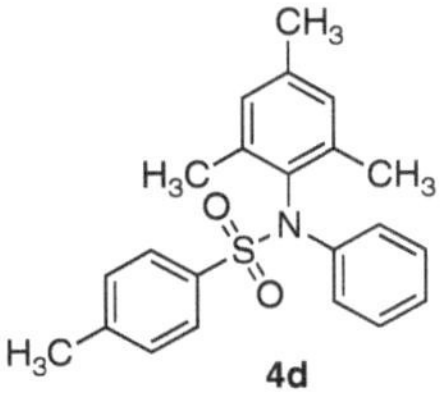

4d

N-Mesityl-4-methyl-*N*-phenylbenzenesulfonamide (**4d**): White solid, reaction time: 16 h; yield: 69%

Experimental procedure

A mixture of *N*-arylsulfonamide (**1**; 0.2 mmol, 1 equiv.), diaryliodonium salt (**2**; 0.24 mmol, 1.2 equiv.), K_3PO_4 (0.4 mmol, 2 equiv.), CuCl (0.04 mmol, 20 mol%) and H_2O (3 mL) in a 10 mL reaction tube was stirred vigorously at room temperature for 10-16 h. After completion of the reaction, as indicated by TLC, the reaction mixture was concentrated under reduced pressure, and the crude compound was purified by chromatography on a silica gel column (ethyl acetate/petroleum ether (1/15)) to afford *N,N*-diarylsulfonamides **3**. Each of the products was characterized by spectral methods.

Characterization data of two representative compounds

N-(4-(Dimethylamino)phenyl)-4-methyl-*N*-phenylbenzenesulfonamide (**4b**): White solid; ^{1}H NMR (400 MHz, $CDCl_3$): δ 7.66-7.61 (m, 2H), 7.33-7.22 (m, 7H), 7.16-7.11 (m, 2H), 2.97 (d, $J=8.8$ Hz, 6H), 2.47 (d, $J=8.0$ Hz, 3H); ^{13}C NMR (100 MHz, $CDCl_3$): δ 149.9, 143.3, 142.5, 137.9, 130.1, 129.7, 129.4, 129.1, 127.9, 127.3, 126.6, 112.4, 40.4, 21.6; HRMS (ESI^+): Calcd. for $C_{21}H_{23}N_2O_2S$ ($[M+H]^+$): 367.1475, found: 367.1464.

N-Mesityl-4-methyl-*N*-phenylbenzenesulfonamide (**4d**): White solid; ^{1}H NMR (400 MHz, $CDCl_3$): δ 7.82 (d, $J=8.4$ Hz, 2H), 7.31 (d, $J=8.4$ Hz, 2H), 7.27-7.21 (m, 4H), 6.96 (s, 1H), 2.44 (s, 3H), 2.33 (s, 3H); ^{13}C NMR (100 MHz, $CDCl_3$): δ 143.73, 141.48, 139.06, 138.48, 138.24, 135.45, 129.99, 129.51, 128.90, 128.01, 124.05, 121.30, 21.56, 21.02, 19.31; HRMS (ESI^+): Calcd. for $C_{22}H_{27}N_2O_2S$ ($[M+NH_4]^+$): 383.1788, found: 383.1700.

Reference

Geng, X., Mao, S., Chen, L., Yu, J., Han, J., Hua, J., and Wang, L. (2014). Copper-catalyzed direct *N*-arylation of N-arylsulfonamides using diaryliodonium salts in water. *Tetrahedron Let.*, **55**, 3856–3859.

C-N bond_11

Type of reaction: C-N bond formation
Reaction conditions: Methanol, room temperature
Synthetic strategy: Chan-Lam coupling forming C-N bond
Catalyst: Cu(I)Cl/air
Keywords: Aryl/heteroaryl boronic acids, sulfonyl azides, Cu(I)Cl, air, methanol, room temperature, Chan-Lam coupling, *N*-aryl/heteroaryl sulfonamides

General reaction scheme

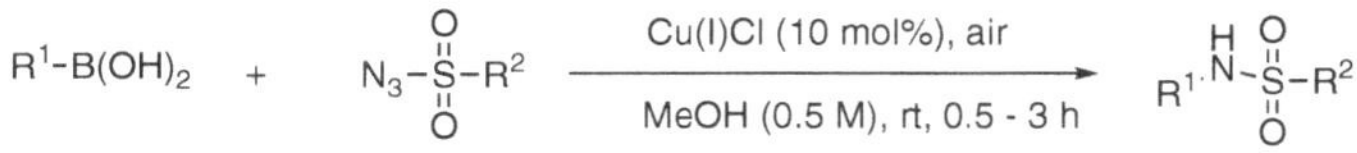

1 (1.2 mmol) **2** (1 mmol)

N-aryl/heteroaryl sulfonamide **3**
37 entries (yield: 28-99%)

R^1 = C_6H_5, naphthyl, 4-ethenylphenyl, 4-MeC_6H_4, 3,5-di-MeC_6H_3, 4-OMeC_6H_4, 3-OMeC_6H_4, 2-OMeC_6H_4, 4-ClC_6H_4, 4-BrC_6H_4, 4-$CF_3C_6H_4$, 4-CO_2MeC_6H_4; 3-(2-methoxypyridyl), 3-(4-methoxypyridyl), 3-furyl, 3-thienyl

R^2 = CH_3, C_2H_5, C_6H_5, 4-ClC_6H_4, 4-MeC_6H_4, 2,4,6-tri-MeC_6H_2, 4-CNC_6H_4, 4-$OCH_3C_6H_4$, 4-$NO_2C_6H_4$,4-(6-methoxypyridyl), naphthyl

Representative entries

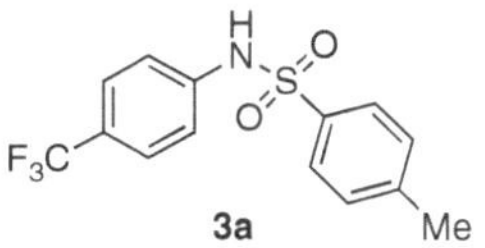

3a

4-Methyl-*N*-(4-(trifluoromethyl)phenyl) benzenesulfonamide (**3a**): White solid; reaction time: 3 h; yield: 92%

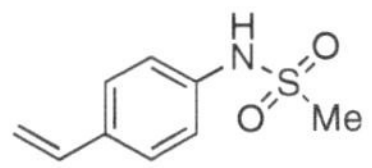

3b

N-(4-Vinylphenyl)methanesulfonamide (**3b**): White solid, mp 90-92 °C; reaction time: 1 h; yield: 99%

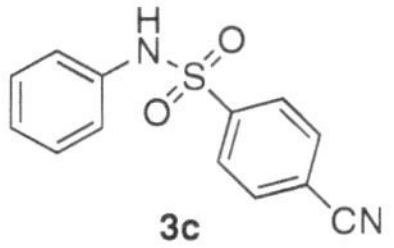

3c

4-Cyano-*N*-phenylbenzenesulfonamide (**3c**): Pale yellow solid, mp 110-112 °C; reaction time: 1 h; yield: 92%

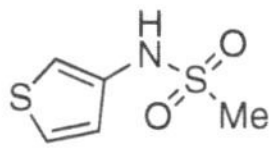

3d

N-(Thiophen-3-yl)methanesulfonamide (**3d**): Pale yellow oil; reaction time: 1 h; yield: 48%

Experimental procedure

A 10 mL round bottom flask was charged with an arylboronic acid (**1**; 1.2 mmol), CuCl (10 mol %) and the corresponding sulfonyl azide (**2**; 1.0 mmol). 2 mL of methanol was then added to the flask. The reaction mixture was stirred (0.5 – 3 h) at room temperature in an open flask. After completion of the reaction, the solvent was removed *in vacuo* and the residue was purified by flash column chromatography on silica gel to obtain pure *N*-aryl/heteroaryl sulfonamide **3**. All the products were characterized by spectral methods.

Characterization data of two representative compounds

4-Methyl-*N*-(4-(trifluoromethyl)phenyl) benzenesulfonamide (**4a**): White solid; ^{1}H NMR (300 MHz, $CDCl_3$): δ 7.93 (s, 1H), 7.78 (d, *J*=8.4 Hz, 2H), 7.48 (d, *J*=8.4 Hz, 2H), 7.25 (dd, *J*=13.8 and 8.1 Hz, 4H), 2.39 (s, 3H); ^{13}C NMR (100 MHz, $CDCl_3$): δ 144.6, 140.0, 135.6, 130.0, 127.2, 126.6, 126.5, 123.9 (q, *J*=270.0 Hz, CF_3), 119.5, 21.5.

N-(Thiophen-3-yl)methanesulfonamide (**4d**): Pale yellow oil; IR (ATR): 3259, 2928, 1692, 1443, 1313, 1144, 1073 cm^{-1}; ^{1}H NMR (300 MHz, $CDCl_3$): δ 7.52 (s, 1H), 7.37 (s, 1H), 6.41 (s, 1H), 6.08 (br, 1H), 3.00 (s, 3H); ^{13}C NMR (125 MHz, $CDCl_3$): δ 143.2, 134.8, 123.1, 107.7, 38.7; HRFABMS: *m*/*z* 162.0225 [$(M+1)^+$; calcd for $C_5H_8NO_3S$: 162.0225].

Reference

Moon, S.-Y., Nam, J., Rathwell, K., and Kim, W.-S. (2014). Copper-catalyzed Chan-Lam coupling between sulfonyl azides and boronic acids at room temperature. *Org. Lett.*, **16**, 338–341.

C-N bond_12

Type of reaction: C-N bond formation
Reaction conditions: Acetone, room temperature
Synthetic strategy: Ru(II)-catalyzed *ortho*-C-H amination
Catalyst: $[Ru^{II}Cl_2(p\text{-cymeme})]_2/K_2CO_3$
Keywords: Substituted aryl/heteroaryl benzamides, cyclic *N*-benzoyloxyamines, $[Ru^{II}Cl_2(p\text{-cymeme})]_2$ complex as catalyst, potassium carbonate, argon atmosphere, acetone, room temperature, *ortho*-C-H amination, substituted aryl/heteroaryl amines

General reaction scheme

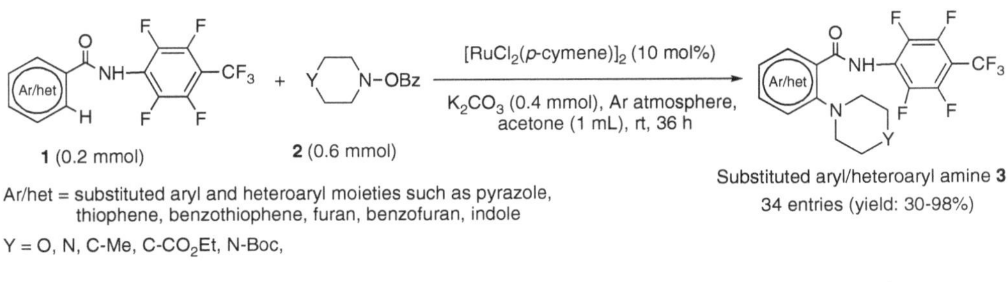

Representative entries

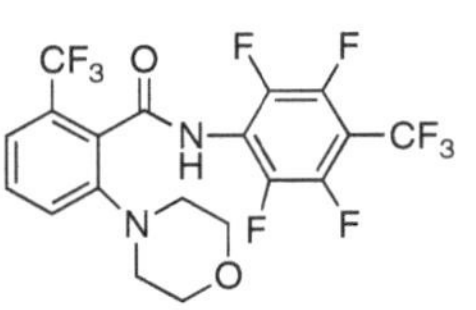

3a

2-Morpholino-*N*-(2,3,5,6-tetrafluoro-4-(trifluoromethyl)phenyl)-6-(trifluoromethyl) benzamide (**3a**): White solid; yield: 71%

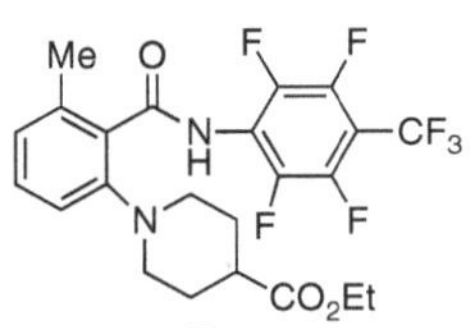

3b

Ethyl 1-(3-methyl-2-((2,3,5,6-tetrafluoro-4-(trifluoromethyl)phenyl)carbamoyl)phenyl) piperidine-4-carboxylate (**3b**): White solid, yield: 98%

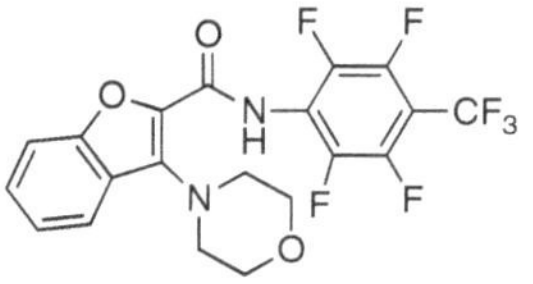

3c

3-Morpholino-*N*-(2,3,5,6-tetrafluoro-4-(trifluoromethyl)phenyl)benzofuran-2-carboxamide (**3c**): White solid, yield: 55%

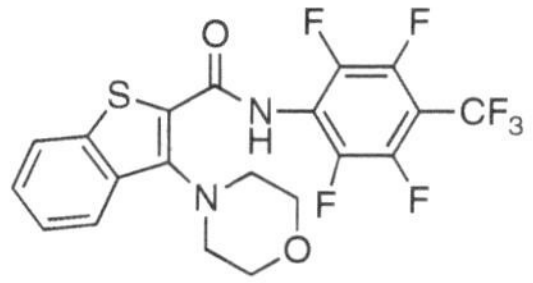

3d

3-Morpholino-*N*-(2,3,5,6-tetrafluoro-4-(trifluoromethyl)phenyl)benzo[*b*]thiophene-2-carboxamide (**3d**): White solid; yield: 39%

Experimental procedure

N-aryl benzamide (**1**; 0.20 mmol), $[RuCl_2(pcymene)]_2$ (0.02 mmol), *O*-benzoyl hydroxylamine (**2**; 0.60 mmol), and K_2CO_3 (0.40 mmol) dissolved in acetone (1 mL) was stirred in a 25 mL Schlenk tube at room temperature under argon atmosphere for 36 h. Upon completion, the reaction mixture was diluted with ethyl acetate and filtered through a small pad of Celite. The filtrate was concentrated *in vacuo* and purified by a silica gel packed flash chromatography column, using ethyl acetate/hexane as the eluent to furnish pure substituted aryl/heteroaryl amine **3**. All the products were characterized by spectral methods.

Characterization data of two representative compounds

Ethyl 1-(3-methyl-2-((2,3,5,6-tetrafluoro-4-(trifluoromethyl)phenyl)carbamoyl)phenyl) piperidine-4-carboxylate (**3b**): White solid, ^{1}H NMR (400 MHz, $CDCl_3$): δ 9.52 (brs, 1H), 7.32 (t, $J=7.8$ Hz, 1H), 7.06 (t, $J=7.9$ Hz, 2H), 4.16 (q, $J=7.1$ Hz, 2H), 3.26 (d, $J=11.9$ Hz, 2H), 2.81 (t, $J=10.4$ Hz, 2H), 2.47 (s, 3H), 2.43-2.39 (m, 1H), 2.05-2.01 (m, 2H), 1.95-1.85 (m, 2H), 1.27 (t, $J=7.1$ Hz, 3H); ^{13}C NMR (100 MHz, $CDCl_3$): δ 174.37, 166.47, 150.54, 139.15, 130.84, 129.15, 127.20, 117.45, 60.47, 53.19, 40.27, 28.27, 20.70, 13.97; HRMS (ESI-TOF): *m*/*z* Calcd for $C_{23}H_{21}F_7N_2O_3$ $[M+H]^+$ 507.1513; Found 507.15.

3-Morpholino-*N*-(2,3,5,6-tetrafluoro-4-(trifluoromethyl)phenyl)benzo[*b*]thiophene-2-carboxamide (**3d**): White solid; ^{1}H NMR (400 MHz, $CDCl_3$): δ 12.58 (brs, 1H), 8.16 (d, $J=7.5$ Hz, 1H), 7.91 (d, $J=7.4$ Hz, 1H), 7.51-7.43 (m, 2H), 3.97 (brs, 4H), 3.50 (brs, 4H); ^{13}C NMR (100 MHz, $CDCl_3$): δ 159.41, 144.34, 139.49, 134.95, 131.78, 126.88, 124.71, 124.23, 124.18, 67.44, 51.05; HRMS (ESI-TOF): *m*/*z* Calcd for $C_{20}H_{13}F_7N_2O_2S$ $[M+H]^+$ 479.0659; Found 479.0647.

Reference

Shang, M., Zeng, S.-H., Sun, S.-Z., Dai, H.-X., and Yu, J.-Q. (2013). Ru(II)-catalyzed *ortho*-C-H amination of arenes and heteroarenes at room temperature. *Org. Lett.*, **15**, 5286–5289.

C-N bond_13

Type of reaction: C-N and C-C bond formation
Reaction conditions: Ethanol, room temperature
Synthetic strategy: One-pot multicomponent
Catalyst: $Bi(NO_3)_3.5H_2O$

Keywords: Aromatic aldehydes, β-keto esters, amines, bismuth nitrate, ethanol, room temperature, homogeneous catalysis, tandem reaction, one-pot synthesis, multicomponent reaction (MCR), Mannich reaction, functionalized piperidines

General reaction scheme

Representative new entries

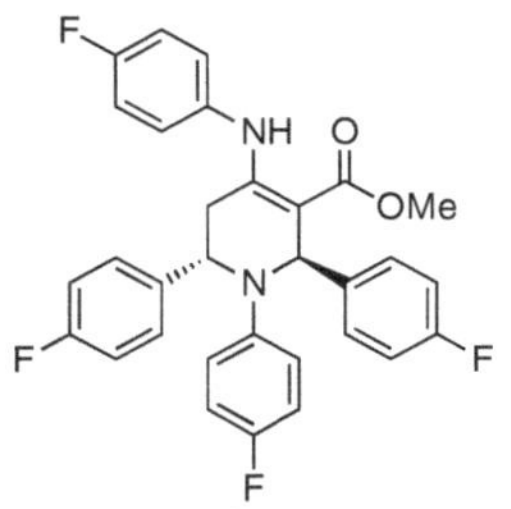

4a

Methyl 2,6-*bis*(4-fluorophenyl)-1-(4-fluorophenyl)-4-(4-fluorophenylamino)-1,2,5,6-tetrahydropyridine-3-carboxylate (**4a**): White crystals, mp 179-181 °C; reaction time: 12 h; yield: 68%

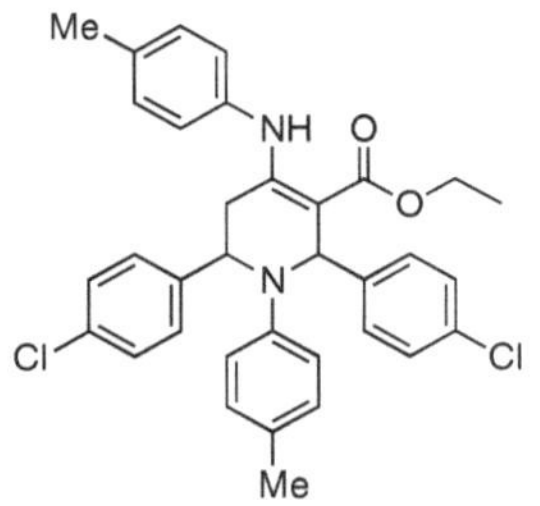

4b

Ethyl 2,6-*bis*(4-chlorophenyl)-1-(4-methylphenyl)-4-(4-methylphenyl amino)-1,2,5,6-tetrahydropyridine-3-carboxylate (**4b**): White crystals, mp 227-229 °C; reaction time: 21 h; yield: 68%

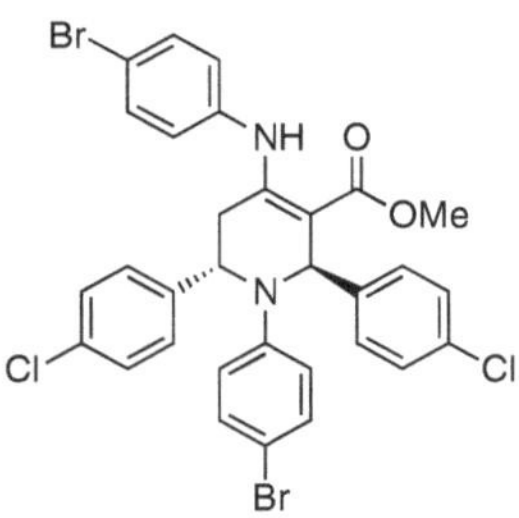

4c

Methyl 2,6-*bis*(4-chlorophenyl)-1-(4-bromophenyl)-4-(4-bromophenylamino)-1,2,5,6-tetrahydropyridine-3-carboxylate (**4c**): White crystals, mp 190-192 °C; reaction time: 21 h; yield: 61%

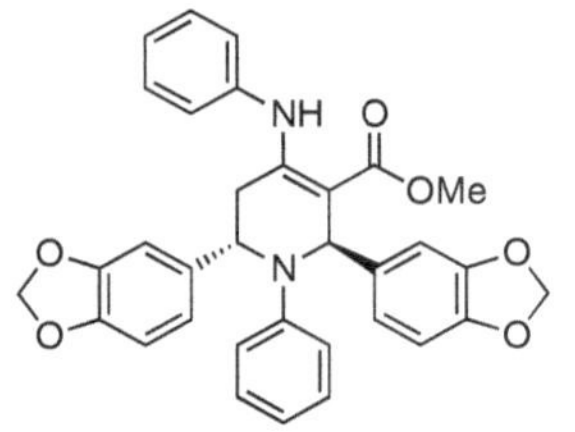

4d

Methyl 2,6-*bis*(3,4-methylenedioxy phenyl)-1-(phenyl)-4-(phenylamino)-1,2,5,6-tetrahydropyridine-3-carboxylate (**4d**): White crystals, mp 193-196 °C; reaction time: 45 h; yield: 76%

Experimental procedure

A dried reaction vessel was charged with a magnetic stir bar, amine (**3**; 2 mmol), β-keto ester (**2**; 1 mmol) and bismuth nitrate (10 mol %) in 4 mL ethanol; the mixture was then started to stir at room temperature. After 20 min, aromatic aldehyde (**1**; 2 mmol) was added to the reaction mixture and stirring was continued up to completion of the reaction (12-55 h for varying entries) as monitored by TLC. After completion of the reaction, a thick precipitate was obtained. The solid residue was filtered off and washed with cold ethanol-water. The solid mass was dissolved in hot ethyl acetate-ethanol mixture and filtered off when bismuth salt separated out; the filtrate on standing afforded crystals of the pure functionalized piperidine (**4**). Each of the products was characterized by detailed spectral studies including IR, ^{1}H NMR, ^{13}C NMR and TOF-MS.

Characterization data of two representative compounds

Ethyl 2,6-*bis*(4-chlorophenyl)-1-(4-methylphenyl)-4-(4-methylphenylamino)-1,2,5,6-tetrahydro pyridine-3-carboxylate (**4b**): White crystals; mp 227-229 °C. IR (KBr): 3242, 3024, 2974, 2915, 2870, 1653, 1589, 1514, 1483, 1256, 1172, 1076, 1012, 804, 678 cm^{-1}; ^{1}H NMR (400 MHz, $CDCl_3$): δ 1.46 (t, $J=6.8$ Hz, 3H), 2.16 (s, 3H), 2.27 (s, 3H), 2.70 (dd, $J=2.4$, 15.2 Hz, 1H), 2.77 (dd, $J=5.2$, 15.2 Hz, 1H), 4.29-4.33 (m, 1H), 4.35-4.44 (m, 1H), 5.05 (brs, 1H), 6.29 (t, $J=8$, 11.2 Hz, 3H), 6.36 (d, $J=8.8$ Hz, 2H), 6.88 (d, $J=8.4$ Hz, 2H), 6.94 (d, $J=8$ Hz, 2H), 7.05(d, $J=8.4$ Hz, 2H), 7.21-7.26 (m, 6H), 10.20 (brs, 1H); ^{13}C NMR (100 MHz, $CDCl_3$): δ 14.82, 20.13, 20.92, 33.62, 54.84, 57.35, 59.74, 97.24, 113.0, 125.77, 125.83, 127.85, 128.07, 128.35, 128.73, 129.59, 132.01, 132.76, 135.01, 135.88, 141.23, 142.80, 144.37, 156.16, 168.03; TOF-MS: cald for $C_{34}H_{32}Cl_2N_2O_2Na$ $[M+Na]^+$: 593.1738; Found: 593.1737.

Methyl 2,6-*bis*(3,4-methylenedioxy phenyl)-1-(phenyl)-4-(phenylamino)-1,2,5,6-tetrahydropyridine-3-carboxylate (**4d**): White crystals, mp 193-196 °C; IR (KBr): 3248, 3007, 2936, 2882, 2345, 1657, 1591, 1493, 1441, 1244, 1074, 1038, 804, 748 678 cm^{-1}; ^{1}H NMR (400 MHz, $CDCl_3$): δ 2.76 (dd, $J=2.4$, 15 Hz, 1H), 2.91 (dd, $J=5.6$, 15.2 Hz, 1H), 3.91 (s, 3H), 5.12 (d, $J=4$ Hz, 1H), 5.91-5.93 (m, 4H), 6.31 (s, 4H), 6.44 (d, $J=7.2$ Hz, 2H), 6.53 (d, $J=8.4$ Hz, 2H), 6.60-6.73 (m, 3H), 6.83 (d, $J=1.2$ Hz, 1H), 7.06-7.18 (m, 5H), 10.32 (br s, 1H); ^{13}C NMR (100 MHz, $CDCl_3$): δ 33.71, 51.10, 54.86, 57.77, 97.77, 100.97, 107.10, 107.34, 107.80, 108.42, 113.01, 116.36, 119.44, 119.49, 125.74, 125.80, 128.89, 128.93, 136.82, 137.80, 137.89, 146.02, 146.59, 146.68, 147.67, 147.84, 156.16, 168.44; TOF-MS: calcd for $C_{33}H_{28}N_2O_6Na$ $[M+Na]^+$: 571.1845; Found: 571.1847.

Reference

Brahmachari, G., and Das, S. (2012). Bismuth nitrate-catalyzed multicomponent reaction for efficient and one-pot synthesis of densely functionalized piperidine scaffolds at room temperature. *Tetrahedron Lett.*, **53**, 1479–1484.

C-N bond_14

Type of reaction: C-N and C-C bond formation
Reaction conditions: Acetonitrile, room temperature
Synthetic strategy: One-pot multicomponent
Catalyst: Cerium ammonium nitrate (CAN)
Keywords: Aromatic aldehydes, amines, β-keto esters, cerium ammonium nitrate (CAN), homogeneous catalysis, acetonitrile, room temperature, one-pot synthesis, multicomponent reaction (MCR), tetrahydropyridines

General reaction scheme

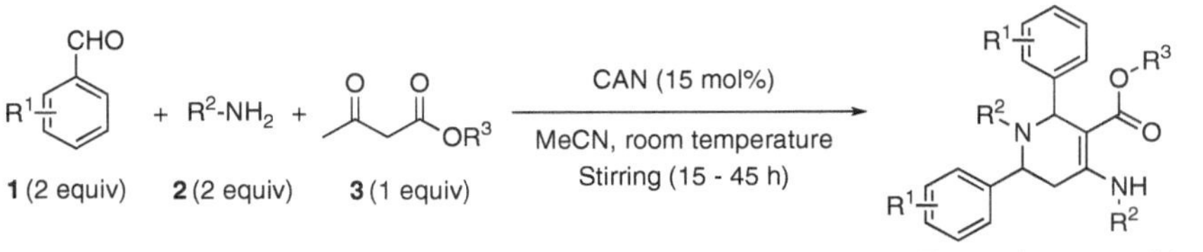

Tetrahydropyridine (**4**)
47 entries (yield: 51-86%)

R^1: H, 3-Me, 4-Me, 2-OMe, 4-OMe, 3,4,5-tri-OMe, 4-F, 3-Cl, 4-Cl, 4-Br, 2-NO_2, 3-NO_2, 4-NO_2,3-CF_3

R^2: C_6H_5, 4-$OCH_3C_6H_4$, 4-BrC_6H_4, 4-ClC_6H_4, $C_6H_5CH_2$, *n*-Bu, *n*-Pr

R^3: Me, Et, ^{t}Bu, $CH_2CH{=}CH_2$, CH_2CH_2OMe

Representative new entries

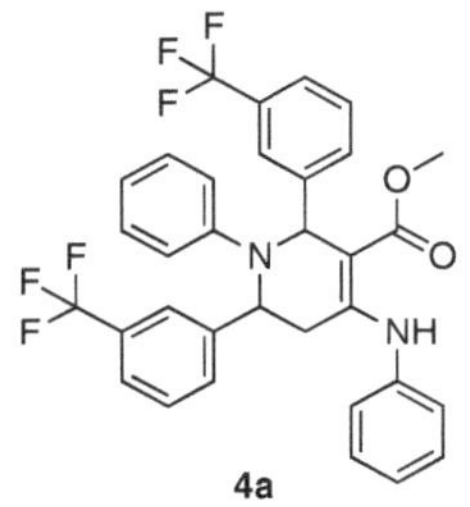

4a

Methyl 1-phenyl-4-(phenylamino)-2,6-bis(3-(trifluoro methyl)phenyl)-1,2,5,6-tetrahydropyridine-3-carboxylate (**4a**): Pale yellow solid, mp 140-142 °C; reaction time: 45 h; yield: 80%

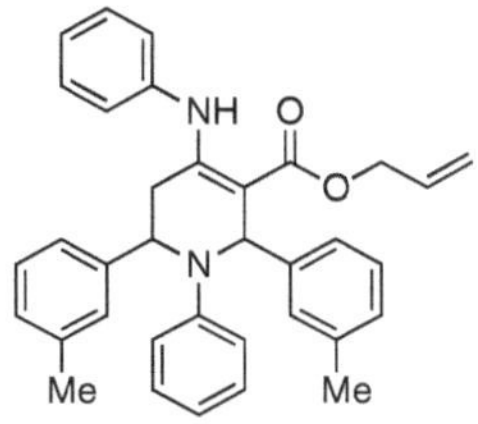

4b

Allyl 1-phenyl-4-(phenylamino)-2,6-di-*m*-tolyl-1,2,5,6-tetrahydropyridine-3-carboxylate (**4b**): White solid, mp 157-159 °C; reaction time: 24 h; yield: 56%

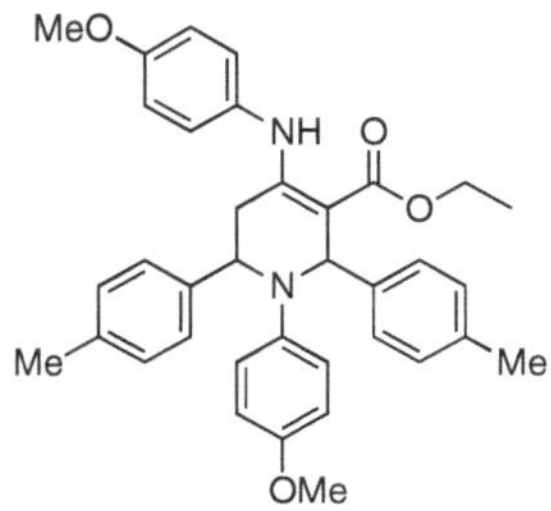

4c

Ethyl 1-(4-methoxyphenyl)-4-((4-methoxyphenyl) amino)-2,6-di-p-tolyl-1,2,5,6-tetrahydropyridine-3-carboxylate (**4c**): Yellow solid, mp 221-224 °C; reaction time: 22 h; yield: 81%

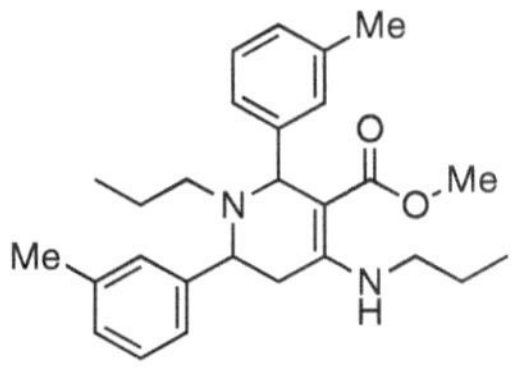

4d

Methyl 1-propyl-4-(propylamino)-2,6-di-*m*-tolyl-1,2,5,6-tetrahydropyridine-3-carboxylate (**4d**): White crystals, mp 151-153 °C; reaction time: 40 h; yield: 53%

Experimental procedure

A mixture of aromatic aldehyde (**1**; 2 mmol), amine (**3**; 2 mmol), β-keto ester (**2**; 1 mmol), and CAN (0.15 mmol) in 5 mL of acetonitrile was stirred at room temperature up to completion of the reaction (15-45 h for varying entries) as monitored by TLC. The crude mixture was diluted with ethyl acetate (20 mL), washed with water and then brine, and dried with anhydrous Na_2SO_4.

The filtrate was concentrated and purified by silica gel column chromatography. The products (**4**) were characterized by elemental analyses and spectral studies including IR, ^{1}H NMR, and ^{13}C NMR.

Characterization data of one representative compound

Allyl 1-phenyl-4-(phenylamino)-2,6-di-*m*-tolyl-1,2,5,6-tetrahydropyridine-3-carboxylate (**4b**): White solid; mp 157-159 °C. IR (KBr): 3445, 2922, 1646, 1593, 1455, 1250, 1064, 1049, 978, 744 cm^{-1}; ^{1}H NMR (500 MHz, $CDCl_3$) δ 2.29 (s, 3H), 2.31 (s, 3H), 2.75 (dd, $J=15.0$, 2.5 Hz, 1H), 2.89 (dd, $J=15.0$, 5.5 Hz, 1H), 4.78–4.82 (m, 1H), 4.88–4.92 (m, 1H), 5.10 (d, $J=4.5$ Hz, 1H), 5.32 (dd, $J=10.5$, 1.5 Hz, 1H), 5.47 (dd, $J=17.5$, 1.5 Hz, 1H), 6.17–6.08 (m, 1H), 6.26–6.28 (m, 2 H), 6.44 (s, 1H), 6.52 (d, $J=8.0$ Hz, 2H), 6.59 (t, $J=7.5$ Hz, 1H), 6.95 (t, $J=7.5$ Hz, 2H), 7.02 (d, $J=7.5$ Hz, 1H), 7.00–7.10 (m, 7 H), 7.12-7.16 (m, 2H), 7.19 (s, 1H), 10.24 (br s, 1 H); ^{13}C NMR (125 MHz, $CDCl_3$) δ 21.6, 21.8, 33.7, 55.2, 58.2, 64.3, 97.8, 113.0, 116.0, 117.7, 123.6, 123.7, 125.8, 126.1, 126.9, 127.1, 127.4, 127.9, 128.0, 128.5, 128.8, 128.9, 133.2, 137.9, 138.1, 142.7, 144.1, 147.1, 156.7, 167.8; Anal. Calcd (%) for $C_{35}H_{34}N_2O_2$: C, 81.68; H, 6.66; N, 5.44; Found: C, 81.84; H, 6.83; N, 5.28.

Reference

Wang, H.-J., Mo, L.-P., and Zhang, Z.-H. (2011). Cerium ammonium nitrate-catalyzed multicomponent reaction for efficient synthesis of functionalized tetrahydropyridines. *ACS Comb. Sci.*, **13**, 181–185.

Further information

Besides $Bi(NO_3)_3.5H_2O$ and CAN, few more catalysts such as bromodimethylsulfonium bromide (BDMS) [1], tetrabutylammonium tribromide (TBATB) [2], molecular iodine (I_2) [3], $InCl_3$ [4], $NiCl_2.6H_2O$ [5], zinc hydrogen sulfate [$Zn(HSO_4)_2$] [6] and $LaCl_3.7H_2O$ [7] have also been reported to accomplish the synthesis of functionalized tetrahydropyridines using ethanol/methanol or acetonitrile medium at room temperature.

[1] Khan, A. T., Parvin, T., and Choudhury, L. H. (2008). Effects of substituents in the β-position of 1,3-dicarbonyl compounds in bromodimethylsulfonium bromide-catalyzed multicomponent reactions: a facile access to functionalized piperidines. *J. Org. Chem.*, **73**, 8398–8402.

[2] Khan, A. T., Lal, M., and Khan M. M. (2010). Synthesis of highly functionalized piperidines by one-pot multicomponent reaction using tetrabutylammonium tribromide (TBATB). *Tetrahedron Lett.*, **51**, 4419–4424.

[3] Khan, A. T., Khan, M. M., and Bannuru, K. K. R. (2010). Iodien catalyzed one-pot five-component reactions for direct synthesis of densely functionalized piperidines. *Tetrahedron*, **66**, 7762–7772.

[4] Clarke, P. A., Zaytzev, A. V., and Whitwood, A. C. (2007). Pot, atom and step economic (PASE) synthesis of highly functionalized piperidines: a five-component condensation. *Tetrahedron Lett.*, **48**, 5209–5212.

[5] Shafiee, M. R. M., Najafabadi, B. H., and Ghashang, M. (2012). Nickel (II) chloride hexahydrate as an efficient catalyst for the preparation of highly functionalised piperidine derivatives. *J. Chem. Res.*, **36**, 336–339.

[6] Ghashang, M. (2012). Zinc hydrogen sulfate promoted multi-component preparation of highly functionalized piperidines. *Lett. Org. Chem.*, **9**, 497–502.

[7] Umamahesh, B., Sathesh, V., Ramachandran, G., Sathishkumar, M., and Sathiyanarayanan, K. (2012). $LaCl_3{\cdot}7H_2O$ as an efficient catalyst for one-pot synthesis of highly functionalized piperidines via multi-component organic reactions. *Catal. Lett.*, **142**, 895–900.

C-N bond_15

Type of reaction: C-N and C-C bond formation
Reaction conditions: Ethanol, room temperature
Synthetic strategy: One-pot multicomponent
Catalyst: Heterogeneous, MCM-41 anchored sulfonic acid (MCM-41-SO_3H)
Keywords: Carbonyl compounds, TMSCN, MCM-41-SO_3H nanocatalyst, solid acid, reusable heterogeneous catalyst, room temperature, one-pot synthesis, multicomponent reaction (MCR), Strecker reaction, α-aminonitriles

General reaction scheme

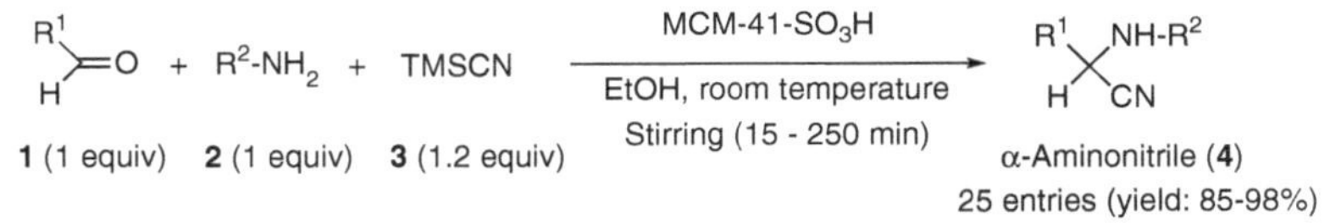

R[1]: C_6H_5, 4-$CH_3C_6H_4$, 4-$OCH_3C_6H_4$, 4-CNC_6H_4, 4-BrC_6H_4, 2-ClC_6H_4, 4-ClC_6H_4, 2-$NO_2C_6H_4$, 4-$NO_2C_6H_4$, 4-OHC_6H_4, $C_6H_5CH{=}CH$, $C_6H_5CH_2CH_2$, *n*-Pr,

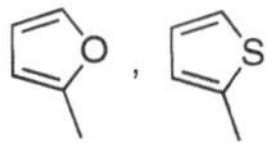

R[2]: C_6H_5, 4-$CH_3C_6H_4$, $C_6H_5CH_2$

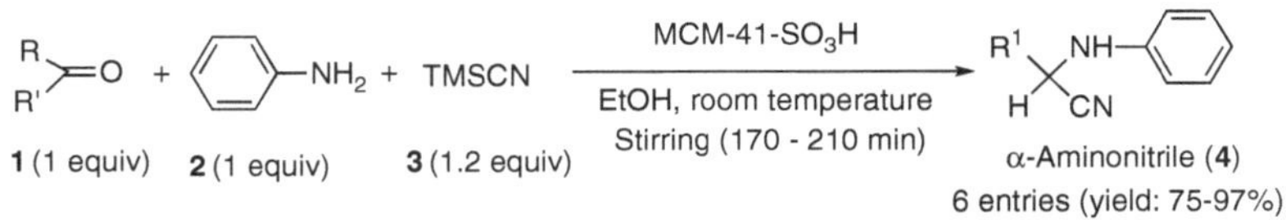

R,R': C_6H_5, CH_3; 4-$NO_2C_6H_4$, CH_3; 4-FC_6H_4, C_6H_5; C_6H_5, *c*-hex; *c*-hex

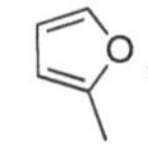 , CH_3

Representative new entries

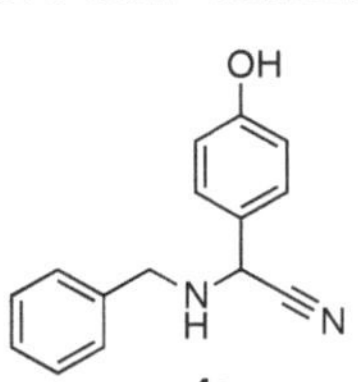

4a

2-(*N*-Benzylamino)-2-(4-hydroxyphenyl) acetonitrile (**4a**): Cream solid, mp 208-211 °C; reaction time: 120 min; yield: 98%

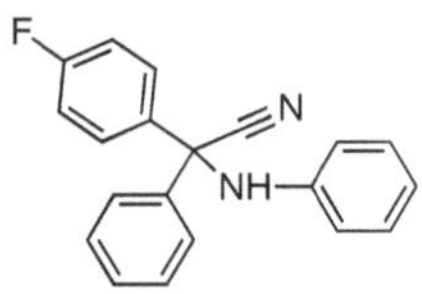

4b

2-(4-Fluorophenyl)-2-phenyl-2-(phenyl amino) acetonitrile (**4b**): Light yellow oil; reaction time: 210 min; yield: 75%

Experimental procedure

To a mixture of carbonyl compound (**1**; 1 mmol), amine (**2**; 1 mmol), and trimethylsilyl cyanide (TMSCN; **3**; 1.2 mmol, 0.15 mL) was added MCM-41-SO_3H (5 mg) in 96% EtOH (1.5 mL) at room temperature. The mixture was then stirred up to completion of the reaction (15-250 min for varying entries) as indicated by TLC. After completion of reaction, 96% EtOH (5 mL) was added to the reaction mixture. The crude mixture was filtered to remove the catalyst, and α-aminonitrile product (**4**) was obtained either on standing the alcoholic filtrate and/or by evaporating the solvent under reduced pressure followed by purification using silica gel column chromatography (EtOAc-hexane, 1:10). The products (**4**) were characterized by elemental analyses and spectral studies including IR, ^{1}H NMR, and ^{13}C NMR. The recovered solid catalyst was washed with fresh aliquot of EtOAc (3 × 5 mL), dried in an oven at 50 °C, and then was subjected to Strecker reaction successfully up to fifth consecutive run.

Characterization data of one representative compound

2-(*N*-Benzylamino)-2-(4-hydroxyphenyl)acetonitrile (**4a**): Cream solid; mp 208-211 °C. IR (KBr): 3443, 3313, 3196, 3063, 3030, 2916, 2362, 2328, 1668, 1600, 1514, 1452, 1283 cm^{-1}; ^{1}H NMR (500 MHz, $CDCl_3$): δ 2.14 (br s, 1H), 3.82 (s, 2H), 4.90 (s, 1H), 5.01 (br s, 1H), 6.52 (d, J=7.2 Hz, 2H), 6.76 (d, J=7.2 Hz, 2H), 7.16 (d, J=8.0 Hz, 2H), 7.23 (t, J=8.0 Hz, 1H), 7.31 (d, J=8.0 Hz, 2H); ^{13}C NMR (75 MHz, $CDCl_3$): δ 44.9, 55.0, 112.8, 115.9, 126.7, 127.6, 127.9, 128.1, 130.0,137.5, 161.9. Anal. Calcd (%) for $C_{15}H_{14}N_2O$: C, 75.61; H, 5.92; N, 11.76; Found: C, 75.56; H, 5.99; N, 12.02.

Reference

Dekamin, M. G., and Mokhtari, Z. (2012). Highly efficient and convenient Strecker reaction of carbonyl compounds and amines with TMSCN catalyzed by MCM-41 anchored sulfonic acid as a recoverable catalyst. *Tetrahedron*, **68**, 922–930.

C-N bond_16

Types of reaction: C-N and C-C bond formation
Reaction conditions: Solvent-free, grinding, room temperature
Synthetic strategy: One-pot multicomponent
Catalyst: Catalyst-free
Keywords: Aldehydes, dimedone, active methylene compounds, ammonium acetate, solvent-free, catalyst-free, room temperature, grinding, one-pot multicomponent synthesis, Hantzsch condensation, polyhydroquinolines

General reaction scheme

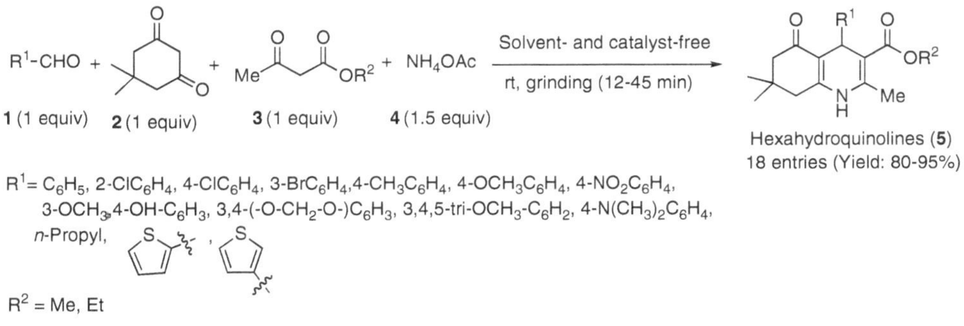

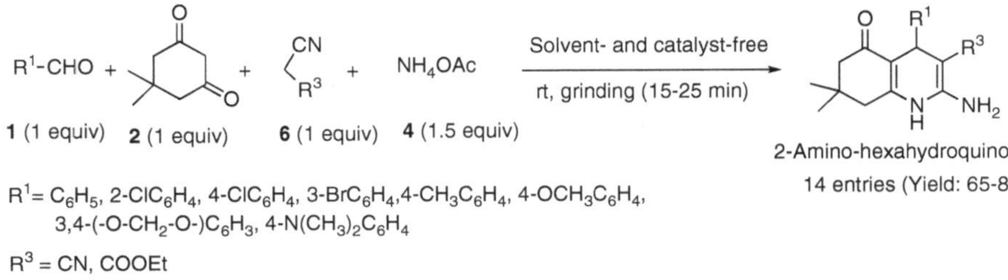

2-Amino-hexahydroquinolines (**7**)
14 entries (Yield: 65-88%)

Representative entries

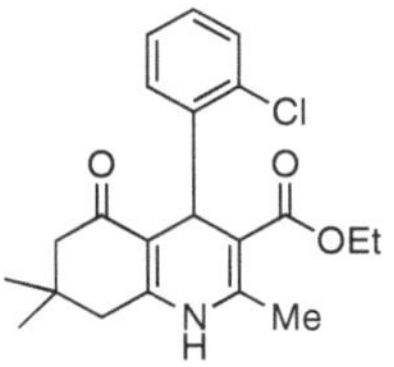

5a

Ethyl-1,4,7,8-tetrahydro-2,7,7-trimethyl-4-(2-chlorolphenyl)-5(6*H*)-oxoquinolin-3-carboxylate (**5a**): Yellow solid, mp 209-211 °C; reaction time: 15 min; yield: 93%

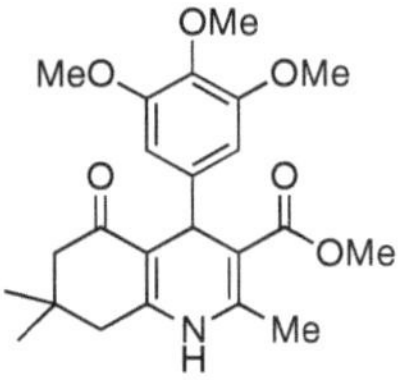

5b

Methyl-1,4,7,8-tetrahydro-2,7,7-trimethyl-4-(3,4,5-trimethoxylphenyl)-5(6*H*)-oxoquinolin-3-carboxylate (**5b**): Yellow solid, mp 203-206 °C; reaction time: 20 min; yield: 88%

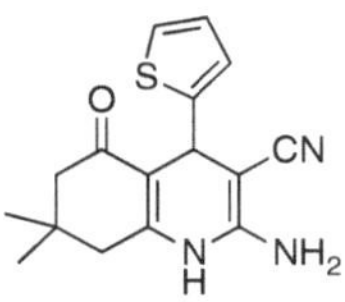

7a

2-Amino-4-(2′-thienyl)-3-cyano-7,7-dimethyl-5-oxo-1,4,5,6,7,8-hexahydroquinoline (**7a**): Yellow solid, mp 253-255 °C; reaction time: 20 min; yield: 74%

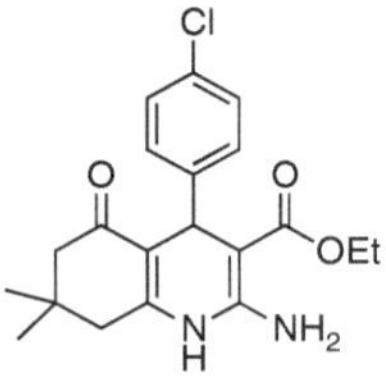

7b

2-Amino-7,7-dimethyl-5-oxo-4-(4′-chlorophenyl)-1,4,5,6,7,8-hexahydroquinoline-3-carboxylic acid ethyl ester (**7b**): Yellow solid, mp 174-176 °C; reaction time: 15 min; yield: 75%

Experimental procedure

*General procedure for the synthesis of 1,4,7,8-tetrahydro-2,7,7-trimethyl-4-aryl-5(6H)-oxo-quinolin-3-carboxylic acid alkyl ester (**5**).* A mixture of aldehyde (**1**; 2 mmol), dimedone (**2**; 2 mmol), β-ketoester (**3**; 2 mmol), and ammonium acetate (**4**; 3 mmol) was thoroughly mixed in a mortar followed by grinding till the completion of reaction as indicated by TLC (12-45 min). The resultant mass was washed with water to remove any unreacted ammonium acetate and was air-dried to give a crude product which on recrystallization from ethanol afforded the pure desired compound (**5**) in excellent yield (80-95%).

*General procedure for the synthesis of 2-amino-4-phenyl-3-cyano-7,7-dimethyl-5-oxo-1,4,5,6,7,8-hexahydroquinoline (**7**).* A mixture of aldehyde (**1**; 2 mmol), dimedone (**2**; 2 mmol), malononitrile or ethyl cyanoacetate (**6**; 2 mmol), and ammonium acetate (**4**; 3 mmol) was mixed thoroughly in a mortar with pestle followed by grinding till the completion of reaction as indicated by TLC (15-25 min). The resultant material was washed with water to remove any unreacted ammonium acetate and was air-dried to furnish the crude product. The pure product (**7**) was obtained by recrystallization from ethyl alcohol.

All the products were characterized from elemental analyses as well as their detailed spectral studies including IR, ^{1}H NMR, ^{13}C NMR and MS.

Characterization data of two representative compounds

Methyl-1,4,7,8-tetrahydro-2,7,7-trimethyl-4-(3,4,5-trimethoxylphenyl)-5(6*H*)-oxoquinolin-3-carboxylate (**5b**): Yellow solid, mp 203-206 °C; IR (KBr): 3280, 3185, 3058, 2931, 1686, 1644, 1603, 1490, 1382, 1333, 1227, 1114, 1004, 787 cm^{-1}; ^{1}H NMR (200 MHz, DMSO-d_6): δ 0.94 (s, 3H, CH_3), 1.03 (s, 3H, CH_3), 2.00-2.51 (m, 4H, 2 × CH_2), 2.27 (s, 3H, CH_3), 3.47 (s, 3H, CO_2Me), 3.59 (s, 3H, OMe), 3.67 (s, 6H, two OMe), 4.83 (s, 1H, CH), 6.41 (s, 2H, ArH), 8.39 (s, 1H, NH); ^{13}C NMR (50 MHz, DMSO-d_6): δ 18.6, 26.7, 29.6, 32.5, 35.9, 50.6, 55.9, 60.2, 103.0, 104.7, 110.0, 136.1, 143.5, 145.4, 150.3, 152.7, 167.8, 194.9; MS (ESI): *m/z* 416.0 $(M+H)^+$. Anal. Calcd for $C_{23}H_{29}NO_6$: C, 66.49; H, 7.04; N, 3.37; Found: C, 66.57; H, 7.11; N, 3.46.

2-Amino-7,7-dimethyl-5-oxo-4-(4′-chlorophenyl)-1,4,5,6,7,8-hexahydroquinoline-3-carboxylic acid ethyl ester (**7b**): Yellow solid, mp 174-176 °C; IR (KBr): 3378, 3319, 3286, 2198, 1679, 1607 cm^{-1}; ^{1}H NMR (200 MHz, DMSO-d_6): δ 0.92 (s, 3H, CH_3), 1.01 (s, 3H, CH_3), 2.06-2.29 (m, 4H, 2 × CH_2), 4.51 (s, 1H, CH), 5.85 (s, 2H, NH_2), 6.85-7.12 (m, 3H, ArH), 8.98 (s, 1H, NH); MS (ESI): *m/z* 300.2 $(M+H)^+$. Anal. Calcd for $C_{16}H_{17}N_3OS$: C, 64.19; H, 5.72; N, 14.04; Found: C, 64.15; H, 5.68; N, 13.99.

Reference

Kumar, S., Sharma, P., Kapoor, K. K., and Hundal, M. S. (2008). An efficient, catalyst- and solvent-free, four-component, and one-pot synthesis of polyhydroquinolines on grinding. *Tetrahedron*, **64**, 536–542.

C-N bond_17

Types of reaction: C-N and C-C bond formation
Reaction conditions: Ethanol, room temperature
Synthetic strategy: One-pot multicomponent
Catalyst: Ceric ammonium nitrate (CAN)
Keywords: Aromatic aldehydes, cyclic 1,3-diketones, β-keto compounds, ammonium acetate, ceric ammonium nitrate (CAN), ethanol, room temperature, one-pot multicomponent synthesis, Hantzsch condensation, polyhydroquinolines

General reaction scheme

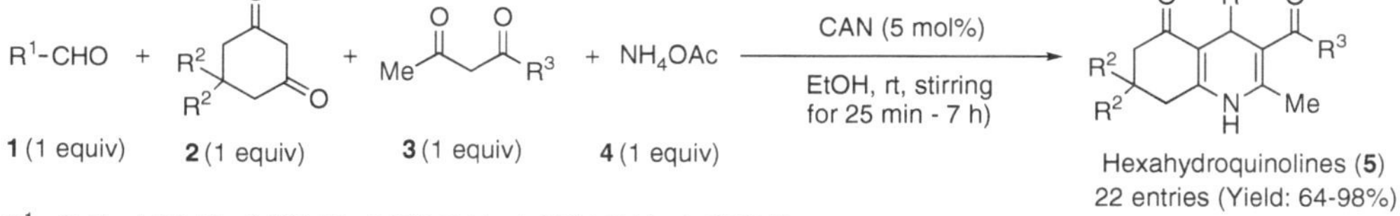

R^1= C_6H_5, 4-FC_6H_4, 4-ClC_6H_4, 4-$CH_3C_6H_4$, 4-$OCH_3C_6H_4$, 4-OHC_6H_4, 2-$NO_2C_6H_4$, 3-$NO_2C_6H_4$, 4-$NO_2C_6H_4$, 2-Furyl, 2-Thienyl

R^2 = H, Me ; R^3 = Me, OMe, OEt, OCH_2CH_2Me

Representative entries

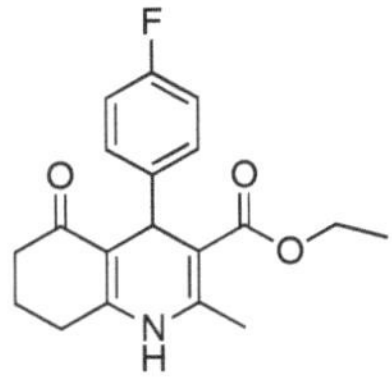

5a

Ethyl 4-(4-fluorophenyl)-2-methyl-5-oxo-1,4,5,6,7,8-hexahydroquinoline-3-carboxylate (**5a**): Solid, mp 243-244 °C; reaction time: 2 h; yield: 98%)

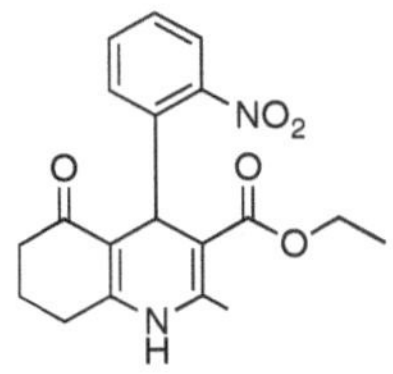

5b

Ethyl 4-(2-nitrophenyl)-2-methyl-5-oxo-1,4,5,6,7,8-hexahydroquinoline-3-carboxylate (**5b**): Solid, mp 190-191 °C; reaction time: 50 min; yield: 93%

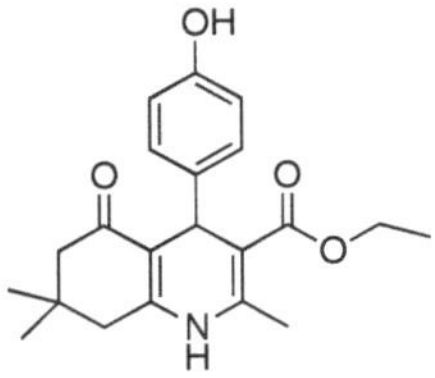

5c

Ethyl 4-(4-hydroxyphenyl)-2,7,7-trimethyl-5-oxo-1,4,5,6,7,8-hexahydroquinoline-3-carboxylate (**5c**): Solid, mp 237-238 °C; reaction time: 2 h; yield: 93%

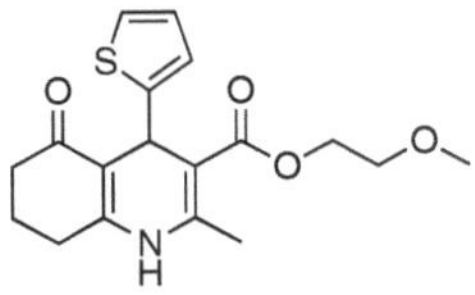

5d

2-Methoxyethyl 2-methyl-5-oxo-4-(thiophene-2-yl)-1,4,5,6,7,8-hexahydroquinoline-3-carboxylate (**7d**): Solid, mp 207-209 °C; reaction time: 7 h; yield: 64%

Experimental procedure

A dried 10 mL round-bottomed flask was charged with aromatic aldehyde (**1**; 1.0 mmol), cyclic 1,3-diketone (**2**; 1,3-cyclohexanedione or 5,5-dimethyl-1,3-cyclohexanedione; 1.0 mmol), β-keto compound (**3**; 2,4-pentadione or acetoacetate derivatives; 1.0 mmol), ammonium acetate (**4**; 1.0 mmol), and ceric ammonium nitrate (CAN) (0.05 mmol; 5 mol%) followed by 0.5 mL of ethanol. The mixture was then stirred at room temperature until the reaction was completed as monitored by TLC (25 min to 7 h). The resulting reaction mixture was treated with brine solution followed by extraction with ethyl acetate (2 × 20 mL). After evaporation of the solvent, the crude yellow product was recrystallized from ethanol to give a yellow or brown solid in good to excellent yield (64-98%). All the products were characterized from their detailed spectral studies including ^{1}H NMR, ^{13}C NMR, EIMS and HRMS.

Characterization data of two representative compounds

Ethyl 4-(4-hydroxyphenyl)-2,7,7-trimethyl-5-oxo-1,4,5,6,7,8-hexahydroquinoline-3-carboxylate (**5c**): Yellow to brown solid, mp 237-238 °C; ^{1}H NMR (Acetone-d_6, 400 MHz): δ 0.91 (s, 3H), 1.04 (s, 3H), 1.21 (t, $J=7.2$ Hz, 3H), 2.13-2.36 (m, 7H), 3.74 (s, 3H), 4.06 (q, $J=7.2$ Hz, 2H), 5.00 (s, 1H), 6.01 (s, 1H), 6.74 (d, $J=8.4$ Hz, 2H), 7.22 (d, $J=8.4$ Hz, 2H); ^{13}C NMR (Acetone-d_6, 100 MHz): δ 15.01, 19.09, 19.16, 27.49, 33.40, 36.70, 41.14, 51.77, 54.98, 60.20, 106.29, 112.65, 115.57, 130.13, 131.38, 140.43, 145.31, 149.71, 156.62, 168.41, 195.36; MS *m/z* (%relative intensity) 356 (M^+, 28), 326 (12), 282 (11), 263 (14), 262 (100), 234 (24); HRMS calcd for $C_{21}H_{25}NO_4$ (M^+) 355.1784; Found 355.1776.

2-Methoxyethyl 2-methyl-5-oxo-4-(thiophene-2-yl)-1,4,5,6,7,8-hexahydroquinoline-3-carboxylate (**7d**): Yellow to brown solid, mp 207-209 °C; ^{1}H NMR ($CDCl_3$, 400 MHz): δ 1.93-2.01 (m, 2H), 2.32-2.46 (m, 7H), 3.33 (s, 3H), 3.58 (t, $J=4.8$ Hz, 2H), 4.22-4.27

(m, 2H), 5.51 (s, 1H), 6.80-6.83 (m, 2H), 7.01-7.07 (m, 2H); ^{13}C NMR ($CDCl_3$, 100 MHz): δ 19.18, 21.05, 27.09, 31.12, 36.99, 58.80, 62.89, 70.47, 104.91, 112.40, 123.09, 123.33, 126.50, 144.69, 150.79, 151.27, 167.14, 195.89; MS *m/z* (%relative intensity) 347 (M^+, 100), 288 (65), 264 (48), 244 (70); HRMS calcd for $C_{18}H_{21}NSO_4$ (M^+) 347.1191; Found 347.1191.

Reference

Ko, S., and Yao, C.-F. (2006). Ceric ammonium nitrate (CAN) catalyzes the one-pot synthesis of polyhydroquinoline via the Hantzsch reaction. *Tetrahedron*, **62**, 7293–7299.

Further information

Applications of few more catalysts such as TMSI (generated *in situ* from TMSCl+NaI) [1], molecular iodine [2], $Yb(OTf)_3$ [3], and organocatalysts [4] have also been reported for the synthesis of polyhydroquinoline derivatives at room temperature under solvent or solvent-free conditions.

[1] Sabitha, G., Kiran Kumar Reddy, G. S., Srinivas Reddy, Ch., and Yadav, J. S. (2003). A novel TMSI-mediated synthesis of Hantzsch1,4-dihydropyridines at ambient temperature. *Tetrahedron Lett.*, **44**, 4129–4131.
[2] Ko, S., Sastry, M. N. V., Lin, C., and Yao, C.-F. (2005). Molecular iodine-catalyzed one-pot synthesis of 4-substituted-1,4-dihydropyridine derivatives via Hantzsch reaction. *Tetrahedron Lett.*, **46**, 5771–5774.
[3] Wang, L.-M., Sheng, J., Zhang, L., Han, J.-W., Fan, Z.-Y., Tian, H., and Qian, C.-T. Facile promoted one-pot synthesis of polyhydroquinoline derivatives through Hantzsch reaction. *Tetrahedron*, **61**, 1539-1543.
[4] Kumar, A., and Maurya, R. A. (2007). Synthesis of polyhydroquinoline derivatives through unsymmetric Hantzsch reaction using organocatalysts. *Tetrahedron*, **63**, 1946–1952.

C-N bond_18

Type of reaction: C-N bond formation
Reaction conditions: Dichloromethane / tetrahydrofuran (THF), room temperature
Synthetic strategy: One-step condensation
Catalyst: Montmorillonite-KSF clay / iodine
Keywords: Amines, 1,4-dicarbonyls, montmorillonite-KSF, iodine, room temperature, Paal-Knorr reaction, condensation, substituted pyrroles

General reaction schemes

Scheme 1.

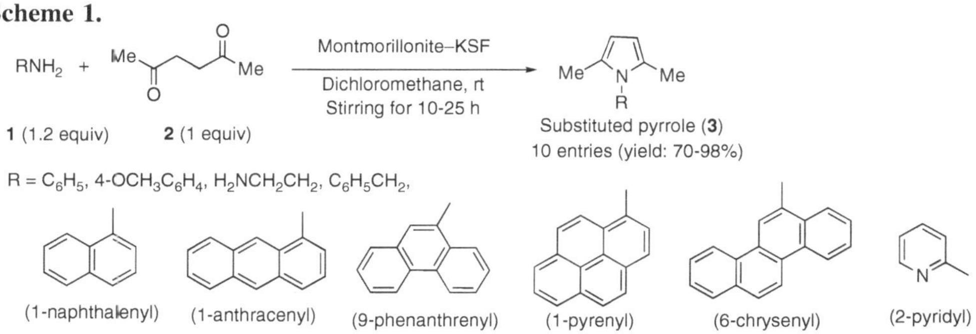

Scheme 2.

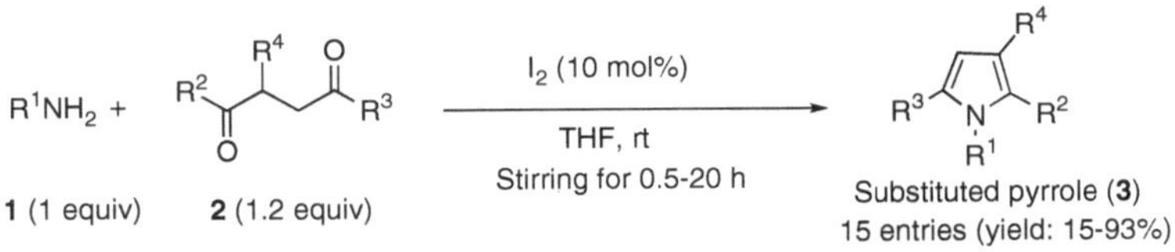

R^1 = C_6H_5, 4-$OCH_3C_6H_4$, $H_2NCH_2CH_2$, $C_6H_5CH_2$, 1-naphthalenyl, 1-anthracenyl, 9-phenanthrenyl, 1-pyrenyl, 6-chrysenyl, 2-pyridyl, *n*-Bu, cycopentyl

R^2 = Me, C_6H_5 ; R^3 = Me, C_6H_5 ; R^4 = H, C_6H_5

Representative entries

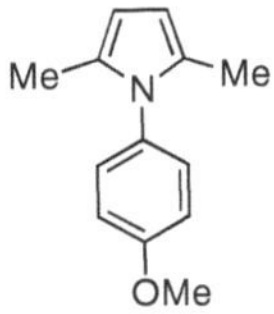

3a

1-(4-Methoxyphenyl)-2,5-dimethyl-1*H*-pyrrole (**3a**): Method 1: reaction time – 15 h, yield - 81%; Method 2: reaction time – 1.5 h, yield - 90%

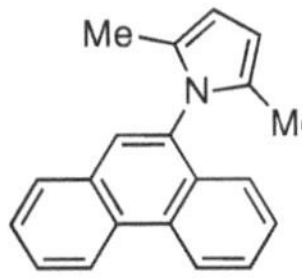

3b

2,5-Dimethyl-1-(phenanthren-9-yl)-1*H*-pyrrole (**3b**): Method 1: reaction time – 18 h, yield - 94%; Method 2: reaction time – 1 h, yield - 91%

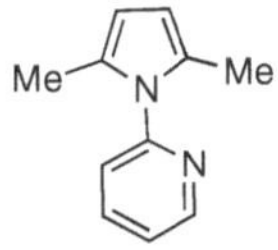

3c

2,5-Dimethyl-1-(2-pyridyl)-1*H*-pyrrole (**3c**): Method 1: reaction time – 25 h, yield - 70%; Method 2: reaction time – 1 h, yield - 78%

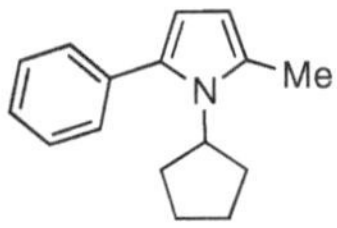

3d

1-Cyclopentyl-2-methyl-5-phenyl-1*H*-pyrrole (**3d**): Method 2: reaction time – 10 h, yield - 76%)

Experimental procedure

Method 1. Montmorillonite-KSF clay (1 g), amine (**1**; 1.2 mmol), and hexane-2,5-dione (**2**; 1 mmol) were mixed together thoroughly using dichloromethane (1 mL). The solvent was then evaporated out, and the mixture was kept at room temperature for specified period (10-25 h for varying entries). On completion of the reaction, the solid mixture was washed with dichloromethane and the washings were evaporated to obtain substituted pyrrole **3** followed by required purification (70-98% yield).

Method 2. To a solution of amine (**1**; 1 mmol) and 1,4-dicarbonyl compound (**2**; 1.2 mmol) in THF (5 mL) was added iodine (0.1 mmol). The mixture was then stirred at room temperature for specified period of 0.5-20 h for varying entries to complete the reaction as monitored by TLC. The crude mixture was extracted with dichloromethane (20 mL), and was washed successively with 5% $Na_2S_2O_3$ solution (2 mL), saturated $NaHCO_3$ solution (2 mL), and brine (2 mL). The organic layer was then dried with sodium sulfate and concentrated to afford pyrrole **3** in sufficiently pure form (>90%). In case of one of the reactants being liquid, the experiment was conducted without using solvent; however, addition of a small amount of ordinary THF or dichloromethane was required for a very thick slurry reaction mixture to ensure a better outcome of the reaction. After completion of the reaction, the crude product was extract with dichloromethane, washed, dried, and the subjected to column chromatography for purification (15-93% yield).

All the products were characterized by ^1H NMR and analytical data.

Characterization data of two representative compounds

2,5-Dimethyl-1-(2-pyridyl)-1*H*-pyrrole (**3c**): Colorless oil; ^{1}H NMR (300 MHz): δ 2.12 (s, 6H), 5.90 (s, 2H), 7.29 (m, 2H), 7.82 (m, 1H), 8.60 (m, 1H). Anal. Calcd for $C_{11}H_{12}N_2$: C, 76.71; H, 7.02; N, 6.95; Found: C, 76.80; H, 6.95; N, 6.83.

1-Cyclopentyl-2-methyl-5-phenyl-1*H*-pyrrole (**3d**): ^{1}H NMR (300 MHz, CDCl3): δ 1.58 (m, 2H), 1.82 (m, 2H), 1.99 (m, 4H), 2.42 (s, 3H), 4.65(m, 1H), 5.96 (d, *J* = 3.3 Hz, 1H), 6.01 (d, *J* = 3.3 Hz, 1H), 7.35 (m, 5H). Anal. Calcd for $C_{16}H_{19}N$: C, 85.28; H, 8.49; N, 6.21; Found: C, 85.24; H, 8.66; N, 6.11

Reference

Banik, B. K., Samajdar, S., and Banik, I. (2004). Simple synthesis of substituted pyrroles. *J. Org. Chem.*, **69**, 213–216.

C-N bond_19

Type of reaction: C-N bond formation
Reaction conditions: Solvent-free, room temperature
Synthetic strategy: One-step condensation
Catalyst: Heterogeneous, sulfamic acid (H_2NSO_3H)
Keywords: Amines, 1,4-dicarbonyls, sulfamic acid, solvent-free, room temperature, Paal-Knorr reaction, condensation, substituted pyrroles

General reaction scheme

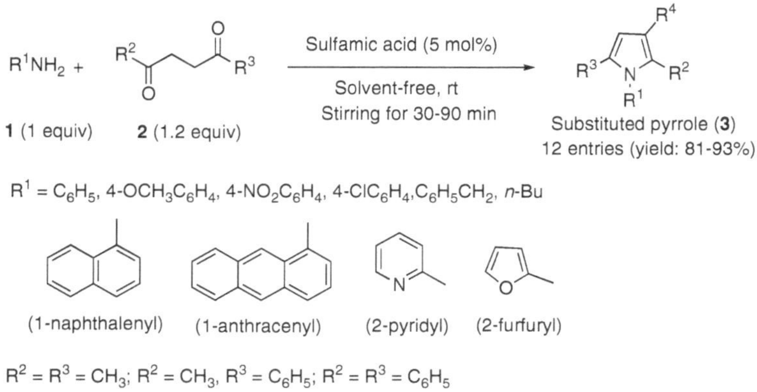

Representative entries

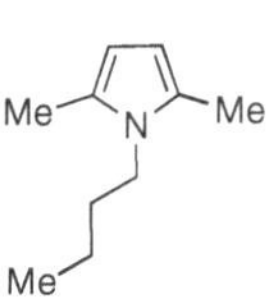

3a

1-Butyl-2,5-dimethyl-1*H*-pyrrole (**3a**): Oil; reaction time: 75 min, yield: 82%

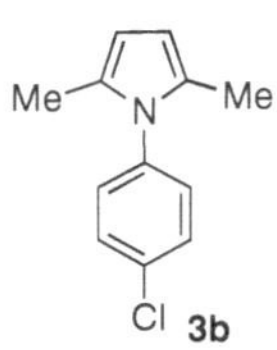

3b

1-(4-Chlorophenyl)-2,5-dimethyl-1*H*-pyrrole (**3b**): Reaction time: 40 min, yield: 91%

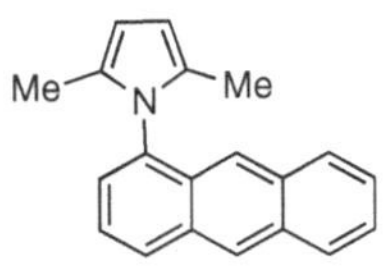

3c

1-(Anthracen-1-yl)-2,5-dimethyl-1*H*-pyrrole (**3c**): Solid, mp 183-184 °C; reaction time: 90 min, yield: 88%

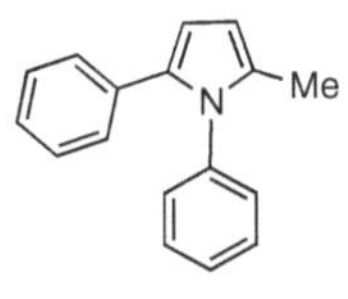

3d

2-Methyl-1,5-diphenyl-1*H*-pyrrole (**3d**): Reaction time: 60 min, yield: 88%

Experimental procedure

A mixture of amine (**1**; 5 mmol), 1,4-dicarbonyl compound (**2**; 6 mmol), and sulfamic acid (25 mg, 5 mol%) was stirred at room temperature under solvent-free conditions for specified period (30-90 min for varying entries). After completion of the reaction, as indicated by thin-layer chromatography (TLC), the reaction mixture was diluted with diethyl ether and filtered to recover the catalyst. The ether layer was washed with brine, dried ($MgSO_4$), and concentrated in vacuo. The residue was purified by silica-gel column chromatography (20% ethyl acetate in hexane) to afford the desired substituted pyrrole (**3**) in excellent yield (81-93%). The recovered catalyst was washed with diethyl ether and activated at 70 °C prior to reuse; it could be reused for the next reaction without losing any significant activity as tested for at least three times. All the products were characterized by ^{1}H NMR, HRMS and analytical data.

Characterization data of two representative compounds

1-Butyl-2,5-dimethyl-1*H*-pyrrole (**3a**): Oil; ^{1}H NMR (300 MHz, $CDCl_3$): δ 0.94 (t, $J=7.2$ Hz, 3H), 1.38 (m, 2H), 1.61 (m, 2H), 2.17 (s, 6H), 3.71 (t, $J=7.2$ Hz, 2H), 5.79 (s, 2H). HRMS calcd. for $C_{10}H_{17}N$ 151.1361; found 151.1360. Anal. calcd. for $C_{10}H_{17}N$: C, 76.41; H, 11.33; N, 9.26. Found: C, 76.43; H, 11.38; N, 9.30.

1-(Anthracen-1-yl)-2,5-dimethyl-1*H*-pyrrole (**3c**): Solid, mp 183–184 °C; ^{1}H NMR (300 MHz, $CDCl_3$): δ 1.95 (s, 6H), 6.04 (s, 2H), 7.35–7.59 (m, 4H), 7.71 (s, 1H), 7.92 (d, $J=8$ Hz, 1H), 8.02 (d, $J=7.9$ Hz, 1H), 8.10 (d, $J=8.4$ Hz, 1H), 8.52 (s, 1H). HRMS calcd. for $C_{20}H_{17}N$ 271.1361; found 271.1362. Anal. calcd. for $C_{20}H_{17}N$: C, 88.52; H, 6.31; N, 5.16. Found: C, 88.58; H, 6.37; N, 5.20.

Reference

De, S. K. (2008). Sulfamic acid as a novel, efficient, cost-effective, and reusable solid acid catalyst for the synthesis of pyrroles under solvent-free conditions. *Syn. Commun.*, **38**, 803–809.

Further information

Sulfamic acid was also reported by Luo *et al.* [1] as an efficient and reusable catalyst for the solvent-free synthesis of substituted pyrroles at room temperature in the year. Besides, some other alternative methods are available in literature for the synthesis of these *N*-heterocycles [2], involving both homogeneous and heterogeneous catalysts such as montmorillonite K10 [3], Fe^{3+}-montmorillonite [4], scandium triflate [$Sc(OTf)_3$] [5], polystyrene sulfonate [6], and acidic styrol resin (D001) [7] in the presence or absence of solvent(s) at room temperature.

[1] Luo, H., Kang, Y., Li, Q., and Yang, L. (2008). Sulfamic acid as efficient and reusable catalytic system for the synthesis of pyrrole, furan, and thiophene derivatives. *Heteroatom Chem.*, **19**, 144–148.

[2] Bellina, F., and Rossi, R. (2006). Synthesis and biological activity of pyrrole, pyrroline and pyrrolidine derivatives with two aryl groups on adjacent positions. *Tetrahedron*, **62**, 7213–7256.
[3] Texier-Boullet, F., Klein, B., and Hamelin, J. (1986). Pyrrole and pyrazole ring closure in heterogeneous media. *Synthesis*, 409–411.
[4] Song, G., Wang, B., Wang, G., Kang, Y., Yang, T., and Yang, L. (2005). Fe^{3+}-montmorillonite as effective, recyclable catalyst for Paal–Knorr pyrrole synthesis under mild conditions. *Syn. Commun.*, **35**, 1051–1057.
[5] Chen, J., Wu, H., Zheng, Z., Jin, C., Zhang, X., and Su, W. (2006). An approach to the Paal-Knorr pyrroles synthesis catalyzed by $Sc(OTf)_3$ under solvent-free conditions. *Tetrahedron Lett.*, **47**, 5383–5387.
[6] Banik, M., Ramirez, B., Reddy, A., Bandyopadhyay, D., and Banik, B. K. (2012). Polystyrenesulfonate-catalyzed synthesis of novel pyrroles through Paal-Knorr reaction. *Org. Med. Chem. Lett.*, **2**, 11.
[7] Yuan, S. Z., Liu, J., and Xu, L. (2010). A convenient synthesis of pyrroles catalyzed by acidic resin under solvent-free condition. *Chin. Chem. Lett.*, **21**, 664–668.

C-N bond_20

Type of reaction: C-N and C-C bond formation
Reaction conditions: Solvent-free, room temperature
Synthetic strategy: One-pot multicomponent
Catalyst: Heterogeneous, 12-tungstophosphoric acid ($H_3PW_{12}O_{40}$)/SiO_2 (40 wt% PW/SiO_2)
Keywords: Aromatic aldehydes, 1,3-dicarbonyls, ammonium acetate, 40 wt% PW/SiO_2, heterogeneous catalysis, solvent-free, room temperature, 1,4-dihydropyridines, Hanztsch reaction

General reaction scheme

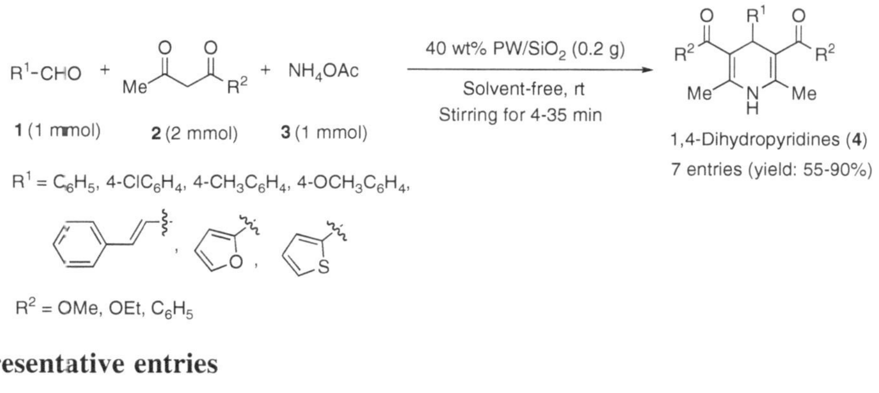

Representative entries

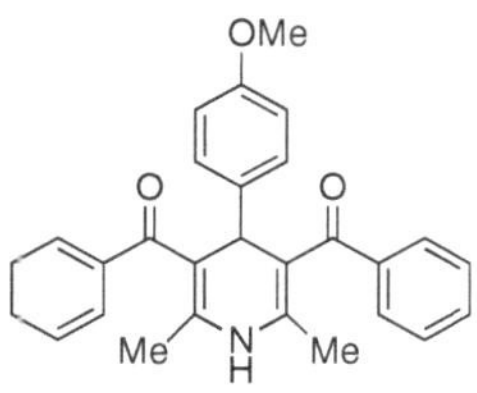

3a

(4-(4-Methoxyphenyl)-2,6-dimethyl-1,4-dihydropyridine-3,5-diyl)bis(phenylmethanone) (**3a**): Reaction time: 32 min; yield: 55%

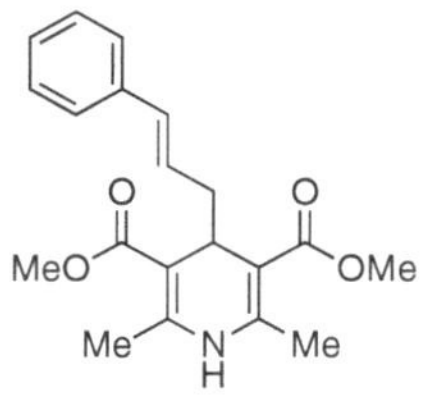

3b

Dimethyl 4-cinnamyl-2,6-dimethyl-1,4-dihydropyridine-3,5-dicarboxylate (**3b**): Reaction time: 20 min; yield: 91%

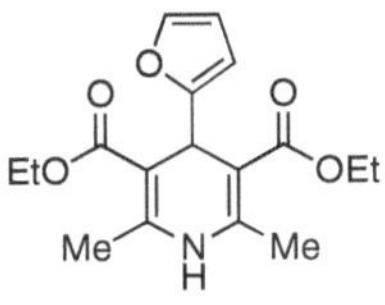

3c

Diethyl 4-(furan-2-yl)-2,6-dimethyl-1,4-dihydropyridine-3,5-dicarboxylate (**3c**): Reaction time: 4 min; yield: 98%

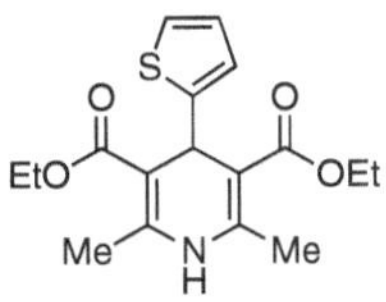

3d

Diethyl 2,6-dimethyl-4-(thiophen-2-yl)-1,4-dihydropyridine-3,5-dicarboxylate (**3d**): Reaction time: 13 min; yield: 94%

Experimental procedure

To a mixture of aromatic aldehyde (**1**; 1 mmol), 1,3-dicarbonyl compound (**2**; 2 mmol), and ammonium acetate (**3**; 1 mmol) was added the solid catalyst 40 wt% PW/SiO_2 (02 g) and stirred at room temperature under solvent-free conditions for specified period (4-35 min for varying entries). After completion of the reaction as monitored by TLC, the resulting solid product was crushed, diluted with acetonitrile (2 × 5 mL), filtered (removed the catalyst and catalyst was washed with ether for reusing) and filtrate was concentrated *in vacuo* to afford the crude product. A pure product was obtained as a yellow solid by further recrystallization in ethanol. All products were characterized and identified by comparing their spectral data with those of authentic samples [1-4].

Reference

Rafiee, E., Eavani, S., Rashidzadeh, S., and Joshaghani, M. (2009). Silica supported 12-tungstophosphoric acid catalysts for synthesis of 1,4-dihydropyridines under solvent-free conditions. *Inorganica Chimica Acta*, **362**, 3555–3562.

References for spectroscopic data

[1] Chari, M. A., and Symasundar, K. (2005). Silica gel/$NaHSO_4$ catalyzed one-pot synthesis of Hantzsch 1,4-dihydropyridines at ambient temperature. *Catal. Commun.*, **6**, 624–626.
[2] Adibi, H., Samimi, H. A., and Beygzadeh, M. (2007). Iron(III) trifluoroacetate and trifluoromethanesulfonate: Recyclable Lewis acid catalysts for one-pot synthesis of 3,4-dihydropyrimidinones or their sulfur analogues and 1,4-dihydropyridines via solvent-free Biginelli and Hantzsch condensation protocols. *Catal. Commun.*, **8**, 2119–2124.
[3] Zolfigol, M. A., Salehi, P., Khorramabadi-Zad, A., and Shayegh, M. (2007). Iodine-catalyzed synthesis of novel Hantzsch *N*-hydroxyethyl 1,4-dihydropyridines under mild conditions. *J. Mol. Catal. A: Chem.*, **261**, 88–92.
[4] Sridharan, V., Perumal, P. T., Avendano, C., and Menéndez, J. C. (2007). A new three-component domino synthesis of 1,4-dihydropyridines. *Tetrahedron*, **63**, 4407–4413.

C-N bond_21

Type of reaction: C-N and C-C bond formation
Reaction conditions: Ethanol, room temperature
Synthetic strategy: One-pot three-component domino synthesis
Catalyst: Ceric ammonium nitrate (CAN)

Keywords: Aromatic amines, α,β-unsaturated aldehydes, acetoacetate ester, ethanol, CAN, homogeneous catalysis, room temperature, one-pot multicomponent domino synthesis, 1,4-dihydropyridines, Hanztsch reaction

General reaction scheme

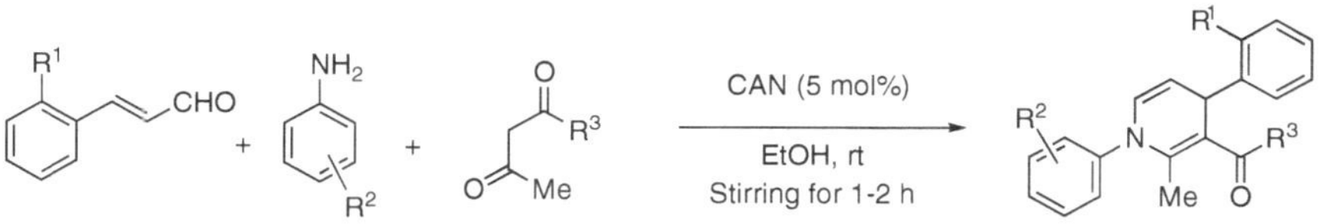

1 (1 equiv) **2** (1 equiv) **3** (1 equiv)

N-aryl-5,6-unsubstituted dihydropyridines (**4**)
19 entries (yield: 50-76%)

R^1 = H, NO_2
R^2 = H, 3-Me, 4-Me, 2,4-di-Me, 3,4-di-Me, 3,5-di-Me, 3-OMe, 4-F, 3-Cl, 4-Cl, 4-Br
R^3 = OEt, $OCMe_3$, $SCMe_3$

Representative entries

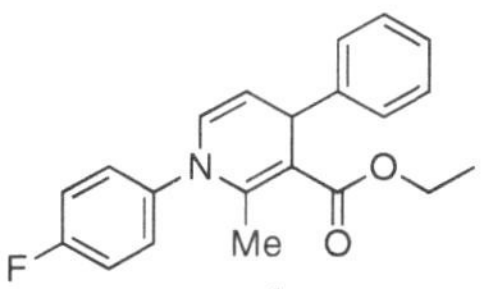

4a

Ethyl 1-(4-fluorophenyl)-2-methyl-4-phenyl-1,4-dihydropyridine-3-carboxylate (**4a**): Semi-solid; reaction time: 1 h; yield: 76%

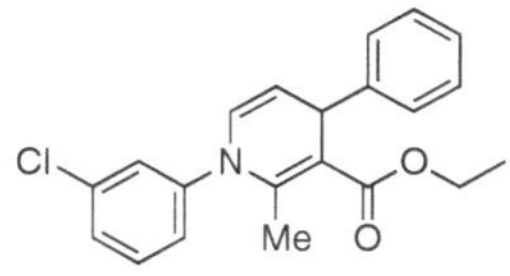

4b

Ethyl 1-(3-chlorophenyl)-2-methyl-4-phenyl-1,4-dihydropyridine -3-carboxylate (**4b**): Semi-solid; reaction time: 1 h; yield: 72%

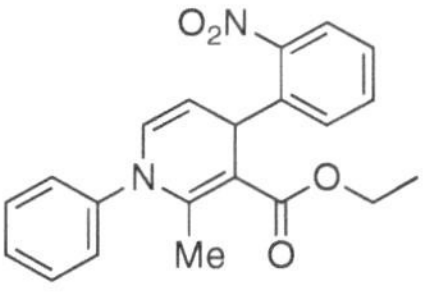

4c

Ethyl 2-methyl-4-(2-nitrophenyl)-1-phenyl-1,4-dihydropyridine-3-carboxylate (**4c**): Semi-solid; reaction time: 1 h; yield: 50%

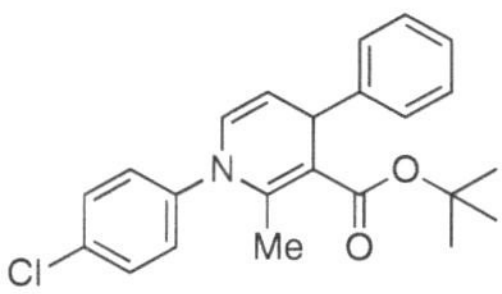

4d

tert-Butyl 1-(4-chlorophenyl)-1,4-dihydro-2-methyl-4-phenylpyridine-3-carboxylate (**4d**): Semi-solid; reaction time: 2 h; yield: 65%

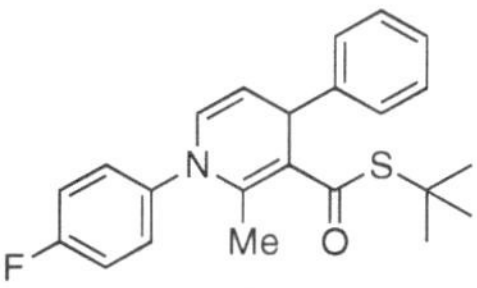

4e

S-tert-butyl 1-(4-fluorophenyl)-2-methyl-4-phenyl-1,4-dihydropyridine-3-carbothioate (**4e**): Solid, mp 138-139 °C; reaction time: 1 h; yield: 62%

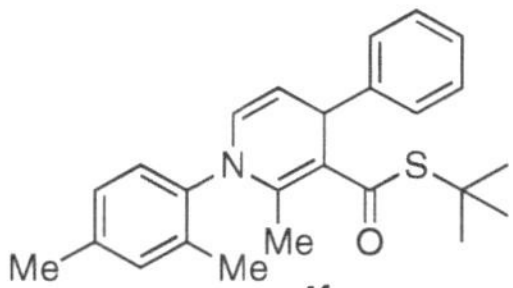

4f

S-tert-Butyl 1-(2,4-dimethylphenyl)-2-methyl-4-phenyl-1,4-dihydropyridine-3-carbothioate (**4f**): Solid as a 1:075 rotameric mixture, mp 132-133 °C; reaction time: 1 h; yield: 61%

Experimental procedure

To a solution of cinnamaldehyde (**1**; 2 mmol) and arylamine (**2**; 2 mmol) in 30 mL of absolute ethanol were added acetoacetate ester (**3**; 2 mmol) and ceric ammonium nitrate (CAN;

5 mol %). The mixture was thoroughly stirred at room temperature under argon atmosphere for 1-2 h. After completion of the reaction, the solvent was evaporated under reduced pressure and extracted with ethyl ether (2 × 20 mL). The ethereal solution was then washed twice with saturated sodium bicarbonate solution (30 mL), dried with anhydrous sodium sulfate, evaporated, and the resultant crude product was purified by flash silica column chromatography using petroleum ether-ethyl acetate as eluant (96:4) to obtain *N*-aryl-5,6-unsubstituted dihydropyridines **4** with 50-76% yield. All the products were characterized from elemental analyses as well as their detailed spectral studies including IR, ^{1}H NMR, and ^{13}C NMR.

Characterization data of representative compounds

tert-Butyl 1-(4-chlorophenyl)-1,4-dihydro-2-methyl-4-phenylpyridine-3-carboxylate (**4d**): Semi-solid; IR (NaCl): 1690.5, 1569.5, 1235.6 cm^{-1}; ^{1}H NMR (250 MHz, $CDCl_3$): δ 1.33 (s, 9H), 2.14 (s, 3H), 4.69 (d, *J*=5.1 Hz, 1H), 4.99 (dd, *J*=5.1, 7.7 Hz, 1H), 6.10 (d, *J*=7.7 Hz, 1H), 7.14- 7.41 (m, 9H); ^{13}C NMR (63 MHz, $CDCl_3$): δ 19.1, 28.6, 41.3, 79.7, 104.5, 108.0, 126.6, 127.9, 128.8, 129.2, 129.3, 130.1, 133.2, 142.8, 146.5, 148.9, 168.6. Anal. Calcd for $C_{23}H_{24}ClNO_2$: C, 72.34; H, 6.33; N, 3.67; Found: C, 72.05; H, 6.10; N, 3.45.

S-tert-Butyl 1-(4-fluorophenyl)-2-methyl-4-phenyl-1,4-dihydropyridine-3-carbothioate (**4e**): Solid, mp 138-139 °C; IR (NaCl): 1672.9, 1628.5, 1508.4, 1153.7 cm^{-1}; ^{1}H NMR (250 MHz, $CDCl_3$): δ 1.43 (s, 9H), 2.08 (s, 3H), 4.83 (d, *J*=5.8 Hz, 1H), 5.10 (dd, J=5.8, 7.5 Hz, 1H,), 6.10 (d, *J*=7.5 Hz, 1H,), 7.12-7.38 (m, 9H); ^{13}C NMR (63 MHz, $CDCl_3$): δ 19.2, 30.4, 41.2, 47.8, 107.2, 111.0, 116.9 (d, J¼22.6 Hz), 126.8, 127.7, 128.9, 129.6, 129.7 (d, *J*=8.2 Hz), 139.8 (d, *J*=3.8 Hz), 144.3, 147.5, 161.9 (d, *J*=247.4 Hz), 194.7. Anal. Calcd for $C_{23}H_{24}FNOS$: C, 72.41; H, 6.34; N, 3.67; Found: C, 72.31; H, 6.41; N, 3.74.

Reference

Sridharan, V., Perumal, P. T., Avendaño, C., and Menéndez, J. C. (2007). A new three-component domino synthesis of 1,4-dihydropyridines. *Tetrahedron*, **63**, 4407–4413.

C-N bond_22

Type of reaction: C-N and C-C bond formation
Reaction conditions: Methanol, room temperature, ultrasound
Synthetic strategy: One-pot condensation
Catalyst: 1-Butylimidazolium tetrafluoroborate ([Hbim][BF_4] ionic liquid, reusable catalyst)
Keywords: *o*-Aminoaryl ketones, α-methylene ketones, [Hbim][BF_4] (IL), methanol, room temperature, ultrasound, condensation, tandem reaction, quinolines, reusable catalyst

General reaction scheme

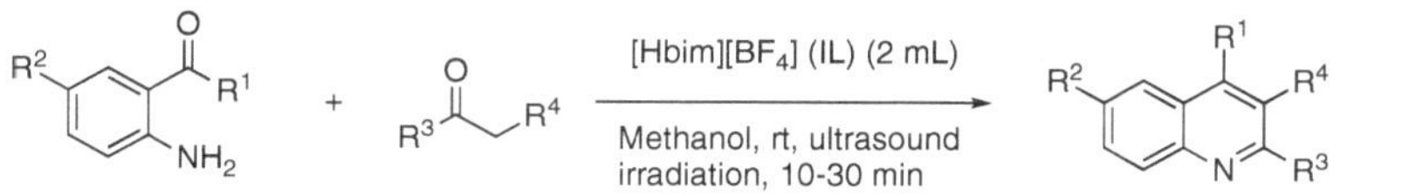

1 (2 mmol) + 2 (2 mmol) → Quinoline (3)

14 entries (yield: 65-93%)

R^1= C_6H_5, R^2= Cl
R^1= CH_3, R^2= H
R^1= C_6H_5, R^2= NO_2

R^3= CH_3, CH_2Cl, CH_2COCH_3, C_6H_5, 4-BrC6H4, 4-ClC_6H_4
R^4= H, CH_3, $COOCH_3$, $COOCH_2CH_3$, $COCH_3$
2: cyclohexanone, cyclooctanone, 4-*tert*-butylcyclohexanone

Representative entries

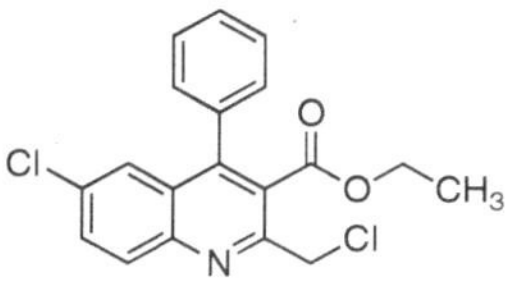

3a

Ethyl 6-chloro-2-(chloromethyl)-4-phenylquinoline-3-carboxylate (**3a**): Solid; mp 103-106 °C, reaction time: 25 min; yield: 74%

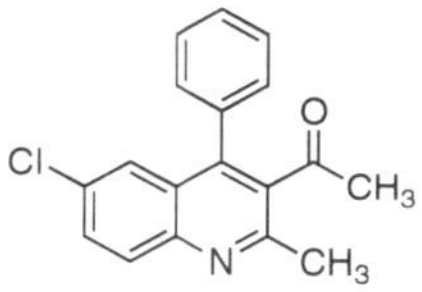

3b

1-(6-Chloro-2-methyl-4-phenylquinolin-3-yl) ethanone (**3b**): Solid; mp 221-223 °C, reaction time: 10 min; yield: 84%

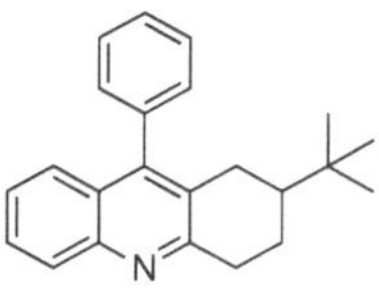

3c

2-(*tert*-Butyl)-9-phenyl-1,2,3,4-tetrahydroacridine (**3c**): Solid; mp 132-133 °C, reaction time: 25 min; yield: 87%

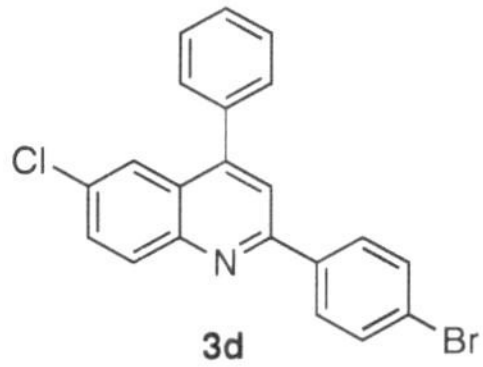

3d

2-(4-Bromophenyl)-6-chloro-4-phenylquinoline (**3d**): Solid; mp 194-196 °C, reaction time: 30 min; yield: 85%

Experimental procedure

A mixture of *o*-aminoaryl ketone (**1**, 2 mmol), α-methylene ketone (**2**, 2 mmol), 2 mL of [Hbim][BF_4] (IL) in 0.5 mL of methanol was sonicated in an atmosphere of argon at ambient conditions in a thermostated (25 ± 1 °C) ultrasonic cleaning bath for a stipulated time (10-35 min for varying entries). After completion of the reaction (as indicated by TLC), the reaction mixture was poured into crushed ice and stirred for about 1 h. The solid separated was filtered through a sintered funnel under suction, washed with ice cold water (20 mL) and then purified by silica gel column chromatography (10% ethyl acetate in hexane) to afford the pure quinoline derivative **3** (65-93% yield). Each of the products was characterized by analytical and spectral studies.

Characterization data of representative compounds

1-(6-Chloro-2-methyl-4-phenylquinolin-3-yl)ethanone (**3b**): Solid; mp 221-223 °C; IR (KBr): 3045, 2983, 1705, 1476, 1367, 823, 711 cm^{-1}; ^1H NMR(300 MHz, $CDCl_3$): δ 1.98 (s, 3H, CH_3), 3.12 (s, 3H, CH_3), 7.62-7.96 (m, 7H, ArH), 9.02 (d, 1H); HRMS (EI): Found: M, 295.0801; $C_{18}H_{14}ClNO$ requires M^+, 295.0712; LRMS *m/z* (EI^+): 295 (100% M^+), 296

(217%); ES$^+$: MNa^+, 319, MH^+, 297. Anal. Calcd for $C_{18}H_{14}ClNO$: C, 73.10; H, 4.77; N, 4.74; found: C, 73.21; H, 4.64; N, 4.66.

2-(4-Bromophenyl)-6-chloro-4-phenylquinoline (**3d**): Solid; mp 194-196 °C; IR (KBr): 3025, 2931, 1655, 1588, 1211, 725 cm^{-1}; ^{1}H NMR(300 MHz, $CDCl_3$): δ 7.55-8.17 (m, 13H); HRMS (EI): Found: M, 392.9907; $C_{21}H_{13}BrClN$ requires M^+, 392.9901; LRMS *m/z* (EI$^+$): 394 (98% M^+); ES$^+$: MNa^+, 415, MH^+, 393. Anal. Calcd for $C_{21}H_{13}BrClN$: C, 63.90; H, 3.32; N, 3.55. Found: C, 63.89; H, 3.71; N, 3.61.

Reference

Heravi, M. R. P. (2009). An efficient synthesis of quinolines derivatives promoted by a room temperature ionic liquid at ambient conditions under ultrasound irradiation via the tandem addition/annulation reaction of *o*-aminoaryl ketones with α-methylene ketones. *Ultrason. Sonochem.*, **16**, 361–366.

C-N bond_23

Type of reaction: C-N and C-C bond formation
Reaction conditions: Dichlorometahne, room temperature
Synthetic strategy: One-pot Lewis acid mediated [3+3]-annulation
Catalyst: Boron trifluoride-etherate ($BF_3.OEt_2$)
Keywords: 3-Ethoxycyclobutanones, aromatic amines, $BF_3.OEt_2$, dichloromethane, room temperature, [3+3]-annulation, 2-alkylquinolines

General reaction scheme

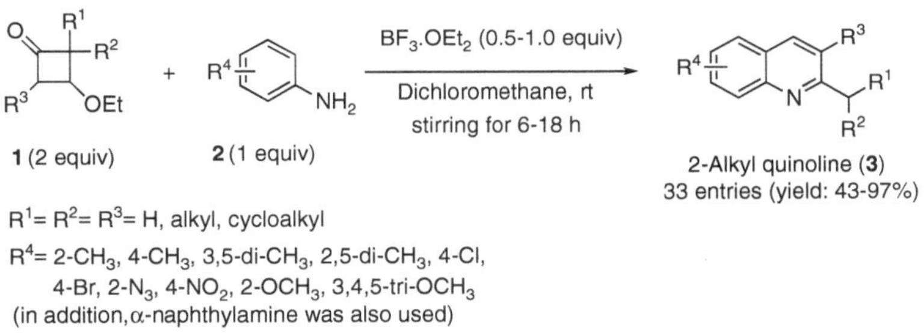

Proposed mechanism

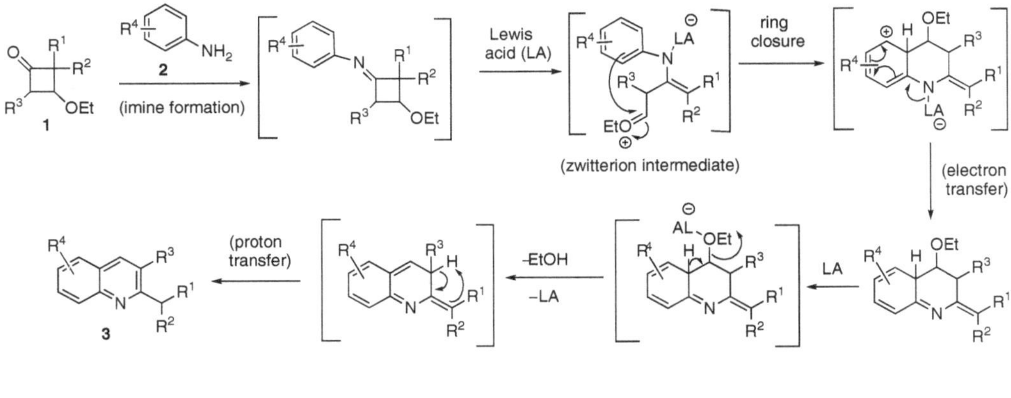

Representative entries

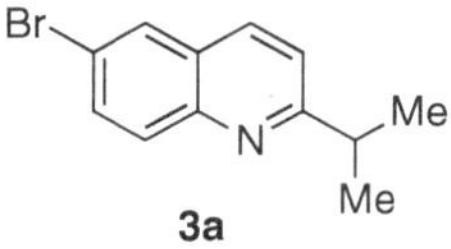

3a

6-Bromo-2-isopropylquinoline (**3a**): Yellow oil; reaction time: 12 h; yield: 74%

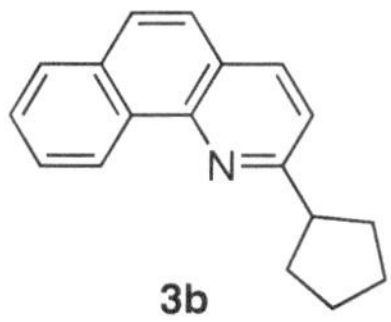

3b

2-Cyclopentylbenzo[*h*]quinoline (**3b**): Yellow oil; reaction time: 6 h; yield: 91%

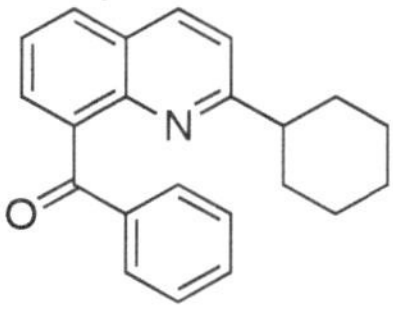

3c

(2-Cyclohexylquinolin-8-yl)(phenyl)methanone (**3c**): Yellow solid; reaction time: 18 h; yield: 64%

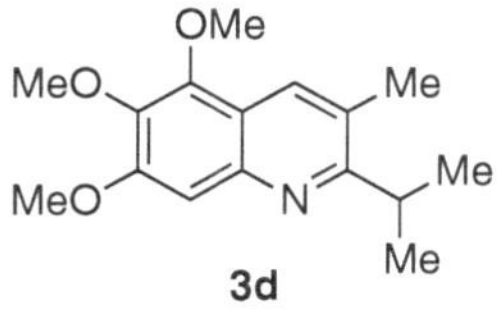

3d

2-Isopropyl-5,6,7-trimethoxy-3-methylquinoline (**3d**): Yellow oil; reaction time: 6 h; yield: 45%

Experimental procedure

To a solution of substituted aryl aniline (**2**; 1.0 equiv) and substituted 3-ethoxycyclobutanone (**1**; 2.0 equiv) in anhydrous dichloromethane in a flame-dried round bottom flask was added BF_3-OEt_2 (0.5-1.0 equiv). Then the reaction mixture was stirred at room temperature under atmosphere of argon for 6-18 h. After completion of the reaction as monitored by TLC and LC-MS, Et_3N was added to quench the reaction. The reaction mixture was diluted with dichloromethane, and washed successively with saturated aqueous $NaHCO_3$ and water. Then organic layer was dried over $MgSO_4$ and concentrated under reduced pressure. Finally the residue was purified by silica gel column chromatography to afford 2-alkyl quinoline **3** with 43-97% yield. All the products were characterized by spectral studies.

Characterization data of representative compounds

6-Bromo-2-isopropylquinoline (**3a**): Yellow oil; ^{1}H-NMR (300 MHz, $CDCl_3$) δ 7.96 - 7.89 (m, 3H), 7.71 (dd, J = 8.91Hz, J = 2.40Hz, 1H), 7.32 (d, J = 8.58Hz, 1H), 3.25 - 3.21 (m, 1H), 1.38 (d, J = 6.87Hz, 6H); ^{13}C-NMR (75 MHz, $CDCl_3$) δ 168.2, 146.4, 135.4, 132.6, 130.8, 129.5, 128.1, 120.2, 119.4, 37.3, 22.5; LRMS(ESI): calcd for $C_{12}H_{13}BrN$ $[M+H]^+$ at *m/z* 250.02; Found 249.88.

2-Isopropyl-5,6,7-trimethoxy-3-methylquinoline (**3d**): Yellow oil; ^{1}H-NMR (300 MHz, $CDCl_3$): δ 8.01 (s, 1H), 7.20 (s, 1H), 4.04 (s, 3H), 4.00 (s, 3H), 3.95 (s, 3H), 3.40 - 3.35 (m, 1H), 2.48 (s, 3H), 1.35 (d, J = 6.51 Hz, 6H); ^{13}C-NMR (75 MHz, $CDCl_3$) δ 165.9, 154.9, 146.6, 144.5, 140.3, 130.3, 126.6, 117.8, 104.1, 61.7, 61.3, 56.2, 32.0, 21.8, 19.3; LRMS (ESI): calcd for $C_{16}H_{22}NO$ $[M+H]^+$ at *m/z* 276.15; Found 276.26.

Reference

Shan, G., Sun, X., Xia, Q., and Rao, Y. (2011). A facile synthesis of substituted2-alkylquinolines through [3+3] annulation between 3-ethoxycyclobutanones and aromatic amines at room temperature. *Org. Lett.*, **13**, 5770–5773.

C-N bond_24

Type of reaction: C-N and C-C bond formation
Reaction conditions: Ethanol, room temperature
Synthetic strategy: One-pot multicomponent
Catalyst: Amberlyst A21
Keywords: Hydrazine hydrate, ethyl acetoacetate, carbonyl compounds, malononitrile, ethanol, room temperature, Amberlyst A21, one-pot multicomponent synthesis, dihydropyrano [2,3-*c*]pyrazoles

General reaction scheme

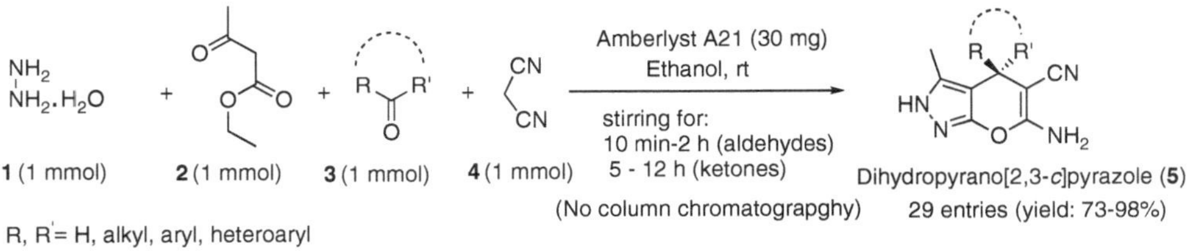

Proposed mechanism

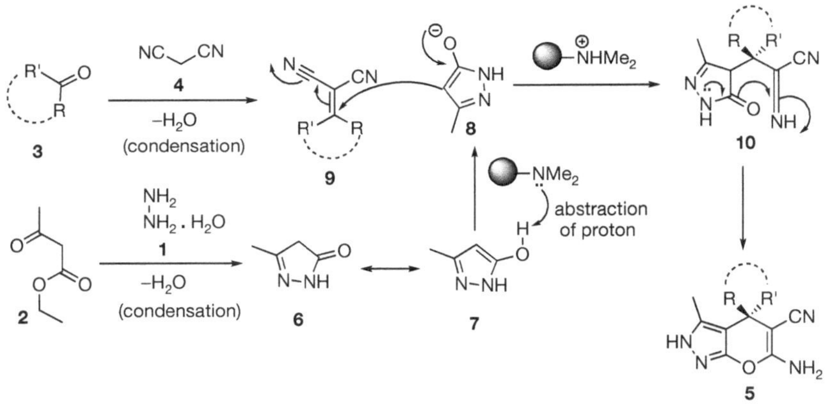

Representative entries

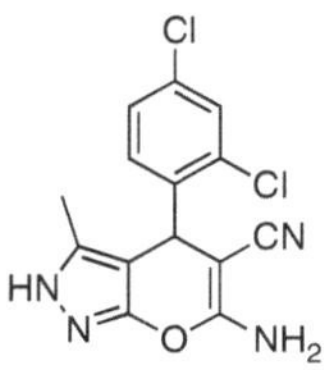

5a

6-Amino-4-(2,4-dichlorophenyl)-3-methyl-2,4-dihydropyrano[2,3-c]pyrazole-5-carbonitrile (**5a**): Solid, mp 209-210 °C; reaction time: 25 min; yield: 97%

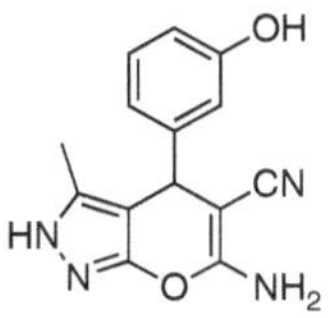

5b

6-Amino-4-(3-hydroxyphenyl)-3-methyl-2,4-dihydropyrano[2,3-*c*]pyrazole-5-carbonitrile (**5b**): Solid, mp 220-223 °C; reaction time: 30 min; yield: 92%

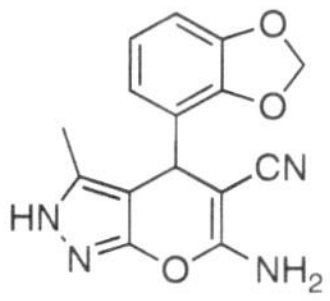

5c

6-Amino-4-(benzo[*d*][1,3]dioxol-4-yl)-3-methyl-2,4-dihydropyrano[2,3-*c*]pyrazole-5-carbonitrile (**5c**): Solid, mp 204-206 °C; reaction time: 65 min; yield: 85%

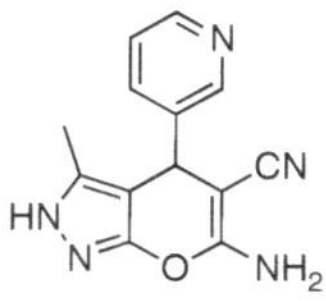

5d

6-Amino-3-methyl-4-(pyridin-3-yl)-2,4-dihydropyrano[2,3-*c*]pyrazole-5-carbonitrile (**5d**): Solid, mp 214-215 °C; reaction time: 15 min; yield: 95%

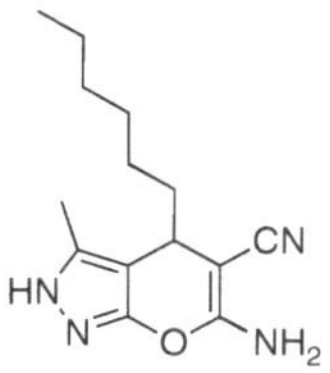

5e

6-Amino-4-hexyl-3-methyl-2,4-dihydropyrano [2,3-*c*]pyrazole-5-carbonitrile (**5e**): Solid, mp 153-154 °C; reaction time: 90 min; yield: 79%

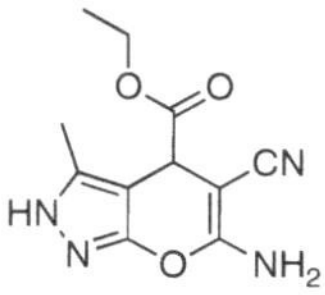

5f

Ethyl 6-amino-5-cyano-3-methyl-2,4-dihydropyrano [2,3-*c*]pyrazole-4-carboxylate (**5f**): Solid, mp 187-189 °C; reaction time: 25 min; yield: 90%

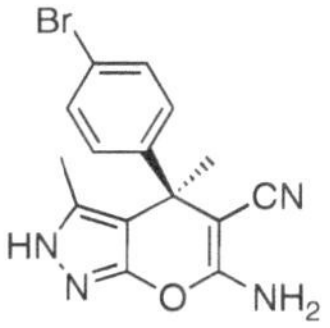

5g

(±)-6-Amino-4-(4-bromophenyl)-3,4-dimethyl-2,4-dihydropyrano[2,3-*c*]pyrazole-5-carbonitrile (**5g**): Solid, mp 178-179 °C; reaction time: 5 h; yield: 88%

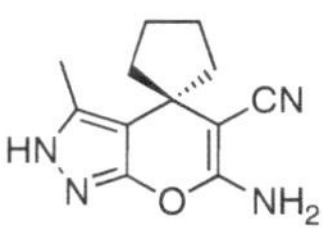

5h

(±)-6′-Amino-3′-methyl-2′*H*-spiro[cyclopentane-1,4′-pyrano[2,3-*c*]pyrazole]-5′-carbonitrile (**5h**): Solid, mp 164-165 °C; reaction time: 12 h; yield: 77%

Experimental procedure

Hydrazine hydrate (**1**; 1 mmol) and ethyl acetoacetate (**2**; 1 mmol) were first mixed together to obtain instantaneously a white solid of 3-methyl-1*H*-pyrazol-5(4*H*)-one which was then dissolved in ethanol (10 mL) followed by addition of carbonyl compound (**3**; 1 mmol) and malononitrile (**4**; 1 mmol). Upon stirring the mixture with Amberlyst A21 (30 mg) at room temperature for the specified time, the desired product of dihydropyrano[2,3-*c*]pyrazole **5** was precipitated out as solid. Warm ethanol (60 °C) was added to dissolve the solid product and filtered. The residue of Amberlyst A21 was washed thoroughly with warm ethanol. The combined ethanolic solution was concentrated and allowed to stand in a refrigerator to afford crystals of **5** in excellent yields. Each of the products was characterized by analytical analyses and spectral studies including IR, ^{1}H NMR, ^{13}C NMR and mass.

Characterization data of representative compounds

6-Amino-4-(benzo[*d*][1,3]dioxol-4-yl)-3-methyl-2,4-dihydropyrano[2,3-*c*]pyrazole-5-carbonitrile (**5c**): Solid; mp 204-206 °C; IR (KBr): 1043, 1248, 1401, 1493, 1600, 1646, 2190, 3184, 3370 cm^{-1}; ^{1}H NMR (400 MHz, DMSO-d_6): δ 1.75 (s, 3H), 4.46 (s, 1H), 5.91 (s, 2H), 6.59 (m, 2H), 6.76 (s, 1H), 6.78 (s, 1H), 12.03 (s, 1H); ^{13}C NMR (100 MHz, DMSO-d_6): δ 9.7, 35.8, 57.3, 97.6, 100.9, 107.6, 107.9, 120.5, 120.7, 135.6, 138.5, 145.9, 147.3, 154.6, 160.7;

MS (ES$^+$) m/z 297.0 $(M+H)^+$, 319.0 $(M+Na)_+$. Anal. analysis for $C_{15}H_{12}N_4O_3$: Calcd C, 60.81; H, 4.08; N, 18.91; Found: C, 60.76; H, 4.04; N, 18.95.

6-Amino-3-methyl-4-(pyridin-3-yl)-2,4-dihydropyrano[2,3-*c*]pyrazole-5-carbonitrile (**5d**): Solid; mp 214-215 °C; IR (KBr): 1049, 1414, 1493, 1606, 1646, 2203, 3177, 3350, 3396 cm^{-1}; ^{1}H NMR (400 MHz, DMSO-d_6): δ 1.72 (s, 3H), 4.63 (s, 1H), 6.93 (s, 1H), 7.28 (dd, $J=3.2$, 12 Hz, 1H), 7.46 (d, $J=8$ Hz, 1H), 8.38 (d, $J=2.4$ Hz, 2H), 12.12 (s, 1H); ^{13}C NMR (100 MHz, DMSO-d_6): δ 9.7, 33.6, 56.2, 96.7, 120.6, 123.8, 135.1, 135.7, 139.7, 148.2, 148.7, 154.7, 161.0; MS (ES$^+$) m/z 254.0. Anal. analysis for $C_{13}H_{11}N_5O$: Calcd C, 61.65; H, 4.38; N, 27.65; Found: C, 61.68; H, 4.34; N, 27.59.

Ethyl 6-amino-5-cyano-3-methyl-2,4-dihydropyrano[2,3-*c*]pyrazole-4-carboxylate (**5f**): Solid: mp 187-189 °C; IR (KBr): 1045, 1175, 1248, 1533, 1732, 2196, 2614, 2939, 3409 cm^{-1}; ^{1}H NMR (400 MHz, DMSO-d_6): δ 1.76 (t, $J=7.2$ Hz, 3H), 4.08-4.13 (q, $J=7.2$, 14.4 Hz, 2H), 4.31 (s, 1H), 7.09 (s, 2H), 12.27 (s, 1H); ^{13}C NMR (100 MHz, DMSO-d_6): δ 9.8, 14, 36.5, 50.8, 61, 92.2, 120.3, 136.3, 154.8, 161.9, 171.5; MS (ES$^+$) m/z 249.0 $(M+H)^+$, 271$(M+Na)^+$. Anal. analysis for $C_{11}H_{12}N_4O_3$: Calcd C, 53.22; H, 4.87; N, 22.57; Found: C, 53.26; H, 4.89; N, 22.50.

6′-Amino-3′-methyl-2′*H*-spiro[cyclopentane-1,4′-pyrano[2,3-*c*]pyrazole]-5′-carbonitrile (**5h**): Solid; mp 164-165 °C; IR (KBr): 1122, 1394, 1493, 1639, 2196, 2952, 3363 cm^{-1}; ^{1}H NMR (400 MHz, DMSO-d_6): δ 1.49-1.61 (m, 8H), 1.93 (s, 3H), 6.31 (s, 2H), 11.83 (s, 1H); ^{13}C NMR (100 MHz, DMSO-d_6): δ 10.9, 26.6, 39.7, 43.4, 64.2, 103.4, 121.3, 134, 154, 159.4; MS (ES$^+$) m/z 231.0 $(M+H)^+$. Anal. analysis for $C_{12}H_{14}N_4O$: Calcd C, 62.59; H, 6.13; N, 24.33; Found: C, 62.38; H, 6.99; N, 24.37.

Reference

Bihani, M., Bora, P. P., Bez, G., and Askari, H. (2013). Amberlyst A21 catalyzed chromatography-free method for multicomponent synthesis of dihydropyrano[2,3-*c*]pyrazoles in ethanol. *ACS Sustainable Chem. Eng.*, **1**, 440–447.

C-N bond_25

Type of reaction: C-N and C-C bond formation
Reaction conditions: Solvent-free, room temperature
Synthetic strategy: One-pot multicomponent
Catalyst: Per-6-amino-β-cyclodextrin (per-6-ABCD) — acts as a supramolecular host and a reusable solid base catalyst
Keywords: Hydrazine hydrate, ethyl acetoacetate, carbonyl compounds, malononitrile, per-6-ABCD, solvent-free, room temperature, mixing, one-pot multicomponent synthesis, dihydropyrano[2,3-*c*]pyrazoles

General reaction scheme

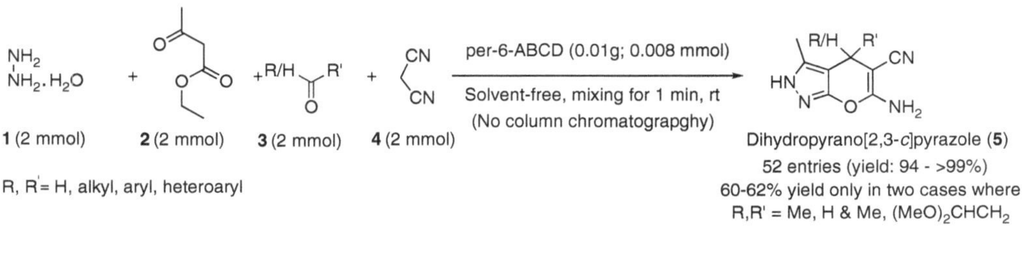

Representative entries

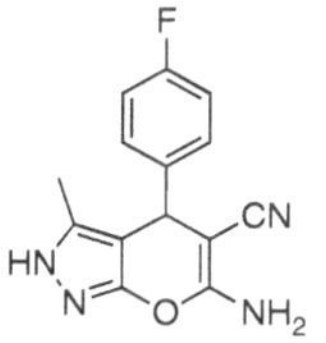

5a

6-Amino-4-(4-fluorophenyl)-2,4-dihydro-3-methylpyrano[2,3-*c*]pyrazole-5-carbonitrile (**5a**): Solid, mp 170-171 °C; yield: >99%

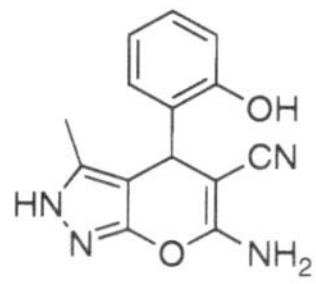

5b

6-Amino-2,4-dihydro-4-(2-hydroxyphenyl)-3-methylpyrano[2,3-*c*]pyrazole-5-carbonitrile (**5b**): Solid, mp 208-210 °C; yield: 99%

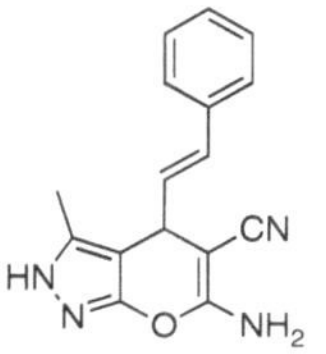

5c

6-Amino-2,4-dihydro-3-methyl-4-styrylpyrano [2,3-*c*]pyrazole-5-carbonitrile (**5c**): Solid, mp 180-181 °C; yield: 98%

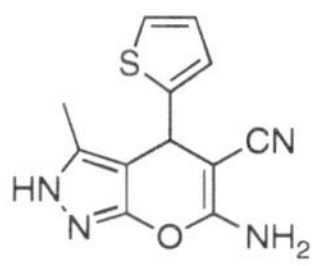

5d

6-Amino-2,4-dihydro-3-methyl-4-(thiophen-2-yl) pyrano [2,3-*c*]pyrazole-5-carbonitrile (**5d**): Solid, mp 190-191 °C; yield: >99%

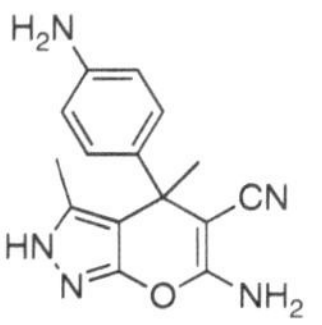

5e

6-Amino-4-(4-aminophenyl)-2,4-dihydro-3,4-dimethylpyrano[2,3-*c*]pyrazole-5-carbonitrile (**5e**): Solid, mp 182-183 °C; yield: >99%

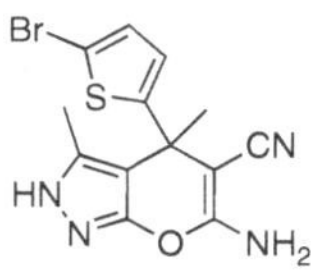

5f

6-Amino-4-(5-bromothiophen-2-yl)-2,4-dihydro-3,4-dimethylpyrano[2,3-*c*]pyrazole-5-carbonitrile (**5f**): Solid, mp 190-192 °C; yield: >99%

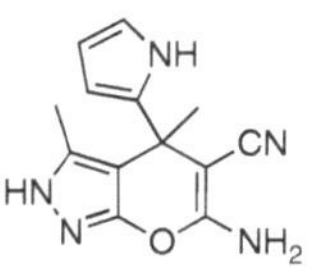

5g

6-Amino-2,4-dihydro-3,4-dimethyl-4-(1*H*-pyrrol-2-yl)pyrano[2,3-*c*]pyrazole-5-carbonitrile (**5g**): Solid, mp 195-197 °C; yield: >99%

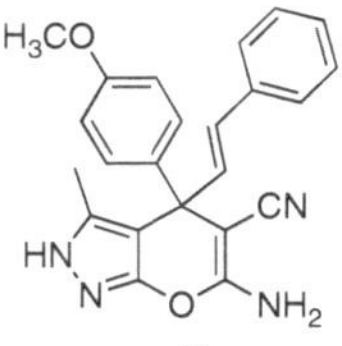

5h

6-Amino-2,4-dihydro-3-methyl-4-phenyl-4-styrylpyrano[2,3-*c*]pyrazole-5-carbonitrile (**5h**): Solid, mp 218-219 °C; yield: 94%

Experimental procedure

Hydrazine hydrate (**1**; 0.1 g, 2 mmol), ethyl acetoacetate (**2**; 0.26 g, 2 mmol), carbonyl compound (**3**; 2 mmol) and malononitrile (**4**; 0.13 g, 2 mmol) were added successively to per-6-ABCD (0.01 g, 0.008 mmol) under solvent-free conditions at room temperature. The reaction mixture was then ground for 1 minute. After completion of the reaction, 1 mL of distilled ethanol was added to the reaction mixture. The precipitated per-6-ABCD was removed by filtration, washed with distilled ethanol (1 mL) for three times, dried in vacuum and reused. The desired product, dihydropyrano[2,3-*c*]pyrazole **5** was obtained by evaporating the combined ethanol portions. Each of the products was characterized by ^{1}H NMR, ^{13}C-NMR and mass spectral studies.

Characterization data of representative compounds

6-Amino-2,4-dihydro-3-methyl-4-styrylpyrano[2,3-*c*]pyrazole-5-carbonitrile (**5c**): Solid; mp 180-181 °C; ^{1}H NMR (300 MHz; $CDCl_3$ & DMSO-*d6*): δ 2.15 (s, 3H), 4.85-4.99 (m, 1H), 6.23 (s br, 2H), 6.85-7.48 (m, 7H); ^{13}C NMR (75 MHz; $CDCl_3$ & DMSO-d_6): δ 11.5, 13.9, 69.8, 119.1, 123.8, 125.6, 126.5, 128.4, 130.0, 130.2, 140.1, 142.0, 153.8, 159; HRMS (ESI): *m/z* Calcd for $C_{16}H_{14}N_4O$ $[M+H]^+$: 279.1245; Found, 279.1246 (rel. int. 100%).

6-Amino-2,4-dihydro-3-methyl-4-(thiophen-2-yl)pyrano[2,3-*c*]pyrazole-5-carbonitrile (**5d**): Solid; mp 190-191 °C; ^{1}H NMR (300 MHz; $CDCl_3$ & DMSO-d_6): δ 2.00 (s, 3H), 4.67 (s,1H), 6.63 (s br, 2H), 6.82-6.95 (m, 2H), 7.79 (t, $J=7.5$ Hz, 1H); ^{13}C NMR (75 MHz; $CDCl_3$ & DMSO-*d6*): δ 11.6, 19.3, 71.3, 113.0, 125.9, 127.8, 129.6, 134.9, 135.5, 136.0, 153.4, 155.3; (HRMS (ESI): *m/z* Calcd for $C_{12}H_{10}N_4OS$ $[M+H]^+$: 259.0653; Found, 259.0653 (rel. int. 100%).

6-Amino-4-(5-bromothiophen-2-yl)-2,4-dihydro-3,4-dimethylpyrano[2,3-*c*]pyrazole-5-carbonitrile (**5f**): Solid; mp 190-192 °C; ^{1}H NMR (300 MHz; $CDCl_3$ & DMSO-d_6): δ 1.65 (s, 3H), 2.34 (s, 3H), 6.28 (s br, 2H), 6.83 (d, $J=8.7$ Hz, 1H), 7.09 (d, $J=8.4$ Hz, 1H); ^{13}C NMR (75 MHz; $CDCl_3$ & DMSO-d_6): δ 11.4, 24.2, 36.6, 76.2, 117.0, 117.8, 125.6, 134.2, 138.2, 139.0, 139.1, 142.0, 152.0, 153.4; HRMS (ESI): *m/z* Calcd for $C_{13}H_{11}BrN_4OS$ $[M+H]^+$: 350.9915; Found, 350.9916 (rel. int. 100%).

6-Amino-2,4-dihydro-3-methyl-4-phenyl-4-styrylpyrano[2,3-*c*]pyrazole-5-carbonitrile (**5h**): Solid; mp 218-219 °C; ^{1}H NMR (300 MHz; $CDCl_3$ & DMSO-d_6): δ 2.12 (s, 3H), 6.08 (s br, 2H), 7.10 (d, $J=14.4$ Hz, 1H), 7.54-7.74 (m, 9H), 7.94 (d, $J=14.4$ Hz, 1H); ^{13}C NMR (75 MHz; $CDCl_3$ & DMSO-d_6): δ 12.0; 37.2; 72.9; 113.6; 123.3; 127.0; 128.3; 129.0; 129.5; 130.5; 131.0; 134.5; HRMS (ESI): *m/z* Calcd for $C_{23}H_{20}N_4O_2$ $[M+H]^+$: 385.1664; Found, 385.1685 (rel. int. 100%).

Reference

Kanagaraj, K., and Pitchumani, K. (2010). Solvent-free multicomponent synthesis of pyranopyrazoles: per-6-amino-β-cyclodextrin as a remarkable catalyst and host. *Tetrahedron Lett.*, **51**, 3312–3316.

C-N bond_26

Type of reaction: C-N and C-C bond formation
Reaction conditions: Ethanol, ultrasound irradiation, room temperature
Synthetic strategy: One-pot multicomponent
Catalyst: Piperidine
Keywords: Hydrazines, isatins, β-keto esters, malononitrile, ethyl cyanoacetate, piperidine, ethanol, room temperature, ultrasound irradiation, one-pot multicomponent synthesis, spiro[indoline-3,4′-pyrano[2,3-*c*]pyrazole] derivatives

General reaction scheme

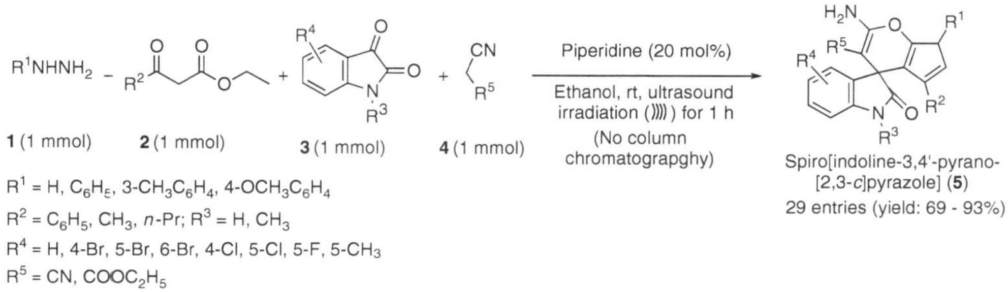

Representative entries

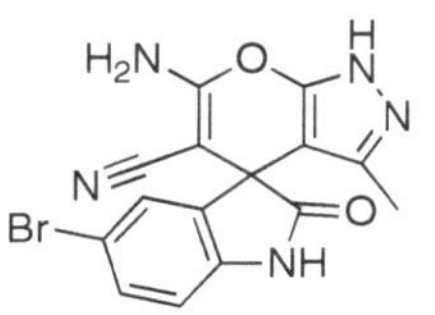

5a

6′-Amino-5-bromo-3′-methyl-2-oxo-1′*H*-spiro[indoline-3,4′-pyrano[2,3-*c*]pyrazole]-5′-carbonitrile (**5a**): White powder, mp 282-283 °C; yield: 80%

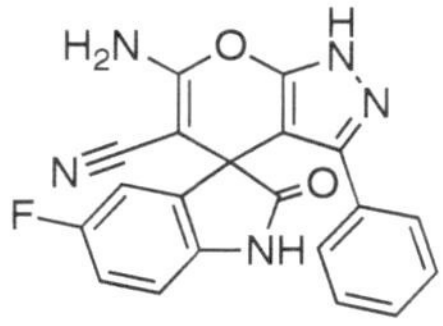

5b

6′-Amino-5-fluoro-2-oxo-3′-phenyl-1′*H*-spiro[indoline-3,4′-pyrano[2,3-*c*]pyrazole]-5′-carbonitrile (**5b**): White powder, mp 255-257 °C; yield: 87%

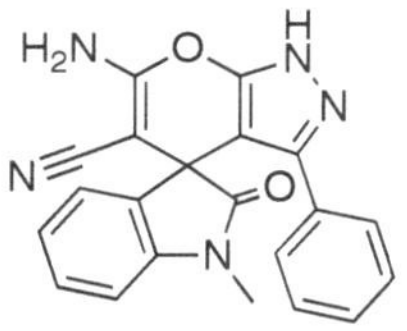

5c

6′-Amino-1-methyl-2-oxo-3′-phenyl-1′*H*-spiro[indoline-3,4′-pyrano[2,3-*c*]pyrazole]-5′-carbonitrile (**5c**): White powder, mp 284-286 °C; yield: 79%

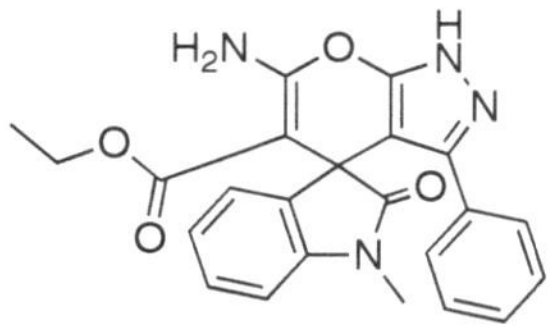

5d

Ethyl 6′-amino-1-methyl-2-oxo-3′-phenyl-1′*H*-spiro[indoline-3,4′-pyrano[2,3-*c*]pyrazole]-5′-carboxylate (**5d**): White powder, mp 224-226 °C; yield: 73%

Experimental procedure

A 100 mL flask was charged with hydrazine (**1**; 1 mmol), β-keto ester (**2**; 1 mmol), isatin (**3**; 1 mmol) and malononitrile/ethyl cyanoacetate (**4**; 1 mmol) and piperdine (20 mol%, 0.2 mmol) in ethanol (15 mL). The mixture was sonicated in the water bath of an ultrasonic cleaner (power 250 W; irradiation frequency 40 kHz) at room temperature (25 ~ 30 °C). After the completion of the reaction (monitored by TLC), the resulting precipitate was filtered and washed with ethanol to afford the pure product of spiro[indoline-3,4′-pyrano[2,3-*c*]pyrazole] **5** as solid in good to excellent yields. Each of the products was characterized by IR, ^{1}H NMR, ^{13}C-NMR and HRMS studies.

Characterization data of representative compounds

6′-Amino-1-methyl-2-oxo-3′-phenyl-1′*H*-spiro[indoline-3,4′-pyrano[2,3-*c*]pyrazole]-5′-carbonitrile (**5c**): White powder, mp 284-286 °C; IR (KBr): 3455, 3297, 3172, 3031, 2926,

2192, 1697, 1637, 1604, 1503, 1399, 1037, 921, 699 cm^{-1}; ^{1}H NMR (400 MHz, DMSO-d_6): δ 2.97 (s, 3H, CH_3), 6.72 (d, $J=7.2$ Hz, 2H, ArH), 6.89 (d, $J=7.6$ Hz, 1H, ArH), 6.97 (t, $J=7.6$ Hz, 1H, ArH), 7.07 (d, $J=7.2$ Hz, 1H, ArH), 7.18 (t, $J=8.0$ Hz, 2H, ArH), 7.24 (q, $J=7.6$ Hz, 2H, ArH), 7.32 (s, 2H, NH_2), 12.91 (s, 1H, NH); HRMS Calcd for $C_{21}H_{15}N_5O_2Na$ $[M+Na]^+$: 392.1118; Found: 392.1138.

Ethyl 6′-amino-1-methyl-2-oxo-3′-phenyl-1′*H*-spiro[indoline-3,4′-pyrano[2,3-*c*]pyrazole]-5′-carboxylate (**5d**): White powder, mp 224-226 °C; IR (KBr): 3463, 3394, 3266, 3186, 1701, 1685, 1611, 1543, 1491, 1475, 1399, 1375, 1287, 1100, 1083, 693 cm^{-1}; ^{1}H NMR (400 MHz, DMSO-d_6): δ 0.45 (s, 3H, CH_3), 3.44-3.47 (m, 2H, CH_2), 6.40-6.45 (m, 3H, ArH), 6.74-7.14 (m, 6H, ArH), 7.97 (s, 2H, NH2), 12.50 (s, 1H, NH); ^{13}C NMR (100 MHz, DMSO-d_6): δ 13.45, 18.66, 25.56, 30.75, 46.92, 56.21, 58.72, 74.36, 98.34, 107.64, 122.15, 122.42, 127.81, 128.64, 128.82, 137.21, 139.19, 143.46, 154.49, 162.87, 168.14, 177.90; HRMS Calcd for $C_{23}H_{21}N_4O_4$ $[M+H]^+$: 417.1558; Found: 417.1554.

Reference

Zou, Y., Hu, Y., Liu, H., and Shi, D. (2012a). A rapid and efficient ultrasound-assisted method for the combinatorial synthesis of spiro[indoline-3,4′-pyrano[2,3-*c*]pyrazole] derivatives. *ACS Comb. Sci.*, **14**, 38–43.

C-N bond_27

Type of reaction: C-N and C-C bond formations
Reaction conditions: Solvent-free, room temperature
Synthetic strategy: Regioselective terminal aziridine opening – aminoacetylative cyclization cascades
Catalyst: [bmim]OH (reusable)/ I_2 (promoter)
Keywords: 2-Phenyl-1,3-oxazolan-5-one, tosylaziridines, [bmim]OH (ionic liquid), molecular iodine (promoter), solvent-free, room temperature, diastereoselective aminoacetylation, pyrrolidine-2-ones

General reaction scheme

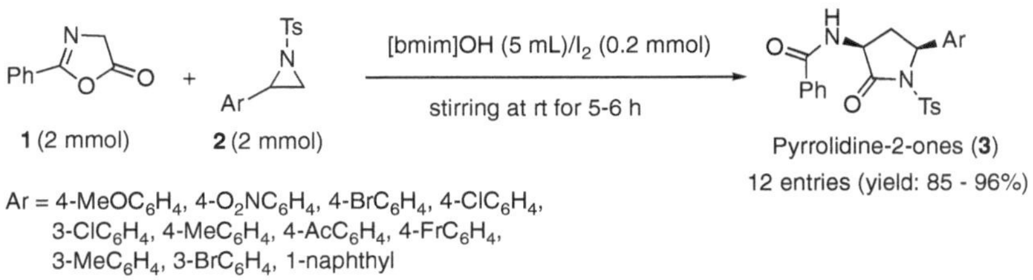

Plausible mechanism

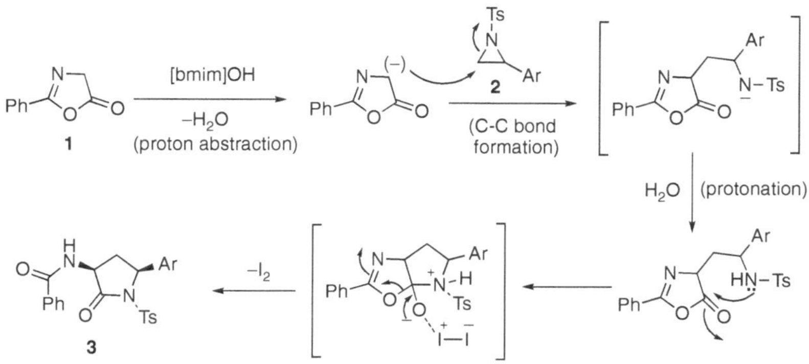

Representative entries

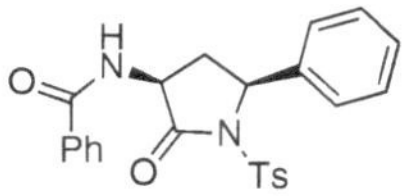

3a

N-((3*S*,5*S*)-2-oxo-5-phenyl-1-tosylpyrrolidin-3-yl)benzamide (**3a**): Colorless solid; mp 103-104°C; reaction time: 5 h; yield: 94%

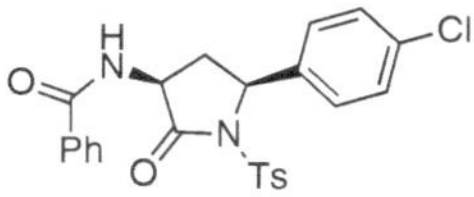

3b

N-((3*S*,5*S*)-5-(4-chlorophenyl)-2-oxo-1-tosylpyrrolidin-3-yl)benzamide (**3b**): Colorless solid; mp 151-153°C; reaction time: 5 h; yield: 92%

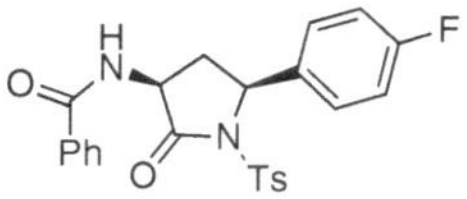

3c

N-((3*S*,5*S*)-5-(4-fluorophenyl)-2-oxo-1-tosylpyrrolidin-3-yl)benzamide (**3c**): Colorless solid; mp 135-137°C; reaction time: 5 h; yield: 85%

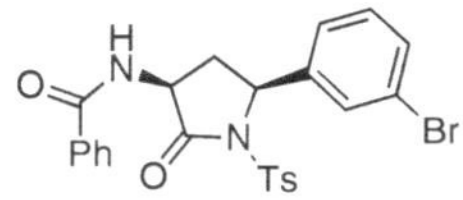

3d

N-((3*S*,5*S*)-5-(3-bromophenyl)-2-oxo-1-tosylpyrrolidin-3-yl)benzamide (**3d**): Reaction time: 5 h; yield: 94%

Experimental procedure

A mixture of 2-phenyl-1,3-oxazol-5-one (**1**, 2.0 mmol), tosylaziridine **2** (2.0 mmol), and a catalytic amount of I_2 (0.2 mmol) in [bmim]OH (5 mL) was stirred at room temperature for 5-6 h. After completion of the reaction as indicated by TLC, distilled water (10 mL) was added, and the mixture was extracted thrice with ethyl acetate (10 mL). The combined organic layer was washed with brine (10 mL), dried over anhyd. Sodium sulfate, filtered, and evaporated under reduced pressure to obtain pure diastereoselectively pure product of pyrrolidine-2-one **3** in excellent yield. The remaining aqueous layer containing the ionic liquid was washed with hexane and dried in *vacuum* resulting in recycled ionic liquid, [bmim]OH and reused. Each of the products **3** was characterized on the basis of detailed spectral as well as elemental analyses.

Characterization data of representative compounds

N-((3*S*,5*S*)-5-(4-chlorophenyl)-2-oxo-1-tosylpyrrolidin-3-yl)benzamide (**3b**): Colorless solid; mp 151-153°C; IR (KBr): 3355, 3032, 2925, 1741, 1701, 1596, 1577, 1451 cm^{-1}; ^1H NMR (400 MHz, $CDCl_3$): δ 2.28 (s, 3 H), 2.78 (ddd, J = 10.7, 7.4, 4.8 Hz, 1 H), 2.90

(ddd, $J = 10.7$, 9.0, 6.6 Hz, 1H), 4.63 (ddd, $J = 9.0$, 7.5, 4.8 Hz, 1 H), 4.85 (dd, $J = 7.4$, 6.6 Hz, 1 H), 7.26–7.53 (m, 9 H), 7.88–7.91 (m, 4 H), 8.15 (br, exch, 1 H); ^{13}C NMR (100 MHz, $CDCl_3$): δ 25.5, 33.2, 43.8, 54.0, 127.2, 127.9, 128.7, 129.5, 130.1, 130.8, 131.4, 132.0, 132.7, 133.3, 134.0, 138.9, 171.2, 178.2; MS (EI): *m/z* 468, 470 [M^+, $M^+ + 2$]. Anal. Calcd for $C_{24}H_{21}ClN_2O_4S$: C 61.47, H 4.51, N 5.97; Found: C 61.69, H 4.13, N 6.19.

N-((3*S*,5*S*)-5-(4-fluorophenyl)-2-oxo-1-tosylpyrrolidin-3-yl)benzamide (**3c**): Colorless solid; mp 135-137°C; IR (KBr): 3350, 3029, 2932, 1743, 1702, 1601, 1578, 1445 cm^{-1}; ^{1}H NMR (400 MHz, $CDCl_3$): δ 2.28 (s, 3 H), 2.82 (ddd, $J = 10.9$, 7.1, 4.3 Hz, 1 H), 2.92 (ddd, $J = 10.9$, 8.9, 6.7 Hz, 1 H), 4.66 (ddd, $J = 8.9$, 7.5, 4.3 Hz, 1 H), 4.86 (dd, $J = 7.1$, 6.7 Hz, 1 H), 7.19–7.55 (m, 9 H), 7.79–7.92 (m, 4 H), 8.17 (br, exch, 1 H); ^{13}C NMR (100 MHz, $CDCl_3$): δ 25.6, 33.1, 43.5, 53.5, 127.7, 128.4, 129.0, 129.6, 130.3, 131.1, 131.8, 132.5, 133.2, 134.0, 134.8, 140.1, 171.0, 178.8; MS (EI): $m/z = 452$ [M^+]. Anal. Calcd for $C_{24}H_{21}FN_2O_4S$: C 63.70, H 4.68, N 6.19; Found: C 63.89, H 4.91, N 5.87.

Reference

Rai, V. K., Sharma, N., and Kumar, A. (2013). The first I_2-promoted efficient aminoacetylation of activated aziridines in ionic liquid. *Synlett*, **24**, 97–101.

C-N bond_28

Type of reaction: C-N and N-S bond formation
Reaction conditions: 1,2-Dichloroethane, room temperature
Synthetic strategy: Oxidative dimerization
Catalyst: 2,3-Dichloro-5,6-dicyanaobenzoquinone (DDQ)
Keywords: Primary thioamides, 1,2-dichloroethane, DDQ, room temperature, oxidative cyclization, 1,2,4-thiadiazoles

General reaction scheme

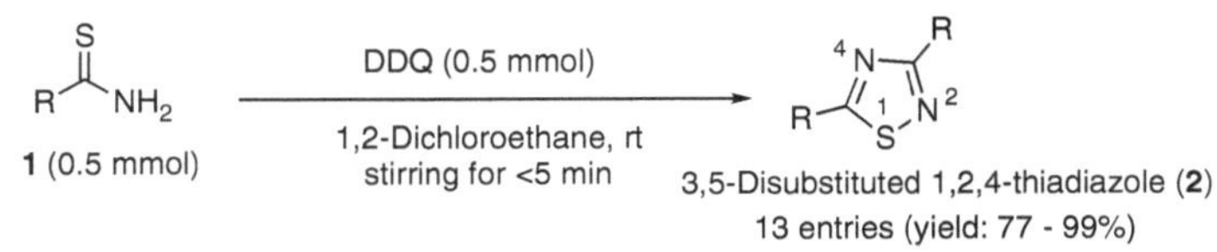

R = C_6H_5, 4-MeC_6H_4, 3-MeC_6H_4, 4-$MeOC_6H_4$, 4-FC_6H_4, 4-ClC_6H_4, 3-ClC_6H_4, 3-F,4-Br-C_6H_3, 4-$NO_2C_6H_4$, 3-$NO_2C_6H_4$, 2-$NO_2C_6H_4$, 4-$NO_2C_6H_4CH_2$, 2-thienyl

Representative entries

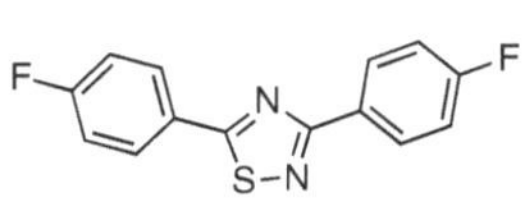

2a

3,5-*Bis*(4-fluorophenyl)-1,2,4-thiadiazole (**2a**): Solid, mp 185-186 °C; reaction time: <5 min; yield: 93%

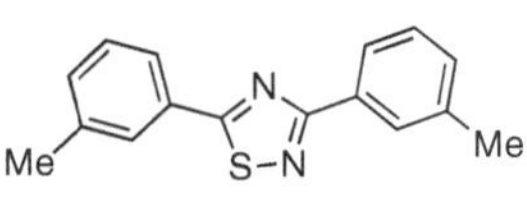

2b

3,5-*Bis*(3-methylphenyl)-1,2,4-thiadiazole (**2b**): Solid, mp 56-58 °C; reaction time: <5 min; yield: 97%

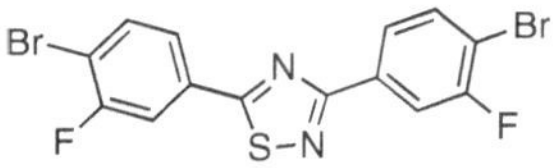

2c

3,5-*Bis*(3-fluoro-4-bromophenyl)-1,2,4-thiadiazole (**2c**): Solid, mp 180-182 °C; reaction time: <5 min; yield: 90%

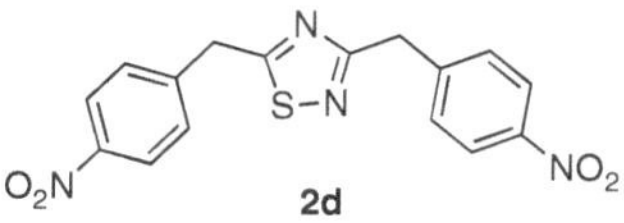

2d

3,5-*Bis*(4-nitrobenzyl)-1,2,4-thiadiazole (**2d**): Solid, mp 73-75 °C; reaction time: <5 min; yield: 77%

Experimental procedure

To a solution of thioamide (**1**; 0.5 mmol) in 1,2-dichloroethane (1 mL) was added DDQ (0.5 mol%), and the mixture was stirred at room temperature. The reaction proceeded rapidly and on its completion within <5 min., the organic solvent was removed under reduced pressure. The residue was then purified through silica gel using petroleum ether-ethyl acetate (10:1) as developer to obtain pure sample of 3,5-di-substituted 1,2,4-thiadiazole **2**. All the products were characterized by means of elemental analyses and detailed spectral studies including IR, ^{1}H-NMR, ^{13}C-NMR and mass.

Characterization data of two representative compounds

3,5-*Bis*(3-fluoro-4-bromophenyl)-1,2,4-thiadiazole (**2c**): Solid; mp 180-182 °C; IR (KBr): 2926, 1504, 1461, 1433, 1384, 841, 739 cm^{-1}; ^{1}H NMR ($CDCl_3$, 500 MHz): δ 8.14-8.12 (m, 1H), 8.06-8.04 (m, 1H), 7.85-7.83 (m, 1H), 7.74-7.66 (m, 3H); ^{13}C NMR ($CDCl_3$,125 MHz): δ 186.3, 171.9, 160.5, 160.3, 158.5, 158.3, 134.7, 134.0, 133.65, 133.59, 131.33, 131.27, 124.99, 124.97, 124.2, 124.1, 116.4, 116.2, 115.2, 115.0, 113.6, 113.4, 111.9, 111.7; MS (EI) *m/z* ((%relative intensity) at 433 (M^+ + 4, 11), 431 (M^+ + 2, 22), 429 (M^+, 11), 232 (97), 230 (100), 152 (32), 120 (24), 94 (11). Anal. Calc. for $C_{14}H_6Br_2F_2N_2S$: C 38.92, H 1.40, N 6.48; Found: C 38.66, H, 1.66, N 6.32%.

3,5-*Bis*(4-nitrobenzyl)-1,2,4-thiadiazole (**2d**): Solid; mp 73-75 °C; IR (KBr): 2925, 1524, 1481, 1347, 843, 716 cm^{-1}; ^{1}H NMR ($CDCl_3$, 500 MHz): δ 8.23 (d, $J=8.5$ Hz, 2H), 8.19 (d, $J=8.5$ Hz, 2H), 7.52-7.49 (m, 4H), 4.49 (s, 2H), 4.41 (s, 2H); ^{13}C NMR (CDC13,125 MHz): δ 189.5, 174.2, 147.5, 147.0, 144.3, 143.0, 130.0, 129.97, 124.2, 123.8, 38.9, 37.1; MS (EI) *m/z* ((%relative intensity) at 356 (M^+, 9), 220(12), 194 (100), 164 (25), 148 (43), 89 (24). Anal. Calc. for $C_{16}H_{12}N_4O_4S$: C 53.93, H 3.39, N 15.72; Found: C 53.75, H 3.48, N 15.65%.

Reference

Cheng, D., Luo, R., Zheng, W., and Yan, J. (2012). Highly efficient oxidative dimerization of thioamides to 3,5-disubstituted 1,2,4-thiadiazoles mediated by DDQ. *Syn. Commun.*, **42**, 2007–2013.

C-N bond_29

Type of reaction: C-N and N-S bond formation
Reaction conditions: *N,N*-Dimethyl formamide, room temperature
Synthetic strategy: Visible-light-driven aerobic oxidative cyclization
Catalyst: Eosin Y (**1**) as a photoredox catalyst

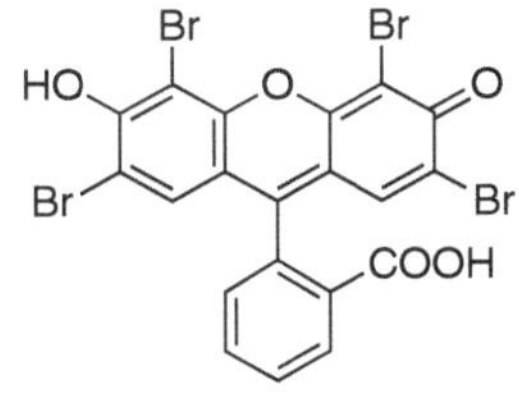

1

Keywords: Primary thioamides, visible light, eosin Y, photoredox, *N,N*-Dimethyl formamide (DMF), room temperature, aerobic oxidative cyclization, 1,2,4-thiadiazoles

General reaction scheme

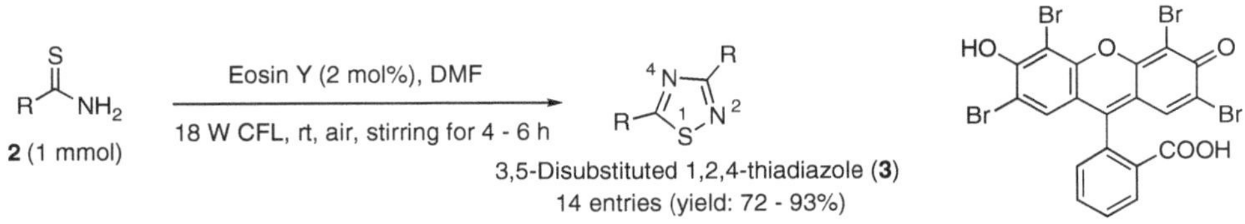

R = C_6H_5, 4-MeC_6H_4, 4-$MeOC_6H_4$, 4-FC_6H_4, 4-ClC_6H_4, 3-ClC_6H_4, 2,5-di-ClC_6H_3, 4-$NO_2C_6H_4$, 3-$NO_2C_6H_4$, 2-$NO_2C_6H_4$, 4-$NO_2C_6H_4CH_2$, Me, 2-thienyl, 4-pyridyl

Plausible mechanism

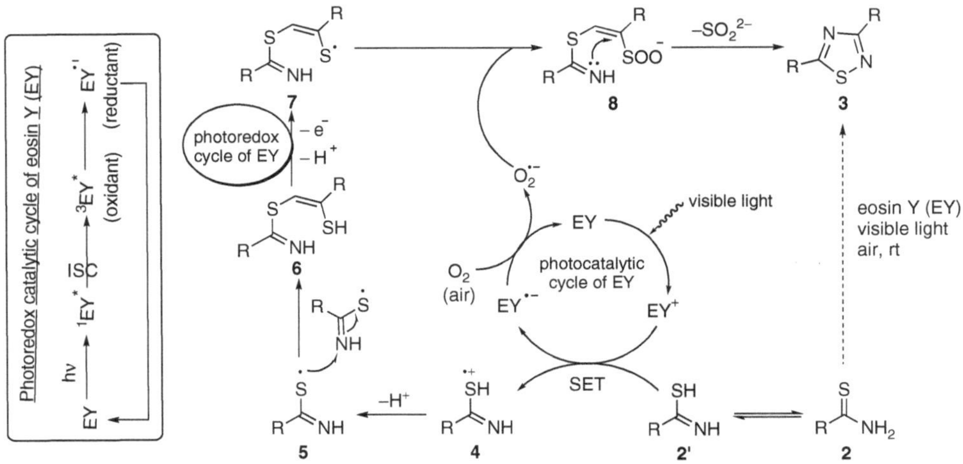

Representative entries

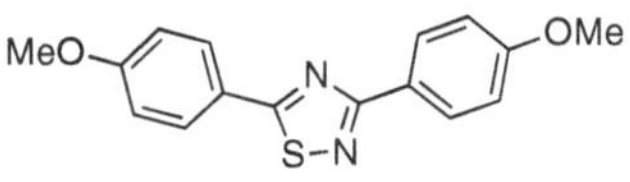

3a

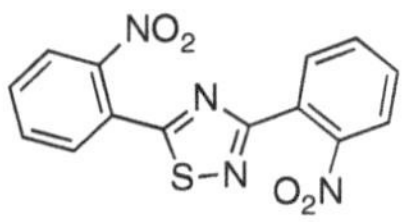

3b

3,5-*Bis*(4-methoxyphenyl)-1,2,4-thiadiazole (**3a**): Reaction time: 4 h; yield: 95%

3,5-*Bis*(2-nitrophenyl)-1,2,4-thiadiazole (**3b**): Reaction time: 6 h; yield: 72%

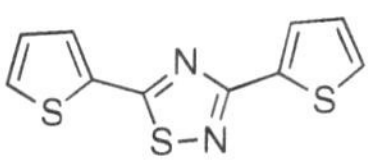

3c

3,5-Di(thiophen-2-yl)-1,2,4-thiadiazole (**3c**):
Reaction time: 4 h; yield: 92%

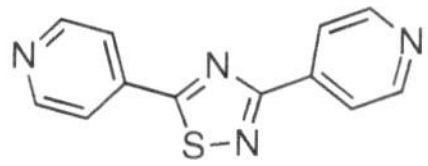

3d

3,5-Di(pyridin-4-yl)-1,2,4-thiadiazole (**3d**):
Reaction time: 6 h; yield: 76%

Experimental procedure

To a solution of thioamide (**2**; 1.0 mmol) in DMF (3 mL) was added eosin Y (**1**; 2 mol%), and the mixture was irradiated with a household 18 W CFL (Philips, 6500 K, 1010 lm, 85 mA) with stirring under an aerobic atmosphere at room temperature for 4-6 h. After completion of the reaction (monitored by TLC), distilled water (3 mL) was added, and the mixture was extracted with EtOAc (3 × 5 mL). The combined organic phases were dried over anhydrous magnesium sulfate, filtered, and evaporated under reduced pressure. The resulting crude product was purified by silica gel column chromatography using a gradient mixture of hexane-EtOAc as eluent to afford pure sample of 3,5-di-substituted 1,2,4-thiadiazole **3**. All the products are known compounds and were characterized by the comparison of their melting points and spectroscopic data with those reported in the literature [1-4].

Reference

Srivastava, V. P., Yadav, A. K., and Yadav, L. D. S. (2013). Eosin Y catalyzed visible-light-driven aerobic oxidative cyclization of thioamides to 1,2,4-thiadiazoles. *Synlett.*, **24**, 465–470.

Further Information

Few more methods for the synthesis of 3,5-di-substituted 1,2,4-thiadiazole derivatives at room temperature have been reported using a variety of catalysts such as iodobenzene diacetate (IBD) or polymer-supported iodobenzene diacetate (PIBD) in dichloromethane [1], phenyliodine(III) diacetate in 1-*n*-butylpyridinium tetrafluoroborate ($BPyBF_4$; ionic liquid) [2], IBX/TEAB in acetonitrile [3], and methyl bromocyanoacetate in methanol [4].

[1] Cheng, D., and Chen, Z. (2002). Hypervalent iodine in synthesis. 84. Facile synthesis of 3,5-disubstituted 1,2,4-thiadiazoles by oxidative dimerization of thioamides using polymer-supported iodobenzene diacetate. *Synth. Commun.*, **32**, 2155–2159.
[2] Yan, M., Chen, Z., and Zheng, Q. (2003). Organic reactions in ionic liquids: oxidative dimerisation of thioamides with phenyliodine(III) diacetate. *J. Chem. Res. (S)*, 618–619.
[3] Patil, P. C., Bhalerao, D. S., Dangate, P. S., and Akamanchi, K. G. (2009). IBX/TEAB- mediated oxidative dimerization of thioamides: Synthesis of 3,5-disubstituted 1,2,4- thiadiazoles. *Tetrahedron Lett.*, **50**, 5820–5822.
[4] Mayhoub, A. S., Kiselev, E., and Cushman, M. (2011). An unexpected synthesis of 3,5- diaryl-1,2,4-thiadiazoles from thiobenzamides and methyl bromocyanoacetate. *Tetrahedron Lett.*, **2011** (52), 4941–4943.

C-N bond_30

Type of reaction: C-N bond formation
Reaction conditions: Acetonitrile, room temperature
Synthetic strategy: One-pot multi-component reaction
Catalyst: Triphenylphospine (Ph_3P)
Keywords: α-Azido esters, isocyanates, Ph_3P, acetonitrile, room temperature, domino reaction, multi-component reaction, *N*-carbamoyl hydantoins, Staudinger reaction

General reaction scheme

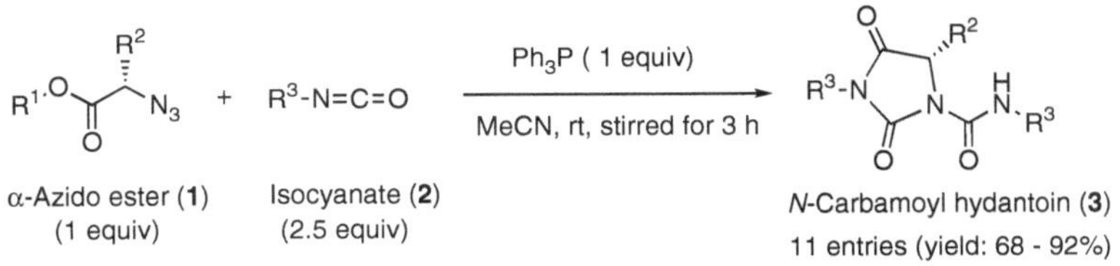

R^1,R^2 = Me, Bn; Me, *s*-Bu; Bn, H; Et, H; Me, CH_2COBn;
Me, CH_2OBn
R^3 = Bn, Et, 4-$MeOC_6H_4CH_2$, *c*-hex

Representative entries

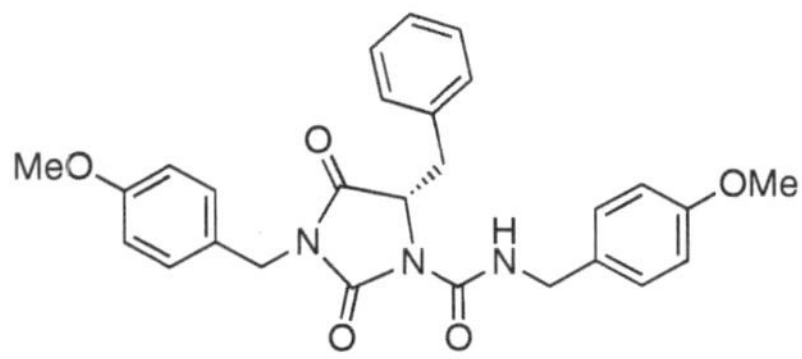

3a

(*S*)-5-Benzyl-*N*,3-*bis*(4-methoxybenzyl)-2,4-dioxoimidazolidine-1-carboxamide (**3a**): Yield: 92%; $[\alpha]^{20}_{D}=+18.5°$ ($CHCl_3$, *c* 0.7, *ee*=*ca.* 60%)

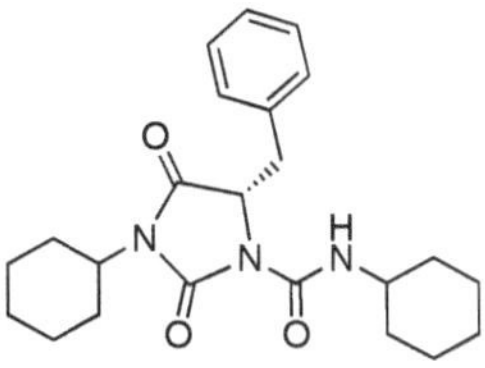

3b

(*S*)-5-Benzyl-*N*,3-dicyclohexyl-2,4-dioxoimidazolidine-1-carboxamide (**3b**): Yield: 76%; $[\alpha]^{20}_{D}=-55.5°$ ($CHCl_3$, *c* 0.4, *ee*=*ca.* 60%)

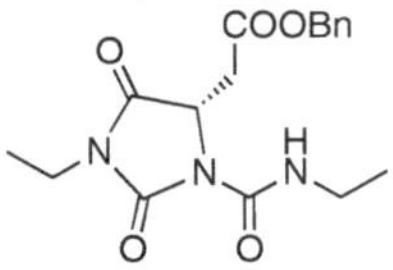

3c

(*S*)-Benzyl 2-(1-ethyl-3-(ethylcarbamoyl)-2,5-dioxoimidazolidin-4-yl)acetate (**3c**): Yield: 68%; $[\alpha]^{20}_{D}=+3.8°$ ($CHCl_3$, *c* 0.7, *ee*=*ca.* 60%)

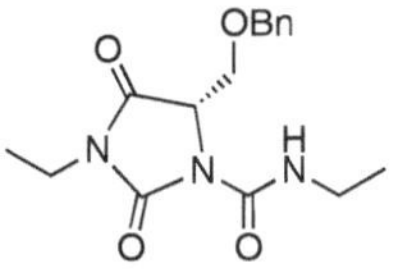

3d

((*S*)-5-(Benzyloxymethyl)-*N*,3-diethyl-2,4-dioxoimidazolidine-1-carboxamide (**3d**): Yield: 76%; $[\alpha]^{20}_{D}=+11.8°$ ($CHCl_3$, *c* 0.5, *ee*=*ca.* 60%)

Experimental procedure

To a stirred solution of α-azido ester (**1**; 1 equiv) and isocyanate (**2**; 2.5 equiv) in acetonitrile (0.1 M solution), solid Ph_3P (1equiv) was added and the solution was stirred at room temperature for 3 h. The organic solvent was removed under reduced pressure and the crude purified using chromatographic technique. All the products were characterized by means of elemental analyses and detailed spectral studies including ^{1}H-NMR, ^{13}C-NMR and mass.

Characterization data of two representative compounds

(*S*)-5-Benzyl-N,3-*bis*(4-methoxybenzyl)-2,4-dioxoimidazolidine-1-carboxamide (**3a**): $R_f=0.37$ (Hex/AcOEt=60:40); $[\alpha]^{20}_{D}=+18.5°$ ($CHCl_3$, *c* 0.7, *ee*=*ca.* 60%); ^{1}H-NMR (400 MHz, $CDCl_3$): δ 8.08 (t, *J*=5.6 Hz, 1H), 7.28 (d, *J*=8.4 Hz, 2H), 7.10-6.90 (m, 9H), 6.76 (d, *J*=8.4 Hz, 2H), 4.78 (dd, *J*=5.6 and 2.4 Hz, 1H), 4.52 (dd, *J*=14.4 and 5.6 Hz, 1H), 4.53-4.30 (m, 3H), 3.82 (s, 3H), 3.79 (s, 3H), 3.67 (dd, *J*=13.6 and 5.2 Hz, 1H), 3.28 (dd, *J*=13.6 and 2.0 Hz, 1H); ^{13}C-NMR (125.7 MHz, $CDCl_3$): δ 170.5, 159.3, 159.1, 155.3, 151.0, 133.2, 129.8, 129.7, 129.6, 129.0, 128.4, 127.3, 126.9, 114.1, 114.0, 113.9, 59.7, 55.3, 55.2, 43.5, 41.9, 34.9; ESI (*m/z*): 474.1 [M^++H; 100%]. Anal. Calcd. for $C_{27}H_{27}N_3O_5$: C 68.48; H 5.75; N 10.16; Found: C 68.49; H, 5.78; N 8.83.

(*S*)-Benzyl 2-(1-ethyl-3-(ethylcarbamoyl)-2,5-dioxoimidazolidin-4-yl)acetate (**3c**): R_f=0.23 (Hex/AcOEt=60:40); $[\alpha]^{20}{}_D$=+3.8° ($CHCl_3$, *c* 0.7, *ee*=*ca.* 60%); ^{1}H-NMR (400 MHz, $CDCl_3$): δ 7.77 (br t, 1H), 7.32 (m, 5H), 5.11 (d, *J*=12.4 Hz, 1H), 5.07 (d, *J*=12.4 Hz, 1H), 4.59 (dd, *J*=4.8 and 3.2 Hz, 1H), 3.56 (m, 3H), 3.33 (m, 2H), 3.15 (dd, *J*=17.6 and 3.2 Hz,1H), 1.25 (t, *J*=7.6 Hz, 3H), 1.20 (t, *J*=7.2 Hz, 3H); ^{13}C-NMR (125.7 MHz, $CDCl_3$): δ 170.6, 169.2, 155.9, 151.0, 135.2, 128.6, 128.4, 128.2, 66.8, 55.4, 34.9, 34.3, 33.8, 14.8, 12.7; ESI (*m*/*z*): 347,1 [M^++H; 57%], 369.2; [M^++Na; 100%)]. Anal. Calcd. for $C_{17}H_{21}N_3O_5$: C 58.78; H 6.09; N 11.50; Found: C 58.80; H 6.06; N 12.12.

Reference

Bellucci, M. C., Frigerio, M., Marcelli, T., and Volonterio, A. (2013). Multicomponent synthesis of *N*-carbamoyl hydantoin derivatives. *Synlett*, **24**, 727–732.

C-N bond_31

Type of reaction: C-N and C-C bond formation
Reaction conditions: Methanol, room temperature
Synthetic strategy: One-pot multi-component click synthesis
Catalyst: Cu_2O
Keywords: Alkyl halides, alkynes, sodium azide, Cu_2O, methanol, room temperature, multi-component reaction, click synthesis, heterocycles, substituted 5-alkynyl 1,2,3-triazoles

General reaction scheme

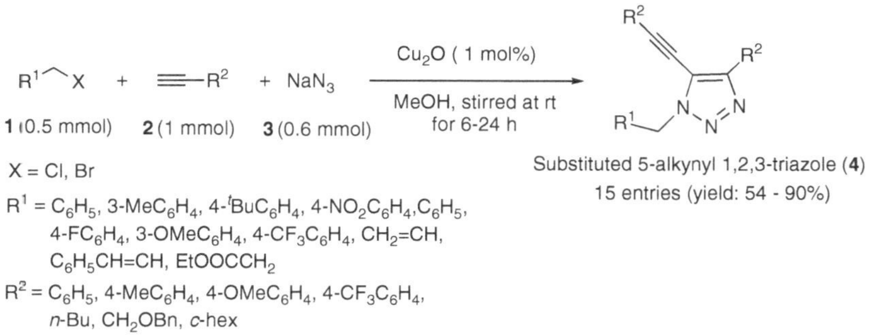

Representative entries

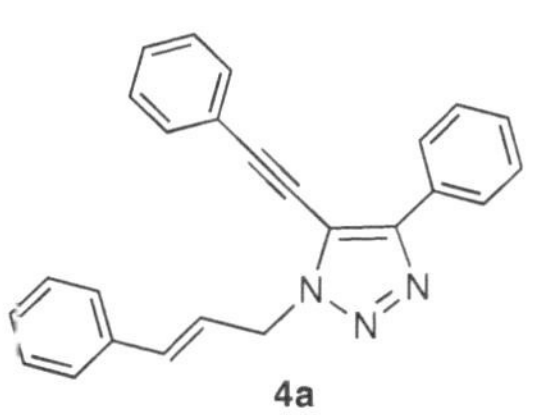

4a

1-Cinnamyl-4-phenyl-5-(phenylethynyl)-1*H*-1,2,3-triazole (**4a**): Pale yellow solid, mp 86-89.4 °C; reaction time: 8 h; yield: 87%

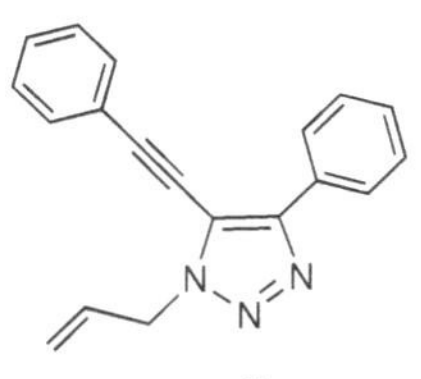

4b

1-Allyl-4-phenyl-5-(phenylethynyl)-1*H*-1,2,3-triazole (**4b**): Pale yellow solid, mp 79.8-83.4 °C; reaction time: 8 h; yield: 90%

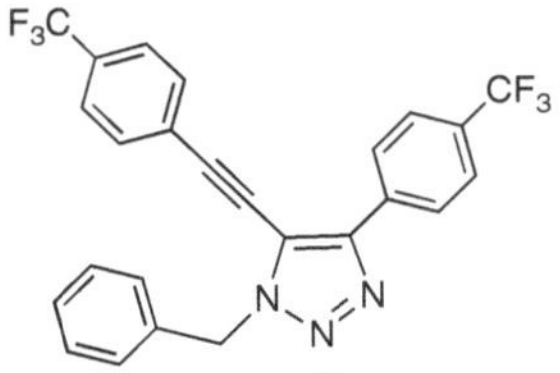

4c

1-Benzyl-4-[4-(trifluoromethyl)phenyl]-5-{[4-(trifluoromethyl)phenyl]ethynyl}-1*H*-1,2,3-triazole (**4c**): White solid, mp 172.3-175.9 °C; reaction time: 12 h; yield: 78%

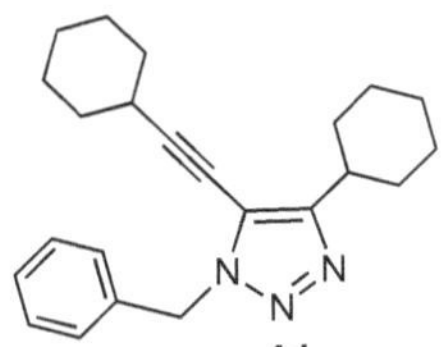

4d

1-Benzyl-4-cyclohexyl-5-(cyclohexylethynyl)-1*H*-1,2,3-triazole (**4d**): Pale yellow solid, mp 79.1-82.1 °C; reaction time: 24 h; yield: 54%

Experimental procedure

Organic halide (**1**; 0.5 mmol), alkyne (**2**; 1.0 mmol), and sodium azide (**3**; 0.6 mmol) were added to a reactor tube containing a suspension of Cu_2O (1.4 mg, 0.01 mmol) in methanol (2 mL). The reaction mixture was then stirred at room temperature for stipulated time. On completion of the reaction as monitored by TLC, GLC, and/or ^{1}H NMR spectroscopy, water (2 mL) was added to the mixture, followed by extraction with EtOAc (3 × 10 mL). The resulting organic phase was evaporated under vacuum, and the residue was purified by column chromatography using silica gel/hexane-EtOAc. All the products were characterized by means of detailed spectral studies including IR, ^{1}H-NMR, ^{13}C-NMR, and mass.

Characterization data of two representative compounds

1-Cinnamyl-4-phenyl-5-(phenylethynyl)-1*H*-1,2,3-triazole (**4a**): Pale yellow solid; mp 86.0-89.4 °C, R_f 0.54 (hexane-EtOAc = 7:3); IR (neat): 3029, 2986, 1602, 1482, 1449, 1353, 1218, 965, 770, 759, 743, 717, 691 cm^{-1}; ^{1}H NMR (400 MHz, $CDCl_3$): δ 8.20 (d, J = 7.3 Hz, 2H), 7.55 (d, J = 7.6 Hz, 2H), 7.49–7.25 (m, 11H), 6.73 (d, J = 15.8 Hz, 1H), 6.43 (dt, J = 15.8, 6.5 Hz, 1H), 5.28 (d, J = 6.5 Hz, 2H); ^{13}C NMR (100 MHz, $CDCl_3$): δ 148.2, 135.8, 135.3, 131.8, 130.5, 129.8, 128.8, 128.7, 128.5, 126.8, 126.4, 122.0, 121.6, 117.3, 102.5, 75.6, 51.5; MS (70 eV): *m/z* 361 [M$^+$, 7%], 332 (24), 217 (29), 216 (100), 117 (53), 115 (32), 113 (17), 91 (15); HRMS calcd. for $C_{25}H_{19}N_3$ 361.1579; Found 361.1569.

1-Benzyl-4-cyclohexyl-5-(cyclohexylethynyl)-1*H*-1,2,3-triazole (**3d**): Pale yellow solid; mp 79.1-82.1 °C, R_f 0.61 (hexane-EtOAc = 7:3); IR (neat): 2927, 1448, 1117, 727, 707, 696 cm^{-1}; ^{1}H NMR (400 MHz, $CDCl_3$): δ 7.33-7.25 (m, 5H), 5.48 (s, 2H), 2.78 (tt, J = 11.9, 3.6 Hz, 1H), 2.71–2.63 (m, 1H), 1.97–1.88 (m, 2H), 1.87–1.77 (m, 4H), 1.74–1.62 (m, 5H), 1.57–1.46 (m, 3H), 1.42-1.22 (m, 6H); ^{13}C NMR (100 MHz, $CDCl_3$): δ 154.4, 135.2, 128.7, 128.2, 127.9, 118.1, 107.2, 66.7, 52.4, 35.9, 32.1, 31.9, 31.7, 29.8, 26.4, 26.3, 26.2, 26.0, 25.8, 24.6; MS (70 eV): *m/z* 347 [M$^+$, 17%], 318 (21), 236 (17), 228 (14), 91 (100); HRMS calcd. for $C_{23}H_{29}N_3$ 347.2361; Found 347.2371.

Reference

Alonso, F., Moglie, Y., Radivoy, G., and Yus, M. (2012). Copper-catalysed multicomponent click synthesis of 5-alkynyl 1,2,3-triazoles under ambient temperature. *Synlett*, **23**, 2179–2182.

C-N bond_32

Type of reaction: C-N bond formation
Reaction conditions: Aqueous micellar, room temperature
Synthetic strategy: One-step Paal-Knorr condensation
Catalyst: Sodium dodecyl sulfate (SDS) as surfactant
Keywords: Amines, 2,5-hexanedione, sodium dodecyl sulfate (SDS), surfactant, water, micelles, room temperature, Paal-Knorr condensation, pyrrole derivatives

General reaction scheme

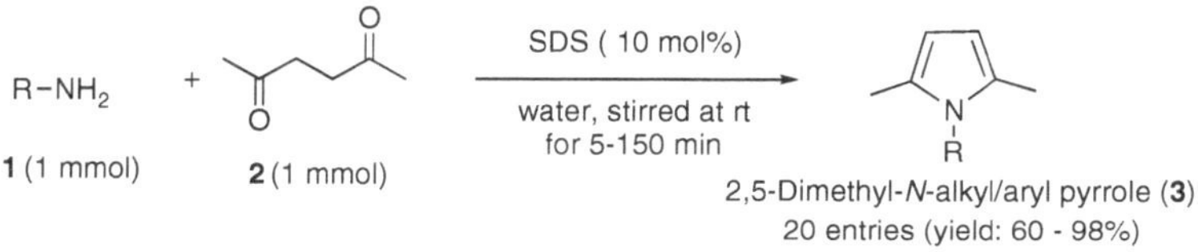

R = C_6H_5, $C_6H_4CH_2$, 3-MeC_6H_4, 4-$OMeC_6H_4CH_2$, 4-$CH_2(NH_2)C_6H_4CH_2$, 4-ClC_6H_4, 4-$OMeC_6H_4$, 4-MeC_6H_4, 3-$CF_3C_6H_4$, 4-$COOHC_6H_4$, 4-$NH_2C_6H_4$, 3-$NH_2C_6H_4$, 2-$NH_2C_6H_4$, naphthyl, 5-aminonaphthyl, 2-pyridyl, 2-(3-indolyl)ethyl, cyclopentyl, $H_2NCH_2CH_2$, $H_2NCH_2CH_2NHCH_2CH_2$, $(H_2NCH_2CH_2)_2NCH_2CH_2$

Representative entries

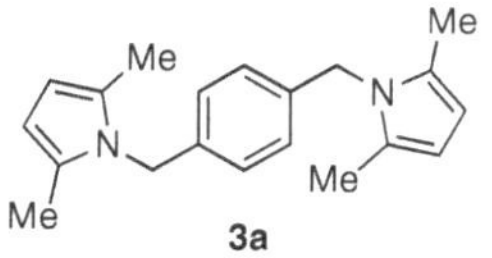

3a

1,4-Bis((2,5-dimethyl-1*H*-pyrrol-1-yl)methyl) benzene (**3a**): Cream solid, mp 197-198 °C; reaction time: 0.5 h; yield: 90%

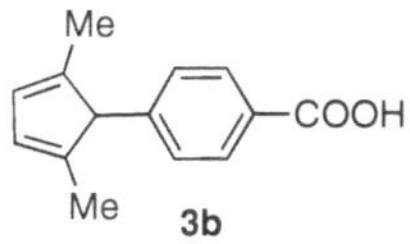

3b

4-(2,5-Dimethylcyclopenta-2,4-dien-1-yl)benzoic acid (**3b**): Pale yellow solid, mp 174-175 °C; reaction time: 1 h; yield: 75%

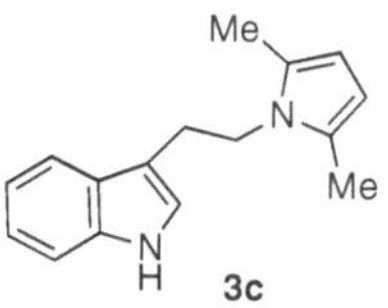

3c

3-(2-(2,5-Dimethyl-1*H*-pyrrol-1-yl)ethyl)-1*H*-indole (**3c**): Brown solid, mp 175-176 °C; reaction time: 0.5 h; yield: 90%

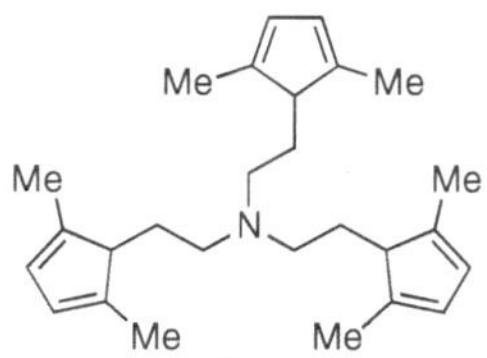

3d

Tris(2-(2,5-dimethylcyclopenta-2,4-dien-1-yl) ethyl)amine (**3d**): Pale yellow solid, mp 110-111 °C; reaction time: 15 min; yield: 96%

Experimental procedure

Amine (**1**; 1 mmol) and 2,5-hexanedione (**2**; 1 mmol) were added to a solution of SDS (10 mol %, 0.03 g) in water (5 mL), and the mixture stirred at room temperature for specified time. On completion of the reaction (monitored by TLC), K_2CO_3 (0.5 mmol, 0.07 g) was added to the reaction mixture, and the resulting precipitate of dodecyl sulfate filtered off. The filtrate was extracted with ethyl acetate (4 × 10 mL), dried over anhydrous $MgSO_4$, and then removed under reduced pressure to give almost pure product. Further purification was performed in certain cases by thin layer chromatography using *n*-hexane/acetone (70:30) as the solvent system to afford the pyrrole derivatives (**3**) with excellent yield. Each of the products was characterized on the basis of analytical and spectral data.

Characterization data of two representative compounds

3-(2-(2,5-Dimethyl-1*H*-pyrrol-1-yl)ethyl)-1*H*-indole (**3c**): Brown solid, mp 175-176 °C; IR (nujol): 3250, 3020, 1605, 1520, 1499, 1455, 1370, 1016 cm^{-1}; ^{1}H-NMR ($CDCl_3$, FT-250 MHz): δ 2.24 (s, CH_3, 6H), 3.06 (t, $J = 15.5$ Hz, CH_2, 2H), 4.04 (t, $J = 15.5$ Hz, CH_2, 2H), 5.83 (s, pyrrolics, 2H), 6.87 (s, CH, 1H), 7.12-7.73 (m, PhH, 4H), 7.94 (s, NH, 1H); HRMS: Found 238.1524 (M^+), $C_{16}H_{18}N_2$ requires M^+ at 238.1536. Anal. Calcd for $C_{16}H_{18}N_2.H_2O$: C 75.00, H 7.03, N 10.93; Found: C 74.49, H 6.98, N 10.39.

Tris(2-(2,5-dimethylcyclopenta-2,4-dien-1-yl)ethyl)amine (**3d**): Pale yellow solid, mp 110-111 °C; IR (nujol): 3100, 2854, 2739, 1571, 1518, 1464, 1407, 1378, 1298, 1166, 1061, 1016, 745 cm^{-1}; ^{1}H-NMR ($CDCl_3$, FT-90 MHz): δ 2.25 (s, CH_3, 18H), 2.74 (t, $J = 19.7$ Hz, CH_2, 6H), 3.75 (t, $J = 19.7$ Hz, CH_2, 6H), 5.78 (s, pyrrolics, 6H). Anal. Calcd for $C_{24}H_{36}N_4.0.5H_2O$: C 74.04, H 9.51, N 14.39; Found: C 74.11, H 9.67, N 14.79.

Reference

Veisi, H., Azadbakht, R., Ezadifar, M., and Hemmati, S. (2013). An efficient and green procedure for synthesis of pyrrole derivatives by Paal-Knorr condensation using sodium dodecyl sulfate in aqueous micellar. *J. Heterocyclic Chem.*, **50**, E241–E246.

C-N bond_33

Type of reaction: C-N and C-C bond formation
Reaction conditions: Toluene, room temperature
Synthetic strategy: Electrocyclic/sigmatropic rearrangement
Catalyst: *p*-Toluenesulfonic acid (PTSA)
Keywords: α,β-Unsaturated aldehydes, (*E*)-1-phenyl-2-(2,2,2-trifluoroethylidene), PTSA, toluene, room temperature, electrocyclic/sigmatropic rearrangement, 3-trifluoromethyl-1,4-dihydropyridazines

General reaction scheme

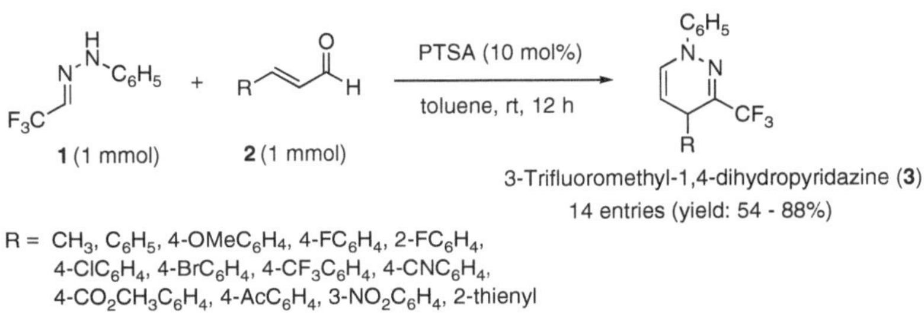

Representative entries

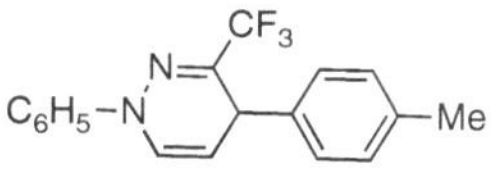

3a

1-Phenyl-4-(*p*-tolyl)-3-(trifluoromethyl)-1,4-dihydropyridazine (**3a**): Yield: 71%

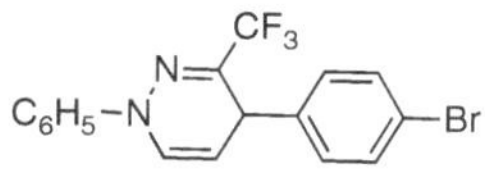

3b

4-(4-Bromophenyl)-1-phenyl-3-(trifluoromethyl)-1,4-dihydropyridazine (**3b**): Yield: 80%

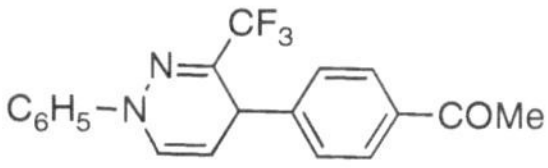

3c

1-(4-(1-Phenyl-3-(trifluoromethyl)-1,4-dihydropyridazin-4-yl)phenyl)ethanone (**3c**): Yield: 78%

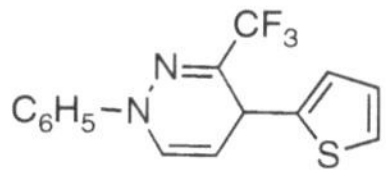

3d

1-Phenyl-4-(thiophen-2-yl)-3-(trifluoromethyl)-1,4-dihydropyridazine (**3d**): Yield: 61%

Experimental procedure

A Schlenk tube was charged with *p*-toluenesulfonic acid (10 mol%), evacuated and backfilled with nitrogen. Toluene (5.0 mL), α,β-unsaturated aldehyde (**2**; 1.0 mmol) and (*E*)-1-Phenyl-2-(2,2,2-trifluoroethylidene)hydrazine (**1**; 1.0 mmol) was successively then added. The reaction mixture was stirred at room temperature for 12 h. The resulting reaction mixture was then partitioned between ethyl acetate and water, the organic layer was washed with brine, dried over $MgSO_4$, and concentrated in *vacuo*. The residue was purified by column chromatography on silica and then recrystallized from petroleum ether (60-90 °C) to afford pure 3-trifluoromethyl-1,4-dihydropyridazine 3 with good yields (54-88%). Each of the products was characterized on the basis of detailed spectral measurements.

Characterization data of two representative compounds

1-(4-(1-Phenyl-3-(trifluoromethyl)-1,4-dihydropyridazin-4-yl)phenyl)ethanone (**3c**): IR(KBr): 3058, 2917, 1683, 1599, 1496, 1415, 1361, 1269, 1242, 1189, 1129, 1056, 836, 754, 692 cm^{-1}; ^{1}HNMR (300 MHz, $CDCl_3$): δ 7.94-7.92 (m, 2H), 7.41-7.40 (m, 4H), 7.34-7.31 (m, 2H), 7.19-7.13 (m, 1H), 6.94 (d, $J=8.1$ Hz, 1H), 5.21 (dd, $J=8.1$ & 5.4 Hz, 1H), 4.50 (d, $J=5.4$ Hz, 1H), 2.58 (s, 3H); ^{19}FNMR (282 MHz, $CDCl_3$): δ −67.9 (s); ^{13}CNMR (100 MHz, $CDCl_3$): δ 197.7, 147.6, 143.4, 136.2, 131.8 (q, $J=33.3$ Hz), 129.3, 129.0, 127.6, 125.1, 124.1, 121.3 (q, $J=274.4$ Hz), 116.3, 103.8, 36.4, 26.4; MS (EI) *m/z* (%relative intensity): 344 (15) [M^+], 225 (100), 77 (25); HRMS Calcd. for $C_{19}H_{15}F_3N_2O$: 344.1136; Found: 344.1134.

1-Phenyl-4-(thiophen-2-yl)-3-(trifluoromethyl)-1,4-dihydropyridazine (**3d**): IR(KBr): 2954, 2917, 2843, 1598, 1495, 1268, 1241, 1190, 1128, 1054, 754, 692 cm^{-1}; 1HNMR (300 MHz, $CDCl_3$): δ 7.40-7.38 (m, 4H), 7.21-7.19 (m, 1H), 7.17-7.12 (m, 1H), 6.96-6.91 (m, 3H), 5.31 (dd, $J=7.6$ & 5.6 Hz, 1H), 4.72 (d, $J=5.6$ Hz, 1H); 19FNMR (282 MHz, $CDCl_3$): δ −67.2 (s); ^{13}CNMR (100 MHz, $CDCl_3$): δ 145.8, 143.6, 131.6 (q, $J=33.7$ Hz), 129.3, 127.1, 125.6, 125.3, 124.9, 124.2, 121.4 (q, $J=273.8$ Hz), 116.6, 103.6, 30.9; MS (EI) *m/z* (%relative intensity): 308 (83) [M^+], 239 (39), 225 (100), 212 (15), 77 (75); HRMS Calcd. For $C_{15}H_{11}F_3N_2S$: 308.0595; Found: 308.0594.

Reference

Xie, H., Zhu, J., Chen, Z., Li, S., and Wu, Y. (2012). Synthesis of 3-trifluoromethyl-1,4-dihydropyridazines by the PTSA-catalyzed reaction of α,β-unsaturated aldehydes with (*E*)-1-phenyl-2-(2,2,2-trifluoroethylidene). *Synlett*, **23**, 935–937.

C-N bond_34

Type of reaction: C-N bond formation
Reaction conditions: Toluene, room temperature
Synthetic strategy: Base-catalyzed cascade reaction
Catalyst: 1,8-Diazabicycl[5.4.0]undec-7-ene (DBU)
Keywords: Sulfonamides, vinyl ketones, toluene, DBU, room temperature, Michael addition, cascade reaction, cyclization, functionalized 3-aroyl-4-hydroxy-4-arylpiperidines, stereoselectivity

General reaction scheme

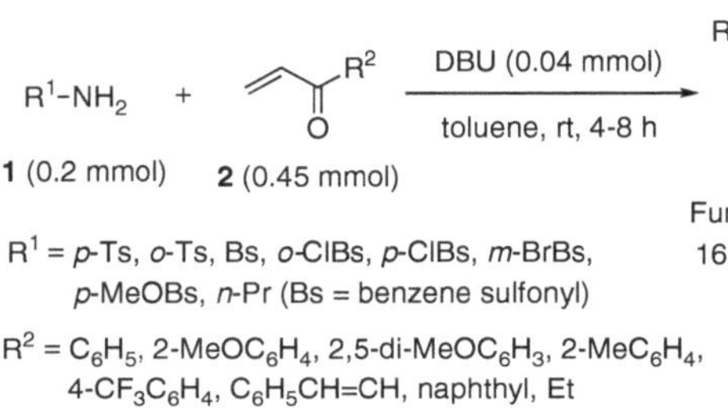

Representative entries

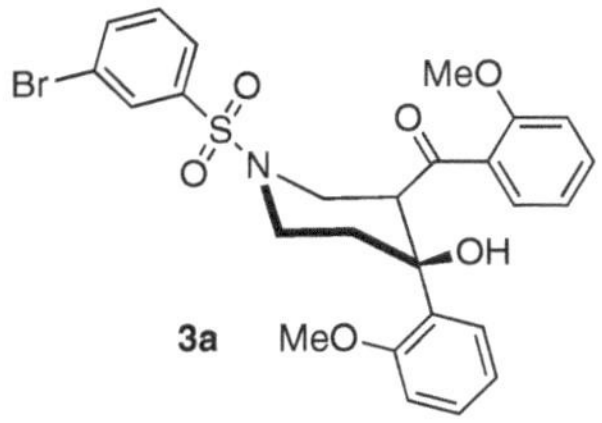

3a

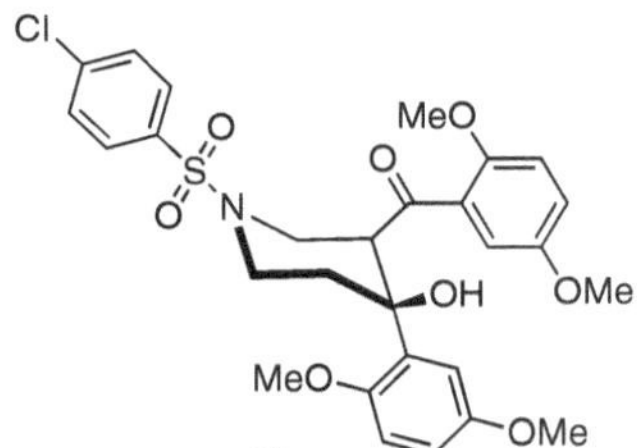

3b

(3*R*,4*S*)-1-((3-Bromophenyl)sulfonyl)-4-hydroxy-4-(2-methoxyphenyl)piperidin-3-yl)(2-methoxyphenyl) methanone (**3a**): White solid, mp 162-163 °C, yield: 95%

((3*R*,4*S*)-1-((4-Chlorophenyl)sulfonyl)-4-(2,5-dimethoxyphenyl)-4-hydroxypiperidin-3-yl)(2,5-dimethoxyphenyl)methanone (**3b**): White solid, mp 178-179 °C, yield: 99%

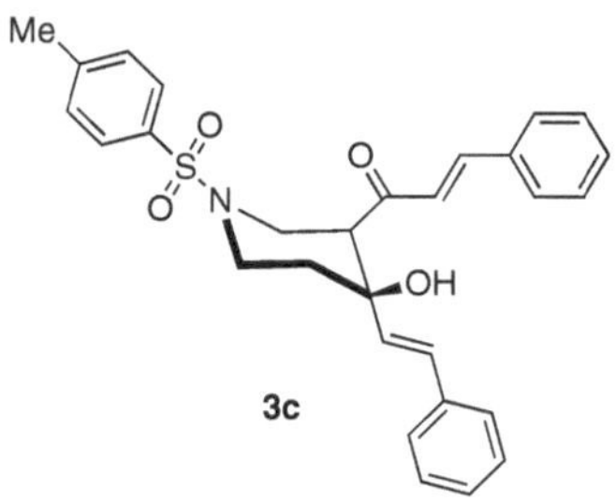

3c

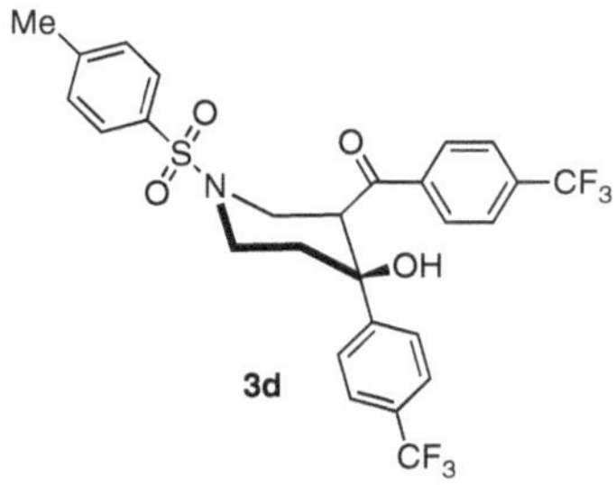

3d

(*E*)-1-((3*R*,4*R*)-4-hydroxy-4-((*E*)-styryl)-1-tosylpiperidin-3-yl)-3-phenylprop-2-en-1-one (**3c**): White solid, mp 159-160 °C, yield: 92%

((3*R*,4*S*)-4-Hydroxy-1-tosyl-4-(4-(trifluoromethyl)phenyl)piperidin-3-yl)(4-(trifluoromethyl)phenyl)methanone (**3d**): White solid, mp 172-173 °C, yield: 52%

Experimental procedure

To a mixture of amine (**1**; 0.2 mmol) and vinyl ketone (**2**; 0.45 mmol) in toluene (1.0 mL) was added DBU (0.04 mmol) and stirred at room temperature for 4-8 h. On completion of the reaction (as monitored by TLC), the organic solvent was evaporated under reduced pressure, and the residue was purified by flash chromatography (petroleum ether/ethyl acetate = 5:1) to afford the functionalized piperidine product **3** with good to excellent yield. The products were characterized with detailed spectral studies.

Characterization data of two representative compounds

(3*R*,4*S*)-1-((3-Bromophenyl)sulfonyl)-4-hydroxy-4-(2-methoxyphenyl)piperidin-3-yl)(2-methoxyphenyl) methanone (**3a**): White solid, mp 162-163 °C; IR ($CDCl_3$): 3432, 2934, 1655, 1597, 1462, 1245, 1166, 1022, 757, 580 cm^{-1}; ^{1}HNMR (300 MHz, $CDCl_3$): δ 7.96 (t, *J* = 1.6 Hz, 1 H), 7.74 (d, *J* = 7.9 Hz, 2 H), 7.58 (dd, *J* = 7.8, 1.6 Hz, 1 H), 7.51–7.31 (m, 2 H), 7.15-7.04 (m, 1 H), 6.91 (t, *J* = 8.4 Hz, 2 H), 6.80-6.62 (m, 2 H), 6.51 (d, *J* = 8.1 Hz, 1 H), 5.08 (dd, *J* = 11.4, 3.8 Hz, 1 H), 4.50 (d, *J* = 2.7 Hz, 1 H), 4.15 (dd, *J* = 11.0, 2.4 Hz, 1 H), 4.04 (s, 3 H), 3.82–3.66 (m, 1 H), 3.42 (s, 3 H), 3.00 (td, *J* = 11.1, 4.8 Hz, 2 H), 2.70-2.51 (m, 1 H), 1.60 (d, *J* = 13.6 Hz, 1 H); ^{13}C NMR (75 MHz, $CDCl_3$): δ 207.4, 158.6, 155.5, 138.7, 135.8, 133.9, 132.2, 130.7, 130.3, 129.2, 128.9, 128.7, 127.5, 126.0, 123.3, 120.7, 120.5, 111.0, 110.7, 72.5, 55.8, 54.6, 51.8, 43.9, 42.3, 34.6; HRMS (ESI): *m/z* calcd. for $C_{26}H_{26}BrNO_6S + H^+$: 560.0737; Found: 560.0738.

(*E*)-1-((3*R*,4*R*)-4-hydroxy-4-((*E*)-styryl)-1-tosylpiperidin-3-yl)-3-phenylprop-2-en-1-one (**3c**): White solid, mp 159-160 °C; IR (CDCl3): 3451, 3029, 2926, 2253, 1598, 1339, 1160, 1091, 919, 733 cm^{-1}; ^{1}HNMR (300 MHz, $CDCl_3$): δ 7.66 (dd, *J* = 12.1, 3.9 Hz, 3 H), 7.60-7.52 (m, 2 H), 7.46-7.36 (m, 3 H), 7.24 (m, 7 H), 6.69 (dd, *J* = 30.5, 16.0 Hz, 2 H), 6.15 (d, *J* = 15.9 Hz, 1 H), 4.54 (d, *J* = 2.4 Hz, 1 H), 3.86 (dd, *J* = 11.5, 2.6 Hz, 1 H), 3.73 (dd, *J* = 9.4, 2.0 Hz, 1 H), 3.48 (dd, *J* = 11.6, 3.9 Hz, 1 H), 2.88-2.71 (m, 2 H), 2.44 (s, 3 H), 2.00-1.84 (m, 1 H), 1.75 (d, *J* = 15.2 Hz, 1 H); ^{13}C NMR (75 MHz, $CDCl_3$): δ 202.3, 146.1, 143.9, 136.3, 133.8, 133.6, 133.0, 131.6, 129.9, 129.5, 129.1, 128.9, 128.5, 127.7, 127.6, 126.6, 125.4, 71.0, 52.0, 44.4, 41.9, 36.6, 21.6; HRMS (ESI): *m/z* calcd for $C_{29}H_{29}NO_4S + Na^+$: 510.1735; Found: 510.1713.

Reference

Liu, X., Chen, Q., Li, W., Liang, Y., and Wang, R. (2012). Synthesis of 3-aroyl-4-hydroxy-4-arylpiperidine derivatives by DBU-catalyzed reactions of amines with vinyl ketones. *Synlett*, **23**, 1691–1695.

C-N bond_35

Type of reaction: C-N bond formation
Reaction conditions: Ethanol-dichloroethane, room temperature
Synthetic strategy: One-step hydroamination
Catalyst: $Au(I)Cl(PPh_3)/AgNTf_2$
Keywords: Alkylnyl carbamates, $AuCl(PPh_3)$, $AgNTf_2$, ethanol-dichloroethane, room temperature, hydroamination, 1,2-dihydroisoquinolines, 1-alkylydeneyl-1,2,3,4-tetrahydroisoquinolines

General reaction schemes

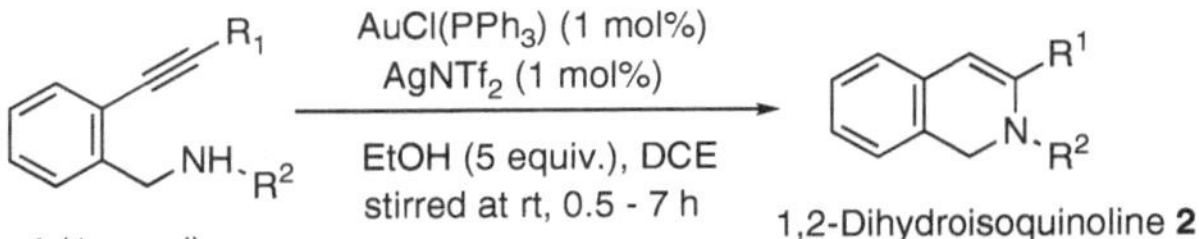

1 (1 mmol)

1,2-Dihydroisoquinoline **2**
8 entries (yield: 28 - 87%)

R^1 = *n*-Pr, C_6H_5, 3,5-di-$CH_3C_6H_3$, 4-ClC_6H_4, 4-$OCH_3C_6H_4$, 4-$CH_2OHC_6H_4$
R^2 = Cbz (benzyloxycarbonyl), Ms, PMP, Boc

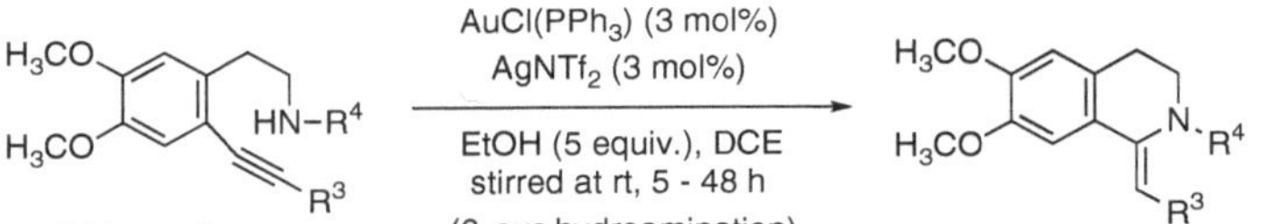

3 (1 mmol)

1-Alkylydenyl-1,2,3,4-tetra-hydroisoquinoline **4**
6 entries (yield: 58 - 87%)

R^1 = *n*-Pr, C_6H_5, 3-ClC_6H_4, 4-^{t}BuC_6H_4
R^2 = Boc, CO_2CH_3

Representative entries

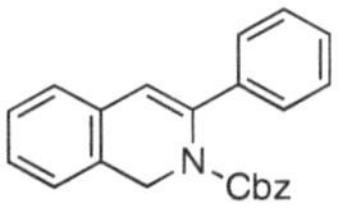

2a

Benzyl 3-phenylisoquinoline-2(1*H*)-carboxylate (**2a**): Colorless crystals, mp 105-106 °C; reaction time: 4 h; yield: 83%

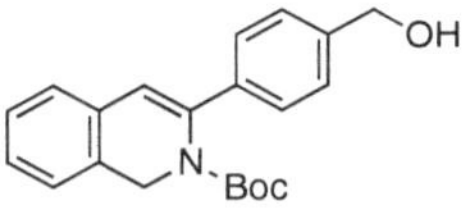

2b

tert-Butyl 3-(4-(hydroxymethyl)phenyl) isoquinoline-2(1*H*)-carboxylate (**2b**): White powder, mp 115-116 °C; reaction time: 2 h; yield: 79%

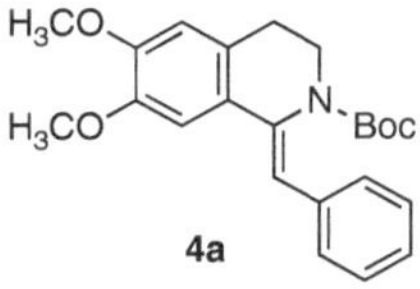

4a

(*Z*)-*tert*-Butyl 1-benzylidene-6,7-dimethoxy-3,4-dihydroisoquinoline-2(1H)-carboxylate (**4a**): Colorless needles, mp 164-165 °C; reaction time: 48 h; yield: 58%

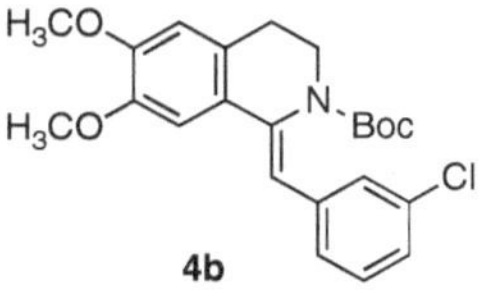

4b

(*Z*)-*tert*-Butyl 1-(3-chlorobenzylidene)-6,7-dimethoxy-3,4-dihydroisoquinoline-2(1*H*)-carboxylate (**4b**): Colorless crystals, mp 146-147 °C; reaction time: 6 h; yield: 86%

Experimental procedure

To a solution of **1** (1 mmol) in dichloroethane (2 mL) were added EtOH (293 μL, 5.02 mmol) and a suspension of AuCl(PPh_3) (1 mmol) and $AgNTf_2$ (1 mmol) in dichloroethane (1 mL), and the reaction mixture was stirred at room temperature for stipulated time (0.5-7 h). On completion of the reaction, saturated aqueous $NaHCO_3$ was added and the product was extracted with $CHCl_3$ (3 × 10 mL). The combined organic extracts were washed with brine, dried over sodium sulfate, filtered, and concentrated. The residue was purified by silica gel column chromatography (hexane-EtOAc) to afford 1,2-dihydroisoquinoline **2** with 28-87% of yield. In a similar fashion, 1-alkylydeneyl-1,2,3,4-tetrahydroisoquinoline **4** were prepared starting from substrate **3** with requisite amount of catalyst as mentioned in the reaction scheme. All the products were characterized with detailed spectral studies and elemental analyses.

Characterization data of two representative compounds

tert-Butyl 3-(4-(hydroxymethyl)phenyl) isoquinoline-2(1*H*)-carboxylate (**2b**): White powder, mp 115-116 °C; IR ($CHCl_3$): 3606, 1693 cm^{-1}; ^1H NMR (500 MHz, $CDCl_3$, TMS): δ 7.46 (d, $J=8.2$ Hz, 2H), 7.34 (d, $J=8.2$ Hz, 2H), 7.15-7.26 (m, 4H), 6.42 (s, 1H), 4.85 (s, 2H), 4.68 (s, 2H), 2.58 (s, 1H), 1.06 (s, 9H); ^{13}C NMR (126 MHz, $CDCl_3$): δ 152.9, 140.6, 140.2, 137.9, 132.5, 131.8, 127.5, 127.2, 126.6,126.2, 125.0, 124.9, 115.2, 81.0, 64.7, 47.4, 27.6; MS (FAB): *m/z* 337 [M^+]. Anal. Calcd for $C_{21}H_{23}NO_3$: C 74.75, H 6.87, N 4.15; Found: C 74.47, H 6.87, N 3.88.

(*Z*)-tert-Butyl 1-(3-chlorobenzylidene)-6,7-dimethoxy-3,4-dihydroisoquinoline-2(1*H*)-carboxylate (**4b**): Colorless crystals, mp 146-147 °C; IR ($CHCl_3$): 1685 cm^{-1}; ^1H NMR (500 MHz, $CDCl_3$, TMS): δ 7.46 (s, 1H), 7.36 (d, $J=7.8$ Hz, 1H), 7.24 (dd, $J_1=J_2=7.8$ Hz, 1H), 7.18 (s, 1H), 7.17 (d, $J=7.8$ Hz, 1H), 6.68 (s, 1H), 6.61 (s, 1H), 4.59 (br, 1H), 3.96 (s, 3H), 3.89 (s, 3H), 3.26 (br, 1H), 3.12 (br, 1H), 2.66 (br, 1H), 1.07 (s, 9H); ^{13}C NMR (126 MHz, $CDCl_3$): δ 153.0, 149.5, 147.7, 139.1, 135.3, 134.2, 129.5, 128.3, 128.2, 126.7, 126.3, 125.0, 116.0, 111.7, 106.4, 80.6, 56.2, 55.9, 43.1, 28.3, 27.8; MS (FAB): *m/z* 415 [M^+]. Anal. Calcd for $C_{23}H_{26}ClNO_4$: C 66.42, H 6.30, N 3.37; Found: C 66.38, H 6.32, N 3.35.

Reference

Enomoto, T., Obika, S., Yasui, Y., and Takemoto, Y. (2008). Gold(I)-catalyzed hydroamination as a general approach toward the synthesis of substituted hydroisoquinolines: remarakable acceleration by ethanol. *Synlett*, 1647–1650.

C-N bond_36

Type of reaction: C-N bond formation
Reaction conditions: Dichloromethane, room temperature
Synthetic strategy: One-pot multicomponent reaction
Catalyst: No catalyst
Keywords: Benzoyl thiocyanate, oxalyl chloride, primary amines, dichloromethane, room temperature, multicomponent reaction, 2-thioparabanic acids

General reaction scheme

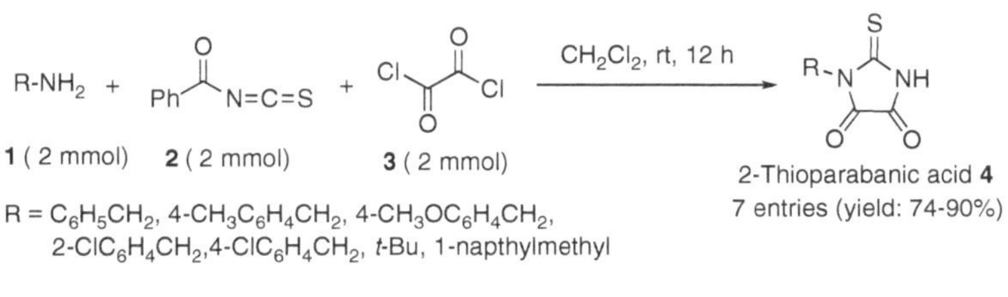

Plausible mechanism

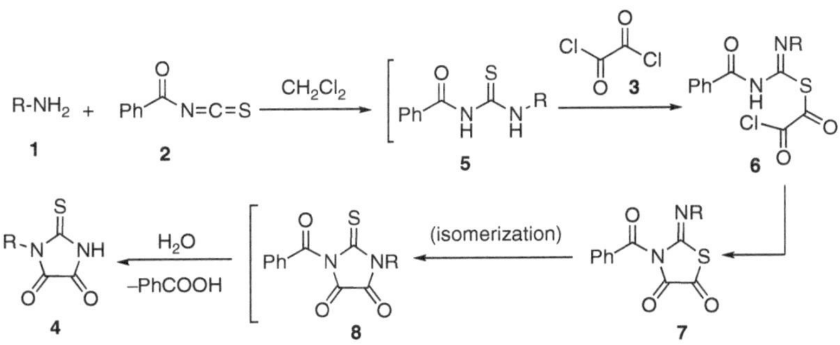

Representative entries

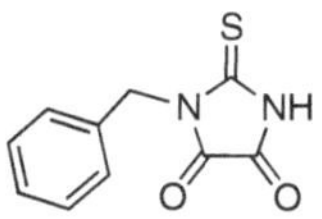

4a

1-Benzyl-2-thioxoimidazolidine-4,5-dione (**4a**): Yellow powder, mp 151-153 °C; yield: 78%

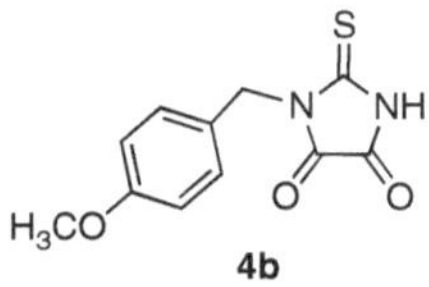

4b

1-(4-Methoxybenzyl)-2-thioxoimidazolidine-4,5-dione (**4b**): Yield: 76%

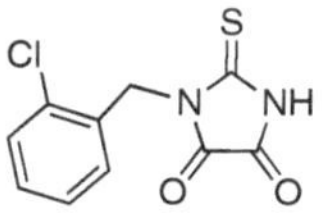

4c

1-(2-Chlorobenzyl)-2-thioxoimidazolidine-4,5-dione (**4c**): Yield: 80%

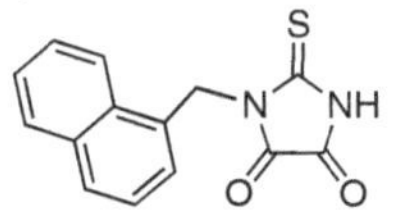

4d

1-(Naphthalen-1-ylmethyl)-2-thioxoimidazolidine-4,5-dione (**4d**): Yield: 90%

Experimental procedure

To a stirred solution of primary amine (**1**; 2 mmol) and benzoyl isothiocyanate (**2**; 0.32 g, 2 mmol) in dichloromethane (10 mL) was added oxalyl chloride (**3**; 0.25 g, 2 mmol) at room temperature, and stirring was continued for about 12 h at rt. On completion of the reaction, the solvent was removed under reduced pressure and the viscous residue was purified by flash column chromatography on SiO_2 (Merck 230-400 mesh) using hexane-EtOAc as eluent to give 2-thioparabanic acid derivatives **4**, characterized on the basis of spectral and elemental analyses.

Characterization data of one representative compound

1-Benzyl-2-thioxoimidazolidine-4,5-dione (**4a**): Yellow powder, mp 151-153 °C; IR (KBr): 3220, 1764, 1735, 1666, 1441, 1340 cm^{-1}; ^{1}H NMR (500 MHz, $CDCl_3$): δ 5.10 (s, 2H, NCH_2), 7.28 (d, $^3J=7.1$ Hz, 1H, C*H*), 7.32 (t, $^3J=7.6$ Hz, 2H, C*H*), 7.41 (d, $^3J=7.3$ Hz, 2H, C*H*), 11.71 (br s, 1H, N*H*); ^{13}C NMR (125 MHz, $CDCl_3$): δ 45.0 (NCH_2), 128.6 (*C*H), 129.0 (*C*H), 129.2 (*C*H), 136.2 (*C*H), 156.8 (C=O), 157.3 (C=O), 182.7 (C=S); MS: *m/z* (%rel) 222 (4) [M^+ + 2], 221 (10) [M^+ + 1], 220 (64) [M^+], 91 (26), 77 (60), 59 (20), 45 (60). Anal. Calcd (%) for $C_{10}H_8N_2O_2S$ (220.24): C 54.53, H 3.66, N 12.72; Found: C, 54.41, H 3.70, N 12.65.

Reference

Yavari, I., Souri, S., and Sirouspour, M. (2008). Efficient one-pot synthesis of unsymmetrical 2-thioparabanic acids from oxalyl chloride, benzoyl isothiocyanate, and primary amines. *Synlett*, 1633–1634.

C-N bond_37

Type of reaction: C-N bond formation
Reaction conditions: Solvent-free, room temperature
Synthetic strategy: Ionic liquid catalyzed oxidative ring-closure
Catalyst: Prolinium nitrate ([Pro]NO_3) IL

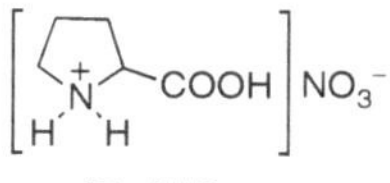

[Pro]NO_3

Keywords: *o*-Phenylenediamines, aromatic aldehydes, prolinium nitrate IL, solvent-free, room temperature, oxidative-cyclization, 2-substituted benzimidazoles, chemoseletivity

General reaction scheme

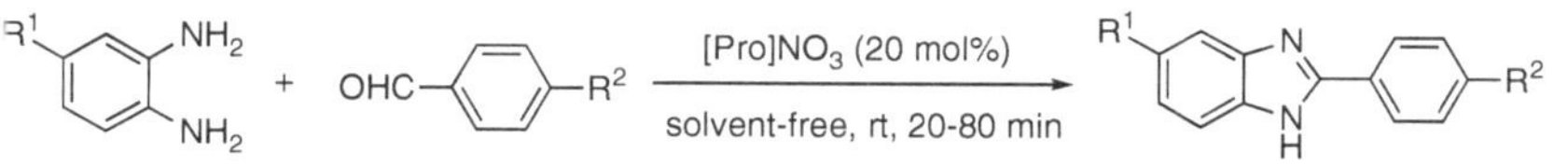

1 (3 mmol) **2** (3 mmol)

R^1 = H, NO_2, COOH
R^2 = H, Me, OMe, Br, Cl, F, NO_2, CN

2-Substituted benzimidazole **3**
17 entries (yield: 76-95%)

Representative entries

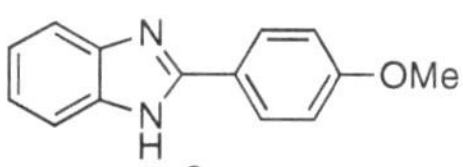

3a

2-(4-Methoxyphenyl)-1*H*-benzo[*d*]imidazole (**3a**): Light yellow solid, mp 217-220 °C; Reaction time: 45 min; yield: 85%

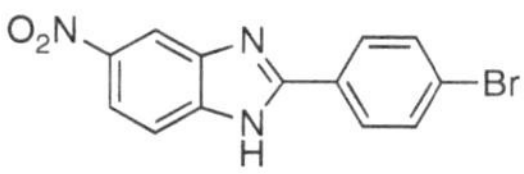

3b

2-(4-Bromophenyl)-5-nitro-1*H*-benzo[*d*]imidazole (**3b**): Yellow solid, mp 161-164 °C; Reaction time: 60 min; yield: 92%

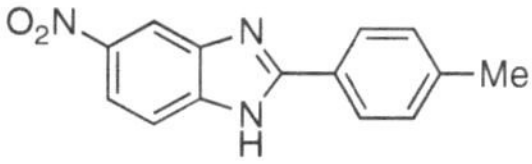

3c

5-Nitro-2-(*p*-tolyl)-1*H*-benzo[*d*]imidazole (**3c**): Light yellow solid, mp 211-214 °C; Reaction time: 80 min; yield: 76%

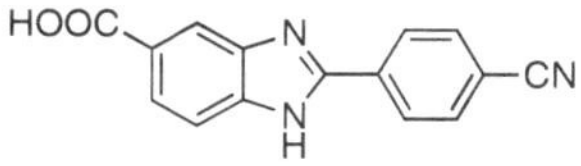

3d

2-(4-Cyanophenyl)-1*H*-benzo[*d*]imidazole-5-carboxylic acid (**3a**): Pale yellow solid, mp 183-185 °C; Reaction time: 30 min; yield: 88%

Experimental procedure

To a mixture of *o*-phenylenediamine (**1**; 3 mmol) and aromatic aldehyde (**2**; 3 mmol) was added prolinium nitrate (0.6 mmol, 20 mol%). The reaction mixture was stirred at room termperature (30 °C) for stipulated time. On completion of reaction (as monitored by TLC), water (10 mL) was added and the mixture was scratched and stirred for at least 30 min. The aqueous layer was filtered and the residues were purified by crystallization from aqueous ethanol to afford pure 2-substituted benzimidazole **3** with good yields (76-95%). Aqueous layer containing prolinium nitrate was concentrated in *vacuo* to recycle and reuse the IL. All the products were characterized with the help of spectral and analytical analyses.

Characterization data of two representative compounds

2-(4-Methoxyphenyl)-1*H*-benzo[*d*]imidazole (**3a**): Light yellow solid, mp 217-220 °C; IR (KBr): 1612(C=N), 3420 (NH) cm^{-1}; ^1H NMR (300 MHz, DMSO-d_6): δ 12.73 (s, 1H), 8.11 (d, *J*=9.0 Hz, 2H), 7.54 (s, 2H), 7.16 (m, 2H), 7.11 (d, *J*=9.0 Hz, 2H), 3.82 (s, 3H).

5-Nitro-2-(*p*-tolyl)-1*H*-benzo[d]imidazole (**3c**): Light yellow solid, mp 211-214 °C; IR (KBr): 1637 (C=N), 3435 (NH) cm^{-1}; ^1H NMR (300 MHz, DMSO-d_6): δ 13.53 (s, 1H), 7.39 (d, *J*=9.0 Hz, 2H), 6.54 (d, *J*=12.0 Hz, 1H), 6.04 (s, 2H), 5.06 (s, 2H), 2.39 (s, 3H); ^{13}C NMR (75 MHz, DMSO-d_6): δ 21.08, 108.03, 111.44, 115.61, 126.98, 129.78, 133.96, 136.79, 143.36, 153.14; MS (EI): *m/z* at 253 (M^+), 153, 107, 80. Anal. Calcd. for $C_{14}H_{11}N_3O_2$: C 66.39, H 4.38, N 16.59; Found: C 66.28, H 4.09, N 16.34.

Reference

Rostamizadeh, S., Aryan, R., Ghaieni, H. R., and Amani, A. M. (2009). Solvent-free chemoselective synthesis of some novel substituted 2-arybenzimidazoles using amino acid-based prolinium nitrate ionic liquid as catalyst. *J. Heterocyclic Chem.*, **46**, 74–78.

C-N bond_38

Type of reaction: C-N bond formation
Reaction conditions: MeOH, room temperature
Synthetic strategy: Magnetically separable nano-material-catalyzed oxidative cyclization
Catalyst: Manganese ferrite ($MnFe_2O_4$) nano-particles (heterogeneous and magnetically separable)
Keywords: *o*-Phenylenediamines, aldehydes, $MnFe_2O_4$-NP, methanol, room temperature, oxidative-cyclization, chemoseletivity, 2-substituted benzimidazoles, quinoxalines

General reaction schemes

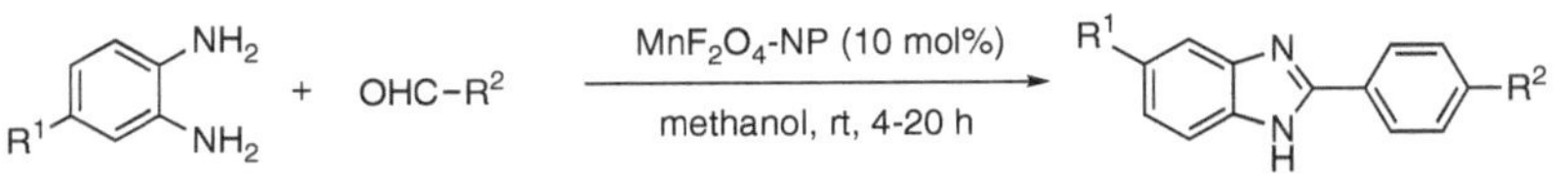

1 (1 mmol) **2** (1 mmol) 2-Substituted benzimidazole **3**

16 entries (yield: 62-94%)

R^1 = H, Cl; R_2 = C_6H_5, 4-MeC_6H_4, 4-OMeC_6H_4, 4-FC_6H_4, 4-ClC_6H_4, 2-ClC_6H_4, 4-BrC_6H_4, 4-CNC_6H_4, 4-OHC_6H_4, 2-furyl, 3,4,5-tri-MeO-C_6H_2, 9-anthracenyl, *n*-Pr

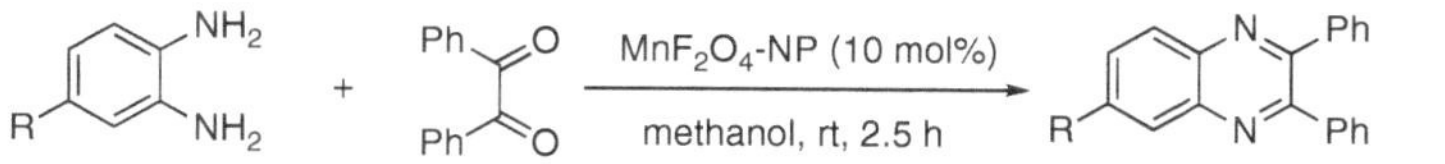

4 (1 mmol) **5** (1 mmol)

R = H, Cl, Me

Quinoxaline **6**
5 entries (yield: 70-91%)

Plausible mechanism

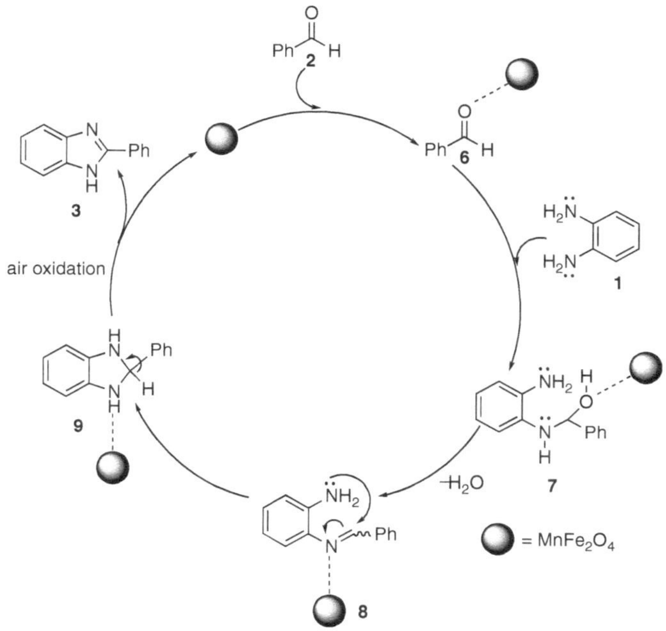

Representative entries

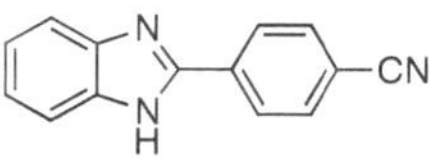

3a

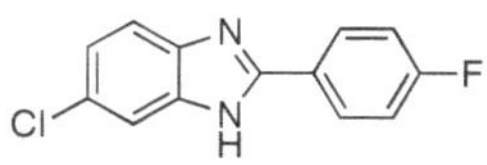

3b

4-(1*H*-Benzo[*d*]imidazol-2-yl)benzonitrile (**3a**): Creamy white solid, mp 235-236 °C; reaction time: 20 h; yield: 67%

6-Chloro-2-(4-fluorophenyl)-1*H*-benzo[*d*]imidazole (**3b**): Off-white solid, mp 204-206 °C; reaction time: 15 min; yield: 75%

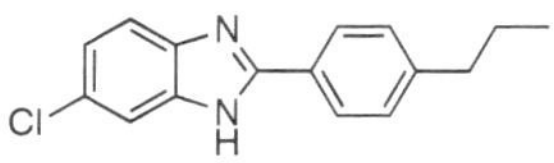

3c

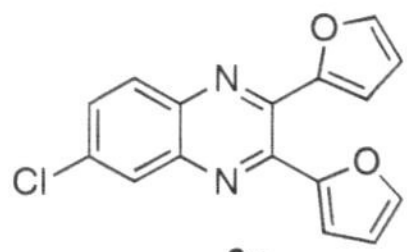

6a

6-Chloro-2-(4-propylphenyl)-1*H*-benzo[*d*]imidazole (**3c**): White amorphous powder; reaction time: 15 h; yield: 65%

6-Chloro-2,3-di(furan-2-yl)quinoxaline (**6a**): Dark brownish solid, mp 120-125 °C; reaction time: 2.5 h; yield: 81%

Experimental procedure

Synthesis of 2-substituted benzimidazole ***3***: To a solution of aldehyde (**2**; 1 mmol) and *o*-phenylenediamine (**1**; 1 mmol) in methanol (3-5 mL), MnFe2O4 (23 mg; 10 mol%) was added and stirred vigorously for stipulated time at room temperature under open atmosphere with the aid of an external bar magnet. The progress of the reaction was monitored by TLC. After the completion of reaction, the solvent was evaporated and the reaction mixture was extracted with dichloromethane/ethyl acetate (3 × 10 mL) for 7-10 h and allowed the mixture to settle the catalyst which was then separated by the simple application of a strong external magnet. The organic extract was dried in sodium sulphate and the product was finally obtained after chromatographic purification. The product was characterized by elemental analyses and spectral studies including FT-IR, ^{1}H-NMR, and ^{13}C-NMR data. The recovered catalyst was further washed with ethanol followed by acetone, dried under vacuum at 100 °C and reused.

Synthesis of quinoxaline derivative ***6***: In a typical procedure, *o*-phenylenediamine (**4**; 1 mmol), benzil (**5**; 1 mmol), and the catalyst $MnFe_2O_4$-NP (10 mol%) were placed in a 10 mL glass tube previously charged with a external bar magnet containing 3 mL of aqueous ethanol. The reaction mixture was stirred for 2.5 h vigorously. The progress of the reaction was monitored by TLC. The mixture was allowed to settle the catalyst which was then removed by the application of a simple magnetic environment; the resulting reaction mixture was extracted with dichloromethane/ethyl acetate (3 × 10 mL). The extract was dried in sodium sulphate and the product was finally obtained after chromatographic purification. The product was characterized by elemental analyses and spectral studies including FT-IR, ^{1}H-NMR, ^{13}C-NMR data. The recovered catalyst was further washed with ethanol followed by acetone, dried under vacuum at 100 °C and reused.

Characterization data of two representative compounds

6-Chloro-2-propyl-1*H*-benzo[*d*]imidazole (**3c**): White amorphous powder; IR (KBr): 3003, 2922, 1612, 1544, 1528, 1450, 1035, 746, 513 cm^{-1}; ^{1}H-NMR (400 MHz, DMSO-d_6): δ 12.39 (br, s, NH), 7.52-7.47 (2H, m, Ar-H), 7.12 (1H, d, J = 8 Hz, Ar-H), 2.78 (2H, t, J = 7.6, 7.2 Hz, –CH_2–), 1.82-1.79 (2H, q, *J* = 7.2, 5.2 Hz, –CH_2–), 1.17 (3H, t, *J* = 7.2Hz, –CH_3); ^{13}C-NMR (100 MHz, DMSO-d_6): δ 170.72, 157.03, 125.85, 121.62, 117.83, 112.39, 30.85, 21.24, 14.44. Anal. Calcd. for $C_{10}H_{11}ClN_2$: C 61.70, H 5.70, N 14.39; Found: C 61.73, H 5.71, N 14.36.

6-Chloro-2,3-di(furan-2-yl)quinoxaline (**6a**): Dark brownish solid, mp 120-125 °C; IR (KBr): 3119, 2959, 1628, 1599, 1507, 1072, 750, 594 cm^{-1}; ^{1}H-NMR (300 MHz, $CDCl_3$): δ 8.24-7.67 (6H, m, Ar-H), 6.71 (3H, br s, Ar-H); ^{13}C-NMR (75 MHz, $CDCl_3$): δ 151.31, 150.46, 150.38, 145.69, 145.51, 143.12, 142.55, 140.58, 138.90, 135.45, 131.62, 130.79, 127.68, 113.98, 113.68, 112.64. Anal. Calcd. for $C_{16}H_9ClN_2O_2$: C 64.77, H 3.06, N 9.44; Found: C 64.72, H 3.09, N 9.43.

Reference

Brahmachari, G., Laskar, S., and Barik, P. (2013). Magnetically separable $MnFe_2O_4$ nano-material: An efficient and reusable heterogeneous catalyst for the synthesis of 2-substituted benzimidazoles and the extended synthesis of quinoxalines at room temperature under aerobic conditions. *RSC Adv.*, **3**, 14245–14253.

Further information

There are few more earlier reports on the synthesis of benzimidazoles and quinoxalines using the catalysts such as $In(OTf)_3$ [1], ammonium bromide [2], silica bonded *S*-sulfonic acid [3], TMSCl [4], 1-methyl-3-pentylimidazolium tetrafluoroborate ([pmim]BF_4) [5], Zn^{2+}-K10-clay (clayzic) [6], and $Co(OH)_2$/CoO [7] in the presence or absence of solvents at room temperature.

[1] Trivedi, R., De, S. K., and Gibbs, R. A. (2006). A convenient one-pot synthesis of 2-substituted benzimidazoles. *J. Mol. Catal. A: Chem.*, **245**, 8–11.
[2] Raju, B. C., Theja, N. D., and Kumar, J. A. (2009). Efficient and inexpensive synthesis of benzimidazoles and quinoxalines. *Syn. Commun.*, **39**, 175–188.
[3] Niknam, K., Saberi, D., and Mohagheghnejad, M. (2009). Silica bonded *S*-sulfonic acid: a recyclable catalyst for the synthesis of quinoxalines at room temperature. *Molecules*, **14**, 1915–1926.
[4] Wan, J.-P., Gan, S.-F., Wu, J.-M., and Pan, Y. (2009). Water mediated chemoselective synthesis of 1,2-disubstituted benzimidazoles using *o*-phenylenediamine and the extended synthesis of quinoxalines. *Green Chem.*, **11**, 1633–1637.
[5] Saha, D., Saha, A., and Ranu, B. C. (2009). Remarkable influence of substituent in ionic. liquid in control of reaction: simple, efficient and hazardous organic solvent free procedure for the synthesis of 2-aryl benzimidazoles promoted by ionic liquid, [pmim]BF_4. *Green Chem.*, **11**, 733–737.
[6] Dhakshinamoorthy, A., Kanagaraj, K., and Pitchumani, K. (2011). Zn^{2+}-K10-clay (clayzic) as an efficient water-tolerant, solid acid catalyst for the synthesis of benzimidazoles and quinoxalines at room temperature. *Tetrahedron Lett.*, **52**, 69–73.
[7] Chari, M. A., Shobha, D., and Sasaki, T. (2011). Room temperature synthesis of benzimidazole derivatives using reusable cobalt hydroxide (II) and cobalt oxide (II) as efficient solid catalysts. *Tetrahedron Lett.*, **52**, 5575–5580.

C-N bond_39

Type of reaction: C-N bond formation
Reaction conditions: Ionic liquid (IL), room temperature
Synthetic strategy: One-pot multicomponent reaction
Catalyst: Task specific IL, [bmim]BF_4
Keywords: Phencyl bromides, amines, diethyl/dimethyl acetylenedicarboxylate, [bmim]BF_4, room temperature, one-pot multicomponent reaction, functionalized pyrroles

General reaction scheme

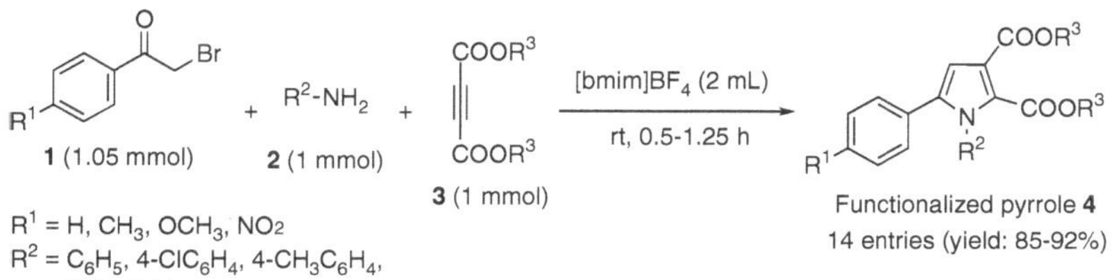

R^1 = H, CH_3, OCH_3, NO_2
R^2 = C_6H_5, 4-ClC_6H_4, 4-$CH_3C_6H_4$, CH_3, C_2H_5
R_3 = CH_3, C_2H_5

Representative entries

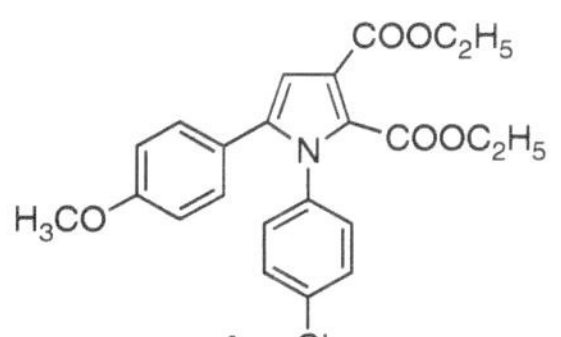

4a

Diethyl 1-(4-chlorophenyl)-5-(4-methoxyphenyl)-1*H*-pyrrole-2,3-dicarboxylate (**4a**): Reaction time: 75 min; yield: 92%

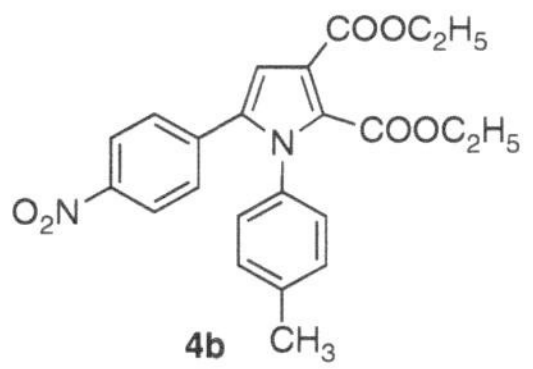

4b

Diethyl 5-(4-nitrophenyl)-1-(*p*-tolyl)-1*H*-pyrrole-2,3-dicarboxylate (**4b**): Reaction time: 75 min; yield: 86%

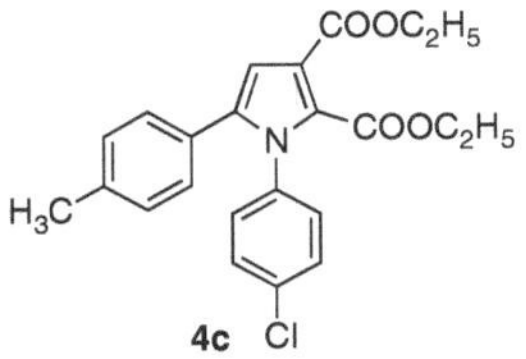

4c

Diethyl 1-(4-chlorophenyl)-5-(*p*-tolyl)-1*H*-pyrrole-2,3-dicarboxylate (**4c**): Reaction time: 65 min; yield: 92%

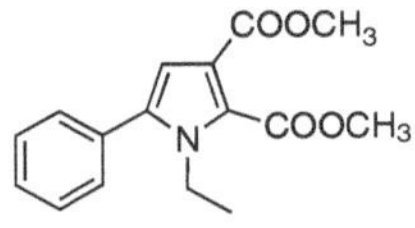

4d

Dimethyl 1-ethyl-5-phenyl-1*H*-pyrrole-2,3-dicarboxylate (**4d**): Reaction time: 55 min; yield: 90%

Experimental procedure

A mixture of phenacyl bromide (**1**; 1.05 mmol) and amine (**2**;1 mmol) in [bmim]BF_4 (2 mL) was added diethyl/dimethyl acetylenedicarboxylate (**3**; 1 mmol) drop-wise and stirred at room temperature for 0.5-1 h. After completion of the reaction, as indicated by TLC, the reaction mixture was extracted with ethyl acetate (3×10 mL). The combined organic extracts were washed with saturated brine solution, dried over anhydrous sodium sulfate, and concentrated under vacuum; the resulting crude product was purified by means of column chromatography (silica gel; 60–120 mesh) using 5-10% ethylacetate/*n*-hexane as eluent to obtain pure product of functionalized pyrroles **4**. The residual ionic liquid was reused after drying under vacuum. All the products were characterized by elemental analyses and spectral studies including ^{1}H-NMR, ^{13}C-NMR and mass.

Characterization data of one representative compound

Diethyl 1-(4-chlorophenyl)-5-(4-methoxyphenyl)-1*H*-pyrrole-2,3-dicarboxylate (**4a**): ^{1}H-NMR (400 MHz, $CDCl_3$): δ 1.25 (t, *J*=8 Hz, 3H), 1.31 (t, *J*=8, 3H), 3.75 (s, 3H), 4.24 (q, *J*=8 Hz, 2H), 4.32 (q, *J*=8 Hz, 2H), 6.88-6.90 (d, *J*=6 Hz, 2H), 7.16 (s, 1H), 7.22-7.37 (m, 6H); ^{13}C NMR (50 MHz, $CDCl_3$): δ 13.5, 13.2, 57.1, 61.5, 61.2, 112.3, 112.9, 116.7, 118.4, 121.3, 124.2, 124.8, 128.7, 129.1, 129.8, 130.1, 140.1, 158, 163, 167.9; MS (ESI): *m*/*z* 428 $[M+H]^+$. Anal. Calcd for $C_{23}H_{22}ClNO_5$: C, 64.56; H, 5.18; N, 3.27; Found: C, 64.68; H, 5.25; N, 6.48.

Reference

Siddiqui, I. R., Kumar, D., and Shamim, S. (2013). Ionic liquid promoted multicomponent reaction: a good strategy for the eco-compatible synthesis of functionalized pyrroles. *J. Heterocyclic Chem.*, **50**, E111–E115.

C-N bond_40

Type of reaction: C-N and C-C bond formation
Reaction conditions: Solvent-free, room temperature
Synthetic strategy: One-pot grinding process
Catalyst: K_2CO_3/Et_2Zn
Keywords: Aromatic aldehydes, glycinemethyl ester hydrochloride, potassium carbonate, diethylzinc, *N*-methylmaleimide, grinding, room temperature, azomethine ylides, 1,3-dipolar cycloaddition, substituted pyrrolodines, stereoselectivity

General reaction scheme

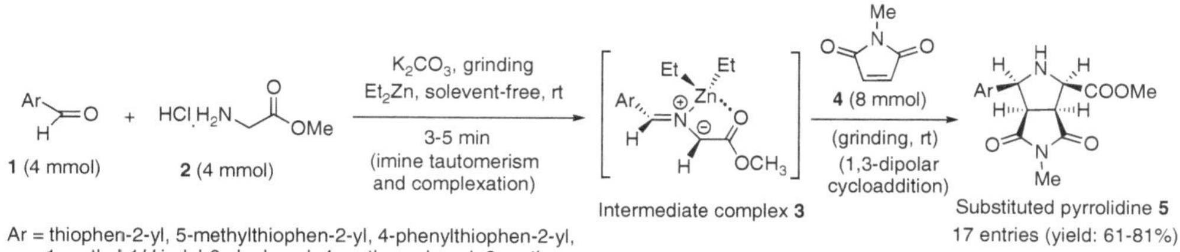

Ar = thiophen-2-yl, 5-methylthiophen-2-yl, 4-phenylthiophen-2-yl, 1-methyl-1*H*-indol-2-yl, phenyl, 4-methoxyphenyl, 2-methoxyphenyl, 4-bromophenyl, 2-bromophenyl, 4-chlorophenyl, 2-chlorophenyl, 4-nitrophenyl, 2-nitrophenyl, 4-cyanophenyl, *p*-tolyl, *o*-tolyl, 4-biphenyl

Representative entries

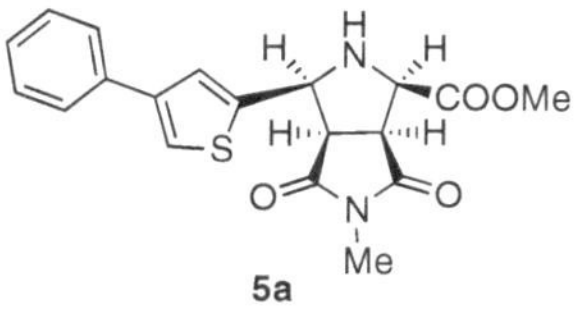

5a

(1*R*,3*S*,3a*R*,6a*S*)-Methyl 5-methyl-4,6-dioxo-3-(4-phenylthiophen-2-yl)octahydropyrrolo[3,4-*c*]pyrrole-1-carboxylate (**5a**): Solid, mp 214-216 °C; reaction time: 3 min; yield: 80%

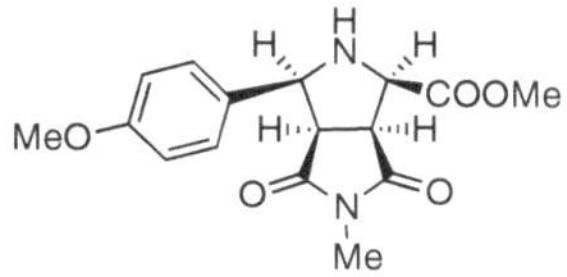

5b

(1*R*,3*S*,3a*R*,6a*S*)-Methyl 3-(4-methoxyphenyl)-5-methyl-4,6-dioxooctahydropyrrolo[3,4-*c*]pyrrole-1-carboxylate (**5b**): Solid, mp 195-197 °C; reaction time: 5 min; yield: 80%

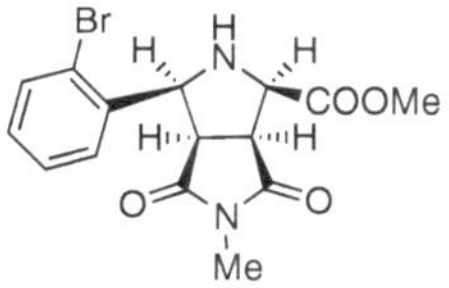

5c

(1*R*,3*S*,3a*R*,6a*S*)-Methyl 3-(2-bromophenyl)-5-methyl-4,6-dioxooctahydropyrrolo[3,4-*c*]pyrrole-1-carboxylate (**5c**): Solid, mp 165-166 °C; reaction time: 3 min; yield: 72%

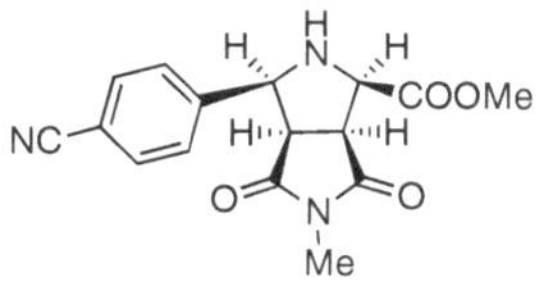

5d

(1*R*,3*S*,3a*R*,6a*S*)-Methyl 3-(4-cyanophenyl)-5-methyl-4,6-dioxooctahydropyrrolo[3,4-*c*]pyrrole-1-carboxylate (**5d**): Solid, mp 242-244 °C; reaction time: 5 min; yield: 79%

Experimental procedure

A mixture of aromatic aldehyde (**1**; 4 mmol), glycinemethylesterhydrochloride (**2**; 4 mmol, 0.490 g) and K_2CO_3 (4.4 mmol, 0.607 g) were added in a mortar and ground effectively with a pestle at room temperature. At the end of this period, diethyl zinc (5 mL, 1.6 M Et_2Zn in hexane) and *N*-methylmaleimide (**4**; 8 mmol, 0.888 g) were added in the same mortar. On completion of the reaction within 3-5 minutes as monitored TLC, the reaction mixture was hydrolyzed with sat. NH_4Cl solution and extracted with ethyl acetate. The combined organic layer was dried over $MgSO_4$, concentrated, and purified by column chromatography on silica gel (Hexane/EtOAc; 1/1 used as the eluent); the crude product was further purified by recrystallization from hexane/EtOAc mixtures to afford substituted pyrrolidines **5** with good yields (71-81%). All the products were characterized by means of spectral studies.

Characterization data of two representative entries

(1*R*,3*S*,3a*R*,6a*S*)-Methyl 5-methyl-4,6-dioxo-3-(4-phenylthiophen-2-yl)octahydropyrrolo[3,4-c]pyrrole-1-carboxylate (**5a**): Solid, mp 214-216 °C; IR (KBr): 1222 (C-O), 1752, 1695

(C=O), 3064 (=C-H), 2819 (CH), 3342 (N-H) cm^{-1}; ^{1}H NMR (300 MHz, $CDCl_3$): δ 2.65 (br, 1H, N-H), 2.93 (s, 3H, *N-CH*$_3$), 3.46 (dd, *J* = 8.3, 8.0 Hz, 1H, H-3a), 3.59 (dd, *J* = 7.2, 6.9 Hz, 1H, H-6a), 3.90 (s, 3H, COOC*H*$_3$), 4.06 (dd, *J* = 6.5, 5.0Hz, 1H, H-1), 4.8 (dd, *J* = 8.7, 5.2Hz, 1H, H-3), 7.32-7.66 (m, 7H, Ar-H); 13 C-NMR (75 MHz, $CDCl_3$): δ 25.2, 48.1, 49.5, 52.3, 59.7, 61.3, 120.0, 124.6, 126.3, 127.1, 128.7, 135.7, 141.1, 142.0, 169.7, 174.3, 175.7; HR-MS (EI): *m/z* calcd for $C_{19}H_{19}N_2O_4S$ $[M+H]^+$: 371.1066; Found: 371.1066.

(1*R*,3*S*,3a*R*,6a*S*)-Methyl 3-(4-cyanophenyl)-5-methyl-4,6-dioxooctahydropyrrolo[3,4-*c*]pyrrole-1-carboxylate (**5d**): Solid, mp 242-244 °C; IR (KBr): 1222 (C-O), 1744, 1691 (C=O), 2226 (CN), 3019 (=C-H), 2935 (C-H), 3338 (N-H) cm^{-1}; ^{1}H NMR (300 MHz, $CDCl_3$): δ 2.44 (br, 1H, N-H), 2.88 (s, 3H, *N-CH*$_3$), 3.49 (dd, *J* = 8.0, 8.0Hz, 1H, H-3a), 3.61 (dd, *J* = 7.4, 7.2 Hz, 1H, H-6a), 3.89 (s, 3H, COOC*H*$_3$), 4.55 (dd, *J* = 8.3, 4.8 Hz, 1H, H-3), 4.99 (dd, *J* = 6.8, 4.5Hz, 1H, H-1), 7.51 (d, *J* = 8.2 Hz, 2H, Ar-H), 7.75 (d, *J* = 8.2 Hz, 2H, Ar-H); ^{13}C NMR (75 MHz, $CDCl_3$): δ 25.1, 47.6, 49.1, 52.4, 61.5, 62.8, 112.2, 118.6, 127.9, 132.2, 142.3, 169.7, 174.2, 175.4; HR-MS (EI): *m/z* calcd for $C_{16}H_{16}N_3O_4$ $[M+H]^+$: 314.1141; Found: 314.1004.

Reference

Yavuz, S., Özkan, H., Tok, G., and Dişli, A. (2013). Facile method for 1,3-dipolar cycloaddition reaction ofazomethineylides: Highly stereoselective synthesis of substituted pyrrolidine derivatives. *J. Heterocyclic Chem.*, **50**, 1437–1440.

C-N bond_41

Type of reaction: C-N and C-O bond formation
Reaction conditions: Methanol, room temperature
Synthetic strategy: One-step condensation
Catalyst: No catalyst
Keywords: 2-Aminophenols, dimethyl-2-oxo-glutarate, methanol, room temperature, condensation, 1,4-benzoxazinones

General reaction scheme

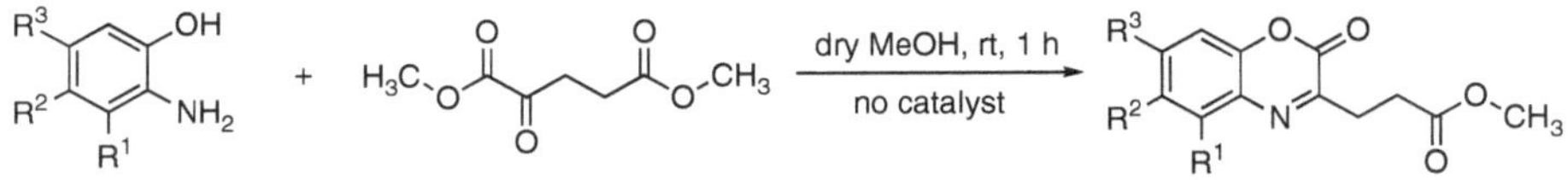

1 (10 mmol) **2** (10 mmol)

R^1,R^2,R^3 = H, H, H; H, H, CH_3; H, CH_3, H; CH_3, H, H; H, Cl, H

Substituted 3-(2-oxo-2*H*-benzo[*b*][1,4]oxazin-3-yl)propanoate **3**
5 entries (yield: 40-60%)

Representative entries

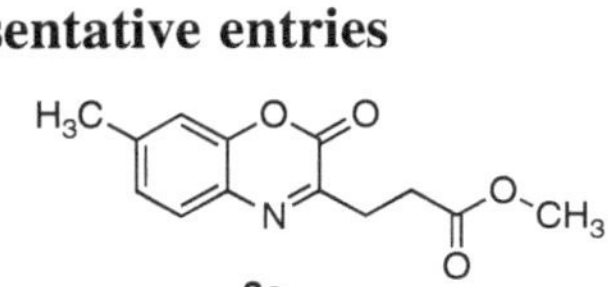

3a

Methyl 3-(7-methyl-2-oxo-2*H*-benzo[*b*][1,4]oxazin-3-yl)propanoate (**3a**): Colorless crystals, mp 99-100 °C; yield: 40%

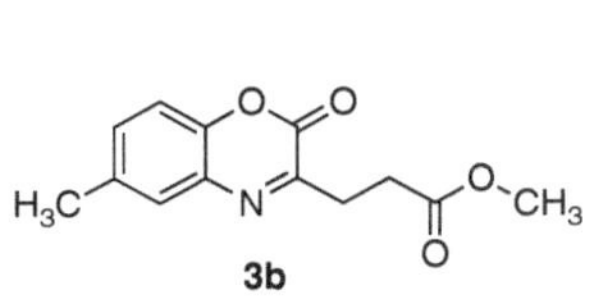

3b

Methyl 3-(6-methyl-2-oxo-2*H*-benzo[*b*][1,4]oxazin-3-yl)propanoate (**3b**): Colorless crystals, mp 100-101°C; yield: 50%

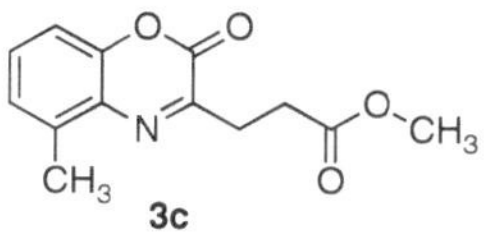

3c

Methyl 3-(5-methyl-2-oxo-2*H*-benzo[*b*][1,4]oxazin-3-yl)propanoate (**3c**): Colorless crystals, mp 132-134°C; yield: 50%

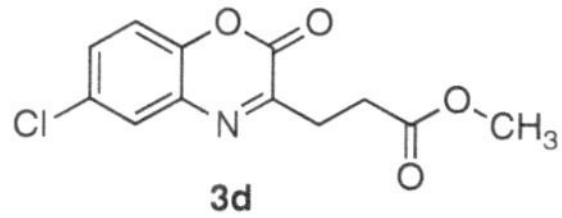

3d

Methyl 3-(6-chloro-2-oxo-2*H*-benzo[*b*][1,4]oxazin-3-yl)propanoate (**3d**): Colorless crystals, mp 114-116°C; yield: 50%

Experimental procedure

A solution of dimethyl-2-oxoglutarate (**2**; 10 mmol, 1.74 g) in 10 mL of dry methanol was added dropwise under stirring to a solution of the corresponding substituted 2-aminophenol (**1**; 10 mmol) in 20 mL of dry methanol at room temperature (20°C). The reaction was almost completed after stirring for 1 h at room temperature; the crystals formed were filtered off and recrystallized from methanol to obtain pure products of substituted 3-(2-oxo-2*H*-benzo[*b*][1,4]oxazin-3-yl)propanoates **3** with 40-60% yield. All the products were characterized by elemental analyses as well as detailed spectral studies.

Characterization data of two representative compounds

Methyl 3-(7-methyl-2-oxo-2H-benzo[*b*][1,4]oxazin-3-yl)propanoate (**3a**): Colorless crystals, mp 99-100 °C; IR(KBr): 1740 (C=O), 1616 (N=C) cm^{-1}; ^{1}H NMR (300/400 MHz, DMSO-d_6): δ 2.38 (s, 3H, CH_3), 2.74 (t, J=7.1 Hz, 2H, CH_2,), 3.01 (t, J=7.1 Hz, 2H, CH_2,), 3.59 (s, 3H, OCH_3), 7.18 (d, J=8.4Hz, 1H, Arom-H), 7.20 (s, 1H, Arom-H), 7.54 (d, J=8.4Hz, 1H, Arom-H); ^{13}C NMR (75/100 MHz, DMSO-d_6): δ 20.98 (CH_3), 28.02 (CH_2), 29.12 (CH_2), 51.44 (OCH_3), 116.18 (C-8), 126.20 (C-6), 127.85 (C-5), 128.55 (C-7), 141.17 (C-4a), 145.93 (C-8a), 152.69 (C=O),155.21 (N=C), 172.66 (COO); EIMS (70 eV): *m/z* (%rel.) 247 (M^+, 88%), 216 (M^+-OCH_3, 51%), 188 ($M^+-CO_2CH_3$, 36%), 160 ($M^+-CH_2-CH_2-CO_2CH_3$, 100%). Anal. Calcd for $C_{13}H_{13}NO_4$: C, 63.15; H, 5.30; N, 5.67; Found: C, 63.08; H, 5.11; N, 5.61%.

Methyl 3-(6-chloro-2-oxo-2*H*-benzo[*b*][1,4]oxazin-3-yl)propanoate (**3b**): Colorless crystals, mp 114-116°C; IR(KBr): 1740 (C=O), 1615 (N=C) cm^{-1}; ^{1}H NMR (300/400 MHz, DMSO-d_6): δ 2.77 (t, J=7.2 Hz, 2H, CH_2), 3.06 (t, J=7.2 Hz, 2H, CH_2), 3.62 (s, 3H, OCH_3), 7.45 (d, J=6.5 Hz, 1H, Arom-H), 7.59 (dd, J=6.5 & 2.1 Hz, 1H, Arom-H,), 7.74 (d, J=2.1 Hz, 1H, Arom-H); ^{13}C NMR (75/100 MHz, DMSO-d_6): δ 28.19 (CH_2), 28.96 (CH_2), 51.46 (OCH_3), 117.94 (C-8), 127.27 (C-5), 128.73 (C-7), 129.97 (C-6), 131.32 (C-4a), 145.02 (C-8a), 152.12 (C=O), 158.19 (N=C), 172.49 (COO); EIMS (70 eV): *m/z* (% rel.) 267 (M^+, 16%), 236 (M^+-OCH_3, 15%), 208 ($M^+-CO_2CH_3$, 5%), 180 ($M^+-CH_2-CH_2-CO_2CH_3$, 100%). Anal. Calcd for $C_{12}H_{10}ClNO_4$: C, 53.85; H, 3.77; N, 5.23; Found: C, 53.83; H, 3.73; N, 5.16%.

Reference

Hachama, K., Khodja, M., Moulay, S., Boutoumi, H., Hennig, L., and Sicker, D. (2013). Novel 3-(2-oxo-2*H*-benzo[*b*][1,4]oxazin-3-yl)propanoates from dimethyl-2-oxoglutarate and test of their biological activity. *J. Heterocyclic Chem.*, **50**, 413–416.

C-N bond_42

Type of reaction: C-N bond formation
Reaction conditions: Acetonitrile-water (9:1), room temperature
Synthetic strategy: One-step condensation
Catalyst: *N*-Bromosuccinimide (NBS)/2,2′-azoisobutyronitrile (AIBN)
Keywords: 5,6-Diamino-1,3-dimethyluracils, aryl/cycloaryl/heteroaryl aldehydes, *N*-bromosuccinimide (NBS), 2,2′-azoisobutyronitrile (AIBN), aqueous acetonitrile, room temperature, condensation, ring closure, 8-substituted xanthines

General reaction scheme

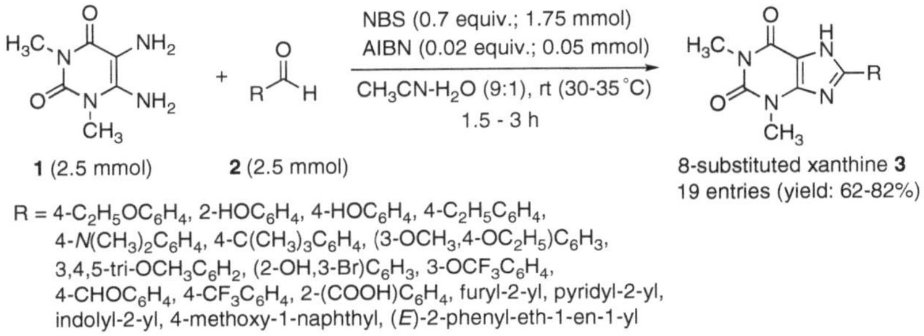

Representative entries

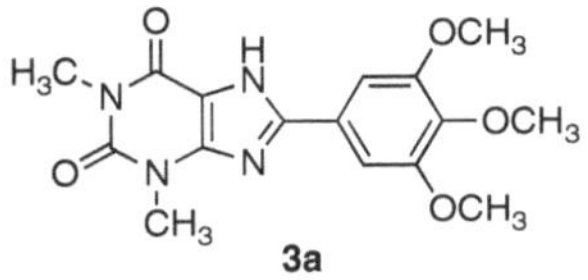

3a

1,3-Dimethyl-8-(3,4,5-trimethoxyphenyl)-1*H*-purine-2,6(3*H*,7*H*)-dione (**3a**): Light brown solid; reaction time: 3 h; yield: 72%

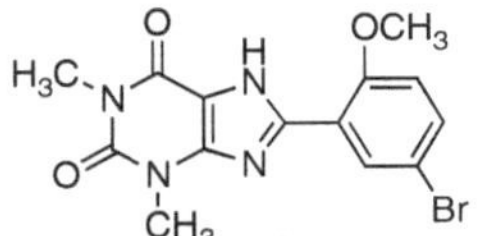

3b

8-(5-Bromo-2-methoxyphenyl)-1,3-dimethyl-1*H*-purine-2,6(3*H*,7*H*)-dione (**3b**): Light brown solid; reaction time: 2.5 h; yield: 78%

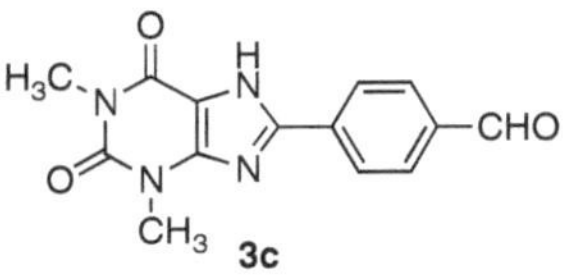

3c

4-(1,3-Dimethyl-2,6-dioxo-2,3,6,7-tetrahydro-1*H*-purin-8-yl)benzaldehyde (**3c**): Yellow solid; reaction time: 1.5 h; yield: 72%

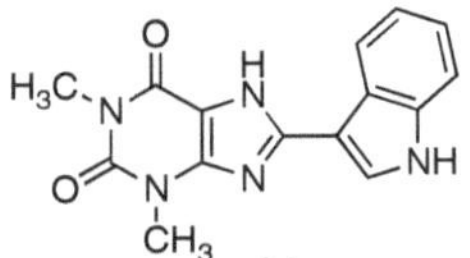

3d

8-(1*H*-Indol-3-yl)-1,3-dimethyl-1*H*-purine-2,6(3*H*,7*H*)-dione (**3d**): Green solid; reaction time: 3 h; yield: 74%

Experimental procedure

In a round bottom flask, 5,6-diamino-1,3-dimethyluracil (**1**; 1 equiv., 2.5 mmol, 0.425 g) and aldehyde (**2**; 1 equiv., 2.5 mmol) was stirred in acetonitrile-water (9:1) (10 mL) for 15 min. To this mixture, catalytic amount of AIBN (0.02 equiv., 0.05 mmol, 0.008 g) was added and after stirring for 5 min, 0.7 equiv. of NBS (0.311 g, 1.75 mmol) was added in two portions, initially 0.5 equiv. and after 30 min additional 0.2 equiv., and the reaction mixture was stirred at room temperature (30-35 °C) for stipulated time (1.5-3 h; monitored by TLC as well as HPLC). On completion of the reaction, the resulting precipitate was filtered under vacuum and the precipitate was washed with ethyl acetate (2-10 mL) and methanol (3-10 mL) to obtain the pure products of 8-substituted xanthines **3** with 62-82% yield. All the products were are fully characterized by physical as well as spectral studies.

Characterization data of two representative compounds

8-(5-Bromo-2-methoxyphenyl)-1,3-dimethyl-1*H*-purine-2,6(3*H*,7*H*)-dione (**3b**): Light brown solid; FT-IR (KBr): 3166, 1693, 1657, 1277, 760 cm^{-1}; ^1H NMR (400 MHz, DMSO-d_6): δ 3.26 (s, 3H, CH_3), 3.50 (s, 3H, CH_3), 3.93 (s, 3H, OCH_3), 7.17-7.19 (m, 1H), 7.63-7.65 (m, 1H), 8.10 (s, 1H), 13.01 (s, br, 1H, -N*H*); ESI-MS: *m/z* 366 $[M+H]^+$.

8-(1*H*-Indol-3-yl)-1,3-dimethyl-1*H*-purine-2,6(3*H*,7*H*)-dione (**3d**): Green solid; FT-IR (KBr): 3294,1688, 1648, 1586; ^1H NMR (400 MHz, DMSO-d_6): δ 3.27 (s, 3H, CH_3), 3.57 (s, 3H, CH_3), 7.17-7.21 (m, 2H), 7.46-7.48 (m, 1H), 8.22 (s, 1H), 8.40-8.42 (m, 1H), 11.70 (s, br, 1H, N*H* of indole), 13.27 (s, br, 1H, -N*H*); ESI-MS: *m/z* 296 $[M+H]^+$.

Reference

Bandyopadhyay, P., Agrawal, S. K., Sathe, M., Sharma, P., and Kaushik, M. P. (2012). A facile and rapid one-step synthesis of 8-substituted xanthine derivatives via tandem ring closure at room temperature. *Tetrahedron*, **68**, 3822–3827.

C-N bond_43

Type of reaction: C-N and C-C bond formation
Reaction conditions: Dichloromethane, room temperature
Synthetic strategy: One-step domino process involving sequential intramolecular cyclization and olefination
Catalyst: $Pd(OAc)_2$, TFA, O_2-atomphere
Keywords: 2-Azido-3-arylacrylates, terminal alkenes, dichloromethane, palladium acetate, trifluoroacetic acid (TFA), oxygen-atmosphere, room temperature, domino reaction, intramolecular cyclization followed by olefination, 3-alkenyl indoles

General reaction scheme

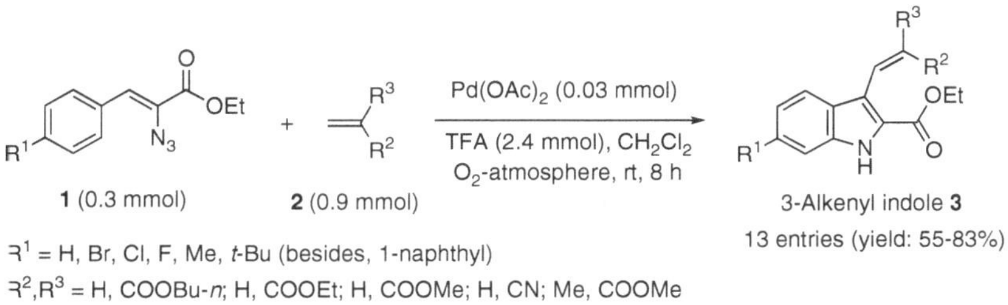

Representative entries

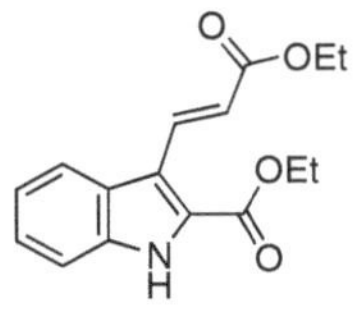

3a

(*E*)-Ethyl 3-(3-ethoxy-3-oxoprop-1-en-1-yl)-1*H*-indole-2-carboxylate (**3a**): Yellow solid, mp 144.2-146.4 °C; yield: 79%

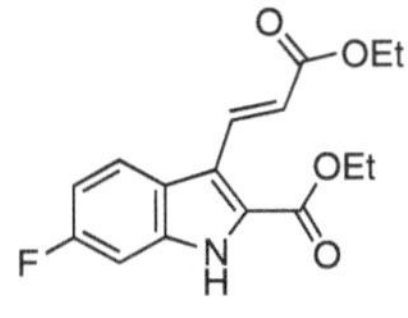

3b

(*E*)-Ethyl 3-(3-ethoxy-3-oxoprop-1-en-1-yl)-6-fluoro-1*H*-indole-2-carboxylate (**3b**): White solid, mp 179.4-181.8 °C; yield: 75%

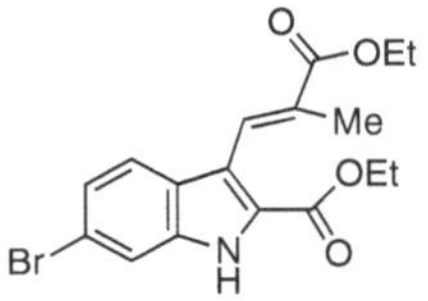

3c

(*E*)-Ethyl 6-bromo-3-(3-ethoxy-2-methyl-3-oxoprop-1-en-1-yl)-1*H*-indole-2-carboxylate (**3c**): White solid, mp >250 °C; yield: 71%

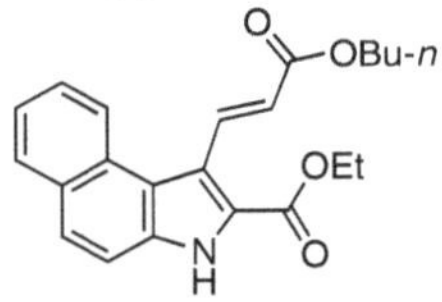

3d

(*E*)-Ethyl 1-(3-*n*-butoxy-3-oxoprop-1-en-1-yl)-3*H*-benzo[e]indole-2-carboxylate (**3d**): Yellow solid, mp 73.5-76.2°C; yield: 55%

Experimental procedure

A mixture of 2-azido-3-arylacrylate (**1**; 0.3 mmol), terminal alkene (**2**; 0.9 mmol), $Pd(OAc)_2$ (0.03 mmol), TFA (2.4 mmol) was stirred in anhydrous dichloromethne (3 mL) under oxygen atmosphere at room temperature for 8 h. The reaction mixture was quenched with water (10 mL) and then extracted with ethyl acetate. The combined organic layers were washed with brine, dried over sodium sulfate and concentrated under vacuum. The concentrated crude residue was purified by column chromatography using petrol ether-ethyl acetate mixture as eluent to furnish 3-alkenyl indoles **3** with 55-83% yield. All the products were are characterized by detailed spectral studies.

Characterization data of two representative compounds

(*E*)-Ethyl 6-bromo-3-(3-ethoxy-2-methyl-3-oxoprop-1-en-1-yl)-1*H*-indole-2-carboxylate (**3c**): White solid, mp >250 °C; IR (KBr) 3309, 2950, 2918, 1722, 1684, 1640, 1564, 1517, 1443, 1377, 1316, 1239, 1191, 1114 cm^{-1}; ^1H NMR (500 MHz, $CDCl_3$): δ 9.04 (s, 1H), 8.00 (d, $J=1.0$ Hz, 1H), 7.60 (d, $J=1.0$ Hz, 1H), 7.37 (d, $J=8.5$ Hz, 1H), 7.30 (dd, $J=8.5$, 1.5 Hz, 1H), 4.41 (q, $J=7.0$ Hz, 2H), 3.85 (s, 3H), 1.91 (d, $J=1.0$ Hz, 3H), 1.41 (t, $J=7.0$ Hz, 3H); ^{13}C NMR (125 MHz, $CDCl_3$): δ 168.8, 161.5, 136.6, 131.8, 130.7, 125.3,124.9,123.9,119.9,118.6,115.2, 61.8, 52.3,15.7,14.5; HRMS (ESI): *m/z* calcd for $C_{16}H_{16}BrNO_4$ $[M+H]^+$: 366.0341; Found: 366.0347.

(*E*)-Ethyl 1-(3-*n*-butoxy-3-oxoprop-1-en-1-yl)-3*H*-benzo[e]indole-2-carboxylate (**3d**): Yellow solid, mp 73.5-76.2°C; IR (KBr): 3289, 3087, 2959, 2923, 2870, 1727, 1675, 1625, 1565, 1506, 1467, 1435, 1379, 1294, 1263, 1169, 1120, 1064, 1015, 983 cm^{-1}; ^1H NMR (500 MHz, $CDCl_3$): δ 9.72 (s, 1H), 8.54 (d, $J=8.5$ Hz, 1H), 8.48 (d, $J=16.0$ Hz, 1H), 7.90 (d, $J=8.0$ Hz, 1H), 7.72 (d, $J=9.0$ Hz, 1H), 7.57 (t, $J=7.5$ Hz, 1H), 7.48 (d, $J=10.0$ Hz, 2H), 6.60 (d, $J=16.5$ Hz, 1H), 4.45 (q, $J=7.0$ Hz, 2H), 4.30 (t, $J=6.5$ Hz, 2H), 1.77-1.73 (m, 2H),

1.53-1.47 (m, 2H), 1.44 (t, $J = 7.0$ Hz, 3H), 1.00 (t, $J = 7.0$ Hz, 3H); ^{13}C NMR (125 MHz, $CDCl_3$): δ 167.4, 162.0, 138.6, 134.0, 130.6, 129.5, 129.4, 128.4, 127.1, 125.0, 124.6, 123.9, 123.0, 120.8, 120.6, 113.0, 64.7, 61.6, 31.1, 19.5, 14.5, 14.0; HRMS (ESI): *m/z* calcd for $C_{22}H_{23}NO_4$ $[M+H]^+$: 366.1705; Found: 366.1706.

Reference

Zhang, Y., Liu, S., Yu, W., Hu, M., Zhang, G., and Yu, Y. (2013). Palladium-catalyzed synthesis of 3-alkenyl indoles from 2-azido-3-arylacrylates and terminal alkenes at room temperature. *Tetrahedron*, **69**, 2070–2074.

C-N bond_44

Type of reaction: C-N and C-C bond formation
Reaction conditions: Ethanol, room temperature
Synthetic strategy: One-pot multicomponent reaction
Catalyst: L-Proline
Keywords: Dimedone, aldehydes, α-naphthylamine, L-proline, ethanol, room temperature, one-pot multicomponent reaction, tetrahydrobenzo[*c*]acridin-8(7*H*)-ones

General reaction scheme

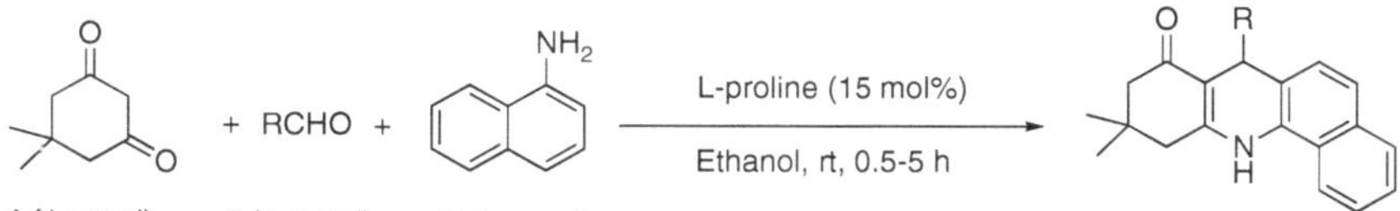

1 (1 mmol) **2** (1 mmol) **3** (1 mmol)

R = C_6H_5, 4-$CH_3C_6H_4$, 4-$CH_3OC_6H_4$, 2-OHC_6H_4, 4-ClC_6H_4, 2-FC_6H_4, 4-FC_6H_4, 4-$CF_3C_6H_4$, 2-$NO_2C_6H_4$, 4-$NO_2C_6H_4$, H, CH_3, $CH(CH_3)_2$

Tetrahydrobenzo[*c*]acridin-8(7*H*)-one **4**
13 entries (yield: 87-97%)

Representative entries

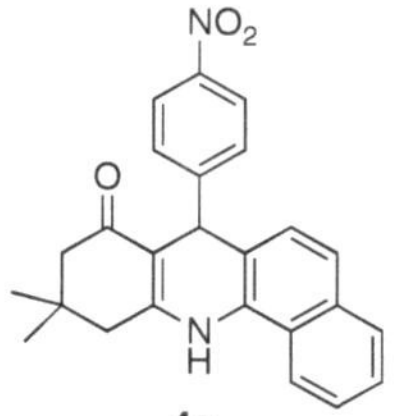

4a

10,10-Dimethyl-7-(4-nitrophenyl)-9,10,11,12-tetrahydrobenzo[*c*]acridin-8(7*H*)-one (**4a**): Yellow solid, mp 281-283 °C; reaction time: 0.5 h; yield: 95%

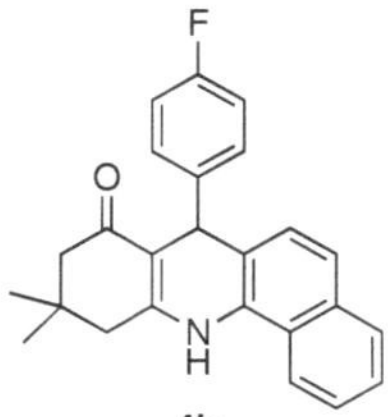

4b

7-(4-Fluorophenyl)-10,10-dimethyl-9,10,11,12-tetrahydrobenzo[*c*]acridin-8(7*H*)-one (**4b**): White solid, mp 218-219 °C; reaction time: 0.5 h; yield: 95%

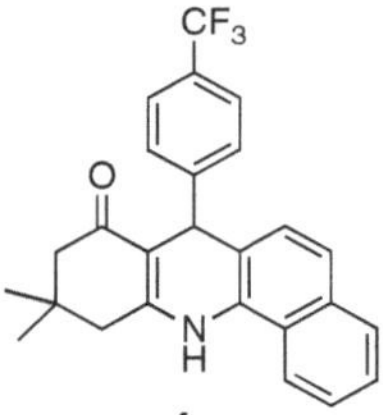

4c

10,10-Dimethyl-7-(4-(trifluoromethyl)phenyl)-9,10,11,12-tetrahydrobenzo[*c*]acridin-8(7*H*)-one (**4c**): White solid, mp 276-277 °C; reaction time: 0.5 h; yield: 97%

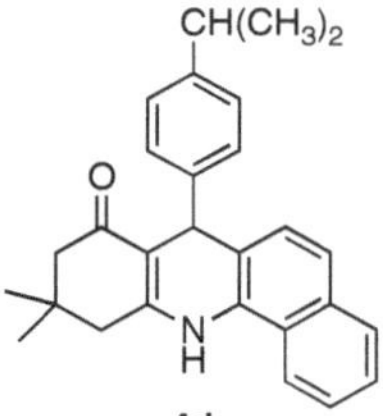

4d

7-(4-Isopropylphenyl)-10,10-dimethyl-9,10,11,12-tetrahydrobenzo[*c*]acridin-8(7*H*)-one (**4d**): Pale yellow solid, mp 195-197 °C; reaction time: 5 h; yield: 89%

Experimental procedure

A mixture of dimedone (**1**; 1 mmol), aldehyde (**2**; 1 mmol) and α-naphthylamine (**3**; 1 mmol) were stirred within a 50 mL round bottom flask in the presence of 15 mol% of L-Proline in ethanol (2 mL) at room temperature for the stipulated time (0.5-5 h). After completion of the reaction (as monitored by TLC), the reaction mixture was diluted with water (5 mL) and extracted with ethyl acetate (3 × 10 mL). The organic layer was dried over anhydrous sodium sulfate, concentrated, and recrystallized from hot ethanol to afford pure tetrahydrobenzo[*c*]acridin-8(7*H*)-ones (**4**) in good yields. All the products were characterized by detailed spectral studies as well as elemental analyses.

Characterization data of two representative compounds

10,10-Dimethyl-7-(4-(trifluoromethyl)phenyl)-9,10,11,12-tetrahydrobenzo[*c*]acridin-8(7*H*)-one (**4c**): White solid, mp 276-277 °C; IR (KBr): 3332, 3076, 2971, 1651, 1522 (C=O), 1525, 1474, 1124, 822 cm^{-1}; ^1H NMR (500 MHz, DMSO-d_6): δ 1.11 (s, 3H, CH_3), 1.23 (s, 3H, CH_3), 2.05-2.32 (dd, 2H, C_9-H), 2.43-2.84 (dd, 2H, C_{11}-H), 5.97 (s, 1H, C_7-H), 7.17-8.12 (m, 9H, Ar-H), 8.73 (d, J=7.5, 1H, C_6-H), 9.45 (s, 1H, NH); ^{13}C NMR (DMSO-d_6, 125 MHz): δ 26.34, 29.42, 31.54, 33.65, 42.76, 55.65, 68.65, 76.54, 85.43, 88.65, 105.43, 119.54, 121.43, 123.98, 124.54, 126.32, 127.08, 128.21, 129.21, 129.76, 131.32, 133.83, 137.96 (q, $^1J_{CF}$=255.76 Hz), 147.34, 156.43, 158.09, 198.56; ^{19} F NMR (DMSO-d_6, 470 MHz): δ −111.2; MS (EI): *m*/*z* (rel %) 421 (M^+, 70), 276 (98); HRMS (EI): *m*/*z* calcd for $C_{26}H_{22}F_3NO$ [M^+]: 421.1709. Found: 421.2106. Anal Calcd for $C_{26}H_{22}F_3NO$: C, 74.10; H, 13.52; N, 3.32; Found: C, 74.54, 13.21; H, 3.45.

7-(4-Isopropylphenyl)-10,10-dimethyl-9,10,11,12-tetrahydrobenzo[*c*]acridin-8(7*H*)-one (**4d**): Pale yellow solid, mp 195-197 °C; IR (KBr): 3318, 3098, 2921, 1651, 1532 (C=O), 1505, 1434, 1154,832 cm^{-1}; ^1H NMR (500 MHz, DMSO-d_6): δ 1.13 (s, 3H, CH_3), 1.31 (s, 3H, CH_3), 2.03-2.25 (dd, 2H, C_9-H), 2.45-2.74 (dd, 2H, C_{11}-H), 3.45 (hep., 1H, CH), 4.56 (d, 6H, 2 × CH_3), 6.21 (s, 1H, C_7-H), 7.16-8.02 (m, 5H, Ar-H), 8.72 (d, 1H, J=7.5, C_6-H), 9.65 (s, 1H, NH); ^{13}C NMR (DMSO-d_6, 125 MHz): δ 26.31, 28.56, 30.54, 32.43, 33.27, 34.65, 35.53, 40.09, 51.43, 64.87, 76.98, 81.21, 89.76, 128.20, 129.43, 129.54, 131.04, 131.23, 143.32, 152.65, 153.43, 194.60. MS (EI): *m*/*z* (rel %) 319 (M^+, 70), 276 (90), 193 (45). HRMS (EI): *m*/*z* calcd for $C_{22}H_{25}NO$ [M^+]: 319.1932. Found: 319.1908. Anal Calcd for $C_{22}H_{25}NO$: C, 82.72; H, 7.89; N, 4.38; Found: C, 82.56; H, 7.65; N, 4.66.

Reference

Heravi, M. R. P., and Aghamohammadi, P. (2012). L-Proline-catalysed one-pot synthesis of tetrahydrobenzo[*c*]acridin-8(7*H*)-ones at room temperature. *Comptes Rendus Chimie*, **15**, 448–453.

C-N bond_45

Type of reaction: C-N and C-C bond formation
Reaction conditions: Diethyl ether, room temperature
Synthetic strategy: One-pot three-component aza-Diels-Alder reaction
Catalyst:

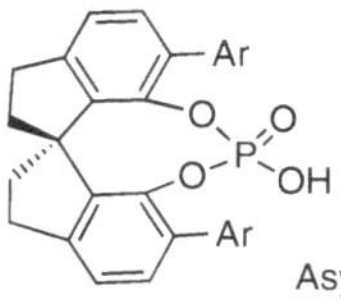

Ar = 2,4,6-(iPr)$_3$C$_6$H$_2$
Asymmetric catalyst [(R)-A]

Keywords: Indoles, aromatic amines, oxetanyl aldehydes, asymmetric catalyst A, diethyl ether, room temperature, asymmetric multicomponent reaction, aza-Diels-Alder reaction, indole-alkaloid-type polycycles, diastereoseltivity

General reaction scheme

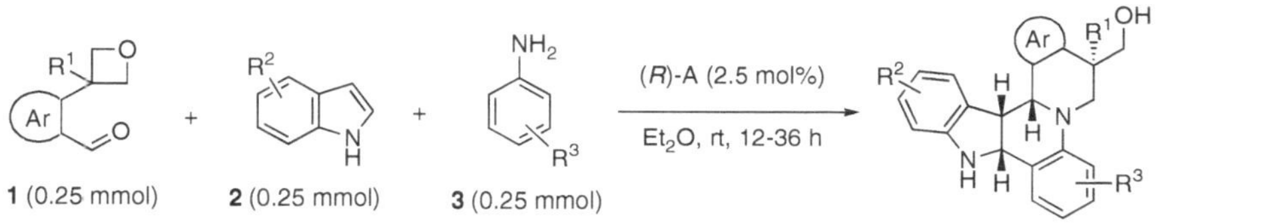

R^1 = H, OAc; R^2 = H, 6-Br, 6-CO$_2$Me
R^3 = 3-OH, 3,5-di-OMe, 3,4,5-tri-OMe
Ar = substituted phenyls [H, 3-FC$_6$H$_4$, 4-FC$_6$H$_4$, 3-OMeC$_6$H$_4$, 3,4-(OMe)$_2$C$_6$H$_3$, 3,4,-(O-CH$_2$-O)C$_6$H$_3$] and thienyl

Indole-alkaloid-type polycyclic derivative **4**
11 entries (yield: 55-93%)

Representative entries

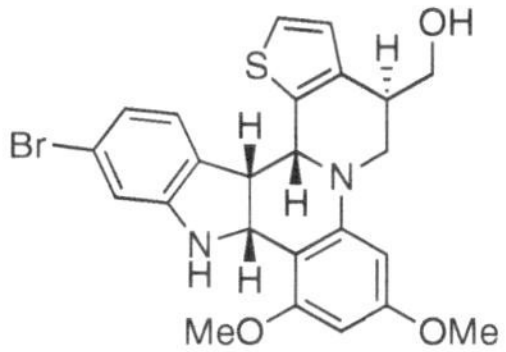

4a

((4R,10bR,15bS,15cR)-13-Bromo-8,10-dimethoxy-4,5,10b,11,15b,15c-hexahydroindolo[3,2-c]thieno[2′,3′:3,4]pyrido[1,2-a]quinolin-4-yl)methanol (**4a**): White solid, yield: 68% (92% ee, d.r. = 80:20), $[\alpha]_D^{25}$ −50.1° (CHCl$_3$, c 0.55)

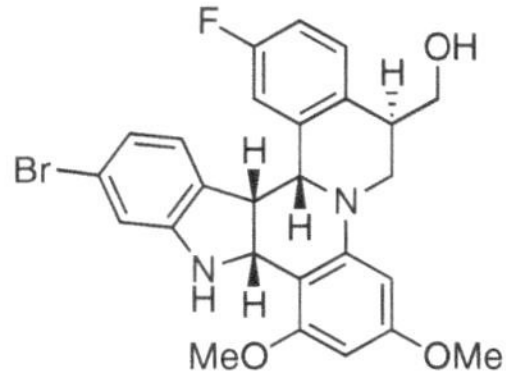

4b

((5R,11bR,16bS,16cR)-14-Bromo-2-fluoro-9,11-dimethoxy-5,6,11b,12,16b,16c-hexahydroindolo[3,2-c]isoquinolino[2,1-a]quinolin-5-yl)methanol (**4b**): White solid; yield: 95% (78% ee, d.r. = 95:5), $[\alpha]_D^{25}$ −88.8° (CHCl$_3$, c 0.34)

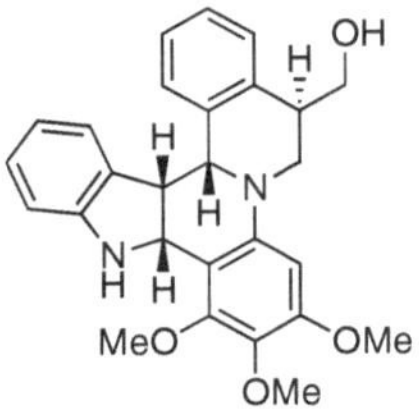

4c

((5*R*,11b*R*,16b*S*,16c*R*)-9,10,11-Trimethoxy-5,6,11b,12,16b,16c-hexahydroindolo[3,2-*c*]isoquinolino[2,1-*a*]quinolin-5-yl)methanol (**4c**): White solid; yield: 45% (65% ee, d.r. = 95:5)

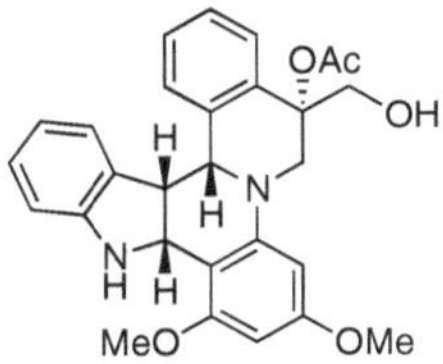

4d

(5*S*,11b*R*,16b*S*,16c*R*)-5-(Hydroxymethyl)-9,11-dimethoxy-5,6,11b,12,16b,16c-hexahydroindolo[3,2-*c*]isoquinolino[2,1-*a*]quinolin-5-yl acetate (**4d**): White solid; yield: 67% (63% ee, d.r. = 90:10)

Experimental procedure

To a 25-mL plastic centrifuge tube charged with oxetanyl aldehyde (**1**; 0.25 mmol), indole (**2**; 0.25 mmol) and aromatic amine (**3**; 0.25 mmol) was added anhydrous diethyl ether (2.5 mL), and the mixture was for about 10 min to get a clear solution. The catalyst (*R*)-A (4.5 mg, 0.63 μmol, 2.5 mol%) was then added to this solution, and the reaction mixture was stirred at room temperature. The progress of the reaction was monitored by TLC; on completion of reaction, hexanes (5 mL) were added to the reaction mixture and the stirring was continued for next 5 min. The reaction mixture was then subjected to centrifugation (5000 rpm for 5 min) and decanted. The solid reside obtained was further centrifuged in 5 mL of hexanes/Et_2O (5:1) mixture for another 5 min. The washing was repeated once and the solid residue was dried under vacuum to afford the desired pure product of indole-alkaloid-type polycycles **4**. All the products were characterized by detailed spectral studies.

Characterization data of two representative compounds

((5*R*,11b*R*,16b*S*,16c*R*)-14-Bromo-2-fluoro-9,11-dimethoxy-5,6,11b,12,16b,16c-hexahydroindolo[3,2-*c*]isoquinolino[2,1-*a*]quinolin-5-yl)methanol (**4b**): White solid, $[\alpha]_D^{25}$ −88.8° ($CHCl_3$, *c* 0.34); IR (neat): 3503, 2963, 2815, 1772, 1603, 1479, 1202, 1145, 1022, 800 cm^{-1}; ^{1}H NMR (400 MHz, DMSO-d_6): δ 7.53 (td, J = 8.0 & 4.5 Hz, 1H), 7.45 (dd, J = 8.4 & 2.0 Hz, 1H), 7.15 (td, J = 8.4 & 2.0 Hz, 1H), 6.51 (d, J = 1.6 Hz, 1H), 6.28 (dd, J = 8.0 & 2.4 Hz, 1H), 6.10 (s, 1H), 6.04 (s, 1H), 5.69 (s, 1H), 5.55 (dd, J = 8.0 Hz, 1H), 5.18 (d, J = 10.2 Hz, 1H), 4.87 (t, J = 9.6 Hz, 1H), 4.52 (s, 1H), 4.35 (d, J = 8.0 Hz, 1H), 3.94-3.89 (m, 1H), 3.82 (s, 3H), 3.73 (s, 3H), 3.68-3.61(m, 2H), 2.77-2.72 (m, 2H); ^{13}C NMR (100 MHz, DMSO-d_6): δ 162.0 (d, J = 244.1 Hz), 160.5, 159.3, 153.3, 149.9, 138.5 (d, J = 6.9 Hz), 131.9, 128.2 (d, J = 7.9 Hz), 126.5, 125.9, 121.4, 120.5, 114.3 (d, J = 21.2 Hz), 113.4 (d, J = 22.0 Hz), 112.3, 108.7, 93.3, 90.2, 64.4, 57.3, 55.8, 55.5, 55.3, 48.0, 47.7, 40.0; ^{19}F NMR (376.5 MHz, $CDCl_3$): δ −114.2; HRMS (ESI): *m*/*z* calcd for $C_{26}H_{25}BrFN_2O_3$ $[M+H]^+$: 511.1027; Found: 511.1023.

(5*S*,11b*R*,16b*S*,16c*R*)-5-(Hydroxymethyl)-9,11-dimethoxy-5,6,11b,12,16b,16c-hexahydroindolo[3,2-*c*]isoquinolino[2,1-*a*]quinolin-5-yl acetate (**4d**): White solid; IR (neat): 3513, 3389, 2992, 2942, 1762, 1606, 1462, 1375, 1242, 1149, 1103, 1051, 753 cm^{-1}; ^{1}H NMR (400 MHz, $CDCl_3$): δ 7.59 (dd, J = 7.2 & 2.0 Hz, 1H), 7.52 (d, J = 6.4 Hz, 1H), 7.47-7.39 (m, 2H), 6.91 (t, J = 7.6 Hz, 1H), 6.66 (d, J = 7.2 Hz, 1H), 6.44 (t, J = 7.6 Hz, 1H), 6.10 (d, J = 2.0 Hz, 1H), 6.06 (d, J = 2.0 Hz, 1H), 5.70 (d, J = 7.6 Hz, 1H), 5.45 (d, J = 10.0 Hz, 1H), 4.71 (d, J = 2.8 Hz, 1H), 4.63 (d, J = 11.6 Hz, 1H), 4.52-4.46 (m, 2H), 3.91(s, 3H), 3.80 (s, 3H), 3.72 (d, J = 11.6 Hz, 1H), 3.16 (d, J = 11.6 Hz, 1H), 2.08 (s, 3H); ^{13}C NMR (100 MHz, $CDCl_3$): δ170.8, 160.4, 159.3, 151.5, 150.2, 136.5, 135.9, 128.9, 128.1, 127.5, 127.1, 126.6, 125.0, 124.8, 118.5, 110.2, 109.2, 92.5, 90.1, 69.0, 66.8, 58.9, 55.5, 55.2, 54.8, 52.1, 48.7, 20.9; HRMS (ESI): *m*/*z* calcd for $C_{28}H_{28}N_2O_5$ $[M^+]$: 472.1998; Found: 472.1999.

Reference

Chen, Z., Wang, B., Wang, Z., Zhu, G., and Sun, J. (2013). Complex bioactive alkaloid-type polycycles through efficient catalytic asymmetric multicomponent aza-Diels-Alder reaction of indoles with oxetane as directing group. *Angew. Chem. Int. Ed.*, **52**, 2027–2031.

C-N bond_46

Type of reaction: C-N and C-C bond formation
Reaction conditions: 1,4-Dioxane, room temperature
Synthetic strategy: One-pot multicomponent reaction
Catalyst: CuBr
Keywords: 2-Ethynylanilines, sulfonyl azide, nitroolefins, CuBr, 1,4-dioxane, room temperature, one-pot three-component reaction, polysubstituted indoles

General reaction scheme

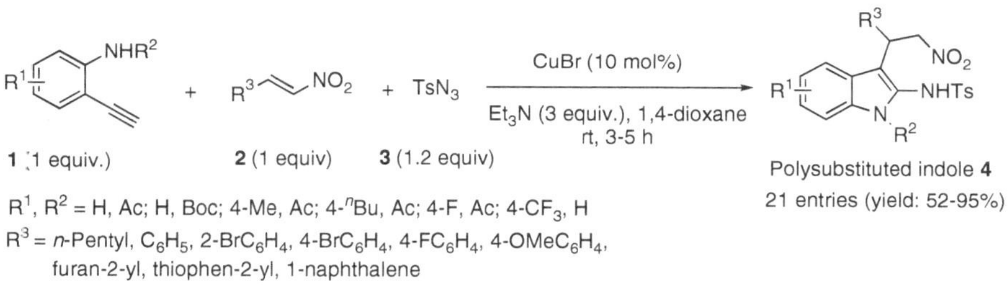

Representative entries

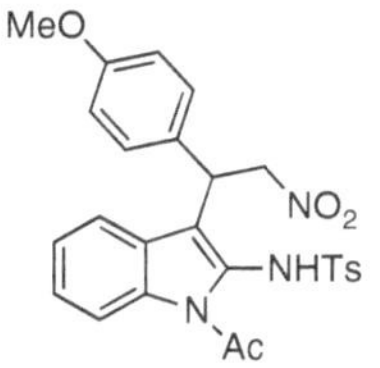

4a

N-(1-Acetyl-3-(1-(4-methoxyphenyl)-2-nitroethyl)-1*H*-indol-2-yl)-4-methylbenzenesulfonamide (**4a**): Yield: 91%

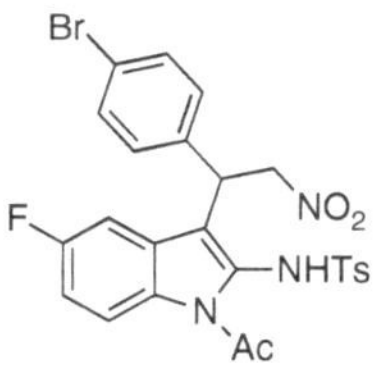

4b

N-(1-Acetyl-3-(1-(4-bromophenyl)-2-nitroethyl)-5-fluoro-1*H*-indol-2-yl)-4-methylbenzenesulfonamide (**4b**): Yield: 71%

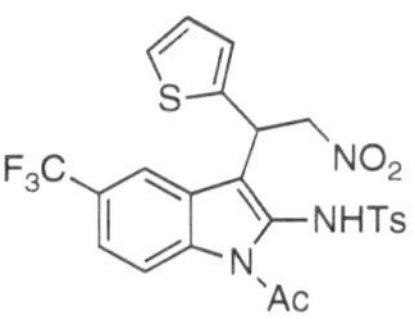

4c

N-(1-Acetyl-3-(2-nitro-1-(thiophen-2-yl)ethyl)-5-(trifluoromethyl)-1*H*-indol-2-yl)-4-methylbenzenesulfonamide (**4c**): Yield: 60%

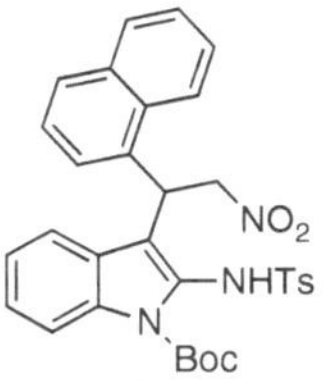

4d

tert-Butyl 2-(4-methylphenylsulfonamido)-3-(1-(naphthalen-1-yl)-2-nitroethyl)-1*H*-indole-1-carboxylate (**4d**): Yield: 85%

Experimental procedure

To a stirred mixture of *p*-toluenesulfonyl azide (**3**; 0.30 mmol, 1.2 equiv), nitroolefin (**2**, 0.25 mmol, 1.0 equiv), and CuBr (0.025 mmol, 10 mol %) in anhydrous 1,4-dioxane (0.5 mL) was added a solution of 2-ethynylaniline (**1**; 0.25 mmol, 1.0 equiv) in 1,4-dioxane (0.5 mL) at room temperature under nitrogen atmosphere. Triethylamine (104 μL, 0.75 mmol, 3.0 equiv) was tehn slowly added by syringe. The whole mixture was thereafter stirred vigorously until completion of the reaction (monitored by TLC, 3-5 h). Upon completion of the reaction, the reaction mixture was quenched with saturated ammonium chloride solution (3.0 mL); the mixture was stirred for additional 10 min and extracted with ethyl acetate (3 × 3.0 mL). The combined organic layers were dried over sodium sulfate, filtered and concentrated in vacuum. The crude product was purified using flash column chromatography (EtOAc/*n*-hexane, 1:3) to obtain polysubstituted indoles **4**. All the products were characterized by their spectral studies.

Characterization data of two representative compounds

N-(1-Acetyl-3-(1-(4-bromophenyl)-2-nitroethyl)-5-fluoro-1*H*-indol-2-yl)-4-methylbenzenesulfonamide (**4b**): ^{1}H NMR (400 MHz, $CDCl_3$): δ δ 7.56-7.48 (m, 3H), 7.41-7.37 (m, 2H), 7.15-7.08 (m, 4H), 7.01-6.94 (m, 1H), 6.81 (dd, *1 J*=2.28 Hz, ^{2} *J*=8.72 Hz, 1H), 5.35-5.26 (m, 1H), 5.23-5.16 (m, 1H), 5.10-5.00 (m, 1H), 2.34 (s, 3H), 2.32 (s, 3H); ^{13}C NMR (100 MHz, $CDCl_3$): δ 21.6, 26.7, 38.7, 76.6, 106.3 (d, ^{2} *J*=24.79 Hz), 113.3 (d, ^{2} *J*=24.79 Hz), 115.5, 115.6, 121.6, 127.2, 127.6, 129.1, 129.7, 129.9, 131.5, 132.0, 135.8, 136.2, 144.6, 159.1 (d, ^{1} *J*=241.22 Hz), 171.0; HRMS: *m*/*z* calcd for $C_{25}H_{21}BrFN_3O_5S$ $[M+Na]^+$: 596.0267; Found: 596.0289.

N-(1-Acetyl-3-(2-nitro-1-(thiophen-2-yl)ethyl)-5-(trifluoromethyl)-1*H*-indol-2-yl)-4-methylbenzenesulfonamide (**4c**): ^{1}H NMR (400 MHz, $CDCl_3$): δ 7.76 (d, *J*=8.72 Hz, 1H), 7.59-7.50 (m, 4H), 7.21-7.14 (m, 3H), 6.94-6.86 (m, 2H), 5.58-5.52 (m, 1H), 5.23-5.15 (m, 1H), 5.00-4.94 (m, 1H), 2.40 (s, 3H), 2.35 (s, 3H); 13C NMR (100 MHz, CDC13): δ 21.6, 26.8, 35.4, 77.9, 114.8, 116.7, 117.9, 122.3, 125.4, 125.6, 125.7 (q, *2 J*=32 Hz), 125.8, 126.9, 1-27.6, 129.9, 130.5, 135.2, 135.4, 140.2, 144.9, 170.8; HRMS: *m*/*z* calcd for $C_{24}H_{20}F_3N_3O_5S_2$ $[M+Na]^+$: 574.0694; Found: 574.0687.

Reference

Chen, Z., Zheng, D., and Wu, J. (2011). A facile route to polysubstituted indoles via three-component reaction of 2-ethynylaniline, sulfonyl azide, and nitroolefin. *Org. Lett.*, **13**, 848–851.

Further information

A similar-type copper-catalyzed C-N coupling reaction for the one-pot multicomponent synthesis of substituted indulines was reported by Wang and his group.

Jin, H., Zhou, B., Wu, Z., Shen, Y., and Wang, Y. (2011). One-pot synthesis of substituted indolines *via* a copper-catalyzed sequential multicomponent/C-N coupling reaction. *Tetrahedron Lett.*, **67**, 1178–1182.

C-N bond_47

Type of reaction: C-N bond formation
Reaction conditions: Solvent-free (neat), room temperature
Synthetic strategy: One-pot multicomponent reaction
Catalyst: Silica nanoparticles (*ca.* 70 nm)

Keywords: 6-Formyl-2,3-dimethoxybenzoic acid, secondary amines, isocyanides, silica NPs, heterogeneous recyclable catalyst, solvent-free, room temperature, one-pot three-component reaction, 1-carboxamido-3-oxoisoindolines

General reaction scheme

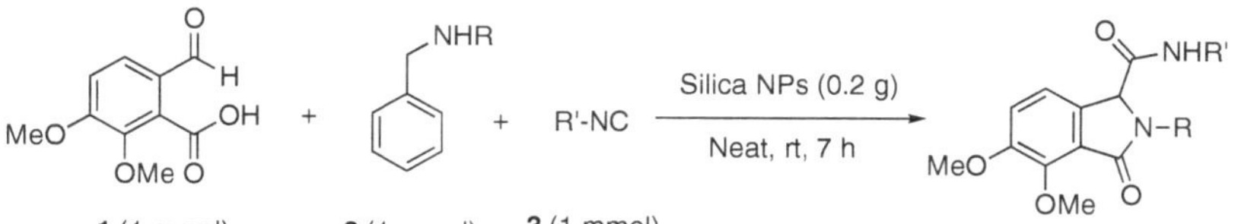

1 (1 mmol) **2** (1 mmol) **3** (1 mmol)

R = $C_6H_5CH_2$, *i*-Pr

R' = *t*-Bu, cyclohexyl, 1,1,3,3-tetramethylbutyl

1-Carboxamido-3-oxoisoindoline **4**

6 entries (yield: 82-94%)

Representative entries

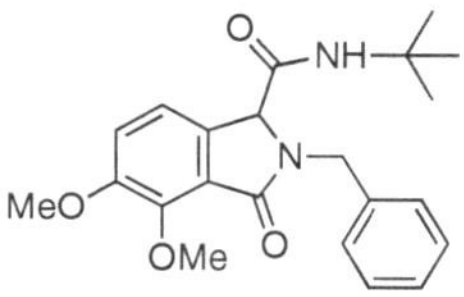

4a

2-Benzyl-*N*-(*tert*-butyl)-4,5-dimethoxy-3-oxoisoindoline-1-carboxamide (**4a**): Colorless oil, yield: 94%

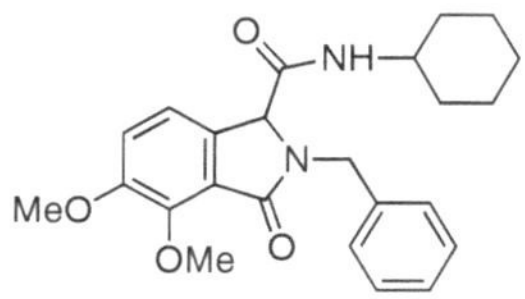

4b

2-Benzyl-*N*-cyclohexyl-4,5-dimethoxy-3-oxoisoindoline-1-carboxamide (**4b**): Colorless oil, yield: 90%

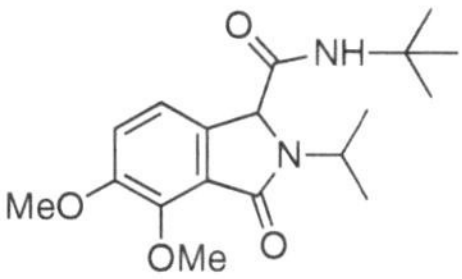

4c

N-(*tert*-Butyl)-2-isopropyl-4,5-dimethoxy-3-oxoisoindoline-1-carboxamide (**4c**): Colorless oil, yield: 92%

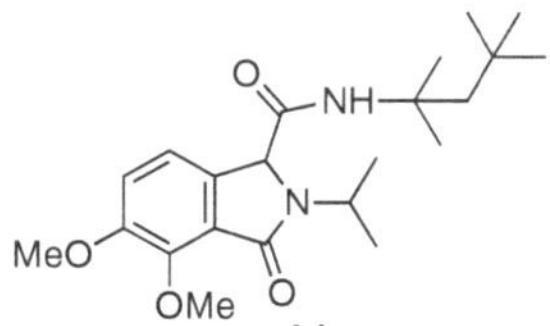

4d

2-Isopropyl-4,5-dimethoxy-3-oxo-*N*-(2,4,4-trimethylpentan-2-yl)isoindoline-1-carboxamide (**4d**): Colorless oil, yield: 88%

Experimental procedure

To a stirred mixture of 6-formyl-2,3-dimethoxybenzoic acid (**1**; 1 mmol), secondary amine (**2**; 1 mmol), and isocyanide **3** (1 mmol) was added silica nanoparticles (0.2 g), and then the whole reaction mixture was stirred vigorously at room temperature for 7 h. After completion of the reaction, flash column chromatography (petroleum ether:EtOAc = 5:2) of the residue furnished purified desired product **4**. All the products were characterized by their elemental analyses and spectral studies.

Characterization data of two representative compounds

2-Benzyl-*N*-cyclohexyl-4,5-dimethoxy-3-oxoisoindoline-1-carboxamide (**4b**): Colorless oil; IR (KBr): 3292, 3084, 2976, 1695, 1677, 1490, 1268 cm^{-1}; ^1H NMR (250 MHz, $CDCl_3$): δ 0.79 - 1.79 (2 m, 5 CH_2 of cyclohexyl), 3.57-3.69 (m, CH of cyclohexyl),

3.86, 4.06 (2 s, 2 × OMe), 4.25, 5.14 (2d, $J = 14.8$, PhCH_2); 4.70 (s, CH); 5.74 (d, $^3J = 8.0$, NH); 7.05, 7.21 (2d, $^3J = 8.2$, 2 × CH of isoindolone), 7.23-7.45 (m, 5 Ar-H); ^{13}C NMR (62.5 MHz, $CDCl_3$): δ 24.8, 24.8 (2 CH_2(β) of cyclohexyl), 25.3 (CH_2(γ) of cyclohexyl), 32.5, 32.7 (2 CH_2(α) of cyclohexyl), 46.0 ($PhCH_2$), 48.6 (CH of cyclohexyl), 56.6, 62.4 (2 MeO), 63.4 (CH), 117.0, 117.8 (2 CH of isoindolone), 122.6, 134.2, 147.1, 153.1 (4 C of isoindolone), 128.0, 128.6, 129.0 (5 CH of Ph), 136.4 (C of Ph), 166.9, 168.0 (2 C=O of amides). Anal. calcd for $C_{24}H_{28}N_2O_4$: C, 70.57; H, 6.91; N, 6.86; Found: C, 70.49; H, 6.95; N, 6.89.

2-Isopropyl-4,5-dimethoxy-3-oxo-*N*-(2,4,4-trimethylpentan-2-yl)isoindoline-1-carboxamide (**4d**): Colorless oil; IR (KBr): 3324, 2955, 1677, 1660, 1480, 1269 cm^{-1}; ^{1}H NMR (250 MHz, $CDCl_3$): δ 0.70 (s, Me_3C), 1.21, 1.35 (m, 2 × Me_2CH), 1.35 (s, Me_2C), 1.68 (s, Me_3CCH_2), 3.85, 4.02 (2 s, 2 × OMe), 4.32 (sept., 3 J = 6.8, Me_2CH), 4.75 (s, CH), 5.59 (s, NH), 7.06, 7.22 (2d, ^{3}J = 8.2, 2 × CH of isoindolone); ^{13}C NMR (62.5 MHz, $CDCl_3$): δ 20.45, 20.50 (Me_2CH); 27.8, 28.9 (Me_2CCH_2), 31.14 (Me_3C), 31.3 (Me_3C), 46.0 (Me_2CH), 52.5 (Me_3CCH2), 55.5 (Me_2CCH_2), 56.7, 62.3 (2 MeO), 63.2 (CH), 117.0, 117.8 (2 CH of isoindolone), 123.4, 134.8, 146.9, 152.9 (4 C of isoindolone), 167.8, 168.5 (2 C=O). Anal. calcd for $C_{22}H_{34}N_2O_4$: C, 67.66; H, 8.78; N, 7.17; Found: C, 67.54; H, 8.73; N, 7.20.

Reference

Ramazani, A., Mahyari, A., Farshadi, A., and Rouhani, M. (2011). Preparation of silica nanoparticles from organic laboratory waste of silica gel HF_{254} and their use as a highly efficient catalyst for the one-pot synthesis of 2,3-dihydro-1*H*-isoindolone derivatives. *Helv. Chim. Acta*, **94**, 1831–1837.

C-N bond_48

Type of reaction: C-N bond formation
Reaction conditions: DMSO-H_2O (1:1 *v/v*), room temperature
Synthetic strategy: One-pot multicomponent reaction
Catalyst: $NaHSO_3$ as an additive
Keywords: *ortho*-Phthalaldehyde/ 3,4-dimethoxy *ortho*-phthalaldehyde, amines, isonitriles, $NaHSO_3$, DMSO-H_2O (1:1), room temperature, one-pot three-component reaction, 1-carboxamido-3-oxoisoindoles

General reaction scheme

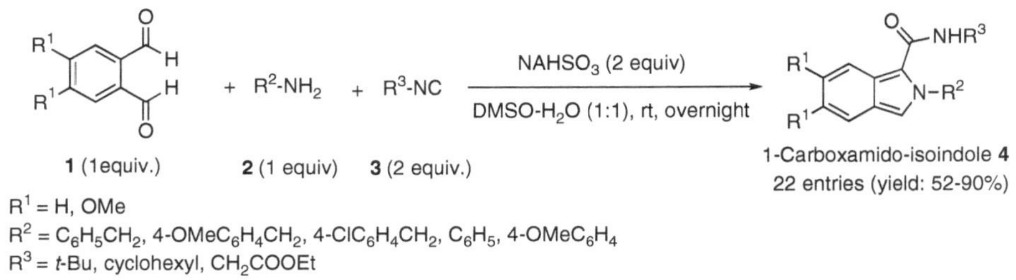

Representative entries

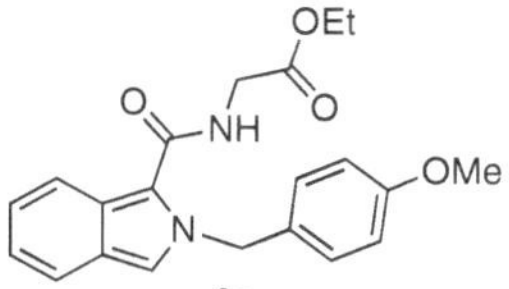

4a

Ethyl 2-(2-(4-methoxybenzyl)-2*H*-isoindole-1-carboxamido)acetate (**4a**): Yield: 76%

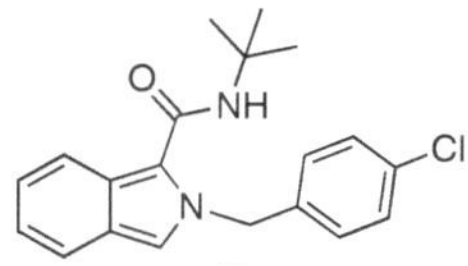

4b

N-(*tert*-Butyl)-2-(4-chlorobenzyl)-2*H*-isoindole-1-carboxamide (**4b**): Yield: 63%

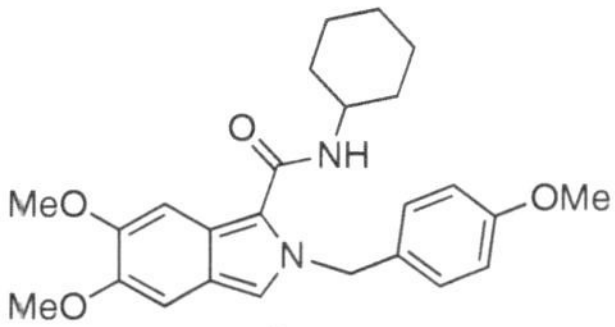

4c

N-Cyclohexyl-5,6-dimethoxy-2-(4-methoxybenzyl)-2*H*-isoindole-1-carboxamide (**4c**): Yield: 90%

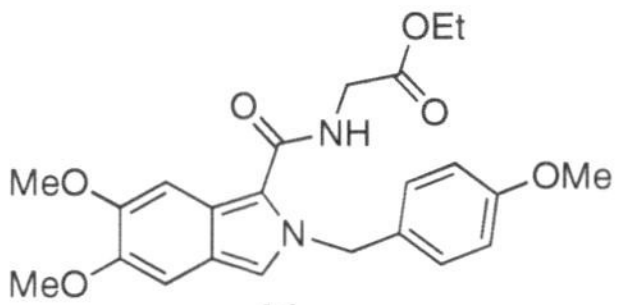

4d

Ethyl 2-(5,6-dimethoxy-2-(4-methoxybenzyl)-2*H*-isoindole-1-carboxamido)acetate (**4d**): Yield: 55%

Experimental procedure

A mixture of *ortho*-phthalaldehyde (**1**; 0.41 mmol; 1 equiv.) and $NaHSO_3$ (0.85 mmol; 2 equiv.) dissolved in DMSO-H_2O (1 mL/1 mL) in the reaction flask with was reacted for 0.5 h at room temperature, followed addition of by 1.0 equiv. of amine **2** and 2.0 equiv. of isonitrile **3**. The whole reaction mixture was then allowed to stir for overnight at room temperature. Thereafter, the reaction mixture was diluted with 30 mL ethyl acetate and extracted with saturated ammonium chloride (3 × 15 mL). The organic layer was dried over sodium sulfate, evaporated under vacuum and the crude product was purified using column chromatography to obtain 10carboxamido-isoindole **4**. In case of 3,4-dimethoxy *ortho*-phthalaldehyde addition of another solvent, acetonitrile is needed (*i.e.* DMSO-H_2O-CH_3CN (4:4:1 *v*/*v*/*v*). All the isolated products of isoindoles were characterized on the basis of spectral studies.

Characterization data of two representative compounds

N-(*tert*-Butyl)-2-(4-chlorobenzyl)-2*H*-isoindole-1-carboxamide (**4b**): IR (KBr): 2966, 2930, 1644, 1502, 1342, 1288, 1208, 1095, 1016, 771 cm^{-1}; ^1H NMR (300 MHz, DMSO-d_6): δ 7.69 (s, 1H), 7.64-7.56 (m, 3H), 7.39-7.28 (m, 4H), 7.05 (t, J=6.61 Hz, 1,), 6.94 (t, J=6.57 Hz, 1H), 5.79 (s, 2H), 1.39 (s, 9H); ^{13}C NMR (75 MHz, DMSO-d_6): δ 161.5, 137.5, 132.1, 129.5 (2C), 128.3 (2C), 124.4, 122.7, 122.6, 120.8, 120.0, 119.8, 117.0, 116.6, 51.0, 50.7, 28.7 (3C); EI-HRMS: Calcd for $C_{20}H_{21}N_2OCl$: *m*/*z* 340.1342; Found: *m*/*z* 340.1334.

Ethyl 2-(5,6-dimethoxy-2-(4-methoxybenzyl)-2*H*-isoindole-1-carboxamido)acetate (**4d**): IR (KBr): 2951, 1638, 1513, 1373, 1229, 1154, 1029 cm^{-1}; ^1H NMR (300 MHz, DMSO-d_6): 8.04 (t, J=5.50 Hz, 1H), 7.38 (s, 1H), 7.28 (d, J=8.83 Hz, 2H), 7.25 (s, 1H), 7.98 (d, J=8.84 Hz, 2H), 6.91 (s, 1H), 4.13-4.08 (q, J=7.12 Hz, 2H), 3.95 (d, J=5.90 Hz, 2H), 3.82 (s, 3H), 3.80 (s, 3H), 3.77 (s, 3H), 1.19 (t, J=7.10 Hz, 3H); ^{13}C NMR (75 MHz, DMSO-d_6): δ 170.3, 161.2, 158.4, 149.3, 147.6, 133.4, 126.3 (2C), 122.1, 118.8, 118.0, 115.4, 114.0 (2C), 98.3, 97.9, 60.5, 55.4, 55.2 (2C), 41.1, 14.1; EI-HRMS: Calcd for $C_{22}H_{24}N_2O_6$: *m*/*z* 412.1634; Found: *m*/*z* 412.1626.

Reference

Zhang, Y., Lee, C. L., Liu, H., and Li, X. (2012). A three-component reaction toward the synthesis of 1-carboxamido-isoindoles. *Org. Lett.*, **14**, 5146–5149.

C-N bond_49

Type of reaction: C-N and C-C bond formation
Reaction conditions: Solvent-free (neat), room temperature
Synthetic strategy: One-pot multicomponent reaction
Catalyst: Ytterbium triflate [$Yb(OTf)_3$]
Keywords: Ethyl pyruvate, aldehydes, amines, $Yb(OTf)_3$, solvent-free, room temperature, one-pot three-component reaction, pyrrol-2-ones, substituted lactums

General reaction scheme

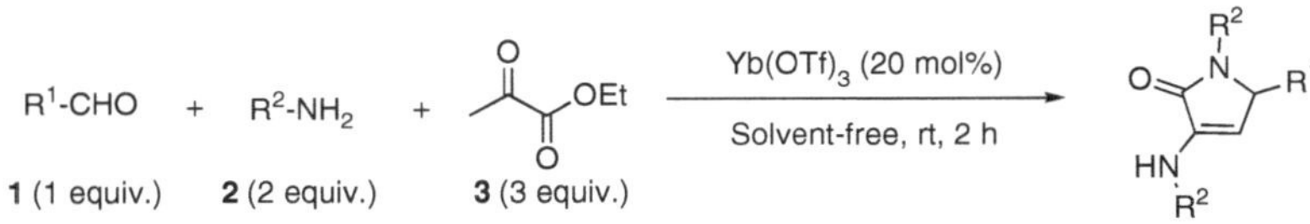

1 (1 equiv.) **2** (2 equiv.) **3** (3 equiv.)

Pyrrol-2-one **4**
13 entries (yield: 57-84%)

R^1 = 2-ClC_6H_4, 4-ClC_6H_4, 4-FC_6H_4, 3-$NO_2C_6H_4$, 4-$NO_2C_6H_4$, 4-OHC_6H_4, 4-$OMeC_6H_4$,3,4-$(OMe)_2C_6H_3$, cyclohexyl, *i*-Pr, *n*-hexyl

R^2 = 4-$OMeC_6H_4$, 2,3-$F_2C_6H_3$, 2-Cl,4-Br-C_6H_3, 2,5-$(C_2H_5)_2$,4-$OMeC_6H_2$, 2,4,6-$(CH_3)_3C_6H_2$, 2,4,6-$Br_3C_6H_2$, 2-Br,4-F-C_6H_3, Et, *n*-Pr

Representative entries

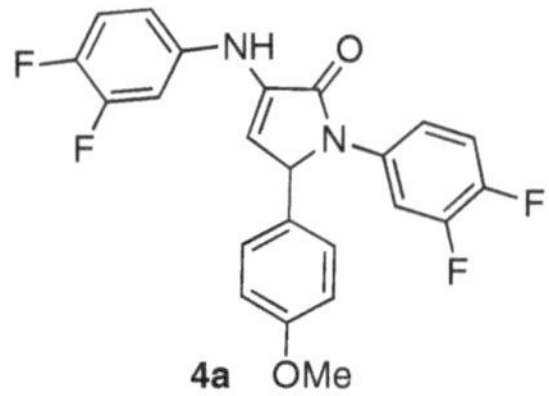

4a

1-(3,4-Difluorophenyl)-3-((3,4-difluorophenyl)amino)-5-(4-methoxyphenyl)-1*H*-pyrrol-2(5*H*)-one (**4a**): Yield: 65%

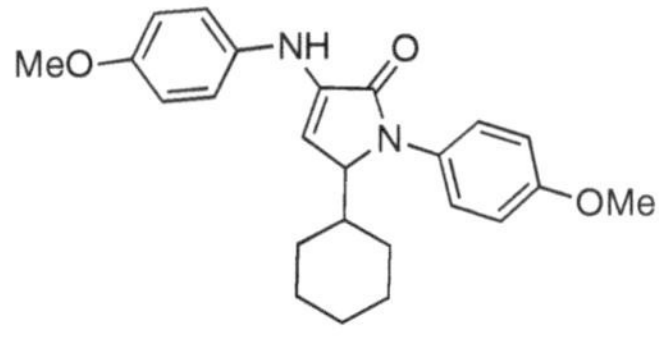

4b

5-Cyclohexyl-1-(4-methoxyphenyl)-3-((4-methoxyphenyl)amino)-1*H*-pyrrol-2(5*H*)-one (**4b**): Yield: 71%

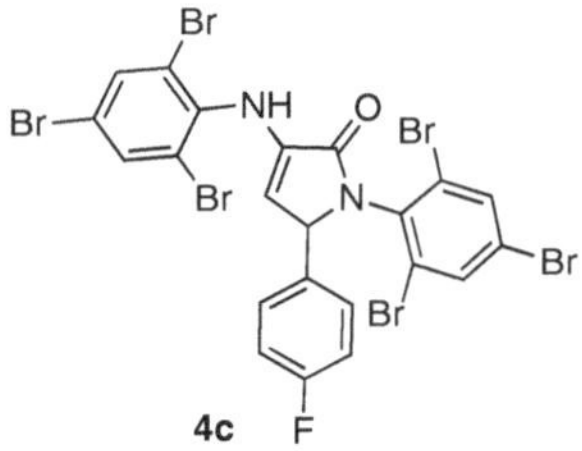

4c

5-(4-Fluorophenyl)-1-(2,4,6-tribromophenyl)-3-((2,4,6-tribromophenyl)amino)-1*H*-pyrrol-2(5*H*)-one (**4c**): Yield 57%

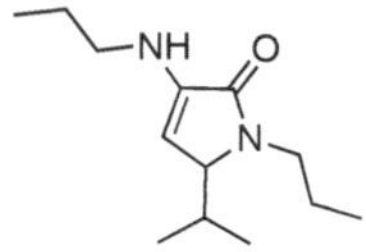

4d

5-Isopropyl-1-propyl-3-(propylamino)-1*H*-pyrrol-2(5*H*)-one (**4d**): Yield: 71%

Experimental procedure

To a stirred mixture of aldehyde (**1**; 0.1 mmol) and amine (**2**; 0.2 mmol) was added ethyl pyruvate (**3**; 0.3 mmol) and ytterbium triflate (20 mol%) successively, and vigorous stirring was continued at room temperature for 1-2 h. The precipitated solid was filtered and washed with a mixture of ethyl acetate-hexane (20:80 v/v), followed by purification by recrystallization from ethyl acetate and ethanol (70:30 v/v). All the products were characterized on the basis of elemental analyses and detailed spectral studies.

Characterization data of two representative compounds

1-(3,4-Difluorophenyl)-3-((3,4-difluorophenyl)amino)-5-(4-methoxyphenyl)-1*H*-pyrrol-2(5*H*)-one (**4a**): IR (KBr): 3357 (N–H), 1677 (C=O) cm^{-1}; ^{1}H NMR (300 MHz, $CDCl_3$): δ 3.75 (s, 3H), 5.72 (d, $J = 2.74$ Hz, 1 H), 6.04 (d, $J = 2.74$ Hz, 1 H), 6.72 (s, 1 H), 6.94 (d, $J = 9.06$ Hz, 2 H), 7.17-7.19 (m, 2 H), 7.25-7.30 (m, 2H), 7.35-7.39 (m, 3 H); ^{13}C NMR (75 MHz, DMSO-d_6): δ 54.7, 64.13, 108.5, 113.4, 118.5, 118.8, 122.5, 126.6, 128.9, 129.5, 131.7, 131.6, 132.4, 136.1, 136.8, 140.9, 54.1, 156.6, 158.1, 159.3, 166.99; LCQ-MS (ESI): *m/z* 428.1. Anal. calcd. for $C_{23}H_{15}F_4N_2O_2$: C, 64.49; H, 3.76; N, 6.54; Found C, 64.41; H, 3.69; N, 6.48.

5-Cyclohexyl-1-(4-methoxyphenyl)-3-((4-methoxyphenyl)amino)-1*H*-pyrrol-2(5*H*)-one (**4b**): IR (KBr): 3327 (s, N–H enamine), 1674 (s, C=O amide) cm^{-1}; ^{1}H NMR (300 MHz, $CDCl_3$): δ 1.97-2.24 (m, 10H), 3.74 (s, 3 H), 3.77 (s, 3 H), 5.96 (d, $J = 2.74$ Hz, 1 H), 6.07 (d, $J = 2.47$ Hz, 1 H), 6.33 (s, 1 H), 6.44 (t, $J = 2.47$ Hz, 2 H,), 6.81-6.86 (m, 2H), 7.03 (d, $J = 9.33$ Hz, 2 H), 7.48 (d, $J = 9.33$ Hz, 2 H); ^{13}C NMR (75 MHz, $CDCl_3$): δ 15.7, 19.3, 21.9, 25.7, 26.6, 27.5, 55.9, 64.2, 98.5, 104.7, 106.8, 117.9, 118.2, 122.2, 127.4, 135.7, 154.1, 159.1, 166.7; MS (ES^+): LCQ-MS (ESI) *m/z* 392.2. Anal. calcd. for $C_{24}H_{28}N_2O_3$: C, 73.44; H, 7.19; N, 7.14; Found C, 73.37; H, 7.23; N, 7.21.

Reference

Nagarajan, A. S., and Reddy, B. S. R. (2013). Efficient one-pot synthesis of substituted lactams via three-component solvent-free reaction catalyzed by ytterbium triflate. *Synth. Commun.*, **43**, 1229–1236.

C-N bond_50

Type of reaction: C-N and C-C bond formation
Reaction conditions: Ethanol, room temperature
Synthetic strategy: One-pot multicomponent reaction
Catalyst: Benzoic acid
Keywords: Ethyl glyoxylate, acetylenedicarboxylates, amines, benzoic acid, ethanol, room temperature, domino reaction, one-pot synthesis, functionalized 2-pyrrolidiones

General reaction schemes

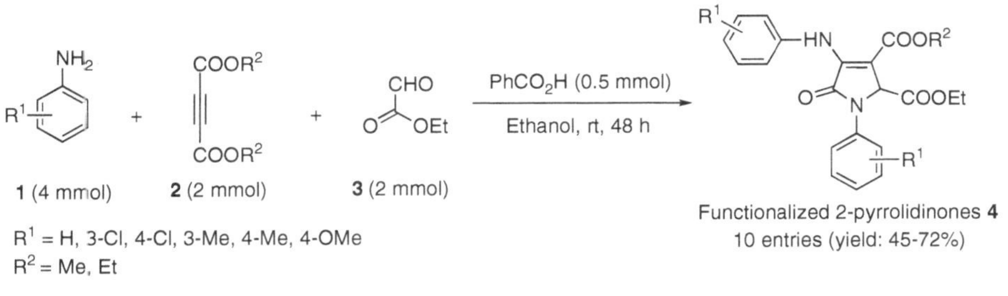

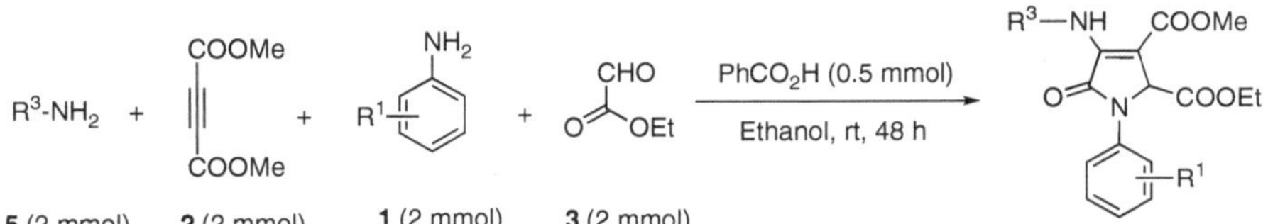

5 (2 mmol) 2 (2 mmol) 1 (2 mmol) 3 (2 mmol)

$R^3 = C_6H_5CH_2, C_6H_5CH_2CH_2, C_6H_5CH(CH_3)$
R^1 = H, Cl, Br, Me, OMe

Functionalized 2-pyrrolidinones **6**
13 entries (yield: 35-75%)

Representative entries

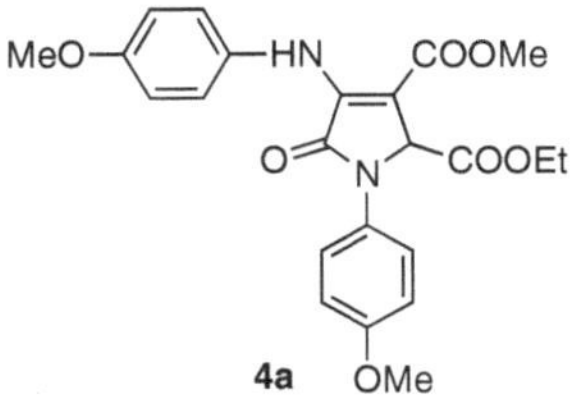

4a

2-Ethyl 3-methyl 1-(4-methoxyphenyl)-4-((4-methoxyphenyl)amino)-5-oxo-2,5-dihydro-1*H*-pyrrole-2,3-dicarboxylate (**4a**): Light yellow solid, mp 122-123 °C; yield: 72%

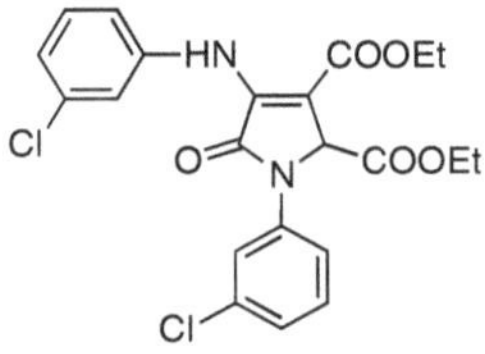

4b

Diethyl 1-(3-chlorophenyl)-4-((3-chlorophenyl) amino)-5-oxo-2,5-dihydro-1*H*-pyrrole-2,3-dicarboxylate (**4b**): White solid, mp 99-101 °C; yield: 47%

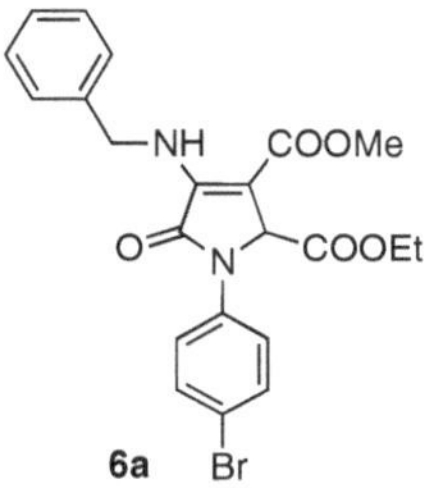

6a

2-Ethyl 3-methyl 4-(benzylamino)-1-(4-bromophenyl)-5-oxo-2,5-dihydro-1*H*-pyrrole-2,3-dicarboxylate (**6a**): White solid, mp 127-128 °C; yield: 40%

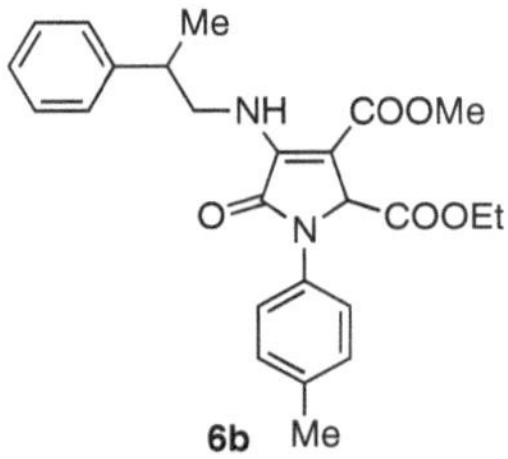

6b

2-Ethyl 3-methyl 5-oxo-4-((2-phenylpropyl) amino)-1-(*p*-tolyl)-2,5-dihydro-1*H*-pyrrole-2,3-dicarboxylate (**6b**): Light orange solid, mp 148-150 °C; yield: 55%

Experimental procedure

Preparation of functionalized 2-pyrrolidiones **4**: In a round bottom flask a mixture of arylamine (**1**: 4.0 mmol) and acetylenedicarboxylate (**2**: 2.0 mmol) in 5 mL ethanol was stirred at room temperature for 10 min. To this solution added ethyl glyoxylate (**3**: 2.0 mmol, 50% soln in toluene, 0.45 g) and benzoic acid (0.5 mmol, 0.060 g), and the resulting mixture was stirred at room temperature for two days. The crude precipitate was filtered, followed by washing with cold alcohol to furnish the pure product **4**.

Preparation of functionalized 2-pyrrolidiones **6**: In a round bottom flask a mixture of aliphatic amine (**5**: 2.0 mmol) and dimethyl acetylenedicarboxylate (**2**: 2.0 mmol) in 5 mL ethanol was stirred at room temperature for 10 min. To this solution added arylamine (**1**: 2.0 mmol), ethyl glyoxylate (**3**: 2.0 mmol, 50% soln in toluene, 0.45 g) and benzoic acid (0.5 mmol,

0.060 g), and the resulting mixture was stirred at room temperature for two days. The crude precipitate was filtered, followed by washing with cold alcohol to furnish the pure product **6**.

All the products (**4** and **6**) were characterized on the basis of elemental analyses and detailed spectral studies.

Characterization data of two representative compounds

Diethyl 1-(3-chlorophenyl)-4-((3-chlorophenyl)amino)-5-oxo-2,5-dihydro-1*H*-pyrrole-2,3-dicarboxylate (**4b**): White solid, mp 99-101 °C; IR (KBr): 3464, 3238, 3121, 3070, 2982, 2913, 1743, 1697, 1636, 1590, 1541, 1480, 1393, 1333, 1296, 1212, 1120, 1034, 929, 872, 781 cm^{-1}; ^1H NMR (600 MHz, $CDCl_3$): δ 8.28 (s, 1H, NH), 7.74 (s, 1H, ArH), 7.54 (t, $J=8.1$ Hz, 1H, ArH), 7.33 (t, $J=8.1$ Hz, 1H, ArH), 7.26 (d, $J=7.8$ Hz, 1H, ArH), 7.21 (t, $J=3.9$ Hz, 1H, ArH), 7.20-7.17 (m, 2H, ArH), 7.10-7.08 (m, 1H, ArH), 5.42 (s, 1H, CH), 4.28-4.15 (m, 4H, CH_2), 1.26 (t, $J=6.9$ Hz, 3H, CH_3), 1.18 (t, $J=7.2$ Hz, 3H, CH_3); ^{13}C NMR (150 MHz, $CDCl_3$): δ 168.0, 163.9, 143.4, 139.1, 137.7, 135.0, 134.1, 130.2, 129.4, 126.3, 123.4, 121.6, 121.5, 119.3, 104.3, 62.3, 61.6, 60.8, 14.1, 14.0; MS (*m*/*z*): 461.25 ($[M-1]^+$) 100%; HRMS (ESI): Calcd for $C_{22}H_{19}Cl_2N_2O_5$ ($[M-H]^+$): 461.0677; Found: 461.0676.

2-Ethyl 3-methyl 5-oxo-4-((2-phenylpropyl)amino)-1-(*p*-tolyl)-2,5-dihydro-1*H*-pyrrole-2,3-dicarboxylate (**6b**): Light orange solid, mp 148-150 °C; IR (KBr): 3457, 3327, 3022, 2974, 2927, 1931, 1740, 1699, 1623, 1515, 1454, 1381, 1311, 1202, 1116, 1021, 906, 807, 760 cm^{-1}; ^1H NMR (600 MHz, $CDCl_3$): δ 7.38 (d, $J=6.6$ Hz, 2H, ArH), 7.34-7.29 (m, 4H, ArH), 7.22 (d, $J=6.6$ Hz, 1H, ArH), 7.16 (d, $J=7.2$ Hz, 2H, ArH), 6.14 (s, 1H, CH), 5.25 (d, $J=21.6$ Hz, 1H, CH), 4.10-4.02 (m, 2H, CH_2), 3.76 (s, 3H, CH_3), 2.31 (s, 3H, CH_3), 1.59 (d, $J=6.6$ Hz, 3H, CH_3), 1.07 (s, 3H, CH_3); ^{13}C NMR (150 MHz, $CDCl_3$): δ 168.7, 136.2, 133.9, 129.7, 128.7, 128.6, 127.1, 125.9, 122.4, 61.7, 51.1, 24.3, 20.9, 14.0; HRMS (ESI) calcd for $C_{24}H_{25}N_2O_5$ ($[M-H]^+$): 421.1769; Found: 421.1768.

Reference

Gao, H., Sun, J., and Yan, C.-G. (2013). Synthesis of functionalized 2-pyrrolidinones via domino reactions of arylamines, ethyl glyoxylate and acetylenedicarboxylates. *Tetrahedron*, **69**, 589–594.

C-N bond_51

Type of reaction: C-N formation
Reaction conditions: Ethanol, room temperature
Synthetic strategy: Sequential three-component one-pot reaction
Catalyst: Catalyst-free
Keywords: 1,2-Diaza-1,3-dienes, amines, isocyanates, ethanol, room temperature, sequential three-component reaction, aza-Michael reaction, one-pot synthesis, functionalized 1,3,5-trisubstituted hydantoins

General reaction scheme

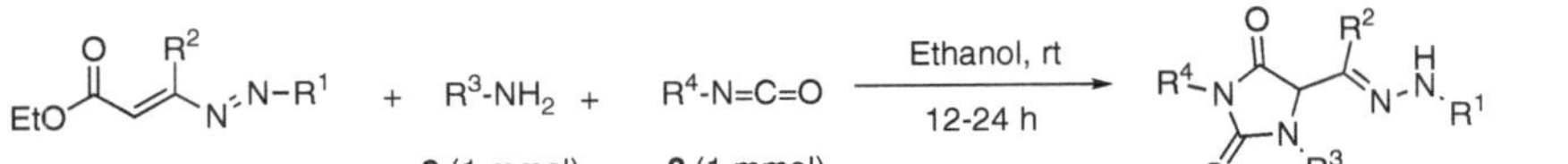

1 (1 mmol) **2** (1 mmol) **3** (1 mmol)

R^1 = COOMe, COOEt, COO*t*-Bu
R^2 = Me, Et, CH_2COOEt, C_6H_5
R^3 = *n*-Pr, *n*-Bu, allyl, propargyl, $CH_2C_6H_5$
R^4 = *n*-Bu, cyclohexyl, CH_2COOEt, C_6H_5, 3-ClC_6H_4, 4-ClC_6H_4

1,3,5-Trisubstituted hydantoin **4**
16 entries (yield: 48-76%)

Plausible mechanism

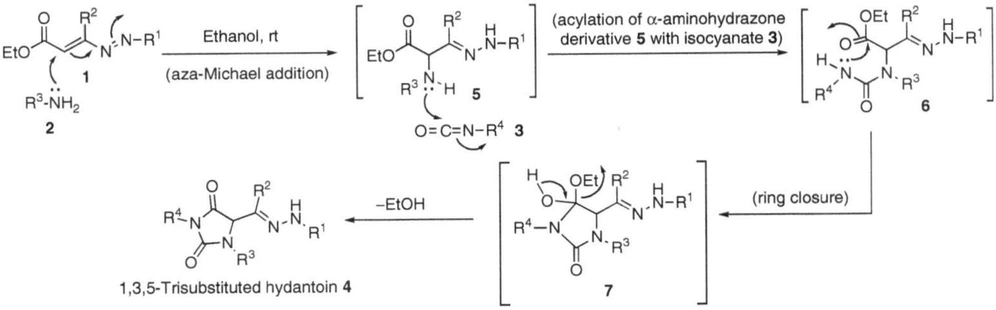

Representative entries

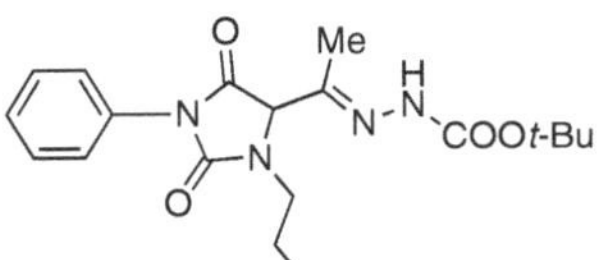

4a

(*E*)-*tert*-Butyl 2-(1-(2,5-dioxo-1-phenyl-3-propylimidazolidin-4-yl)ethylidene) hydrazinecarboxylate (**4a**): White powder, mp 179-180 °C; yield: 76%

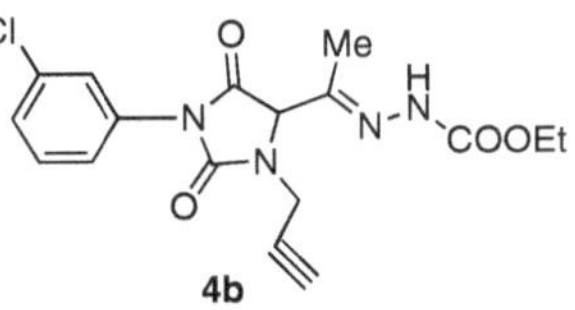

4b

(*E*)-Ethyl 2-(1-(1-(3-chlorophenyl)-2,5-dioxo-3-(prop-2-yn-1-yl)imidazolidin-4-yl)ethylidene) hydrazinecarboxylate (**4b**): White powder, mp 151-153 °C; yield: 63%

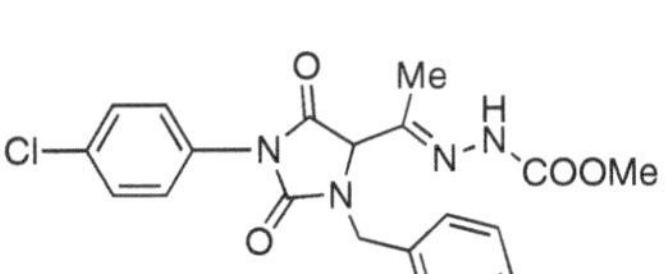

4c

(*E*)-Methyl 2-(1-(3-benzyl-1-(4-chlorophenyl)-2,5-dioxoimidazolidin-4-yl)ethylidene) hydrazinecarboxylate (**4c**): White powder, mp 192-194 °C; yield: 73%

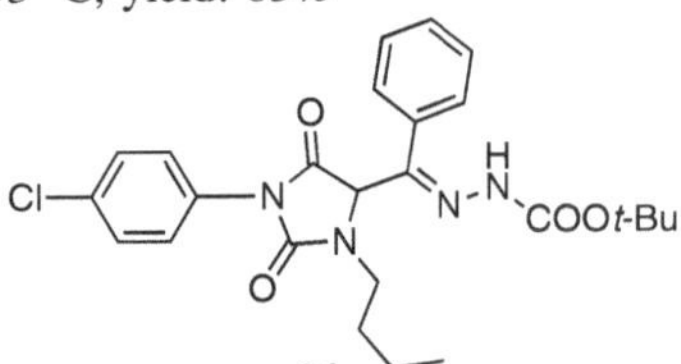

4d

(*E*)-*tert*-Butyl 2-((3-butyl-1-(4-chlorophenyl)-2,5-dioxoimidazolidin-4-yl)(phenyl)methylene) hydrazinecarboxylate (**4d**): White powder, mp 172-176 °C; yield: 48%

Experimental procedure

To a stirred solution of 1,2-diaza-1,3-diene (**1**: 1 mmol) in EtOH was added amine (**2**; 1 mmol), and stirred at room temperature. On dissolution of **1** (within 0.1-0.5 h as checked by TLC), isocyanate **3** was added and the reaction mixture was allowed to stand at room temperature (12-24 h). In some cases solid hydantoin derivative **4** was obtained by precipitation from the reaction medium or by crystallization from appropriate solvents, while, in other cases, it was obtained after removal of ethanol *in vacuo* and subsequent column chromatographic purification. All the products were characterized on the basis of elemental analyses and detailed spectral studies.

Characterization data of two representative compounds

(*E*)-Ethyl 2-(1-(1-(3-chlorophenyl)-2,5-dioxo-3-(prop-2-yn-1-yl)imidazolidin-4-yl)ethylidene)hydrazinecarboxylate (**4b**): White powder, mp 151-153 °C; IR (nujol): 3244, 3209, 3038, 1780, 1722, 1709, 1645, 1595, 1583, 1538 cm^{-1}; ^{1}H NMR (400 MHz, DMSO-d_6): δ 1.85 (s, 3H, CH_3), 3.36 (s, 1H, CH), 3.70 (s, 3H, OCH_3), 4.21 (s, 2H, CH_2), 4.88 (s, 1H, CH), 7.42 (s, 1H, ArH), 7.51-7.57 (m, 3H, ArH), 10.37 (s, 1H, NH); ^{13}C NMR (100 MHz, DMSO-d_6): δ 12.4 (q), 31.4 (t), 52.0 (q), 66.5 (d), 75.6 (d), 77.6 (s), 125.4 (d), 126.5 (d), 128.2 (d), 130.5 (d), 132.9 (s), 133.0 (s), 144.5 (s), 154.2 (s), 154.4 (s), 168.0 (s); MS (EI): *m*/*z* (%) = 364 [M^+ + 2 (1)], 362 (M+, 3), 288 (10), 153 (22), 135 (27), 111 (100). Anal. calcd. for $C_{16}H_{15}ClN_4O_4$: C, 52.97; H, 4.17; N, 15.44; Found: C, 52.84; H, 4.26; N, 15.35.

(*E*)-Methyl 2-(1-(3-benzyl-1-(4-chlorophenyl)-2,5-dioxoimidazolidin-4-yl)ethylidene) hydrazinecarboxylate (**4c**): White powder, mp 192-194 °C; IR (nujol): 3211, 1780, 1732, 1713, 1704, 1637, 1542 cm^{-1}; ^{1}H NMR (400 MHz, DMSO-d_6): δ 1.84 (s, 3H, CH_3), 3.69 (s, 3H, OCH_3), 4.44 (d, *J* = 15.2 Hz, 1H, NCH_aH_bPh), 4.54 (d, *J* = 15.2 Hz, 1H, NCH_aH_bPh), 4.79 (s, 1H, CH), 7.27-7.36 (m, 5H, ArH), 7.47-7.50 (m, 2H, ArH), 7.57-7.60 (m, 2H, ArH), 10.22 (s, 1H, NH); ^{13}C NMR (100 MHz, DMSO-d_6): δ 12.1 (q), 45.4 (t), 52.0 (q), 66.8 (d), 127.6 (d), 128.3 (d), 128.5 (d), 128.6 (d), 128.9 (d), 130.8 (s), 132.6 (s), 135.9 (s), 144.6 (s), 154.3 (s), 154.9 (s), 168.3 (s); MS (EI): *m*/*z* (%) = 416 [M^+ + 2 (1)], 414 (M+, 3), 340 (10), 324 (89), 209 (4), 153 (57), 132 (100). Anal. calcd. for $C_{20}H_{19}ClN_4O_4$: C, 57.90; H, 4.62; N, 13.51; Found: C, 58.02; H, 4.72; N, 13.47.

Reference

Attanasi, O. A., De Crescentini, L., Favi, G., Nicolini, S., Perrulli, F. R., and Santeusanio, S. (2011). 1,3,5-Trisubstituted and 5-acyl-1,3-disubstituted hydantoin derivatives via novel sequential three-component reaction. *Org. Lett.*, **13**, 353–355.

C-N bond_52

Type of reaction: C-N bond formation
Reaction conditions: Chloroform, room temperature
Synthetic strategy: Sequential three-component one-pot reaction
Catalyst: Catalyst-free
Keywords: 1,2-Diaza-1,3-dienes, amines, isothiocyanates, chloroform, room temperature, sequential three-component reaction, one-pot synthesis, 2-thiohydantions

General reaction scheme

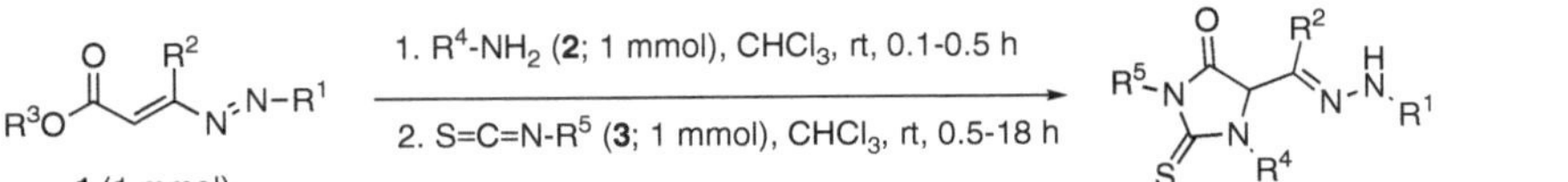

1 (1 mmol)

1,3,5-Trisubstituted 2-thiohydantoin **4**
12 entries (yield: 54-97%)

R^1 = COOMe,COO*t*-Bu, $COONH_2$, $COONHC_6H_5$
R^2 = Me, C_6H_5; R^3 = Me, Et
R^4 = *n*-Pr, *n*-Bu, allyl, $CH_2C_6H_5$
R^5 = *n*-Bu, allyl, C_6H_5, $CH_2C_6H_5$, 4-ClC_6H_4, 4-$OMeC_6H_4$

Plausible mechanism

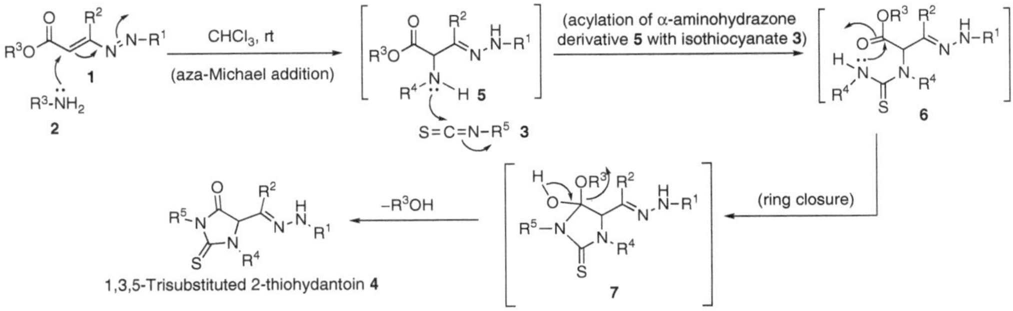

Representative entries

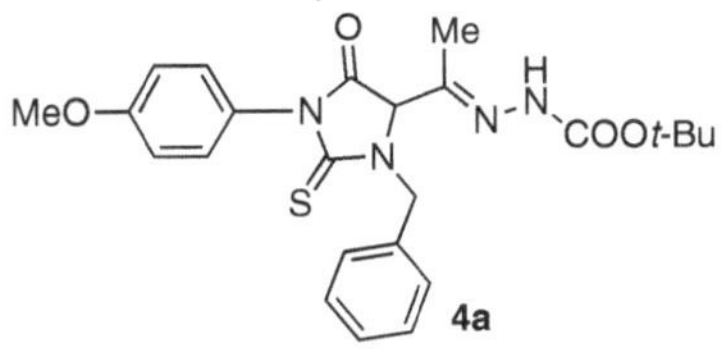

4a

(*E*)-*tert*-Butyl 2-(1-(3-benzyl-1-(4-methoxyphenyl)-5-oxo-2-thioxoimidazolidin-4-yl)ethylidene)hydrazinecarboxylate (**4a**): White powder, mp 193-195 °C; yield: 86%

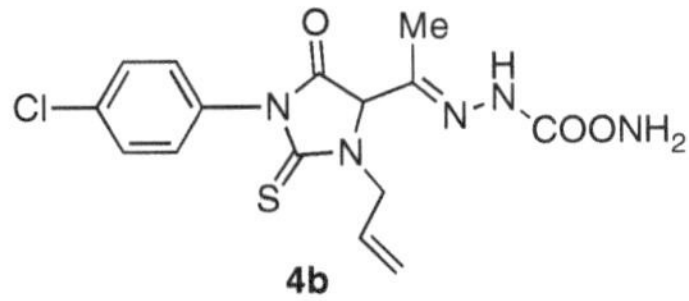

4b

(*E*)-1-Allyl-5-(1-(2-((aminooxy)carbonyl) hydrazono)ethyl)-3-(4-chlorophenyl)-2-thioxoimidazolidin-4-one (**4b**): White powder, mp 182-184 °C; yield: 97%

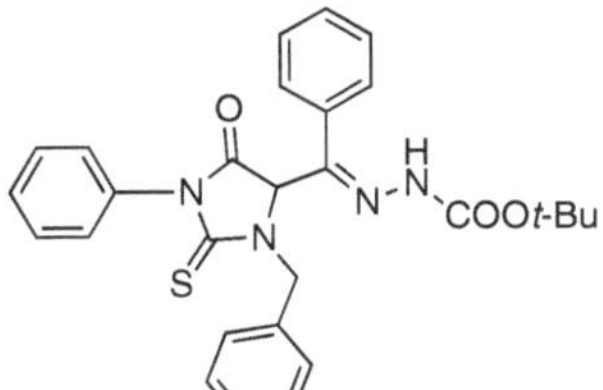

4c

(*E*)-*tert*-Butyl 2-((3-benzyl-5-oxo-1-phenyl-2-thioxoimidazolidin-4-yl)(phenyl)methylene) hydrazinecarboxylate (**4c**): White powder, mp 181-183 °C; yield: 75%

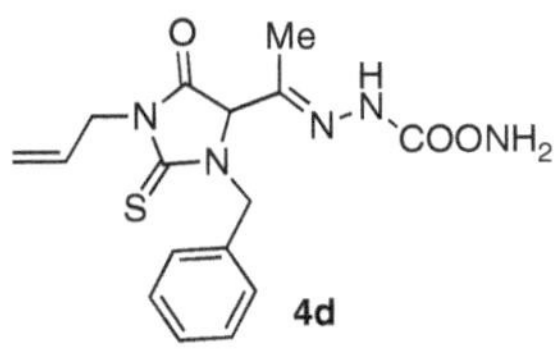

4d

(*E*)-3-Allyl-5-(1-(2-((aminooxy)carbonyl) hydrazono)ethyl)-1-benzyl-2-thioxoimidazolidin-4-one (**4d**): White powder, mp 163-166 °C; yield: 54%

Experimental procedure

To a stirred solution of 1,2-diaza-1,3-diene (**1**: 1 mmol) in chloroform was added amine (**2**; 1 mmol), and stirred at room temperature. On disappearance of the reagents (within 0.1-0.5 h as checked by TLC), isothiocyanate **3** was added to the reaction vessel and the reaction mixture was allowed to stand at room temperature until completion of the reaction (0.5-18 h). In some cases solid thiohydantoin derivative **4** was obtained by precipitation from the reaction medium or by crystallization from appropriate solvents, while, in other cases, it was obtained after removal of chloroform *in vacuo* and subsequent column chromatographic purification (cyclohexane/ethyl acetate mixtures). All the products were characterized on the basis of elemental analyses and detailed spectral studies.

Characterization data of two representative compounds

(*E*)-*tert*-Butyl 2-(1-(3-benzyl-1-(4-methoxyphenyl)-5-oxo-2-thioxoimidazolidin-4-yl)ethylidene)hydrazinecarboxylate (**4a**): White powder, mp 193-195 °C; IR (nujol): 3230, 1755, 1715, 1638, 1620, 1592 cm^{-1}; ^{1}H NMR (400 MHz, DMSO-d_6): δ 1.48 (s, 9H), 1.55 (s, 3H), 3.80 (s, 3H), 4.84 (d, *J* = 15.2 Hz, 1H), 5.01 (s, 1H), 5.11 (d, *J* = 15.2 Hz, 1H), 7.04 (d, *J* = 6.8 Hz, 2H), 7.29-7.39 (m, 7H), 9.98 (s, 1H);); ^{13}C NMR (100 MHz, DMSO-d_6): δ 12.5 (q), 28.0 (q), 48.8 (t), 55.4 (q), 69.1 (d), 79.8 (s), 114.1 (d), 126.3 (s), 127.7 (d), 128.4 (d), 128.5 (d), 129.9 (d), 135.3 (s), 142.9 (s), 152.7 (s), 159.4 (s), 170.1(s), 183.1 (s). MS (EI): *m/z* at 468 (M^+, 5), 412 (15), 368 (17), 352 (25), 311 (9), 277 (12), 165 (100), 106 (40). Anal. Calcd for $C_{24}H_{28}N_4O_4S$ (468.56): C, 61.52; H, 6.02; N, 11.96; Found: C, 61.38; H, 6.14; N, 11.89.

(*E*)-3-Allyl-5-(1-(2-((aminooxy)carbonyl)hydrazono)ethyl)-1-benzyl-2-thioxoimidazolidin-4-one (**4d**): White powder, mp 163-166 °C; IR (nujol): 3414, 3287, 3285, 3157, 1750, 1702, 1671, 1628, 1584 cm^{-1}; ^{1}H NMR (400 MHz, DMSO-d_6): δ 1.58 (s, 3H), 4.09 (dd, *J* = 15.4, 6.0 Hz, 1H), 4.49 (dd, J = 15.4, 6.0 Hz, 1H), 4.89-5.01 (m, 3H), 5.16-5.20 (m, 2H), 5.75-5.82 (m, 1H), 6.39 (br s, 2H), 7.26-7.35 (m, 5H), 9.59 (s, 1H); ^{13}C NMR (100 MHz, DMSO-d_6): δ 12.5 (q), 44.7 (t), 47.8 (t), 68.3 (d), 119.0 (t), 127.4 (d), 128.4 (d), 130.9 (d), 136.0 (s), 138.3 (s), 156.5 (s), 170.6 (s), 182.0 (s); MS (EI): *m/z* (%) at 345 (M^+, 2),328 (7), 286 (100), 270 (43), 245 (17), 205 (52), 106 (27), 100 (58). Anal. Calcd for $C_{16}H_{19}N_5O_2S$ (345.42): C, 55.63; H, 5.54; N, 20.27; Found: C, 55.50; H, 5.66; N, 20.34.

Reference

Attanasi, O. A., Bartoccini, S., Favi, G., Giorgi, G., Perrulli, F. R., and Santeusanio, S. (2012a). Powerful approach to heterocyclic skeletal diversity by sequential three-component reaction of amines, isothiocyanates, and 1,2-diaza-1,3-dienes. *J. Org. Chem.*, **77**, 1161–1167.

C-N bond_53

Type of reaction: C-N bond formation
Reaction conditions: DMF/Et_3N, room temperature
Synthetic strategy: Sequential three-component one-pot reaction
Catalyst: $Cu(OAc)_2.H_2O$, 1,10-phenanthroline (as ligand)

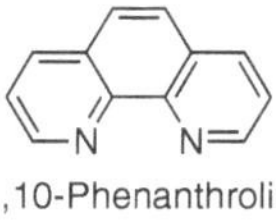

1,10-Phenanthroline

Keywords: 2-Bromoanilines, isothiocyanates, sodium azide, copper acetate, potassium phosphate, 1,10-phenanthroline, DMF, Et_3N, room temperature, sequential three-component reaction, tandem addition, electrocyclization, *N*-arylation, 1-substituted-2-azido-1*H*-benzo[*d*]imidazoles

General reaction scheme

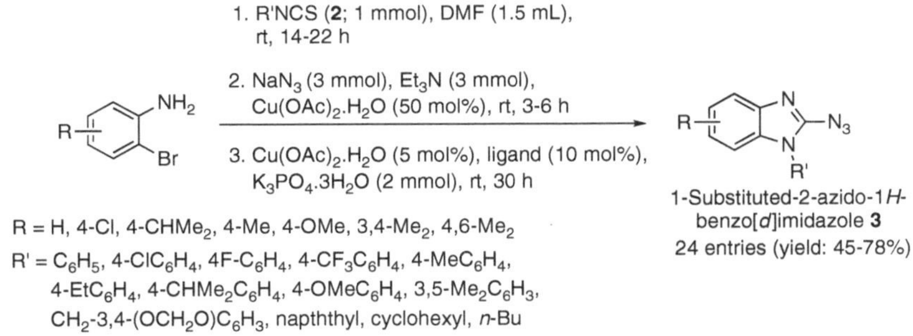

Representative entries

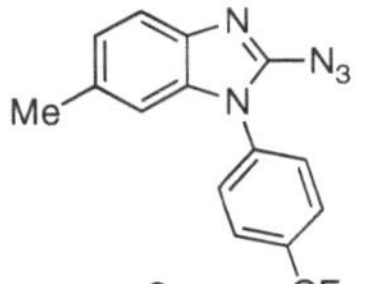

3a

2-Azido-6-methyl-1-(4-(trifluoromethyl)phenyl)-1*H*-benzo[*d*]imidazole (**3a**): White solid, mp 135-136 °C; reaction time: 30 h; yield: 45%

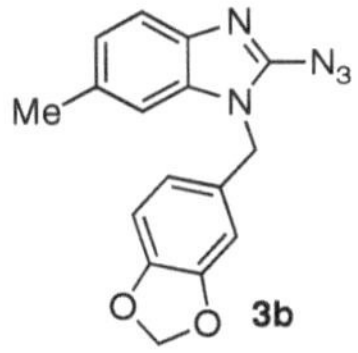

3b

2-Azido-1-(benzo[*d*][1,3]dioxol-5-ylmethyl)-6-methyl-1*H*-benzo[*d*]imidazole (**3b**): White solid, mp 101-102 °C; reaction time: 30 h; yield: 49%

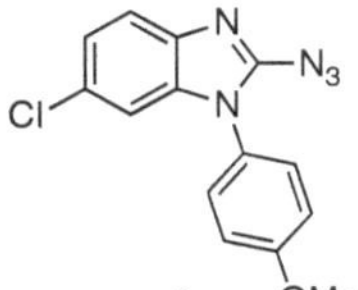

3c

2-Azido-6-chloro-1-(4-methoxyphenyl)-1*H*-benzo[*d*]imidazole (**3c**): White solid, mp 104-105 °C; reaction time: 25 h; yield: 70%

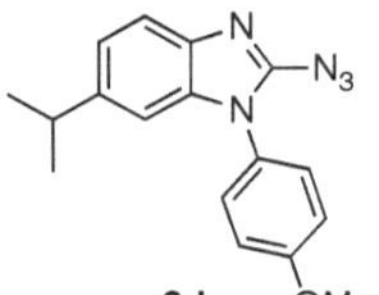

3d

2-Azido-6-isopropyl-1-(4-methoxyphenyl)-1*H*-benzo[*d*]imidazole (**3d**): White solid, mp 106-106 °C; reaction time: 27 h; yield: 70%

Experimental procedure

Initially, 2-bromoaniline (**1**; 1 mmol) and isothiocyanate (**2**; 1 mmol) were stirred at room temperature in DMF (1.5 mL) for 14-22 h. Then, sodium azide (3 mmol), $Cu(OAc)_2 \cdot H_2O$ (50 mol %) and triethyl amine (3 mmol) were added. The resulting mixture was stirred for 4 h at room temperature, and the black colored CuS precipitate formed was removed from the reaction mixture by centrifugation. To this solution was added $K_3PO_4 \cdot 3H_2O$ (2 mmol), $Cu(OAc)_2 \cdot H_2O$ (5 mol %) and 1,10-phenanthroline (10 mol %), and the stirring was further continued for 25-30 h at room temperature. On completion of the reaction (as monitored by TLC), the reaction mixture was extracted with ethyl acetate (2×10 mL) and the organic layer was washed successively with brine (1×5 mL) and water (2×5 mL). Drying and evaporation of the solvent provided a residue which on purification by using silica gel chromatography (ethyl acetate-hexane mixture as eluent) afforded the desired product 1-substituted-2-azido-1*H*-benzo[*d*]imidazole (**3**). All the products were characterized on the basis of elemental analyses and spectral studies.

Characterization data of two representative compounds

2-Azido-6-methyl-1-(4-(trifluoromethyl)phenyl)-1*H*-benzo[d]imidazole (**3a**): White solid, mp 135-136 °C; IR (KBr): 3072, 2923, 2416, 2218, 2141, 1614, 1507, 1434, 1167, 1005 cm^{-1}; ^{1}H NMR (400 MHz, $CDCl_3$): δ 7.81 (d, $J=8.4$ Hz, 2H), 7.56 (dd, $J=2.4$, 8.4 Hz, 3H), 7.11 (d, $J=8.0$ Hz, 1H), 7.00 (s, 1H), 2.41 (s, 3H); 13C NMR (100 MHz, $CDCl_3$): δ 146.6, 139.7, 137.7, 135.1, 133.4, 127.0 (q, $^{1}J_{CF}=99.9$ Hz), 125.1, 123.1, 118.3, 113.3, 109.9, 21.9. Anal. Calcd. for $C_{15}H_{10}F_3N_5$: C, 56.78; H, 3.18; N, 22.07; Found: C, 56.91; H, 3.17; N, 22.00.

2-Azido-1-(benzo[*d*][1,3]dioxol-5-ylmethyl)-6-methyl-1*H*-benzo[d]imidazole (**3b**): White solid, mp 101-102 °C; IR (KBr): 3275, 2962, 2149, 1610, 1500, 1448, 1260, 1097, 1038 cm^{-1}; ^{1}H NMR (400 MHz, $CDCl_3$): δ 7.57 (s, 1H), 7.49 (d, $J=8.4$ Hz, 1H), 7.04 (d, $J=8.4$ Hz, 1H), 6.98 (s, 1H), 6.75-6.73 (m, 2H), 5.93 (s, 2H), 5.02 (s, 2H), 2.42 (s, 3H); ^{13}C NMR (100 MHz, $CDCl_3$): δ 155.5, 148.1, 147.5, 134.7, 131.4, 129.9, 129.3, 124.1, 121.5, 121.2, 120.7, 117.9, 101.3, 48.4, 21.2. Anal. Calcd. for $C_{16}H_{13}N_5O_2$: C, 62.53; H, 4.26; N, 22.79; Found: C, 62.64; H, 4.24; N, 22.72.

Reference

Ramana, T., and Punniyamurthy, T. (2012). Preparation of 2-azido-1-substituted-1*H*-benzo[*d*]imidazoles using a copper-promoted three-component reaction and their further conversion into 2-amino and 2-triazolyl derivatives. *Chem. Eur. J.*, **18**, 13279–13283.

C-N bond_54

Type of reaction: C-N bond formation
Reaction conditions: Dichloromethane, room temperature
Synthetic strategy: Sequential four-component one-pot reaction
Catalyst: $Cu(OAc)_2.H_2O$
Keywords: 1,2-Diaza-1,3-dienes, trimethylsililazide, copper(II) acetate, triphenyl phosphine, isocyanates/isothiocyanates, dichloromethane, room temperature, sequential four-component aza-Michael, Staudinger and aza-Wittig reactions, one-pot synthesis, 1,2-diaminoimidazoles

General reaction scheme

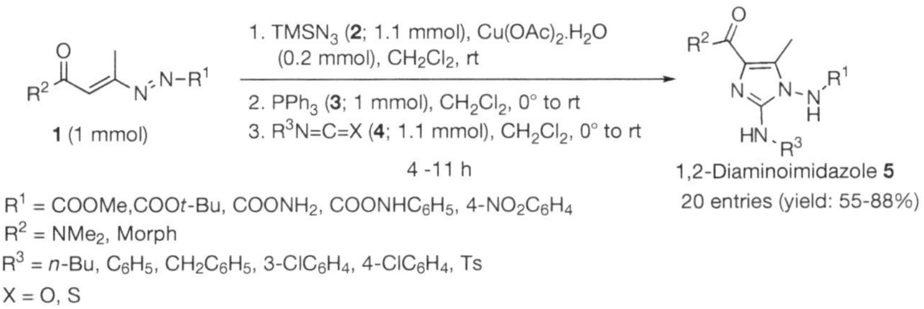

Plausible mechanism

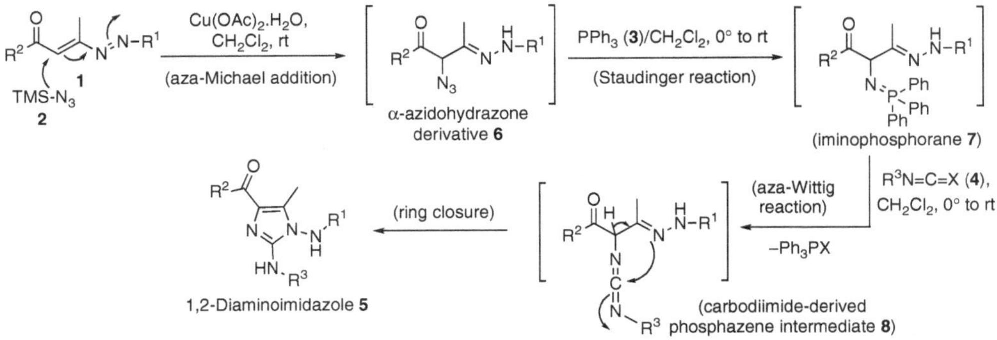

Representative entries

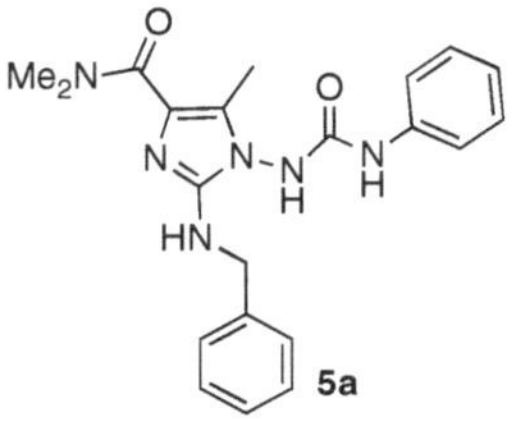

5a

2-(Benzylamino)-*N*,*N*,5-trimethyl-1-(3-phenylureido)-1*H*-imidazole-4-carboxamide (**5a**): White powder, mp 192-195 °C; reaction time: 4 h; yield: 73%

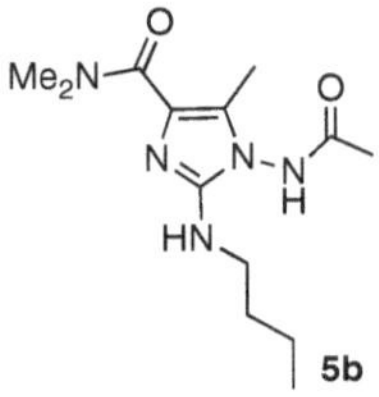

5b

1-Acetamido-2-(butylamino)-*N*,*N*,5-trimethyl-1*H*-imidazole-4-carboxamide (**5b**): Pale yellow crystals, mp 79-82 °C; reaction time: 7 h; yield: 55%

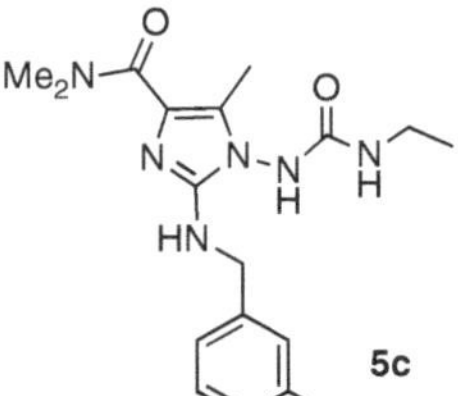

5c

2-((3-Chlorobenzyl)amino)-1-(3-ethylureido)-*N*,*N*,5-trimethyl-1*H*-imidazole-4-carboxamide (**5c**): White powder, mp 192-194 °C; reaction time: 7 h; yield: 88%

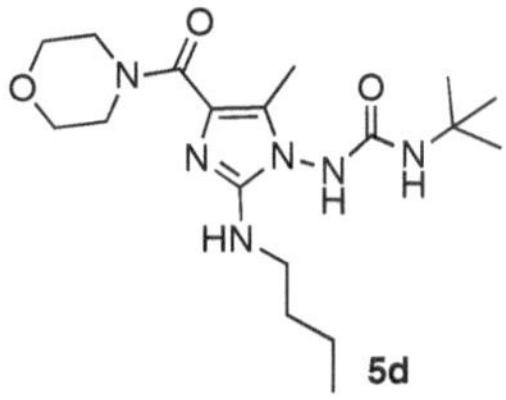

5d

1-(*tert*-Butyl)-3-(2-(butylamino)-5-methyl-4-(morpholine-4-carbonyl)-1*H*-imidazol-1-yl)urea (**5d**): Beige powder, mp 139-141 °C; reaction time: 6 h; yield: 68%

Experimental procedure

To a stirred solution of 1,2-diaza-1,3-diene (**1**: 1 mmol) in dichloromethane (4 mL) taken in a 25 mL flask was added trimethylsilyl azide (**2**; 1.1 mmol), followed by the addition of $Cu(OAc)_2 \cdot H_2O$ (0.2 mmol) and the reaction mixture was stirred at room temperature until the complete formation of organic azide (TLC check). The reaction flask was then immersed in an ice bath (0 °C), and a cooled solution of triphenylphosphine (**3**; 1.0 mmol) in CH_2Cl_2 (1 mL) was added drop-wise; the reaction mixture was then stirred at room temperature until the disappeareance of organic azide (monitored by TLC). The reaction mixture was cooled again to 0 °C, and isocyanate or isothiocyanate derivative (**4**; 1.1 mmol) was slowly added. Within 5 min, the reaction mixture was taken at room temperature and stirred until the complete formation of the triphenylphosphine oxide or triphenylphosphine thioxide as byproduct (TLC check). After removal of the solvent in *vacuo*, the residue was purified by column chromatography on silica gel (elution mixture: EtOAc/cyclohexane or EtOAc/MeOH) to give 1,2-diaminosubstituted imidazoles **5**. All the products were characterized on the basis of elemental analyses and detailed spectral studies.

Characterization data of two representative compounds

2-(Benzylamino)-*N*,*N*,5-trimethyl-1-(3-phenylureido)-1*H*-imidazole-4-carboxamide (**5a**): White powder, mp 192-195 °C; IR (nujol): 3262, 3208, 3145, 1736, 1680, 1668, 1652, 1612, 1571 cm^{-1}; ^1H NMR (400 MHz, DMSO-d_6): δ 2.11 (3H, s), 2.85 and 3.19 (2 br s, 6H), 4.36 (br s, 2H) 6.53 (t, $J=6.0$ Hz, 1H), 6.98 (t, $J=7.2$ Hz, 1H), 7.17-7.50 (m, 9H), 9.05 (s, 1H), 9.28 (s, 1H); ^{13}C NMR (100 MHz, DMSO-d_6): δ 8.9 (q), 35.6 (q), 36.4 (q), 45.5 (t), 118.6 (d), 122.2 (d), 124.5 (d), 126.3 (d), 127.3 (s), 127.6 (d), 128.6 (d), 129.3 (s), 139.3 (s), 140.7 (s), 147.1 (s), 153.6 (s), 165.1 (s); MS (%rel.): *m/z* 392 (M^+, 50), 299 (39), 228 (40), 200 (100), 185 (27), 106 (13). Anal. Calcd for $C_{21}H_{24}N_6O_2$ (392.45): C, 64.27; H, 6.16; N, 21.41; Found: C, 64.38; H, 6.28; N,21.33.

1-(*tert*-Butyl)-3-(2-(butylamino)-5-methyl-4-(morpholine-4-carbonyl)-1*H*-imidazol-1-yl) urea (**5d**): Beige powder, mp 139-141 °C; IR (nujol): 3373, 3206, 1748, 1703, 1605, 1596, 1510 cm^{-1}; ^1H NMR (400 MHz, DMSO-d_6): δ 0.86 (t, $J=7.2$ Hz, 3H), 1.25-1.49 (m, 13H), 2.07 (s, 3H), 3.10 (br s, 2H), 3.57-3.77 (m, 8H), 5.87 (br s, 1H), 9.80 (s, 1H); ^{13}C NMR (100 MHz, DMSO-d_6): δ 8.7 (q), 13.7 (q), 19.4 (t), 27.8 (q), 31.2 (t), 41.7 (t), 46.5 (t), 66.4 (t), 80.7 (s), 124.1 (s), 129.0 (s), 147.1 (s), 154.0 (s), 163.7 (s); MS (%rel.): *m/z* 381 (M^+, 18), 325 (39), 281 (19), 265 (22), 240 (41), 212 (100), 195 (30), 168 (53). Anal. Calcd for $C_{18}H_{31}N_5O_4$ (381.47): C, 56.67; H, 8.19; N, 18.36; Found: C, 56.79; H, 8.28; N, 18.28.

Reference

Attanasi, O. A., Bartoccini, S., Favi, G., Filippone, P., Perrulli, F. R., and Santeusanio, S. (2012b). Tandem aza-Wittig/carbodiimide-mediated annulation applicable to 1,2-diaza-1,3-dienes for the one-pot synthesis of fully substituted 1,2-diaminoimidazoles. *J. Org. Chem.*, **77**, 9338–9343.

C-N bond_55

Type of reaction: C-N and C-C bond formation
Reaction conditions: Methanol, room temperature
Synthetic strategy: One-pot multicomponent reaction
Catalyst: KOH/MeOH
Keywords: Aldehydes, active methylene compounds, Bestmann-Ohira reagent [dimethyl(diazomethyl)phosphonate], potassium hydroxide, methanol, room temperature, Knoevenagel condensation, one-pot multicomponent reaction, click chemistry, cycloaddition, phosphonates, regioselective synthesis, phosphonyl pyrazoles

General reaction scheme

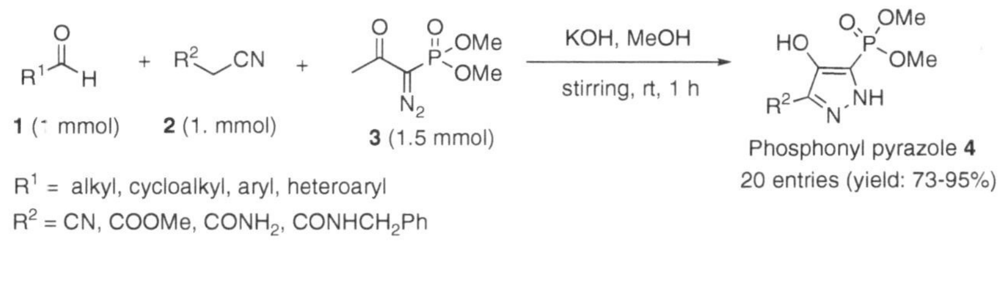

Proposed mechanism

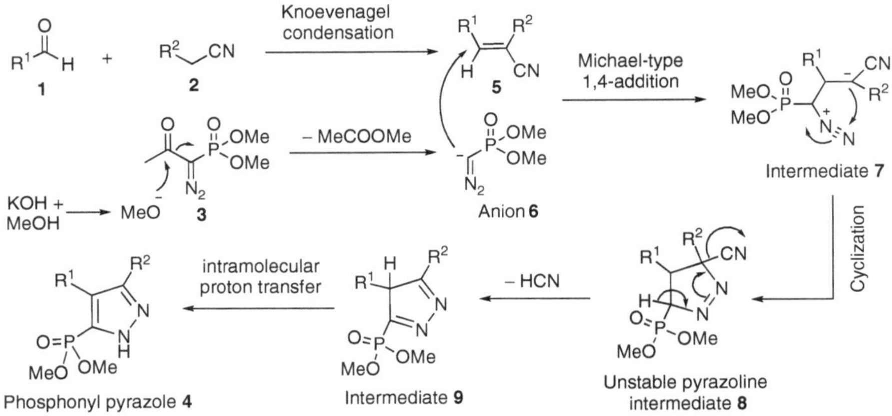

Representative entries

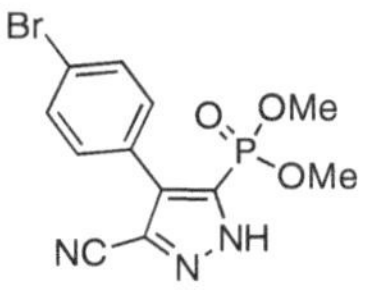

4a

Dimethyl (4-(4-bromophenyl)-3-cyano-1*H*-pyrazol-5-yl)phosphonate (**4a**): Colorless crystalline solid, mp 164-167 °C; yield: 95%

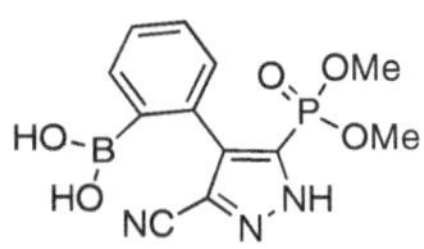

4b

(2-(3-Cyano-5-(dimethoxyphosphoryl)-1*H*-pyrazol-4-yl)phenyl)boronic acid (**4b**): Colorless amorphous solid; yield: 79%

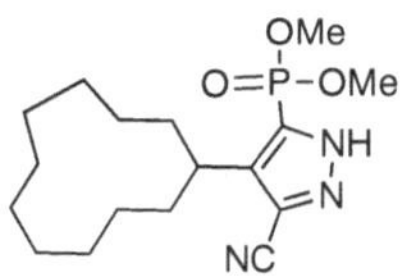

4c

Dimethyl (3-cyano-4-cycloundecyl-1*H*-pyrazol-5-yl)phosphonate (**4c**): Colorless oil; yield: 78%

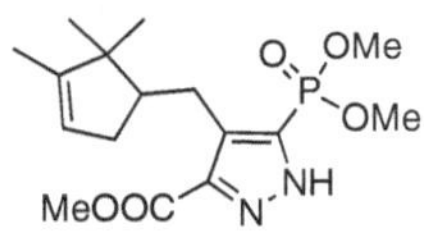

4d

Methyl 5-(dimethoxyphosphoryl)-4-((2,2,3-trimethylcyclopent-3-en-1-yl)methyl)-1*H*-pyrazole-3-carboxylate (**4d**): Colorless oil; yield: 78%

Experimental procedure

To a stirred solution of aldehyde (**1**; 1 mmol), active methylene cyano derivative (**2**; 1.2 mmol) and Bestmann-Ohira reagent (**3**; 1.5 mmol) in distilled methanol (4 mL), was added molecular sieves followed by powdered KOH (2 mmol) and the reaction mixture was stirred at room temperature for one hour. Upon completion of the reaction (TLC monitoring), methanol was distilled off under reduced pressure. The crude residue was dissolved in ethyl acetate (50 mL) and washed with saturated ammonium chloride (2 × 20 mL). Finally the organic layer was washed with saturated brine (20 mL) and dried over anhydrous sodium sulphate. After removal of the solvent, column chromatographic purification was carried out using 0-30% acetone-dichloromethane as eluent to afford phosphonyl pyrazole **4**. All the products were characterized based on spectral studies.

Characterization data of two representative compounds

(2-(3-Cyano-5-(dimethoxyphosphoryl)-1*H*-pyrazol-4-yl)phenyl)boronic acid (**4b**): Colorless amorphous solid; ^{1}H NMR (300 MHz, DMSO-d_6): δ 14.7 (bs, 1H), 7.84-7.82 (m, 1H), 7.46-7.42 (m, 2H), 7.24 (d, J = 6.3 Hz, 1H), 3.54 (s, 3H), 3.51 (s, 3H); ^{13}C NMR (75 MHz, MeOD): δ 134.2, 134.1, 133.8, 132.0, 131.7, 130.4, 130.2, 129.2, 129.1, 113.8, 54.1, 54.0; ^{31}P NMR (121 MHz, DMSO-d_6): δ 7.11; HRMS: Calcd. for $C_{12}H_{14}BN_3O_5P$: 322.0765 $(M+H)^+$; Found: 322.0764.

Methyl 5-(dimethoxyphosphoryl)-4-((2,2,3-trimethylcyclopent-3-en-1-yl)methyl)-1*H*-pyrazole-3-carboxylate (**4d**): Colorless oil; ^{1}H NMR (300 MHz, $CDCl_3$): δ 5.19 (s, 1H), 3.94 (s, 3H), 3.86 (s, 3H), 3.80 (s, 3H), 3.02-2.93 (m, 2H), 2.12-2.04 (m, 2H), 1.83-1.81 (m, 1H), 1.62 (s, 3H), 1.11 (s, 3H), 0.91 (s, 3H); ^{13}C NMR (75 MHz, $CDCl_3$): δ 161.7, 148.3, 131.8, 131.6, 121.8, 53.2, 53.0, 52.0, 51.3, 46.9, 34.0, 29.6, 25.2, 23.3, 19.5, 12.6; ^{31}P NMR (121 MHz, $CDCl_3$): δ 10.60; HRMS: Calcd. for $C_{16}H_{26}N_2O_5P$: 357.1579 $(M+H)^+$; Found: 357.1579.

Reference

Mohanan, K., Martin, A. R., Toupet, L., Smietana, M., and Vasseur, J.-J. (2010). Three-component reaction using the Bestmann-Ohira reagent: A regioselective synthesis of phosphonyl pyrazole rings. *Angew. Chem. Int. Ed.*, **49**, 3196–3199.

C-N bond_56

Type of reaction: C-N and C-C bond formation
Reaction conditions: Ethanol, room temperature
Synthetic strategy: One-pot, two-step three-component reaction
Catalyst: Cs_2CO_3/EtOH and KOH/CuI
Keywords: Aldehydes, Bestmann-Ohira reagent [diethyl(diazomethyl)phosphonate], diazomethyl sulfone, cesium carbonate, potassium hydroxide, copper(I) iodide, ethanol, room temperature, Horner-Wadsworth-Emmons (HWE) reaction, click chemistry, cycloaddition, phosphonates, one-pot synthesis, phosphonylpyrazoles, sulfonylpyrazols

General reaction scheme

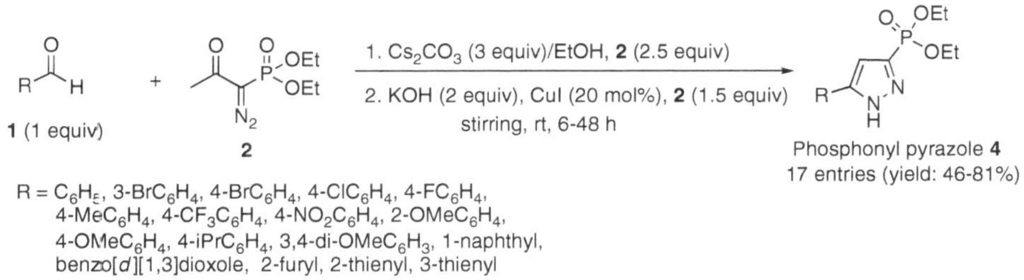

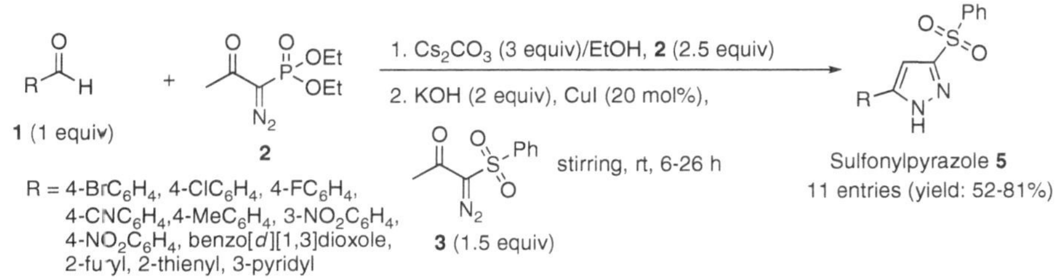

Representative entries

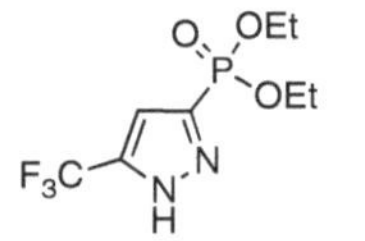

4a

Diethyl (5-(trifluoromethyl)-1*H*-pyrazol-3-yl) phosphonate (**4a**): White solid, mp 76-78 °C; yield: 75%

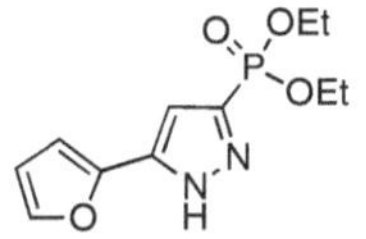

4b

Diethyl (5-(furan-2-yl)-1*H*-pyrazol-3-yl) phosphonate (**4b**): Yellow liquid; yield: 60%

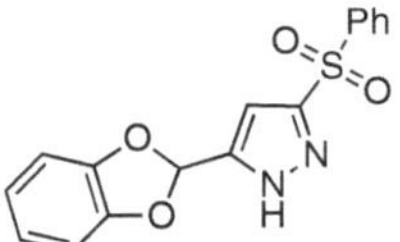

5a

5-(Benzo[*d*][1,3]dioxol-2-yl)-3-(phenylsulfonyl)-1*H*-pyrazole (**5a**): White solid, mp 206-208 °C; yield: 77%

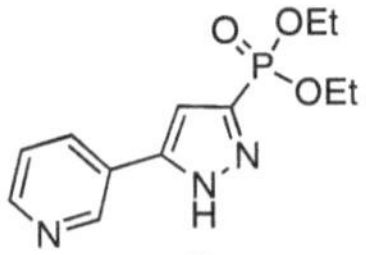

5b

Diethyl (5-(pyridin-3-yl)-1*H*-pyrazol-3-yl) phosphonate (**5b**): White solid, mp 175-177 °C; yield: 52%

Experimental procedure

Phosphonylpyrazole **4**: To a stirred solution of aldehyde (**1**; 0.5 mmol) and diethyl(diazomethyl)phosphonate (BOR; **2**; 275 mg, 1.25 mmol) in dry EtOH (10 mL) was added Cs_2CO_3 (480 mg, 1.5 mmol) at 0 °C and the resulting mixture was stirred at room temperature until complete consumption of aldehyde **1** (monitored by TLC). To the reaction mixture containing alkyne intermediate were then added CuI (19 mg, 0.10 mmol), BOR (**2**;165 mg, 0.75 mmol) and KOH (56 mg, 1 mmol) and stirring continued until all the alkyne intermediate was consumed (monitored by TLC). Then the reaction mixture was concentrated in *vacuo* and the crude residue was directly subjected to silica gel column chromatography (*n*-hexane/ethyl acetate: 3/7) to afford pure phosphonylpyrazole **4**, characterized by spectral studies.

Sulfonylpyrazole **5**: To a stirred solution of aldehyde (**1**; 0.5 mmol) and diethyl(diazomethyl)phosphonate (BOR; **2**; 275 mg, 1.25 mmol) in dry EtOH (10 mL) was added Cs_2CO_3 (480 mg, 1.5 mmol) at 0 °C and the resulting mixture was stirred at room temperature until complete consumption of aldehyde **1** (monitored by TLC). To the reaction mixture containing alkyne intermediate were then added CuI (19 mg, 0.10 mmol), sulfone **3** (168 mg, 0.75 mmol) and KOH (56 mg, 1 mmol) and stirring continued until all the alkyne intermediate was consumed (monitored by TLC). Then the reaction mixture was concentrated in *vacuo* and the crude residue was directly subjected to silica gel column chromatography (*n*-hexane/ethyl acetate: 3/7) to afford pure sulfonylpyrazole **5**, characterized by spectral studies.

Characterization data of two representative compounds

Diethyl (5-(furan-2-yl)-1*H*-pyrazol-3-yl)phosphonate (**4b**): Yellow liquid; IR (neat): 3460, 3134, 2987, 2930 (m), 1665, 159, 1477, 1442, 1394, 1377, 1326, 1236, 1165, 1052, 1026, 976, 705, 602 cm^{-1}; ^{1}H NMR (400 MHz, $CDCl_3$): δ 1.34 (td, $J=7.0$, 0.4 Hz, 6H), 4.12-4.24 (m, 4H), 6.48 (dd, $J=3.3$, 1.8 Hz, 1H), 6.75 (dd, $J=3.3$, 0.6 Hz, 1H), 6.90 (d, $J=2.1$ Hz, 1H), 7.44 (dd, $J=0.6$, 1.8 Hz, 1H); ^{13}C NMR (100 MHz, $CDCl_3$): δ 16.2 (d, $J_{C\text{-}P}=6.0$ Hz), 63.2 (d, $J_{C\text{-}P}=5.0$ Hz), 106.9, 108.0 (d, $J_{C\text{-}P}=21.0$ Hz), 111.5, 135.6 (d, $J_{C\text{-}P}=228.0$ Hz), 141.4, 142.3, 142.4 (d, $J_{C\text{-}P}=18.0$ Hz), 146.6; MS (ES$^+$): *m*/*z* (%rel intensity) 272 ($[MH+1]^+$, 14), 271 (MH^+, 100), 243 (42), 215 (13); HRMS (ES$^+$): Calcd for $C_{11}H_{16}N_2O_4P$ (MH^+) 271.0848; Found 271.0855.

5-(Benzo[*d*][1,3]dioxol-2-yl)-3-(phenylsulfonyl)-1*H*-pyrazole (**5a**): White solid, mp 206-208 °C; IR (KBr): 3292, 3054, 2927, 1610, 1498, 1462, 1360, 1306, 1240, 1181, 1146, 1081, 1036, 936, 862, 808, 733, 666, 617 cm^{-1}; ^1H NMR (400 MHz, DMSO-d_6): δ 6.07 (s, 2H), 7.02 (d, $J=8.0$ Hz, 1H), 7.27 (s, 1H), 7.32 (d, $J=8.0$ Hz, 1H), 7.40 (s, 1H), 7.62-7.66 (m, 2H), 7.69-7.73 (m, 1H), 7.95 (d, $J=7.6$ Hz, 2H), 14.0 (br s, 1H); ^{13}C NMR (100 MHz, DMSO-d_6): δ 101.5, 103.1, 106.1, 108.8, 119.7, 121.7, 127.2, 129.5, 133.7, 140.9, 144.3, 147.9, 152.3; MS (ES^+): *m/z* (%rel intensity) 351 ($[M+Na]^+$, 36), 331 ($[M+H+2]^+$, 13), 330 ($[M+H+1]^+$, 38), 329 ($[M+H]^+$, 100); HRMS (ES^+): Calcd. for $C_{16}H_{13}N_2O_4S$ ($[M+H]^+$) 329.0596; Found 329.0595.

Reference

Kumar, R., Verma, D., Mobin, S. M., and Namboothiri, I. N. N. (2012). One-pot, two-step conversion of aldehydes to phosphonyl- and sulfonylpyrazoles using Bestmann-Ohira reagent. *Org. Lett.*, **14**, 4070–4073.

C-N bond_57

Type of reaction: C-N and C-C bond formation
Reaction conditions: Ethanol, room temperature
Synthetic strategy: One-pot four-component ultrasound-promoted reaction
Catalyst: Piperidine
Keywords: Hydrazines, β-keto esters, isatins, malononitrile, ethyl acetoacetate, piperidine, ethanol, room temperature, ultrasound irradiation, one-pot multicomponent reaction, heterocycles, spiroindoline, spiro[indoline-3,4′-pyrano[2,3-*c*]pyrazole]

General reaction scheme

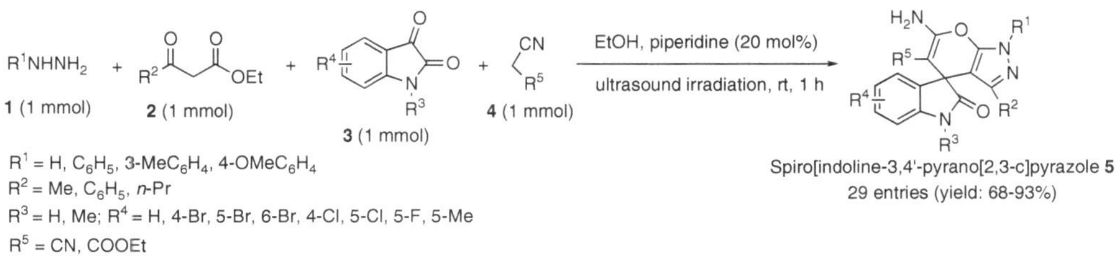

Spiro[indoline-3,4'-pyrano[2,3-c]pyrazole **5**
29 entries (yield: 68-93%)

Representative entries

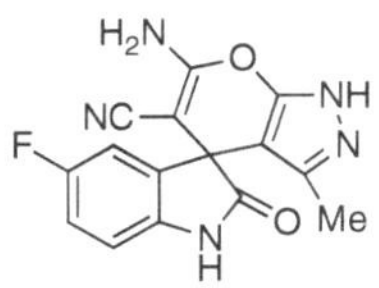

5a

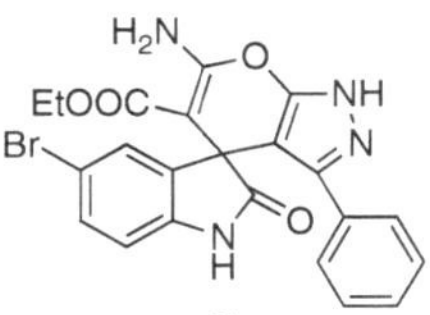

5b

6′-Amino-5-fluoro-3′-methyl-2-oxo-1′*H*-spiro[indoline-3,4′-pyrano[2,3-*c*]pyrazole]-5′-carbonitrile (**5a**): Light yellow powder, mp 274-275 °C; yield: 86%

Ethyl 6′-amino-5-bromo-2-oxo-3′-phenyl-1′*H*-spiro[indoline-3,4′-pyrano[2,3-*c*]pyrazole]-5′-carboxylate (**5b**): White powder, mp 257-259 °C; yield: 81%

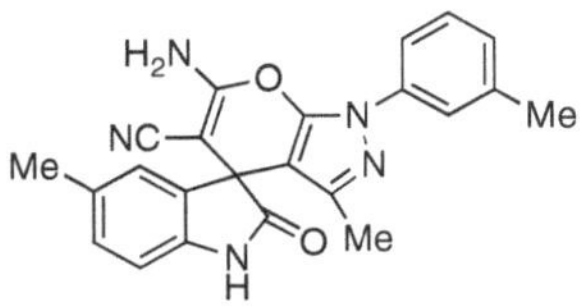

5c

6′-Amino-3′,5-dimethyl-2-oxo-1′-(*m*-tolyl)-1′*H*-spiro[indoline-3,4′-pyrano[2,3-*c*]pyrazole]-5′-carbonitrile (**5c**): White powder, mp 220-221 °C; yield: 89%

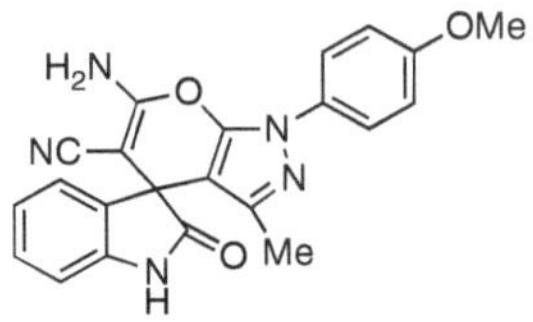

5d

6′-Amino-1′-(4-methoxyphenyl)-3′-methyl-2-oxo-1′*H*-spiro[indoline-3,4′-pyrano[2,3-*c*]pyrazole]-5′-carbonitrile (**5b**): Yellow powder, mp 270-272 °C; yield: 87%

Experimental procedure

A 100 mL flask was charged with hydrazine (1;1 mmol), β-keto ester (**2**; 1 mmol), isatin (**3**; 1 mmol), malononitrile or ethyl cyanoacetate (**4**; 1 mmol) and piperdine (20 mol %) in ethanol (15 mL). The mixture was sonicated in the water bath of an ultrasonic cleaner at 25 ~ 30 °C for an hour. After the completion of the reaction (monitored by TLC), the resulting precipitate was filtered and washed with ethanol to afford the pure product of spiroindoline **5** as solid in good to excellent yields. All the products were characterized by spectral studies.

Characterization data of two representative compounds

Ethyl 6′-amino-5-bromo-2-oxo-3′-phenyl-1′*H*-spiro[indoline-3,4′-pyrano[2,3-*c*]pyrazole]-5′-carboxylate (**5b**): White powder, mp 257-259 °C; IR (KBr): 3421, 3260, 1718, 1677, 1618, 1543, 1491, 1475, 1380, 960 cm^{-1}; ^{1}H NMR (400 MHz, DMSO-d_6): δ 0.70 (t, J=6.0 Hz, 3H, CH_3), 3.70 (q, J=6.8 Hz, 2H, CH_2O), 6.42 (d, J=8.0 Hz, 1H, Ar-H), 6.67 (d, J=6.4 Hz, 2H, Ar-H), 7.11 (s, 1H, Ar-H), 7.21 (d, J=7.2 Hz, 3H, Ar-H), 7.32 (s, 1H, Ar-H), 8.12 (s, 2H, NH_2), 10.11 (s, 1H, NH), 12.68 (s, 1H, NH); ^{13}C NMR (100 MHz, DMSO-d_6): δ 13.25, 58.93, 74.27, 97.51, 110.94, 112.92, 125.66, 127.98, 128.63, 128.96, 130.15, 139.35, 140.52, 141.99, 154.54, 162.78, 165.94, 168.04, 179.46; HRMS calculated for $C_{22}H_{17}^{79}BrN_4O_4$ [M]+: 480.0433; Found: 480.0455.

6′-Amino-3′,5-dimethyl-2-oxo-1′-(*m*-tolyl)-1′*H*-spiro[indoline-3,4′-pyrano[2,3-*c*]pyrazole]-5′-carbonitrile (**5c**): White powder, mp 220-221 °C; IR (KBr): 3360, 3314, 3185, 3038, 2970, 2205, 1702, 1650, 1611, 1516, 1493, 1396, 1313, 1137, 808, 696 cm^{-1}; ^{1}H NMR (400 MHz, DMSO-d_6): δ 1.55 (s, 3H, CH_3), 2.24 (s, 3H, CH_3), 2.40 (s, 3H, CH_3), 6.83 (d, J=8.0 Hz, 1H, Ar-H), 7.00 (s, 1H, Ar-H), 7.08 (d, J=8.0 Hz, 1H, Ar-H), 7.16 (d, J=7.6 Hz, 1H, Ar-H), 7.39 (t, J=7.6 Hz, 1H, Ar-H), 7.57-7.60 (m, 4H, Ar-H and NH_2), 10.66 (s, 1H, NH); HRMS calculated for $C_{23}H_{19}N_5O_2Na$ [M+Na]+: 420.1431; Found: 420.1432.

Reference

Zou, Y., Hu, Y., Liu, H., and Shi, D. (2012). Rapid and efficient ultrasound-assisted method for the combinatorial synthesis of spiro[indoline-3,4′-pyrano[2,3-*c*]pyrazole] derivatives. *ACS Comb. Sci.*, **14**, 38–43.

C-N bond_58

Type of reaction: C-N and C-C bond formation
Reaction conditions: Methanol-dichloromethane, room temperature
Synthetic strategy: One-pot two-step synthesis

Catalyst: Trifluoroacetic acid
Keywords: Methyl levulinate, primary amines, isocyanides, azidotrimethylsilane, methanol, dichloromethane, room temperature, one-pot, two-step synthesis, Ugi reaction, 1,5-disubstituted tetrazoles, *bis*-pyrrolidinone tetrazoles

General reaction scheme

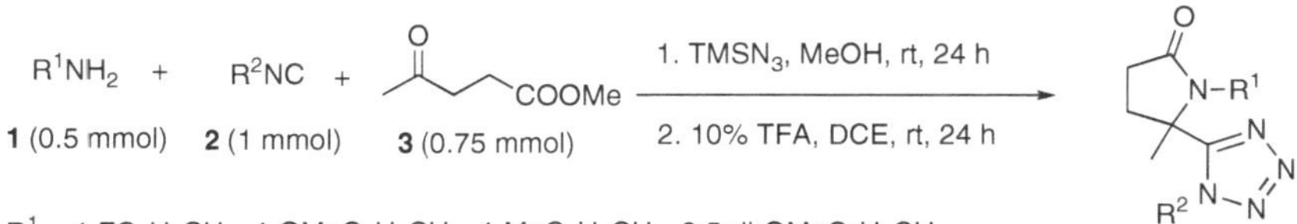

1,5-Disubstituted tetrazole **4**
24 entries (yield: 2-83%)

R^1 = 4-$FC_6H_4CH_2$, 4-$OMeC_6H_4CH_2$, 4-$MeC_6H_4CH_2$, 2,5-di-$OMeC_6H_3CH_2$, 3,4-di-$OMeC_6H_3CH_2$, 2,6-di-Cl-$C_6H_3CH_2$, 2-furylmethyl, 2-thienylmethyl, 3-pyridylmethyl, 4-morpholinophenyl, 3-$CF_3C_6H_4CH_2$, 3,4-di-$OMeC_6H_3CH_2CH_2$, 2-$FC_6H_4CH_2CH_2$, 4-$OMeC_6H_4CH_2CH_2$, 3-indolylethyl, F_3CCH_2, $MeOCH_2CH_2$, cyclohexyl, cyclohexylmethyl, cyclopropyl, 3-morpholinopropyl, 3-(1*H*-imidazol-1-yl)propyl, 1-benzyl-4-piperidyl

R^2 = 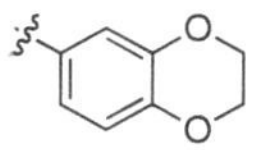, 2,6-diemthylphenyl, benzyl, *n*-butyl

Representative entries

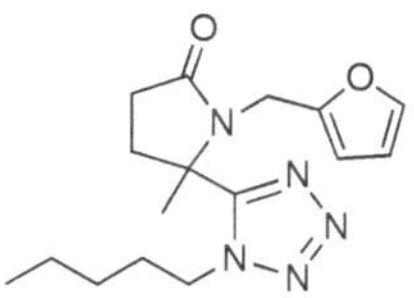

4a

1-(Furan-2-ylmethyl)-5-methyl-5-(1-pentyl-1*H*-tetrazol-5-yl)pyrrolidin-2-one (**4a**): Viscous liquid; yield: 78%

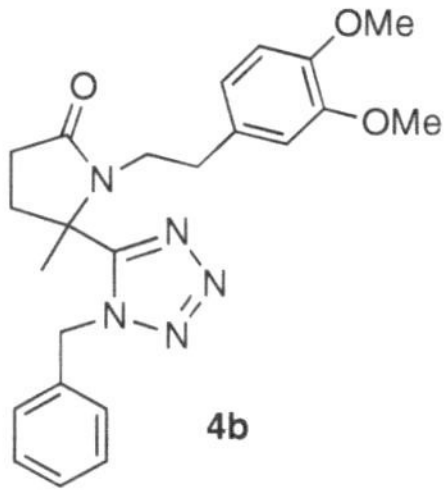

4b

5-(1-Benzyl-1*H*-tetrazol-5-yl)-1-(3,4-dimethoxyphenethyl)-5-methylpyrrolidin-2-one (**4b**): White solid, mp 161-163 °C; yield: 59%

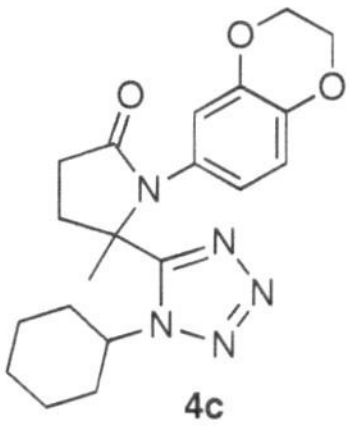

4c

5-(1-Cyclohexyl-1*H*-tetrazol-5-yl)-1-(2,3-dihydrobenzo[*b*][1,4]dioxin-6-yl)-5-methylpyrrolidin-2-one (**4c**): Viscous liquid; yield: 48%

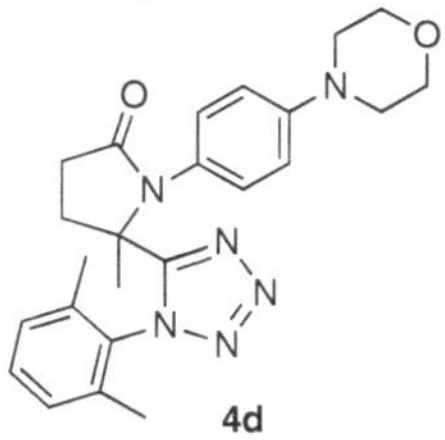

4d

5-(1-(2,6-Dimethylphenyl)-1*H*-tetrazol-5-yl)-5-methyl-1-(4-morpholinophenyl)pyrrolidin-2-one (**4d**): Viscous liquid; yield: 40%

Experimental procedure

A mixture of primary amine (**1**; 0.5 mmol), methyl levulinate (**3**; 0.093 mL, 0.75 mmol) and $TMSN_3$ (0.066 mL, 0.5 mmol) in MeOH (1.0 mL) in a 10 mL vial. Isocyanide (**2**; 1.0 mmol) was then added and the reaction was allowed to run at room temperature for 24 h. Subsequently,

10% TFA solution in DCE (2.0 mL) was added without prior removal of MeOH and the reaction mixture was further stirred for another 24 h (to expedite the process the second step can also be irradiated in the CEM microwave initiator at 100 °C for 10 min). Upon completion of the reaction, the mixture was diluted in EtOAc (10 mL) and washed with saturated sodium bicarbonate solution. The organic layer was dried over anhydrous sodium sulfate and then concentrated *in vacuo*. The crude mixture was purified by using flash column chromatographic technique to afford bis-pyrrolidinone tetrazole **4**. All the products were characterized by spectral studies.

Characterization data of two representative compounds

1-(Furan-2-ylmethyl)-5-methyl-5-(1-pentyl-1*H*-tetrazol-5-yl)pyrrolidin-2-one (**4a**): Viscous liquid; ^{1}H NMR (400 MHz, $CDCl_3$): δ 7.19 (dd, J=1.8, 0.8 Hz, 1H), 6.20 (dd, J=3.2, 1.9 Hz, 1H), 5.98 (d, J=3.2 Hz, 1H), 4.53 (d, J=15.7 Hz, 1H), 4.18 (d, J=15.7 Hz, 1H), 3.90 (ddd, J =13.8, 8.5, 7.1 Hz, 1H), 3.73 (ddd, J=13.8, 8.6, 6.9 Hz, 1H), 2.74-2.57 (m, 2H), 2.33-2.21 (m, 2H), 1.92 (s, 3H), 1.91-1.83 (m, 2H), 1.42-1.26 (m, 4H), 0.91 (t, J=8.0 Hz, 3H); ^{13}C NMR (100 MHz, $CDCl_3$): δ 173.8, 156.0, 149.0, 142.4, 110.8, 109.4, 59.9, 48.2, 36.6, 31.9, 29.1, 28.9, 28.6, 24.9, 22.1, 13.8; $[M+H]^+$=318.4; HRMS (ESI): *m/z* calcd for $C_{16}H_{24}N_5O_2$: 318.1925; Found: 318.1925.

5-(1-Cyclohexyl-1*H*-tetrazol-5-yl)-1-(2,3-dihydrobenzo[*b*][1,4]dioxin-6-yl)-5-methylpyrrolidin-2-one (**4c**): Viscous liquid; ^{1}H NMR (400 MHz, $CDCl_3$): δ 6.77 (d, J=8.6 Hz, 1H), 6.51 (d, J=2.4 Hz, 1H), 6.38 (dd, J=8.7, 2.4 Hz, 1H), 4.70 (p, J=6.4 Hz, 1H), 4.21 (s, 4H), 2.81 (t, J=8.0 Hz, 2H), 2.63 (dt, J=13.7, 8.7 Hz, 1H), 2.38 (dt, J=13.7, 8.7 Hz, 1H), 2.17-1.96 (m, 4H), 1.95 (s, 3H), 1.80-1.55 (m, 4H); ^{13}C NMR (100 MHz, $CDCl_3$); δ 174.3, 160.7, 156.1, 143.8, 143.3, 128.5, 119.9, 117.8, 116.2, 64.2, 64.1, 62.2, 60.3, 49.9, 34.6, 34.1, 34.0, 33.5, 33.1, 29.6, 26.1, 25.0, 23.6, 23.4; $[M+H]^+$=370.4; HRMS (ESI): *m/z* calcd for $C_{19}H_{24}N_5O_3$: 370.1874; Found: 370.1873.

Reference

Gunawan, S., Petit, J., and Hulme, C. (2012). Concise one-pot preparation of unique bis-pyrrolidinone tetrazoles. *ACS Comb. Sci.*, **14**, 160–163.

C-N bond_59

Type of reaction: C-N and C-C bond formation
Reaction conditions: Methanol, room temperature
Synthetic strategy: One-pot multicomponent reaction
Catalyst: No catalyst
Keywords: Isocyanides, carbodiimides, trimethylsilyl azide, methanol, room temperature, one-pot multicomponent synthesis, 1,5-disubstituted 1*H*-tetrazoles

General reaction scheme

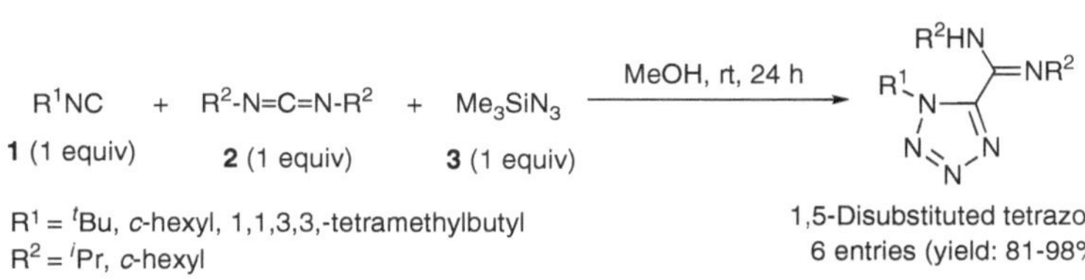

1,5-Disubstituted tetrazole **4**
6 entries (yield: 81-98%)

Proposed mechanism

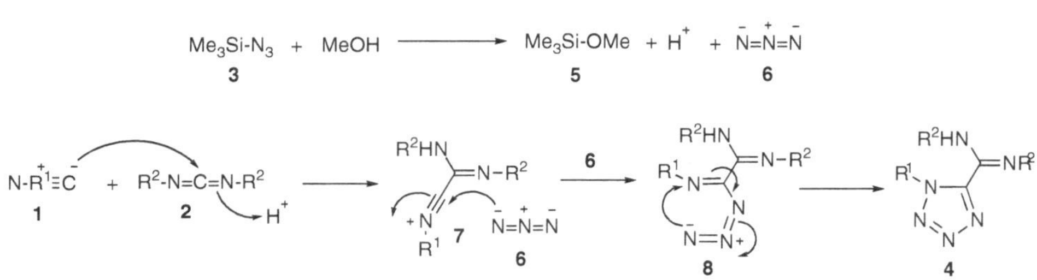

Representative entries

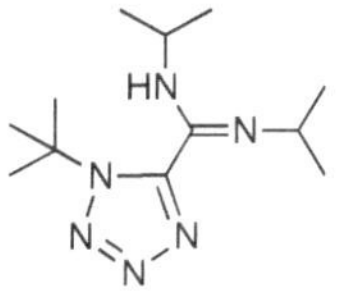

4a

1-(*tert*-Butyl)-*N*,*N*′-diisopropyl-1*H*-tetrazole-5-carboximidamide (**4a**): White solid, mp 105.4-107.9 °C; yield: 93%

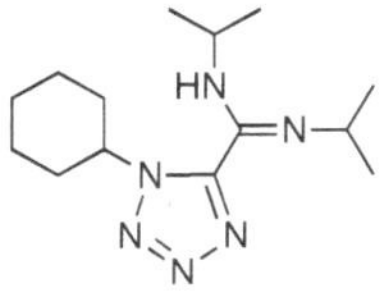

4b

1-Cyclohexyl-*N*,*N*′-diisopropyl-1*H*-tetrazole-5-carboximidamide (**4b**): White solid, mp 111.9-113.4 °C; yield: 82%

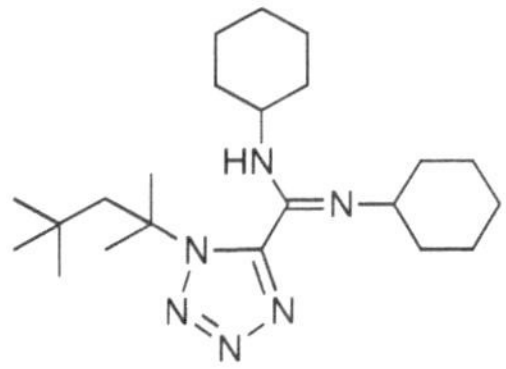

4c

N,*N*′-Dicyclohexyl-1-(2,4,4-trimethylpentan-2-yl)-1*H*-tetrazole-5-carboximidamide (**4c**): White solid, mp 65.8-66.2 °C; yield: 97%

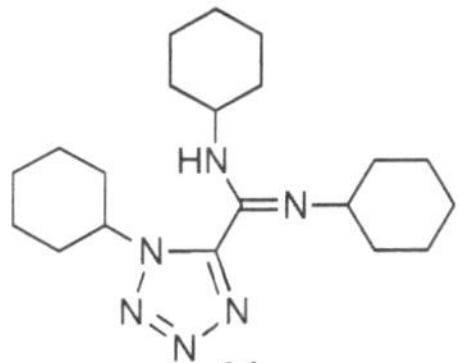

4d

N,*N*′,1-Tricyclohexyl-1*H*-tetrazole-5-carboximidamide (**4d**): White solid, mp 72.4-74.1 °C; yield: 81%

Experimental procedure

To a magnetically stirred solution of isocyanide (**1**; 0.2 mmol) and carbodiimide (**2**; 0.2 mmol) in methanol (3 mL) was added trimethylsilyl azide (**3**; 0.2 mmol), and the resulting mixture was then stirred at room temperature for 24 h. The solvent was evaporated to give the pure product of 1,5-disubstituted tetrazole **4**, characterized on the basis of spectral and analytical studies.

Characterization data of one representative compound

N,*N*′-Dicyclohexyl-1-(2,4,4-trimethylpentan-2-yl)-1*H*-tetrazole-5-carboximidamide (**4c**): White solid, mp 65.8-66.2 °C; IR (KBr): 3313, 2926, 1594, 1445, 1100 cm^{-1}; ^{1}H NMR (250 MHz, $CDCl_3$): δ 0.69 (s, 9H, Me_3C), 1.07-2.04 (m, 20 H, 10 CH_2 of *c*-hex), 1.71 (s, 6H, Me_2C), 1.99 (s, 2H, CH_2), 3.51-3.68 (m, 1H, CH of *c*-hex), 3.93-4.08 (m, 1H, CH of *c*-hex), 5.15 (d, *J* = 7.5, NH); ^{13}C NMR (62.5 MHz, $CDCl_3$): δ 24.88, 24.98, 25.02, 25.47, 31.86, 33.14 (10 CH_2 of *c*-hex); 29.82 (Me_2C); 30.50 (Me_3C); 31.46 (Me_3C); 53.52, 55.17 (2 CH of *c*-hex); 54.11 (CH_2); 62.90 (Me_2C); 140.74, 154.11 (2 C¼N). Anal. calc. for $C_{22}H_{40}N_6$ (388.59): C 68.00, H 10.38, N 21.63; Found: C 68.07, H 10.34, N 21.60.

Reference

Kazemizadeh, A. R., Hajaliakbari, N., Hajian, R., Shajari, N., and Ramazani, A. (2012). Synthesis of 1,5-disubstituted 1*H*-tetrazole derivatives via a three-component reaction of carbodiimides, isocyanides, and trimethylsilyl azide. *Helv. Chim. Acta*, **95**, 594–597.

C-N bond_60

Type of reaction: C-N bond formation
Reaction conditions: Ionic liquid, room temperature
Synthetic strategy: Aziridination of alkenes
Catalyst: *N*-bromosuccinimide (NBS)/[bmim]BF_4
Keywords: Alkenes, Chloramine-T, *N*-bromosuccinimide (NBS), [bmim]BF_4, room temperature, ionic liquid-promoted aziridination of alkenes, recyclable reaction media, aziridines

General reaction scheme

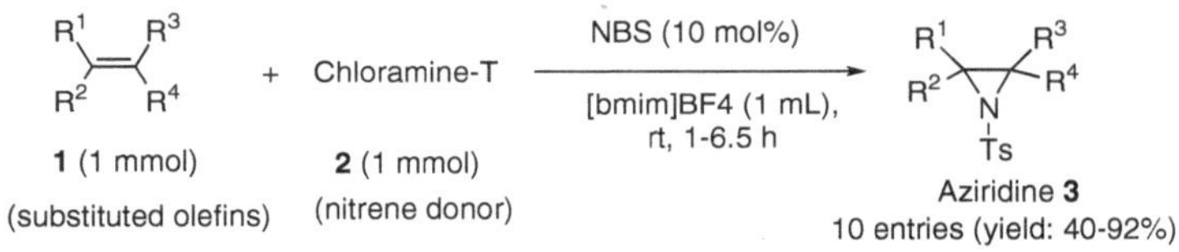

Representative entries

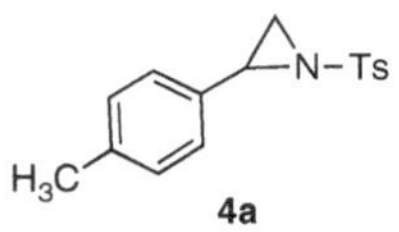

4a

2-(*p*-Tolyl)-1-tosylaziridine (**4a**): Reaction time: 6.5 h; yield: 40%

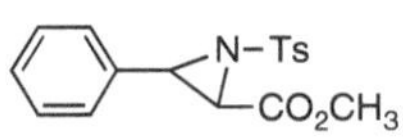

4b

Methyl 3-phenyl-1-tosylaziridine-2-carboxylate (**4b**): Reaction time: 3 h; solid, mp 43.5 °C; yield: 65%

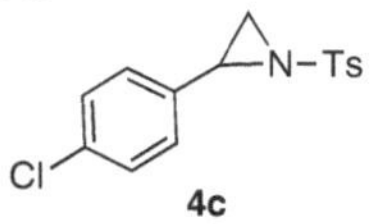

4c

2-(4-Chlorophenyl)-1-tosylaziridine (**4c**): Reaction time: 2.5 h; solid, mp 111-112 °C; yield: 80%

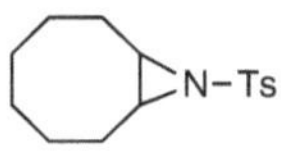

4d

9-Tosyl-9-azabicyclo[6.1.0]nonane (**4d**): Reaction time: 3 h; solid, mp 87-88 °C; yield: 67%

Experimental procedure

A mixture of an alkene (**1**; 1 mmol), Chloramine-T (**2**; 1 mmol, 228 mg), *N*-bromosuccinimide (10 mol%) and [bmim]BF_4 (1 mL) in a 25 mL round bottomed flask was stirred at room temperature for stipulated time (1-6.5 h). On completion of the reaction (monitored by TLC), the product was extracted by diethyl ether and purified by passing through a small column of silica gel using ethyl acetate/hexane (4:6) as eluent. Evaporation of the solvent under vacuum yielded pure aziridine **3** with good yield. Each of the products was identified by means of spectral methods. The ionic liquid layer containing succinimide and NaCl was diluted with ethyl acetate and insoluble solids were separated by filtration through a small Buckner funnel. The ethyl acetate was evaporated under reduced pressure and ionic liquid was used for the next run.

Characterization data of two representative compounds

Methyl 3-phenyl-1-tosylaziridine-2-carboxylate (**4b**): Solid, mp 43.5 °C; IR (KBr): 3068, 2960, 1750, 1600, 1290, 1161 cm^{-1}; ^1H NMR (300 MHz, $CDCl_3$): δ 2.40 (s, 3H), 3.50 (d, $J=4.0$ Hz, 1H), 3.87 (s, 3H), 4.46 (d, $J=3.9$ Hz, 1H), 7.25-7.30 (m, 7H), 7.75 (d, 2H).

2-(4-Chlorophenyl)-1-tosylaziridine (**4c**): Solid, mp 111-112 °C; IR (KBr): 3052, 1596, 1370, 1182 cm^{-1}; ^1H NMR (300 MHz, $CDCl_3$): δ 2.34 (d, $J=4.2$Hz, 1H), 2.45 (s, 3H), 2.85 (d, $J=7.0$ Hz, 1H), 3.77 (dd, $J=7.0$ & 4.2 Hz, 1H), 7.15-7.37 (m, 6H), 7.83 (d, 2H).

Reference

Jain, S. L., and Sain, B. (2006). Ionic liquid promoted highly efficient aziridination of alkenes with chloramine-T using *N*-bromosuccinimide as catalyst. *Green Chem.*, **8**, 943–946.

C-N bond_61

Type of reaction: C-N and C-C bond formation

Reaction conditions: Ethyl lactate-water system (as a bio-based green solvent), room temperature

Synthetic strategy: One-pot multicomponent reaction

Catalyst: Ethyl lactate-water media

Keywords: Isatins, naphthaquinone, proline, ethyl lactate-water system, bio-based green solvent, room temperature, 1,3-dipolar cycloaddition, one-pot three-component reaction, spirooxindoles, spiro[benzo[*f*]pyrrolo[2,1-*a*]isoindole-5,3′-indoline]-2′,6,11-triones

General reaction scheme

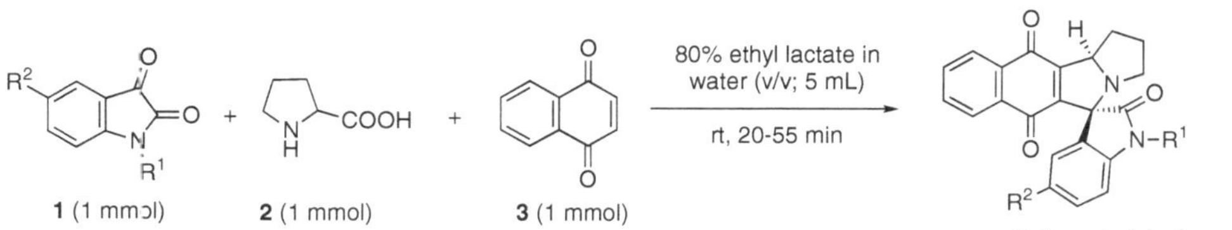

1 (1 mmol) **2** (1 mmol) **3** (1 mmol)

R^1 = H, CH_3, C_2H_5, $C_6H_5CH_2$
R^2 = H, CH_3, Cl, Br, NO_2

Spiro-oxindole **4**
8 entries (yield: 88-93%)

Representative entries

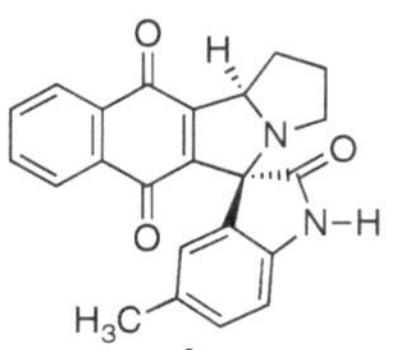

4a

5′-Methyl-1,2,3,11b-tetrahydrospiro[benzo[*f*]pyrrolo[2,1-*a*]isoindole-5,3′-indoline]-2′,6,11-trione (**4a**): Red solid, mp 234-236 °C; reaction time: 45 min; yield: 90%

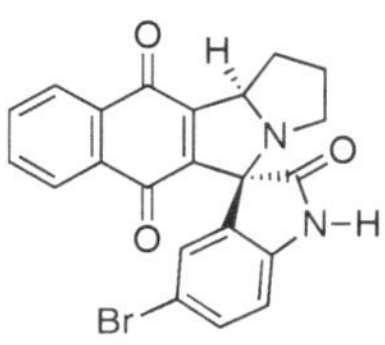

4b

5′-Bromo-1,2,3,11b-tetrahydrospiro[benzo[*f*]pyrrolo[2,1-*a*]isoindole-5,3′-indoline]-2′,6,11-trione (**4b**): Yellow solid, mp 225-227 °C; reaction time: 55 min; yield: 90%

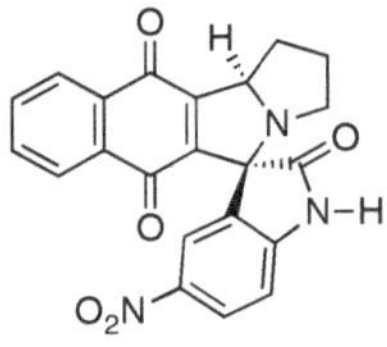

4c

5′-Nitro-1,2,3,11b-tetrahydrospiro[benzo[*f*]pyrrolo[2,1-*a*]isoindole-5,3′-indoline]-2′,6,11-trione (**4c**): Brown solid, mp 229-231 °C; reaction time: 42 min; yield: 92%

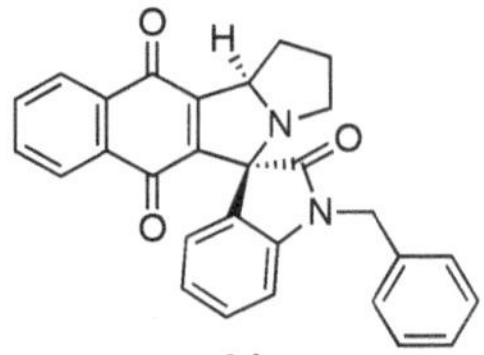

4d

1′-Benzyl-1,2,3,11b-tetrahydrospiro[benzo[*f*]pyrrolo[2,1-a]isoindole-5,3′-indoline]-2′,6,11-trione (**4d**): Red solid, mp 180-182 °C; reaction time: 20 min; yield: 93%

Experimental procedure

A mixture of isatin (**1**; 1 mmol), proline (**2**; 1 mmol) and 1,4-naphthoquinone (**3**; 1 mmol) in ethyl lactate-water system (5 mL; 80% ethyl lactate in water v/v) was stirred magnetically at room temperature for stipulated time (20-55 min). On completion of the reaction (as indicated by TLC), the reaction mixture was left overnight. The precipitate was isolated by filtration and washed with water to furnish pure spiro[benzo[f]pyrrolo[2,1-a]isoindole-5,3′-indoline]-2′,6,11-trione **4**. Each of the products was characterized by analytical and spectral studies.

Characterization data of two representative compounds

5′-Methyl-1,2,3,11b-tetrahydrospiro[benzo[*f*]pyrrolo[2,1-*a*]isoindole-5,3′-indoline]-2′,6,11-trione (**4a**): Red solid, mp 234-236 °C; IR (KBr): 3259, 1735, 1660 cm^{-1}; ^1H NMR (400 MHz, DMSO-d_6): δ 1.77-2.01 (m, 3H, CH and CH_2), 2.24 (s, 3H, CH3), 2.34-2.38 (m, 1H, CH), 2.65-2.77 (m, 2H, CH2), 4.87-4.91 (m, 1H, CH), 6.79-8.07 (m, 7H, Ar–H), 10.40 (s, 1H, NH); ^{13}C NMR (100 MHz, DMSO-d_6): δ 20.62, 27.50, 30.95, 47.13, 70.32, 76.82, 110.00, 121.28, 125.32, 125.68, 125.78, 126.43, 130.00, 130.18, 132.22, 132.42, 133.63, 140.17, 145.87, 151.52, 177.52, 180.66, 181.93; MS (*m/z*): 371 $[M+H]^+$. Anal. Calcd for $C_{23}H_{18}N_2O_3$: C, 74.58; H, 4.90; N, 7.56; Found: C, 74.72; H, 4.77; N, 7.76.

1′-Benzyl-1,2,3,11b-tetrahydrospiro[benzo[*f*]pyrrolo[2,1-a]isoindole-5,3′-indoline]-2′,6,11-trione (**4d**): Red solid, mp 180-182 °C; IR (KBr): 1728, 1679 cm^{-1}; ^1H NMR (400 MHz, DMSO-d_6): δ 1.82-2.03 (m, 3H, CH and CH_2), 2.42-2.50 (m, 1H, CH), 2.62-2.68 (m, 2H, CH_2), 4.78-4.85 (t, $J=8.2$ Hz, 1H, CH), 4.90-4.97 (m, 2H, CH_2), 6.99-8.24 (m, 13H, Ar–H); ^{13}C NMR (100 MHz, DMSO-d_6): δ 27.39, 30.52, 44.12, 49.15, 70.22, 76.57, 110.28, 122.37, 125.54, 125.80, 125.99, 126.61, 129.91, 130.84, 132.13, 132.62, 134.29, 135.01, 135.26, 142.66, 144.26, 145.69, 152.16, 177.71, 181.13, 182.93; MS (m/z): 447 [M+H]+. Anal. Calcd for $C_{29}H_{22}N_2O_3$: C, 78.01; H, 4.97; N, 6.27; Found: C, 78.19; H, 4.82; N, 6.10.

Reference

Dandia, A., Jain, A. K., and Laxkar, A. K. (2013). Ethyl lactate lactate as a promising bio based green solvent for the synthesis of spiro-oxindole derivatives via 1,3 dipolar cycloaddtion reaction. *Tetrahedron Lett.*, **54**, 3929–3932.

C-N bond_62

Type of reaction: C-N bond formation
Reaction conditions: Acetonitrile, room temperature
Synthetic strategy: Recyclable fluorous organocatalyst-promoted multi-component reaction
Catalyst: Fluorous imine-1,2-*bis*(carbothioate) (**I**), an organocatalyst

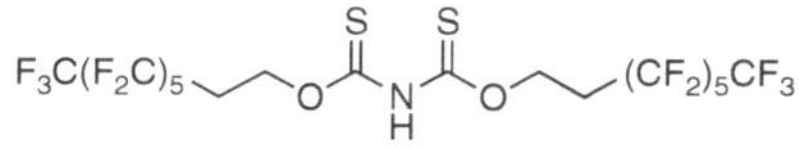

I

Keywords: Pyruvates, aromatic amines, aldehydes, fluorous imine-1,2-*bis*(carbothioate) (**I**), recyclable organocatalyst, acetonitrile, room temperature, one-pot three-component condensation reaction, pyrrole-2-ones, 1,5-dihydro-2*H*-pyrrole-2-ones

General reaction scheme

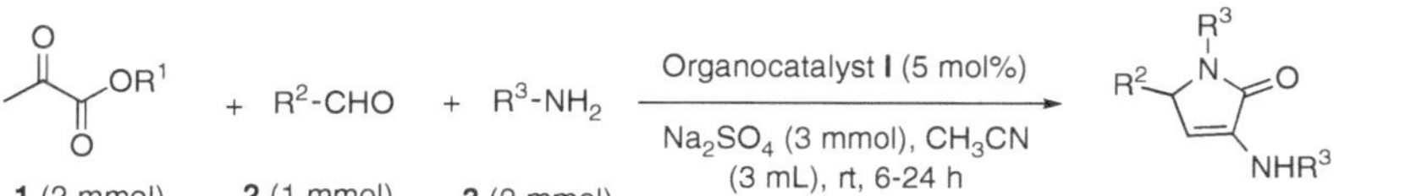

1 (2 mmol) **2** (1 mmol) **3** (2 mmol)

Pyrrole-2-one **4**
10 entries (yield: 41-93%)

R^1 = CH_3, C_2H_5
R^2 = C_6H_5, 4-ClC_6H_4, 4-$OCH_3C_6H_4$, 4-$NO_2C_6H_4$, *n*-butyl
R^3 = C_6H_5, 4-$CH_3C_6H_4$, 4-$OCH_3C_6H_4$

Representative entries

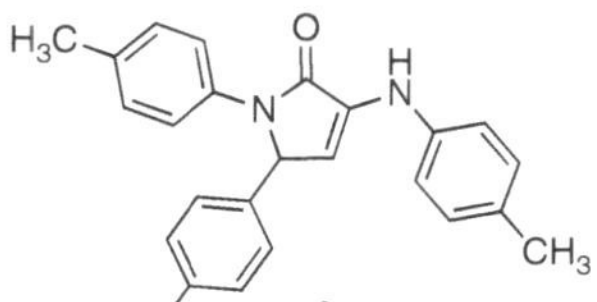

4a

5-(4-Chlorophenyl)-1-(*p*-tolyl)-3-(p-tolylamino)-1*H*-pyrrol-2(5H)-one (**4a**): White solid; yield: 87%

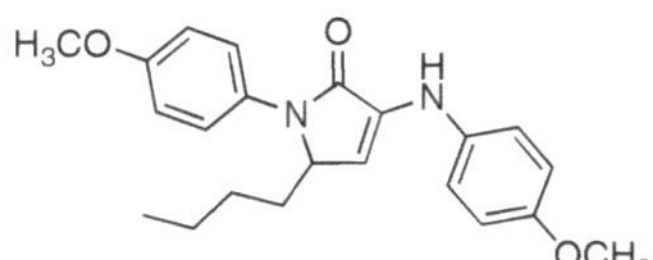

4b

5-Butyl-1-(4-methoxyphenyl)-3-((4-methoxyphenyl)amino)-1*H*-pyrrol-2(5*H*)-one (**4b**): White solid; yield: 53%

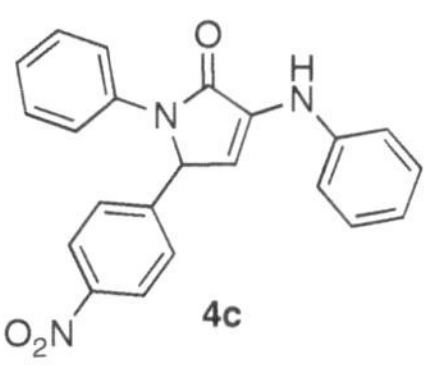

4c

5-(4-Nitrophenyl)-1-phenyl-3-(phenylamino)-1*H*-pyrrol-2(5*H*)-one (**4c**): Brown solid, mp 229-231 °C; reaction time: 42 min; yield: 92%

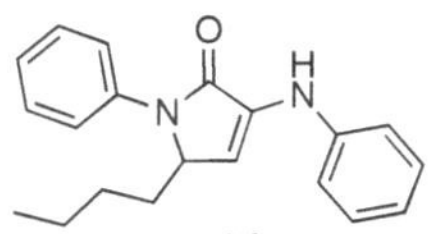

4d

5-Butyl-1-phenyl-3-(phenylamino)-1*H*-pyrrol-2 (5*H*)-one (**4d**): White solid; yield: 41%

Experimental procedure

A mixture of pyruvate (**1**; 2 mmol), aldehyde (**2**; 1 mmol), arylamine (**3**; 2 mmol), fluorous organocatalyst **I** (5 mol %), Na_2SO_4 (3 mmol) and acetonitrile (3 mL) was stirred at room temperature for stipulated time (6-24 h). After completion of the reaction (as indicated by TLC), the mixture was concentrated and then loaded onto a FluoroFlash® silica gel (5 g) cartridge for F-SPE, eluted by 80% methanol at first for non-fluorous components. Methanol was then added onto the fluorous gel column to wash out the fluorous imine-1,2-bis(carbothioate) **I**.

On removal of the solvent under reduced pressure a crude residue was obtained, followed by purification by recrystallization with 50% ethyl acetate in petroleum ether to give the desired product **4**. On the other hand, fluorous imine-1,2-bis(carbothioate) **I** obtained on removal of methanol was dried *in vacuo* at 40 °C for 12 h for its direct use in the next run. Each of the products was characterized by spectral studies.

Characterization data of one representative compound

5-(4-Chlorophenyl)-1-(*p*-tolyl)-3-(*p*-tolylamino)-1*H*-pyrrol-2(5H)-one (**4a**): White solid; ^{1}H NMR (500 MHz, DMSO-d_6): δ 2.22 (s, 6H), 6.03 (d, 1H), 6.23 (d, 1H), 7.05 (d, 2H), 7.12 (d, 2H), 7.17 (d, 2H), 7.26 (d, 2H), 7.33 (d, 2H), 7.47 (d, 2H), 8.02 (s, 1H); ^{13}C NMR (125 MHz, DMSO-d_6): δ 20.8 (d), 108.6, 117.3, 122.1, 129.2(d), 129.6, 129.7, 129.9, 132.6, 132.8, 134.3, 135.0, 137.8, 140.0, 166.8.

Reference

Qian, J.-L., Yi, W.-B., and Cai, C. (2013). Recyclable fluorous organocatalysts promoted three-component reactions of pyruvate, aldehyde and amine under room temperature. *Tetrahedron Lett.*, **54**, 7100–7102.

C-N bond_63

Type of reaction: C-N bond formation
Reaction conditions: Acetonitrile, room temperature
Synthetic strategy: Reductive-cyclization
Catalyst: $BF_3.OEt_2$/NaI
Keywords: Aromatic azides, $BF_3.OEt_2$/NaI, MeCN, room temperature, selective reduction, amines, reductive-cyclization, pyrrolo[2,1-*c*][1,4]benzodiazepines

General reaction schemes

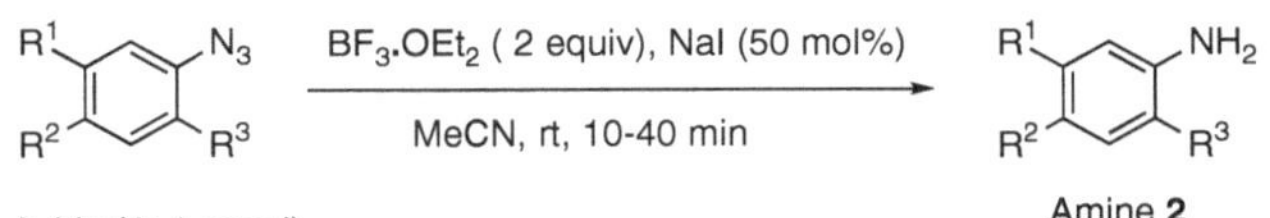

Azide (**1**; 1 mmol)
R^1 = H, OMe, OBn; R^2 = H, Me, OMe, Cl
R^3 = COOH, COOMe

Amine **2**
10 entries (yield: 84-96%)

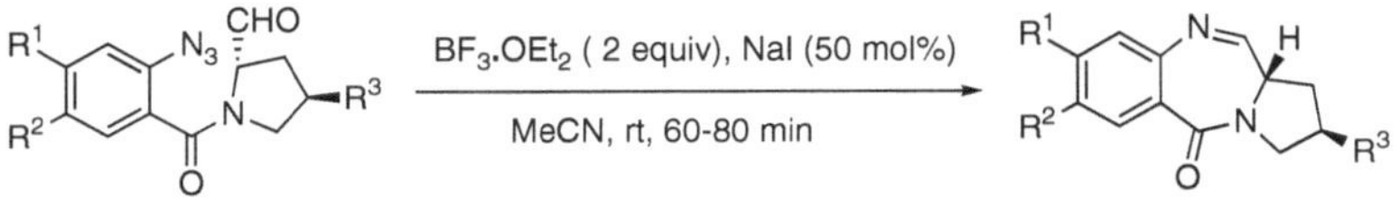

Azide (**3**; 1 mmol)
R^1 = H, OMe, OBn; R^2 = H, Me, OMe,
R^3 = H, OH

Pyrrolo[2,1-*c*][1,4]benzodiazepine **4**
8 entries (yield: 75-92%)

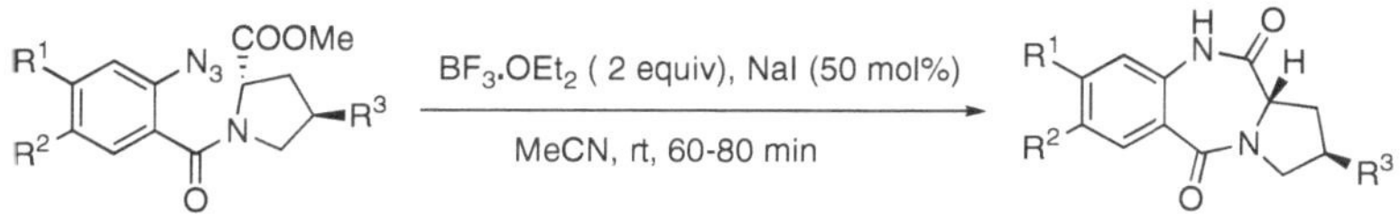

Pyrrolo[2,1-*c*][1,4]benzodiazepine
5,11-dione **6**
10 entries (yield: 78-96%)

R^1 = H, OMe, OBn; R^2 = H, Me, OMe, ClAzide (**5**; 1 mmol)
R^3 = H, OH

Representative entries

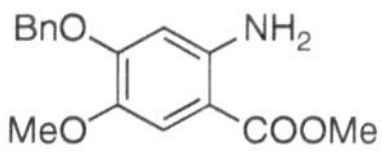

2a

Methyl 2-amino-4-(benzyloxy)-5-methoxybenzoate (**2a**): Reaction time: 30 min; yield: 96%

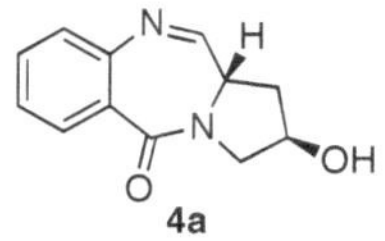

4a

(2*R*,11a*S*)-2-Hydroxy-2,3-dihydro-1*H*-benzo[e]pyrrolo[1,2-*a*][1,4]diazepin-5(11aH)-one (**4a**): Reaction time: 75 min; yield: 75%

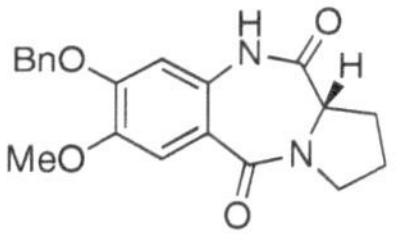

6a

(*S*)-8-(Benzyloxy)-7-methoxy-2,3-dihydro-1*H*-benzo[*e*]pyrrolo[1,2-*a*][1,4]diazepine-5,11(10*H*,11a*H*)-dione (**6a**): Reaction time: 70 min; yield: 96%

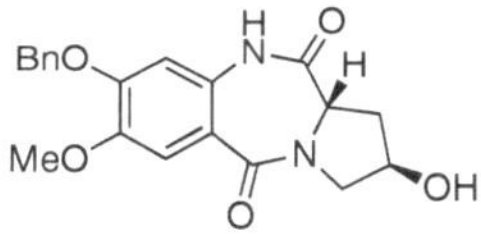

6b

(2*R*,11a*S*)-8-(Benzyloxy)-2-hydroxy-7-methoxy-2,3-dihydro-1*H*-benzo[*e*]pyrrolo[1,2-*a*][1,4] diazepine-5,11(10*H*,11a*H*)-dione (**6b**): Reaction time: 80 min; yield: 85%

Experimental procedure

Substituted aromatic azide (**1/3/5**; 1 mmol) was dissolved in MeCN (15 mL) containg $BF_3 \cdot OEt_2$ (2.0 equiv) and NaI (50 mol%) and the reaction mixture was then stirred at room temperature for 10-80 min. It was quenched with aq. $Na_2S_2O_3$ followed by neutralization with aq. $NaHCO_3$ solution, and extracted with EtOAc (3 × 35 mL). The combined organic portion was dried over anhydrous Na_2SO_4, and the solvent was evaporated under vacuum to afford crude product of respective compound **2/4/6**. The desired product was purified by column chromatography on SiO_2 (60-120 mesh) eluting with EtOAc-hexane with good to moderate yields. Each of the products was characterized on the basis of ^{1}H-NMR, ^{13}C-NMR and mass spectral analyses.

Characterization data of two representative compounds

(2*R*,11a*S*)-2-Hydroxy-2,3-dihydro-1*H*-benzo[e]pyrrolo[1,2-*a*][1,4]diazepin-5(11a*H*)-one (**4a**): ^{1}H NMR (200 MHz, DMSO-d_6): δ 7.84 (d, *J* = 7.32, 1H), 7.12–7.26 (m, 1H), 6.95 (d, *J* = 5.01 Hz, 1H,), 6.70-7.80 (m, 2H), 4.92 (d, *J* = 4.81 Hz, 1H), 4.17-4.26 (m, 1H), 4.02 (m, 1H), 3.50-3.75 (m, 2H), 1.95-2.10 (m, 1H), 1.70-1.80 (m, 1H); ^{13}C NMR (50 MHz, DMSOd_6): δ 166.5, 150.2, 144.6, 133.5, 131.7, 131.4, 128.6, 125.2, 71.0, 56.6, 52.0, 39.5; MS (EI): *m/z* 216 [M^+]; HRMS: *m/z* calcd for $C_{12}H_{12}N_2O_2$: 216.2358; Found: 216.2361.

(2*R*,11a*S*)-8-(Benzyloxy)-2-hydroxy-7-methoxy-2,3-dihydro-1*H*-benzo[e]pyrrolo[1,2-*a*] [1,4] diazepine-5,11(10*H*,11a*H*)-dione (**6b**): ^{1}H NMR (200 MHz, $CDCl_3$): δ 10.10 (s, 1H), 7.26-7.40 (m, 6H), 6.49 (s, 1H), 5.10 (s, 2H), 3.98-4.01 (d, *J* = 6.44 Hz, 1H), 3.91 (s, 3H), 3.65-3.75 (m, 1H), 3.42-3.50 (m, 2H), 2.65-2.75 (m, 1H), 1.90-2.00 (m, 2H); ^{13}C NMR

(50 MHz, DMSO-d_6): δ 171.2, 166.5, 152.2, 146.6, 137.2, 131.4, 129.5, 128.6, 118.3, 113.4, 106.8, 71.0, 68.5, 56.6, 54.0, 35.8; MS (EI): m/z 368 [M^+]; HRMS: m/z calcd for $C_{20}H_{20}N_2O_5$: 368.3824; Found: 368.3831.

Reference

Kamal, A., Shankaraiah, N., Markandeya, N., and Reddy, C. S. (2008). An efficient selective rduction of aromatic azides to amines employing $BF_3.OEt_2$/NaI: Synthesis of pyrrolobenzodiazepines. *Synlett*, 1297–1300.

C-N bond_64

Type of reaction: C-N bond formation
Reaction conditions: Water, room temperature
Synthetic strategy: One-step condensation
Catalyst: Polystyrene supported sulfonic acid (PSSA)
Keywords: 1,3-Diketones, diamines, hydrazines, hydrazides, PSSA, water, room temperature, condensation, pyrazoles, diazepines

General reaction scheme

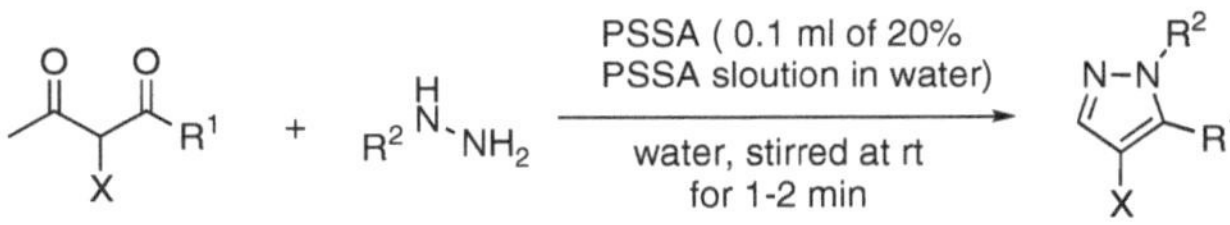

1 (1 mmol) **2** (1.1 mmol) Pyrazole (**3**)
17 entries (yield: 72 - 92%)

X = H, Et, Cl
R^1 = Me, OEt
R^2 = C_6H_5, 4-ClC_6H_4, COC_6H_5, CO-furyl, CO-thienyl

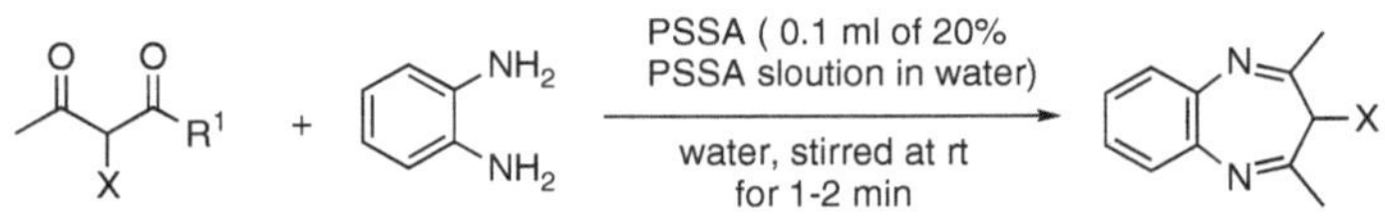

4 (1 mmol) **5** (1.1 mmol) Diazepine **6**
4 entries (yield: 85 - 90%)

X = H, Me, Et, Cl

Representative entries

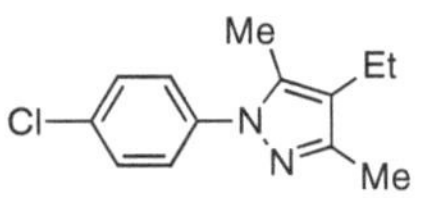

3a

1-(4-Chlorophenyl)-4-ethyl-3,5-dimethyl-1*H*-pyrazole (**3a**): Yield: 80%

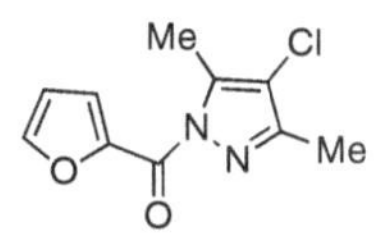

3b

(4-Chloro-3,5-dimethyl-1*H*-pyrazol-1-yl)(furan-2-yl)methanone (**3b**): Yield: 85%

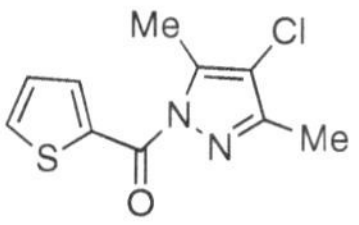

3c

(4-Chloro-3,5-dimethyl-1*H*-pyrazol-1-yl)(thiophen-2-yl)methanone (**3c**): Yield: 85%

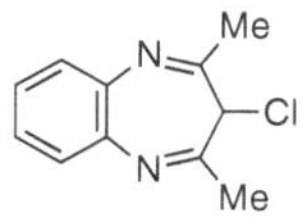

6a

3-Chloro-2,4-dimethyl-3*H*-benzo[b][1,4]diazepine (**6a**): Yield: 85%

Experimental procedure

In a reaction tube, a mixture of 1,3-diketone (**1**; 1 mmol) and hydrazines/hydrazides (**2**; 1.1 mmol) was added with 0.1 mL of 20% PSSA solution in water. The whole mixture was stirred for 1-2 min to complete the reaction. The crude product of pyrazole **3** was then extracted with ethyl acetate followed by washing with dilute sodium bicarbonate solution. After drying the organic layer over sodium sulfate and removal of the solvent, pure product of **3** was isolated by passing the crude reaction mixture over short silica gel column (Silica gel: 20%EtOAc/hexane). Following the same procedure, diazepine **6** was synthesized from 1.0 equiv of 1,3-diketone and 1.1 equiv. of diamines. All the compounds were characterized from detailed spectral studies.

Reference

Polshettiwar, V., and Varma, R. S. (2008). Greener and rapid access to bio-active heterocycles: Room temperature synthesis of pyrazoles and diazepines in aqueous medium. *Tetrahedron Lett.*, **49**, 397–400.

C-N bond_65

Type of reaction: C-N bond formation
Reaction conditions: Solvent-free, room temperature
Synthetic strategy: One-step condensation
Catalyst: Ytterbium triflate [Yb(OTf)$_3$]
Keywords: 1,5-Benzodiazepines, *o*-phenylenediamine, ketones, condensation, Yb(OTf)$_3$, solvent-free, room temperature

General reaction scheme

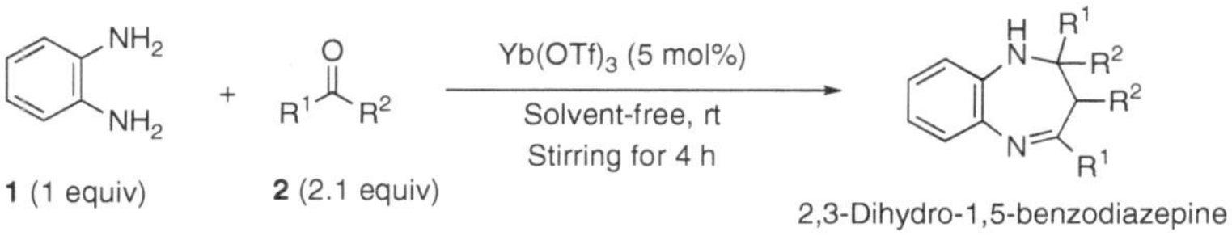

2,3-Dihydro-1,5-benzodiazepine (**3**)
8 entries (yield: 88-99%)

Ketones: acetone, acetophenone, 3-pentanone, 2-butanone, 6-methyl-5-hepten-2-one, cyclopentanone, cyclohexanone, cycloheptanone

Representative entries

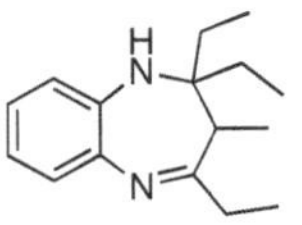

3a

2,2,4-Triethyl-3-methyl-2,3-dihydro-1*H*-1,5-benzodiazepine (**3a**; yellow solid; mp 144-145 °C; yield: 95%)

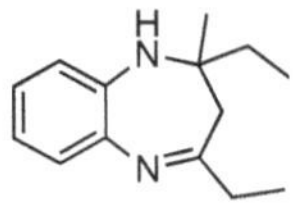

3b

2,4-Diethyl-2-methyl-2,3-dihydro-1*H*-1,5-benzodiazepine (**3b**; yellow solid; mp 137-138 °C; yield: 93%)

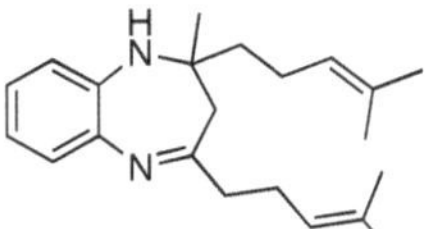

3c

2-Methyl-2,4-di(4-methyl-3-pentenyl)-2,3-dihydro-1*H*-1,5-benzodiazepine (**3c**; yellow solid; mp 158-160 °C; yield: 99%)

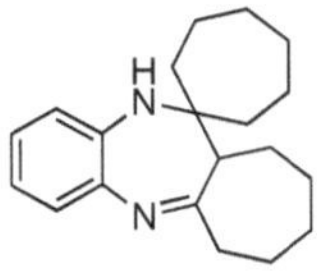

3d

7,8,9,10,10a,12-Hexahydro-6*H*-spiro[benzo[*b*]cyclohepta[*e*][1,4]diazepine-11,1′-cycloheptane (**3d**; yellow solid; mp 136-137 °C; yield: 88%)

Experimental procedure

A mixture of *o*-phenylenediamine (**1**; 1 mmol) and ketone (**2**; 2.1 mmol) was well stirred with $Yb(OTf)_3$ (0.05 mmol; 5 mol%) at room temperature for 4 h so as to complete the reaction. Then 2 mL of CH_2Cl_2 was added to separate out $Yb(OTf)_3$. After removal of the catalyst, the solvent was removed under reduced pressure and the residue was purified by SiO_2 gel column chromatography using CH_2Cl_2-MeOH (95:5) as eluent to obtain pure 2,3-dihydro-1,5-benzodiazepine (**3**) as yellow solid. Each of the products was characterized by IR, ^{1}H NMR, ^{13}C NMR and GC-MS studies.

Characterization data of two representative compounds

2,2,4-Triethyl-3-methyl-2,3-dihydro-1*H*-1,5-benzodiazepine (**3a**): Yellow solid; mp 144-145 °C. IR (KBr): 3330 (s br),1638 (s) cm^{-1}; ^{1}H NMR (200 MHz, $CDCl_3$): δ 0.76-1.48 (m, 13H), 1.61 (s, 3H), 2.87 (q, $J=7.0$ Hz, 1H), 3.75 (s br, 1H), 6.65-7.40 (m, 4H); ^{13}C NMR (50 MHz, $CDCl_3$): δ 173.8, 142.4, 139.0, 132.7, 126.7, 118.0, 117.5, 68.7, 46.2, 35.6, 28.4, 28.0, 12.2, 11.5, 7.8, 7.3; GC-MS: $M^+=244$.

2-Methyl-2,4-di(4-methyl-3-pentenyl)-2,3-dihydro-1*H*-1,5-benzodiazepine (**3c**): Yellow solid; mp 158–160 °C. IR (KBr): 3330 (s br), 1638 (s) cm^{-1}; ^{1}H NMR (200 MHz, $CDCl_3$): δ 1.34 (s, 3H), 1.66 (s, 3H), 1.68 (s, 3H), 1.73 (s, 3H), 1.74 (s, 3H), 2.01-2.70 (m, 10H), 3.15 (s br, 1H), 5.03-5.24 (m, 2H), 6.67-7.19 (m, 4H); ^{13}C NMR (50 MHz, $CDCl_3$): δ 174.7, 140.6, 137.9, 132.2, 131.9, 127.0, 125.48, 123.8, 123.4, 121.8, 121.7, 70.6, 43.0, 42.9, 42.7, 27.5, 25.7, 25.7, 25.1, 23.0, 17.7, 17.7; GC-MS: $M^+=324$.

Reference

Curini, M., Epifano, F., Marcotullio, M. C., and Rosati, O. (2001). Ytterbium triflate promoted synthesis of 1,5-benzodiazepine derivatives. *Tetrahedron Lett.*, **42**, 3193–3195.

C-N bond_66

Type of reaction: C-N bond formation
Reaction conditions: Solvent-free, room temperature
Synthetic strategy: One-step condensation
Catalyst: (Bromodimethyl)sulfonium bromide [$Me_2S^+BrBr^-$]
Keywords: *o*-Phenylenediamines, ketones, (bromodimethyl)sulfonium bromide, solvent-free, room temperature, condensation, 1,5-benzodiazepines

General reaction scheme

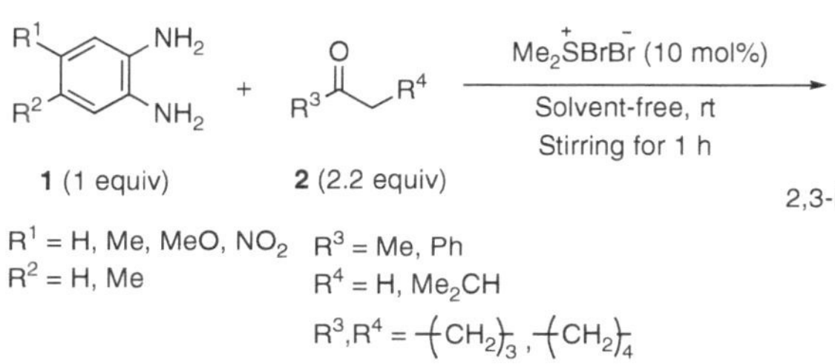

2,3-Dihydro-1,5-benzodiazepine (**3**)
11 entries (yield: 79-96%)

Representative entries

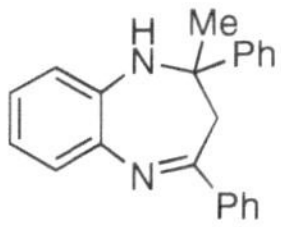

3a

2-Methyl-2,4-diphenyl-2,3-dihydro-1*H*-1,5-benzodiazepine (**3a**): Yield: 91%

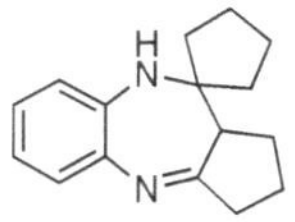

3b

2,3,9,10a-Tetrahydro-1*H*-spiro[benzo[*b*] cyclopenta [*e*][1,4]diazepine-10,1′-cyclopentane (**3b**): Yield: 85%

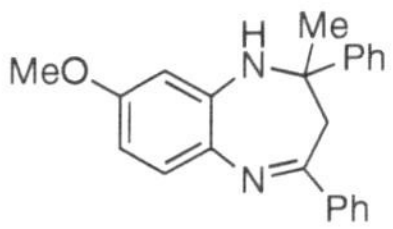

3c

8-Methoxy-2-methyl-2,4-diphenyl-2,3-dihydro-1*H*-1,5-benzodiazepine (**3c**): Yield: 84%

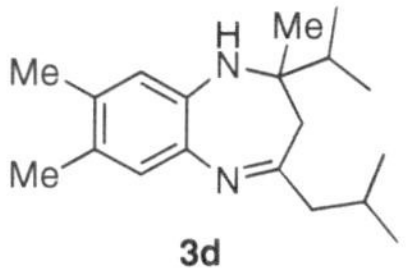

3d

4-Isobutyl-2-isopropyl-2,7,8-trimethyl-2,3-dihydro-1*H*-1,5-benzodiazepine (**3d**): Yield: 95%

Experimental procedure

A mixture of *o*-phenylenediamine (**1**; 1 mmol) and ketone (**2**; 2.2 mmol) was well stirred with (bromodimethyl) sulfonium bromide (10 mol%) at room temperature for 1 h so as to complete the reaction as monitored by TLC. The reaction mixture was then quenched with 5 mL of water and extracted with EtOAc (3 × 5 mL). The extract was concentrated and the residue was subjected to column chromatography over silica gel using EtOAc-hexane (1:4) as eluent to afford pure 1,5-benzodiazepine (**3**). Physical and spectral data of all the isolated products were found to be in accordance with the known compounds reported in literature.

Reference

Das, B., Ramu, R., Ravikanth, B., and Saidi Reddy, V. (2006). (Bromodimethyl)sulfonium bromide: An efficient catalyst for solvent-free synthesis of 1,5-benzodiazepines. *J. Mol. Catal. A: Chem.*, **246**, 76–78.

Further information

A good number of alternative methods are also available in literature for the synthesis of 1,5-benzodiazepine derivatives involving the use of a variety of catalysts such magnesium oxide/ phosphorus oxychloride (MOPO) [1], ionic liquid (1,3-di-*n*-butylimidazolium bromide) [2], silica gel-supported cerium(III) chloride/sodium iodide [3], $Sc(OTf)_3$ [4], $InCl_3$ [5], $InBr_3$ [6], ceric ammonium nitrate (CAN) [7], 2,4,6-trichloro-1,3,5-triazine (TCT) [8,9], $GaCl_3$ [10], and $ZrOC_2.8H_2O$ [11] in the presence or absence of solvent(s) at room temperature.

[1] Balakrishna, M. S., and Kaboudin, B. (2001). A simple and new method for the synthesis of 1,5-benzodiazepine derivatives on a solid surface. *Tetrahedron Lett.*, **42**, 1127–1129.

[2] Jarikote, D. V., Siddiqui, S. A., Rajagopal, R., Daniel, T., Lahoti, R. J., and Srinivasan, K. V. (2003). Room temperature ionic liquid promoted synthesis of 1,5-benzodiazepine derivatives under ambient conditions. *Tetrahedron Lett.*, **44**, 1835–1838.

[3] Sabitha, G., Kiran Kumar Reddy, G. S., Bhaskar Reddy, K., Mallikarjuna Reddy, N., and Yadav, J. S. (2004). A new, efficient and environmentally benign protocol for the synthesis of 1,5-benzodiazepines using cerium(III) chloride/sodium iodide supported on silica gel. *Adv. Syn. Catal.*, **346**, 921–923.

[4] De, S. K., and Gibbs, R. A. (2005). Scandium(III) triflate as an efficient and reusable catalyst for synthesis of 1,5-benzodiazepine derivatives. *Tetrahedron Lett.*, **46**, 1811–1813.

[5] Yadav, J. S., Reddy, B. V. S., Satheesh, G., Srinivasulu, G., and Kunwar, A. C. (2005). $InCl_3$-catalyzed stereoselective synthesis of optically pure 1,5-benzodiazepines. *ARKIVOC*, 221–227.

[6] Yadav, J. S., Reddy, B. V. S., Praveenkumar, S., and Nagaiah, K. (2005). Indium(III) bromide: A novel and efficient reagent for the rapid synthesis of 1,5-benzodiazepines under solvent-free conditions. *Synthesis*, 480–484.

[7] Varala, R., Enugala, R., Nuvula, S., and Adapa, S. R. (2006). Ceric ammonium nitrate (CAN) promoted efficient synthesis of 1,5-benzodiazepine derivatives. *Synlett*, 1009–1014.

[8] Khodaei, M. M., Bahrami, K., and Nazarian, Z. (2008). TCT as a rapid and efficient catalyst for the synthesis of 1,5-benzodiazepines. *Bull. Korean Chem. Soc.*, **29**, 1280–1282.

[9] Kuo, C.-W., Wang, C.-C., Kavala, V., and Yao, C.-F. (2008). Efficient TCT-catalyzed synthesis of 1,5-benzodiazepine derivatives under mild conditions. *Molecules*, **13**, 2313–2325.

[10] Kumar, S., and Sandhu, J. S. (2008). An efficient synthesis of 1,5-benzodiazepines catalyzed by $GaCl_3$ under solvent free conditions. *Indian J. Chem.*, **47B**, 1463–1466.

[11] Baseer, M. A., and Khan, A. J. (2011). ZOC-Catalyzed an efficient synthesis of 1,5-benzodiazepines under mild conditions. *Der Chemica Sinica*, **2**, 84–87.

Chapter 3

Carbon – Oxygen Bond Forming Reactions at Room Temperature

C-O bond_1

Types of reaction: C-O and C-C bond formation
Reaction conditions: Toluene, room temperature, oxygen (balloon)
Synthetic strategy: Aerobic oxidation of alcohols, and coupling between primary alcohols and ketones
Catalyst: Gold nano-particles (immobilized in aluminum oxyhydroxide support) [Au/AlO(OH)]
Keywords: Alcohols, toluene, oxygen (ballon), room temperature, Au/AlO(OH), oxidation, ketones, coupling

Reaction scheme

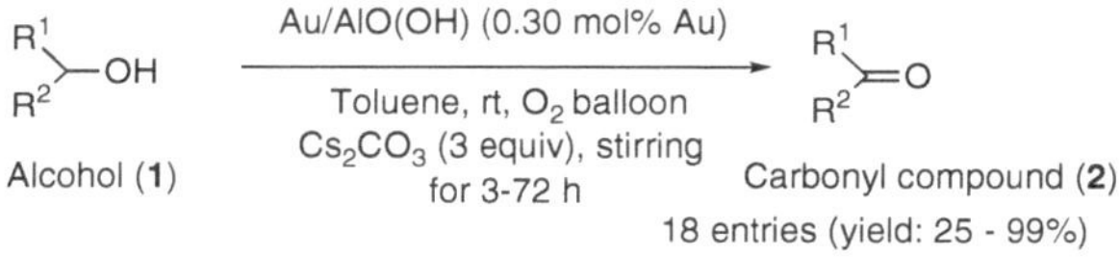

R^1 = C_6H_5, 4-ClC_6H_4, 4-OMeC_6H_4, $C_6H_5CH_2$, $C_6H_5CH_2CH_2$, $C_6H_5CH{=}CH_2$, *n*-hexyl, cyclohexyl, 2-furyl, 4-pyridyl, naphthyl

R^2 = H, Me, CF_3, *tert*-butyl, cyclopropyl, phenyl

1:

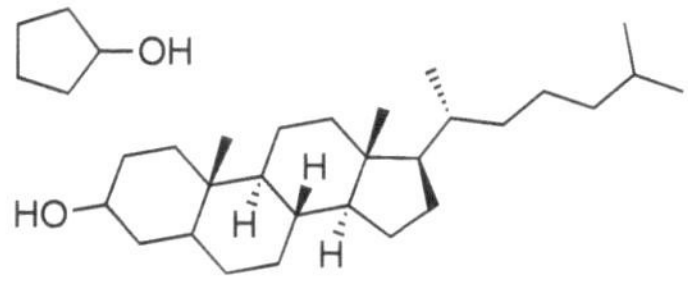

(two more alcohol substartes **1**)

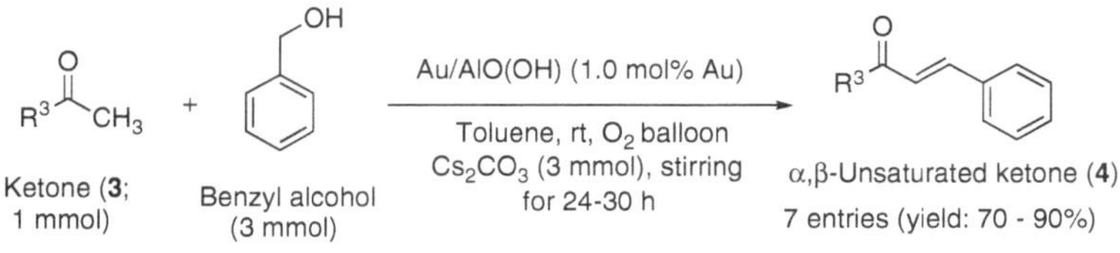

R^3 = C_6H_5, 4-ClC_6H_4, 4-OMeC_6H_4, isopropyl, *n*-pentyl

Representative entries

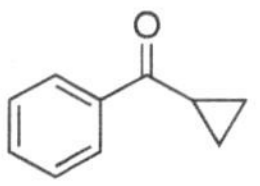

2a

Cyclopropyl(phenyl)methanone (**2a**): Reaction time: 7 h; yield: 96%

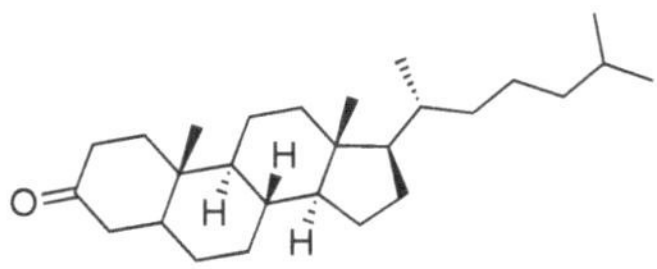

2b

(8*R*,9*S*,10*S*,13*R*,14*S*,17*R*)-10,13-Dimethyl-17-((*R*)-6-methylheptan-2-yl) tetradecahydro-1*H*-cyclopenta [a]phenanthren-3(2*H*)-one (**2b**): Reaction time: 72 h; yield: 83%

Room Temperature Organic Synthesis. http://dx.doi.org/10.1016/B978-0-12-801025-9.00003-6

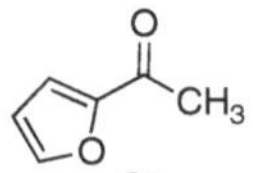

2c

1-(Furan-2-yl)ethanone (**2c**): Reaction time: 24 h; yield: 50%

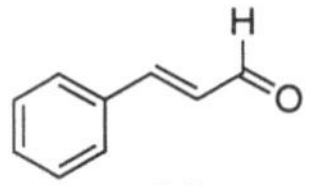

2d

Cinnamaldehyde (**2d**): Colorless crystals; mp 118-119 °C; reaction time: 24 h; yield: 65%

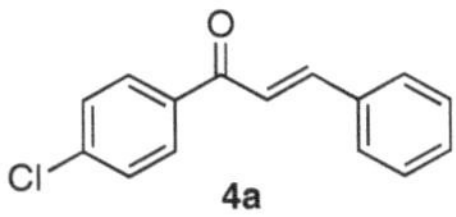

4a

(*E*)-1-(4-Chlorophenyl)-3-phenylprop-2-en-1-one (**4a**): Reaction time: 24 h; yield: 86%

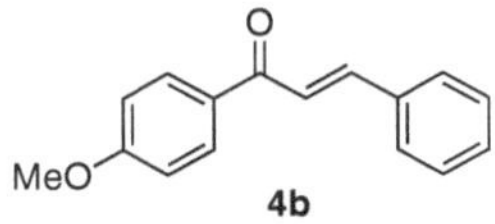

4b

(*E*)-1-(4-Methoxyphenyl)-3-phenylprop-2-en-1-one (**4b**): Reaction time: 24 h; yield: 70%

Experimental procedure

Aerobic oxidation of alcohols: To a solution of alcohol (**1**; 0.5 mmol) in toluene was added Cs_2CO_3 (490 mg, 1.5 mmol, 3 equiv) and the catalyst, Au/AlO(OH) (12 mg, 0.30 mol % of Au), and the reaction mixture was stirred at room temperature for stipulated hours (3-72 h) under O_2 balloon. On completion of the reaction, the crude oxidized product **2** was obtained by removal of the solvent from the filtrate followed by proper purification. Characterization of each product was done on the basis of GC and spectral data.

Reactions of benzyl alcohol with ketones: Cs_2CO_3 (980 mg, 3 mmol) and the catalyst, Au/AlO(OH) (69 mg, 1.0 mol % of Au) were added to a solution of ketonic compound **3** (1 mmol) and benzyl alcohol (324 mg, 3 mmol) in toluene (3 mL) in a 10 mL tube. The tube with a rubber stopper was connected with an oxygen balloon through a needle, and the mixture was stirred at room temperature for 24-30 h. On completion of the reaction, the crude α,β-unsaturated carbonyl compound product **4** was obtained by removal of the solvent from the filtrate followed by proper purification. Characterization of each product was done on the basis of GC and spectral data.

Reference

Kim, S., Bae, S. W., Lee, J. S., and Park, J. (2009). Recyclable gold nanoparticle catalyst for the aerobic alcohol oxidation and C–C bond forming reaction between primary alcohols and ketones under ambient conditions. *Tetrahedron*, **65**, 1461–1466.

C-O bond_2

Type of reaction: C-O bond formation
Reaction conditions: Acetonitrile, room temperature
Synthetic strategy: Esterification of alcohols
Catalyst: *In situ* generated $Ph_3P(OAc)_2$ from Ph_3P, Br_2 and NH_4OAc
Keywords: Alcohols, triphenylphosphine, bromine, ammonium acetate, $Ph_3P(OAc)_2$ (*in situ* generated), acetonitrile, room temperature, acetylation

General reaction scheme

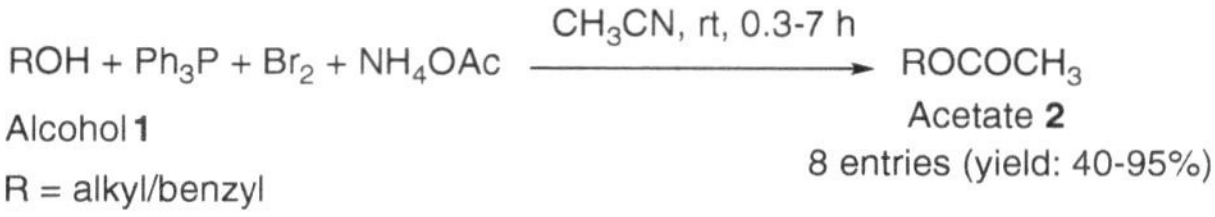

Plausible mechanism

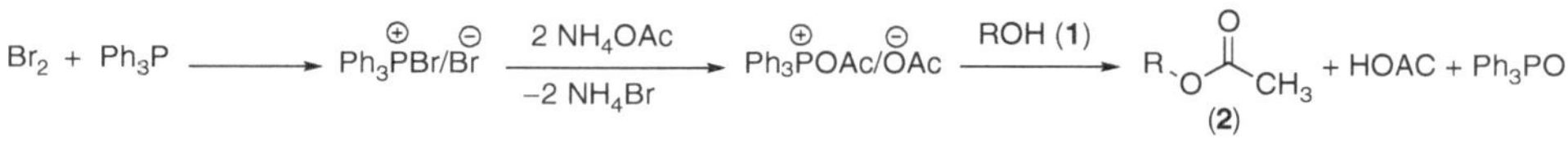

Representative entries

$PhCH_2CH_2CH_2OCOCH_3$

2a

3-Phenylpropyl acetate (**2a**): Reaction time: 0.3 h; yield: 90%

$CH_3(CH_2)_5CH(CH_3)OCOCH_3$

2b

Octan-2-yl acetate (**2b**): Reaction time: 4.5 h; yield: 87%

Experimental procedure

To a solution of $Ph_3P(OAc)_2$ was added alcohol (**1**; 1 mmol), and the progress of the reaction was monitored by TLC. After completion of the reaction (0.3-7 h) the reaction mixture was filtered to remove the precipitated NH_4Br followed by evaporation of the solvent. Column chromatography of the crude mixture on silica gel using *n*-hexane-ethyl acetate mixture as eluent to afford pure acetate derivative **2**. Each of the products was characterized on the basis of spectral properties.

Characterization data of one representative compound

3-Phenylpropyl acetate (**2a**): ^1H NMR (250 MHz, $CDCl_3$): δ 1.88 (q, $J=7.5$ Hz, 2H), 1.97 (s, 3H), 2.6 (t, $J=7.6$ Hz, 2H,), 4.0 (t, $J=6.5$ Hz, 2H,), 7.0-7.2 (m, 5H); ^{13}C NMR (62.5 MHz, $CDCl_3$): δ 20.9, 30.1, 32.1, 63.8, 126.0, 128.3, 128.4, 141.1, 171.1

Reference

Iranpoor, N., Firouzabadi, H., and Davan, E. E. (2013). In situ generated $Ph_3P(OAc)_2$ as a novel reagent for the efficient acetylation of alcohols and thiols at room temperature. *Tetrahedron Lett.*, **54**, 1813–1816.

C-O bond_3

Type of reaction: C-O bond formation

Reaction conditions: Acetic acid, room temperature

Synthetic strategy: One-step conversion of a *tert*-butyl protecting group to an acetyl protecting group for thiols

Catalyst: Bromine in AcCl/AcOH

Keywords: *tert*-Butyl protected thiols, bromine, acetyl chloride (AcCl), acetic acid, room temperature, acetyl protected thiols

General reaction scheme

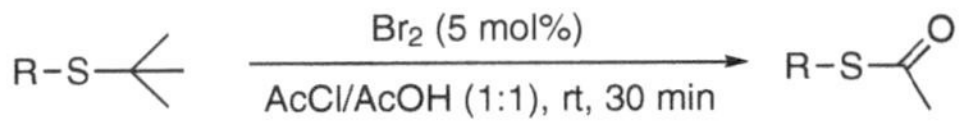

1 (0.01 mmol)

Acetyl protected thiol **2**
9 entries (yield: 29-97%)

R = substituted phenyls, phenyloxy, benzyl, propargyl,

Representative entries

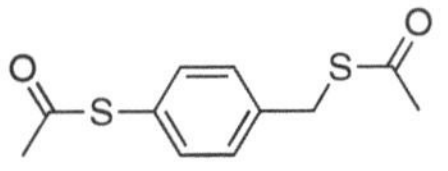

2a

S-4-(Acetylthio)benzyl ethanethioate (**2a**): Yield: 97%

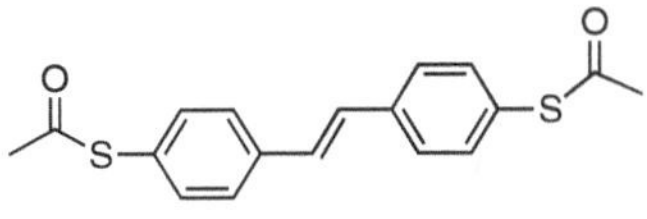

2b

(*E*)-*S*,*S*′-(Ethene-1,2-diylbis(4,1-phenylene)) diethanethioate (**2b**): Solid, mp. 168.5-169.5 °C; yield: 96%

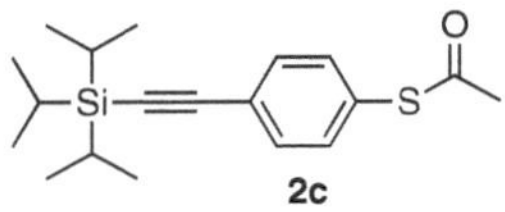

2c

S-(4-((Triisopropylsilyl)ethynyl) phenyl) ethanethioate (**2c**): Yield: 29%

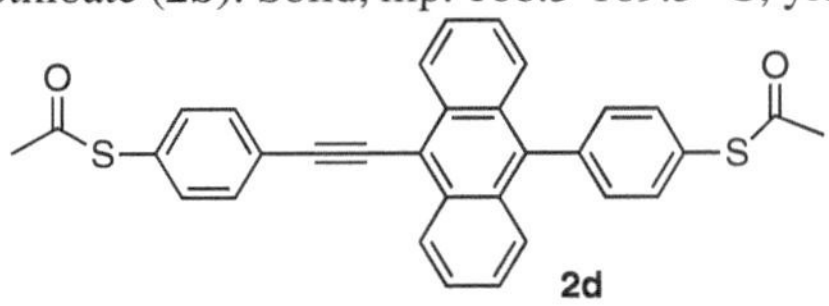

2d

S-(4-((10-(4-(Acetylthio)phenyl)anthracen-9-yl)ethynyl) phenyl) ethanethioate (**2d**): Solid, mp. 248-249 °C; yield: 42%

Experimental procedure

To a well-stirred solution of *tert*-butyl protected thiol (**1**; 0.01 mmol) in acetyl chloride (10 mL), a solution of bromine (5 mol%) in acetyl chloride-acetic acid (1 : 1) was added in 30 minutes at room temperature. After completion of the reaction (1-30 minutes after the bromine addition as monitored by TLC), all solvents were removed by evaporation and the crude residues were purified by silica gel chromatography to obtain pure *S*-acetyl protected thiols 2. The products were characterized on the basis of spectral studies.

Characterization data for two representative compounds

(*E*)-*S*,*S*′-(Ethene-1,2-diylbis(4,1-phenylene)) diethanethioate (**2b**): Solid, mp. 168.5-169.5 °C; ^{1}H NMR (300 MHz, $CDCl_3$): δ 2.44 (s, 3H, CH_3), 7.14 (s, 2H, CH=CH), 7.41 (d, $J = 8.1$ Hz, 4H, ArH), 7.55 (d, $J = 8.1$ Hz, 4H, ArH); ^{13}C NMR (75 MHz, $CDCl_3$): δ 30.7 (CH_3), 127.6, 127.7, 129.5, 135.1, 138.5, 194.5 (C=O).

S-(4-((10-(4-(Acetylthio)phenyl)anthracen-9-yl)ethynyl)phenyl) ethanethioate (**2d**): Solid, mp. 248-249 °C; ^{1}H NMR (300 MHz, $CDCl_3$): δ 2.48 (s, 6H, CH_3), 7.51 (d, $J = 8.1$, 4H, ArH), 7.65-7.68 (m, 4H, ArH), 7.81 (d, $J = 8.2$, 4H, ArH), 8.65-8.69 (m, 4H, ArH); ^{13}C NMR (75 MHz, $CDCl_3$): δ 30.4 (CH_3), 88.1 (C≡C), 101.74 (C≡C), 118.4, 124.6, 127.1, 127.2, 128.6, 132.3, 134.5, 193.4 (C=O); MALDI-TOF-MS: *m/z* 525.91 [M^+], 452.80 [$M^+ - SAc$].

Reference

Błaszczyk, A., Elbing, M., and Mayor, M. (2004). Bromine catalyzed conversion of *S*-*tert*-butyl groups into versatile and, for self-assembly processes accessible, acetyl-protected thiols. *Org. Biomol. Chem.*, **2**, 2722–2724.

C-O bond_4

Type of reaction: C-O bond formation
Reaction conditions: Acetonitrile-water (1:1), room temperature
Synthetic strategy: Oxidation of alcohols to aldehydes and ketones
Catalyst: *tert*-Butylhydrperoxide (TBHP) as oxidant, and vanadyl sulfate ($VOSO_4$) as catalyst
Keywords: Primary and secondary alcohols, *tert*-butylhydrperoxide (TBHP), vanadyl sulfate, acetonitrile-water (1:1), room temperature, oxidation, aldehydes/ketones

General reaction scheme

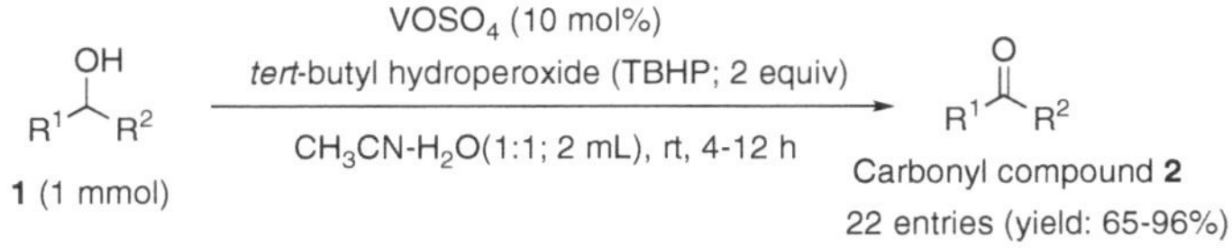

R^1, R^2 = C_6H_5, CH_3; 4-ClC_6H_4, CH_3; 4-BrC_6H_4, CH_3; 4-$OCH_3C_6H_4$, CH_3; C_6H_5, C_6H_5CO; C_6H_5, C_6H_5; C_6H_5, H; 4-$NO_2C_6H_4$, H; 4-ClC_6H_4, H; 4-$OCH_3C_6H_4$, H; 2-OHC_6H_4, H; octyl, H; hexyl, H; furyl, H

In addition, cyclohexanol, propylene glycol, isomayl alcohol, hexane-1,2-diol, hexane-1,4-diol, hydrobenzoin, 2-(benzyloxy)ethan-1-ol, hex-5-en-1-ol, chloesterol

Representative entries

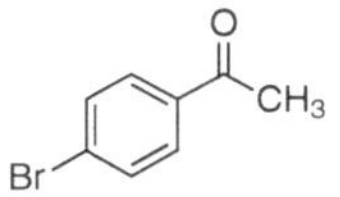

2a

1-(4-Bromophenyl)ethanone (**2a**): Yield: 89%

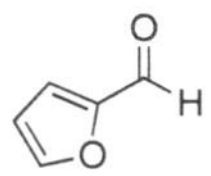

2b

Furan-2-carbaldehyde (**2b**): Yield: 80%

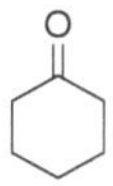

2c

Cyclohexanone (**2c**): Yield: 94%

2d

Hexanal (**2d**): Yield: 70%

Experimental procedure

To a mixture of alcohol (**1**; 1 mmol) and 10 mol % (25.30 mg) $VOSO_4.5H_2O$ in 2 mL CH_3CN-H_2O (1:1) was added 2 equiv 70% TBHP (aqueous), and the resulting mixture was stirred at room temperature for the time specified time (4-12 h). Upon completion of the reaction (as indicated by TLC), the mixture was extracted with ethyl acetate (3 × 20 mL). The combined ethyl acetate layer was dried with anhydrous sodium sulfate and evaporated under reduced pressure. The crude product was purified by column chromatography using ethyl acetate-hexane as the eluent. The products were identified by means of spectral methods and GC-MS as well from the comparison with authentic samples.

Reference

Sarmah, G., Bharadwaj, S. K., Dewan, A., Gogoi, A., and Bora, U. (2014). An efficient and reusable vanadium based catalytic system for room temperature oxidation of alcohols to aldehydes and ketones. *Tetrahedron Lett.*, **55**, 5029–5032.

C-O bond_5

Type of reaction: C-O bond formation
Reaction conditions: Acetonitrile, room temperature
Synthetic strategy: Aerobic oxidation of secondary alcohols
Catalyst: *N*-Hydroxyphthalimide (NHPI)/$Fe(NO_3)_3.9H_2O$
Keywords: Secondary alcohols, *N*-hydroxyphthalimide (NHPI), $Fe(NO_3)_3.9H_2O$, acetonitrile, room temperature, aerobic oxidation, ketones

General reaction scheme

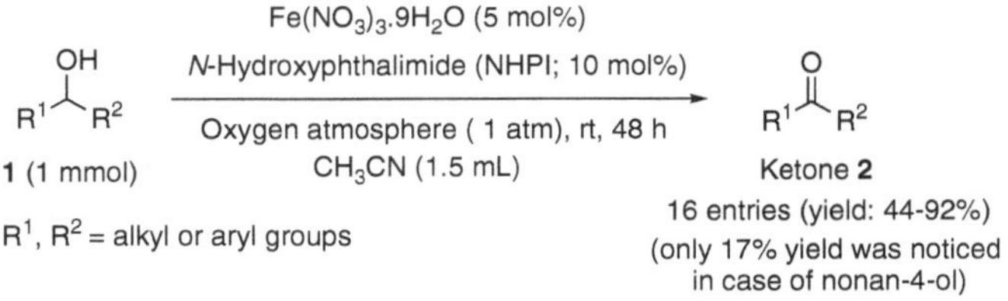

Representative entries

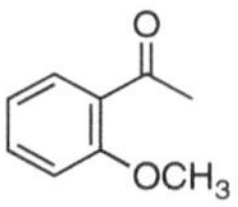

2a

1-(2-Methoxyphenyl)ethanone (**2a**): Yield: 45%

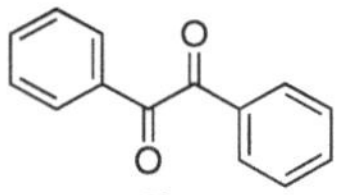

2b

Benzil (**2b**): Yield: 57%

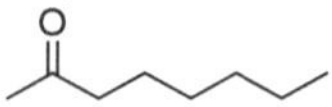

2c

Octan-2-one (**2c**): Yield: 74%

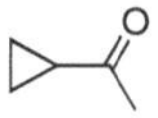

2d

1-Cyclopropylethanone (**2d**): Yield: 60%

Experimental procedure

A mixture of secondary alcohol (**1**; 1 mmol), $Fe(NO_3)_3 \cdot 9H_2O$ (5 mol%), *N*-hydroxyphthalimide (10 mol%), and 1.5 mL of acetonitrile in a 15 mL test tube was maintained for 48 h under an atmospheric pressure of O_2 at room temperature. Upon completion of the reaction the crude product was processed, followed by purification using column chromatography using silica gel (200-300 mesh). The product was characterized using GC-MS and NMR studies.

Characterization data for one representative compound

1-(2-Methoxyphenyl)ethanone (**5a**): ^1H NMR (400 MHz, $CDCl_3$): δ 7.80-7.67 (m, 1H), 7.53-7.38 (m, 1H), 7.06-6.90 (m, 2H), 3.90 (d, $J=0.8$ Hz, 3H), 2.61 (s, 3H); ^{13}C NMR (100 MHz, $CDCl_3$): δ 199.8, 158.9, 133.7, 130.3, 128.3, 120.5, 111.6, 55.5, 31.8; GC-MS: *m/z* 150.1.

Reference

Zhao, H., Sun, W., Miao, C., and Zhao, Q. (2014). Aerobic oxidation of secondary alcohols using NHPI and iron salt as catalyst at room temperature. *J. Mol. Catal. A: Chem.*, **393**, 62–67.

C-O bond_6

Types of reaction: C-O and C-C bond formation
Reaction conditions: Dichloromethane, room temperature
Synthetic strategy: One-pot multicomponent stereospecific synthesis
Catalyst: Triethylamine (TEA)
Keywords: Isocyanides, *syn*-chlorooximes, carboxylic acids, TEA, dichloromethane, Alcohols, room temperature, *syn*-α-oximinoamides, stereospecific synthesis, Passerini-type reaction, Mumm-type rearrangement

Reaction scheme

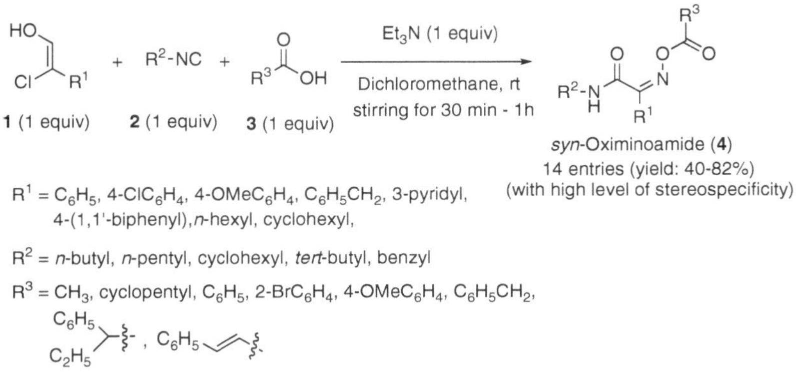

Proposed mechanism

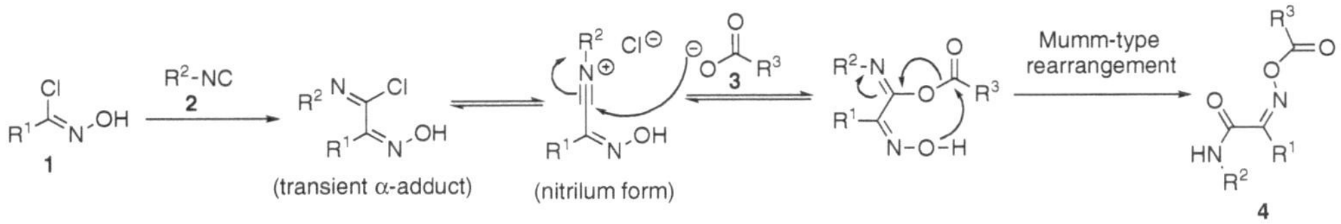

Representative entries

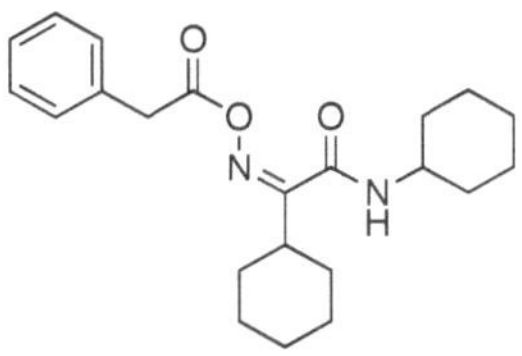

4a

(*Z*)-*N*,2-Dicyclohexyl-2-(2-phenylacetoxyimino) acetamide [**4a**; light yellow amorphous solid; yield: 68% (*syn*/*anti*: 99.5/0.5)]

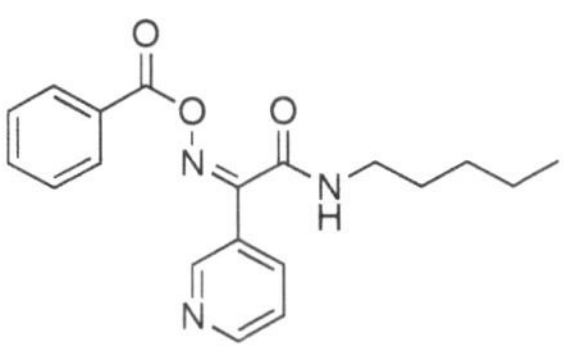

4b

(*Z*)-2-(Benzoyloxyimino)-*N*-pentyl-2-(pyridin-3-yl)acetamide [**4b**; white solid; mp 98-99 °C; yield: 55% (*syn*/*anti*: 96.9/3.1)]

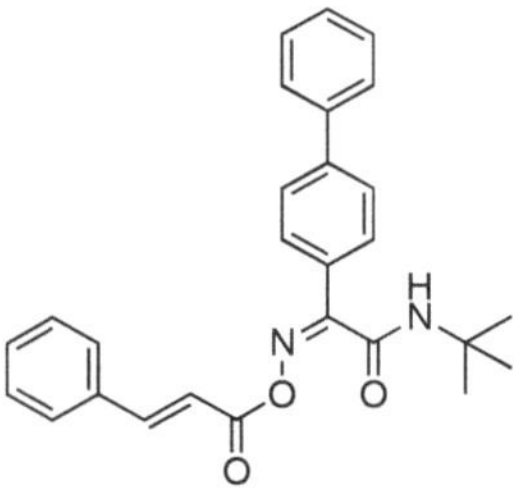

4c

(*Z*)-2-(Biphenyl-4-yl)-*N*-*tert*-butyl-2-(cinnamoyloxyimino)acetamide [**4c**; white solid; mp 198-199 °C; yield: 70% (*syn*/*anti*: 100/0)]

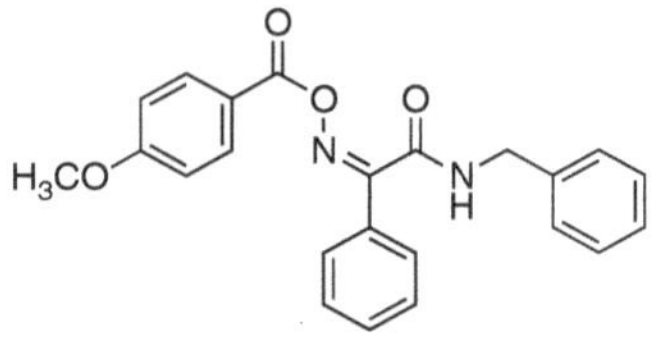

4d

(*Z*)-*N*-Benzyl-2-(4-methoxybenzoyloxy imino)-2-phenylacetamide [**4d**; white solid; mp 165-166 °C; yield: 56% (*syn*/*anti*: 100/0)]

Experimental procedure

To a solution of carboxylic acid (**3**; 1 equiv) in dry dichloromethane, TEA (1 equiv), isocyanide (**2**; 1 equiv), and *syn*-chlorooxime (**1**; 1 equiv) were added and the reaction mixture was then stirred at room temperature under a nitrogen atmosphere until all the chlorooxime was disappeared (typically 30 minutes to1 hour as monitored by TLC). On completion of reaction, the mixture was diluted with dichloromethane and washed with saturated Na_2CO_3. The organic phase is concentrated under reduced pressure. The crude material was purified by column chromatography using petroleum ether and ethyl acetate mixture as eluent. All the products were characterized by analytical and spectral studies including IR, ^{1}H NMR, ^{13}C NMR and mass.

Characterization data of representative compounds

(*Z*)-2-(Biphenyl-4-yl)-*N*-*tert*-butyl-2-(cinnamoyloxyimino)acetamide (**4c**): White solid; mp 198-199 °C; IR ν_{max} (KBr): 1736, 1673, 1636, 1118, 905, 762, 602 cm^{-1}; 1H NMR (300 MHz, $CDCl_3$): δ 7.86 (d, *J*=8.5 Hz, 2 H), 7.70 (d, *J*=15.9 Hz, 1H), 7.59 (d, *J*=8.5 Hz, 2 H), 7.50 (dd, *J*=9.6, 1.9 Hz, 2 H), 7.43-7.30 (m, 8 H), 6.32 (d, *J*=15.9 Hz, 1H), 6.31 (br s, 1 H), 1.59 (s, 9 H); ^{13}C NMR (75 MHz, $CDCl_3$): δ 163.9, 161.5, 160.0, 149.8, 146.5, 144.4, 139.7, 133.9, 130.8, 129.0 (2C), 128.3, 128.2, 128.0, 127.4, 127.0, 114.9, 52.9, 28.9; MS (ESI): *m*/*z* 449 $(M+Na)^+$. Anal. calcd. for $C_{27}H_{26}N_2O_3$: C, 76.03; H, 6.14; N, 6.57%; found: C, 76.13; H, 6.21; N, 6.62.

(*Z*)-*N*-Benzyl-2-(4-methoxybenzoyloxy imino)-2-phenylacetamide (**4d**): White solid; mp 165-166 °C ; IR ν_{max} (KBr): 1768, 1641, 1220, 1105, 898, 710, 690 cm^{-1}; ^{1}H NMR (300 MHz, $CDCl_3$): δ 7.68 (m, 4 H), 7.41-7.21 (m, 8 H), 7.03 (t, *J*=5.2 Hz, 1 H), 6.76 (d, *J*=6.9 Hz, 2 H), 4.72 (d, *J*=2 H), 3.85 (s, 3 H); ^{13}C NMR (75 MHz, $CDCl_3$): δ 164.4, 163.5, 162.3, 160.3, 137.5, 132.2, 131.9, 129.5, 129.3, 129.0, 128.5, 128.0, 127.9, 120.1, 114.0, 55.7, 43.8; MS (ESI): *m*/*z* 389 $(M+H)^+$. Anal. calcd. for $C_{23}H_{20}N_2O_4$: C, 71.12; H, 5.19; N, 7.21%; found: C, 71.19; H, 5.25; N, 7.29.

Reference

Pirali, T., Mossetti, R., Galli, S., and Tron, G. C. (2011). Stereospecific synthesis of *syn*-α-oximinoamides by a three-component reaction of isocyanides, *syn*-chlorooximes, and carboxylic acids. *Org. Lett.*, **13**, 3734–3737.

C-O bond_7

Types of reaction: C-O bond formation
Reaction conditions: Aqueous H_2O_2, room temperature
Synthetic strategy: *ipso*-Hydroxylation of arylboronic acids
Catalyst: Molecular iodine
Keywords: Arylboronic acids, aqueous hydrogen peroxide, iodine, room temperature, *ipso* hydroxylation, phenols

Reaction scheme

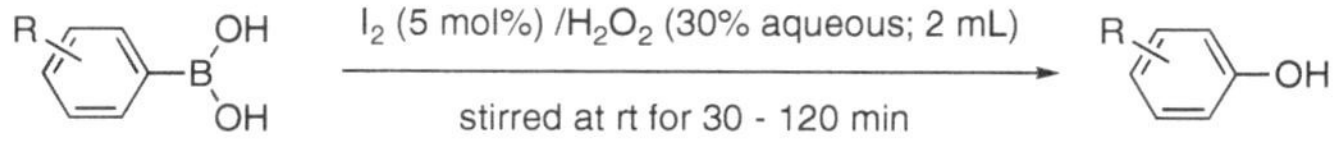

1 (1 mmol)

Phenol (**2**)
13 entries (yield: 80-93%)

R = H, *p*-OMe, *o*-OMe, *m*-OMe, *o*-Me, *p*-Me, *p*-Cl, *p*-COMe, *o*-NO_2, *p*-NH_2, *p*-OH
(besides, thiophene-2-boronic acid was also used as substrate)

Representative entries

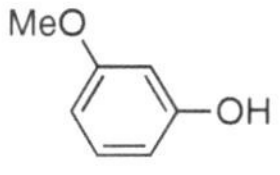

2a

3-Methoxyphenol (**2a**): Reaction time: 45 min; yield: 91%

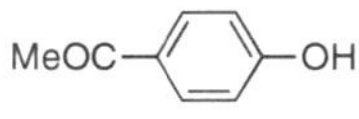

2b

1-(4-Hydroxyphenyl)ethanone (**2b**): Reaction time: 1 h; yield: 91%

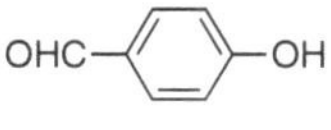

2c

4-Hydroxybenzaldehyde (**2c**): Reaction time: 50 min; yield: 88%

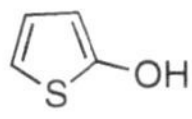

2d

Thiophen-2-ol (**2d**): Yield: 87%

Experimental procedure

A mixture of arylboronic acid (**1**; 1 mmol) and iodine (5 mol%) in 30% aqueous hydrogen peroxide (2 mL) was stirred in a reaction vessel at room temperature for 30-120 min. on completion of reaction (as monitored by TLC), the reaction mixture was diluted with water and extracted with diethyl ether. The organic layer was then washed with sodium thiosulfate solution, and dried over anhydrous sodium sulfate and filtered. The organic solvent was evaporated out under reduced pressure to furnish the crude phenolic product **2** (80-93% yield) which was purified by column chromatography on silica gel (ethyl acetate/hexane) followed by crystallization. Physical and spectral properties of all the products were found to be in close agreement with those reported for authentic samples.

Reference

Gogoi, A., and Bora, U. (2012). An iodine-promoted, mild, and efficient method for the synthesis of phenols from arylboronic acids. *Synlett.*, **23**, 1079–1081.

C-O bond_8

Type of reaction: C-O bond formation
Reaction conditions: Tetrahydrofuran (TMF), room temperature
Synthetic strategy: Methylhydrazine-mediated aerobic oxidative cleavage of arylboronic acids to phenols
Catalyst: Air
Keywords: Arylboronic acids, methylhydrazine, tetrahydrofuran, air, room temperature, aerobic oxidative C-B bond cleavage, phenols

General reaction scheme

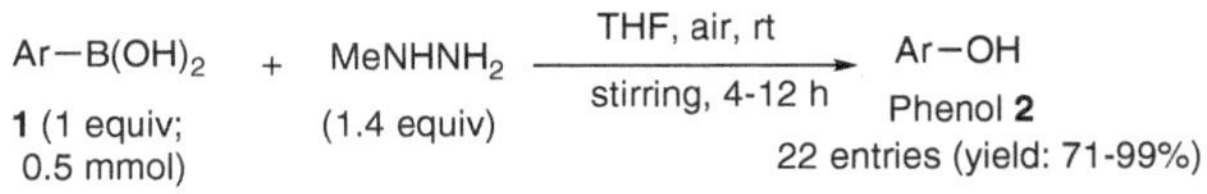

Ar = substituted phenyls, heteroaryl, napthyl, pyrenyl

Representative entries

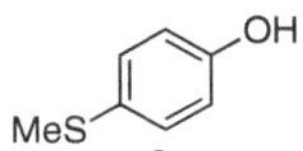

2a

4-(Methylthio)phenol (**3a**): White solid; reaction time: 8 h; yield: 90%

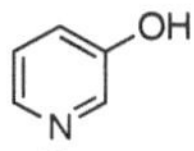

2b

Pyridin-3-ol (**3b**): White solid; reaction time: 24 h; yield: 71%

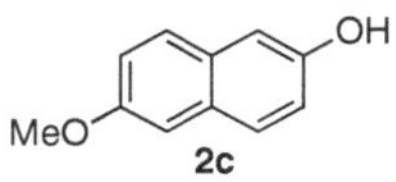

2c

6-Methoxynaphthalen-2-ol (**3c**): Brown solid; reaction time: 6 h; yield: 80%

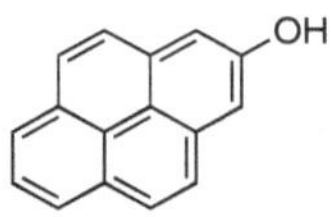

2d

Pyren-2-ol (**3d**): Brown solid; reaction time: 4 h; yield: 90%

Experimental procedure

A 50 mL round-bottom flask was charged with aryl boronic acid (**1**; 0.5 mmol), methylhydrazine (40% aq. 1.4 equiv, 80 uL) (*potentially dangerous to handle, be careful!!!*) and THF (0.017 M, 30 mL). Then, the reaction mixture was stirred at room temperature in an open flask. After the reaction was completed (as monitored by TLC), the mixture was concentrated under reduced pressure and purified by flash chromatography on silica gel (eluting with ethyl acetate / petroleum ether = 1:5) to pure product of phenol **2**. All the products were characterized by spectral analysis.

Characterization data of two representative entries

4-(Methylthio)phenol (**3a**): White solid; ^{1}H NMR (600 MHz, $CDCl_3$): δ 7.22 (d, J = 8.6 Hz, 2 H), 6.78 (d, J = 8.6 Hz, 2 H), 4.97 (s, 1H), 2.44 (s, 3 H); ^{13}C NMR (100 MHz, $CDCl_3$): δ 153.96, 130.30, 128.66, 116.06, 17.95; MS: *m/z* 140.2 (M^+)

Pyren-2-ol (**3d**): Brown solid; ^{1}H NMR (600 MHz, $CDCl_3$): δ 10.66 (s, 1 H), 8.33 (d, J = 9.1, 1H,), 8.13 (dd, J = 12.2 & 4.7 Hz, 3 H,), 8.03 (dd, J = 17.4 & 9.0 Hz, 2 H), 7.97 (t, J = 7.6 Hz, 1 H), 7.90 (d, J = 8.9 Hz, 1 H), 7.60 (d, J = 8.2, 1 H); ^{13}C NMR (100 MHz, $CDCl_3$): δ 152.17,

131.33, 127.39, 126.10, 125.45, 124.46, 123.87, 123.77, 123.59, 121.42, 118.08, 113.23; MS: *m/z* 218.2 (M^+)

Reference

Ding, W., Chen, J.-R., Zou, Y.-Q., Duan, S.-W., Lu, L.-Q., and Xiao, W.-J. (2014). Aerobic oxidative C–B bond cleavage of arylboronic acids mediated by methylhydrazines. *Org. Chem. Front.*, **1**, 151–154.

C-O bond_9

Types of reaction: C-O and C-S bond formation
Reaction conditions: Methanol, room temperature
Synthetic strategy: Oxidative desulfurization
Catalyst: Iodobenzene/Oxone®/Et_3N
Keywords: Thiosemicarbazides, iodobenzene, Oxone®, triethylamine, hypervalent iodine, methanol, room temperature, oxidative sulfurization, oxadiazoles, bisdiarylthiourea, thiadiazoles, azole derivatives

Reaction scheme

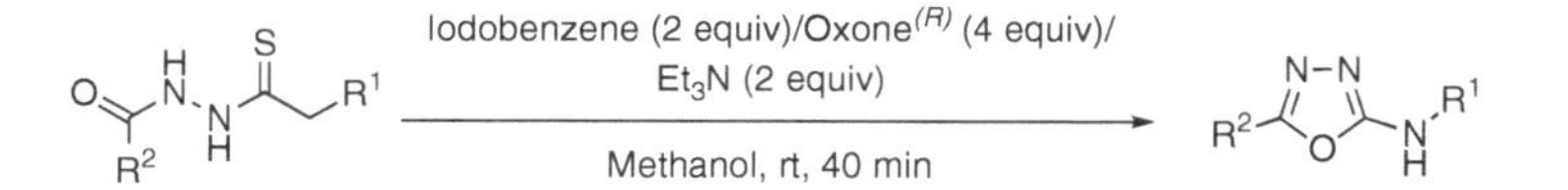

Thiosemicarbazide (**1**; 1 equiv)

1,3,4-Oxadiazole (**2**)
9 entries (yield: 72-91%)

R^1, R^2= 4-BrC_6H_4, 4-ClC_6H_4; 4-BrC_6H_4, 4-BrC_6H_4; 4-BrC_6H_4, 3,4-di-$OMeC_6H_3$; 4-$OMeC_6H_4$,4-ClC_6H_4; 4-BrC_6H_4, 4-$OMeC_6H_4$; 4-$OMeC_6H_4$, 3,4-di-$OMeC_6H_3$; Et, 2-$NO_2C_6H_4$; $C_6H_5CH_2$, C_6H_5; C_6H_5,C_6H_5

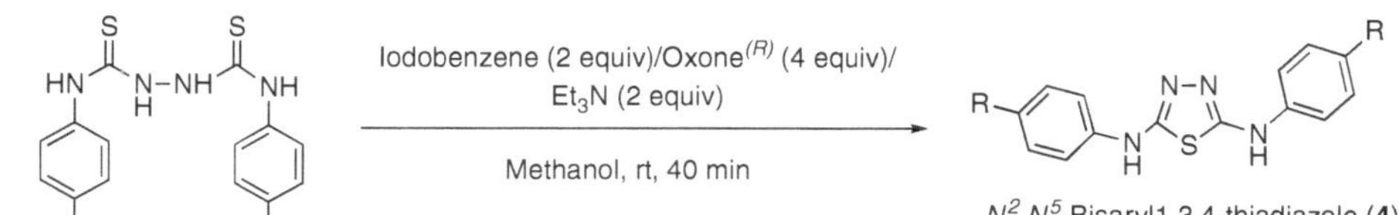

N²,N⁵-Bisaryl1,3,4-thiadiazole (**4**)
4 entries (yield: 78-90%)

Bisdiarylthiourea (**3**; 1 equiv)
R = H, Br, Cl, MeO

Representative entries

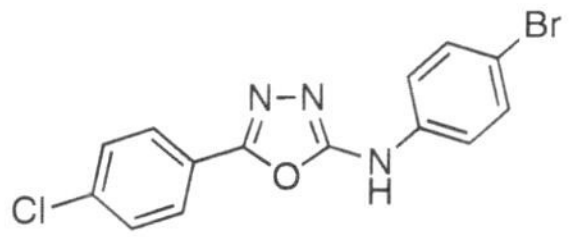

2a

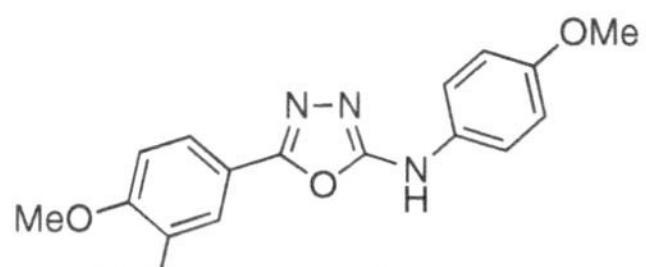

2b

N-(4-Bromophenyl)-5-(4-chlorophenyl)-1,3,4-oxadiazol-2-amine (**2a**): White solid, mp 273-274 °C; yield: 91%

5-(3,4-Dimethoxyphenyl)-*N*-(4-methoxyphenyl)-1,3,4-oxadiazol-2-amine (**2b**): Solid; yield: 84%

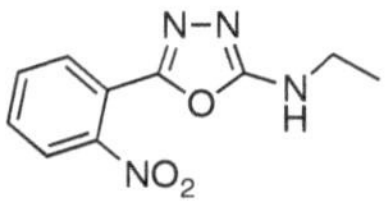

2c

N-Ethyl-5-(2-nitrophenyl)-1,3,4-oxadiazol-2-amine (**2c**): Yield: 85%

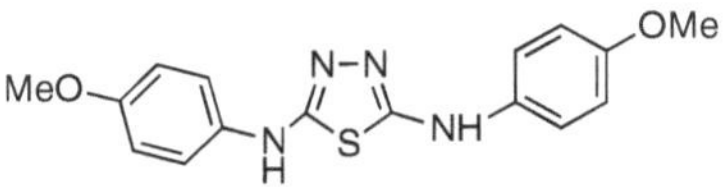

4a

(N^2,N^5-Bis(4-methoxyphenyl)-1,3,4-thiadiazole-2,5-amine (**4a**): White solid, mp 234-236 °C; yield: 90%

Experimental procedure

A mixture of Oxone® (4 equiv) and iodobenzene (2 equiv.) in methanol was stirred at room temperature for 20 min followed by addition of triethylamine (2 equiv.) and the substrate (**1** or **3**; 1 equiv.); the resulting mixture was stirred for another 40 min. On completion of the reaction, the mixture was diluted with water and then extracted twice with EtOAc. The organic layer was washed successively with 10% $NaHCO_3$ (2 × 20 mL) and H_2O (2 × 20 mL), and then dried over anhyd. Na_2SO_4, filtered, and concentrated under reduced pressure to yield the crude product. The crude material was purified by column chromatography using petroleum ether and ethyl acetate mixture (70:30) as eluent. All the products were characterized by spectral studies.

Characterization data of representative compounds

N-(4-Bromophenyl)-5-(4-chlorophenyl)-2-amino-1,3,4-oxadiazole (**2a**): White solid, mp 273-274 °C; IR (KBr): 3318, 3032, 1551, 1435 cm^{-1}; ^1H NMR (300 MHz, $CDCl_3$): δ 7.49-7.53 (m, 4 H),7.61 (d, 2H), 7.89 (d, 2H), 4.10 (s, 1H).

N^2,N^5-Bis(4-methoxyphenyl)-1,3,4-thiadiazole-2,5-diamine (**4a**): White solid, mp 234-236 °C; IR (KBr): 3328, 1543, 1436, 1053 cm^{-1}; ^1H NMR (300 MHz, DMSO-d_6): δ 3.74 (s, 6H), 7.04 (d, 4 H), 7.38 (d, 4H), 8.21 (s, 2H).

Reference

Patel, K. N., Jadhav, N. C., Jagadhane, P. B., and Telvekar, V. N. (2012). A novel approach strategy for the construction of azole heterocycles via an oxidative dwsulfurization approach using iodobenzene and Oxone®. *Synlett*, **23**, 1970–1972.

C-O bond_10

Types of reaction: C-O bond formation
Reaction conditions: Dichloromethane, room temperature
Synthetic strategy: One-step protection of carbonyl group
Catalyst: *N*-Hydroxybenzenesulfonamide ($PhSO_2NHOH$)/Et_3N
Keywords: Aldehydes, ketones, ethylene glycol, *N*-hydroxybenzenesulfonamide, triethylamine, dichloromethane, room temperature, ketal, 1,3-dioxolanes

Reaction scheme

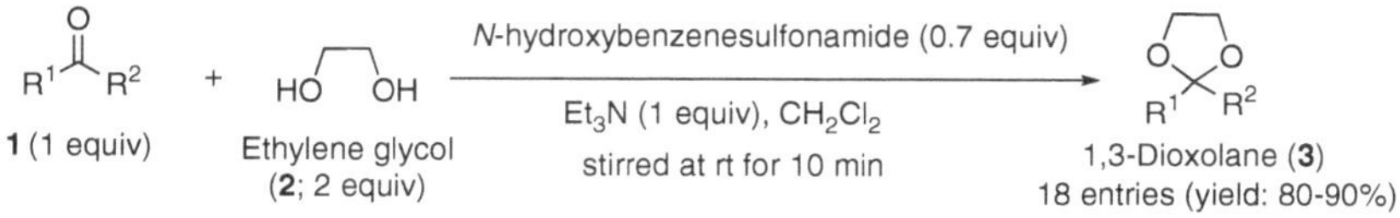

R^1, R^2= 2-carboethoxycyclohexanone; C_6H_5, CH_3; $(CH_3)_2CH$; C_6H_5, H; 4-ClC_6H_4,H; 4-$NO_2C_6H_4$, H; cyclopentaone; cyclohexanone; cycloheptanone; adamantan-2-one; 2-naphthyl, H; *N*-Boc-cyclohexanone; PhCHMe, H; Ph_2CH, H; 4-$TBSOC_6H_4$, H; 4-$THPOC_6H_4$, H; 4-$TBDPSOC_6H_4$, H; 4-$CH_3C_6H_4$, H

Representative entries

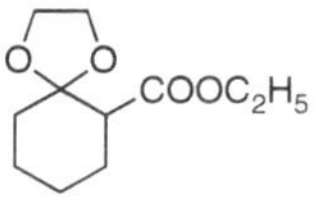

3a

Ethyl 1,4-dioxaspiro[4.5]decane-6-carboxylate (**3a**): Yield: 84%

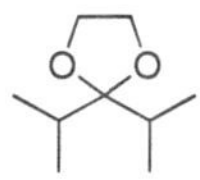

3b

2,2-Diisopropyl-1,3-dioxolane (**2b**): Yield: 90%

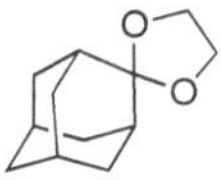

3c

Spiro[adamantane-2,2′-[1,3]dioxolane] (**2c**): Yield: 89%

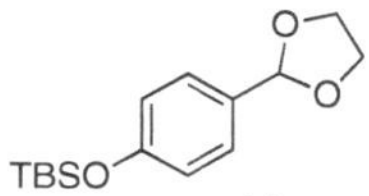

4d

(4-(1,3-Dioxolan-2-yl)phenoxy)(*tert*-butyl)dimethylsilane (**2d**): Yield: 87%

Experimental procedure

A solution of a carbonyl compound (ketone or aldehyde **1**; 1 mmol), commercially available *N*-hydroxybenzenesulfonamide (0.7 mmol), and ethylene glycol (**2**; 2 mmol) in dichloromethane (4 mL) was stirred at room temperature for 10 min. The solution was washed with water and brine, and dried over anhydrous magnesium sulfate. On concentration under reduced pressure the resulting solution yielded crude product which was purified by flash chromatography (basic alumina; EtOAc-hexane, 1:9) to obtain pure 1,3-dioxolane 3 in 80–90% yield. The analytical data of the products were compared with literature values and found to be in good agreement with those of authentic compounds.

Reference

Hassner, A., Reddi, B. C., and Panchgalle, S. (2012). A mild, room-temperature protection of ketones and aldehydes as 1,3-dioxolanes under basic conditions. *Synlett*, **23**, 2773–2776.

C-O bond_11

Type of reaction: C-O and C-C bond formation
Reaction conditions: *N*,*N*-Dimethylformamide (DMF), room temperature
Synthetic strategy: One-pot tandem reaction involving transesterification and subsequent Dieckmann cyclization

Catalyst: Potassium *tert*-butoxide (KO*t*-Bu)

Keywords: Esters, KO*t*-Bu, DMF, room temperature, tandem process, transesterification, Dieckmann cyclization, ring-closure, heterocycles, tetronic acids

General reaction scheme

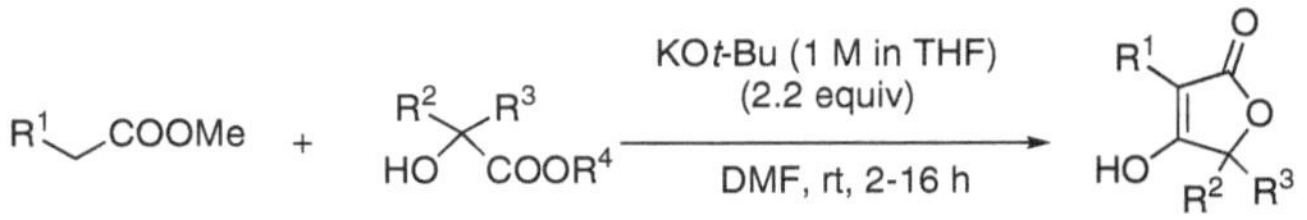

1 (2 mmol) **2** (2.2 mmol)

Tetronic acid **3**
7 entries (yield: 39-96%)

R^1= C_6H_5, 4-BrC_6H_4, 4-OMeC_6H_4, thien-2-yl
R^2 = H, Me; R_3 = H, Me, *n*-Bu, Ph; R^4 = Me, Et

Representative entries

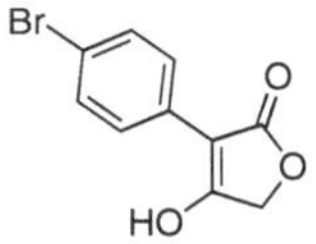

3a

3-(4-Bromophenyl)-4-hydroxyfuran-2(5*H*)-one (**3a**): Yield: 59%

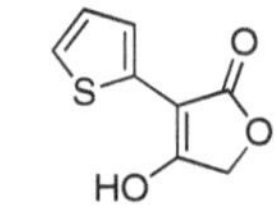

3b

4-Hydroxy-3-(thiophen-2-yl)furan-2(5*H*)-one (**3b**): Yield: 70%

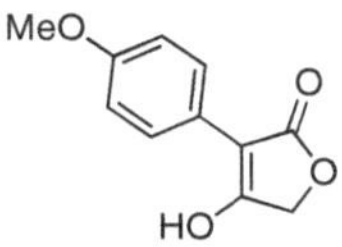

3c

4-Hydroxy-3-(4-methoxyphenyl)furan-2(5*H*)-one (**3c**): Yield: 60%

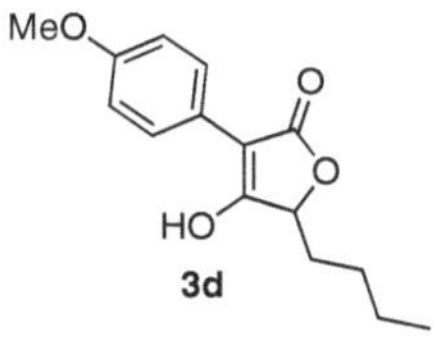

3d

5-Butyl-4-hydroxy-3-(4-methoxyphenyl)furan-2(5*H*)-one (**3d**): Yield: 39%

Experimental procedure

To a solution of alkyl aryllacetate (**1**; 2.0 mmol) and alkyl hydroxyacetate (**2**; 2.2 mmol) in DMF (10 mL) was added a 1 M solution of KO*t*-Bu in THF (4.4 mL, 4.4 mmol). The mixture was then stirred under argon at room temperature for 2-16 h. On completion of the reaction, the reaction mixture was poured into cooled 1 N HCl (15 mL). The aqueous phase was extracted with ethyl acetate (3 × 10 mL), and the combined organic layers was washed with brine, dried over Na_2SO_4. After filtration and concentration in *vacuo*, the residue was purified by column chromatography to furnish tetronic acid **3** with 39-96% yield. All the products were characterized from spectral analyses.

Characterization data of one representative compound

4-Hydroxy-3-(4-methoxyphenyl)furan-2(5*H*)-one (**3c**): White solid, mp 228 °C; IR (neat): 2955, 1695, 1641, 1610, 1514, 1426, 1398, 1351, 1297, 1256, 1169, 1053, 1030, 1022, 960, 836, 736 cm^{-1}; ^{1}H NMR (400 MHz, acetone-d_6): δ 3.81 (s, 3H, OCH_3), 4.76 (s, 2H, CH_2), 6.95 (d, J=9.0 Hz, 2H, CH), 7.94 (d, J=9.0 Hz, 2H, CH), 10.94 (br s, 1H, OH); ^{13}C NMR

(100 MHz, acetone-d_6): δ 55.5 (OCH_3), 66.6 (C-5), 100.3 (C-3), 114.3(C-3′), 123.8 (C-1′), 129.2 (C-2′), 159.4 (C-4′), 172.1, 173.5 (C-2, C-4); MS (ESI-TOF): *m/z* 207 $[M+H]^+$.

Reference

Mallinger, A., Gall, T. L., and Mioskowski, C. (2008). One-pot synthesis of tetronic acids from esters. *Synlett*, 386–388.

C-O bond_12

Type of reaction: C-O and C-N bond formation
Reaction conditions: Solvent-free, room temperature
Synthetic strategy: One-pot multicomponet reaction
Catalyst: *N*-Methylimidazole
Keywords: Ammonium thiocyanate, benzoyl chlorides, ethyl bromopyruvate, ethyl 2-chloroacetoacetate, *N*-methylimidazole, solvent-free, room temperature, one-pot multicomponent reaction, condensation, 1,3-oxazoline-2-thiones

General reaction schemes

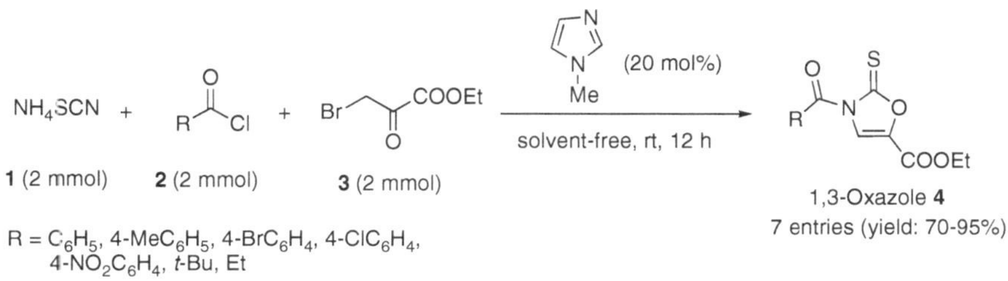

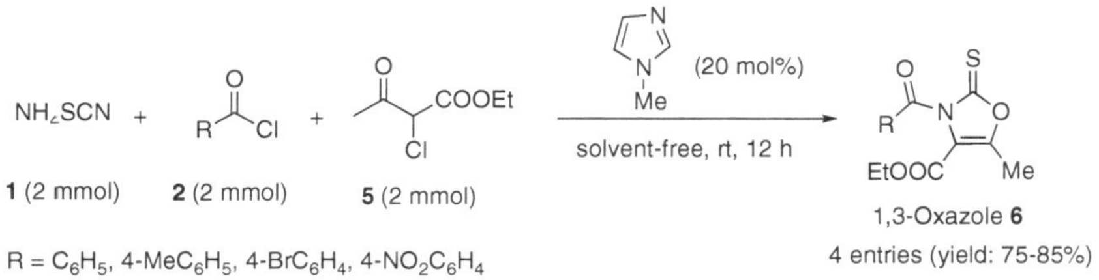

Representative entries

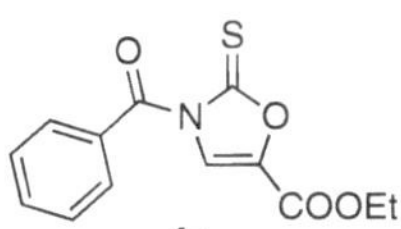

4a

Ethyl 3-benzoyl-2-thioxo-2,3-dihydrooxazole-5-carboxylate (**4a**): Pale yellow crystals, mp 129-131 °C; yield: 85%

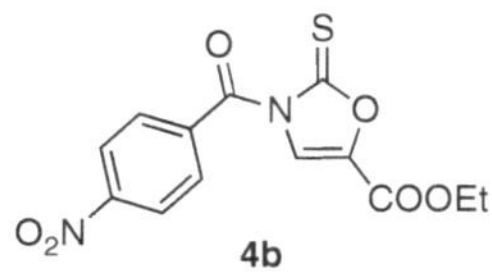

4b

Ethyl 3-(4-nitrobenzoyl)-2-thioxo-2,3-dihydrooxazole-5-carboxylate (**4b**): Yield: 85%

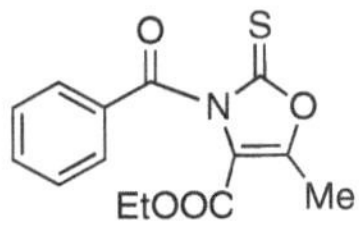

6a

Ethyl 3-benzoyl-5-methyl-2-thioxo-2,3-dihydrooxazole-4-carboxylate (**6a**): White powder, mp 140-142 °C; yield: 85%

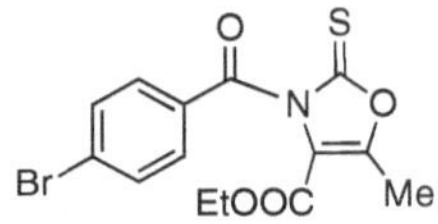

6b

Ethyl 3-(4-bromobenzoyl)-5-methyl-2-thioxo-2,3-dihydrooxazole-4-carboxylate (**9a**): Yield: 76%

Experimental procedure

A stirred mixture of ammonium isothiocyanate (**1**; 2 mmol) and acid chloride (**2**; 2 mmol) was warmed at about 90 °C in a water bath for 5 min; after then ethyl bromopyruvate (**3**; 2 mmol) or ethyl 2-chloroaceto acetate (**5**; 2 mmol) was added to the mixture slowly. The resulting mixture was allowed to cool to room temperature and *N*-methylimidazole (20 mol%) was added. The reaction mixture was now stirred for 12 h to complete the conversion. Extraction with Et_2O (2 mL) afforded the pure 1,3-oxazoline-2-thione **4** or **6**. The products were characterized by means of detailed spectral and elemental studies.

Characterization data of representative compounds

Ethyl 3-benzoyl-2-thioxo-2,3-dihydrooxazole-5-carboxylate (**4a**): Pale yellow crystals, mp 129-131 °C; IR (KBr): 1724, 1631, 1585, 1518, 1470 cm^{-1}; 1H NMR: δ 1.45 (t, $^3J=7.2$ Hz, 3H, Me), 4.46 (q, $^3J=7.2$ Hz, 2H, OCH_2), 7.52 (t, $^3J=7.8$ Hz, 2H, 2 × CH), 7.61 (t, $^3J=6.1$ Hz, 1H, CH), 7.65 (s, 1H, CH), 7.52 (d, $^3J=6.1$ Hz, 2H, 2 × CH); ^{13}C NMR: δ 14.6 (Me), 63.0 (OCH_2), 118.4 (CH), 128.9 (2 × CH), 130.5 (2 × CH), 133.8 (CH), 134.9 (C), 139.8 (C), 156.6 (C═O), 176.7 (C═O), 178.1 (C═S); EI-MS: *m/z* 227 (10) [M^+], 121 (20), 105 (100), 77 (90), 57 (30), 51 (64), 45 (36). Anal. Calcd for $C_{13}H_{11}NO_4S$ (277.29): C 56.31, H 4.00, N 5.05; Found: C 56.30, H 4.03, N 5.00.

Ethyl 3-benzoyl-5-methyl-2-thioxo-2,3-dihydrooxazole-4-carboxylate (**6a**): White powder, mp 140-142 °C; IR (KBr): 1720, 1635, 1582, 1510, 1475 cm^{-1}; ^1H NMR: δ 1.38 (t, $^3J=7.2$ Hz, 3H, Me), 2.68 (s, 3H, Me), 4.34 (q, $^3J=7.2$ Hz, 2H, OCH_2), 7.48 (t, $^3J=7.5$ Hz, 2H, 2 × CH), 7.58 (t, $^3J=7.5$ Hz, 1H, CH), 8.31 (d, $^3J=7.5$ Hz, 2H, 2 × CH); ^{13}C NMR: δ 13.8 (Me), 14.1 (Me), 62.0 (OCH_2), 110.4 (C), 128.4 (2 × CH), 129.9 (2 × CH), 133.1 (CH), 134.8 (C), 155.5 (C), 160.6 (C═O), 176.1 (C═O), 177.5 (C═S); EI-MS: *m/z* 291 (15) [M^+], 186 (78), 105 (100), 77 (48), 45 (48). Anal. Calcd for $C_{14}H_{13}NO_4S$ (291.32): C 57.72, H 4.50, N 4.81; Found: C 57.76, H 4.54, N 4.80.

Reference

Yavari, I., Hossaini, Z., Souri, S., and Sabbaghan, M. (2008). N-Methylimidazole-promoted efficient synthesis of functionalized 1,3-oxazoline-2-thiones under solvent-free conditions. *Synlett*, 1287–1288.

C-O bond_13

Type of reaction: C-O bond formation
Reaction conditions: Dichloromethane, room temperature
Synthetic strategy: Cyclization
Catalyst: Silver hexafluoroantimonate ($AgSbF_6$)
Keywords: *N*-Propargylamides, $AgSbF_6$, dichloromethane, room temperature, oxazolines

General reaction scheme

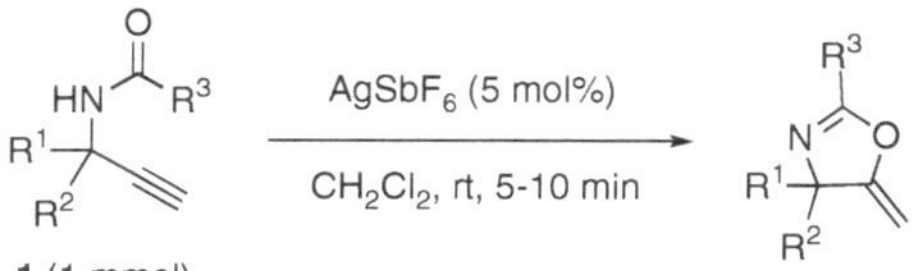

1 (1 mmol)

Oxazoline **2**
11 entries (yield: 89-98%)

Extension

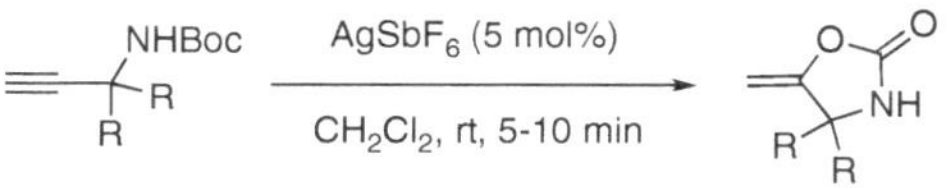

3a: R,R = Me, Me
3b: R,R = $(\text{-}CH_2\text{-})_5$

4a: R,R = Me, Me; yield: 92%
4b: R,R = $(\text{-}CH_2\text{-})_5$; yield: 91%

Representative entries

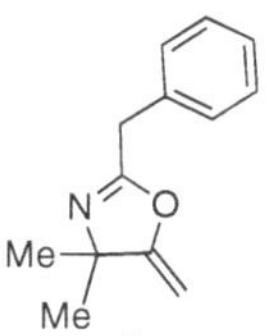

2a

2-Benzyl-4,4-dimethyl-5-methylene-4,5-dihydrooxazole (**2a**): Solid, mp 192-193 °C; yield: 97%

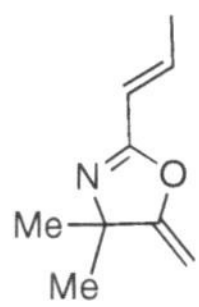

2b

(*E*)-4,4-Dimethyl-5-methylene-2-(prop-1-en-1-yl)-4,5-dihydrooxazole (**2b**): Oil; yield: 95%

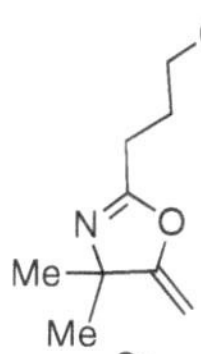

2c

2-(3-Chloropropyl)-4,4-dimethyl-5-methylene-4,5-dihydrooxazole (**2c**): Oil; yield: 89%

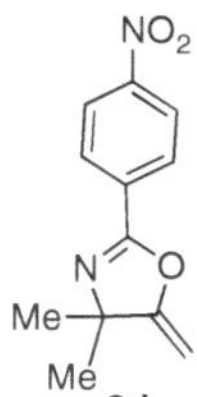

2d

4,4-Dimethyl-5-methylene-2-(4-nitrophenyl)-4,5-dihydrooxazole (**2d**): Oil; yield: 94%

Experimental procedure

To a solution of *N*-propargylamide (**1**; 1 mmol) in dichloromethane (2 mL) within a dry flask was added $AgSbF_6$ (5 mol%) in one portion; the contents were then stirred at room temperature just for 5-10 min to complete the reaction (TLC monitored). The reaction mixture was filtered through Celite and washed with dichloromethane. The filtrate was evaporated to afford the pure product of oxazolines (**2**). The starting amides were prepared by the reaction of amines with acid chlorides. All products were characterized by means of detailed spectral studies.

Characterization data of representative compounds

(*E*)-4,4-Dimethyl-5-methylene-2-(prop-1-en-1-yl)-4,5-dihydrooxazole (**2b**): Oil; IR (CH_2Cl_2): 2970, 1674, 1307, 1184, 972, 661 cm^{-1}; ^1H NMR (500 MHz, $CDCl_3$): δ 6.84 (dt, J=6.0, 15.5 Hz, 1H), 6.07 (d, J=15.5 Hz, 1H), 4.72 (d, J=3.0 Hz, 1H), 4.27 (d, J=3.0 Hz, 1H), 1.94 (d, J=6.0 Hz, 3H), 1.41 (s, 6H); ^{13}C NMR (125.8 MHz, $CDCl_3$): δ 165.9, 161.3, 143.3, 117.6, 83.5, 68.1, 29.8, 18.5; HRMS: m/z calcd for $C_9H_{13}NONa$: 174.0889; Found: 174.0886.

4,4-Dimethyl-5-methylene-2-(4-nitrophenyl)-4,5-dihydrooxazole (**2d**): Oil; IR (CH_2Cl_2): 3427, 3105, 2982, 1646, 1597, 1519, 1307, 1066 cm^{-1}; ^1H NMR (500 MHz, $CDCl_3$): δ 8.30 (d, J=8.0 Hz, 2H), 8.18 (d, J=8.0 Hz, 2H), 4.80 (d, J=3.0 Hz, 1H), 4.33 (d, J=3.0 Hz, 1H), 1.48 (s, 6H); ^{13}C NMR (125.8 MHz, $CDCl_3$): δ 167.2, 158.2, 149.6, 132.6, 129.1, 123.6, 83.5, 69.6, 29.5; HRMS: m/z calcd for $C_{12}H_{12}N_2O_3Na$: 255.0740; Found: 255.0739.

Reference

Harmata, M., and Huang, C. (2008). Silver-catalyzed preparation of oxazolines from *N*-propargylamides. *Synlett*, 1399–1401.

C-O bond_14

Type of reaction: C-O and C-N bond formation
Reaction conditions: Dichloromethane, room temperature
Synthetic strategy: One-pot multicomponent reaction
Catalyst: No catalyst
Keywords: *N*-Isocyaniminotriphenylphosphorane, biacetal, (*E*)-cinnamic acids, dichloromethane, room temperature, one-pot multicomponent reaction, *aza*-Wittig reaction, 1,3,4-oxadiazoles

General reaction scheme

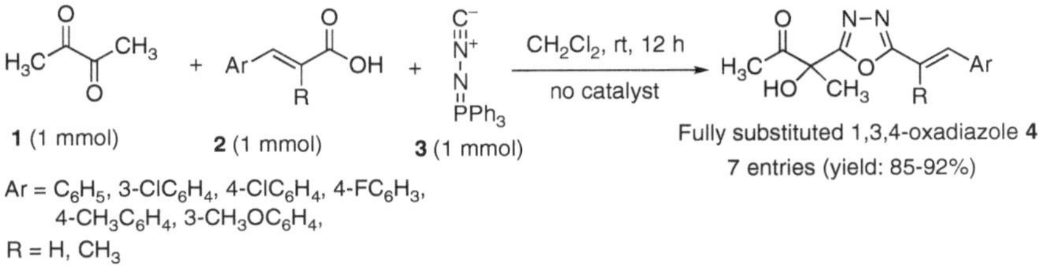

Plausible mechanism

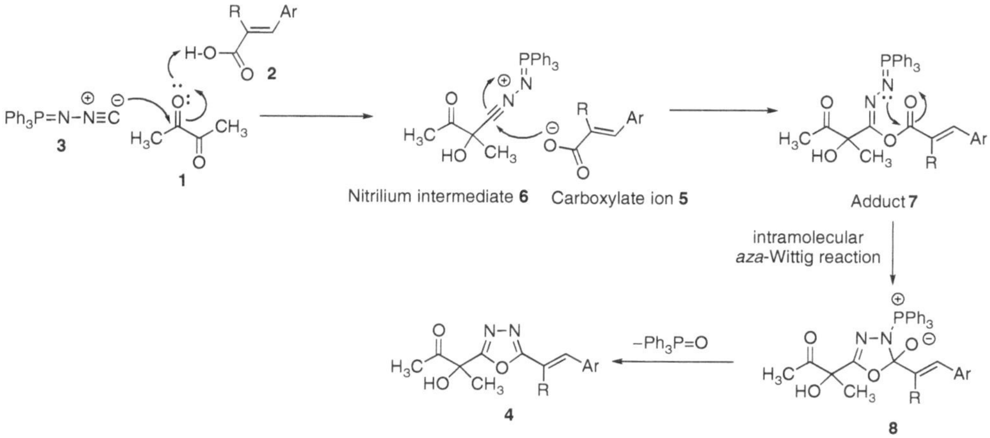

Representative entries

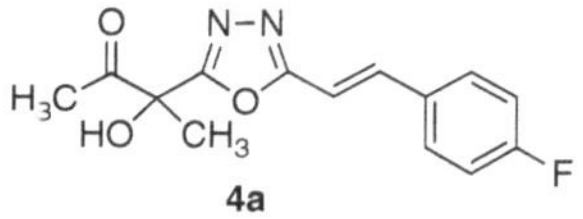

4a

(*E*)-3-(5-(4-Fluorostyryl)-1,3,4-oxadiazol-2-yl)-3-hydroxybutan-2-one (**4a**): White powder, mp 84-86 °C; yield: 90%

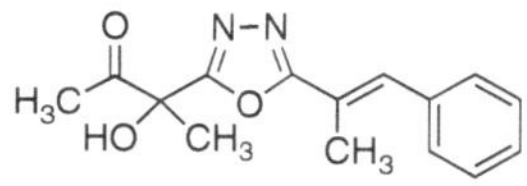

4b

(*E*)-3-Hydroxy-3-(5-(1-phenylprop-1-en-2-yl)-1,3,4-oxadiazol-2-yl)butan-2-one (**4b**): Yellow viscous oil; yield: 86%

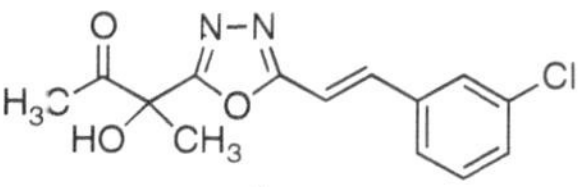

4c

(*E*)-3-(5-(3-Chlorostyryl)-1,3,4-oxadiazol-2-yl)-3-hydroxybutan-2-one (**4c**): White powder, mp 62-64 °C; yield: 86%

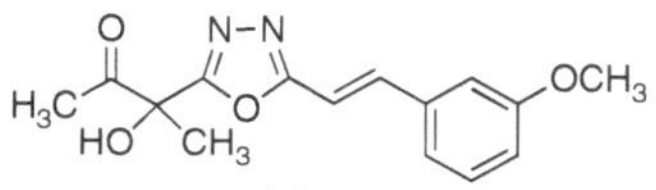

4d

(*E*)-3-Hydroxy-3-(5-(3-methoxystyryl)-1,3,4-oxadiazol-2-yl)butan-2-one (**4d**): White powder, mp 70-72 °C; yield: 87%

Experimental procedure

A mixture of biacetyl (**1**; 1.0 mmol, 0.10 mL), (*E*)-cinnamic acid (**2**; 1.0 mmol), and *N*-isocyaniminotriphenylphosphorane (**3**; 1.0 mmol, 0.300 g) in dichloromethane (5 mL) was stirred at room temperature for 12 h to complete the reaction. The solvent was removed under reduced pressure and the viscous residue was purified by PLC (silica gel (F_{254}) powder; petroleum ether-ethyl acetate = 4:1) to obtain pre fully substituted 1,3,4-oxadiazoles **4**. All the products were characterized from spectral and analytical measurements.

Characterization data of two representative compounds

(*E*)-3-(5-(4-Fluorostyryl)-1,3,4-oxadiazol-2-yl)-3-hydroxybutan-2-one (**4a**): White powder, mp 84-86 °C; IR (neat): 3372, 2923, 1731, 1670, 1529, 1505, 1163, 818 cm^{-1}; ^{1}H NMR (250 MHz, $CDCl_3$): δ 1.87 (s, 3H, CH_3), 2.35 (s, 3H, CH_3), 4.70 (s, 1H, O*H*), 6.92 (d, $^3J_{HH}$ = 16.5 Hz, 1H, =C*H*), 7.01-7.55 (m, 5H, Ar-*H* and = C*H*); ^{13}C NMR (62.5 MHz, $CDCl_3$): δ = 23.3 (CH_3), 23.7 (CH_3), 75.7 (*C*-OH), 109.0 (*C*H=), 138.9 (*C*H=), 116.2

(d, $^2J_{CF}$ = 28.3 Hz, Ar-*C*H), 129.4 (d, $^3J_{CF}$ = 8.2 Hz, Ar-*C*H), 130.7 (C), 162.5 (d, $^1J_{CF}$ = 440.3 Hz, C), 164.9 (C=N), 165.2 (C=N), 204.6 (C=O); MS: *m/z* (%) = 276 (M^+, 12), 234 (100), 191 (56), 164 (32), 147 (60), 121 (20), 101 (24), 43 (68). Anal. calcd. for $C_{14}H_{13}FN_2O_3$ (276.09): C, 60.87; H, 4.74; N, 10.14; Found: C, 60.81; H, 4.80; N, 10.08%.

(*E*)-3-Hydroxy-3-(5-(3-methoxystyryl)-1,3,4-oxadiazol-2-yl)butan-2-one (**4d**): White powder, mp 70-72 °C; IR (neat): 3236, 3004, 2922, 1730, 1650, 1580, 1466, 1108, 876, 795, 687 cm^{-1}; ^1H NMR (250 MHz, $CDCl_3$): 1.93 (s, 3H, CH_3), 2.35 (s, 3H, CH_3), 3.83 (s, 3H, OCH_3), 4.73 (s, 1H, *OH*), 6.91-7.35 (m, 5H, Ar-*H* and *CH*), 7.52 (d, $^3J_{HH}$ = 16.5 Hz, 1H, *CH*); ^{13}C NMR (62.5 MHz, $CDCl_3$): δ 23.3 (CH_3), 23.8 (CH_3), 55.3 (s, 3H, OCH_3), 75.8 (*C*-OH), 109.6 (*C*H), 140.1 (*C*H), 112.5 (*C*H), 116.1 (*C*H), 120.3 (*C*H), 130.0 (*C*H), 135.8 (C), 160.0 (C), 164.9 (*C*=N), 165.3 (*C*=N), 204.6 (*C*=O). Anal. calcd. for $C_{15}H_{16}N_2O_4$ (288.11): C, 62.49; H, 5.59; N, 9.72; Found: C, 62.44; H, 5.64; N, 9.67%.

Reference

Ramazani, A., Abdian, B., Nasrabadi, F. Z., and Rouhani, M. (2013). The reaction of *N*-isocyaniminotriphenylphosphorane with biacetyl in the presence of (*E*)-cinnamic Acids: Synthesis of fully substituted 1,3,4-oxadiazole derivatives *via* intramolecular *aza*-Wittig reactions of *in situ* generated iminophosphoranes. *Phosphorus Sulfur Silicon Relat. Elem.*, **188**, 642–648.

C-O bond_15

Type of reaction: C-O and C-N bond formation
Reaction conditions: Acetonitrile, room temperature
Synthetic strategy: One-pot synthesis
Catalyst: Phosphorus pentoxide (P_2O_5)
Keywords: Acid hydrazides, acyl halides, phosphorus pentoxide, acetonitrile, room temperature, one-pot synthesis, 2,5-di-substituted 1,3,4-oxadiazoles

General reaction scheme

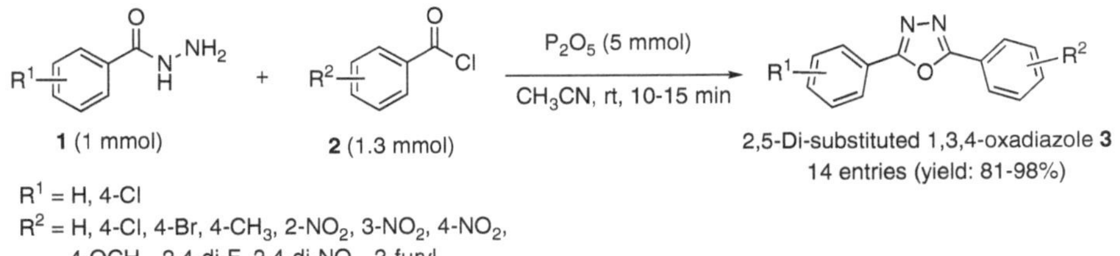

Representative entries

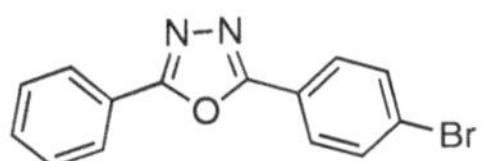

3a

2-(4-Bromophenyl)-5-phenyl-1,3,4-oxadiazole (**3a**): Solid, mp 167-168 °C; reaction time: 10 min; yield: 94%

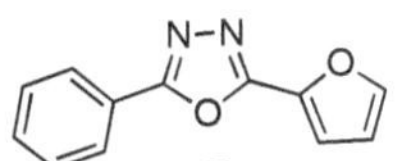

3b

2-(Furan-2-yl)-5-phenyl-1,3,4-oxadiazole (**3b**): Solid, mp 103-104 °C; reaction time: 15 min; yield: 81%

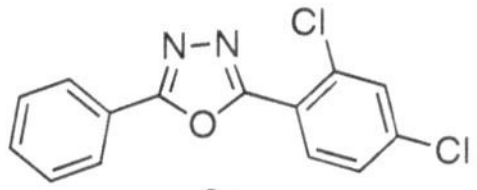
3c

2-(2,4-Dichlorophenyl)-5-phenyl-1,3,4-oxadiazole (**3c**): Solid, mp 100-101 °C; reaction time: 12 min; yield: 89%

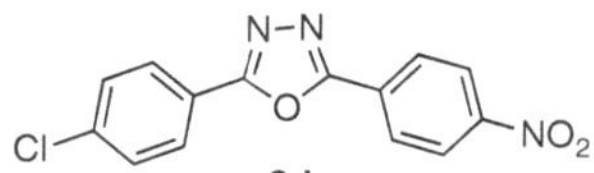
3d

2-(4-Chlorophenyl)-5-(4-nitrophenyl)-1,3,4-oxadiazole (**3d**): Solid, mp 218-219 °C; reaction time: 12 min; yield: 87%

Experimental procedure

A mixture of acid hydrazide (**1**; 1 mmol), acyl chloride (**2**; 1.3 mmol), and phosphorous pentoxide (5 mmol) dissolved in acetonitrile (10 mL) was stirred at room temperature for 10-15 min to complete the reaction (as indicated by TLC). The solvent was then removed under vacuo and the mixture was dissolved in dichloromethane and washed with water (3 × 50 mL). The organic layer was then dried over sodium sulfate and evaporated under vacuo. The solid precipitates were further recrystallized from ethanol to afford pure 2,5-di-substituted 1,3,4-oxadiazoles (**3**) in good yields. The products were characterized by comparison of their spectral and physical data with those of known compounds reported in the literature.

Reference

Rostamizadeh, S., and Ghamkhar, S. (2008). A mild and facile method for one pot synthesis of 2,5-di-substituted 1,3,4-oxadiazoles at room temperature. *Chinese Chem. Lett.*, **19**, 639–642.

C-O bond_16

Type of reaction: C-O bond formation
Reaction conditions: Dichloromethane, room temperature
Synthetic strategy: 5-*endo-dig* cycloisomerization
Catalyst: $ZnCl_2$
Keywords: 1,4-Di- and 1,2,4-trisubstituted but-3-yn-1-ones, dichloromethane, $ZnCl_2$, room temperature, 5-*endo-dig* cycloisomerization, 2,5- and 2,3,5-trisubstituted furans

General reaction scheme

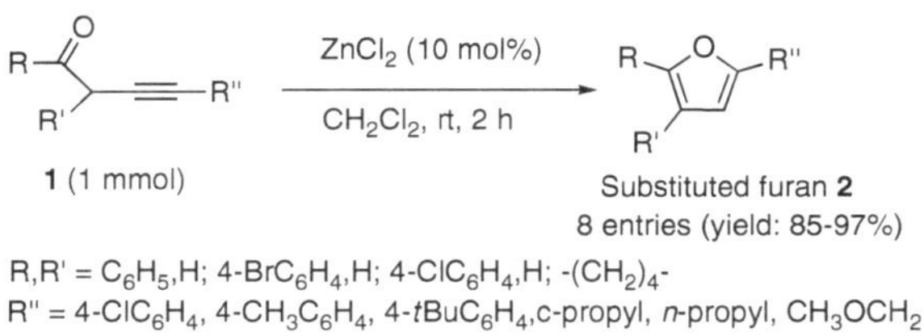

Representative entries

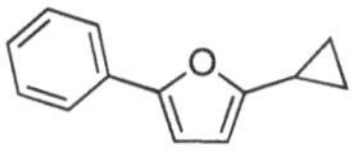

2a

2-Cyclopropyl-5-phenylfuran (**2a**): Yield: 85%

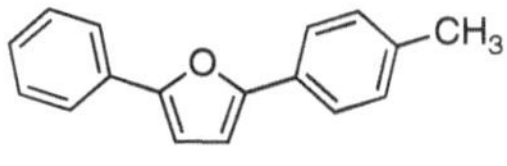

2b

2-Phenyl-5-(*p*-tolyl)furan (**2b**): White solid, mp. 97-99 °C; yield: 97%

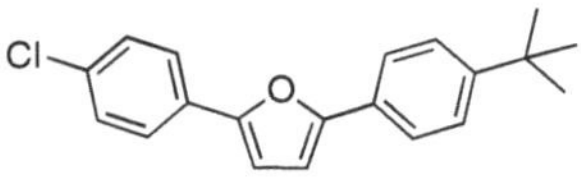

2c

2-(4-(*tert*-Butyl)phenyl)-5-(4-chlorophenyl) furan (**2c**): Yield: 85%

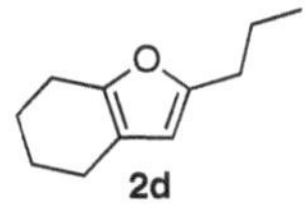

2d

2-Propyl-4,5,6,7-tetrahydrobenzofuran (**2d**): Yield: 89%

Experimental procedure

A round-bottom flask was charged with substituted but-3-yn-1-one (**1**; 1.00 mmol) and dichloromethane (10 mL). Zinc chloride (1.0 M in ether, 0.10 mL, 0.10 mmol) was then added dropwise with a syringe. The solution was stirred at room temperature (22 °C) for 2 h (monitored by ^{1}H NMR). The reaction mixture was passed through a short path silica gel column (2.5 × 15 cm; CH_2Cl_2). The solvent was removed by rotary evaporation and the residue was dried by an oil pump vacuum for 3 h to furnish the desired product **2**, characterized based on spectral and elemental studies.

Characterization data for one representative compound

2-Phenyl-5-(*p*-tolyl)furan (**2b**): White solid, mp. 97-99 °C; UV-vis (ether): λ_{max} 227, 326 nm; IR (KBr): 1498, 1482, 821, 794, 757, 691 cm^{-1}; ^{1}H NMR ($CDCl_3$): δ 7.83-7.64 (m, 4H), 7.50-7.20 (m, 5H), 6.77 (d, J=3.3 Hz, 1H), 6.72 (d, J=3.3 Hz, 1H), 2.42 (s, 3H); ^{13}C NMR ($CDCl_3$): δ 153.7, 153.1, 137.4, 131.0, 129.5, 128.8, 128.2, 127.3, 123.8, 107.3, 106.6, 21.5; MS: 234 (M^+, 100%). Anal. Calcd for $C_{17}H_{14}O$: C, 87.15; H, 6.02; Found: C, 87.05; H, 6.03.

Reference

Sniady, A., Durham, A., Morreale, M. S., Wheeler, K. A., and Dembinski, R. (2007). Room temperature zinc chloride-catalyzed cycloisomerization of alk-3-yn-1-ones: synthesis of substituted furans. *Org. Lett.*, **9**, 1175–1178.

C-O bond_17

Type of reaction: C-O, C-N and C-C bond formation
Reaction conditions: Ethanol, room temperature
Synthetic strategy: One-pot multicomponent reaction
Catalyst: No catalyst
Keywords: *N*-(Isocyanoimino)triphenylphosphorane, 2-aminobenzoic acid, 2-hydroxybenzaldehydes, isocyanides, ethanol, catalyst-free, room temperature, cyclocondensation, one-pot multicomponent, intramolecular aza-Wittig reaction, N^2-alkyl-N^3-[2-(1,3,4-oxadiazol-2-yl)aryl]benzofuran-2,3-diamines

General reaction scheme

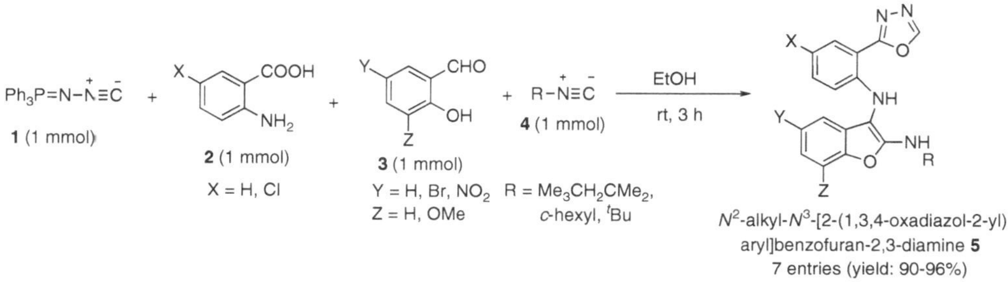

Proposed Mechanism

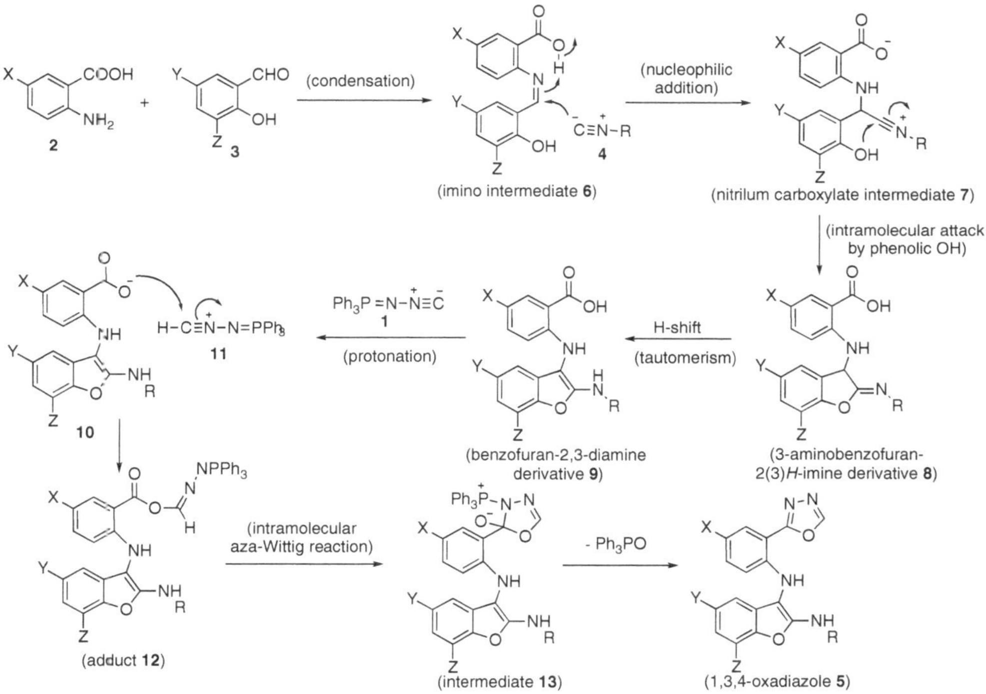

Representative entries

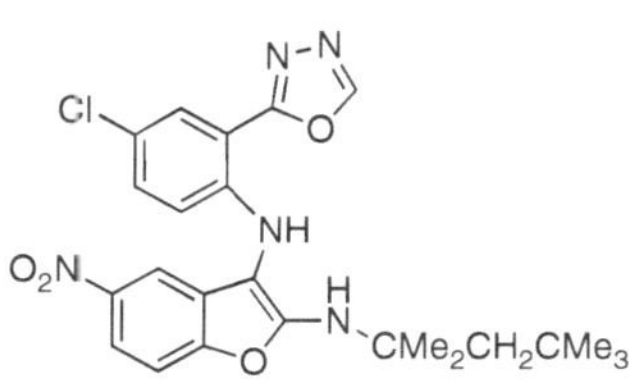

5a

N^3-(4-Chloro-2-(1,3,4-oxadiazol-2-yl)phenyl)-5-nitro-N^2-(2,4,4-trimethylpentan-2-yl)benzofuran-2,3-diamine (**5a**): Orange crystals, mp 195-197 °C; yield: 90%

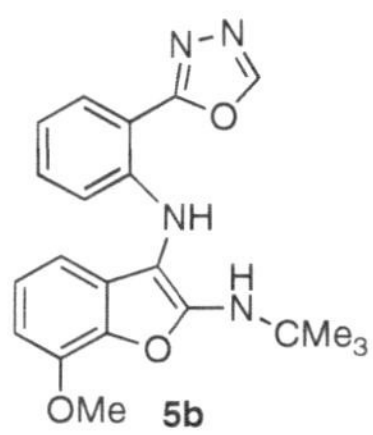

5b

N^3-(2-(1,3,4-oxadiazol-2-yl)phenyl)-N^2-(*tert*-butyl)-7-methoxybenzofuran-2,3-diamine (**5b**): Yellow crystals, mp 223 °C; yield: 91%

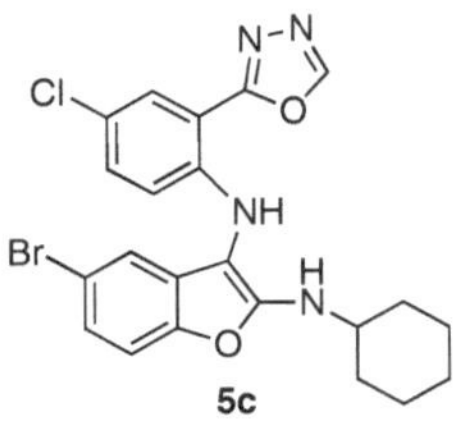

5c

5-Bromo-N^3-(4-chloro-2-(1,3,4-oxadiazol-2-yl)phenyl)-N^2-cyclohexylbenzofuran-2,3-diamine (**5c**): Yellow crystals, mp 194-196 °C; yield: 93%

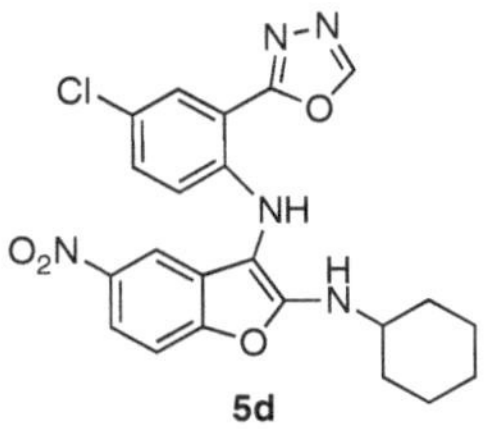

5d

N^3-(4-Chloro-2-(1,3,4-oxadiazol-2-yl)phenyl)-N^2-cyclohexyl-5-nitrobenzofuran-2,3-diamine (**5d**): Orange crystals, mp 174-176 °C; yield: 96%

Experimental procedure

A mixture of 2-aminobenzoic acid derivative (**2**; 1 mmol), 2-hydroxy-5-nitrobenzaldehyde derivative (**3**; 1 mmol), and isocyanide (**4**; 1 mmol) in EtOH (2 ml) was stirred at room temperature for 1 h. Next, *N*-(isocyanoimino)triphenylphosphorane (**1**; 0.302 g, 1 mmol) was added to the mixture, and the resulting mixture was continued to stir at room temperature further 2 h. Then, the solvent was evaporated, and the residue purified using column chromatography (SiO_2, hexane/AcOEt 3 : 1) and recrystallization (hexane/AcOEt 1 : 1) to afford the derised product **5**. Each of the products was characterized by spectral and analytical analyses.

Characterization data for two representative compounds

N^3-(2-(1,3,4-oxadiazol-2-yl)phenyl)-N^2-(*tert*-butyl)-7-methoxybenzofuran-2,3-diamine (**5b**): Yellow crystals, mp 223 °C; IR (KBr): 3315 and 3167 (NH), 1629, 1574, 1507, 1445, 1379, 1333, 1273, 1168, 1068, 950, 813, 717 cm^{-1}; ^{1}H-NMR (500.1 MHz, $CDCl_3$): δ 1.40 (s, 9 H), 3.88 (s, 1 H), 4.04 (s, 3H), 6.67 (d, $J=8.1$, 1 H), 6.71 (d, $J=7.7$, 1 H), 6.75 (d, $J=8.5$, 1 H), 6.83 (t, $J=7.7$, 1 H), 7.02 (t, $J=7.9$, 1 H), 7.27 (t, $J=7.7$, 1 H), 7.89 (dd, $J=1.3$, 7.9, 1 H), 8.47 (s, 1 H), 8.58 (br s, 1H); ^{13}C NMR (125.8 MHz, $CDCl_3$): δ 30.5, 53.8, 56.4, 99.5, 105.3, 105.9, 109.7, 113.9, 117.1, 123.4, 128.2, 129.6, 133.2, 138.4, 144.6, 147.1, 151.0, 154.5, 164.8; EI-MS : *m*/*z* (%rel. intensity) 378 (<1,M^+), 277 (100), 236 (8), 199 (15), 183 (10), 151 (17), 109 (12), 91 (8), 77 (14), 65 (5), 51 (11). Anal. calc. for $C_{21}H_{22}N_4O_3$ (378.43): C, 66.65; H, 5.86; N, 14.81; Found: C, 66.60; H, 5.78; N, 14.74.

N^3-(4-Chloro-2-(1,3,4-oxadiazol-2-yl)phenyl)-N^2-cyclohexyl-5-nitrobenzofuran-2,3-diamine (**5d**): Orange crystals, mp 174-176 °C; IR (KBr): 3368, 3136, 1651, 1590, 1512, 1474, 1390, 1336, 1232, 1164, 1097, 1044, 1000, 952, 876, 808, 742 cm^{-1}; ^{1}H NMR (500.1 MHz, $CDCl_3$): δ 1.14-2.05 (m, 10 H), 3.62-3.67 (m, 1 H), 3.98 (br s, 1 H), 6.61 (d, $J=9.0$, 1 H), 7.23 (dd, $J=2.0$, 8.9, 1 H), 7.33 (d, $J=8.8$, 1 H), 7.82 (s, 1H), 7.87 (s, 1H), 7.94 (dd, $J=2.0$, 8.8, 1 H), 8.50 (s, 1 H), 8.52 (br s, 1H); ^{13}C NMR (125.8 MHz, $CDCl_3$): δ 24.9, 25.4, 34.4, 52.6, 91.9, 107.1, 110.0, 111.1, 114.9, 116.1, 122.4, 127.7, 129.9, 133.3, 144.7, 145.4, 151.3, 151.8, 156.9, 164.7; EI-MS: *m*/*z* (%rel intensity) 455 (5, $M^+(^{37}Cl)$), 453 (11, $M^+(^{35}Cl)$), 370 (5), 354 (7), 337 (6), 279 (8), 167 (19), 149 (57), 97 (17), 81 (33), 69 (74), 57 (61), 43 (100). Anal. calc. for $C_{22}H_{20}ClN_5O_4$ (453.88): C, 58.22; H, 4.44; N, 15.43; Found: C, 58.31; H, 4.51; N, 15.37.

Reference

Adib, M., Sheikhi, E., Karimzadeh, M., Bijanzadeh, H. R., and Amanlou, M. (2012). One-pot four-component synthesis of N^2-alkyl-N^3-[2-(1,3,4-oxadiazol-2-yl)aryl]benzofuran-2,3-diamines. *Helv. Chim. Acta*, **95**, 788–794.

C-O bond_18

Type of reaction: C-O, C-N and C-C bond formation
Reaction conditions: Acetonitrile, room temperature
Synthetic strategy: One-pot multicomponent reaction
Catalyst: No catalyst
Keywords: *N*-(Isocyanoimino)triphenylphosphorane, aromatic bis-aldehydes, aromatic carboxylic acids, acetonitrile, catalyst-free, room temperature, cyclocondensation, one-pot multicomponent, intramoleculer aza-Witting reaction, 1,3,4-oxadiazoles

General reaction scheme

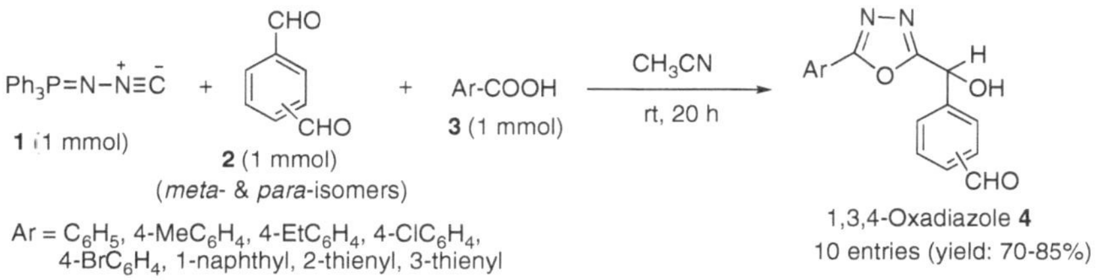

Proposed Mechanism

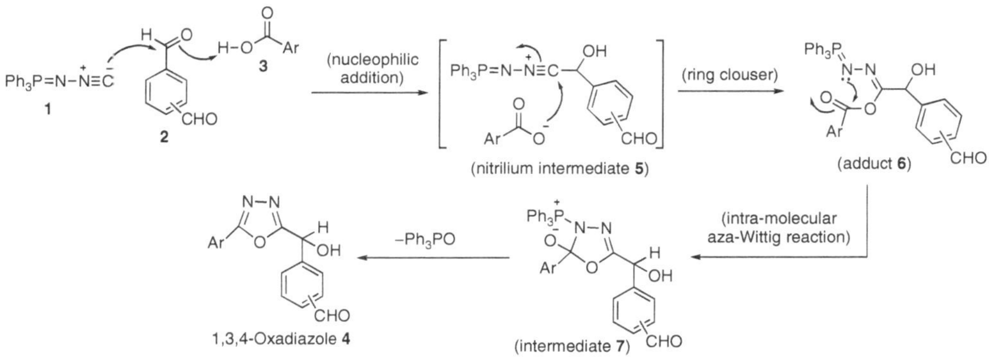

Representative entries

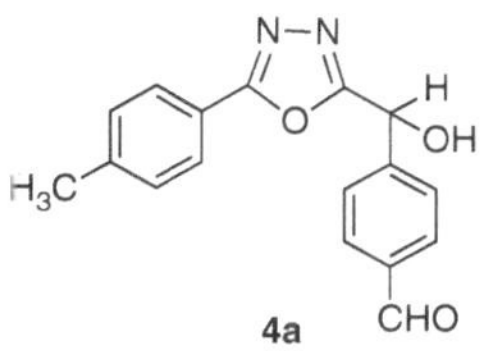

4a

4-(Hydroxy(5-(*p*-tolyl)-1,3,4-oxadiazol-2-yl)methyl)benzaldehyde (**4a**): Yellow crystals, mp 91-93 °C; yield: 85%

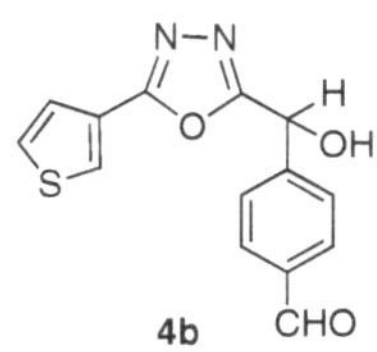

4b

4-(Hydroxy(5-(thiophen-3-yl)-1,3,4-oxadiazol-2-yl)methyl)benzaldehyde (**4b**): Yellow oil; yield: 72%

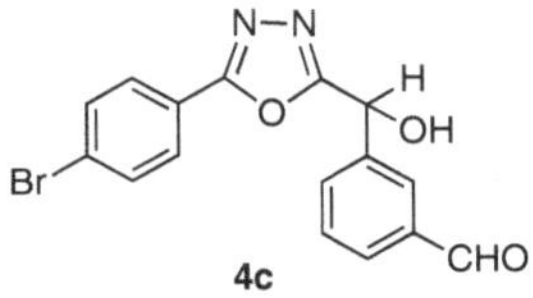

4c

3-((5-(4-Bromophenyl)-1,3,4-oxadiazol-2-yl)(hydroxy)methyl)benzaldehyde (**4c**): Yellow crystals, mp 133-135 °C; yield: 80%

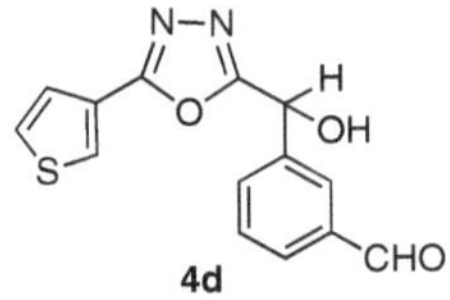

4d

3-(Hydroxy(5-(thiophen-3-yl)-1,3,4-oxadiazol-2-yl)methyl)benzaldehyde (**4d**): Yellow oil; yield: 72%

Experimental procedure

A mixture of (*N*-isocyanimino)triphenylphosphorane (**1**; 1.0 mmol), aromatic bis-aldehyde (**2**; 1.0 mmol), and aromatic carboxylic acid (**3**; 1.0 mmol) in acetonitrile (5 mL) was stirred at room temperature for 20 h. Then, the solvent was removed under reduced pressure, and the viscous residue was purified by flash column chromatography using silica-gel (petroleum ether-ethyl acetate (4:1)). The solvent was removed under reduced pressure to give the sterically congested 1,3,4-oxadiazole **4**. Each of the products was characterized by spectral and analytical analyses.

Characterization data for two representative compounds

4-(Hydroxy(5-(*p*-tolyl)-1,3,4-oxadiazol-2-yl)methyl)benzaldehyde (**4a**): Yellow crystals, mp 91-93 °C; IR (KBr): 3389, 2922, 1700, 1607, 1480, 1202, 1068 cm^{-1}; ^{1}H NMR (250.13 MHz, $CDCl_3$): δ 2.39 (3H, s, CH_3), 4.25 (1H, br s, OH), 6.23 (1H, s, CH), 7.25-7.92 (8H, m, H-Ar), 10.02 (1H, s, CHO); ^{13}C NMR (62.53 MHz, $CDCl_3$): δ 21.66, 67.79, 126.99, 127.15, 129.78, 130.21, 129.18, 136.61, 142.86, 144.03, 164.85, 166.20 ; 191.72; MS: *m/z* (%rel. int.) 294 (M^+, 60), 159 (100), 133 (27), 119 (54), 105 (15), 91 (67), 77 (47), 57 (20), 43 (24). Anal. calcd. for $C_{17}H_{14}N_2O_3$: C, 69.38; H, 4.79; N, 9.52%; Found: C, 69.45; H, 4.74; N, 9.45.

3-(Hydroxy(5-(thiophen-3-yl)-1,3,4-oxadiazol-2-yl)methyl)benzaldehyde (**4d**): Yellow oil; IR (KBr): 3399, 2920, 1698, 1605, 1437, 1057 cm^{-1}; ^{1}H NMR (250.13 MHz, $CDCl_3$): δ 2.70 (1H, br s, OH), 6.20 (1H, s, CH), 7.25–8.35 (7H, m, H-Ar), 9.95 (1H, s, CHO); ^{13}C NMR (62.53 MHz, $CDCl_3$): δ 66.96,125.87, 127.13, 12.66, 130.19, 129.18, 136.61, 144.03, 160.01, 169.80, 192.05. Anal. calcd. for $C_{14}H_{10}N_2O_3S$: C, 58.73; H, 3.52; N, 9.78%; Found: C, 58.67; H, 3.45; N, 9.75.

Reference

Ramazani, A., Ahmadi, Y., and Mahyar, A. (2011). One-pot efficient synthesis of fully substituted 1,3,4-oxadiazole derivatives from (*N*-isocyanimino)triphenylphosphorane, carboxylic acids, and aromatic bis-aldehydes. *Synth. Commun.*, **41**, 2273–2282.

C-O bond_19

Type of reaction: C-O, C-N and C-C bond formation
Reaction conditions: Acetonitrile, room temperature
Synthetic strategy: One-pot multicomponent reaction
Catalyst: No catalyst

Keywords: *N*-(Isocyanoimino)triphenylphosphorane, aromatic carboxylic acids, acenaphthaquinone, acetonitrile, catalyst-free, room temperature, cyclocondensation, one-pot multicomponent, intramolecular aza-Wittig reaction, 1,3,4-oxadiazoles

General reaction scheme

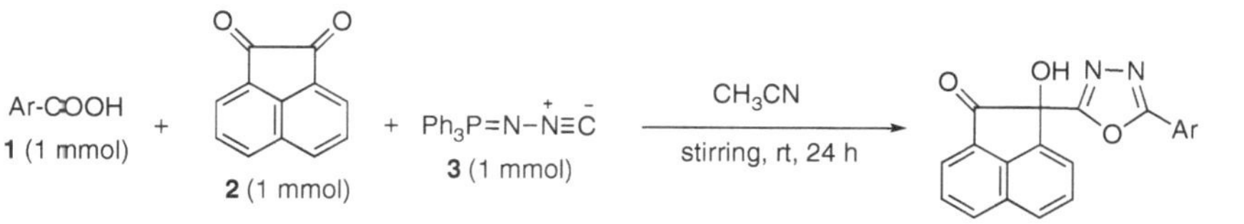

Ar = C_6H_5, 2-MeC_6H_4, 4-MeC_6H_4, 3,4-di-MeC_6H_3, 3-ClC_6H_4, 4-ClC_6H_4, 4-BrC_6H_4, 4-CNC_6H_4, 4-OMeC_6H_4, 4-BrCH_2, 4-FC_6H_4, 3,5-di-OMeC_6H_3, 3-OPhC_6H_4, 4-^{t}BuC_6H_4, 1-naphthyl

1,3,4-Oxadiazole **4**
16 entries (yield: 67-85%)

Representative entries

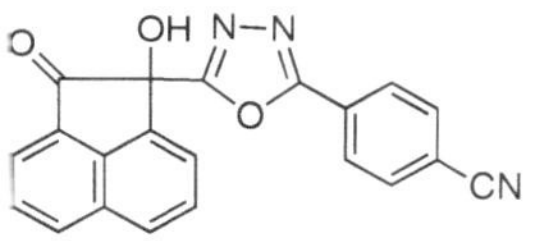

4a

4-(5-(1-Hydroxy-2-oxo-1,2-dihydroacenaphthylen-1-yl)-1,3,4-oxadiazol-2-yl) benzonitrile (**4a**): Yellow powder, mp 161-163 °C; yield: 73%

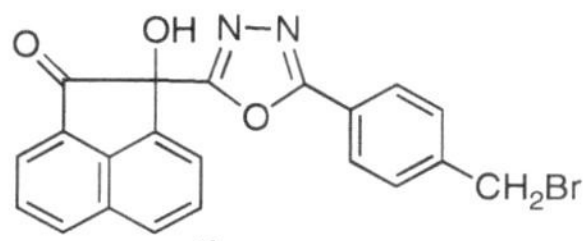

4b

2-(5-(4-(Bromomethyl)phenyl)-1,3,4-oxadiazol-2-yl)-2-hydroxyacenaphthylen-1(2*H*)-one (**4b**): Yellow powder, mp 163-165 °C; yield: 82%

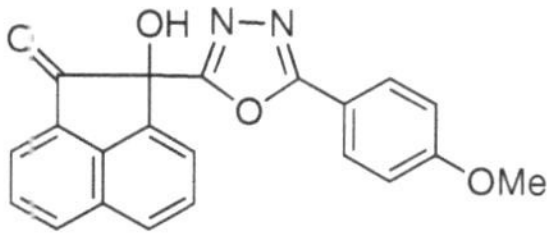

4c

2-Hydroxy-2-(5-(4-methoxyphenyl)-1,3,4-oxadiazol-2-yl)acenaphthylen-1(2*H*)-one (**4c**): Yellow powder, mp 195-197 °C; yield: 68%

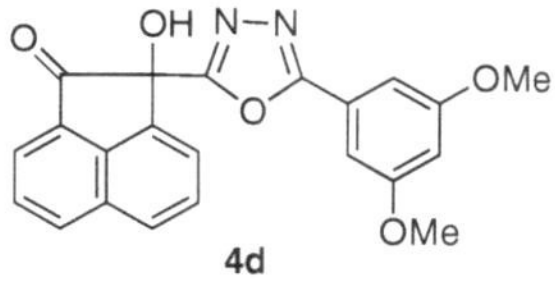

4d

2-(5-(3,5-Dimethoxyphenyl)-1,3,4-oxadiazol-2-yl)-2-hydroxyacenaphthylen-1(2*H*)-one (**4d**): Yellow powder, mp 168-170 °C; yield: 67%

Experimental procedure

A mixture of (*N*-isocyanimino)triphenylphosphorane (**3**; 1.0 mmol), acenaphthoquinone (**2**; 1.0 mmol), and aromatic carboxylic acid (**1**; 1.0 mmol) in 5 mL acetonitrile was stirred at room temperature for 24 h. The solvent was removed under reduced pressure, and the viscous residue was purified using chromatographic technique (silica gel (F_{254}) powder; petroleum ether/ethyl acetate 4:1) to furnish 1,3,4-oxadiazole **4**. Each of the products was characterized by spectral and analytical analyses.

Characterization data for two representative compounds

2-(5-(4-(Bromomethyl)phenyl)-1,3,4-oxadiazol-2-yl)-2-hydroxyacenaphthylen-1(2*H*)-one (**4b**): Yellow powder, mp 163-165 °C; IR (KBr): 3,343, 3,072, 1,717, 1,603, 1,486, 1,012 cm^{-1}; ^{1}H NMR (250.13 MHz, $CDCl_3$): δ 4.46 (2H, s, CH_2), 4.98 (1H, br s, OH), 7.43 (2H, d, $^3J_{HH}$ = 8.3 Hz, Ar-H), 7.90 (2H, d, $^3J_{HH}$ = 8.3 Hz, Ar-H), 7.70-7.86 (3H, m, Ar-H),

7.99-8.29 (3H, m, Ar-H); ^{13}C NMR (62.53 MHz, $CDCl_3$): δ 32.09, 76.77, 122.56, 123.89, 126.95, 127.55, 129.07, 129.61, 132.62, 132.87, 123.10, 128.44, 128.79, 130.92, 135.24, 141.80, 165.12, 166.22, 198.15.

2-(5-(3,5-Dimethoxyphenyl)-1,3,4-oxadiazol-2-yl)-2-hydroxyacenaphthylen-1(2*H*)-one (**4d**): Yellow powder, mp 168-170 °C; IR (KBr): 3,380, 3,094, 1,725, 1,604, 1,463, 1,162 cm^{-1}; ^{1}H NMR (250.13 MHz, $CDCl_3$): δ 3.80 (6H, s, 2 x OCH_3), 4.43 (1H, s, OH), 6.56-8.30 (9H, m, Ar-H); ^{13}C NMR (62.53 MHz, $CDCl_3$): δ 55.64 (2 x OCH_3), 87.00 (*C*-OH), 105.00, 107.05, 122.04, 123.94, 128.43, 132.59, 132.84, 124.75, 127.00, 131.04, 135.03, 142.10, 162.15, 164.02, 166.43, 189.51.

Reference

Ramazani, A., Nasrabadi, F. Z., Malekzadeh, A. M., and Ahmadi, Y. (2011). Synthesis of sterically congested 1,3,4-oxadiazole derivatives from aromatic carboxylic acids, acenaphthoquinone, and (*N*-isocyanimino)triphenylphosphorane. *Monatsh. Chem.*, **142**, 625–630.

C-O bond_20

Type of reaction: C-O, C-N and C-C bond formation
Reaction conditions: Dichloromethane, room temperature
Synthetic strategy: One-pot multicomponent reaction
Catalyst: No catalyst
Keywords: *N*-(Isocyanoimino)triphenylphosphorane, primary amine, chloroacetone, aromatic carboxylic acids, dichloromethane, catalyst-free, room temperature, cyclocondensation, one-pot multicomponent, intramolecular aza-Wittig reaction, 1,3,4-oxadiazoles

General reaction scheme

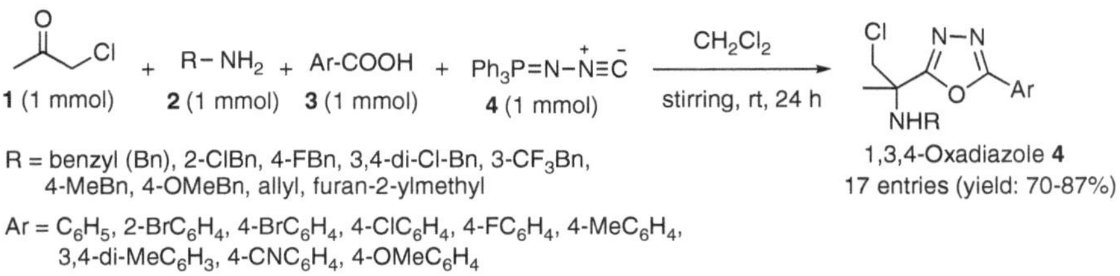

Representative entries

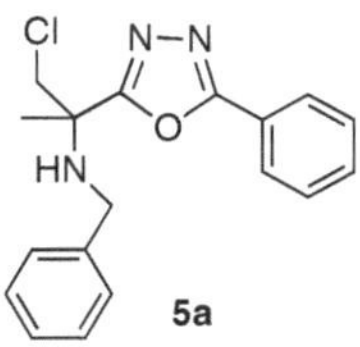

5a

N-Benzyl-1-chloro-2-(5-phenyl-1,3,4-oxadiazol-2-yl)propan-2-amine (**5a**): Yellow powder, mp 82-84 °C; yield: 85%

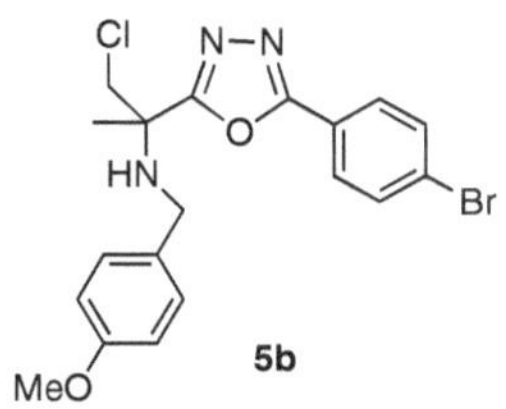

5b

2-(5-(4-Bromophenyl)-1,3,4-oxadiazol-2-yl)-1-chloro-N-(4-methoxybenzyl)propan-2-amine (**5b**): Yellow powder, mp 95-97 °C; yield: 75%

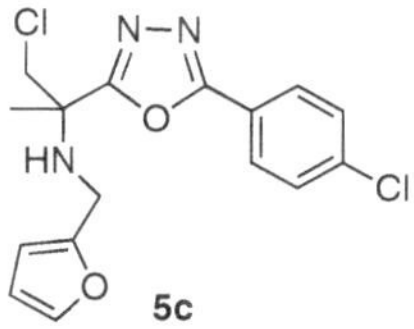

5c

1-Chloro-2-(5-(4-chlorophenyl)-1,3,4-oxadiazol-2-yl)-*N*-(furan-2-ylmethyl)propan-2-amine (**5c**): Yellow powder, mp 79-81 °C; yield: 80%

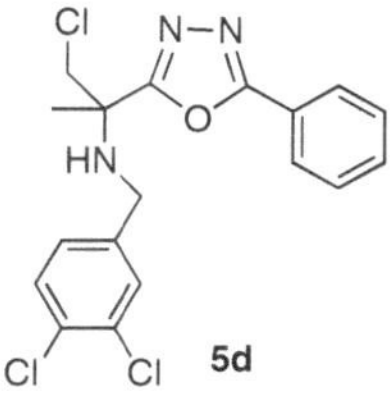

5d

1-Chloro-*N*-(3,4-dichlorobenzyl)-2-(5-phenyl-1,3,4-oxadiazol-2-yl)propan-2-amine (**5d**): Yellow powder, mp 73-75 °C; yield: 85%

Experimental procedure

To a magnetically stirred soln. of chloroacetone (**1**; 1 mmol), primary amine (**2**; 1 mm ol), and (*N*-isocyanoimino)triphenylphosphorane (**4**; 1 mmol) in dichloromethane (5 mL) was added dropwise a solution of benzoic acid derivative (**3**; 1 mmol) in dichloromethane (5 mL) at room temperature over 15 min. The mixture was then well stirred for 12 h. The solvent was evaporated and the viscous residue purified by preparative TLC (silica gel F254, petroleum ether/AcOEt 4 : 1) to obtain sterically congested a,3,4-oxadiazole **5**. Each of the products was characterized by spectral and analytical analyses.

Characterization data for one representative compound

N-Benzyl-1-chloro-2-(5-phenyl-1,3,4-oxadiazol-2-yl)propan-2-amine (**5a**): Yellow powder, mp 82-84 °C; IR (KBr):3456, 3269, 2927, 1608, 1449, 1069 cm^{-1}; ^1H NMR (250 MHz, $CDCl_3$): δ 1.77 (3H, s, Me); 2.23 (1H, s, NH), 3.69 (1H, d, $J=12.3$ Hz), 3.76 (1H, d, $J=12.3$ Hz), 3.96 (1H, d, $J=11.5$ Hz), 4.07 (1H, d, $J=11.5$ Hz), 7.24 - 8.07 (10H, m, Ar-H); ^{13}C NMR (62.5 MHz, $CDCl_3$): δ 22.53, 47.82, 50.24, 57.58, 127.00, 127.31, 128.27, 128.51, 129.07, 131.8, 123.70, 139.33, 165.24, 167.77; EI-MS: *m/z* (%rel. int.) 327 (M^+), 187 (16), 149 (33), 105 (70), 91 (100), 76 (29), 43 (54). Anal. calc. for $C_{18}H_{18}ClN_3O$ (327.81): C, 65.95; H, 5.53; N, 12.82; Found: C, 65.83; H, 5.49; N, 12.77.

Reference

Ramazani, A., Nasrabadi, F. Z., and Ahmadi, Y. (2011). One-pot, four-component synthesis of fully substituted 1,3,4-oxadiazole derivatives from (isocyanoimino)triphenylphosphorane, a primary amine, an aromatic carboxylic acid, and chloroacetone. *Helv. Chim. Acta*, **94**, 1024–1029.

C-O bond_21

Type of reaction: C-O and C-C bond formation

Reaction conditions: *N,N*-Dimethylformamide (DMF), room temperature

Synthetic strategy: Sequential carboxylation with carbon dioxide and intramolecular cyclization

Catalyst: $AgBF_4$ / MTBD (MTBD = 7-methyl-1,5,7-triazabicyclo[4.4.0]dec-5-ene)

Keywords: *ortho*-Alkynyl acetophenones, carbon dioxide (balloon), silver(I) tetrafluoroborate ($AgBF_4$), MTBD, anhydrous DMF, methyl iodide, room temperature, sequential carboxylation/intramolecular cyclization, selective 5-*exo*-oxygen cyclization, 1(3*H*)-isobenzofuranylidene acetic acids/methyl esters

General reaction scheme

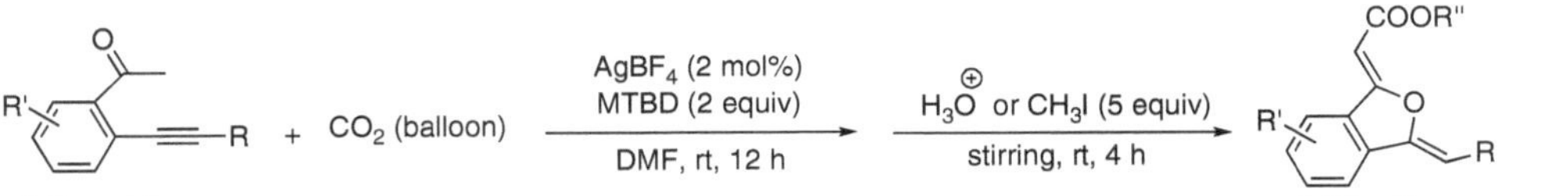

1 (1 equiv)

2: R'' = H
3: R'' = CH_3
1(3*H*)-isobenzofuranylidene acetic acid **2** /methyl ester **3**
16 entries (yield: 50-89%)

R,R',R'' = 4-$CH_3C_6H_4$, H, H; 4-$CH_3C_6H_4$, H, CH_3; 4-^{t}BuC_6H_4, H, H; 4-^{t}BuC_6H_4, H, CH_3; 4-FC_6H_4, H, CH_3; 2-ClC_6H_4, H, CH_3; 4-O$CH_3C_6H_4$, H, CH_3; 4-$CF_3C_6H_4$, H, CH_3; 3-OHC_6H_4, H, CH_3; 3-NHCO_2MeC_6H_4, H, CH_3; cyclopentyl, H, CH_3; $CH_2CH_2C_6H_5$, H, CH_3; C_6H_5, 4-OMe, CH_3; 4-OMeC_6H_4, 4-OMe, CH_3; 4-$CF_3C_6H_4$, 4-OMe, CH3; 4-^{t}BuC_6H_4, 6-F, CH_3

Representative entries

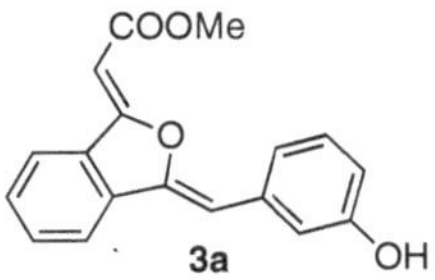

3a

(*Z*)-Methyl 2-((*Z*)-3-(3-hydroxybenzylidene) isobenzofuran-1(3*H*)-ylidene)acetate (**3a**): Yield: 75%

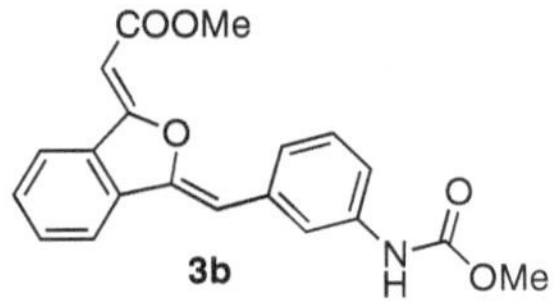

3b

(*Z*)-Methyl 2-((*Z*)-3-(3-((methoxycarbonyl)amino) benzylidene)isobenzofuran-1(3*H*)-ylidene)acetate (**3b**): Yield: 71%

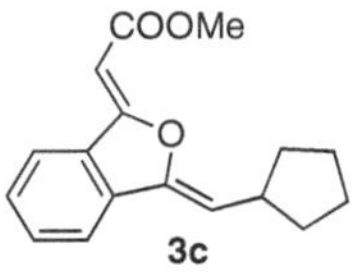

3c

(*Z*)-Methyl 2-((*Z*)-3-(cyclopentylmethylene) isobenzofuran-1(3*H*)-ylidene)acetate (**3c**): Yield: 80%

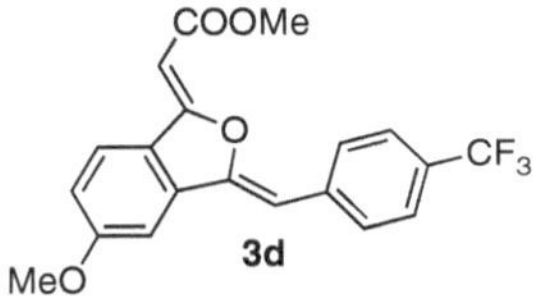

3d

(*Z*)-Methyl 2-((*Z*)-5-methoxy-3-(4-(trifluoromethyl)benzylidene)isobenzofuran-1 (3*H*)-ylidene)acetate (**3d**): Yield: 50%

Experimental procedure

A 50 mL oven dried Schlenk flask containing a stir bar was charged with *ortho*-alkynyl acetophenone (1; 1 equiv, 0.20 mmol), MTBD (61 mg, 2 equiv; 0.4 mmol), $AgBF_4$ (mol%; 0.8 mg, 0.004 mmol) and dry DMF (3.0 mL) under argon. After purging the flask with carbon dioxide two times, the reaction mixture was stirred at room temperature (25 °C) with a CO_2 balloon attached. After 12 h, methyl iodide (142 mg, 1.0 mmol) was injected into the flask and the reaction mixture was stirred at room temperature for another 4 h. The reaction mixture was then diluted with water (30 mL) and extracted with ethyl acetate (3 × 30 mL). The combined organic layers were washed with water and brine, dried over sodium sulfate and filtered. The solvent was removed under vacuum. The crude product was purified by column chromatography on silica gel (ethyl acetate/petroleum ether) to furnish methyl 1(3*H*)-isobenzofuranylidene acetate **3**. All the products were characterized by means of spectral studies.

Characterization data of two representative entries

(*Z*)-Methyl 2-((*Z*)-3-(3-((methoxycarbonyl)amino)benzylidene)isobenzofuran-1(3*H*)-ylidene) acetate (**3b**). IR (neat): 3332, 2952, 2926, 1708, 1638, 1609, 1547, 1439, 1230, 1153, 1086,

1039, 909, 767, 733, 691, 648 cm^{-1}; ^{1}H NMR (400 MHz, $CDCl_3$): δ 8.00 (s, 1H), 7.71-7.34 (m, 7H), 7.01 (s, 1H), 6.25 (s, 1H), 5.67 (s, 1H), 3.86 (s, 3H), 3.79 (s, 3H); ^{13}C NMR (100 MHz, $CDCl_3$): δ 165.94, 162.20, 154.33, 151.10, 138.50, 135.46, 134.85, 131.95, 131.84, 129.62, 129.56, 124.66, 121.52, 120.18, 103.74, 88.47, 52.49, 51.54; HRMS (ESI, *m/z*): Calcd for $C_{20}H_{17}O_5NNa$ $[M+Na]^+$: 374.1004, Found: 374.1014.

(*Z*)-Methyl 2-((*Z*)-5-methoxy-3-(4-(trifluoromethyl)benzylidene)isobenzofuran-1(3*H*)-ylidene)acetate (**3d**): IR (neat) 2947, 2920, 1686, 1637, 1613, 1490, 1435, 1325, 1268, 1229, 1070, 1035, 870, 812, 805, 758, 736 cm^{-1}; ^{1}H NMR (400 MHz, $CDCl_3$): δ 8.08 (d, $J=8.2$ Hz, 2H), 7.66 (d, $J=8.3$ Hz, 2H), 7.49 (d, $J=8.5$ Hz, 1H), 7.04 (d, $J=2.1$ Hz, 1H), 7.00 (dd, $J=8.6$, 2.2 Hz, 1H), 6.21 (s, 1H), 5.54 (s, 1H), 3.90 (s, 3H), 3.85 (s, 3H); ^{13}C NMR (100 MHz, $CDCl_3$): δ 165.69, 162.89, 161.63, 152.13, 137.56, 136.93, 129.30, 128.65, 125.46, 124.98, 122.61, 118.69, 102.76, 102.00, 87.58, 55.71, 51.31; HRMS (ESI, *m/z*): Calcd for $C_{20}H_{15}O_4F_3Na$ $[M+Na]^+$: 399.0820, Found: 399.0826.

Reference

Zhang, W.-Z., Shi, L.-L., Liu, C., Yang, X.-T., Wang, Y.-B., Luo, Y., and Lu, X.-B. (2014). Sequential carboxylation/intramolecular cyclization reaction of *o*-alkynyl acetophenone with CO_2. *Org. Chem. Front.*, 1, 275–283.

C-O bond_22

Type of reaction: C-O and C-C bond formation
Reaction conditions: Ethanol, room temperature
Synthetic strategy: One-pot multicomponent reaction
Catalyst: Piperidine
Keywords: Dimedone/4-hydroxycoumarin, aromatic aldehydes, (*E*)-*N*-methyl-1-(methylthio)-2-nitroethenamine (NMSM), piperidine, ethanol, room temperature, one-pot multicomponent reaction, tandem Knoevenagel-Michael cyclization, functionalized chromen-5-ones, pyrano[3,2-*c*]chromen-5-ones

General reaction schemes

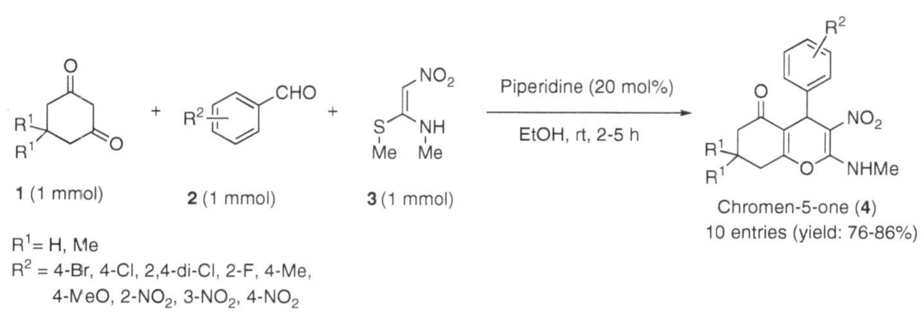

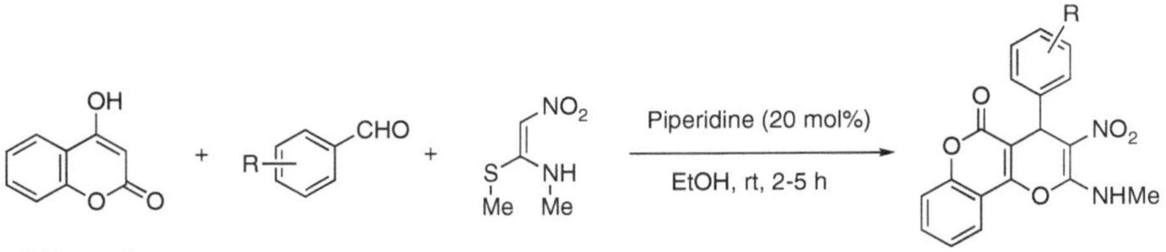

5 (1 mmol) **2** (1 mmol) **3** (1 mmol)

R = 4-Br, 4-Cl, 2-F, 4-Me, 4-MeO, 3-NO_2

Pyrano[3,2-*c*]chromen-5-one (**6**)
6 entries (yield: 76-86%)

Representative entries

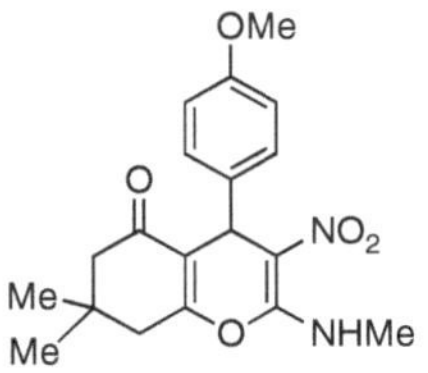

4a

4-(4-Methoxyphenyl)-7,7-dimethyl-2-(methylamino)-3-nitro-7,8-dihydro-4*H*-chromen-5(6*H*)-one (**4a**): White solid, mp 254-256 °C ; reaction time: 5 h, yield: 78%

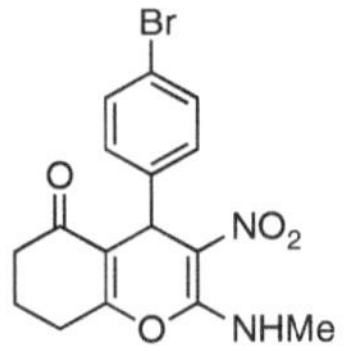

4b

4-(4-Bromophenyl)-2-(methylamino)-3-nitro-7,8-dihydro-4*H*-chromen-5(6*H*)-one (**4b**): White solid, mp 254-256 °C ; reaction time: 3.2 h, yield: 85%

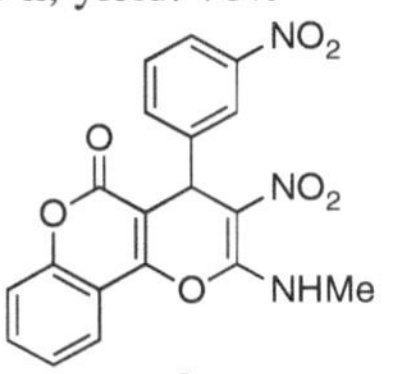

6a

2-(Methylamino)-3-nitro-4-(3-nitrophenyl) pyrano [3,2-*c*]chromen-5(4*H*)-one (**6a**): White solid, mp 264-266 °C ; reaction time: 2.2 h, yield: 76%

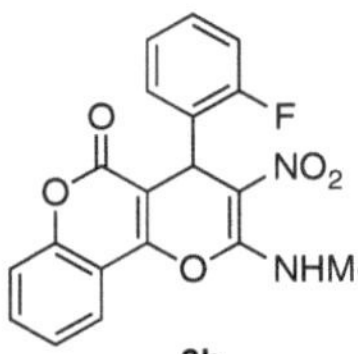

6b

4-(2-fluorophenyl)-2-(methylamino)-3-nitropyrano[3,2-*c*]chromen-5(4H)-one (**6b**): White solid, mp 280-282 °C ; reaction time: 2.7 h, yield: 80%

Experimental procedure

A solution of cyclic 1,3-dicarbonyl compound (**1**; 1.0 mmol)/ 4-hydroxycoumarin (**5**; 1.0 mmol), arylaldehyde (**2**; 1.0 mmol), NMSM (**3**; 1.0 mmol), and piperidine (20 mol%) in etahnol (2 mL) was stirred at room temperature for the appropriate time (2-5 h). On completion of the reaction (as indicated by TLC), the product (**4/6**) was filtered and washed with etahnol (2 mL) to remove the excess base and other impurities. Pure chromen-5-ones (**4**)/pyrano[3,2-*c*] chromen-5-ones (**6**) were obtained when recrystallized from EtOH. All the products were characterized from spectral (IR, ^{1}H NMR, ^{13}C NMR, MS) and elemental analyses.

Characterization data of representative compounds

4-(4-Methoxyphenyl)-7,7-dimethyl-2-(methylamino)-3-nitro-7,8-dihydro-4*H*-chromen-5(6*H*)-one (**4a**): White solid, mp 254-256 °C; IR (KBr): 3197, 2960, 1683, 1511, 1463, 1366, 1298, 1126, 1050, 920, 869, 758, 698, 528,422 cm^{-1}; ^{1}H NMR (500 MHz, DMSO-d_6): δ 0.91 (s, 3H, $-CH_3$), 1.05 (s, 3H, $-CH_3$), 2.11 (d, $J=16$Hz, 1H, -CH), 2.26 (d, $J=16$Hz, -CH), 2.61

(dd, $J = 17.5$Hz, 2H, $-CH_2$), 3.08 (d, $J = 5$Hz, 3H, $-NCH_3$), 3.68 (s, 3H, $-OCH_3$), 4.83 (s, 1H, -CH), 6.79 (d, $J = 8.5$Hz, 2H, Ar-H), 7.10 (d, $J = 9$Hz, 2H, Ar-H), 10.20 (d, $J = 4$Hz, 1H, D_2O exchangeable, -NH); ^{13}C NMR (125 MHz, DMSO-d_6): δ 27.1, 28.8, 28.9, 32.4, 35.2, 39.5, 50.3, 55.4, 109.2, 113.8, 116.1, 129.5, 134.9, 157.9, 158.3, 161.3, 195.9; ESI-MS: *m/z* 359 [M^++1]. Anal. Calcd (%) for $C_{19}H_{22}N_2O_5$: C 63.67, H 6.19, N 7.82; Found: C 63.72, H 6.23, N 7.87.

4-(2-Fluorophenyl)-2-(methylamino)-3-nitropyrano[3,2-*c*]chromen-5(4*H*)-one (**6b**): White solid, mp 280-282 °C ; IR (KBr): 3201, 3074, 2947, 1729, 1676, 1628, 1459, 1360, 1262, 1160, 947, 847, 764, 696, 575, 436 cm^{-1}; ^{1}H NMR (500 MHz, DMSO-d_6): δ 3.32 (d, $J = 5.5$Hz, 3H, $-NCH_3$), 5.25 (s, 1H, -CH), 7.06-7.13 (m, 1H, Ar-H), 7.42 (t, $J = 7.5$Hz, Ar-H), 7.48-7.53 (m, 2H, Ar-H), 7.74 (t, $J = 7.5$Hz, Ar-H), 8.02 (d, $J = 8$Hz, 1H, Ar-H), 10.45 (d, $J = 5.5$Hz, 1H, , D_2O exchangeable, -NH); ^{13}C NMR (125 MHz, DMSO-d_6): δ 29.2, 33.2, 105.5, 107.2, 112.9, 115.7, 115.9, 117.1, 123.3, 124.5, 125.6, 128.0, 128.1, 129.6, 129.7, 132.2, 133.8, 152.5, 152.6, 157.3, 159.6, 160.2, 162.2. ESI-MS: *m/z* 369 [M^++1]. Anal. Calcd (%) for $C_{19}H_{13}FN_2O_5$: C 61.96, H 3.56, N 7.61; Found: C 61.91, H 3.61, N 7.57.

Reference

Kamalraja, J., Muralidharan, D., and Perumal, P. T. (2012). An efficient, one-pot regioselective synthesis of highly functionalized chrome-5-ones and pyrano[3,2-*c*]chromen-5-ones via a tandem Knoevenagel-Michael-cyclization sequence. *Synlett*, **23**, 2894–2898.

C-O bond_23

Type of reaction: C-O and C-C bond formation
Reaction conditions: Dichloromethane, room temperature
Synthetic strategy: Oxidative coupling of Baylis-Hillman adducts with silyl enol ethers
Catalyst: Dess-Martin periodinane (DMP)/Pyridine
Keywords: Baylis-Hillman adducts, silyl enol ethers, Dess-Martin periodinane (DMP), pyridine, dichloromethane, room temperature, oxidative Mukaiyama-Michael addition, diastereoselective cyclization, *cis*-fused dihydrpyrans

General reaction scheme

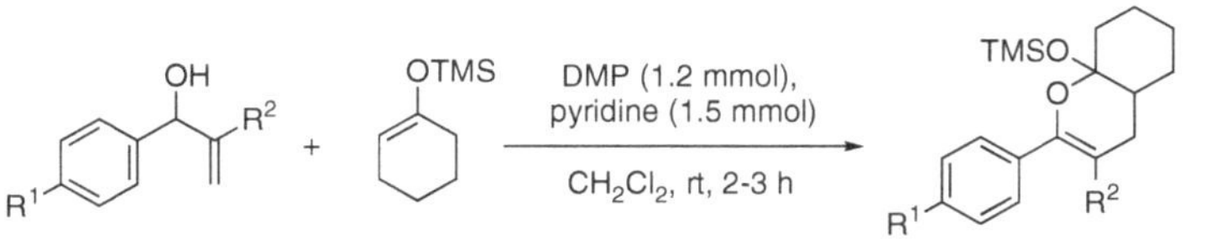

1 (1 mmol) **2** (1.5 mmol) *cis*-fused dihydropyran **3**

6 entries (yield: 78-86%)

R^1= H, Me, MeO, Cl; R^2 = COOMe, COOEt, CN

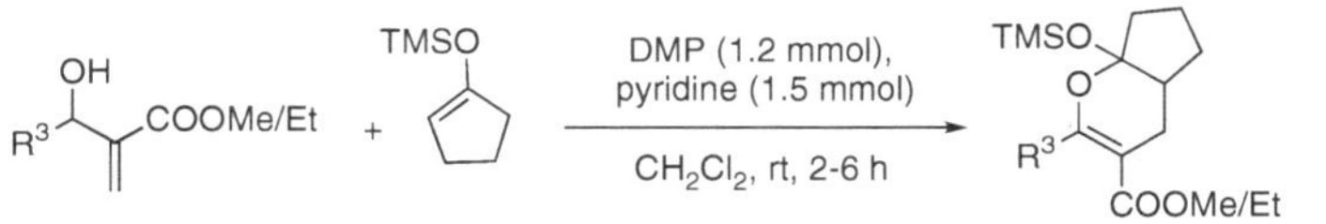

4 (1 mmol) **5** (1.5 mmol) *cis*-fused dihydropyran **6**

3 entries (yield: 78-80%)

R^3= 4-MeC_6H_4, *n*-pentyl, (*E*)-C_6H_5CH=CH-

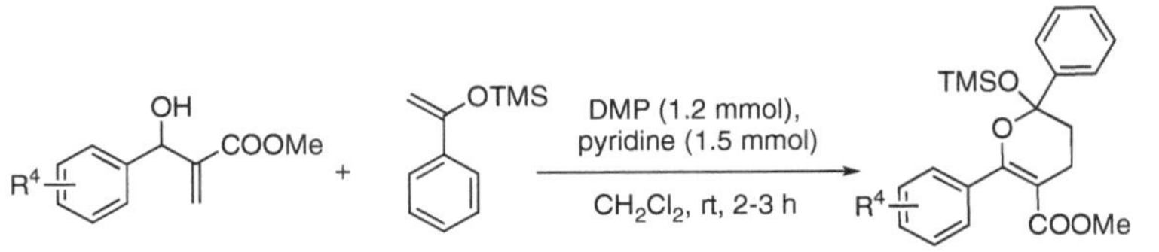

6 (1 mmol) **7** (1.5 mmol)

R^4= 4-Cl, 4-F, 3-OPh

cis-fused dihydropyran **9**

3 entries (yield: 81-84%)

Representative entries

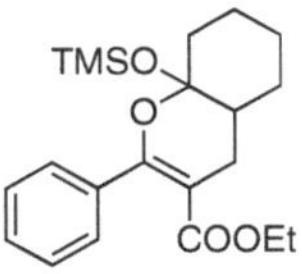

3a

Ethyl 2-phenyl-8a-((trimethylsilyl)oxy)-4a,5,6,7,8,8a-hexahydro-4*H*-chromene-3-carboxylate (**3a**): Reaction time: 2 h; yield: 80%

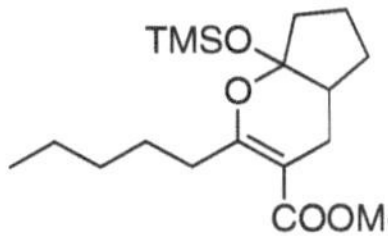

6a

Methyl 2-pentyl-7a-((trimethylsilyl)oxy) - 4,4a,5,6,7,7a-hexahydrocyclopenta[*b*]pyran-3-carboxylate (**6a**): Colorless liquid; reaction time: 6 h; yield: 78%

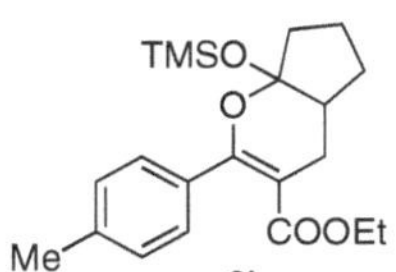

6b

Ethyl 2-(*p*-tolyl)-7a-((trimethylsilyl)oxy)-4,4a,5,6,7,7a-hexahydrocyclopenta[*b*]pyran-3-carboxylate (**6b**): Reaction time: 3 h; yield: 78%

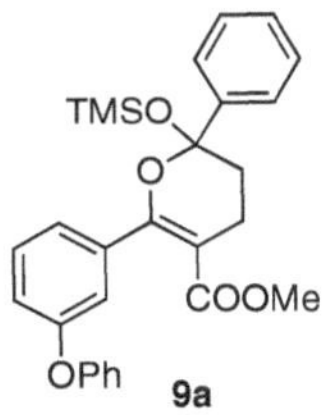

9a

Methyl 6-(3-phenoxyphenyl)-2-phenyl-2-((trimethylsilyl)oxy)-3,4-dihydro-2*H*-pyran-5-carboxylate (**9a**): Colorless liquid; reaction time: 3 h; yield: 82%

Experimental procedure

A mixture of Baylis-Hillman adduct (**1**, **4** or **6**;1 mmol), Dess-Martin periodinane (DMP; 1.2 mmol), and pyridine (1.5 mmol) in anhydrous dichlorometahne (10 mL) was stirred at room temperature to oxidize the substrate completely into the α,β-unsaturated ketone first. To this resulting reactionmixture, trimethylsilyl enol ether (**2**, **5** or **7**;1.5 mmol) was added and stirring was continued up to completion of the overall reaction (as indicated by TLC). After then the reaction mixture was diluted with water (50 mL) and extracted with Et_2O (3 × 15 mL). The combined etherlayer was washed with sat. aq. $NaHCO_3$ solution (1 × 15 mL), brine (1 × 10 mL), dried over Na_2SO_4, and evaporated. The crude product was purified by silica gel column chromatography using a gradient mixture of hexane-EtOAc (9:1) as eluent to afford pure substituted dihydropyran derivative **3**, **6** or **9**. The products were characterized by means of spectral studies.

Characterization data of representative compounds

Ethyl 2-phenyl-8a-(tosyloxy)-4a,5,6,7,8,8a-hexahydro-4H-chromene-3-carboxylate (**3a**): ^{1}H NMR (600 MHz, $CDCl_3$): δ 7.33 (m, 5H, Ph), 3.92 (q, 2H, J=7.2 Hz, OCH_2), 2.72 (dd,1H, J=16.7, 6.2 Hz, H-6), 2.23 (dd, 1H, J=16.7, 2.0 Hz, H-6′), 2.12 (dt, 1H, J=13.0, ca.

3.6 Hz, H-1e), 1.85 (dddd, 1H, $J=10.5$, 6.2, 4.2, 2.0 Hz, H-5a), 1.65 (m, 1H, H-3e), 1.62 (m, 1H, H-2e), 1.57 (m, 1H, H-4e), 1.55 (dt, 1H, $J=3.8$, ca. 12.8 Hz, H-1a), 1.44 (tq, 1H, $J=$ca. 3.5, ca. 12.6 Hz, H-2a), 1.34 (dq, 1H, $J=3.2$, ca. 12.3 Hz, H-4a), 1.28 (m, 1 H, H-3a), 0.91 (t, $J=7.2$ Hz, 1H, CH_3), 0.15 (s, 9H, $3\times CH_3$).

Methyl 2-pentyl-7a-((trimethylsilyl)oxy)-4,4a,5,6,7,7a-hexahydrocyclopenta[*b*]pyran-3-carboxylate (**6a**): Colorless liquid; IR (KBr): 2928, 2857, 1715, 1504, 1433, 1151, 1039, 754 cm^{-1}; ^1H NMR (300 MHz, $CDCl_3$): δ 0.0 (s, 9H), 0.78 (t, $J=6.8$ Hz, 3H), 1.10-1.27 (m, 6H), 1.33-1.72 (m, 6H), 1.86-2.02 (m, 1H), 2.07 (dd, $J=6.0$, 12.0 Hz, 1H), 2.16 (dd, 1H, $J=6.0$, 12.0 Hz), 3.55 (s, 3H).

Methyl 6-(3-phenoxyphenyl)-2-phenyl-2-((trimethylsilyl)oxy)-3,4-dihydro-2*H*-pyran-5-carboxylate (**9a**): Colorless liquid; IR (KBr): 3029, 2948, 2865, 1718, 1495, 1265, 1217, 1137, 1037, 854 cm^{-1}; ^1H NMR (300 MHz, $CDCl_3$): δ 0.0 (s, 9H), 1.80 (ddd, $J=1.8$, 5.6, 13.4 Hz, 1H), 2.21 (ddd, $J=3.5$, 5.4, 13.4 Hz, 1H,), 2.52 (ddd, $J=3.5$, 5.6, 16.9 Hz, 1H), 2.66 (ddd, $J=5.4$, 11.3, 16.8 Hz, 1H), 3.59 (s, 3H), 7.07-7.13 (m, 5H), 7.22 (td, $J=1.1$, 7.9 Hz, 1H), 7.3-7.43 (m, 6H), 7.50 (dd, $J=1.5$, 7.7 Hz, 2H); HRMS: *m/z* calcd for $C_{28}H_{31}O_5Si$: 475.1940; Found: 475.1954.

Reference

Yadav, J. S., Subba Reddy, B. V., Mandal, S. S., Basak, A. K., Madavi, C., and Kunwar, A. C. (2008). Dess-Martin periodinane promoited oxidative coupling of Baylis-Hillman adducts with silyl enol ethers: A novel synthesis of *cis*-fused dihydropyrans. *Synlett*, 1175–1178.

C-O bond_24

Type of reaction: C-O and C-N bond formation
Reaction conditions: Water, room temperature
Synthetic strategy: One-pot multicomponent reaction
Catalyst: No catalyst
Keywords: Acetylenic esters, benzoyl cyanide, quinoline, isoquinoline, water, room temperature, one-pot multicomponent reaction, oxazines

General reaction scheme

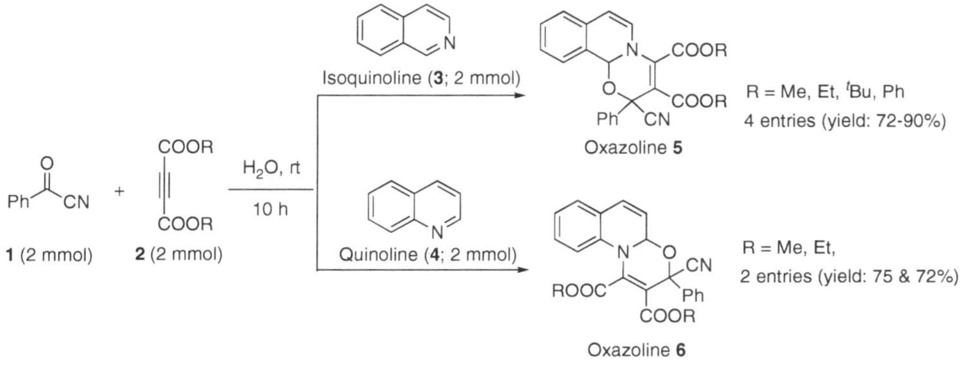

Representative entries

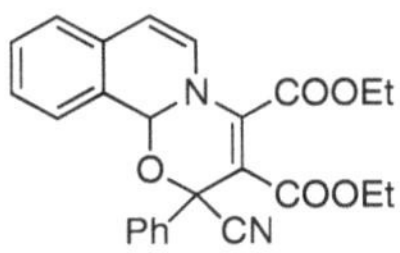

5a

Diethyl 2-cyano-2-phenyl-2,11b-dihydro-[1,3] oxazino[2,3-*a*]isoquinoline-3,4-dicarboxylate (**5a**): White powder, mp 148-150 °C; yield: 72%

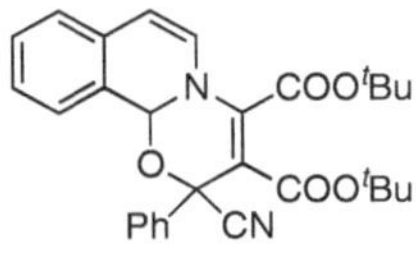

5b

Di-*tert*-butyl 2-cyano-2-phenyl-2,11b-dihydro-[1,3]oxazino[2,3-*a*]isoquinoline-3,4-dicarboxylate (**5b**): Yellow powder, mp 69-71 °C; yield: 75%

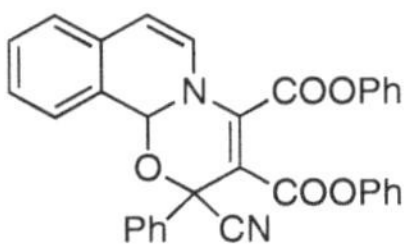

5c

Diphenyl 2-cyano-2-phenyl-2,11b-dihydro-[1,3] oxazino[2,3-*a*]isoquinoline-3,4-dicarboxylate (**5c**): Yellow powder, mp 100-102 °C; yield: 82%

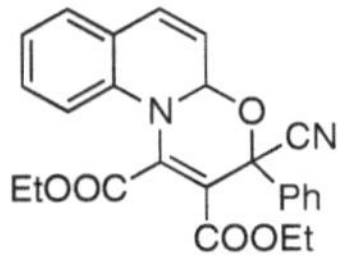

6a

Diethyl 3-cyano-3-phenyl-3,4a-dihydro-[1,3] oxazino[3,2-*a*]quinoline-1,2-dicarboxylate (**6a**): Yellow powder, mp 150-152 °C; yield: 72%

Experimental procedure

To a stirred solution of benzoyl cyanide (**1**; 2 mmol) and dialkyl acetylenedicarboxylate (**2**; 2 mmol) in water (10 mL) was added isoquinoline (**3**; 2 mmol) or quinoline (**4**; 2 mmol), and the reaction was continued to stir for 10 h at room temperature. The completion of reaction was confirmed by TLC (EtOAc-hexane 6:1). The resulting precipitate was separated by filtration and recrystallized from ethanol to afford pure oxazine derivatives (**5/6**). All the compounds were characterized on the basis of analytical and spectral studies.

Characterization data of representative compounds

Di-tert-butyl 2-cyano-2-phenyl-2,11b-dihydro-[1,3]oxazino[2,3-*a*]isoquinoline-3,4-dicarboxylate (**5b**): Yellow powder, mp 69-71 °C; IR (KBr): 1726, 1696 (2 C=O), 1280, 1230, 1138, 763 cm^{-1}; ^{1}H NMR (500 MHz, $CDCl_3$): δ 1.29 (s, 9H, $C(CH_3)_3$), 1.67 (s, 9H, $C(CH_3)_3$), 5.78 (d, $^{3}J=7.4$, 1H, *CH*), 5.80 (s, 1H, *CH*), 6.43 (d, $^{3}J=7.0$, 1H, *CH*), 6.82 (d, $^{3}J=6.7$, 1H, *CH*), 7.05 (d, $^{3}J=6.7$, 1H, *CH*), 7.12 (t, $^{3}J=6.7$, 1H, *CH*), 7.25-7.26 (m, 1H, *CH*), 7.35-7.36 (m, 1H, *CH*), 7.48-7.50 (m, 3H, 3 × *CH*), 7.71-7.72 (m, 1H, *CH*); ^{13}C NMR (125 MHz, $CDCl_3$): δ 27.7 ($C(CH_3)_3$), 27.8 ($C(CH_3)_3$), 75.1 (C), 78.0 (CH), 82.5 ($C(CH_3)_3$), 84.9 ($C(CH_3)_3$), 103.3 (C), 105.0 (CH), 119.0 (CN), 122.8 (CH), 125.1 (C), 125.2 (CH), 126.6 (C), 127.1 (CH), 127.4 (CH), 128.2 (2 CH), 128.7 (2 CH), 129.6 (CH), 129.7 (CH), 139.4 (C), 146.2 (C), 161.5 (C=O), 162.2 (C=O). Anal. Calcd for $C_{29}H_{30}N_2O_5$ (486.56): C, 71.59; H, 6.21; N, 5.76%; Found: C, 71.68; H, 6.24; N, 5.80%.

Diethyl 3-cyano-3-phenyl-3,4a-dihydro-[1,3]oxazino[3,2-*a*]quinoline-1,2-dicarboxylate (**6a**): Yellow powder, mp 150-152 °C; IR (KBr): 1738, 1725, 1715, 1658, 1458, 1325, 1245 cm^{-1}; ^{1}H NMR (500 MHz, $CDCl_3$): δ 1.25 (t, $^{3}J=7.3$Hz, 3H, CH_3), 1.32 (t, $^{3}J=7.3$Hz, 3H, CH_3), 4.20 (q, $^{3}J=7.3$Hz, 2H, CH_2O), 4.40 (q, $^{3}J=7.3$ Hz, 2H, OCH_2), 5.83 (d, $^{3}J_{HH}=7.2$Hz, 1H, *CH*), 7.52 (d, $^{3}J_{HH}=7.4$ Hz, 2H, 2 × *CH*), 7.62 (t, $^{3}J_{HH}=7.2$Hz, 2H, 2 × *CH*), 7.64 (t, $^{3}J_{HH}=7.3$Hz, 2H, 2 × *CH*), 7.75 (2 H, d, $^{3}J_{HH}=7.5$ Hz, 2 × *CH*), 8.22 (d, $^{3}J_{HH}=7.2$Hz, 2H, 2 × *CH*), 8.25 (d, $^{3}J_{HH}=7.7$Hz, 1H, *CH*); ^{13}C NMR (125 MHz, $CDCl_3$): δ 14.1 (CH_3), 14.2 (CH_3), 62.5 (OCH_2), 62.7 (OCH_2), 76.2 (C), 83.5 (CH), 117.2 (C), 118.0 (2 CH), 118.6 (CN),

120.1 (2 × CH), 124.5 (C), 125.6 (2 × CH), 126.7 (C), 129.2 (2 × CH), 129.5 (2 × CH), 130.4 (CH), 132.5 (C), 133.4 (C), 164.8 (C=O), 172.7 (C=O). Anal. Calcd for $C_{25}H_{22}N_2O_5$ (430.46): C, 69.76; H, 5.15; N, 6.51%; Found: C, 69.68; H, 5.07; N, 6.47%.

Reference

Rostami-Charati, F., Hossaini, Z., Gharaee, E., and Khalilzadeh, M. A. (2013). One-pot three-component synthesis of oxazine derivatives in water. *J. Hetercyclic Chem.*, **50**, E174–E177.

C-O bond_25

Type of reaction: C-O bond formation
Reaction conditions: Ethanol, room temperature, ultrasound
Synthetic strategy: One-pot multicomponent reaction
Catalyst: Piperidine
Keywords: Aldehydes, malononitrile, 3-cyanoacetylindoles, piperidine, ethanol, ultrasonic irradiation, room temperature, one-pot multicomponent reaction, indol-3-yl substituted pyrans

General reaction scheme

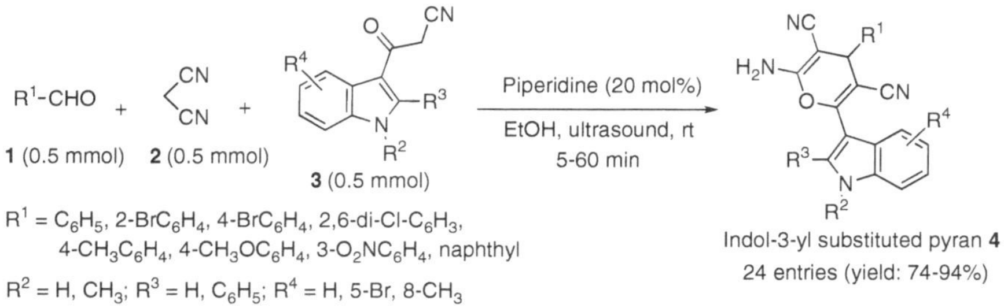

Representative entries

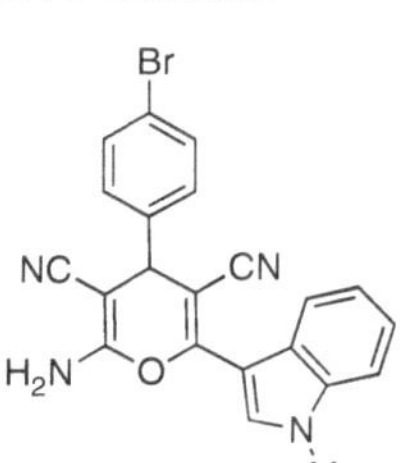

4a

2-Amino-4-(4-bromophenyl)-6-(1-methyl-1*H*-indol-3-yl)-4*H*-pyran-3,5-dicarbonitrile (**4a**): Solid, mp 236-238 °C; reaction time: 6 min; yield: 92%

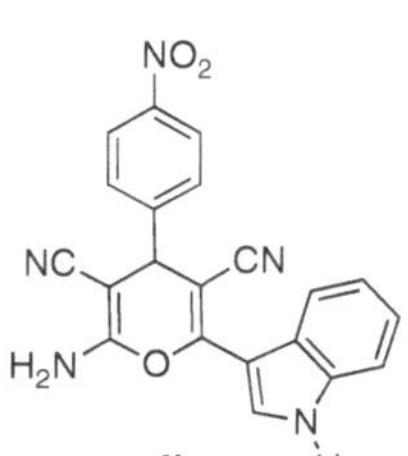

4b

2-Amino-6-(1*H*-indol-3-yl)-4-(4-nitrophenyl)-4*H*-pyran-3,5-dicarbonitrile (**4b**): Solid, mp 238-240 °C; reaction time: 11 min; yield: 86%

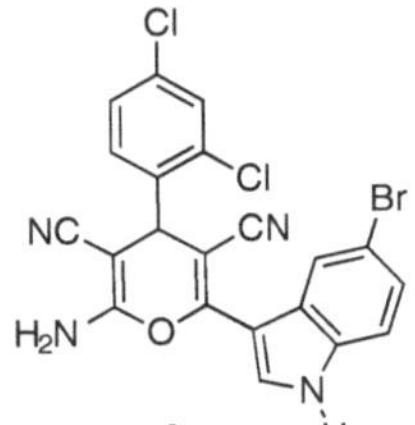

4c

2-Amino-6-(5-bromo-1*H*-indol-3-yl)-4-(2,4-dichlorophenyl)-4*H*-pyran-3,5-dicarbonitrile (**4c**): Solid, mp 298-300 °C; reaction time: 17 min; yield: 85%

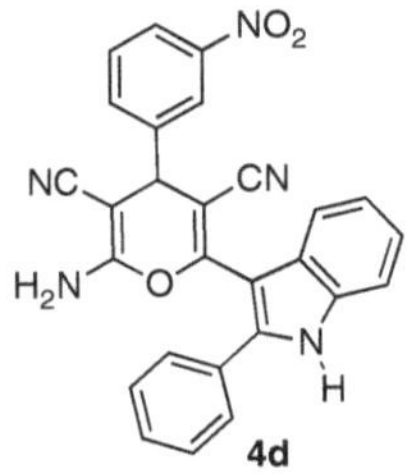

4d

2-Amino-4-(3-nitrophenyl)-6-(2-phenyl-1*H*-indol-3-yl)-4H-pyran-3,5-dicarbonitrile (**4d**): Solid, mp 230-232 °C; reaction time: 13 min; yield: 82%

Experimental procedure

A mixture of aldehyde (**1**; 0.5 mmol), malononitrile (**2**; 0.5 mmol), 3-cyanoacetylindole (**3**; 0.5 mmol), and piperidine (20 mol%) in ethanol (3 mL) was placed in a round bottom flask. The reaction was performed at room temperature under ultrasonic irradiation (using a KQ-250E ultrasonic cleaner with a frequency of 40 KHz and a normal power of 250 W) for requisite time (5-60 min). After completion of reaction, as indicated by TLLC, the reaction mixture was filtered; the solid precipitate was washed with water (5 mL) and ethanol (2 mL) to afford the pure product of indol-3-yl substituted pyrans **4,** characterized by elemental analyses as well as detailed spectral studies.

Characterization data of two representative compounds

2-Amino-4-(4-bromophenyl)-6-(1-methyl-1*H*-indol-3-yl)-4*H*-pyran-3,5-dicarbonitrile (**4a**): Solid, mp 236-238 °C; IR (KBr): 3346.6, 3335.7, 3058.1, 2933.7, 2197.0, 1672.6, 1614.5, 1587.7, 1528.1, 1474.5, 1404.5, 1364.7, 1121.4, 738.1 cm^{-1}; ^{1}H NMR (400 MHz, DMSO-d_6): δ 8.15 (s, 1H, Ar*H*), 7.97 (d, 1H, J=8.0 Hz, Ar*H*), 7.63 (d, 2H, J=8.4 Hz, Ar*H*), 7.56 (d, 1H, J=8.4 Hz, Ar*H*), 7.29-7.35 (m, 5H, NH_2+Ar*H*), 7.19-7.23 (m, 1H, Ar*H*), 4.48 (s, 1H, C*H*), 3.88 (s, 3H, CH_3); ^{13}C NMR (100 MHz, DMSO-d_6): δ 158.7, 155.4, 142.5, 136.5, 132.9, 131.8, 129.9, 124.9, 122.8, 121.8, 121.3, 120.9, 119.1, 118.7, 110.6, 104.3, 83.3, 55.4, 39.2, 33.2. Anal. calcd for $C_{22}H_{15}BrN_4O$: C, 61.27; H, 3.51; N, 12.99. Found: C, 61.45; H, 3.77; N, 13.05%.

2-Amino-6-(5-bromo-1*H*-indol-3-yl)-4-(2,4-dichlorophenyl)-4*H*-pyran-3,5-dicarbonitrile (**4c**): Solid, mp 298-300 °C; IR (KBr): 3483.0, 3215.4, 3083.8, 2201.5, 1672.5, 1587.8, 1519.0, 1361.1, 1402.3, 1102.8, 887.0, 825.6 cm^{-1}; ^{1}H NMR (400 MHz, DMSO-d_6): δ 12.17 (d, 2H, J=2.0 Hz, N*H*), 8.17 (d, 1H, J=3.2 Hz, Ar*H*), 8.03 (d, 1H, J=1.6 Hz, Ar*H*), 7.69 (d, 1H, J=1.6 Hz, Ar*H*), 7.47-7.57 (m, 5H, NH_2+Ar*H*), 7.35-7.38 (m, 1H, Ar*H*), 4.95 (s, 1H, C*H*); ^{13}C NMR (75 MHz, DMSO-d_6): δ 159.1, 155.9, 138.3, 134.8, 133.5, 133.4, 132.6, 130.8, 129.4, 128.4, 126.1, 125.5, 123.3, 118.9, 118.3, 114.4, 114.0, 105.0, 83.0, 53.9, 37.0. Anal. calcd for $C_{21}H_{11}BrCl_2N_4O$: C, 51.88; H, 2.28; N, 11.52. Found: C, 52.07; H, 2.41; N, 11.53%.

Reference

Chen, T., Xu, X.-P., and Ji, S.-J. (2013). Facile and efficient synthesis of indol-3-yl substituted pyran derivatives via one-pot multicomponent reactions under ultrasonic irradiation. *J. Heterocyclic Chem.*, **50**, 244–251.

C-O bond_26

Type of reaction: C-O and C-C bond formation
Reaction conditions: Ethanol-water (1:1), room temperature
Synthetic strategy: One-pot multicomponent reaction
Catalyst: Urea
Keywords: Aldehydes, malononitrile, C-H activated acids, urea, organocatalyst, ethanol-water (1:1), room temperature, tandem Knoevenagel-cyclocondensation, one-pot three-component reaction, chemoselectivity, functionalized 2-amino-3-cyano-4*H*-pyrans and pyran annulated heterocycles

General reaction scheme

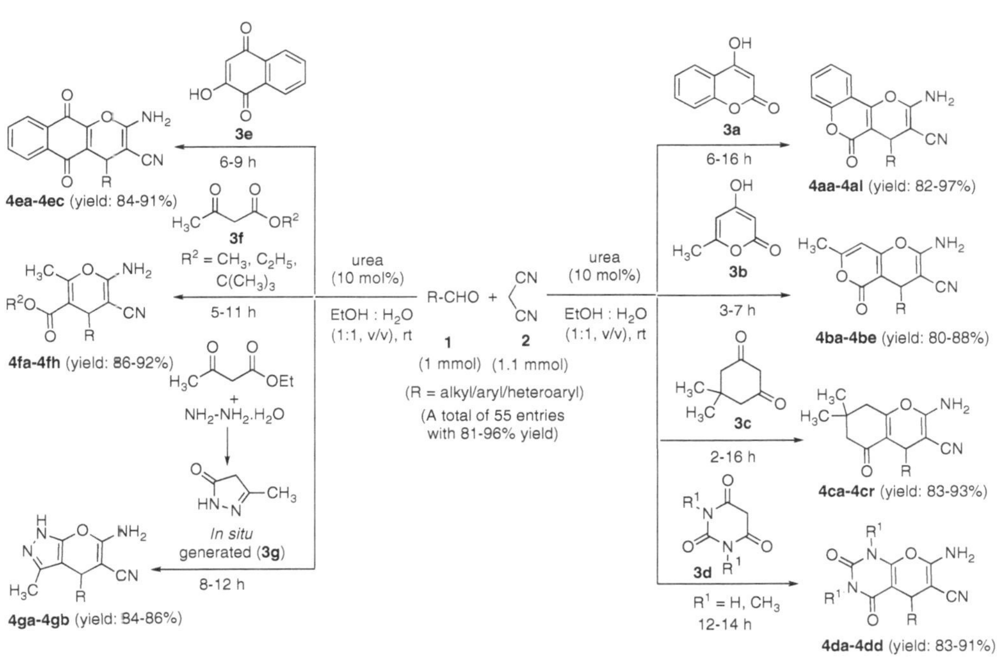

Proposed mechanism

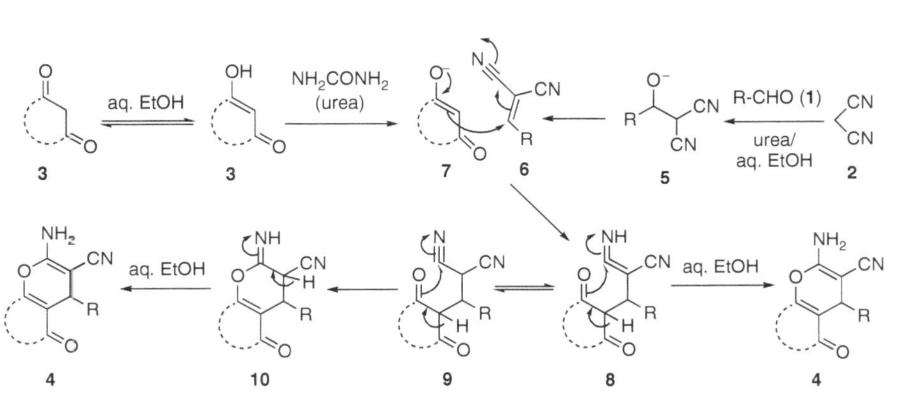

Representative entries

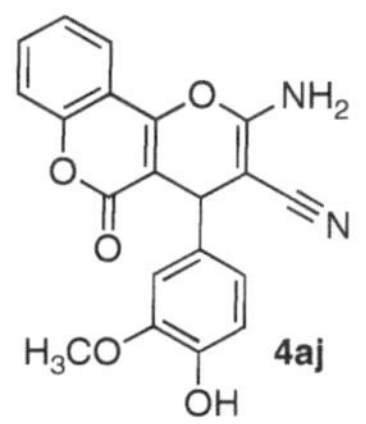

4aj

2-Amino-4-(4-hydroxy-3-methoxyphenyl)-5-oxo-4,5-dihydropyrano[3,2-*c*]chromene-3-carbonitrile (**4aj**). Yellow solid, mp 253-255 °C; yield: 87%

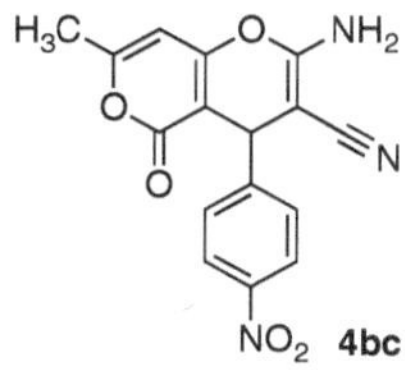

4bc

2-Amino-7-methyl-4-(4-nitrophenyl)-5-oxo-4,5-dihydropyrano[4,3-*b*]pyran-3-carbonitrile (**4bc**). Yellowish white solid, mp 208-210 °C; yield 83%

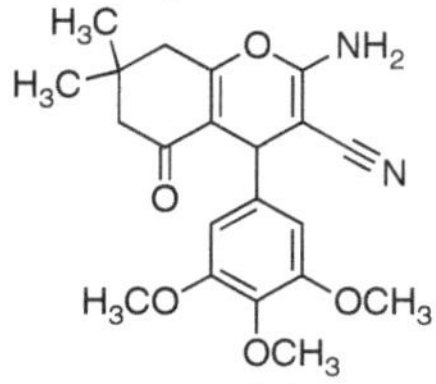

4ck

2-Amino-7,7-dimethyl-5-oxo-4-(3,4,5-trimethoxyphenyl)-5,6,7,8-tetrahydro-4*H*-chromene-3-carbonitrile (**4ck**). White solid, mp 208-210 °C; yield 83%

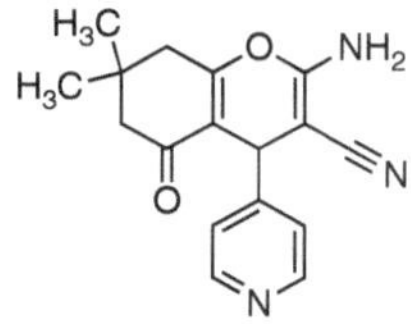

4cl

2-Amino-7,7-dimethyl-5-oxo-4-(pyridin-4-yl)-5,6,7,8-tetrahydro-4*H*-chromene-3-carbonitrile (**4 cl**). White solid, mp 214-216 °C; yield: 92%

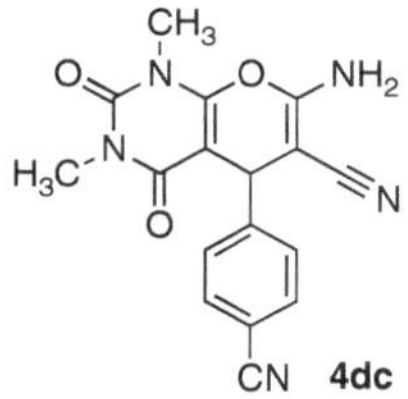

4dc

7-Amino-5-(4-cyanophenyl)-1,3-dimethyl-2,4-dioxo-2,3,4,5-tetrahydro-1*H*-pyrano[2,3-*d*] pyrimidine-6-carbonitrile (**4 dc**). White solid, mp 202-204 °C; yield 91%

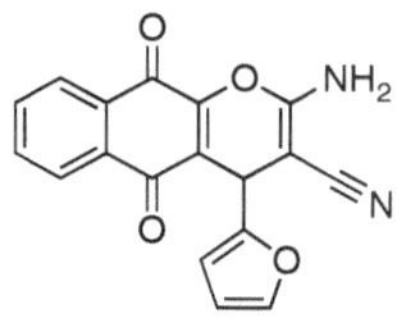

4eb

2-Amino-4-(furan-2-yl)-5,10-dioxo-5,10-dihydro-4*H*-benzo[*g*]chromene-3-carbonitrile (**4eb**). Blackish, mp 266-268 °C; yield 91%

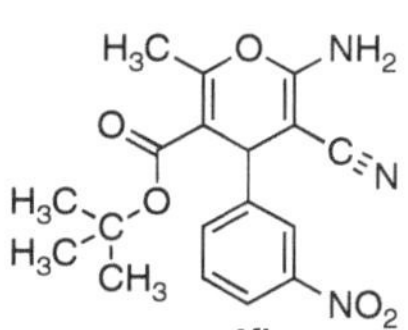

4fh

tert-Butyl 6-amino-5-cyano-2-methyl-4-(3-nitrophenyl)-4*H*-pyran-3-carboxylate (**4fh**). White solid, mp 188-190 °C; yield 92%

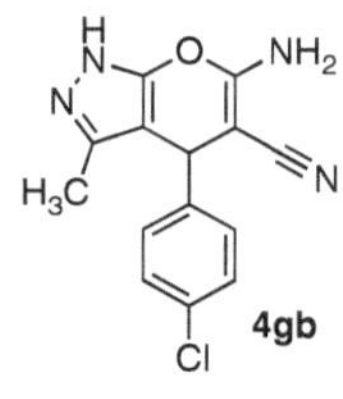

4gb

6-Amino-4-(4-chlorophenyl)-3-methyl-1,4-dihydropyrano[2,3-*c*]pyrazole-5-carbonitrile (**4gb**). White solid, mp 233-235 °C; yield: 84%

Experimental procedure

An oven-dried screw cap test tube was charged with a magnetic stir bar, aldehyde (**1**; 1 mmol), malononitrile (**2**; 1.1 mmol), urea (10 mol% as organo-catalyst), EtOH: H_2O (1:1 v/v; 4 mL) in a sequential manner; the reaction mixture was then stirred vigorously at room temperature for about 20 min. After then C−H activated acid (**3**) (1 mmol) was added to the stirred reaction mixture, and the stirring was continued for appropriate range of time as indicated in respective tables in the text. The progress of the reaction was monitored by TLC. On completion of reaction, a solid mass precipitated out which was filtered off followed by washing with aqueous ethanol to obtain crude product (**4**) purified just by re-crystallization from ethanol without carrying out column chromatography. The filtrate containing residual solvent, catalyst and substrates obtained upon filtration of the reaction mixture after completion of reaction could be successfully reused for a particular entry up to thrice without appreciable loss of catalytic activity. The structure of each purified pyran annulated heterocyclic scaffold was confirmed by analytical as well as spectral studies including FT-IR, ^{1}H NMR, ^{13}C NMR and TOF-MS.

Characterization data of representative entries

2-Amino-4-(4-hydroxy-3-methoxyphenyl)-5-oxo-4,5-dihydropyrano[3,2-*c*]chromene-3-carbonitrile (**4aj**). Yellow solid, mp 253-255 °C; yield 87%; IR (KBr) cm^{-1}: 3365, 3309, 3042, 3003, 2942, 2326, 2201, 1709, 1682, 1595, 1508, 1371, 1234, 1176, 1040, 965, 949, 858, 773, 633 cm^{-1}; ^{1}H NMR (400 MHz, DMSO-d_6): δ (1H, s, aromatic OH), 7.89 (1H, d, $J=7.2$ Hz, aromatic H), 7.70 (1H, t, $J=7.2$ & 7.6 Hz, aromatic H), 7.49 (1H, d, $J=7.6$ Hz, aromatic H), 7.46 (1H, d, $J=8.0$ Hz, aromatic H), 7.35 (2H, s, NH_2), 6.81 (1H, d, $J=1.2$ Hz, aromatic H), 6.70 (1H, d, $J=8.0$ Hz, aromatic H), 6.61 (1H, dd, $J=8.0$ & 2.0 Hz, Ar-*H*), 4.36 (1H, s, CH), 3.73 (3H, s, CH_3); ^{13}C NMR (100 MHz, DMSO-d_6): δ 160.00, 158.32, 153.46, 152.49, 147.71, 146.14, 134.73, 133.21, 125.04, 122.85, 120.28, 119.80, 116.96, 115.90, 113.47, 112.45, 104.74, 58.71, 56.07, 36.87; TOF-MS: 385.0798 $[M+Na]^+$. Elemental analysis: Calcd. (%) for $C_{20}H_{14}N_2O_5$: C, 66.30; H, 3.89; N, 7.73; Found: C, 66.28; H, 3.91; N, 7.69.

2-Amino-7-methyl-4-(4-nitrophenyl)-5-oxo-4,5-dihydropyrano[4,3-*b*]pyran-3-carbonitrile (**4bc**). Yellowish white solid, mp 208-210 °C; yield 83%; IR (KBr) cm^{-1}: 3402, 3296, 3283, 3084, 2937, 2849, 2314, 2197, 1707, 1676, 1595, 1514, 1377, 1344, 1257, 1142, 1041, 974, 825, 694, 544 cm^{-1}. ^{1}H NMR (400 MHz, DMSO-d_6): δ 8.18 (2H, d, $J=8.8$ Hz, aromatic H), 7.51 (2H, d, $J=8.8$ Hz, aromatic H), 7.37 (2H, s, NH_2), 6.31 (1H, s, CH), 4.51 (1H, s, CH), 2.23 (3H, s, CH_3); ^{13}C NMR (100 MHz, DMSO-d_6): δ 163.89, 161.75, 159.04, 158.60, 151.43, 146.94, 129.44 (2C), 124.08 (2C), 119.40, 99.97, 98.43, 57.07, 36.57, 19.74; TOF-MS: 349.0671 $[M+Na]^+$. Elemental analysis: Calcd. (%) for $C_{16}H_{11}N_3O_5$: C, 59.08; H, 3.41; N, 12.92; Found: C, 59.06; H, 3.43; N, 12.90.

2-Amino-7,7-dimethyl-5-oxo-4-(3,4,5-trimethoxyphenyl)-5,6,7,8-tetrahydro-4*H*-chromene-3-carbonitrile (**4ck**). White solid, mp 208-210 °C; yield 83%; IR (KBr) cm^{-1}: 3377, 3300, 31787, 2947, 2324, 2183, 1663, 1639, 1603, 1587, 1346, 1231, 1119, 1113, 1020, 856, 716, 561 cm^{-1}; ^{1}H NMR (400 MHz, DMSO-d_6): δ 6.98 (2H, s, aromatic H), 6.38 (2H, s, NH_2), 4.14 (1H, s, CH), 3.73 (6H, s, OCH_3), 3.63 (3H, s, OCH_3), 2.54 (2H, br s, CH_2), 2.29 (1H, d, $J=16.0$ Hz), 2.15 (1H, d, $J=16.0$ Hz), 1.06 (3H, s, CH_3), 1.03 (3H, s, CH_3); ^{13}C NMR (100 MHz, DMSO-d_6) : δ 195.71, 162.75, 158.32, 152.71 (3C), 140.43, 119.69, 112.30, 104.11 (2C), 59.88, 58.29, 55.74 (2C), 49.93, 35.56, 31.71 (2C), 28.56, 26.53; TOF-MS: 407.1586 $[M+Na]^+$. Elemental analysis: Calcd. (%) for $C_{21}H_{24}N_2O_5$: C, 65.61; H, 6.29; N, 7.29; Found: C, 65.64; H, 6.33; N, 7.25.

2-Amino-7,7-dimethyl-5-oxo-4-(pyridin-4-yl)-5,6,7,8-tetrahydro-4*H*-chromene-3-carbonitrile (**4 cl**). White solid, mp 214-216 °C; yield 92%; IR (KBr) cm^{-1}: 3389, 3312, 3038, 2962, 2878, 2341, 2330, 2183, 1674, 1653, 1599, 1529, 1479, 1360, 1213, 1149, 1041, 922, 852, 787, 658 cm^{-1}; ^{1}H NMR (400 MHz, DMSO-d_6): δ 8.49 (2H, d, $J=5.6$ Hz, aromatic H), 7.19 (2H, s, NH_2), 7.18 (2H, d, $J=1.2$ Hz, aromatic H), 4.23 (1H, s, CH), 2.54 (2H, s, CH_2), 2.27 (1H, d, $J=16.0$ Hz), 2.14 (1H, d, $J=16.4$ Hz), 1.04 (3H, s, CH_3), 0.97 (3H, s, CH_3); ^{13}C NMR (100 MHz, DMSO-d_6) : δ 196.11, 163.69, 159.11, 153.45, 150.12 (2C), 122.93

(2C), 119.78, 111.84, 57.17, 50.27, 35.58, 32.23 (2C), 28.68, 27.36; TOF-MS: 296.0589 [M +H]$^+$. Elemental analysis: Calcd. (%) for $C_{17}H_{17}N_3O_2$: C, 69.14; H, 5.80; N, 14.23; Found: C, 69.18; H, 5.81; N, 14.21.

7-Amino-5-(4-cyanophenyl)-1,3-dimethyl-2,4-dioxo-2,3,4,5-tetrahydro-1*H*-pyrano[2,3-*d*] pyrimidine-6-carbonitrile (**4 dc**). White solid, mp 202-204 °C; yield 91%; IR (KBr): 3389, 3286, 3198, 3067, 2951, 2330, 2206, 1678, 1663, 1483, 1375, 1259, 1194, 1051, 957, 945, 864, 831, 762, 638, 613 cm^{-1}; ^{1}H NMR (400 MHz, DMSO-d_6): δ 7.77 (2H, d, J=8.0 Hz, aromatic H), 7.48 (2H, d, J=8.4 Hz, aromatic H), 7.45 (2H, s, NH_2), 4.45 (1H, s, CH), 3.35 (3H, s, CH_3), 3.07 (3H, s, CH_3); ^{13}C NMR (100 MHz, DMSO-d_6) : δ 160.91, 158.12, 151.87, 150.41, 150.09, 132.73 (2C), 128.97 (2C), 119.24, 119.16, 110.05, 88.13, 57.88, 37.04, 29.58, 28.08; TOF-MS: 358.0915 [M+Na]$^+$. Elemental analysis: Calcd. (%) for $C_{17}H_{13}N_5O_3$: C, 60.89; H, 3.91; N, 20.89; Found: C, 60.93; H, 3.89; N, 20.88.

2-Amino-4-(furan-2-yl)-5,10-dioxo-5,10-dihydro-4*H*-benzo[*g*]chromene-3-carbonitrile (**4eb**). Blackish, mp 266-268 °C; yield 91%; IR (KBr): 3406, 3198, 3070, 2332, 2197, 1794, 1663, 1643, 1580, 1516, 1441, 1352, 1279, 1051, 997, 872, 787, 764, 723, 662, 573 cm^{-1}; ^{1}H NMR (400 MHz, DMSO-d_6): δ 8.27 (1H, d, J=8.4 Hz, aromatic H), 8.13 (1H, d, J=7.2 Hz, aromatic H), 8.06 (1H, d, J=7.2 Hz, aromatic H), 7.99 (1H, d, J=7.2 Hz, aromatic H), 7.93 (2H, d, J=7.2 Hz, aromatic H), 7.85 (2H, br s, NH_2), 7.79 (1H, d, J=8.0 Hz, aromatic H), 6.17 (1H, s, CH); ^{13}C NMR (100 MHz, DMSO-d_6): δ 185.08, 181.67, 160.02, 135.28, 134.88, 133.72, 133.63, 132.28, 130.95, 127.06, 126.48, 126.35, 125.82, 124.25, 124.19, 121.64, 111.40, 60.78; TOF-MS: 341.0541 [M+Na]$^+$. Elemental analysis: Calcd. (%) for $C_{18}H_{10}N_2O_4$: C, 67.92; H, 3.17; N, 8.80; Found: C, 67.98; H, 3.13; N, 8.82.

tert-Butyl 6-amino-5-cyano-2-methyl-4-(3-nitrophenyl)-4*H*-pyran-3-carboxylate (**4fh**). White solid, mp 188-190 °C; yield 92%; IR (KBr): 3435, 3329, 3192, 3184, 3072, 2986, 2345, 2193, 1680, 1670, 1584, 1593, 1521, 1265, 1057, 839, 706, 690 cm^{-1}; ^{1}H NMR (400 MHz, DMSO-d_6): δ 8.15-8.12 (1H, m, aromatic H), 7.98 (1H, s, aromatic H), 7.66 (2H, dd, J=4.8 & 1.6 Hz, aromatic H), 7.05 (2H, s, NH_2), 4.47 (1H, s, CH), 2.31 (3H, s, CH_3), 1.21 (9H, s, $C(CH_3)_3$); ^{13}C NMR (100 MHz, DMSO-d_6): δ164.74, 158.96, 157.38, 148.07, 147.83, 134.64, 130.63, 122.34, 122.18, 119.86, 107.64, 81.31, 56.61, 39.15, 27.83 (3C), 18.50; TOF-MS: 380.1228 [M+Na]$^+$. Elemental analysis: Calcd. (%) for $C_{18}H_{19}N_3O_5$: C, 60.50; H, 5.36; N, 11.76; Found: C, 60.53; H, 5.39; N, 11.74.

6-Amino-4-(4-chlorophenyl)-3-methyl-1,4-dihydropyrano[2,3-c]pyrazole-5-carbonitrile (**4gb**). White solid, mp 233-235 °C; IR (KBr): 3392, 3345, 3288, 3196, 3182, 3018, 2960, 2338, 2185, 1612, 1595, 1493, 1398, 1389, 1308, 1186, 1171, 1055, 1043, 945, 874, 746, 735, 609 cm^{-1}; ^{1}H NMR (400 MHz, DMSO-d_6): δ 12.14 (1H, s, NH), 7.38 (2H, d, J=8.4 Hz aromatic H), 7.20 (2H, d, J=8.4 Hz, aromatic H), 6.93 (2H, s, NH_2), 4.64 (1H, s, CH), 1.79 (3H, s, CH_3); ^{13}C NMR (100 MHz, DMSO-d_6): δ 161.33, 155.13, 143.89, 136.10, 131.65, 129.77 (2C), 128.87 (2C), 121.05, 97.61, 57.20, 35.99, 10.14; TOF-MS: 309.0513 [M+Na]$^+$. Elemental analysis: Calcd. (%) for $C_{14}H_{11}ClN_4O$: C, 58.65; H, 3.87; N, 19.54; Found C, 58.61; H, 3.89; N, 19.52.

Reference

Brahmachari, G., and Banerjee, B. (2014). Facile and one-pot access to diverse and densely functionalized 2-amino-3-cyano-4*H*-pyrans and pyran annulated heterocyclic scaffolds *via* an eco-friendly multicomponent reaction at room temperature using urea as a novel organo-catalyst. *ACS Sustainable Chem. Eng.*, **2**, 411–422.

C-O bond_27

Type of reaction: C-O, C-N, C-C and C-S bond formation
Reaction conditions: Ethanol, room temperature
Synthetic strategy: One-pot multicomponent reaction
Catalyst: Sodium formate
Keywords: Salicylic aldehydes, malononitrile, cyclic secondary amines/thiols, sodium formate, ethanol, room temperature, one-pot multicomponent reaction, tandem Knoevenagel-cyclocondensation, benzopyrano[2,3-*d*]pyrimidines/2-amino-4-arylsulfanyl-4*H*-chromene-3-corbonitriles

General reaction scheme

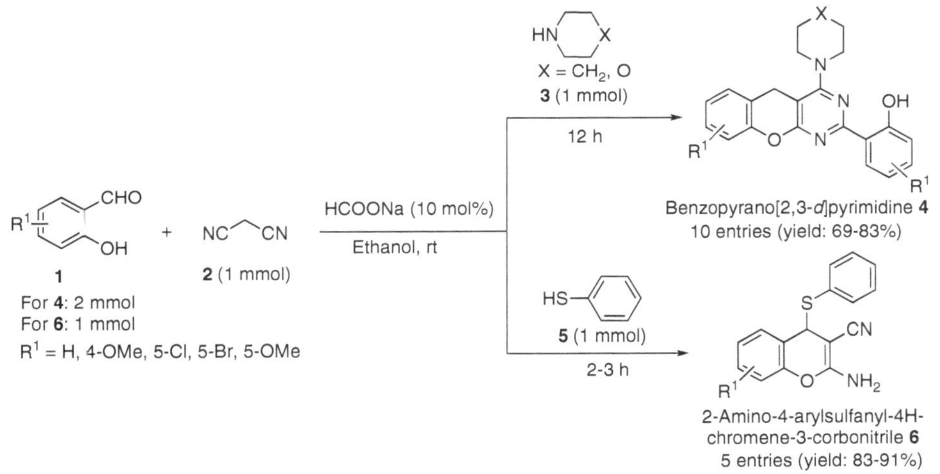

Proposed mechanism

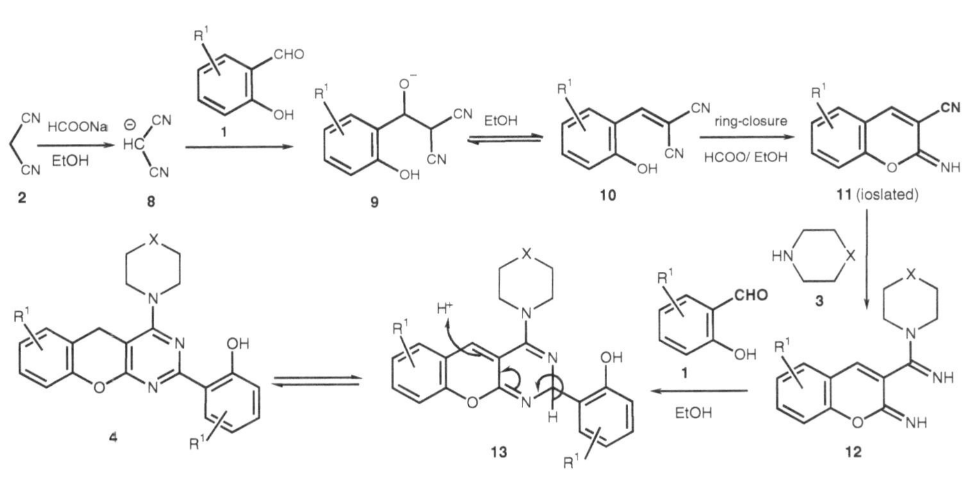

Representative entries

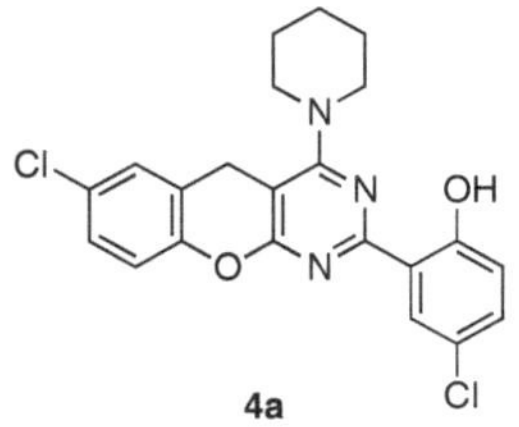

4a

4-Chloro-2-(7-chloro-4-(piperidin-1-yl)-5*H*-chromeno[2,3-*d*]pyrimidin-2-yl)phenol (**4a**): White powder, mp 210-2013 °C; yield: 93%

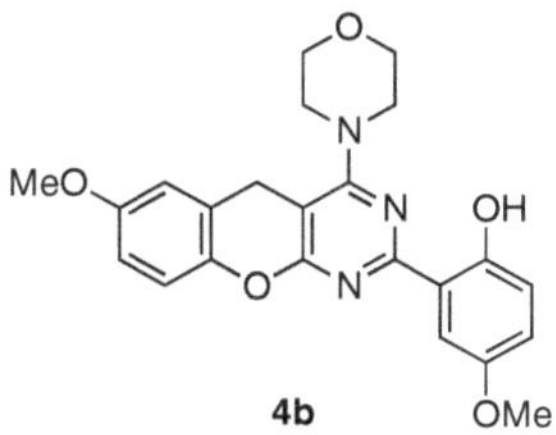

4b

4-Methoxy-2-(7-methoxy-4-morpholino-5*H*-chromeno[2,3-*d*]pyrimidin-2-yl)phenol (**4b**). Yellowish white powder, mp 209-211 °C; yield 69%

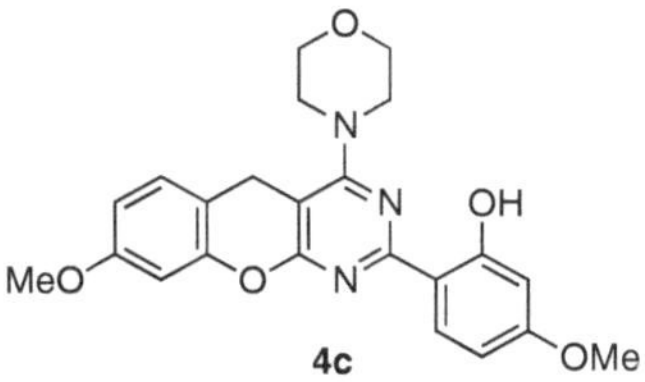

4c

5-Methoxy-2-(8-methoxy-4-morpholino-5*H*-chromeno[2,3-*d*]pyrimidin-2-yl)phenol (**4c**). White powder, mp 214-216 °C; yield 81%

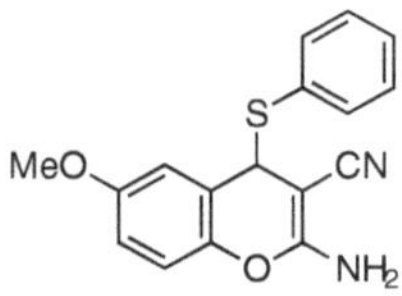

6a

2-Amino-6-methoxy-4-(phenylthio)-4*H*-chromene-3-carbonitrile (**6a**): Yellowish white powder, mp 150-152 °C; yield: 87%

Experimental procedures

General procedure for preparation of benzopyrano[2,3-*d*]pyrimidine **4:** An oven-dried screw cap test tube was charged with a magnetic stir bar, salicylic aldehyde **1** (2 mmol), malononitrile **2** (1 mmol), secondary amine **3** (1 mmol), and sodium formate (10 mol%) in 4 mL ethanol. The reaction mixture was stirred at room temperature for 12 h. On completion of the reaction as monitored by TLC, the product was precipitated out and filtered; the filtrate was preserved for reuse. The crude residue was washed with water followed by ethanol to obtain pure product **4**, characterized by analytical and conventional spectroscopic methods.

General procedure for preparation of 2-amino-4-(phenylthio)-4*H*-chromene-3-carbonitrile **6:** In an oven-dried screw cap test tube, a mixture of salicylic aldehyde **1** (1 mmol), malononitrile **2** (1 mmol), thiophenol **5** (1 mmol) and sodium formate (10 mol%) was stirred with 4 mL of ethanol at room temperature for 2-3 h. On completion of the reaction as monitored by TLC, the product was precipitated out and filtered; the filtrate was preserved for reuse. The crude residue was subjected to column chromatography using silica gel (60-120 mesh) and petrol ether-ethyl acetate mixture to obtain pure product **6**, characterized by analytical and conventional spectroscopic methods.

Characterization data of two representative entries

4-Methoxy-2-(7-methoxy-4-morpholino-5*H*-chromeno[2,3-*d*]pyrimidin-2-yl)phenol (**4b**): Yellowish white powder, mp 209-211 °C; IR (KBr): 3485, 3045, 2904, 2844, 1614, 1579, 1114, 1026, 823 cm^{-1}; ^1H NMR (400 MHz, DMSO-d_6): δ 3.50 (t, $J=4$, 4.8 Hz, 4H, 2 × $-CH_2-$), 3.76 (s, 3H, $-OCH_3$), 3.77 (s, 3H, $-OCH_3$), 3.81 (t, $J=4.8$, 4.4 Hz, 4H, 2 × $-CH_2-$), 4.02 (s, 2H, $-CH_2-$), 6.86-6.88 (m, 2H, Ar-H), 6.95 (d, $J=2.8$ Hz, 1H, Ar-H), 7.02 (dd, $J=3.2$, 9.2 Hz, 1H, Ar-H), 7.16 (d, $J=9.2$ Hz, 1H, Ar-H), 7.77 (d, $J=3.2$ Hz, 1H, Ar-H),

12.67 (s, 1H, −OH). Anal. Calcd for $C_{23}H_{23}N_3O_5$: C, 65.55; H, 5.50; N, 9.97; Found: C, 65.57; H, 5.48; N, 9.98.

2-Amino-6-methoxy-4-(phenylthio)-4*H*-chromene-3-carbonitrile (**6a**): Yellowish white powder, mp 150-152 °C; IR (KBr): 3385, 3227, 3061, 2933, 2831, 2180, 1636 cm^{-1}; ^{1}H NMR (400 MHz, $CDCl_3$): δ 3.83 (s, 3H, $-OCH_3$), 4.57 (s, 2H, $-NH_2$), 5.01 (s, 1H, –CH–), 6.66 (d, $J = 12$ Hz, 2H, Ar-H), 7.09 (s, 1H, Ar-H), 7.23-7.35 (m, 5H, Ar-H); ^{13}C NMR (100 MHz, $CDCl_3$): δ 46.96, 55.93, 57.84, 112.26, 115.88, 116.95, 119.44, 121.78, 128.67, 129.30, 129.63, 130.84, 137.04, 143.56, 156.71, 161.40. Anal. Calcd for $C_{17}H_{14}N_2O_2S$: C, 65.79; H, 4.55; N, 9.03; Found: C, 65.77; H, 4.56; N, 9.04.

Reference

Brahmachari, G., and Das, S. (2014). Sodium formate-catalyzed one-pot synthesis of benzopyranopyrimidines and 4-thio-substituted 4*H*-chromenes *via* multicomponent reaction at room temperature. *J. Heterocyclic Chem.*, http://dx.doi.org/10.1002/jhet.2123.

C-O bond_28

Types of reaction: C-O and C-C bond formation
Reaction conditions: Solvent-free, room temperature
Synthetic strategy: One-pot condensation
Catalyst: Titanium(IV) chloride ($TiCl_4$)
Keywords: Phenols, β-keto esters, $TiCl_4$, solvent-free, room temperature, Pechmann condensation, coumarins

General reaction scheme

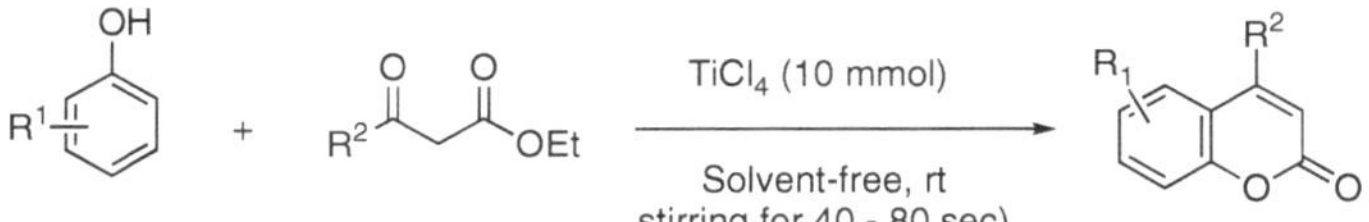

1 (20 mmol) **2** (30 mmol) Coumarins (**3**)

11 entries (Yield: 60-97%)

R^1= H; 4-OH; 4-OMe; 4-Me; 4-OCH_2CO_2Et; 4-OCH_2COPh; 4-OCHMeCOMe; 4,6-di-OH; 4-OH,6-Me
R^2 = Me; CF_3; CH_2Cl

[no reaction took place for the reactants where R^1 = 4-NO_2; 4-Cl, and R^2 = Me]

Representative entries

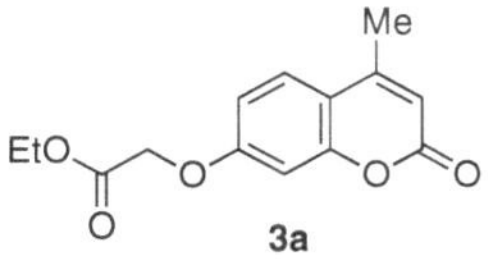

3a

Ethyl 2-((4-methyl-2-oxo-2*H*-chromen-7-yl)oxy) acetate (**3a**): Solid, mp 98-100 °C; reaction time: 55 sec; yield: 89%

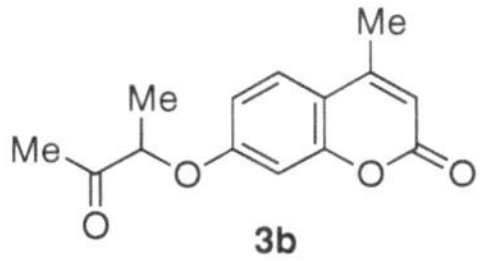

3b

4-Methyl-7-((3-oxobutan-2-yl)oxy)-2*H*-chromen-2-one (**3b**): Solid, mp 231-233 °C; reaction time: 80 sec; yield: 90%

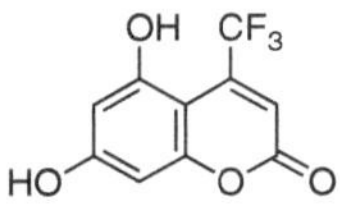

3c

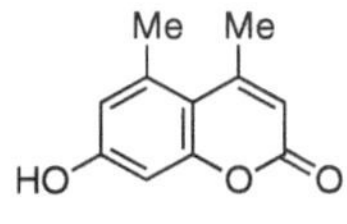

3d

5,7-Dihydroxy-4-(trifluoromethyl)-2*H*-chromen-2-one (**3c**): Solid, mp 250-252 °C; reaction time: 60 sec; yield: 97%

7-Hydroxy-4,5-dimethyl-2H-chromen-2-one (**3d**): Solid, mp 243-245 °C; reaction time: 50 sec; yield: 92%

Experimental procedure

To a thorough mixture of phenol (**1**; 20 mmol) and β-keto ester (**2**; 30 mmol) was added with titanium(IV) chloride (10 mmol) under solvent-free conditions and the resulting mixture was stirred at room temperature. The reaction was found to be very rapid and was completed within the range of 40-80 sec. The resulting reaction mixture was poured onto crushed ice and the crude product was filtered off followed by purification through recrystallization from hot ethanol-water (9:1) to furnish pure coumarin **3** with good to excellent yield (60-97%). All the products were characterized from their spectral studies including FT-IR, ^{1}H NMR and EIMS.

Characterization data of two representative compounds

5,7-Dihydroxy-4-(trifluoromethyl)-2*H*-chromen-2-one (**3c**): Solid, mp 250-252 °C; IR (KBr): 3158, 1665 cm^{-1}; ^{1}H NMR (DMSO-d_6): δ 6.24 (d, J=2.62, 1H, ArH), 6.34 (d, J=2.62, 1H, ArH), 6.50 (s, 1H, C=CH), OH not observed; EIMS: *m*/*z* 246 (M^+).

7-Hydroxy-4,5-dimethyl-2*H*-chromen-2-one (**3d**): Solid, mp 243-245 °C; IR (KBr): 3159, 1665 cm^{-1}; ^{1}H NMR (DMSO-d_6): δ 2.65 (s, 3H, Me), 2.95 (s, 3H, Me), 6.25 (d, J=2.60, 1H, ArH), 6.35 (d, J=2.60, 1H, ArH), 6.41 (s, 1H, C=CH), OH not observed; EIMS: *m*/*z* 190 (M^+).

Reference

Valizadeh, H., and Shockravi, A. (2005). An efficient procedure for the synthesis of coumarin derivatives using TiCl4 as catalyst under solvent-free conditions. *Tetrahedron Lett.*, **46**, 3501–3503.

C-O bond_29

Types of reaction: C-O and C-C bond formation
Reaction conditions: Ethanol/methanol/dichlorometahne, room temperature
Synthetic strategy: One-pot multicomponent
Catalyst: No catalyst
Keywords: 2-Hydroxybenzaldehydes, Meldrum's acid, isocyanides, aromatic/aliphatic alcohols, dichloromethane, room temperature, one-pot multicomponent reaction, Ugi reaction, substituted 3,4-dihydrocoumarins

Reaction scheme

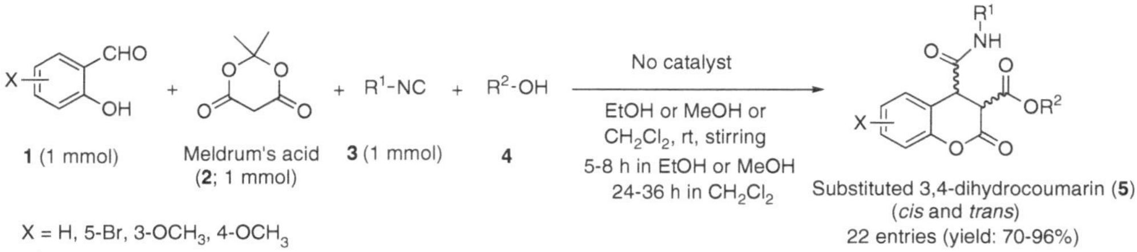

X = H, 5-Br, 3-OCH_3, 4-OCH_3

R^1 = CH_2COOEt, *t*-butyl, cyclohexyl, benzyl, 4-toluene-methanesulfonyl, 2,6-dimethylphenyl

R^2 = Et, Me, isobutyl, 4-$ClC_6H_4CH_2$, 4-$FC_6H_4CH_2$, *n*-heptyl, cycloheptyl, cyclododecyl

Proposed mechanism

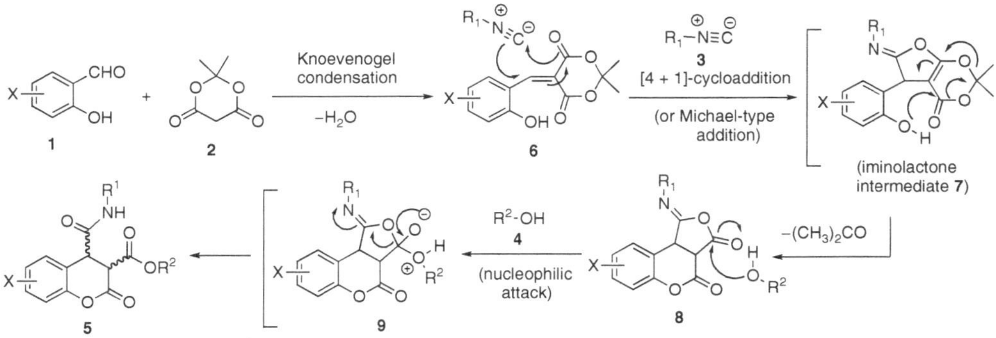

Representative entries

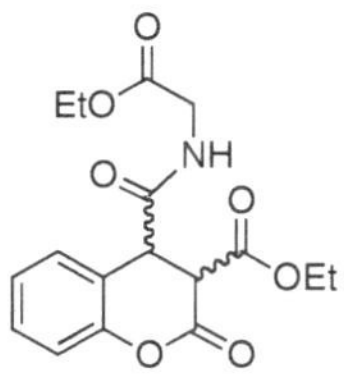

5a

Ethyl-4-((2-ethoxy-2-oxyethyl)carbamoyl)-3,4-dihydro-2-oxo-2*H*-chromene-3-carboxylate (**5a**): White powder, mp 126-129 °C; reaction time: 8 h; yield: 75% (*cis*:*trans* = 72:28)

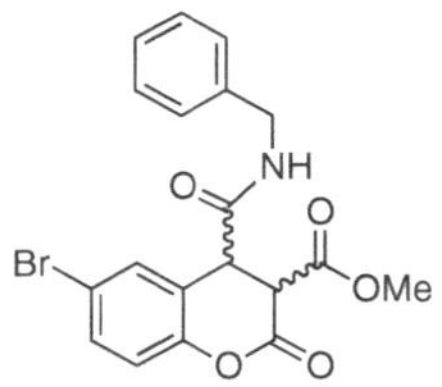

5b

Methyl 4-(benzylcarbamoyl)-6-bromo-3,4-dihydro-2-oxo-2*H*-chromene-3-carboxylate (**5b**): White powder; mp 177-180 °C; reaction time: 8 h; yield: 90% (*cis*:*trans* = 61:39)

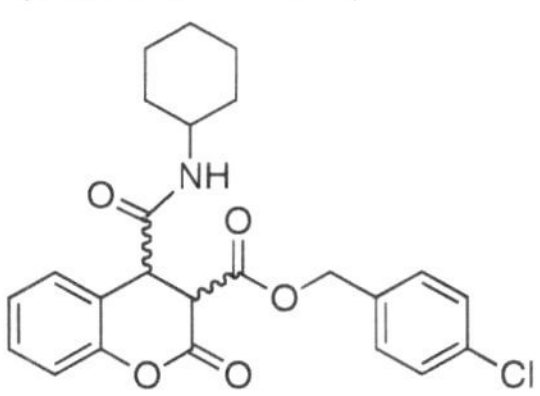

5c

4-Chlorobenzyl 4-(cyclohexylcarbamoyl)-3,4-dihydro-2-oxo-2*H*-chromene-3-carboxylate (**5c**): White powder; mp 148-152 °C; reaction time: 36 h; yield: 78% (*cis*:*trans* = 63:37)

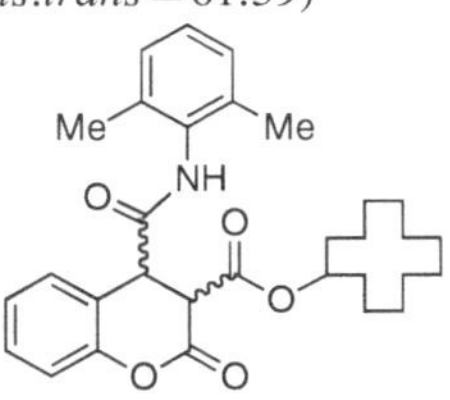

5d

Cyclododecyl 4-((2,6-dimethylphenyl) carbamoyl)-3,4-dihydro-2-oxo-2*H*-chromene-3-carboxylate (**5d**): White powder; mp 154-158 °C; reaction time: 36 h; yield: 75% (*cis*:*trans* = 100:00)

Experimental procedure

To a magnetically stirred solution of 2-hydroxybenzaldehyde (**1**, 1 mmol) and Meldrum's acid (**2**, 1 mmol) in ethanol or methanol (5 mL) (and alcohol **4** other than ethanol or methanol in dichloromethane) was added isocyanide (**3**, 1 mmol) and the reaction mixture was stirred at room temperature for stipulated period (5-8 h in ethanol or methanol and 24-36 h in dichloromethane). After completion of the reaction, as indicated by TLC (ethyl acetate/*n*-hexane, 2:1), the precipitate (in case of dichloromethane medium, the solvent was removed under vacuum to have the crude mass) was washed with ethanol or diethyl ether to afford substituted 3,4-dihydrocoumarin 5 as white powder in good to excellent yield (70-96%). Each of the products was characterized by analytical as well as detailed spectral studies including IR, ^{1}H NMR, ^{13}C NMR, and mass.

Characterization data of representative entries

Methyl 4-(benzylcarbamoyl)-6-bromo-3,4-dihydro-2-oxo-2*H*-chromene-3-carboxylate (**5b**: *cis*:*trans* = 61:39): White powder; mp 177-180 °C; IR (KBr): 3360, 2930, 2851, 1782, 1654, 1540 cm^{-1}; ^{1}H NMR (300 MHz, DMSO-d_6) *Cis* and *Trans*: δ 1.08-1.78 (20H, m, 10CH_2 of two cyclohexyls, *cis* and *trans*), 3.43 (2H, brs, 2CH-N of cyclohexyls), 3.61 (OCH_3, *cis*), 3.70 (OCH_3, *trans*), 4.01 (1H, d, $^{3}J_{HH}$ = 4.1 Hz, CH-4, *cis*), 4.26 (1H, d, $^{3}J_{HH}$ = 3.45 Hz, CH-3, *trans*), 4.32 (1H, d, $^{3}J_{HH}$ = 3.4 Hz, CH-3, *trans*), 4.36 (1H, d, $^{3}J_{HH}$ = 4.1 Hz, CH-4 *cis*), 7.04-7.10 (2H, m, H-Ar, *cis* and *trans*), 7.48-7.52 (2H, m, H-Ar, *cis* and *trans*), 7.65 (2H, brs, H-Ar, *cis* and *trans*), 8.17 (1H, brs, NH, *cis*), 8.39 (1H, brs, NH, *trans*); ^{13}C NMR (75 MHz, DMSO-d_6) *Cis*: δ 24.69, 24.72, 25.52, 32.49, 32.55 (*C*-cyclohexyls), 44.84 (CH-4), 47.90 (CH–N of cyclohexyl), 48.22 (CH-3), 53.59 (OCH3), 116.68, 119.48, 123.42, 130.87, 132.56, 150.45 (C-Ar), 162.93, 167.41, 168.19 (3C=O); 13C NMR (75 MHz, DMSO-*d6*) *Trans*: δ 24.60, 24.65, 25.60, 32.36, 32.73 (*C*-cyclohexyls), 44.37 (CH-4), 46.52 (CH-N of cyclohexyl), 48.07 (CH-3), 52.78 (OCH_3), 116.11, 119.24, 125.47, 130.69, 132.16, 150.94 (C-Ar), 163.57, 167.24, 169.17 (3C=O); MS at *m*/*z* (%): 419 (M^+, ^{81}Br, 5), 417 (M^+, ^{79}Br, 5), 387 (20), 385 (20), 353 (10), 351 (10), 227 (25), 226 (20), 225 (25), 224 (20), 198 (20), 196 (20), 91 (100), 65 (20), 31 (15). Anal. Calcd for $C_{19}H_{16}BrNO_5$: C, 54.56; H, 3.86; N, 3.35; Found: C, 54.34; H, 3.95; N, 3.47.

Cyclododecyl 4-((2,6-dimethylphenyl) carbamoyl)-3,4-dihydro-2-oxo-2*H*-chromene-3-carboxylate (**5d**; *cis*:*trans* = 100:00): White powder; mp 154-158 °C; IR (KBr): 3256, 2940, 2859, 1780, 1743, 1651, 1590 cm^{-1}; ^{1}H NMR (300 MHz, DMSO-d_6) *Cis* and *Trans*: δ 1.31-1.72 (22H, m, 11CH_2 of cyclododecyl), 2.09 (3H, s, CH_3), 2.17 (3H, s, CH_3), 4.09 (1H, d, $^{3}J_{HH}$ = 6.0 Hz, CH-4), 4.54 (1H, d, $^{3}J_{HH}$ = 6.0 Hz, CH-3), 5.05 (1H, brs, OCH), 6.79-7.32 (7H, m, H-Ar), 10.28 (1H, brs, NH); ^{13}C NMR (75 MHz, DMSO-d_6): δ 17.52, 17.82 (2CH_3), 20.75, 20.85, 23.16, 23.22, 23.39, 23.45, 23.74, 23.91, 24.08, 28.99, 29.05, (11CH_2 of cyclododecyl), 48.71 (CH-4), 54.14 (CH-3), 73.98 (OCH), 115.82, 119.66, 122.64, 128.67, 128.86, 129.76, 130.06, 130.92, 132.11, 139.03, 136.97, 155.39 (126C-Ar), 168.06, 171.22, 175.65 (3C=O); MS at *m*/*z* (%): 505 (M^++1, 1), 339 (10), 321 (50), 295 (10), 147 (100), 118 (40), 82 (30), 57 (30), 41 (50). Anal. Calcd for $C_{31}H_{39}NO_5$: C, 73.63; H, 7.77; N, 2.77; Found: C, 73.54; H, 7.87; N, 2.88.

Reference

Shaabani, A., Soleimani, E., Rezayan, A. H., Sarvary, A., and Khavasi, H. R. (2008). Novel Isocyanide-based four-component reaction: A facile synthesis of fully substituted 3,4-dihydrocoumarin derivatives. *Org. Lett.*, **10**, 2581–2584.

C-O bond_30

Type of reaction: C-O, C-C and C-N bond formation
Reaction conditions: Water, room temperature
Synthetic strategy: One-pot multicomponent
Catalyst: Catalyst-free
Keywords: Hydrazine hydrate, ethyl acetoacetate, 2-hydroxy benzaldehydes, malononitrile, water, room temperature, one-pot multicomponent synthesis, 2-amino-4-(5-hydroxy-3-methyl-1*H*-pyrazol-4-yl)-4*H*-chromene-3-carbonitrile derivatives

General reaction scheme

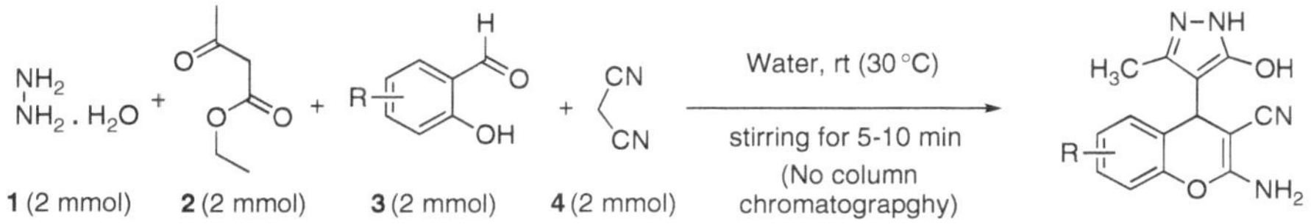

R = H; 5-CH_3; 5-C_2H_5; 4,6-di-CH_3; 5-OCH_3; 5-Br; 3,5-di-Br; 5-Cl; 5-OH; 5-NO_2; 4-CH_3,5-Cl

2-Amino-4-(5-hydroxy-3-methyl-1*H*-pyrazol-4-yl)-4*H*-chromene-3-carbonitrile (**5**)
11 entries (yield: 74 - 92%)

Representative entries

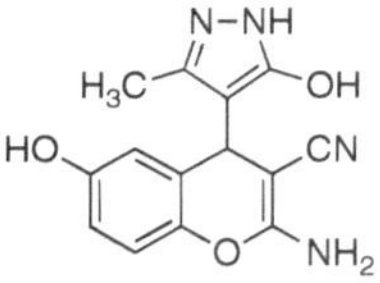

5a

2-Amino-6-hydroxy-4-(5-hydroxy-3-methyl-1*H*-pyrazol-4-yl)-4*H*-chromene-3-carbonitrile (**5a**): White solid; yield: 79%

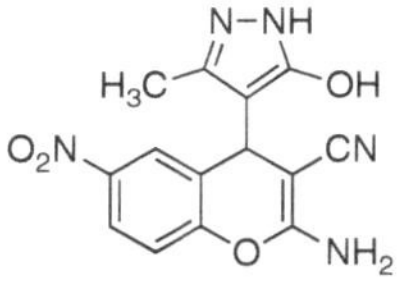

5b

2-Amino-6-nitro-4-(5-hydroxy-3-methyl-1*H*-pyrazol-4-yl)-4*H*-chromene-3-carbonitrile (**5a**): White solid; yield: 86%

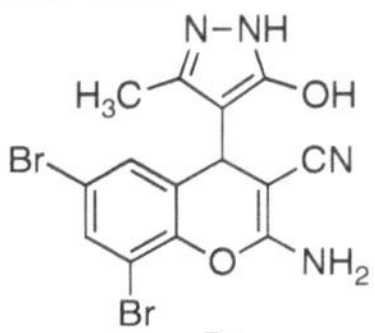

5c

2-Amino-6,8-dibromo-4-(5-hydroxy-3-methyl-1*H*-pyrazol-4-yl)-4*H*-chromene-3-carbonitrile (**5c**): White solid; yield: 74%

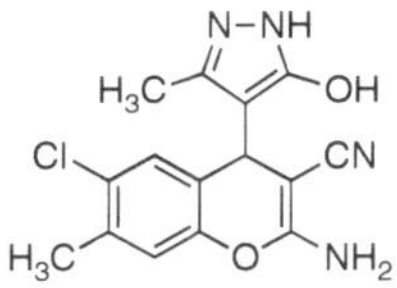

5d

2-Amino-6-chloro-4-(5-hydroxy-3-methyl-1*H*-pyrazol-4-yl)-7-methyl-4*H*-chromene-3-carbonitrile (**5d**): White solid; yield: 79%

Experimental procedure

To a stirred aqueous mixture of hydrazine hydrate 98% (**1**; 0.107 g, 2 mmol) and ethyl acetoacetate (**2**; 0.260 g, 2 mmol), 2-hydroxy benzaldehyde (**3**; 2 mmol), malononitrile (**4**; 0.132 g, 2 mmol) were added successively at ambient temperature (30 °C) under an open atmosphere with vigorous stirring for 5-10 min. On completion of the reaction (as monitored by TLC), a solid precipitate was obtained. The precipitate was filtered, washed with water and then with 5 mL of ethyl acetate/hexane mixture (1:1) affording pure 2-amino-4-(5-hydroxy-3-methyl-1*H*-pyrazol-4-yl)-4*H*-chromene-3-carbonitrile (**5**), characterized by IR, ^{1}H NMR, ^{13}C-NMR and mass spectral studies.

Characterization data of representative compounds

2-Amino-6-hydroxy-4-(5-hydroxy-3-methyl-1*H*-pyrazol-4-yl)-4*H*-chromene-3-carbonitrile (**5a**): White solid; IR (KBr): 3448, 3344, 2191, 1652, 1596, 1508, 1462, 1409 cm^{-1}; ^1H NMR (400 MHz; DMSO-d_6; TMS): δ 1.93 (s, 3H, -CH_3), 4.56 (s, 1H, 4-CH), 6.39 (d, 1H, Ar-H), 6.52 (s, 2H, -NH_2), 6.54-6.57 (m, 1H, Ar-H), 6.77 (d, 1H, Ar-H), 9.17 (s, 1H, -OH), 10.43 (br s, 2H, -OH & -NH); ^{13}C NMR (100 MHz; DMSO-d_6; TMS): δ 9.9, 28.8, 54.6, 104.8, 114.2, 114.5, 116.2, 121.0, 124.3, 136.6, 141.3, 153.6, 159.0, 160.4; LCMS (*m/z*) 285.4 $(M+1)^+$.

2-Amino-6-chloro-4-(5-hydroxy-3-methyl-1*H*-pyrazol-4-yl)-7-methyl-4*H*-chromene-3-carbonitrile (**5d**): White solid; IR (KBr): 3449, 3372, 3174, 2186, 1659,1610, 1530, 1487, 1404 cm^{-1}; ^1H NMR (400 MHz; DMSO-d_6; TMS): δ 1.99 (s, 3H, -CH_3), 2.25 (s, 3H, -CH_3), 4.58 (s, 1H, 4-CH), 6.74 (s, 2H, -NH2), 6.96 (s, 2H, Ar-H), 10.02 (br s, 1H, -OH), 10. 80 (br s, 1H, -NH); ^{13}C NMR (100 MHz; DMSO-d_6; TMS): δ 9.8, 19.1, 54.7, 104.5, 117.8, 120.6, 123.0, 128.0, 128.4, 134.9, 136.6, 147.0, 159.0, 159.8; LCMS (*m/z*) 317.1 $(M+1)^+$.

Reference

Kumaravel, K., and Vasuki, G. (2009). Four-component catalyst-free reaction in water: Combinatorial library synthesis of novel 2-amino-4-(5-hydroxy-3-methyl-1*H*-pyrazol-4-yl)-4*H*-chromene-3-carbonitrile derivatives. *Green Chem.*, **11**, 1945–1947.

C-O bond_31

Type of reaction: C-O and C-C bond formations
Reaction conditions: Acetonitrile, room temperature
Synthetic strategy: One-pot multicomponent reaction
Catalyst: No catalyst
Keywords: Isocyanides, dialkyl acetylenedicarboxylates, 2,5-dihydroxycyclohexa-2,5-diene-1,4-dione, 2-hydroxynaphthalene-1,4-dione, acetonitrile, room temperature, one-pot multicomponent synthesis, functionalized *bis*(4*H*-chromeme) and 4*H*-benzo[*g*]chromene derivatives

General reaction scheme

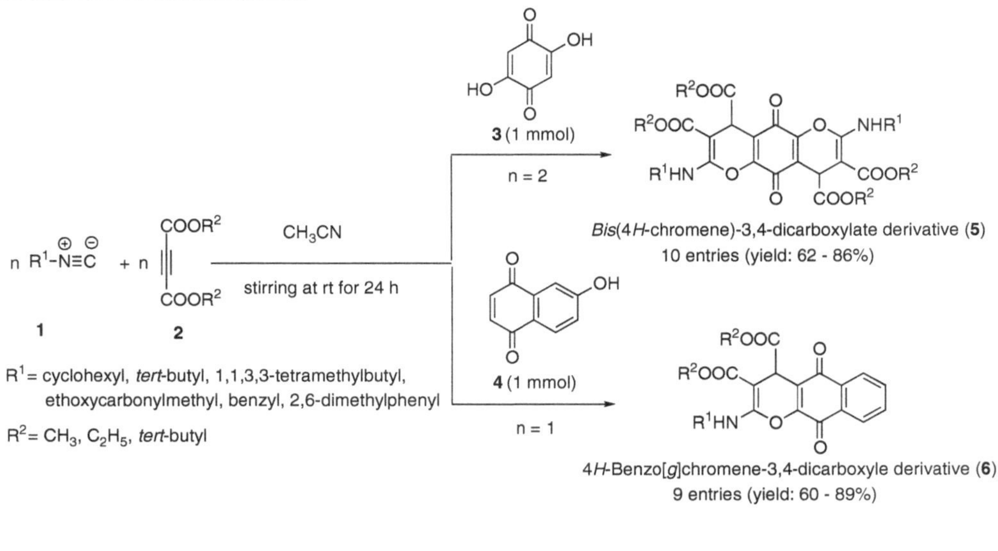

Proposed mechanism

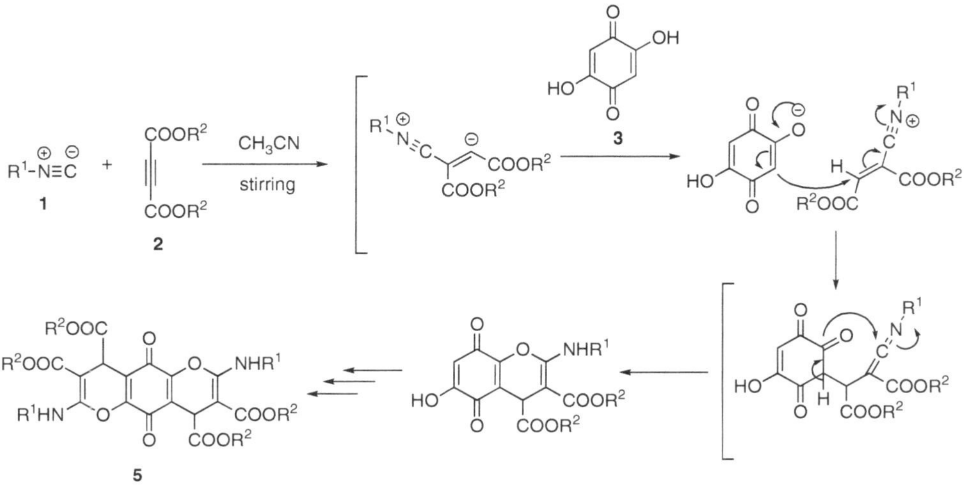

Representative entries

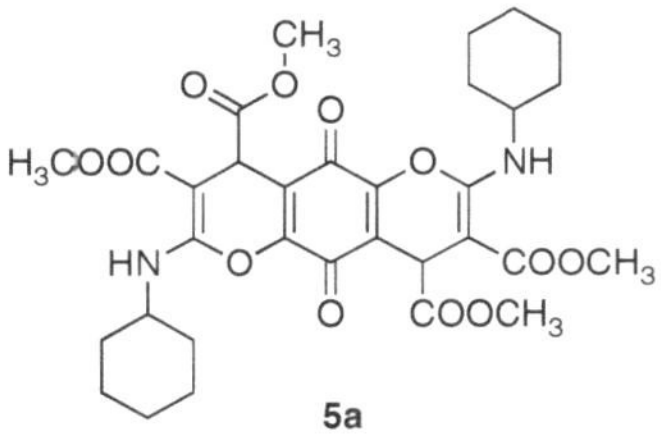

5a

Tetramethyl 2,7-*bis*(cyclohexylamino)-5,10-dioxo-4,5,9,10-tetrahydropyrano[2,3-*g*]chromene-3,4,8,9-tetracarboxylate (**5a**): Brownish red powder, mp 257-258 °C; reaction time: 24 h; yield: 86%

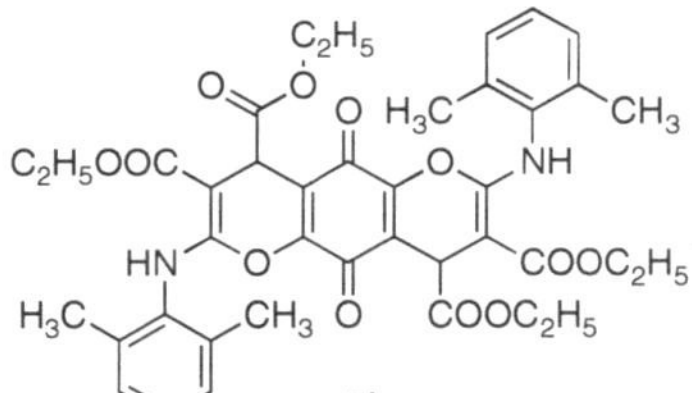

5b

Tetraethyl 2,7-*bis*((2,6-dimethylphenyl) amino)-5,10-dioxo-4,5,9,10-tetrahydropyrano[2,3-*g*] chromene-3,4,8,9-tetracarboxylate (**5b**): Dark red powder, mp 245-248 °C; reaction time: 24 h; yield: 65%

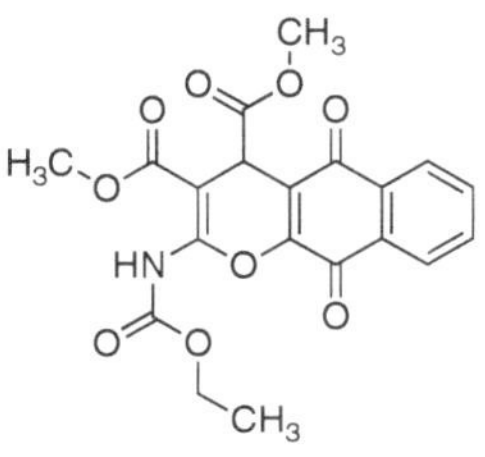

6a

Dimethyl 2-((ethoxycarbonyl)amino)-5,10-dioxo-5,10-dihydro-4*H*-benzo[*g*]chromene-3,4-dicarboxylate (**6a**): Yellow powder, mp 167-169 ° C; reaction time: 24 h; yield: 60%

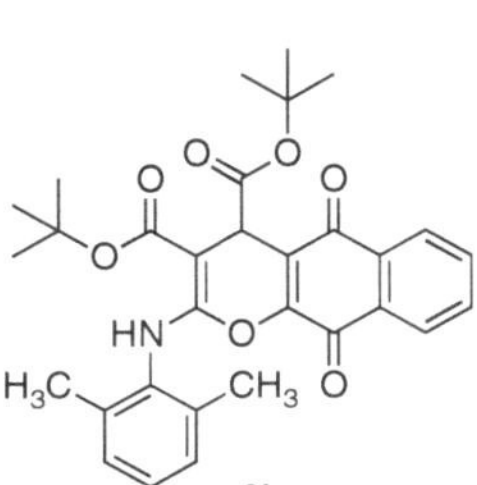

6b

Di-*tert*-butyl 2-((2,6-dimethylphenyl)amino)-5,10-dioxo-5,10-dihydro-4*H*-benzo[*g*] chromene-3,4-dicarboxylate (**6b**): Dark yellow powder, mp 171-173 °C; reaction time: 24 h; yield: 64%

Experimental procedure

Functionalized bis(4H-chromeme) derivative (**5**): To a magnetically stirred solution of 2,5-dihydroxycyclohexa-2,5-diene-1,4-dione (**3**; 1 mmol) and dialkyl acetylenedicarboxylate (**2**; 2 mmol) in acetonitrile (5 mL) was added a solution of isocyanide (**1**; 2 mmol) in acetonitrile (2 mL) at room temperature over 5 min. The mixture was then stirred for 24 h. After completion of the reaction, the precipitated product was separated out from the reaction mixture by filtration and was washed with 5 mL of *n*-hexane to afford bis(4H-chromene)-3,4-dicarboxylate derivative **5** with reasonable good yield (62-86%). Each of the products was characterized by analytical analyses as well as detailed spectral studies including IR, ^{1}H-NMR, ^{13}C NMR, MS.

4H-Benzo[g]chromene derivative (**6**): To a magnetically stirred solution of 2-hydroxy-naphthalene-1,4-dione (**4**, 1 mmol) and dialkyl acetylenedicarboxylate (**2**; 1 mmol) in acetonitrile (5 mL) was added a solution of isocyanide (**1**; 1 mmol) in acetonitrile (2 mL) at room temperature over 5 min. The mixture was then stirred for 24 h. After completion of the reaction, the precipitated product was separated out from the reaction mixture by filtration and was washed with 5 mL of *n*-hexane to afford 4*H*-Benzo[g]chromene derivative **6** with reasonable good yield (60-89%). Each of the products was characterized by analytical analyses as well as detailed spectral studies including IR, ^{1}H-NMR, ^{13}C NMR, MS.

Characterization data of representative compounds

Tetramethyl 2,7-*bis*(cyclohexylamino)-5,10-dioxo-4,5,9,10-tetrahydropyrano[2,3-*g*]chromene-3,4,8,9-tetracarboxylate (**5a**): Brownish red powder, mp 257-258 °C; IR (KBr): 2933, 2859, 1751, 1673, 1602,1446 cm^{-1}; ^{1}H NMR (300 MHz, $CDCl_3$): δ 1.35-1.96 (m, 20H, 10CH_2 of 2 cyclohexyl), 3.66 (s, 6H, 2 × O-C*H*$_3$), 3.71 (s, 6H, 2 × O-C*H*$_3$), 3.79 (bs, 2H, C*H*-NH), 4.64 (s, 2H, 2 × C*H*-CO_2Me), 8.57 (bs, 2H, 2 × CH-N*H*); ^{13}C NMR (75 MHz, $CDCl_3$): δ 24.3, 25.4, 33.5, (*C*-cyclohexyl), 35.2 (*C*H-CO_2Me), 50.3 (*C*HNH), 51.2, 52.8 (2 × O-*C*H$_3$), 70.7, 117.7, 147.8, 158.3 (*C*-alkene), 169.1, 172.2, 176.9 (3C=O); MS at *m*/*z* (rel. %): 643 (M^++1, 2), 611 (3), 585 (5), 553 (3), 525 (4), 486 (5), 402 (5), 327 (5), 297 (5), 269 (5), 170 (5), 96 (20), 83 (30), 55 (80), 41 (80), 31 (100). Anal. Calcd for $C_{32}H_{38}N_2O_{12}$: C, 59.81; H, 5.96; N, 4.36; Found: C, 59.76; H, 5.84; N, 4.28.

Di-*tert*-butyl 2-((2,6-dimethylphenyl)amino)-5,10-dioxo-5,10-dihydro-4*H*-benzo[*g*] chromene-3,4-dicarboxylate (**6b**): Dark yellow powder, mp 171-173 °C; IR (KBr): 2972, 2957, 1722, 1681, 1578 cm^{-1}; ^{1}H NMR (300 MHz, $CDCl_3$): δ 1.44 (s, 9H, C(C*H*$_3$)$_3$), 1.74 (s, 9H, C(C*H*$_3$)$_3$), 2.17 (s, 6H, 2 × C*H*$_3$), 4.76 (s, 1H, C*H*-CO_2*t*-Bu), 7.13 (bs, 3H, Ar-*H*), 7.72-7.76 (m, 2H, Ar-*H*), 8.02-8.12 (m, 2H, Ar-*H*), 9.92 (bs, 1H, N*H*); ^{13}C NMR (75 MHz, $CDCl_3$): δ 18.4 (*C*H$_3$), 27.8, 28.5 (2 × OC(*C*H$_3$)$_3$), 37.5 (*C*H-CO_2*t*-Bu), 74.0 (*C*-alkene), 80.5, 81.5 (2 × O*C*(CH$_3$)$_3$), 121.4, 122.5, 127.5, 128.0, 130.6, 131.5, 133.7, 133.9, 134.0, 134.3, 135.6, 149.5, 157.7 (*C*-alkene, *C*-arom), 168.6, 170.7, 176.8, 182.7 (4C=O); MS at *m*/*z* (rel. %): 430 (M^+−101, 8), 374 (100), 105 (20), 77 (20), 57 (30), 41 (25). Anal. Calcd for $C_{31}H_{33}NO_7$: C, 70.04; H, 6.26; N, 2.63; Found: C, 70.11; H, 6.28; N, 2.71.

Reference

Shaabani, A., Ghadari, R., Sarvary, A., and Rezayan, A. H. (2009). Synthesis of highly functionalized bis(4*H*-chromene) and 4*H*-benzo[*g*]chromene derivatives *via* an isocyanide-based pseudo-five-component reaction. *J. Org. Chem.*, **69**, 4372–4374.

Chapter 4

Carbon – Sulfur Bond Forming Reactions at Room Temperature

C-S bond_1

Type of reaction: C-S bond formation
Reaction conditions: Acetonitrile, room temperature
Synthetic strategy: Esterification of alcohols
Catalyst: *In situ* generated $Ph_3P(OAc)_2$ from Ph_3P, Br_2 and NH_4OAc
Keywords: Thiols, triphenylphosphine, bromine, ammonium acetate, $Ph_3P(OAc)_2$ (*in situ* generated), acetonitrile, room temperature, acetylation

General reaction scheme

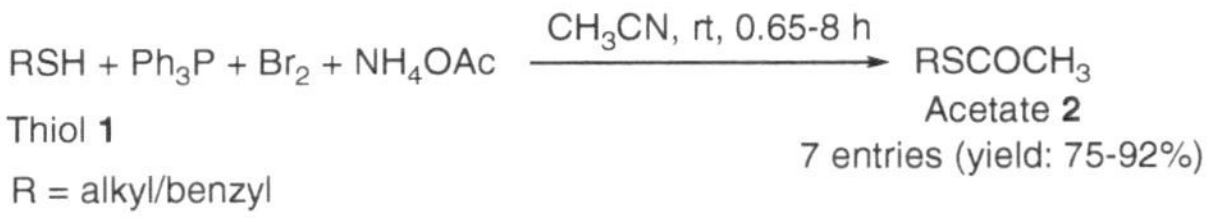

Plausible mechanism

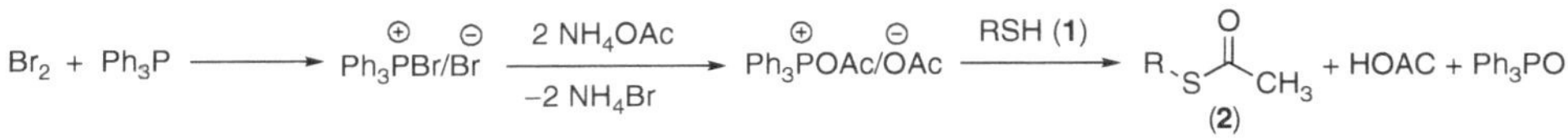

Representative entries

$PhCH_2SCOCH_3$

2a

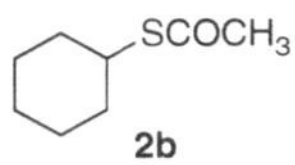

2b

S-Benzyl ethanethioate (**2a**): Reaction time: 0.8 h; yield: 88%

S-Cyclohexyl ethanethioate (**2b**): Reaction time: 4 h; yield: 80%

Experimental procedure

To a solution of $Ph_3P(OAc)_2$ was added thiol (**1**; 1 mmol), and the progress of the reaction was monitored by TLC. After completion of the reaction (0.3-7 h) the reaction mixture was filtered to remove the precipitated NH_4Br followed by evaporation of the solvent. Column chromatography of the crude mixture on silica gel using *n*-hexane-ethyl acetate mixture as eluent to afford pure acetate derivative **2**. Each of the products was characterized on the basis of spectral properties.

Characterization data of one representative compound

3-Phenylpropyl acetate (**2a**): ^{1}H NMR (250 MHz, $CDCl_3$): δ 2.3 (s, 3H), 4.1 (s, 2H), 7.26 (m, 5H); ^{13}C NMR (62.5 MHz, $CDCl_3$): δ 30.3, 32.1, 127.2, 128.4, 128.6, 137.5, 193.1.

Reference

Iranpoor, N., Firouzabadi, H., and Davan, E. E. (2013). In situ generated $Ph_3P(OAc)_2$ as a novel reagent for the efficient acetylation of alcohols and thiols at room temperature. *Tetrahedron Lett.*, **54**, 1813–1816.

Room Temperature Organic Synthesis. http://dx.doi.org/10.1016/B978-0-12-801025-9.00004-8

C-S bond_2

Type of reaction: C-S bond formation
Reaction conditions: Solvent-free (neat condition), room temperature
Synthetic strategy: One-pot synthesis for *S*-alkyl thiocarbamates
Catalyst: Catalyst-free
Keywords: Secondary amines, carbon disulphide, alkyl halides, solvent-free, catalyst-free, room temperature, dithiocarbamates

General reaction scheme

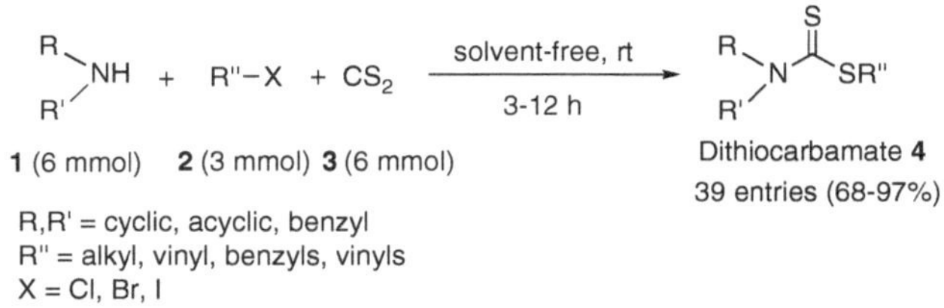

Representative entries

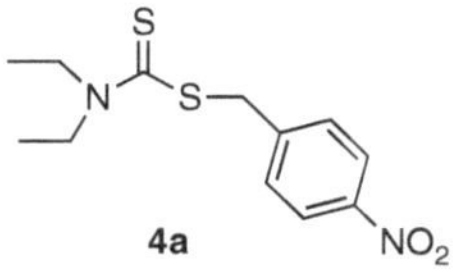

4a

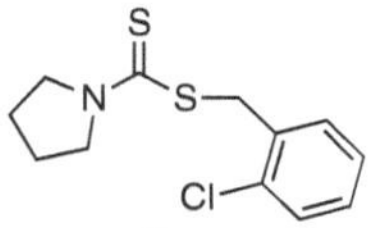

4b

4-Nitrobenzyl diethylcarbamodithioate (**4a**): Yield 76%

2-Chlorobenzyl pyrrolidine-1-carbodithioate (**4b**): Yield 92%

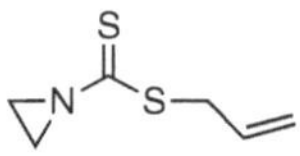

4c

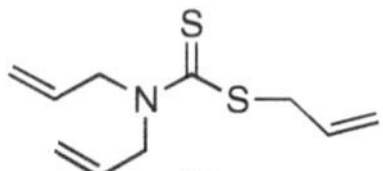

4d

Allyl aziridine-1-carbodithioate (**4c**): Yield 68%

Allyl diallylcarbamodithioate (**4d**): Yield 70%

Experimental procedure

To a ice cold mixture of organic halide (**2**; 3 mmol) and carbon disulfide (**3**; 6 mmol) was added secondary amine (**1**: 6 mmol) and stirred at 0 °C for 30 min, then warmed to room temperate and stirring was continued until the reaction was complete (monitored by TLC). The reaction was quenched by addition of water and was extracted with ethyl acetate, dried over anhydrous sodium sulfate, and evaporated to give almost pure product. Further purification in some cases was done by silica gel column chromatography eluted with petroleum ether and ethyl acetate or recrystallized to afford the pure dithiocarbamate **4**. All the products were characterized by spectral studies and data were compared with literature values.

Reference

Azizi, N., Aryanasab, F., and Saidi, M. R. (2006). Straightforward and highly efficient catalyst-free one-pot synthesis of dithiocarbamates under solvent-free conditions. *Org. Lett.*, **8**, 5275–5277.

C-S bond_3

Type of reaction: C-S and C-N bond formation
Reaction conditions: Solvent-free (neat condition), room temperature
Synthetic strategy: One-pot multicomponent reaction
Catalyst: Catalyst-free
Keywords: Michael acceptors, amines, carbon disulfide (CS_2), neat condition, room temperature, one-pot multicomponent reaction, dithiocarbamates

General reaction scheme

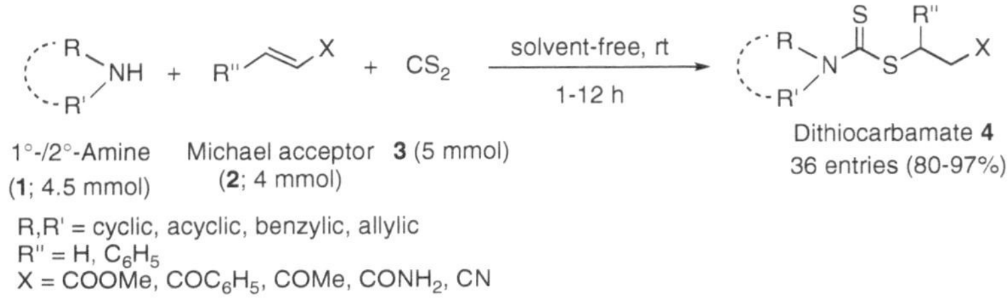

Representative entries

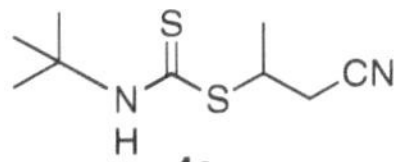

4a

1-Cyanopropan-2-yl *tert*-butylcarbamodithioate (**4a**): Yield 84%

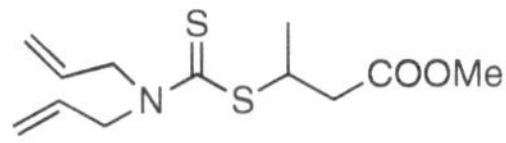

4b

Methyl 3-((diallylcarbamothioyl)thio) butanoate (**4b**): Yield 78%

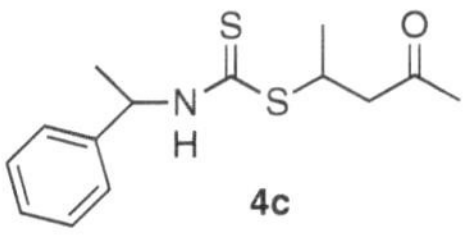

4c

4-Oxopentan-2-yl (1-phenylethyl) carbamodithioate (**4c**): Yield 94%

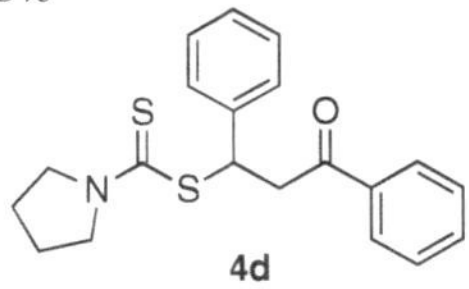

4d

3-Oxo-1,3-diphenylpropyl pyrrolidine-1-carbodithioate (**4d**): Yield 92%

Experimental procedure

Amine (**1**; 4.5 mmol) was added slowly to the mixture of carbon disulfide (CS_2, 5 mmol) and the Michael acceptor (**2**; 4 mmol) into a test tube placed in an ice bath. The reaction mixture was stirred at 0 °C for 30 min, and then warmed to room temperature with continued stirring for stirred for another 1–12 h to complete the reaction. On competition of the reaction, excess of CS_2 and amine was removed under reduced pressure to afford dithiocarbamate derivative **4** in the almost pure form; in some cases, further purification was carried out by recrystallization or short-column chromatography on silica gel (EtOAc-petrol ether). All compounds were characterized on the basis of NMR spectral data.

Characteristic data for two compounds

4-Oxopentan-2-yl (1-phenylethyl)carbamodithioate (**4c**): ^{1}H NMR (500 MHz, $CDCl_3$-CCl_4,): δ 1.58 (d, J=6.8 Hz, 3H), 2.14 (s, 3H), 2.89 (t, J=6.1 Hz, 2H), 3.40 (t, J=6.1 Hz, 2H), 5.81

(q, $J=6.8$ Hz, 1H,), 7.26-7.35 (m, 5H), 8.18 (br s, NH, 1H); ^{13}C NMR (125 MHz $CDCl_3$-CCl_4,): δ 21.3, 29.0, 30.4, 43.3, 56.4, 127.0, 128.2, 129.3, 141.7, 196.8, 207.6.

3-Oxo-1,3-diphenylpropyl pyrrolidine-1-carbodithioate (**4d**): ^{1}H NMR (500 MHz, $CDCl_3$): δ 1.93-2.05 (m, 4H), 3.58-4.02 (m, 6H), 5.83 (dd, $J=4.1$, 9.8 Hz, 1H), 7.27-7.56 (m, 8H), 8.00 (d, $J=7.3$ Hz, 2H); ^{13}C NMR (125 MHz, $CDCl_3$): δ 24.6, 26.5, 51.0, 54.6, 55.3, 128.0, 128.6, 128.8, 129.0, 129.4, 133.6, 137.0, 140.0, 191.5, 197.4.

Reference

Azizi, N., Ebrahimi, F., Aakbari, E., Aryanasab, F., and Saidi, M. R. (2007). Waste-free and environment-friendly uncatalyzed synthesis of dithiocarbamates under solvent-free conditions. *Synlett*, 2797–2800.

C-S bond_4

Type of reaction: C-S bond formation
Reaction conditions: Triethylamine/dichloromethane, room temperature
Synthetic strategy: *S*-Acylation using *N*-acylbenzotriazoles
Catalyst: No other additive
Keywords: *N*-Acylbenzotriazoles, thiols, triethylamine, dichloromethane, room temperature, *S*-acylation, thiol esters

General reaction scheme

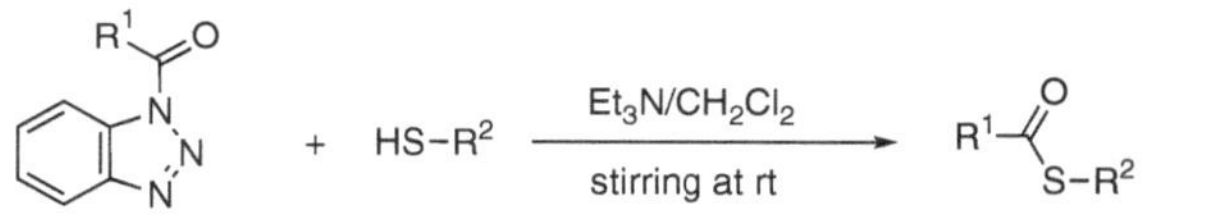

1 (5 mmol) **2** (5.1 mmol) Thiol ester **3** 27 entries (85-99%)

R^1 = C_6H_5, 2-$OMeC_6H_4$, 4-$NEt_2C_6H_4$, 2-pyridyl, 2-indolyl, 2-furyl
R^2 = C_6H_5, $CH_2C_6H_5$, CH_2COOEt, CH_2COOH

R^2	Amt. of Et_3N	Time
Ph	5 mmol	1 h
$PhCH_2$	5 mmol	12 h
$EtOOCCH_2$	10-20 mmol	1.5 h
$HOOCCH_2$	10 mmol	1 h

Representative entries

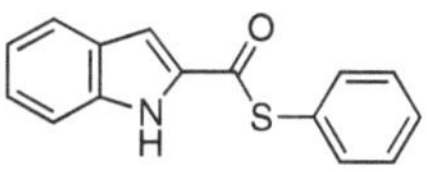

3a

S-Phenyl 1*H*-indole-2-carbothioate (**3a**): White crystals, mp. 197–198 °C; yield 93%

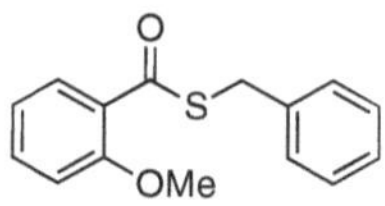

3b

S-Benzyl 2-methoxybenzothioate (**3b**): Colorless oil; yield 99%

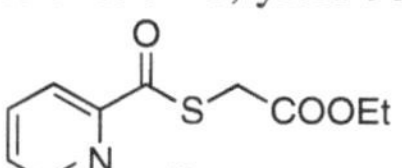

3c

Ethyl 2-(picolinoylthio)acetate (**3c**): White crystals, mp. 31–35 °C; yield 90%

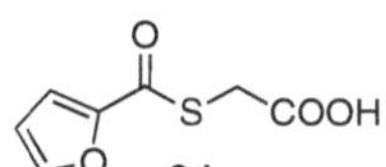

3d

2-((Furan-2-carbonyl)thio)acetic acid (**3d**): White microcrystals, mp. 112–113 °C; yield 88%

Experimental procedure

To a solution of *N*-acylbenzotriazole (**1**; 5.0 mmol) in dichloromethane (30 mL) in the presence of triethylamine (5.0-20 mmol as the case may be), was added thiophenol (**2**; 5.1 mmol), and the mixture was stirred at room temperature (25 °C) for stipulated time (1–12 h). On completion of the reaction, 10 mL of dichloromethane was added, shaken well, and separated organic layer was washed with aq. 10% Na_2CO_3, aq sat. NH_4Cl, and dried over $MgSO_4$. After evaporation of the solvents, the thiol ester **3** was obtained in almost pure form. In some cases, the crude product was purified by means of recrystallization from dichloromethane-hexanes, or by using column chromatographic technique (eluent: EtOAc-hexanes, 1:2; silica gel). All the products were characterized by spectral and elemental analyses.

Characteristic data for the four types of compounds

S-Phenyl 1*H*-indole-2-carbothioate (**3a**): White crystals, mp. 197–198 °C; ^{1}H NMR (300 MHz, DMSO-d_6): δ 7.14 (t, J=7.4 Hz, 1 H), 7.31-7.36 (m, 1H), 7.47-7.60 (m, 7 H), 7.75 (d, J=8.1 Hz, 1 H), 12.11 (br s, 1 H); ^{13}C NMR (75 MHz, DMSO-d_6): δ 108.4, 112.9, 120.7, 122.6, 125.8, 126.4, 126.7, 129.4, 129.7, 133.0, 135.0, 137.9, 180.8. Anal. Calcd for $C_{15}H_{11}NOS$: C, 71.12; H, 4.38; N, 5.53; Found: C, 70.83; H, 4.26; N, 5.34.

S-Benzyl 2-methoxybenzothioate (**3b**): Colorless oil; ^{1}H NMR (300 MHz, $CDCl_3$): δ 3.91 (s, 3 H), 4.28 (s, 2 H), 6.96–7.02 (m, 2 H), 7.23-7.40 (m, 5 H), 7.43-7.50 (m, 1 H), 7.81 (dd, J=8.0, 1.5 Hz, 1 H); ^{13}C NMR (75 MHz, $CDCl_3$): δ 33.9, 55.7, 111.9, 120.4, 127.1, 128.5, 129.0, 129.8, 133.7, 137.6, 157.9, 190.3.

Ethyl 2-(picolinoylthio)acetate (**3c**): White crystals, mp. 31–35 °C; ^{1}H NMR (300 MHz, $CDCl_3$): δ 1.30 (t, J=7.1 Hz, 3 H), 3.84 (s, 2 H), 4.22 (q, J=7.1 Hz, 2 H), 7.55 (t, J=6.3 Hz, 1 H), 7.87 (t, J=7.7 Hz, 1 H), 7.96 (d, J=7.7 Hz, 1 H), 8.70 (d, J=4.1 Hz, 1 H); ^{13}C NMR (75 MHz, $CDCl_3$): δ 14.2, 31.4, 61.8, 120.6, 128.3, 137.4, 149.3, 151.2, 169.0, 192.4. Anal. Calcd for $C_{10}H_{11}NO_3S$: C, 53.32; H, 4.92; N, 6.22; Found: C, 52.93; H, 5.15; N, 5.91.

2-((Furan-2-carbonyl)thio)acetic acid (**3d**): White microcrystals, mp. 112–113 °C; ^{1}H NMR (300 MHz, $CDCl_3$): δ 3.90 (s, 2 H), 6.58 (apparent s, 1 H), 7.22-7.28 (m, 1 H), 7.62 (s, 1 H), 10.43 (br s, 1 H); ^{13}C NMR (75 MHz, $CDCl_3$): δ 30.5, 112.8, 115.3, 116.8, 126.6, 147.0, 174.3. Anal. Calcd for $C_7H_6O_4S$: C, 45.16; H, 3.25; Found: C, 44.77; H, 3.03.

Reference

Katritzky, A. R., Shestopalov, A. A., and Suzuki, K. (2004). A new convenient preparation of thiol esters utilizing *N*-acylbenzotriazoles. *Synthesis*, 1806–1813.

C-S bond_5

Type of reaction: C-S bond formation
Reaction conditions: Acetonitrile, room temperature
Synthetic strategy: One-step sulfination
Catalyst: Acetic acid/KI-$NaIO_4$
Keywords: Alkenes, aryl sulfinates, acetic acid/KI-$NaIO_4$, acetonitrile, room temperature, one-step sulfination, vinyl sulfones

General reaction scheme

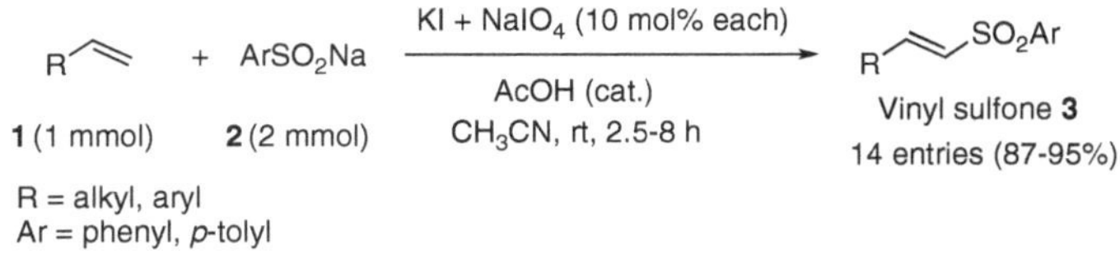

Proposed mechanism

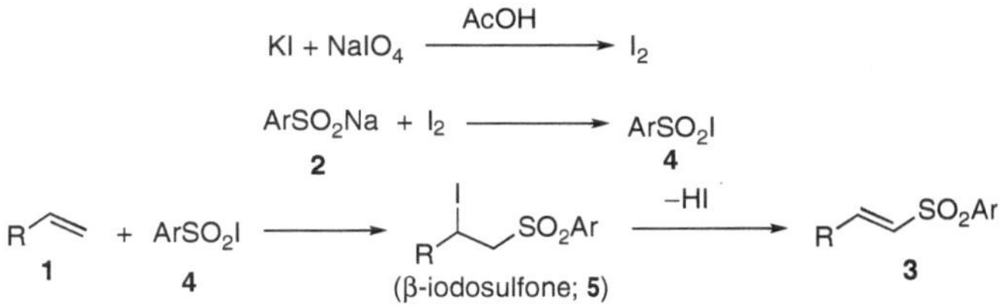

Representative entries

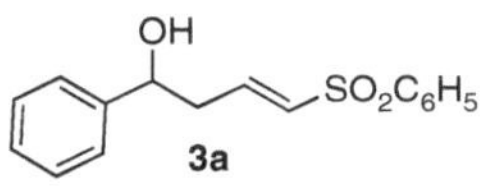

(*E*)-1-Phenyl-4-(phenylsulfonyl)but-3-en-1-ol (**3a**): White solid, mp. 101–103 °C; reaction time: 7 h; yield 91%

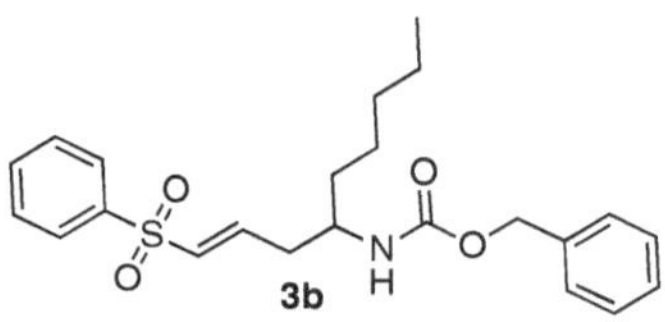

(*E*)-Benzyl (1-(phenylsulfonyl)non-1-en-4-yl) carbamate (**3b**): White solid, mp. 90–92 °C; reaction time: 6 h; yield 90%

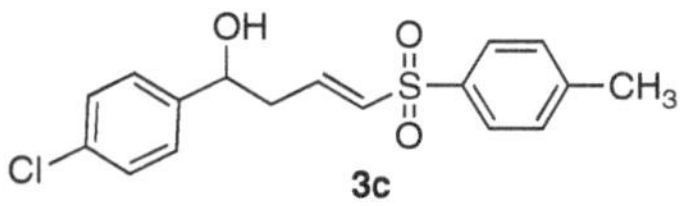

(*E*)-1-(4-Chlorophenyl)-4-tosylbut-3-en-1-ol (**3c**): White solid, mp. 117–119 °C; reaction time: 7 h; yield 89%

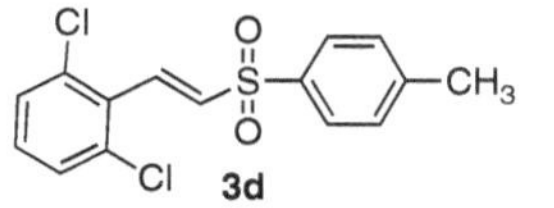

(*E*)-1,3-Dichloro-2-(2-tosylvinyl)benzene (**3d**): White solid, mp. 101–103 °C; reaction time: 3 h; yield 91%

Experimental procedure

To a stirred solution of alkene (**1**; 1 mmol) and sodium aryl sulfinate (**2**; 2 mmol) in acetonitrile (10 mL) was added with KI (10 mol%), $NaIO_4$ (10 mol%) and catalytic amount of acetic acid. The resulting reaction mixture was stirred at room temperature for the appropriate time (2.5-7 h as monitored by TLC). After completion of the reaction, the mixture was quenched by the addition of a saturated aqueous solution of sodium thiosulfate (5 mL) and extracted with ethyl acetate (2×10 mL). The organic layer was separated and concentrated, and the residue was subjected to column chromatography (silica gel; hexane–EtOAc) to obtain the pure product of vinyl sulfone **3**. All compounds were characterized based on detailed spectral studies and elemental analyses.

Characteristic data for two compounds

(*E*)-1-Phenyl-4-(phenylsulfonyl)but-3-en-1-ol (**3a**): White solid, mp. 101–103 °C; IR (KBr): 3522, 1626, 1449, 1285 cm^{-1}; ^1H NMR (200 MHz, $CDCl_3$): δ 7.82 (d, J=8.0 Hz, 2H), 7.61-7.46 (m, 3H), 7.39-7.20 (m, 5 H), 6.96 (m, 1H), 6.31 (d, J=14.0 Hz, 1H), 4.82

(m, 1H), 3.22 (m, 1H), 2.61 (m, 1H); ^{13}C NMR (50 MHz, $CDCl_3$): δ 143.0, 142.9, 140.4, 133.2, 132.5, 129.2, 128.6, 128.0, 127.7, 125.3, 72.8, 40.9; MS (ESI): *m/z* 311 $[M+Na]^+$. Anal. Calcd for $C_{16}H_{16}O_3S$: C, 66.67; H, 5.56. Found: C, 66.78; H, 5.61.

(*E*)-1-(4-Chlorophenyl)-4-tosylbut-3-en-1-ol (**3a**): White solid, mp. 117–119 °C; IR (KBr): 3502, 1638, 1595, 1488, 1310 cm^{-1}; ^{1}H NMR (200 MHz, $CDCl_3$): δ 7.69 (d, $J=8.0$ Hz, 2 H), 7.38-7.19 (m, 6 H), 6.90 (m, 1 H), 6.29 (d, $J=14$ Hz, 1 H), 4.83 (t, $J=7.0$ Hz, 1 H), 2.68-2.53 (m, 2 H), 2.48 (s, 3 H); ^{13}C NMR (50 MHz, $CDCl_3$): δ 144.3, 141.7, 141.4, 137.2, 133.2, 129.8, 127.5, 127.2, 127.0, 71.7, 40.7, 21.6; MS (ESI): *m/z* 354 $[M+NH_4]^+$, 359 $[M+Na]^+$. Anal. Calcd for $C_{17}H_{17}ClO_3S$: C, 60.62; H, 5.05. Found: C, 60.73; H, 5.01.

Reference

Das, B., Lingaiah, M., Damodar, K., and Bhunia, N. (2011). An efficient synthesis of vinyl sulfones from alkenes and aryl sulfinates. *Synthesis*, 2941–2944.

C-S bond_6

Type of reaction: C-S bond formation
Reaction conditions: Acetonitrile, room temperature
Synthetic strategy: Alcohols and thiols to thiocyanate
Catalyst: PPh_3/DEAD
Keywords: Alcohols and thiols, ammonium thiocyanate, triphenylphosphine (PPh_3), diethylazodicarboxylate (DEAD), acetonitrile, room temperature, thiocyanates

General reaction scheme

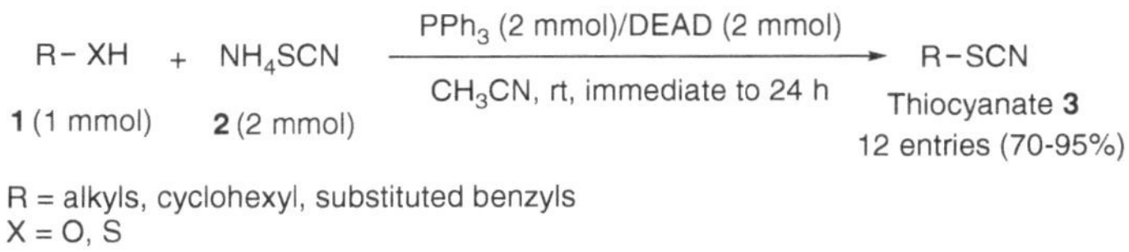

Representative entries

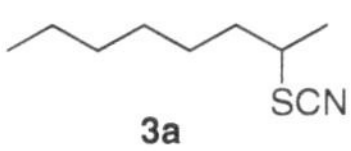

3a

2-Thiocyanatooctane (**3a**): Reaction time: 2 h; yield 70%

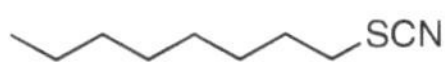

3b

1-Thiocyanatooctane (**3b**): Reaction time: immediate; yield 94%

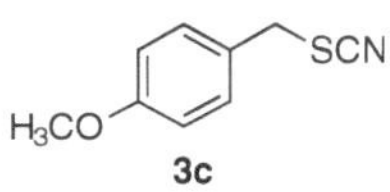

3c

1-Methoxy-4-(thiocyanatomethyl)benzene (**3c**): Reaction time: 24 h; yield 82%

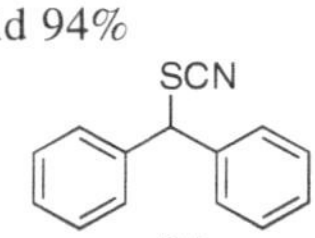

3d

(Thiocyanatomethylene)dibenzene (**3d**): Reaction time: 45 min; yield 75%

Experimental procedure

To a mixture of PPh_3 (2 mmol) and DEAD (2 mmol) in anhydrous acetonitrile (5–7 mL) was added ammonium thiocyanate (**2**; 2 mmol). Alcohol/thiol (**1**; 1 mmol) was then added to this mixture, and the whole mixture was stirred at room temperature for stipulated time. On completion of the reaction, the solvent was evaporated under reduced pressure. Column chromatography of the crude mixture on silica-gel (*n*-hexane) afforded pure thiocyanate **3** in moderate to good yield. All the products were identified by comparing physical and spectral data with the respective literature values.

Reference

Iranpoor, N., Firouzabadi, H., Akhlaghinia, B., and Azadi, R. (2004). Conversion of alcohols, thiols, carboxylic acids, trimethylsilyl ethers, and carboxylates to thiocyanates with triphenylphosphine/diethylazodicarboxylate/NH_4SCN. *Synthesis*, 92–96.

C-S bond_7

Type of reaction: C-S bond formation
Reaction conditions: Acetonitrile, room temperature
Synthetic strategy: Thiocyanation of aromatic compounds
Catalyst: Bromodimethylsulfonium bromide (BDMS)
Keywords: Aromatic substrates, ammonium thiocyanate, bromodimethylsulfonium bromide (BDMS), acetonitrile, room temperature, thiocyanation, aryl thiocyanate

General reaction scheme

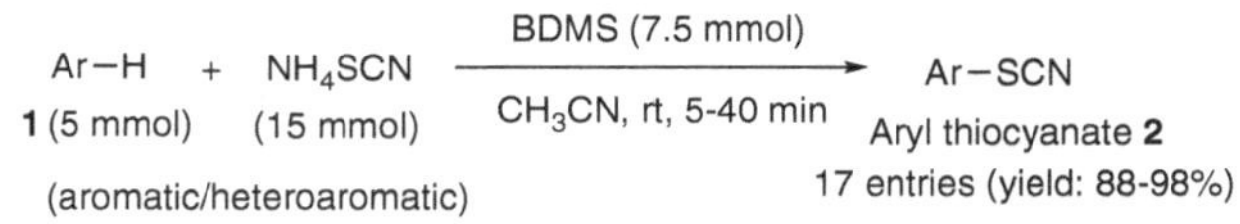

Representative entries

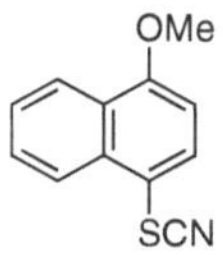

2a

1-methoxy-4-thiocyanatonaphthalene (**2a**): Solid, mp 104–106 °C; reaction time: 30 min; yield 95%

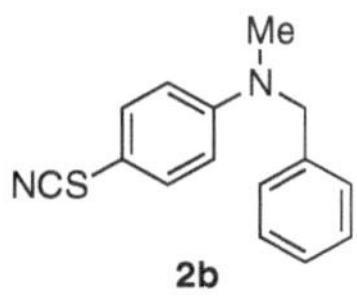

2b

N-Benzyl-*N*-methyl-4-thiocyanatoaniline (**2b**): Solid, mp 68–70 °C; reaction time: 15 min; yield 95%

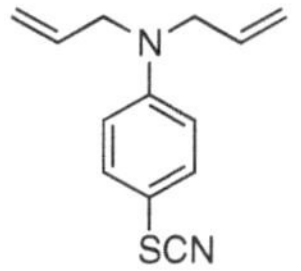

2c

N,N-Diallyl-4-thiocyanatoaniline (**2c**): Oil; reaction time: 15 min; yield 92%

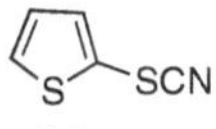

2d

2-Thiocyanatothiophene (**2d**): Oil; reaction time: 10 min; yield 92%

Experimental procedure

To a suspension of BDMS (1.66 g, 7.5 mmol) and ammonium thiocyanate (1.14 g, 15 mmol) in 50 mL of anhydrous acetonitrile was added aryl substrate (1; 5 mmol) in one portion, and the reaction was stirred at room temperature for stipulated time. Upon completion of reaction (monitored by TLC), it was quenched with 50 ml of saturated solution of sodium bicarbonate and the solid residue was removed by filtration and the residue was washed with ethyl acetate. The organic layer was separated, washed with water (50 mL), brine (50 mL) and dried over anhydrous sodium sulfate and finally concentrated under reduced pressure to give the crude thiocyanato followed by purification using column chromatography (silica gel, mesh size 60–120, eluent EtOAc-hexane, 10:90) to afford **2**. All the products were characterized by spectral and analytical studies.

Characterization data of two representative compounds

N-Benzyl-*N*-methyl-4-thiocyanatoaniline (**2b**): Solid, mp 68–70 °C; IR (KBr): 3014, 2144 (SCN), 1599, 1460, 1023, 824 cm^{-1}; ^1H NMR (60 MHz, $CDCl_3$): δ 3.07 (s, 3 H), 4.57 (s, 2 H), 6.69 (d, $J=9.0$ Hz, 2 H), 7.05 (s, 5 H), 7.39 (d, $J=9.0$ Hz, 2 H). Anal. Calcd for $C_{15}H_{14}N_2$: C, 70.83; H, 5.55; N, 11.01; Found: C, 70.85; H, 5.54; N, 10.98.

N,N-Diallyl-4-thiocyanatoaniline (**2c**): Oil; IR (KBr): 3082, 2916, 2148 (SCN), 1641, 1591, 1504, 1238, 811 cm^{-1}; ^1H NMR (60 MHz, $CDCl_3$): δ 3.78–3.97 (m, 4 H), 4.96-5.26 (m, 4 H), 5.55-6.06 (m, 2 H), 6.63 (d, $J=9.0$ Hz, 2 H), 7.35 (d, $J=9.0$ Hz, 2 H). Anal. Calcd for $C_{13}H_{14}N_2S$: C, 67.79; H, 6.13; N, 12.16; Found: C, 67.82; H, 6.15; N, 12.18.

Reference

Bhalerao, D. S., and Akamanchi, K. G. (2007). Efficient and novel method for thiocyanation of aromatic and heteroaromatic compounds using bromodimethylsulfonium bromide and ammonium thiocyanate. *Synlett*, 2952–2956.

C-S bond_8

Types of reaction: C-S bond formation
Reaction conditions: Ionic liquid (IL), room temperature
Synthetic strategy: Sulfonylation of arylboronic acid
Catalyst: Cupric acetate [Cu(OAc)$_2$]
Keywords: Arylboronic acids, sulfinic acid salts, [bim]OTf (ionic liquid), Cu(OAc)$_2$, room temperature, sulfonylation, aryl sulfones

General reaction scheme

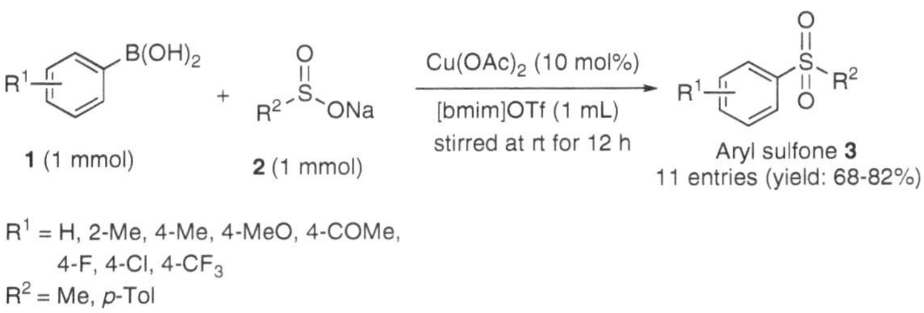

Representative entries

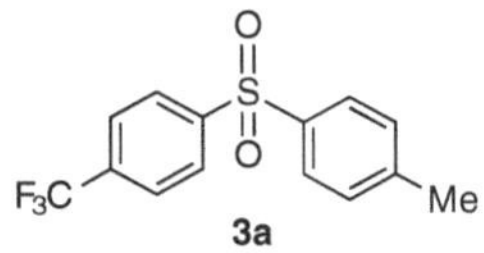

3a

1-Methyl-4-((4-(trifluoromethyl)phenyl)sulfonyl)benzene (**3a**): Yield: 80%

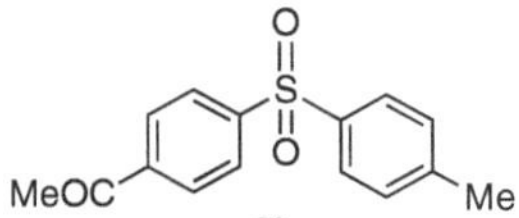

3b

(1-(4-Tosylphenyl)ethanone (**3b**): Yield: 75%

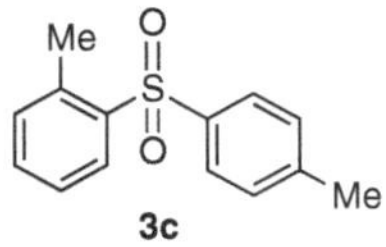

3c

1-Methyl-2-tosylbenzene (**3c**): Yield: 70%

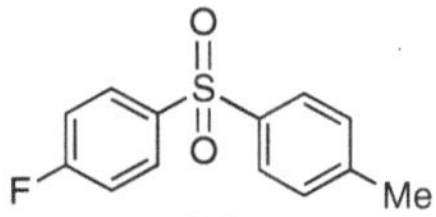

3d

1-Fluoro-4-tosylbenzene (**3d**): Yield: 72%

Experimental procedure

To a stirred solution of $Cu(OAc)_2$ (10 mol%) in ionic liquid ([bmim]OTf; 1 mL) and arylboronic acid (**1**; 1 mmol), sodium sulfinate salt (**2**; 1 mmol) was added and stirring was continued at room temperature for 12 h. After completion of the reaction, as monitored by TLC, the crude product was extracted with Et_2O (3×10 mL). The combined Et_2O extracts were concentrated in *vacuo* and the resulting product was purified by column chromatography on silica gel using EtOAc-*n*-hexane (2:8) as eluent to afford the pure product of aryl sulfone **3** (68-82% yield). The ionic liquid containing $Cu(OAc)_2$ was dried under vacuum and preserved for the next run. All the compounds were characterized from comparison of their physical and spectral properties with those reported for authentic samples in literature.

Reference

Lakshmi Kantam, M., Neelima, B., Sreedhar, B., and Chakravarti, R. (2008). Copper-catalyzed sulfonylation of arylboronic acids in ionic liquids. *Synlett*, 1455–1458.

C-S bond_9

Types of reaction: C-S bond formation
Reaction conditions: Dichloromethane, room temperature
Synthetic strategy: One-step β-sulfonylation of α,β-enones
Catalyst: $FeCl_3$/TMSCl
Keywords: α,β-Unsaturated ketones, sodium *p*-toluenesulfinate, iron(III) chloride, trimethysilyl chloride, dichloromethane, room temperature, conjugate addition, β-sulfonyl ketones

General reaction scheme

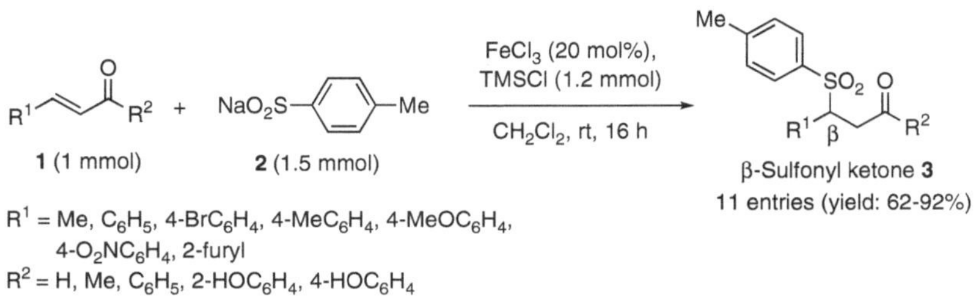

Representative entries

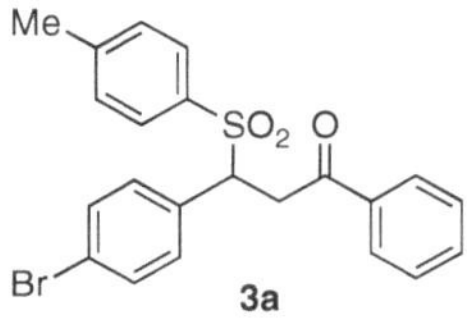

3a

3-(4-Bromophenyl)-1-phenyl-3-tosylpropan-1-one (**3a**): Yield: 78%

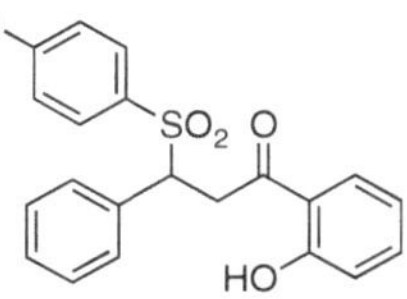

3b

1-(2-Hydroxyphenyl)-3-phenyl-3-tosylpropan-1-one (**3b**): Yield: 75%

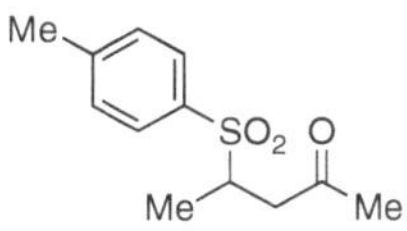

3c

4-Tosylpentan-2-one (**3c**): Yield: 72%

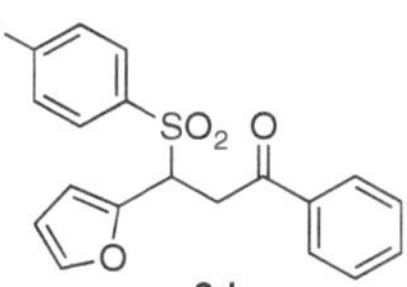

3d

3-(Furan-2-yl)-1-phenyl-3-tosylpropan-1-one (**3d**): Yield: 62%

Experimental procedure

To a stirred solution of α,β-enone (**1**; 1 mmol), TMSCl (1.2 mmol), and sodium sulfinate (**2**; 1.5 mmol) in dichloromethane (3 mL), $FeCl_3$ (20 mol%) was added, and stirring was continued at room temperature for 16 h to complete the reaction (monitored by TLC). After completion of the reaction, the reaction mixture was quenched with saturated solution of sodium bicarbonate. The aqueous layer was extracted with EtOAc (3 × 20 mL), and the combined organics were dried over anhydrous Na_2SO_4, concentrated in *vacuo*, and purified by column chromatography on SiO_2 to furnish the pure product of β-sulfonyl ketone **3**. All products were characterized by IR, ^{1}H NMR, ^{13}C NMR, and MS spectral analyses.

Characterization data of two representative entries

3-(4-Bromophenyl)-1-phenyl-3-tosylpropan-1-one (**3a**): IR (KBr): 2884, 1671, 1341, 1332, 1152, 1064, 810 cm^{-1}; ^{1}H NMR (300 MHz, $CDCl_3$): δ 2.40 (s, 3H), 3.78 (dd, J_1 = 9.3 Hz, J_2 = 18.0 Hz, 1H), 4.05 (dd, J_1 = 3.9 Hz, J_2 = 18.0 Hz, 1H), 4.77 (dd, J_1 = 3.9 Hz, J_2 = 9.3 Hz, 1H), 7.10-7.24 (m, 7H), 7.37 (d, J = 8.5 Hz, 2H), 7.57 (d, J = 8.5 Hz, 2H), 7.82 (d, J = 8.5 Hz, 2H); ^{13}C NMR (75 MHz, $CDCl_3$): δ 21.5, 37.1, 65.9, 121.3, 127.4, 128.3, 128.7, 128.8, 129.2, 130.0, 130.6, 131.2, 133.2, 133.5, 144.7, 198.3; ESI-MS: *m/z* 455 $[M+1]^+$, 467 $[M+Na]^+$.

4-Tosylpentan-2-one (**3c**): IR (KBr): 2924, 1710, 1596, 1455, 1282, 1134, 1080, 812, 658 cm^{-1}; ^{1}H NMR (300 MHz, $CDCl_3$): δ 1.20 (s, 3H), 2.47 (s, 3H), 2.82 (dd, J_1 = 9.2 Hz, J_2 = 17.6 Hz, 1H,), 3.21 (dd, J_1 = 3.8 Hz, J_2 = 17.6 Hz, 1H), 3.61-3.66 (m, 1H), 7.34 (d, J = 8.1 Hz, 2H), 7.70 (d, J = 8.1 Hz, 2H); ^{13}C NMR (75 MHz, $CDCl_3$): δ 18.9, 21.4, 29.0, 45.8, 64.1, 129.6, 130.1, 131.2, 145.1, 206.8; LC-MS: *m/z* 263 $[M+Na]^+$.

Reference

Seedhar, B., Reddy, M. A., and Reddy, P. S. (2008). $FeCl_3$/TMSCl: An efficient catalytic system for the conjugate addition of sodium *p*-toluenesulfinate to α,β-enones. *Synlett*, 1949–1952.

C-S bond_10

Type of reaction: C-S bond formation
Reaction conditions: Acetonitrile, room temperature
Synthetic strategy: One-step thiocyanation
Catalyst: Anhydrous Fe(III)Cl_3
Keywords: Arylalkenes (styrenes), potassium thiocyanate, anhydrous ferric chloride, acetonitrile, room temperature, thiocyanation, oxidative radical addition, dithiocyanates

General reaction scheme

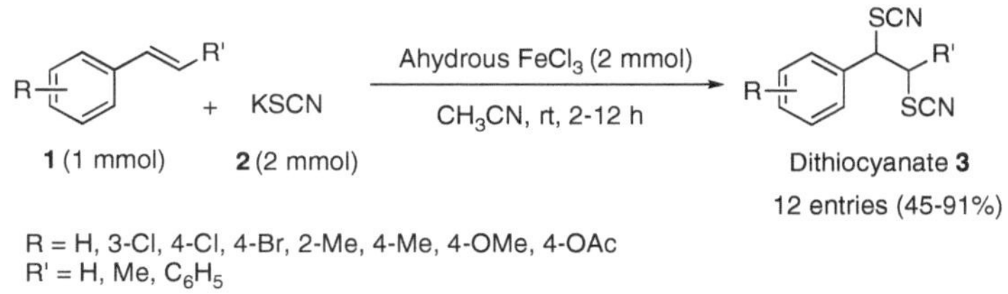

Representative entries

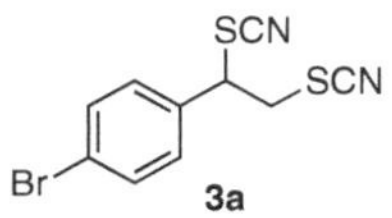

3a

1-Bromo-4-(1,2-dithiocyanatoethyl)benzene (**3a**): White crystalline solid, mp. 87 °C; reaction time: 3.5 h; yield 82%

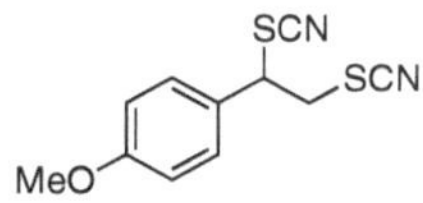

3b

1-(1,2-Dithiocyanatoethyl)-4-methoxybenzene (**3b**): Oil; reaction time: 4 h; yield 85%

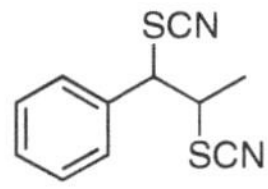

3c

(1,2-Dithiocyanatopropyl)benzene (**3c**): White crystalline solid, mp. 83 °C; reaction time: 5 h; yield 85%

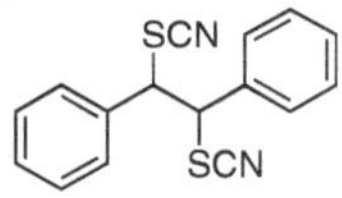

3d

1,2-Diphenyl-1,2-dithiocyanatoethane (**3d**): Oil; reaction time: 12 h; yield 45%

Experimental procedure

A mixture of arylalkene (**1**;1 mmol), potassium thiocyanate (**2**; KSCN; 2 mmol) and anhydrous $FeCl_3$ (2 mmol) in acetonitrile (10 mL) was stirred at room temperature until completion of the reaction (2–12 h; monitored by TLC). After completion of the reaction, the reaction mixture was diluted with distilled water and extracted with ethyl acetate (2 × 10 mL). The combined organic layers were dried over anhyd. sodium sulfate, concentrated in *vacuo* and purified by column chromatography on silica gel (Merck, 100–200 mesh, EtOAc-hexane, 1:9) to afford pure dithiocyanate **3**, characterized on the basis of spectral studies.

Characteristic data for two compounds

1-(1,2-Dithiocyanatoethyl)-4-methoxybenzene (**3b**): Oil; IR (KBr): 3015, 2065, 1610, 1513, 1252, 1219, 1178, 1032, 832, 786, 541 cm^{-1}; ^1H NMR (300 MHz, $CDCl_3$): δ 7.25

(d, $J=8.0$ Hz, 2 H), 6.90 (d, $J=8.0$ Hz, 2 H), 5.05 (dd, $J=6.1, 9.6$ Hz, 1 H), 3.82 (s, 3 H), 3.35 (dd, $J=6.1$, 13.7 Hz, 1 H), 3.30 (dd, $J=6.7$, 13.7 Hz, 1 H); FABMS: *m/z* (rel %) at 251 (M+1, 15), (10), 207 (5), 192 (75), 178 (50), 165 (35), 147 (45), 134 (100), 121 (80), 109 (85).

(1,2-Dithiocyanatopropyl)benzene (**3c**): White crystalline solid, mp. 83 °C; IR (KBr): 2921, 2151, 1450, 1219, 772, 693 cm^{-1}; 1H NMR (200 MHz, $CDCl_3$): δ 7.45-7.30 (m, 5 H), 4.30 (d, $J=9.7$ Hz, 1 H), 3.89-3.70 (m, 1 H), 1.89 (d, $J=7.0$ Hz, 3 H); FABMS: *m/z* (rel %) at 234 (M^+, 10), 176 (M^+ – SCN, 20), 154 (35), 95 (50), 69 (100).

Reference

Yadav, J. S., Reddy, B. V. S., and Gupta, M. K. (2004). Ferric(III) chloride-promoted efficient thiocyanation of arylalkenes: a facile synthesis of dithiocyanates. *Synthesis*, 1983–1986.

C-S bond_11

Type of reaction: C-S bond formation
Reaction conditions: Toluene, room temperature
Synthetic strategy: Metal-free direct C-S coupling from thiols and diaryliodonium salts
Catalyst: Cesium carbonate (Cs_2CO_3)
Keywords: Arylthiols, diaryliodonium salts, cesium carbonate, toluene, room temperature, direct C-S coupling from thiols and diaryliodonium salts, aryl sulfides

General reaction scheme

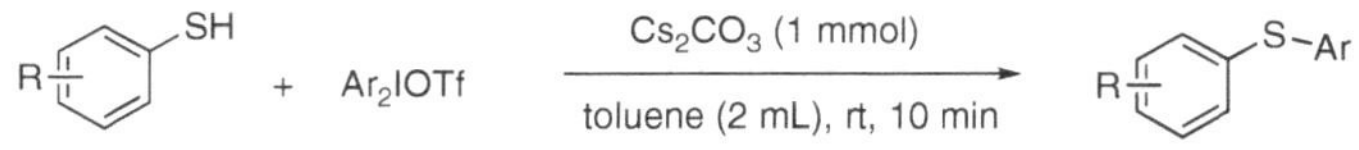

1 (1 mmol) **2** (1.1 mmol) Aryl sulfide **3**
12 entries (yield: 80-96%)

R = 4-Br, 4-Cl, 4-F, 3-OMe, 4-Me
Ar = C_6H_5, 4-BrC_6H_4, 4-ClC_6H_4, 4-FC_6H_4, 4-MeC_6H_4

Representative entries

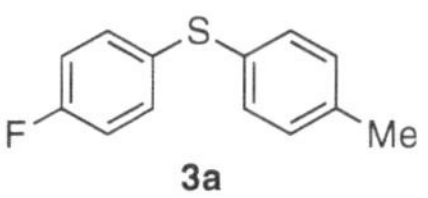

3a

(4-Fluorophenyl)(*p*-tolyl)sulfane (**2a**): Yield 84%

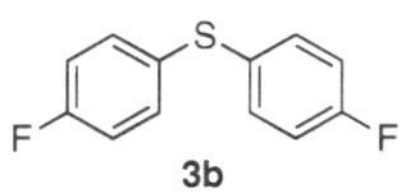

3b

Bis(4-fluorophenyl)sulfane (**2b**): Yield 85%

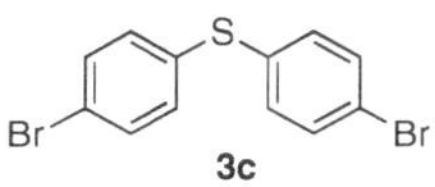

3c

Bis(4-bromophenyl)sulfane (**2c**): Yield 81%

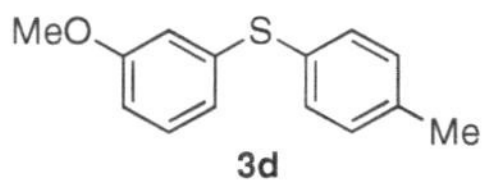

3d

(3-Methoxyphenyl)(*p*-tolyl)sulfane (**2d**): Yield 80%

Experimental procedure

A mixture of arylthiol (**1**, 1 mmol), diaryliodonium triflate (**2**, 1.1 mmol), cesium carbonate (326 mg, 1 mmol), in toluene (2.0 mL) was stirred at room temperature for 10 min. Upon completion, the reaction mixture was dissolved in water, then extracted with ethyl acetate (3 × 15 mL), dried over anhydrous $MgSO_4$ and evaporated to give the crude product which was then purified by column chromatography on silica gel (ethyl acetate / petroleum ether = 1:20) to afford pure aryl sulfide **3** with good yield. All the products were identified by means of spectral measurements.

Characterization data of two representative compounds

(4-Fluorophenyl)(*p*-tolyl)sulfane (**3a**). ^1H NMR (400 MHz, $CDCl_3$): δ 7.47-7.37 (m, 2H), 7.32-7.26 (m, 2H), 7.25-7.20 (m, 2H), 7.11 (d, J = 8.0 Hz, 2H), 2.32 (s, 3H).

(4-Methoxyphenyl)(*p*-tolyl)sulfane (**3d**). ^1H NMR (400 MHz, $CDCl_3$): δ 7.32 (d, J = 8.1 Hz, 2H), 7.20-7.09 (m, 3H), 6.85- 6.77 (m, 2H), 6.75-6.70 (m, 1H), 3.75 (d, J = 11.0 Hz, 3H).

Reference

Wang, D., Yu, X., Zhao, K., Li, L., and Ding, Y. (2014). Rapid synthesis of aryl sulfides through metal-free C-S coupling of thioalcohols with diaryliodonium salts. *Tetrahedron Lett.*, **55**, 5739–5741.

C-S bond_12

Types of reaction: C-S bond formation
Reaction conditions: *tert*-Butyl alcohol, room temperature
Synthetic strategy: One-pot multicomponent
Catalyst: Potassium *tert*-butoxide (^{t}BuOK)
Keywords: Acetphenones, aromatic aldehydes, thiols, ^{t}BuOK, ^{t}BuOH, room temperature, β-aryl-β-mercapto ketones, one-pot multicomponent synthesis, thio-Michael addition

General reaction scheme

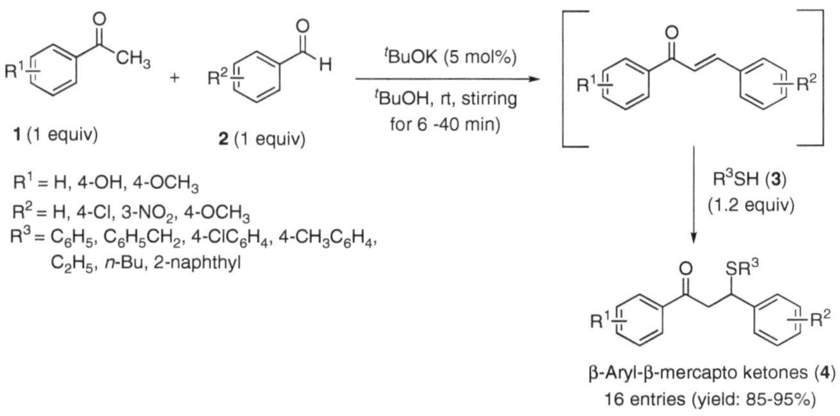

Representative entries

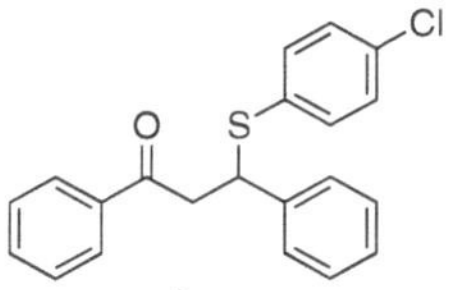

4a

3-((4-Chlorophenyl)thio)-1,3-diphenylpropan-1-one (**4a**): Reaction time: 12 min; yield: 92%

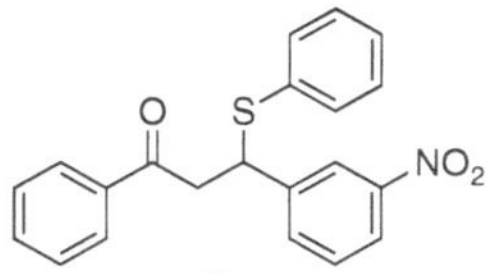

4b

3-(3-Nitrophenyl)-1-phenyl-3-(phenylthio)propan-1-one (**4b**): Solid, mp 112–115 °C; reaction time: 12 min; yield: 85%

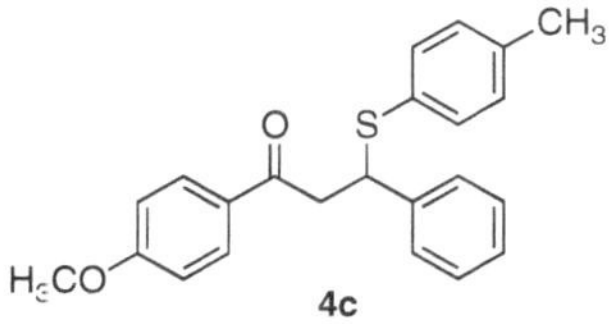

4c

1-(4-Methoxyphenyl)-3-phenyl-3-(*p*-tolylthio)propan-1-one (**4c**): Solid, mp 98–100 °C; reaction time: 6 min; yield: 95%

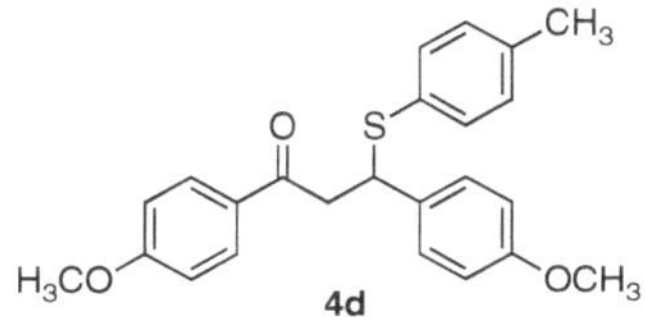

4d

1,3-bis(4-Methoxyphenyl)-3-(*p*-tolylthio)propan-1-one (**4d**): Reaction time: 11 min; yield: 89%

Experimental procedure

To a solution of potassium *tert*-butoxide (0.05 mmol; 5 mol%) in *tert*-butyl alcohol (1 mL) in a test tube, a mixture of acetophenone (**1**; 1.0 mmol) and aldehyde (**2**; 1.0 mmol) was added followed by continuous stirring for several minutes at room temperature until the starting materials were completely disappeared (monitored by TLC); thiol (**3**; 1.2 mmol) was then added to the reaction mixture, and was stirred to completion of the reaction. The solvent was evaporated, and chloroform (1 mL) was added and the crude product was subjected to preparative TLC (silica gel, eluent petroleum ether/Et_2O = 7:4) to afford a pure β-aryl-β-mercapto ketone **4** in excellent yield (85-95%). The structures of all the products were established on the basis of physical and spectral properties.

Characterization data of two representative compounds

3-(3-Nitrophenyl)-1-phenyl-3-(phenylthio)propan-1-one (**4b**): Solid, mp 112–115 °C; IR (KBr): 1680, 1530, 1345 cm^{-1}; ^1H NMR (300 MHz, $CDCl_3$): δ 3.68 (d, J = 7.1 Hz, 2H,), 5.01 (t, J = 7.1 Hz, 1H,), 7.20-7.31 (m, 6H), 7.35-7.67 (m, 4H), 7.87-8.10 (m, 3H), 8.18 (s, 1H); ^{13}C NMR (75 MHz, $CDCl_3$): δ 29.69, 47.73, 122.34, 122.64, 128.05, 128.31, 128.77, 129.11, 129.24, 133.46, 133.64, 134.27, 197.21.

1-(4-Methoxyphenyl)-3-phenyl-3-(*p*-tolylthio)propan-1-one (**4c**): Solid, mp 98–100 °C; IR (KBr): 1671, 1595 cm^{-1}; ^1H NMR (300 MHz, $CDCl_3$): δ 2.30 (s, 3H), 3.50 (dd, J = 16.9, 6.2 Hz, 1H), 3.59 (dd, J = 16.9, 8.1 Hz, 1H), 3.86 (s, 3H), 4.88 (dd, J = 8.1, 6.2 Hz, 1H), 6.90 (d, J = 9.0 Hz, 2H), 7.04 (d, J = 8.3 Hz, 2H), 7.20-7.33 (m, 7H), 7.87 (d, J = 9.0 Hz, 2H); ^{13}C NMR (75 MHz, $CDCl_3$): δ 21.12, 44.13, 48.75, 55.47, 113.71, 127.23, 127.81, 128.37, 129.60, 129.89, 130.37, 130.47, 133.38, 137.74, 141.42, 163.56, 195.59.

Reference

Movassagh, B., and Rakhshani, A. (2011). A mild and highly efficient one-pot three-component reaction for carbon-sulfur bond formation catalyzed by potassium *tert*-butoxide. *Chinese Chem Lett.*, **22**, 1179–1182.

C-S bond_13

Types of reaction: C-S, C-N, C-O and C-C bond formation
Reaction conditions: Ethanol, room temperature
Synthetic strategy: One-pot multicomponent
Catalyst: Triethylamine (Et_3N)
Keywords: Ninhydrin, malononitrile, aryl isothiocyanates, primary aliphatic amines, ethanol, triethylamine, room temperature, one-pot multicomponent synthesis, thio-Michael and Knoevenagel reactions, sulfur-nitrogen heterocycles, oxathiaaza[3.3.3]-propellanes, chemoselectivity, regioselectivity

General reaction scheme

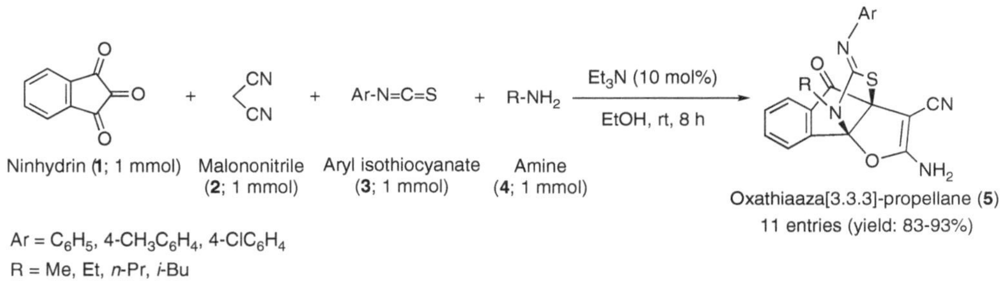

Plausible mechanism

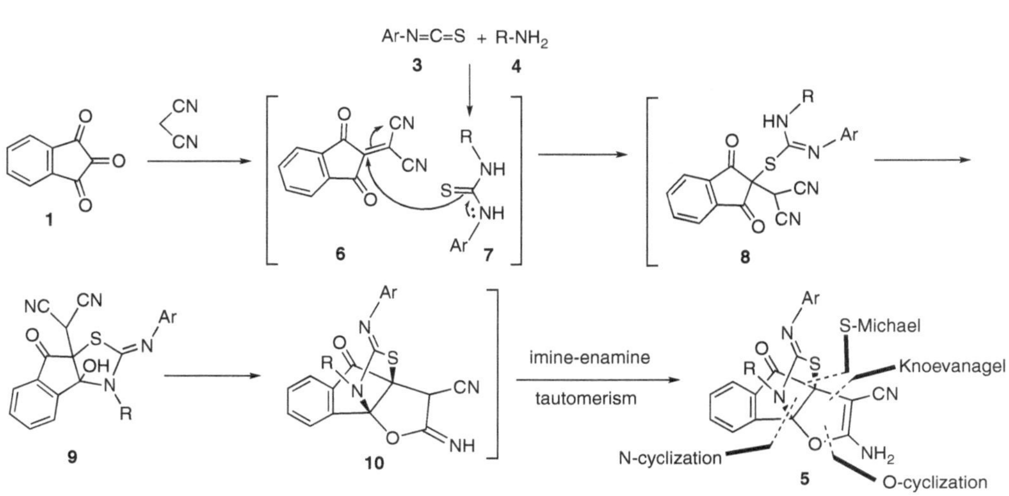

Representative entries

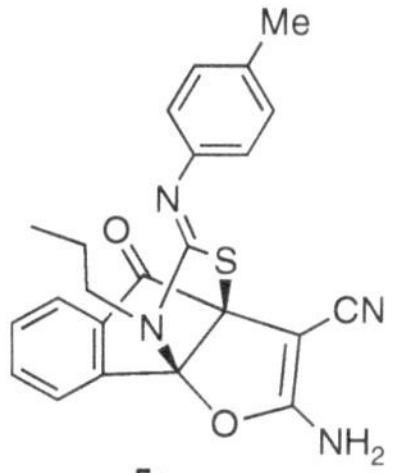

5a

(3a*S*,8b*R*,*Z*)-2-Amino-9-propyl-4-oxo-10-(*p*-tolylimino)-4*H*-3a,8b-(epithiomethanoimino) indeno[1,2-*b*]furan-3-carbonitrile (**5a**): White powder, mp 191–193 °C, yield: 90%

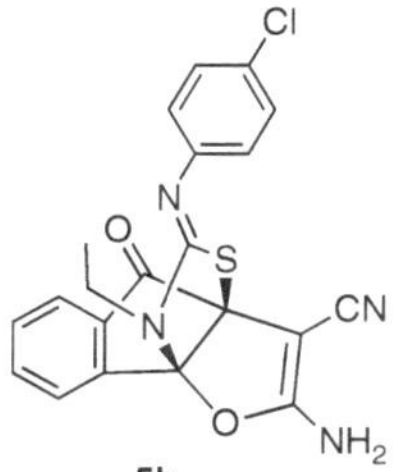

5b

(3a*S*,8b*R*,*Z*)-2-Amino-10-((4-chlorophenyl)imino)-4-oxo-9-ethyl-4*H*-3a,8b-(epithiomethanoimino) indeno[1,2-*b*]furan-3-carbonitrile (**5b**): White powder, mp 178–180 °C, yield: 88%

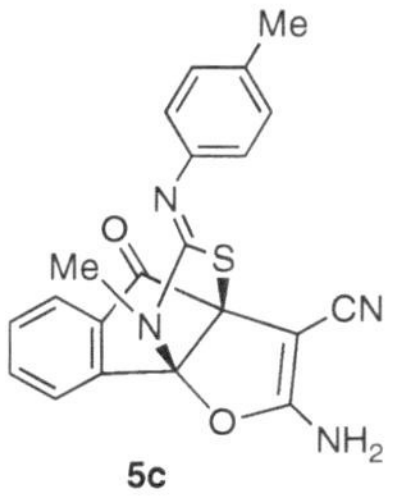

5c

(3a*S*,8b*R*,*Z*)-2-Amino-9-methyl-4-oxo-10-(*p*-tolylimino)-4*H*-3a,8b-(epithiomethanoimino) indeno[1,2-*b*]furan-3-carbonitrile (**5c**): White powder, mp 189–191 °C, yield: 84%

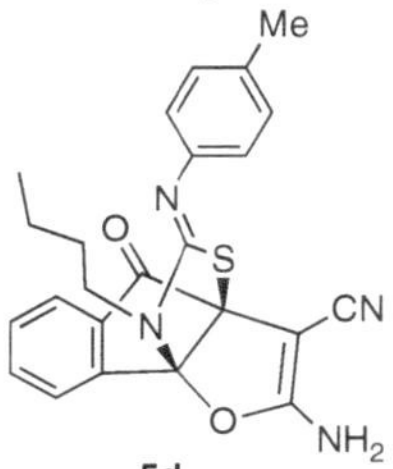

5d

(3a*S*,8b*R*,*Z*)-2-Amino-9-butyl-4-oxo-10-(*p*-tolylimino)-4*H*-3a,8b-(epithiomethanoimino) indeno[1,2-*b*]furan-3-carbonitrile (**5d**): White powder, mp 202–204 °C, yield: 90%

Experimental procedure

To a solution of ninhydrin (**1**;1 mmol) and malononitrile (**2**; 1 mmol) in ethanol (4 mL) was added triethylamine (0.1 mmol), and the solution was stirred at room temperature for 1 h. To this stirred solution was added aryl isothiocyanate (**3**; 1 mmol) and primary amine (**4**; 1 mmol) at once and drop-wise over 15 min, respectively. After then the whole mixture was stirred again for another 7 h to complete the reaction (monitored by TLC). On completion, the mixture was filtered and the precipitate was washed with ethanol (4 mL) to afford the pure product of oxathiaaza[3.3.3]propellane **5** with good yield (83-93%). Each of the products was characterized on the basis of analytical analyses and detailed spectral studies.

Characterization data of two representative compounds

(3a*S*,8b*R*,*Z*)-2-Amino-9-propyl-4-oxo-10-(p-tolylimino)-4*H*-3a,8b-(epithiomethanoimino)indeno [1,2-*b*]furan-3-carbonitrile (**5a**): White powder, mp 191–193 °C, IR (KBr): 3320, 3296, 2190, 1739, 1664, 1593 cm^{-1}; ^1H NMR (500 MHz, DMSO): δ 0.86 (s, 3H, Me), 1.51 (s, 1H, CH_2), 1.75 (s, 1H, CH_2), 2.23 (s, 3H, Me), 3.38 (s, 1H, CH_2), 3.69 (s, 1H, CH_2), 6.72 (s, 2H, 2 × CH of Ar), 7.07 (s, 2H, 2 × CH of Ar), 7.81-7.82 (m, 1H, CH of Ar), 7.94-8.16 (m, 5H, 3 × CH of Ar, NH_2); ^{13}C NMR (125 MHz, DMSO): δ10.95, 20.41, 20.77, 45.45, 53.27, 70.74, 91.81, 107.35, 116.09, 121.03, 123.69, 125.28, 125.58, 127.93, 129.53, 130.01, 132.44, 132.67, 137.23, 144.32, 147.35, 154.12, 166.04, 192.86; MS: *m*/*z*=416 [M^+], 373, 344, 331, 271, 257, 236, 210, 174, 146, 127, 104, 91, 76, 57. Anal. Calcd for $C_{23}H_{20}N_4O_2S$: C 66.33, H, 4.84, N 13.45; Found: C 66.28, H 13.4.86, N, 13.88. Crystal data available for **5a**: CCDC 869053.

(3a*S*,8b*R*,*Z*)-2-Amino-9-butyl-4-oxo-10-(*p*-tolylimino)-4*H*-3a,8b-(epithiomethanoimino) indeno [1,2-*b*]furan-3-carbonitrile (**5d**): White powder, mp 202–204 °C; IR (KBr): 3350, 3199, 2924, 2357, 2227, 1717, 1649, 1593 cm^{-1}; 1H NMR (300 MHz, DMSO): δ 0.78 (d, $J=6.5$ Hz, 3H, Me), 0.91 (d, $J=6.2$ Hz, 3H, Me), 2.23 (s, 3H, Me), 2.24-2.47 (m, 1H, CH), 3.44-3.75 (m, 2 H, CH_2), 6.68 (d, $J=7.4$ Hz, 2H, 2 × CH of Ar), 7.06 (d, $J=7.7$ Hz, 2H, 2 × CH of Ar), 7.83 (d, $J=6.5$ Hz, 1 H, CH of Ar), 7.92-8.10 (m, 4 H, CH of Ar, NH_2), 8.31 (t, $J=6.2$ Hz, 1 H, CH of Ar); ^{13}C NMR (75 MHz, DMSO): δ 20.12, 20.46, 20.92, 27.44, 51.20, 58.0, 75.10, 108.50, 115.93, 121.41, 124.14, 128.40, 130.05, 130.54, 133.0, 137.59, 137.59, 147.80, 153.73, 155.28, 166.53, 193.20; MS: $m/z=430$ [M^+], 374, 331, 242, 210, 155, 127, 91, 57. Anal. Calcd for $C_{24}H_{22}N_4O_2S$: C 66.33, H 4.84, N 13.45; Found: C 66.26;, H 4.88, N13.49.

Reference

Rezvanian, A., Alizadeh, A., and Zhu, L.-G. (2012). Chemo- and regioselective 4CR synthesis of oxathiaaza[3.3.3]-propellanes via sequential C-S, C-N and C-O bond formation in a single pot. *Synlett*, **23**, 2526–2530.

C-S bond_14

Types of reaction: C-S and C-N bond formation
Reaction conditions: Acetone, room temperature
Synthetic strategy: One-pot multicomponent reaction
Catalyst: No catalyst
Keywords: Ammonium thiocyanate, benzylamine, ethyl bromopyruvate, benzoyl chloride, acetone, room temperature, multicomponent reaction, benzoyl thiocyanate, 1,3-thiazoles

General reaction scheme

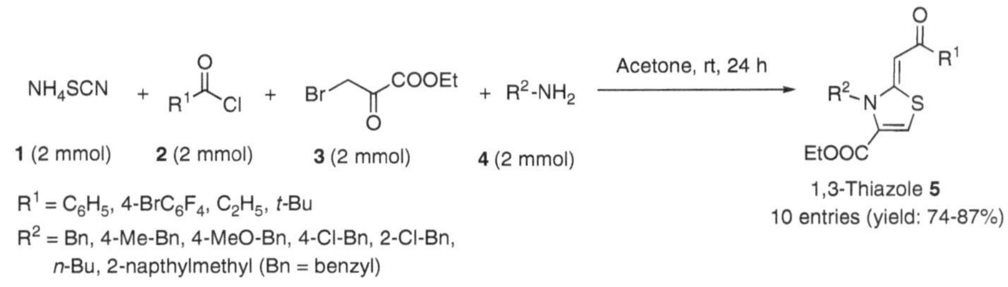

1,3-Thiazole **5**
10 entries (yield: 74-87%)

Plausible mechanism

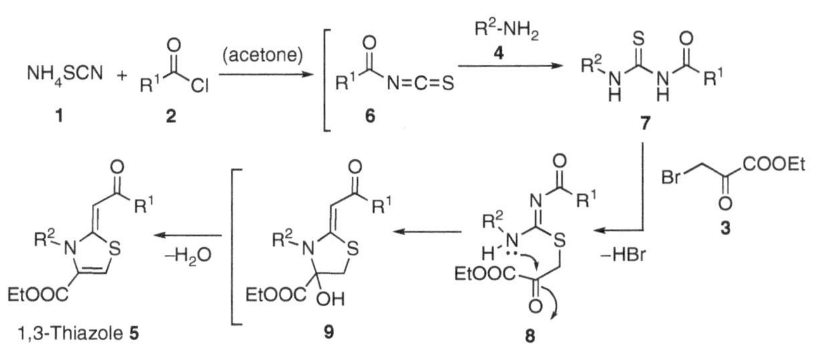

Representative entries

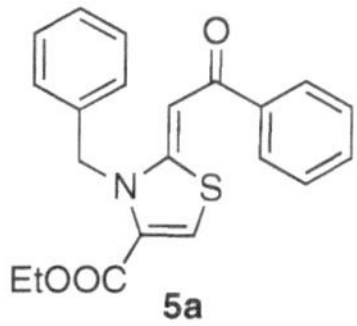

5a

(*Z*)-Ethyl 3-benzyl-2-(2-oxo-2-phenylethylidene)-2,3-dihydrothiazole-4-carboxylate (**5a**): Colorless crystals, mp 119–121 °C; yield: 85%

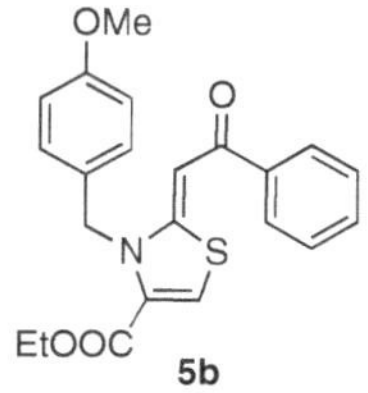

5b

(*Z*)-Ethyl 3-(4-methoxybenzyl)-2-(2-oxo-2-phenylethylidene)-2,3-dihydrothiazole-4-carboxylate (**5b**): Yield: 78%

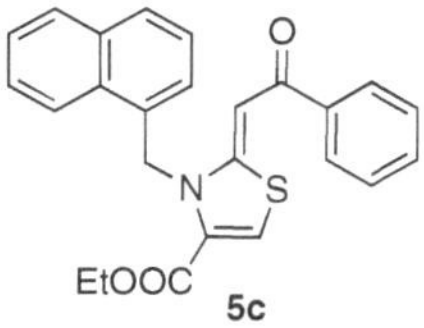

5c

(*Z*)-Ethyl 3-(naphthalen-1-ylmethyl)-2-(2-oxo-2-phenylethylidene)-2,3-dihydrothiazole-4-carboxylate (**5c**): Yield: 75%

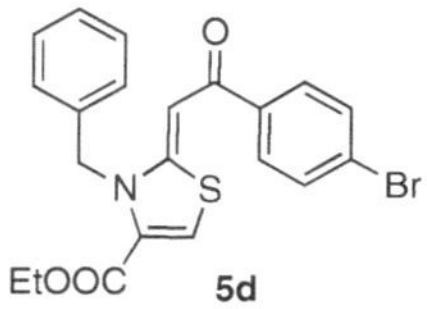

5d

(*Z*)-Ethyl 3-benzyl-2-(2-(4-bromophenyl)-2-oxoethylidene)-2,3-dihydrothiazole-4-carboxylate (**5d**): Yield: 78%

Experimental procedure

To a stirred solution of ammonium thiocyanate (**1**; 0.15 g, 2 mmol) in acetone (15 mL) was added benzoyl chloride (**2**; 0.28 g, 2 mmol), and the mixture was refluxed for 5 min and then a solution of ethyl bromopyruvate (**3**; 0.39 g, 2 mmol) in acetone (10 mL) was added drop-wise. After addition of the primary amine (**4**; 2 mmol) at room temperature, the resulting reaction mixture was then stirred for 24 h, the solvent was removed under reduced pressure, and the residue was purified by column chromatography (SiO_2;hexane-EtOAc) to give 1,3-thiazole derivatives 5 with 74-97% of yield. All the products were characterized on the basis of detailed spectral and elemental analyses.

Characterization data of one representative compound

(*Z*)-Ethyl 3-benzyl-2-(2-oxo-2-phenylethylidene)-2,3-dihydrothiazole-4-carboxylate (**5a**): Colorless crystals, mp 119–121 °C; IR (KBr): 1716, 1658, 1587, 1477, 1429, 1369, 1369, 1304, 1231, 1105, 1054 cm^{-1}; ^{1}H NMR (500.1 MHz, $CDCl_3$): δ 1.33 (t, $^{3}J=7.2$ Hz, 3H, Me), 4.31 (q, $^{3}J=7.2$ Hz, 2H, OCH_2), 6.07 (s, 2H, NCH_2), 7.24-7.32 (m, 3H, 3x*CH*), 7.38 (d, $^{3}J=7.4$ Hz, 2H, 2 x CH), 7.45 (t, $^{3}J=7.2$ Hz, 2H, *CH*), 7.51 (d, $^{3}J=7.2$ Hz, 1H, CH), 7.63 (s, 1H, CH), 8.34 (d, $^{3}J=7.2$ Hz, 2H, 2 x CH); ^{13}C NMR (500.1 MHz, $CDCl_3$): δ 14.0 (Me), 50.4 (CH_2), 61.9 (OCH_2), 120.8 (*C*H), 127.7 (2 x *C*H), 128.1 (3 x *C*H), 128.6 (2 x *C*H), 129.5 (3 x *C*H), 131.8 (C), 136.5 (C), 136.8 (C), 158.3 (C=O), 169.3 (C=N), 174.8 (C=O); MS: *m/z* (%rel) 366 (10) [M^+], 293 (75), 261 (62), 105 (100), 91 (82), 45 (84). Anal. Calcd (%) for $C_{20}H_{18}N_2O_3S$ (366.43): C 65.56, H 4.95, N 7.64; Found: C 65.63, H 4.99, N, 7.68.

Reference

Yavari, I., Hossaini, Z., Shirgahi-Talari, F., and Seyfi, S. (2008). Synthesis of functionalized 1,3-thiazoles from acid chlorides, primary amines, ethyl bromopyruvate, and ammonium thiocyanate. *Synlett*, 1631–1632.

C-S bond_15

Types of reaction: C-S bond formation
Reaction conditions: Dichloromethane, room temperature
Synthetic strategy: One-pot multicompoment reaction
Catalyst: No catalyst
Keywords: Diethyl/dimethyl acetylenedicarboxylate, phenacyl halides, tetramethylthiourea, dichloromethane, room temperature, one-pot multicomponent reaction, functionalized 2-aminothiophenes

General reaction scheme

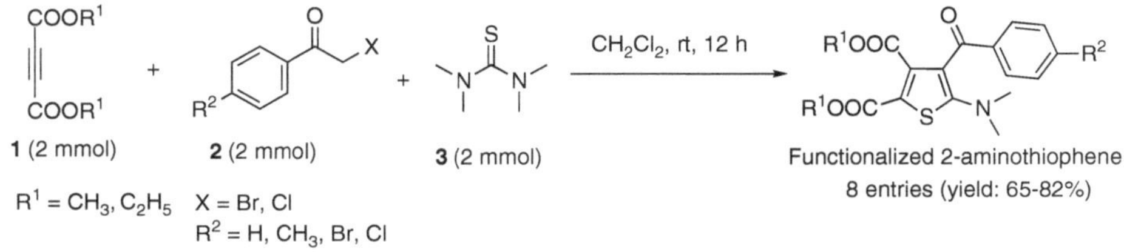

Representative entries

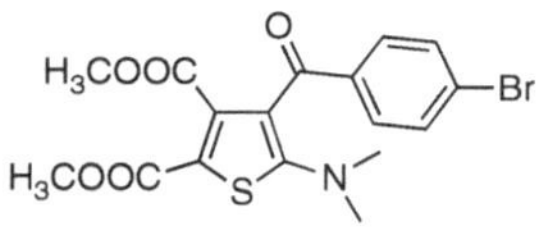

4a

Dimethyl 4-(4-bromobenzoyl)-5-(dimethylamino) thiophene-2,3-dicarboxylate (**4a**): Yellow oil; yield: 70%

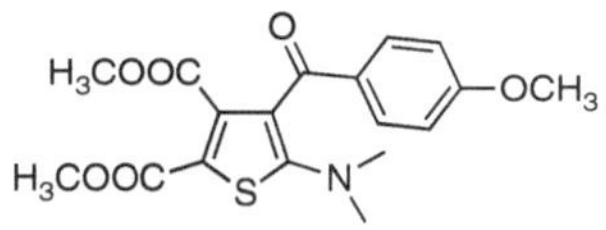

4b

Dimethyl 5-(dimethylamino)-4-(4-methoxybenzoyl)thiophene-2,3-dicarboxylate (**4b**): Yellow oil; yield: 73%

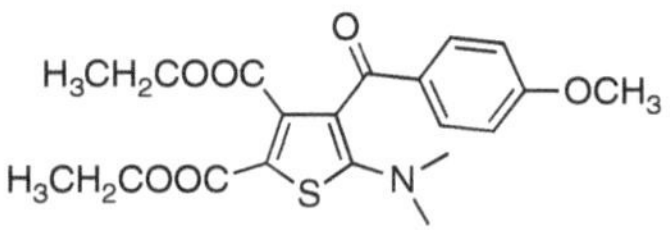

4c

Diethyl 5-(dimethylamino)-4-(4-methoxybenzoyl) thiophene-2,3-dicarboxylate (**4c**): Yellow oil; yield: 70%

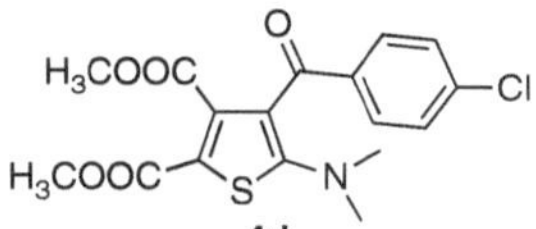

4d

Dimethyl 4-(4-chlorobenzoyl)-5-(dimethylamino) thiophene-2,3-dicarboxylate (**4d**): Yellow oil; yield: 82%

Experimental procedure

To a stirred solution of acetylenic dicarboxylate (**1**; 2 mmol) and phenacyl halide (**2**; 2 mmol) in dichloromethane (15 mL) was added tetramethylthiourea (3; 0.26 g; 2 mmol) at room temperature, and the reaction mixture was stirred for 12 h. On completion of reaction, the solvent was evaporated out under reduced pressure to produce crude residue that was purified by column chromatography (SiO_2; *n*-hexane/EtOAc 6:1) to afford pure products of functionalized 2-aminothiphenes (**4**). All the products were characterized by IR, ^{1}H NMR, ^{13}C NMR, and mass as well as elemental analyses.

Characterization data of two representative entries

Dimethyl 4-(4-bromobenzoyl)-5-(dimethylamino)thiophene-2,3-dicarboxylate (**4a**): Yellow oil; IR (KBr): 1734, 1630, 1542, 1436, 1260 cm^{-1}; ^{1}H NMR (500 MHz, $CDCl_3$): δ 2.81 (s, 6H, $(CH_3)_2$ N), 3.86 (s, 3H, CH_3O), 3.97 (s, 3H, CH_3O), 7.62 (d, $^{3}J=8.6$ Hz, 2H CH), 7.70 (d, $^{3}J=8.6$ Hz, 2H, CH); ^{13}C NMR (125 MHz, $CDCl_3$): δ 43.3 ($(CH_3)_2$ N), 52.8 (CH_3O), 53.1 (CH_3O), 124.9 (C), 127.9 (C), 130.6 (CH), 131.7 (CH), 133.3 (C), 134.4 (C), 138.1 (C). 153.2 (C), 162.2 (C=O), 164.2 (C=O), 186.0 (C=O); EI-MS: *m/z* (rel. int.%) at 426 (M^-, 7), 184 (100), 170 (54), 156 (60), 59 (47), 44 (23). Anal. Calcd for $C_{17}H_{16}BrNO_5S$ (426.28): C, 47.90; H, 3.78; N, 3.29; Found: C, 48.1; H, 3.9; N, 3.4%.

Diethyl 5-(dimethylamino)-4-(4-methoxybenzoyl)thiophene-2,3-dicarboxylate (**4c**): Yellow oil; IR (KBr): 1732, 1598, 1541, 1418, 1260 cm^{-1}; ^{1}H NMR (500 MHz, $CDCl_3$):δ 1.34 (t, $^{3}J=7.1$ Hz, 3H, CH_3), 1.41 (t, 3 J 7.1 Hz, 3H, CH_3), 2.88 (s, 6H, $(CH_3)_2$ N), 3.89 (s, 3H, CH_3O), 4.32 (q, $^{3}J=7.1$ Hz, 2H, CH_2O), 4.43 (q, $^{3}J=7.1$ Hz, 2H, CH_2O), 6.96 (d, $^{3}J=8.8$ Hz, 2H,CH), 7.86 (d, $^{3}J=8.8$ Hz, 2H, CH); ^{13}C NMR (125 MHz, $CDCl_3$): δ 13.9 (CH_3), 14.1 (CH_3), 43.3 ($(CH_3)_2$ N), 55.5 (CH_3O), 61.9 (CH_2O), 62.2 (CH_2O), 113.7 (CH), 125.3 (C), 131.4 (C), 131.7 (CH), 133.7 (C), 134.7 (C), 153.0 (C), 160.7 (C), 163.6 (C=O), 165.7 (C=O), 186.0 (C=O); EI-MS: *m/z* (rel. int.%) at 405 (M^+, 8), 135 (100), 121 (47), 107 (55), 73 (38), 45 (41), 44 (19); Anal. Calcd for $C_{20}H_{23}NO_6S$ (405.46): C, 59.24; H, 5.72; N, 3.45; Found: C, 59.0; H, 5.9; N, 3.3%.

Reference

Yavari, I., and Malekafzali, A. (2013). Formation of highly functionalized thiophenes by reaction of tetramethylthiourea, acetylenic esters, and α-haloketones. *J. Heterocyclic Chem.*, **50**, 992–994.

C-S bond_16

Type of reaction: C-S and C-N bond formation
Reaction conditions: Chloroform, room temperature
Synthetic strategy: Sequential three-component one-pot reaction
Catalyst: Catalyst-free
Keywords: 1,2-Diaza-1,3-dienes, amines, isothiocyanates, chloroform, room temperature, sequential three-component reaction, one-pot synthesis, 5-hydrazinoethylidene-2-iminothiazolidinones

General reaction scheme

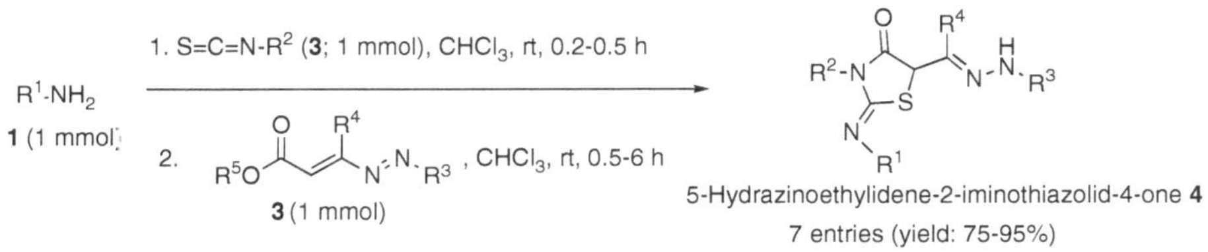

$R^1 = C_6H_5$, $4\text{-}ClC_6H_4$, $4\text{-}OMeC_6H_4$
R^2 = *n*-Bu, allyl, $CH_2C_6H_5$
R^3 = COOMe,COO*t*-Bu, $COONH_2$, $COONHC_6H_5$
R^4 = Me, C_6H_5; R^5 = Me, Et

Plausible mechanism

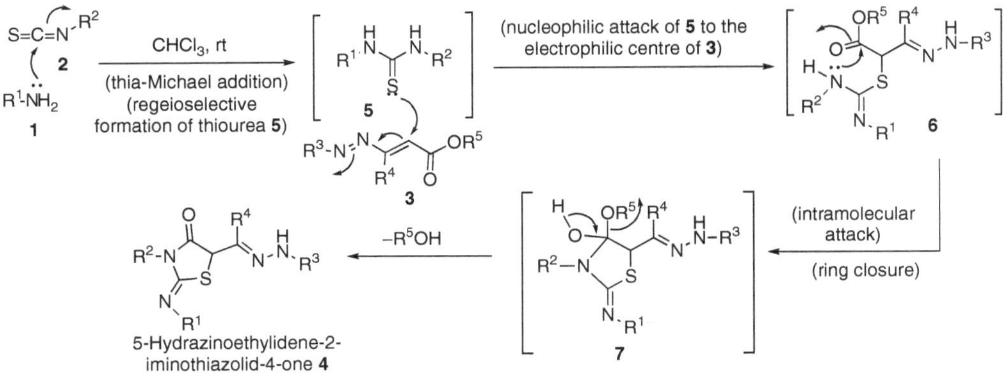

Representative entries

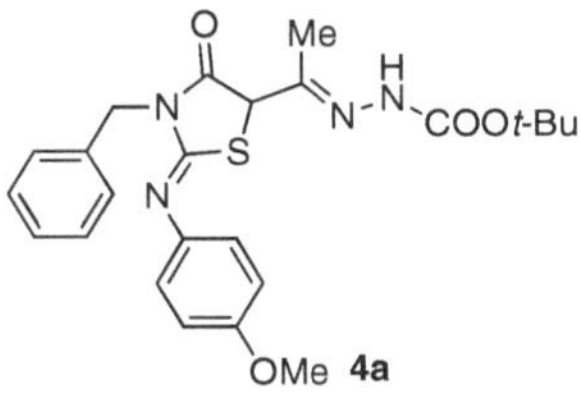

tert-Butyl 2-(1-(3-benzyl-2-((4-methoxyphenyl) imino)-4-oxothiazolidin-5-yl)ethylidene) hydrazinecarboxylate (**4a**): Light yellow foam; yield 95%

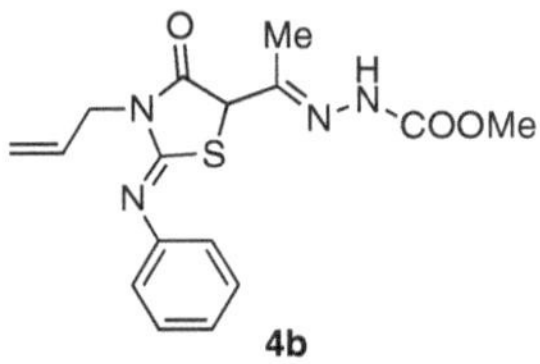

Methyl 2-(1-(3-allyl-4-oxo-2-(phenylimino) thiazolidin-5-yl)ethylidene)hydrazinecarboxylate (**4b**): White powder; yield: 90%

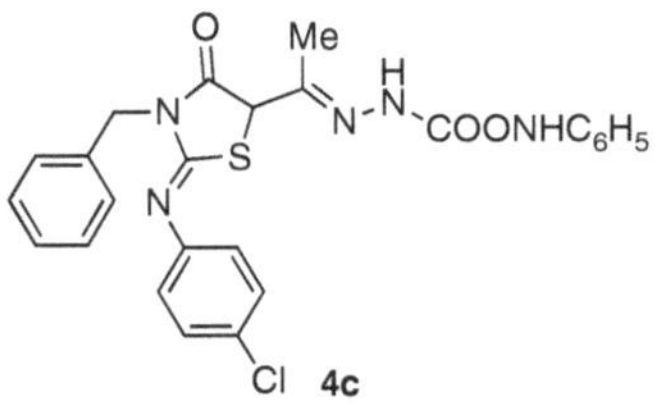

3-Benzyl-2-((4-chlorophenyl)imino)-5-(1-(2-(((phenylamino)oxy)carbonyl)hydrazono)ethyl) thiazolidin-4-one (**4c**): White crystals, yield: 84%

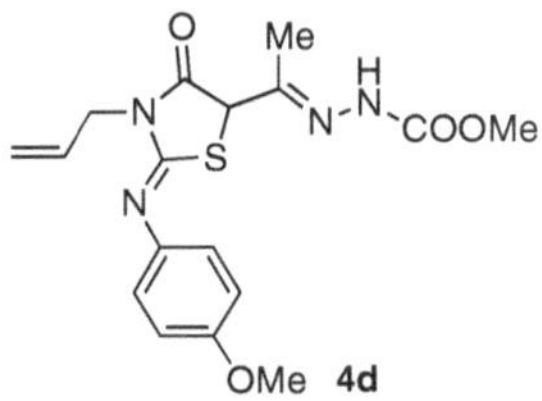

Methyl 2-(1-(3-allyl-2-((4-methoxyphenyl) imino)-4-oxothiazolidin-5-yl)ethylidene) hydrazinecarboxylate (**4d**): Light gray powder; yield: 85%

Experimental procedure

To a stirred solution of primary aromatic amine (**1**: 1 mmol) in chloroform (4 mL) was added alkylthiocyanate (**2**; 1 mmol), and stirred at room temperature. On disappearance of the reagents forming the thiourea intermediate (within 0.2-0.5 h as checked by TLC), 1,2-diaza-1,3-diene **3** was added solid or dissolved in $CHCl_3$ (1 mL) to the reaction vessel and the reaction mixture was allowed to stand at room temperature until completion of the reaction (0.5-6 h). The solvent was then removed under reduced pressure and the crude reaction mixture was purified by column chromatography (cyclohexane/ethyl acetate mixtures) to obtain 5-hydrazinoethylidene-2-thiazolidin-4-one (**4**) as mixture of C_5 *E*/*Z* isomers (approximately ratio 3:1). In the case of solid product was obtained by precipitation from the reaction medium and subsequent recrystallization from the appropriate solvents. All the products were characterized on the basis of elemental analyses and detailed spectral studies.

Characterization data of two representative compounds

3-Benzyl-2-((4-chlorophenyl)imino)-5-(1-(2-(((phenylamino)oxy)carbonyl)hydrazono) ethyl) thiazolidin-4-one (**4c**): White crystals; IR (nujol): 3276, 3201, 3099, 1711, 1698, 1652, 1629, 1603, 1573, 1547 cm^{-1}; ^1H NMR (400 MHz, DMSO-d_6): δ 1.82 and 2.41 (2 s, 3H), 4.94 and 4.99 (2 s, 2H), 6.96 – 7.01 (m, 2H), 7.25 – 7.55 (m, 12H), 8.48 (br s, 1H), 8.85 (s, 1H), 8.94 and 9.99 (br s and s, 1H); ^{13}C NMR (100 MHz, DMSO-d_6): δ 12.5 (q), 16.9 (q), 44.3 (t), 44.6 (t), 84.0 (s), 85.9 (s), 118.7 (d), 119.2 (d), 122.3 (d), 122.7 (d), 123.0 (d), 123.1 (d), 127.3 (d), 127.4 (d), 127.5 (s), 127.6 (d), 127.9 (s), 128.3 (d), 128.4 (d), 128.5 (d), 128.7 (d), 129.0 (d), 129.3 (d), 136.6 (s), 137.1 (s), 139.1 (s), 139.3 (s), 147.6 (s), 147.9 (s), 152.9 (s), 153.4 (s), 155.1 (s), 155.4 (s), 165.3 (s), 166.0 (s); MS (EI): *m*/*z* at 493 [M^+ + 2 (5)], 491 (M^+, 13), 398 (52), 274 (100), 241 (57), 179 (79), 119 (83). Anal. Calcd for $C_{25}H_{22}ClN_5O_2S$ (491.99): C, 61.03; H, 4.51; N, 14.23; Found: C, 61.12; H, 4.39; N, 14.32.

Methyl 2-(1-(3-allyl-2-((4-methoxyphenyl)imino)-4-oxothiazolidin-5-yl)ethylidene) hydrazinecarboxylate (**4d**): Light gray powder; IR (Nujol): 3317, 3290, 1742, 1653, 1616, 1585 cm^{-1}; ^1H NMR (400 MHz, DMSO-d_6): δ 1.86 and 2.34 (2 s, 3H), 3.58 and 3.62 (2 s, 3H), 3.73 (s, 3H), 4.33-4.38 (m, 2H), 5.09-5.17 (m, 2H), 5.84-5.94 (m, 1H), 6.83-6.92 (m, 4H), 8.75 and 9.82 (br and s, 1H), 9.52 (br s, 1H); ^{13}C NMR (100 MHz, DMSO-d_6): δ 16.2 (q), 16.7 (q), 43.4 (t), 43.5 (t), 52.4 (q), 55.2 (q), 84.7 (s), 86.7 (s), 114.4 (d), 114.6 (d), 116.5 (t), 116.7 (t), 122.1 (d), 122.3 (d), 132.0 (d), 132.4 (d), 141.9 (s), 142.1 (s), 150.8 (s), 155.7 (s), 155.9 (s), 156.7 (s), 156.8 (s), 165.4 (s), 165.8 (s); MS (EI): *m*/*z* at 376 (M^+, 46), 344 (8), 287 (2), 189 (36), 147 (100), 101(53). Anal. Calcd for $C_{17}H_{20}N_4O_4S$ (376.43): C, 54.24; H, 5.36; N, 14.88; Found: C, 54.16; H, 5.49; N, 14.79.

Reference

Attanasi, O. A., Bartoccini, S., Favi, G., Giorgi, G., Perrulli, F. R., and Santeusanio, S. (2012). Powerful approach to heterocyclic skeletal diversity by sequential three-component reaction of amines, isothiocyanates, and 1,2-diaza-1,3-dienes. *J. Org. Chem.*, **77**, 1161–1167.

C-S bond_17

Type of reaction: C-S and C-C bond formation
Reaction conditions: Dichloromethane, room temperature
Synthetic strategy: One-pot two-component [3+2]-cycloaddition
Catalyst: 4-Dimethylaminopyridine (DMAP)
Keywords: Dialkyl acetylenedicarboxylates, β-oxodithioesters, 4-dimethylaminopyridine (DMAP), dichloromethane, room temperature, one-pot two-component [3+2]-cycloaddition, 2,3-dicarboalkoxy-4-aroyl/heteroaroyl/alkanoyl thiophenes

General reaction scheme

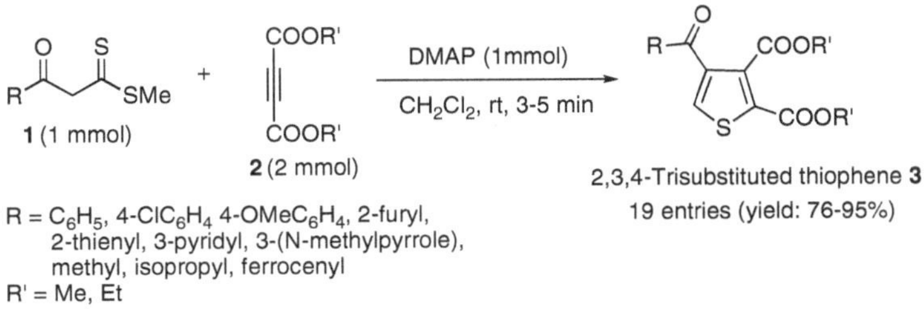

Proposed Mechanism

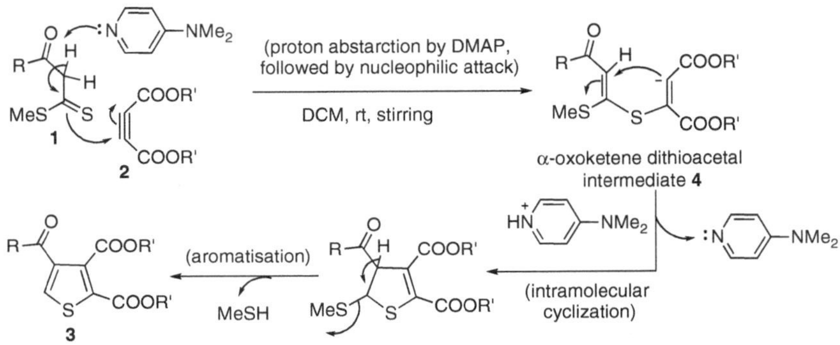

Representative entries

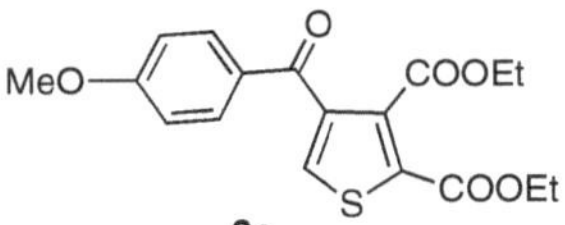

3a

Diethyl 4-(4-methoxybenzoyl)thiophene-2,3-dicarboxylate (**3a**): White solid, mp 75–76 °C; reaction time: 3 min; yield 93%

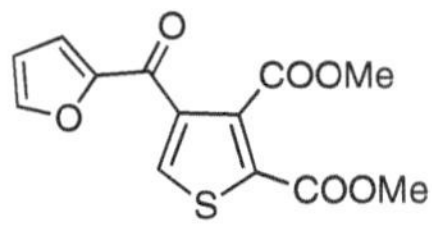

3b

Dimethyl 4-(furan-2-carbonyl)thiophene-2,3-dicarboxylate (**3b**): White solid, mp 110–111 °C; reaction time: 4 min; yield 95%

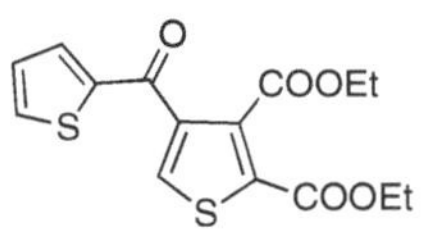

3c

Diethyl 4-(thiophene-2-carbonyl)thiophene-2,3-dicarboxylate (**3c**): White solid, mp 110–112 °C; reaction time: 5 min; yield 88%

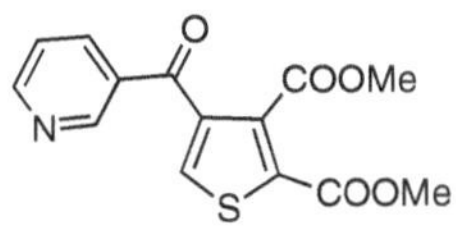

3d

Dimethyl 4-nicotinoylthiophene-2,3-dicarboxylate (**3d**): White solid, mp 94–95 °C; reaction time: 5 min; yield 85%

Experimental procedure

To a mixture of β-ketodithioester (**1**; 1.0 mmol) and dialkyl acetylenedicarboxylate (**2**; 1.0 mmol) in dichloromethane (3 mL) was added 4-dimethylaminopyridine (DMAP; 1.0 mmol), and the reaction mixture was then stirred for the stipulated period of time at room temperature. Upon completion of the reaction (monitored by TLC), water (20 mL) was added to the reaction mixture followed by extraction with dichloromethane (2 x 10 mL). The combined organic layer was dried over anhydrous sodium sulfate and then evaporated in *vacuo*. The crude residue was purified by column chromatography over silica gel using ethyl acetate/hexane (1:10) as eluent to afford pure thiophene **3**. All the products were characterized by spectral and analytical studies.

Characterization data of two representative compounds

Diethyl 4-(4-methoxybenzoyl)thiophene-2,3-dicarboxylate (**3a**): White solid, mp 75–76 °C; IR (KBr): 3020, 1721, 1600, 1526, 1215 cm^{-1}; ^1H NMR (300 MHz, $CDCl_3$): δ 7.84-7.81 (m, 3H), 6.95 (d, J=8.7 Hz, 2H), 4.40-4.34 (m, 4H), 3.88 (s, 3H), 1.39-1.33 (m, 6H); ^{13}C NMR(75 MHz, $CDCl_3$): δ 187.2, 164.6, 163.6, 160.3, 140.5, 139.7, 136.2, 132.9, 131.9, 129.9, 113.8, 62.1, 62.0, 55.5, 14.1, 13.9; MS: m/z at 362 (M^+). Anal. Calcd for $C_{18}H_{18}O_6S$: C, 59.66; H, 5.01; Found: C, 59.82; H, 4.91.

Dimethyl 4-nicotinoylthiophene-2,3-dicarboxylate (**3d**): White solid, mp 94–95 °C; IR (KBr): 3081, 2957, 1731, 1727, 1622, 1276 cm^{-1}; ^1H NMR (300 MHz, $CDCl_3$): δ 9.03 (s, 1H), 8.83 (s, 1H), 8.12 (d, J=6.6 Hz, 1H), 7.95 (s, 1H), 7.47 (t, J=5.1 Hz, 1H), 3.96 (s, 3H), 3.93 (s, 3H); ^{13}C NMR (75 MHz, $CDCl_3$): δ 186.6, 164.8, 160.4, 153.4, 150.1, 139.5, 139.1, 138.1, 136.7, 132.9, 132.8, 123.6, 53.2, 53.0; MS: m/z at 305 (M^+). Anal. Calcd for $C_{14}H_{11}NO_5S$: C, 55.08; H, 3.63; N, 4.59; Found: C,55.25; H, 3.48; N, 4.78.

Reference

Nandi, G. C., Samai, S., and Singh, M. S. (2011). One-pot two-component [3+2] cycloaddition/annulation protocol for the synthesis of highly functionalized thiophene derivatives. *J. Org. Chem.*, **76**, 8009–8014.

C-S bond_18

Type of reaction: C-S and C-O bond formation
Reaction conditions: Methanol, room temperature
Synthetic strategy: Methoxide ion-promoted one-step reaction
Catalyst: Sodium hydride (NaH)
Keywords: Carbon disulfide, oxiranes, sodium hydride, methanol, room temperature, methoxide ion-promoted reaction, oxathiolanes, 1,3-oxathiolane-2-thiones

General reaction scheme

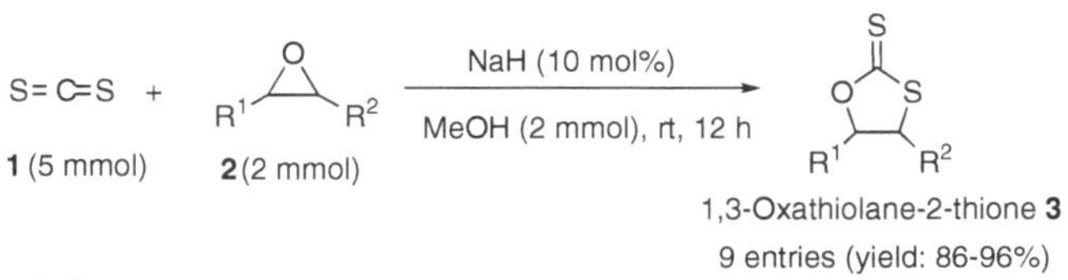

R^1,R^2 = Me,H; Et,H; -$(CH_2)_4$-; C_6H_5,H; $C_6H_5OCH_2$,H; $H_2C{=}CHCH_2O$,H; Me_2CHOCH_2,H; C_6H_5,C_6H_5 (*cis*); C_6H_5,C_6H_5 (*trans*)

Proposed mechanism

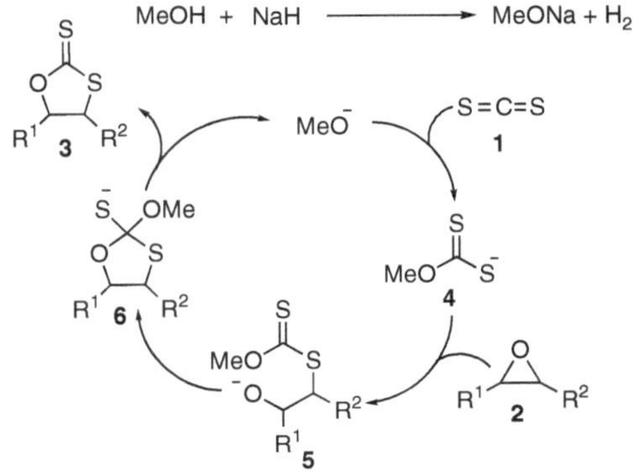

Representative entries

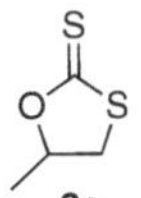

3a

5-Methyl-1,3-oxathiolane-2-thione (**2a**): Yellow oil; yield 96%

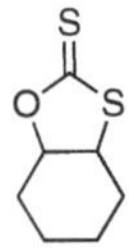

3b

Hexahydrobenzo[*d*][1,3]oxathiole-2-thione (**2b**): Yellow crystals, mp 176–178 °C; yield 96%

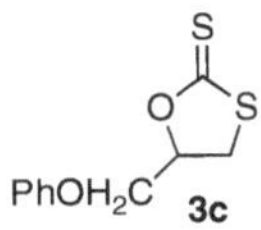

3c

5-(Phenoxymethyl)-1,3-oxathiolane-2-thione (**2c**): Yellow crystals, mp 55–57 °C; yield 94%

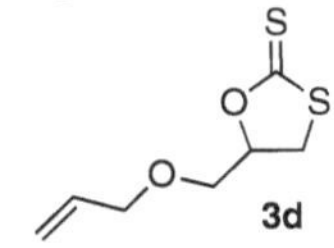

3d

5-((Allyloxy)methyl)-1,3-oxathiolane-2-thione (**2d**): Yellow oil; yield 94%

Experimental procedure

To a stirred solution of carbon disulfide (CS_2, **1**; 0.35 g, 5 mmol) in methanol (0.064 g, 2 mmol) containing NaH (10 mol%), was added the oxirane derivative (**2**; 2 mmol) at room temperature. The mixture was then stirred for 12 h, and filtered to remove the white precipitates (presumably, NaOMe). The residue was purified by extraction with diethyl ether to afford pure 1,3-oxathiolane-2-thione **3**. All the compounds were characterized by spectral and analytical studies.

Characterization data of two representative compounds

Hexahydrobenzo[*d*][1,3]oxathiole-2-thione (**2b**): Yellow crystals, mp 176–178 °C; IR (KBr): 1628, 1431, 1326, 1272, 1094 cm^{-1}; ^1H NMR (500 MHz, $CDCl_3$): δ 1.43-1.48 (2 H, m, CH_2), 1.68-1.75 (2 H, m, CH_2), 1.93-1.97 (2 H, m, CH_2), 2.17-2.22 (2 H, m, CH_2), 4.08-4.09 (1 H, m, CH), 4.09-4.11 (1 H, m, CH); ^{13}C NMR (125.7 MHz, $CDCl_3$): δ 25.5 (2 CH_2), 29.5 (2 CH_2), 65.0 (2 CH), 227.6 (C=S); MS (EI): *m/z* (%) 174 (5) [M^+], 118 (78), 92 (100), 82 (64), 76 (48), 56 (45). Anal. Calcd (%) for $C_7H_{10}OS_2$ (174.27): C, 48.24; H, 5.78; Found: C, 48.18; H, 5.79.

5-((Allyloxy)methyl)-1,3-oxathiolane-2-thione (**2d**): Yellow oil; IR (KBr): 1702, 1630, 1451, 1414, 1348 cm^{-1}; ^1H NMR (500 MHz, $CDCl_3$): δ 3.64 (1 H, dd, $^2J=9.8$ Hz, $^3J=5.8$ Hz, CH), 3.79-3.81 (1 H, m, CH), 3.94 (1 H, dd, $^2J=12.1$ Hz, $^3J=4.7$ Hz, CH), 4.04 (2 H, d, $^3J=5.6$ Hz, 2 CH), 4.07 (1 H, dd, $^2J=12.1$ Hz, $^3J=5.7$ Hz, CH), 4.44-4.49

(1 H, m, CH), 5.20-5.30 (2 H, m, 2 CH), 5.83–5.91 (1 H, m, CH); ^{13}C NMR (125.7 MHz, $CDCl_3$): δ 45.4 (CH_2), 58.7 (CH), 69.4 (CH_2), 72.8 (CH_2), 118.4 (CH_2), 134.3 (CH), 227.9 (C=S); MS (EI): *m/z* (%) 190 (15) [M^+], 133 (74), 114 (58), 92 (46), 57 (100). Anal. Calcd (%) for $C_7H_{10}O_2S_2$ (190.27): C, 44.19; H, 5.30; Found: C, 44.15; H, 5.36.

Reference

Yavari, I., Ghazanfarpour-Darjani, M., Hossaini, Z., Sabbaghan, M., and Hosseini, N. (2008). Methoxide ion promoted efficient synthesis of 1,3-oxathiolane-2-thiones by reaction of oxiranes and carbon disulfide. *Synlett*, 889–891.

C-S bond_19

Type of reaction: C-S and C-N bond formation
Reaction conditions: Dichloromethane, room temperature
Synthetic strategy: One-pot DMAP-promoted cascade C-S/C-N bonds formation
Catalyst: DCC/DMAP
Keywords: α-Halocarboxylic acid, β-ketothioamides, *N,N′*-dicyclohexylcarbodiimide (DCC), 4-dimethylaminopyridine (DMAP), dichloromethane, room temperature, one-pot annulations, thiazolidin-4-ones

General reaction scheme

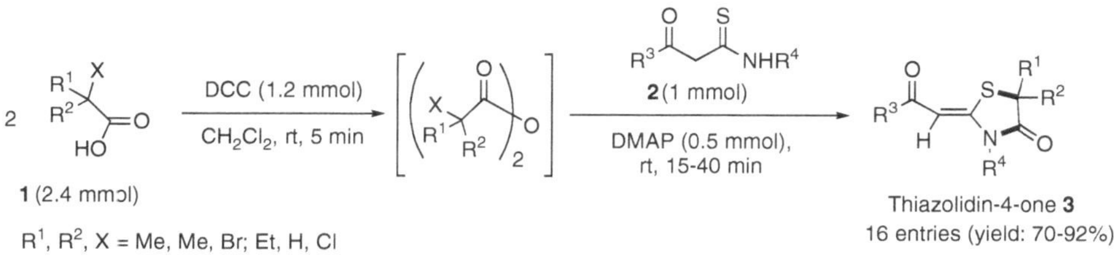

R^3, R^4 = 4-BrC_6H_4, Me; C_6H_5, C_6H_5; 4-BrC_6H_4, C_6H_5; 3,4-OCH_2O-C_6H_3, Me; 2-thienyl, Me; 2-furyl, C_6H_5; 2-naphthyl, C_6H_5; 4-Ph-C_6H_4, C_6H_5; ^{t}Bu, C_6H_5; 4-MeC_6H_4, Me; 4-OMeC_6H_4, 2-Cl$C_6H_4CH_2$;

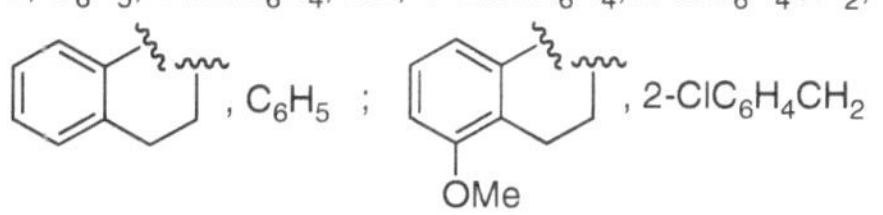

Representative entries

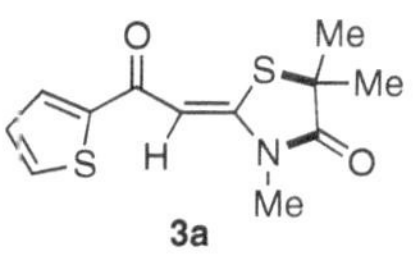

3a

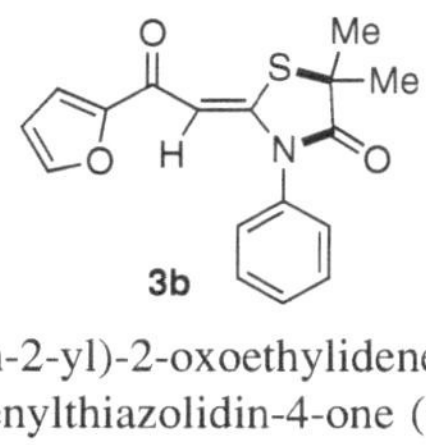

3b

(*Z*)-3,5,5-Trimethyl-2-(2-oxo-2-(thiophen-2-yl)ethylidene)thiazolidin-4-one (**3a**): Yellow solid, mp. 160–162 °C; reaction time: 15 min; yield: 92%

(*Z*)-2-(2-(Furan-2-yl)-2-oxoethylidene)-5,5-dimethyl-3-phenylthiazolidin-4-one (**3b**): Yellow solid, mp. 141–142 °C; reaction time: 25 min; yield: 88%

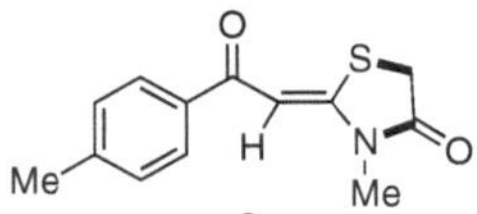

3c

(*Z*)-3-Methyl-2-(2-oxo-2-(*p*-tolyl)ethylidene) thiazolidin-4-one (**3c**): Yellow solid, mp. 171–173 °C; reaction time: 30 min; yield: 80%

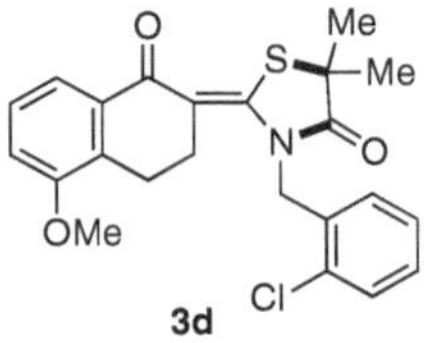

3d

(*Z*)-3-(2-Chlorobenzyl)-2-(5-methoxy-1-oxo-3,4-dihydronaphthalen-2(1*H*)-ylidene)-5,5-dimethylthiazolidin-4-one (**3d**): Yellow solid, mp. 230–231 °C; reaction time: 40 min; yield: 70%

Experimental procedure

To a solution of α-halocarboxylic acid (**1**; 2.4 mmol) in 5 mL of dichloromethane was added 1.2 mmol of DCC, and the mixture was stirred at room temperature for 5 min. After formation of the white precipitate, β-ketothioamide (**2**; 1.0 mmol) was added followed by addition of 0.5 mmol of DMAP. Reaction mixture was stirred at room temperature for the stipulated period of time (15–40 min). Upon completion of the reaction (monitored by TLC), the white precipitate was filtered out followed by washing with dichloromethane (2 × 5 mL). The combined organic layer was washed with water (2 × 15 mL) and dried over anhydrous sodium sulfate. On removal of solvent evaporated under reduced pressure, a crude product was obtained which was then purified by column chromatography over silica gel (ethyl acetate/hexane as eluent) to afford the pure thiazolidin-4-one **3**. The products were identified by means of spectral and analytical measurements.

Characterization data of two representative compounds

(*Z*)-3,5,5-Trimethyl-2-(2-oxo-2-(thiophen-2-yl)ethylidene)thiazolidin-4-one (**3a**): Yellow solid, mp. 160–162 °C; IR (KBr): 2961, 1712, 1614, 1528, 1416, 1344, 1241, 1153, 1068 cm^{-1}; ^{1}H NMR (300 MHz, $CDCl_3$): δ 7.69 (d, *J* = 3.6 Hz, 1H), 7.57 (d, *J* = 4.5 Hz, 1H), 7.11 (t, *J* = 3.6 Hz, 1H), 6.50 (s, 1H), 3.31 (s, 3H), 1.61 (s, 6H); ^{13}C NMR (75 MHz, $CDCl_3$): δ 181.0, 178.4, 158.6, 145.5, 132.1, 129.5, 127.9, 94.9, 48.8, 30.3, 27.7; HRMS [ESI]: calcd for $C_{12}H_{13}NO_2S_2$ $(M+H)^+$: *m*/*z* 268.0466, Found: 268.0460.

((*Z*)-3-(2-Chlorobenzyl)-2-(5-methoxy-1-oxo-3,4-dihydronaphthalen-2(1*H*)-ylidene)-5,5-dimethylthiazolidin-4-one (**3d**): Yellow solid, mp. 230–231 °C; IR (KBr): 2930, 2840, 1712, 1629, 1582, 1493, 1298, 1263, 1175, 1037 cm^{-1}; ^{1}H NMR (300 MHz, $CDCl_3$): δ 7.32-7.31 (m, 5H), 6.92-6.83 (m, 2H), 4.85 (d, *J* = 5.1 Hz, 2H), 3.83 (s, 3H), 2.93-2.84 (m, 4H), 1.95 (s, 6H); ^{13}C NMR (75 MHz, $CDCl_3$): δ 197.2, 170.6, 156.0, 139.4, 134.4, 134.0, 130.5, 129.7, 129.4, 129.0, 127.0, 124.6, 119.5, 114.5, 113.6, 111.6, 55.6, 54.7, 48.8, 30.7, 28.0, 27.9, 20.1.

Reference

Verma, G. K., Shukla, G., Nagaraju, A., Srivastava, A., and Singh, M. S. (2014). DMAP-promoted cascade C-S/C-N bonds formation approach to 1,3-thiazolidin-4-ones via annulation of β-ketothioamides with α-halocarboxylic acids at room temperature. *Tetrahedron*, **70**, 6980–6984.

C-S bond_20

Types of reaction: C-S bond formation
Reaction conditions: DMSO, room temperature
Synthetic strategy: One-step S-arylation
Catalyst: DDQ
Keywords: *o*-Halobenzothiaoureas, DDQ, DMSO, room temperature, S-arylation, 2-aminobenzothiazoles

General reaction scheme

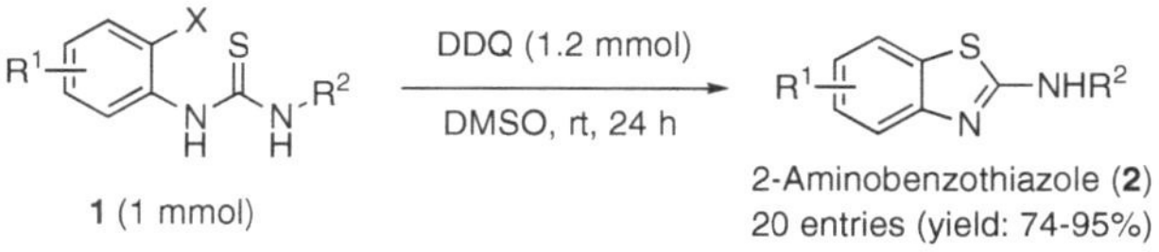

X = I, Br
R^1 = H, 4-Cl, 4-F, 4-CF_3, 4-Me, 4-OMe, 4,5-di-Me
R^2 = C_6H_5, 4-MeC_6H_4, 4-ClC_6H_4, 4-BrC_6H_4, 4-OMeC_6H_4, 4-FC_6H_4, 4-$NO_2C_6H_4$, *c*-hex

Representative entries

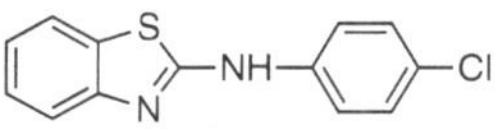

2a

N-(4-Chlorophenyl)benzo[*d*]thiazol-2-amine (**2a**): Yield: 89%

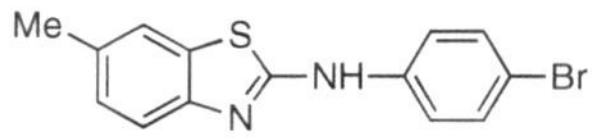

2b

N-(4-Bromophenyl)-6-methylbenzo[*d*]thiazol-2-amine (**2b**): Yield: 95%

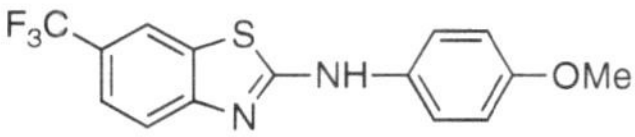

2c

N-(4-Methoxyphenyl)-6-(trifluoromethyl)benzo[*d*]thiazol-2-amine (**2c**): Yield: 77%

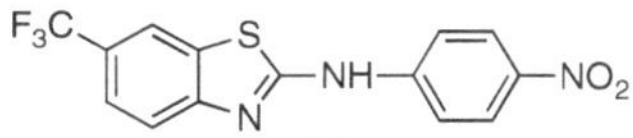

2d

N-(4-Nitrophenyl)-6-(trifluoromethyl)benzo[*d*]thiazol-2-amine (**2a**): Yield: 86%

Experimental procedure

To a solution of (*o*-iodo/bromo)thiaourea (**1**; 1.0 mmol) in DMSO (3 mL) was added with DDQ (1.2 mmol) in five equal portions. The mixture was then stirred at room temperature for 24 h. On completion of the reaction (TLC monitoring), water (10 mL) and a saturated aqueous $NaHCO_3$ solution (5 mL) were added, and then the aqueous solution was extracted with ethylacetate (3 × 15 mL). The combined organic extracts were dried over anhydrous Na_2SO_4 and concentrated, and then the crude residue was purified by column chromatography [eluent: PE–EtOAc (5:1 to 7:1)] on silica gel to afford pure 2-aminobenzothiazole 2 in good to excellent yield. Each of the products was characterized on the basis of spectral studies.

Characterization data of two representative compounds

N-(4-Chlorophenyl)benzo[*d*]thiazol-2-amine (**2a**): ^{1}H NMR (400 MHz, DMSO-d_6): δ 7.14 (t, *J* = 8.4 Hz, 1H), 7.31 (t, *J* = 7.6 Hz, 1H), 7.40 (d, *J* = 8.8 Hz, 2H), 7.57 (d, *J* = 8 Hz, 2H), 7.80 (d, *J* = 10 Hz, 2H), 10.62 (br, 1H); ^{13}C NMR (100 MHz, DMSO-d_6): δ 119.6, 119.8, 121.6, 123.0, 125.8, 126.4, 129.3, 130.5, 140.0, 152.4, 161.7; ESI-HRMS: *m/z* calcd. for $[C_{13}H_9ClN_2S+H]^+$: 260.0175; Found: 260.0183.

N-(4-Bromophenyl)-6-methylbenzo[*d*]thiazol-2-amine (**2b**): ^{1}H NMR (400 MHz, DMSO-d_6): δ 2.35 (s, 3H), 7.14 (d, *J* = 8.2 Hz, 1H), 7.49-7.54 (m, 3H), 7.62 (s, 1H), 7.79 (d, *J* = 8.8 Hz, 2H), 10.53 (br, 1H); ^{13}C NMR (100 MHz, DMSO-d_6): δ 21.3, 113.5, 119.5, 119.9, 121.5, 127.5, 130.5, 132.1, 132.3; 10.5, 150.3, 160.9; ESI-HRMS: *m*/*z* calcd. for $[C_{14}H_{11}BrN_2S+H]^+$: 318.9905; Found: 318.9937.

Reference

Wang, R., Yang, W.-J., Yue, L., Pan, W., and Zeng, H.-Y. (2012). DDQ-promoted C-S bond formation: Synthesis of 2-aminobenzothiazole derivatives under transition-metal-, ligand-, and base-free conditions. *Synlett*, **23**, 1643–1648.

C-S bond_21

Type of reaction: C-S bond formation
Reaction conditions: *N,N*-Dimethylformamide, room temperature
Synthetic strategy: Ligand-free Pd/C-catalyzed C-S bond formation
Catalyst: Pd/C
Keywords: *o*-Iodobenzanilides, Pd/C catalyst, *N,N*-dimethylformamide, room temperature, cyclization reaction, ligard-free and additive-free synthesis, 2-alkyl/aryl-substituted benzothiazoles

General reaction scheme

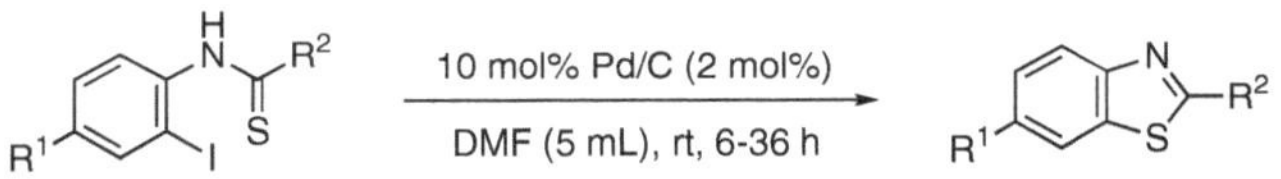

1 (0.5 mmol)

2-Alkyl/aryl-substituted benzothiazole **2**
24 entries (yield: 71-94%)

R^1 = H, Me, Cl, F (in addition, a pyridine nucleus)
R^2 = Alkyl: Me, *n*-Pr, *n*-Bu, Me_3CCH_2, cyclopropyl, cyclohexyl, *t*-Bu, adamantyl, CH_2NHcbz, PhCH=CH
Aryl: C_6H_5, 4-ClC_6H_4, 4-OMeC_6H_4, 3-OMeC_6H_4, 4-CNC_6H_4

Representative entries

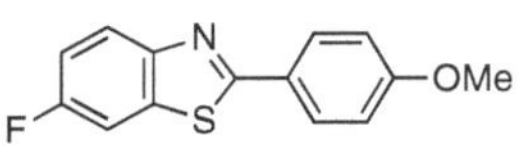

2a

6-Fluoro-2-(4-methoxyphenyl)benzo[*d*]thiazole (**2a**): White solid, mp. 132.9-133.7 °C; reaction time: 12 h; yield: 91%

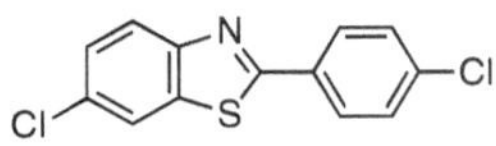

2b

6-Chloro-2-(4-chlorophenyl)benzo[*d*]thiazole (**2b**): White solid, mp. 200.9-203.1 °C; reaction time: 6 h; yield: 88%

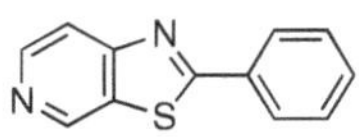

2c

2-Phenylthiazolo[5,4-*c*]pyridine (**2c**): White solid, mp. 137.8-138.4 °C; reaction time: 12 h; yield: 78%

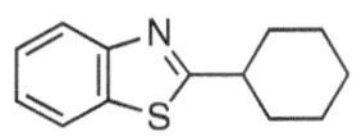

2d

2-Cyclohexylbenzo[d]thiazole (**2d**): Colorless oil; reaction time: 6 h; yield: 86%

Experimental procedure

A solution of *o*-iodothiobenzanilide (**1**; 0.5 mmol) in DMF (5 mL) was stirred in the presence of 10% Pd/C (2 mol %) at room temperature under a nitrogen atmosphere. After completion of the reaction as monitored by TLC, the reaction mixture was diluted with ethyl acetate (10 mL) and filtered through a Celite pad to remove Pd/C. To the filtrate was added water (15 mL). The layers were separated, and the aqueous layer was extracted with ethyl acetate (3 × 10 mL). The combined organic layers were washed with brine (10 mL) and dried over magnesium sulfate. Flash silica gel column chromatography rinsed with petroleum ether/ethyl acetate afforded the desired benzothiazole product **2**. The products were identified by means of spectral measurements.

Characterization data of two representative compounds

6-Fluoro-2-(4-methoxyphenyl)benzo[*d*]thiazole (**2a**): White solid, mp. 132.9-133.7 °C; IR 3080, 2836, 2357, 1601, 1453, 1250, 968, 813, 697, 575 cm^{-1}; ^{1}H NMR (400 MHz, $CDCl_3$): δ 7.92-7.97 (m, 3H), 7.51 (dd, $J=8.0$ and 1.9 Hz, 1H), 7.18 (td, $J=8.0$ and 2.0 Hz, 1H), 6.96 (d, $J=8.6$ Hz, 2H), 3.85 (s, 3H); ^{13}C NMR (100 MHz, $CDCl_3$): δ 167.50, 161.87, 160.2 (d, $J=244.0$ Hz), 150.77, 135.74 (d, $J=11$ Hz), 128.92, 126.08, 123.55 (d, $J=9$ Hz), 114.64 (d, $J=24.0$ Hz), 114.34, 107.70 (d, $J=27.0$ Hz) 55.39; HRMS (ESI) calcd for $[M+H]^+$ $C_{14}H_{11}FNOS$ 260.0545; Found 260.0534.

2-Phenylthiazolo[5,4-*c*]pyridine (**2c**): White solid, mp. 137.8-138.4 °C; IR 3036, 2363, 1578, 1474, 1405, 1230, 965, 843, 772, 688, 569 cm^{-1}; ^{1}H NMR (400 MHz, $CDCl_3$): δ 9.20 (d, $J=0.7$Hz, 1H), 8.65 (d, $J=5.6$ Hz, 1H), 8.13 (dd, $J=8.0$ and 1.5 Hz, 2H), 7.93 (dd, $J=5.6$ and 0.7 Hz, 1H), 7.59-7.49 (m, 3H); ^{13}C NMR (101 MHz, $CDCl_3$): δ 173.3, 158.8, 145.9, 144.2, 132.7, 132.1, 132.0, 129.2, 128.00, 117.4; HRMS (ESI) calcd for $[M+H]^+$ $C_{12}H_9N_2S$ 213.0490; Found 213.0481.

Reference

Cheng, Y., Peng, Q., Fan, W., and Li, P. (2014). Room-temperature ligand-free Pd/C-catalyzed C – S bond formation: synthesis of 2-substituted benzothiazoles. *J. Org. Chem.*, **79**, 5812–5819.

Chapter 5

Carbon – Phosphorus Bond Forming Reactions at Room Temperature

C-P bond_1

Types of reaction: C-P bond formation
Reaction conditions: Dichloromethane, room temperature
Synthetic strategy: One-step transformation
Catalyst: Zinc bromide ($ZnBr_2$)
Keywords: Arylmethyl and heteroarylmethyl halides/alcohols, zinc bromide, dichloromethane, room temperature, Michaelis-Arbuzov reaction, arylmethyl and heteroarylmethyl phosphonate esters, aryl aldehydes, diethyl [hydroxyl(arylmethyl)]phosphonate

General reaction scheme

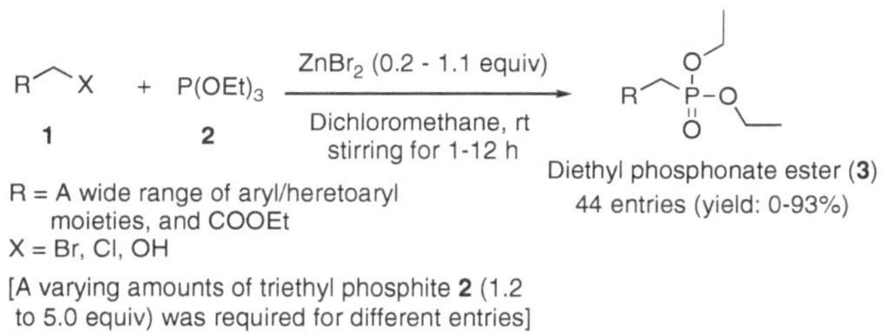

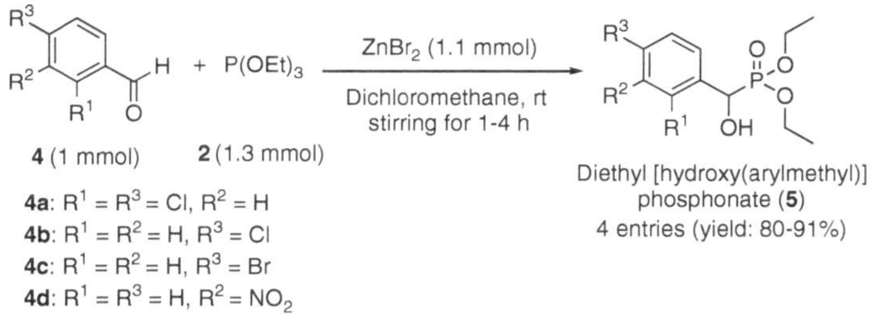

Representative entries

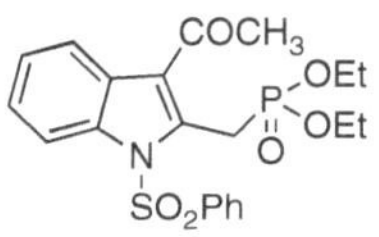

3a

Diethyl ((3-acetyl-1-(phenylsulfonyl)-1*H*-indol-2-yl)methyl)phosphonate [**3a**; colorless solid, mp 88-90 °C, $P(OEt)_3$: 1.2 equiv; $ZnBr_2$: 0.2 equiv; reaction time: 5 h; yield: 85%]

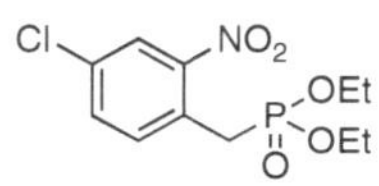

3b

Diethyl 4-chloro-2-nitrobenzyl phosphonate [**3b**; thick yellow liquid, $P(OEt)_3$: 1.2 equiv; $ZnBr_2$: 0.2 equiv; reaction time: 5 h; yield: 88%]

Room Temperature Organic Synthesis. http://dx.doi.org/10.1016/B978-0-12-801025-9.00005-X

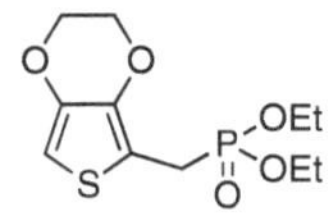

3c

Diethyl ((2,3-dihydrothieno[3,4-*b*][1,4]dioxin-5-yl)methyl)phosphonate [**3c**; colorless oil, $P(OEt)_3$: 5 equiv; $ZnBr_2$: 1.1 equiv; reaction time: 1 h; yield: 90%]

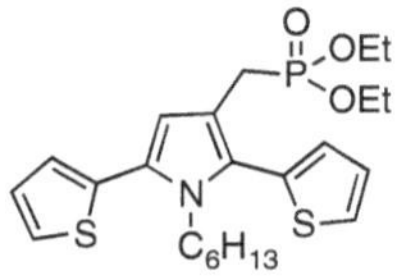

3d

Diethyl ((1-hexyl-2,5-di(thiophen-2-yl)-1*H*-pyrrol-3-yl)methyl)phosphonate [**3d**; colorless liquid, $P(OEt)_3$: 5 equiv; $ZnBr_2$: 1.1 equiv; reaction time: 2 h; yield: 80%]

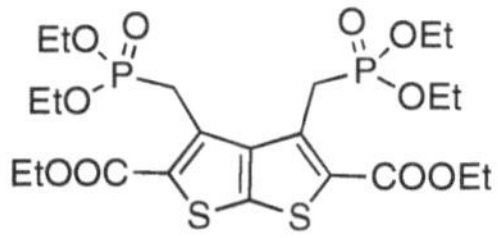

3e

Diethyl 3,4-bis((diethoxyphosphoryl) methyl) thieno[2,3-b]thiophene-2,5-dicarboxylate [**3e**; colorless oil, $P(OEt)_3$: 2.5 equiv; $ZnBr_2$: 0.4 equiv; reaction time: 7 h; yield: 85%]

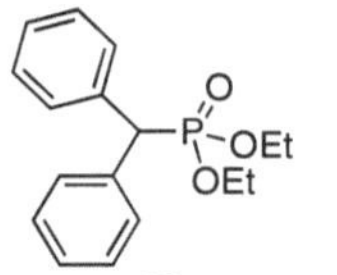

3f

Diethyl benzhydrylphosphonate [**3f**; colorless solid, mp 40-42 °C, $P(OEt)_3$: 5 equiv; $ZnBr_2$: 1.1 equiv; reaction time: 4 h; yield: 89%]

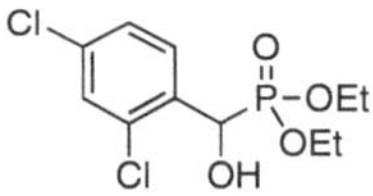

5a

Diethyl ((2,4-dichlorophenyl)(hydroxy)methyl) phosphonate [**5a**; colorless oil, $P(OEt)_3$: 1.3 equiv; $ZnBr_2$: 1.1 equiv; reaction time: 1 h; yield: 91%]

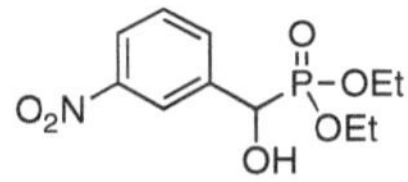

5b

Diethyl (hydroxy(3-nitrophenyl)methyl) phosphonate [**5b**; yellow solid, mp 92-94 °C, P $(OEt)_3$: 1.3 equiv; $ZnBr_2$: 1.1 equiv; reaction time: 4 h; yield: 80%]

Experimental procedure

To a solution of arylmethyl/heteroarylmethyl halide/alcohol (**1**; 1 mmol) or arylaldehyde (**4**; 1 mmol) and triethylphosphite (**2**; 1.2 to 5 mmol for varying substrates) in dry dichloromethane at room temperature, $ZnBr_2$ (0.2 to 1.1 mmol for varying substrates) was added and allowed to stir for required time span (1-12 h) under nitrogen atmosphere. After consumption of the substrate compound (monitored by TLC) volatile components were removed under *vacuo*. The residual mass then poured over crushed ice containing conc. HCl. The precipitated solid was filtered, washed with water and dried to give crude phosphonate ester duly purified by flash column chromatography to afford pure product **3** and/or **5**. All the products were characterized from their analytical and detailed spectral studies.

Characterization data of two representative compounds

Diethyl ((1-hexyl-2,5-di(thiophen-2-yl)-1*H*-pyrrol-3-yl)methyl)phosphonate (**3d**): Colorless liquid; ^{1}H NMR (300 MHz, $CDCl_3$): δ 7.44-7.42 (1 H, m, Ar-H), 7.30-7.27 (1 H, m, Ar-H), 7.15-7.10 (2 H, m, Ar-H), 7.08-7.05 (2 H, m, Ar-H), 6.45 (1 H, s, Ar-H), 4.09-3.95 (6H, m, -N-CH_2 & -OCH_2), 2.99 (2 H, d, $J = 20.4$ Hz, CH_2-P), 1.49-1.42 (2 H, m, -CH_2-), 1.27 (6 H, t, $J = 7.1$ Hz, -CH_3), 1.17-1.05 (6 H, m, -CH_2-), 0.78 (3 H, t, $J = 6.9$ Hz, -CH_3); ^{13}C NMR (75.4 MHz, $CDCl_3$): δ 134.83, 132.33, 129.73, 127.50, 127.24, 127.18, 127.16, 125.92 (d, $J = 10.9$ Hz), 125.73, 125.07, 113.54 (d, $J = 8.5$ Hz), 111.64 (d, $J = 2.9$ Hz), 61.92 (d, $J = 6.7$ Hz), 45.32, 31.15, 31.03, 26.04, 24.46 (d, $J = 142.6$ Hz), 22.34, 16.43 (d, $J = 6.1$ Hz), 13.9; HRMS (FAB^+): Calc for $C_{23}H_{32}NO_3PS_2$ $(M)^+$ at *m/z* 465.1561; found: 465.1731. Anal calcd for $C_{23}H_{32}NO_3PS_2$: C, 59.33; H, 6.93; N, 3.01; S, 13.77; Found: C, 59.51; H, 7.22; N, 3.24; S, 13.96.

Diethyl (hydroxy(3-nitrophenyl)methyl) phosphonate (**5b**): Yellow solid, mp 92-94 °C; 1H NMR (300 MHz, $CDCl_3$): δ 8.41 (1 H, s, ArH), 8.16 (1 H, d, $J = 8.1$ Hz, Ar-H), 7.82 (1 H, d, $J = 8.1$ Hz, Ar-H), 7.53 (1 H, t, $J = 8.0$ Hz, Ar-H), 5.71 (1 H, s, -OH), 5.18 (1 H, d, $J = 11.7$ Hz, -CH-P), 4.19-4.09 (4 H, m, -OCH_2-), 1.32-1.24 (6 H, m, -CH_3); ^{13}C NMR (75.4 MHz, $CDCl_3$): δ 148.19, 139.40, 133.05 (d, $J = 5.3$ Hz), 129.0 (d, $J = 2.26$ Hz), 122.77 (d, $J = 3.0$ Hz), 122.03 (d, $J = 5.3$ Hz), 69.78 (d, $J = 159.8$ Hz), 63.97 (d, $J = 7.20$ Hz), 63.32 (d, $J = 7.54$ Hz), 16.39 (d, $J = 2.26$ Hz), 16.31 (d, $J = 3$ Hz).

Reference

Rajeshwaran, G. G., Nandakumar, M., Sureshbabu, R., and Mohanakrishnan, A. K. (2011). Lewis acid-mediated Michaelis-Arbuzov reaction at room temperature: A facile preparation of arylmethyl/heteroarylmethyl phosphonates. *Org. Lett.*, **13**, 1270–1273.

C-P bond_2

Types of reaction: C-P bond formation
Reaction conditions: Dichloromethane, room temperature
Synthetic strategy: Multi-component one-pot protocol for the Wolff-Kishner type reductive deoxygenation of acyl phosphonates
Reagents: Oxalyl chloride [$(COCl)_2$]/DMF, triethylphosphite [$P(OEt)_3$], hydrazine/THF/benzoic acid, ^tBuOK
Keywords: Carboxylic acids, acyl chlorides, dichloromethane, room temperature, Arbuzov reaction, acyl phosphonates, Wolff-Kishner type reductive deoxygenation, alkyl phosphonates

General reaction scheme

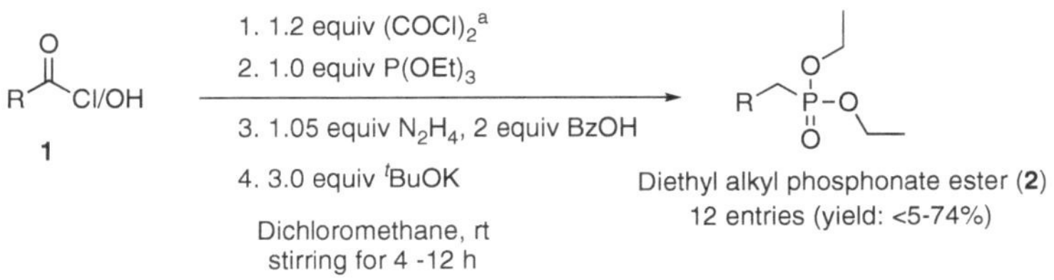

R = CH_3, C_2H_5, $(CH_3)_2CH$, *tert*-butyl, C_6H_5, $C_6H_5CH_2$, $C_6H_5CH_2CH_2$, 4-$OCH_3C_6H_4CH_2CH_2$, 4-$CNC_6H_4CH_2CH_2$, 4-$OAcC_6H_4CH_2CH_2$, 4-$NHAcC_6H_4CH_2CH_2$,

([a]Oxalyl chloride used in case of starting carboxylic acid derivatives)

Proposed mechanism

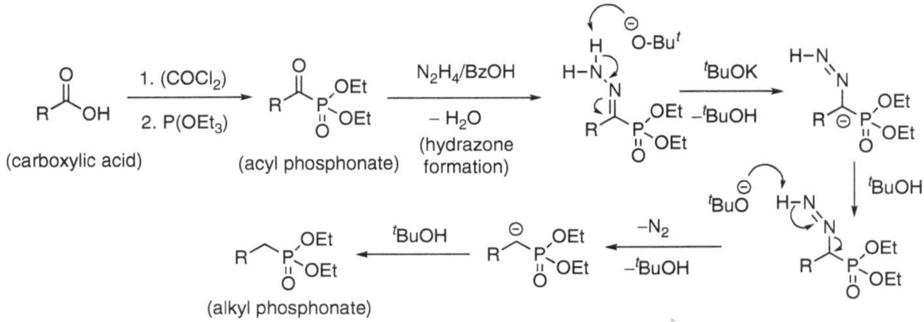

Representative entries

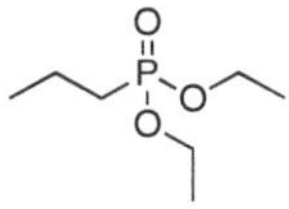

2a

Diethyl propylphosphonate (**2a**): Colorless liquid, reaction time: 6 h; yield: 74%

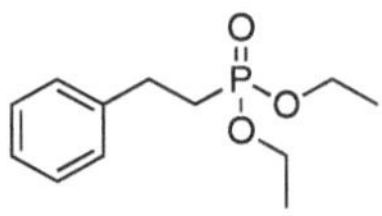

2b

Diethyl 2-phenylethylphosphonate (**2b**): Yellow liquid; reaction time: 6 h; yield: 65%

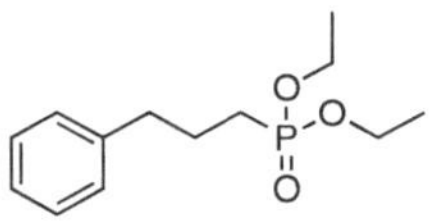

2c

Diethyl 3-phenylpropyl-1-phosphonate (**2c**): Faintly yellow liquid; reaction time: 6 h; yield: 74%

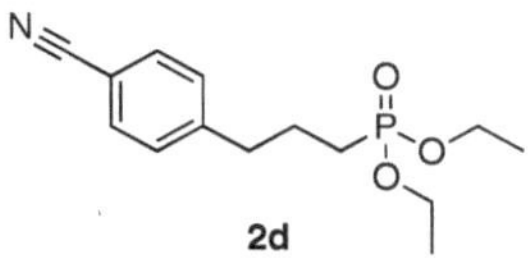

2d

Diethyl 3-(4-cyanophenyl)propyl-1-phosphonate (**2d**): Colorless liquid, reaction time: 6 h; yield: 58%

Experimental procedure

In case of carboxylic acid starting materials, they were first converted into their corresponding acyl chlorides by reacting with oxalyl chloride in dichloromethane. A cooled dichloromethane solution of an acyl chloride (**1**; 1 mmol) in a 25 mL round bottomed flask was then added with triethylphosphite (1 mmol) drop-wise under stirring, and the reaction mixture was allowed to attain room temperature. On completion of the formation of acyl phosphonate, the reaction mixture was reduced *in vacuo*, and benzoic acid (2 mmol) was added directly to the residue, followed by benzene. The mixture was stirred to dissolve completely, then hydrazine solution in THF (1.05 mmol) was added drop-wise under rapid stirring. On completion of hydrazone formation, the reaction was flash-frozen and lyophilized. After lyophilization, 1:1 v/v tetrahydrofuran: *tert*-butanol was added to the flask and stirred to dissolve the solid. Potassium *tert*-butoxide solution (3 mmol) in 1:1 v/v tetrahydrofuran : *tert*-butanol was then added to the stirring solution in one portion. After stirring at room temperature for several hours, the reaction was diluted with ethyl acetate (50 mL), quenched with 1 N HCl (20 mL), washed with saturated sodium bicarbonate (2 × 15 mL), then with brine (15 mL), dried on sodium sulfate, reduced *in vacuo*, then purified by flash chromatography on silica (15% acetone in dichloromethane) to give the desired product, alkyl phosphonate (**2**). All the products were characterized from their spectral studies including ^{1}H-, ^{13}C- and ^{31}P- NMR, and HRMS.

Characterization data of two representative compounds

Diethyl 2-phenylethylphosphonate (**2b**): Yellow liquid; ^{1}H NMR (300 MHz, $CDCl_3$): δ 7.16-7.30 (m, 5H), 4.03-4.13 (m, 4H), 2.85-2.94 (m, 2H), 1.98-2.10 (m, 2H), 1.30 (t, $J=7.0$ Hz, 6H); ^{13}C NMR (75 MHz, $CDCl_3$): δ 141.1, 140.8, 128.5, 128.0, 126.3, 61.6, 61.5, 28.6, 28.5, 28.5, 26.6, 16.5, 16.4; ^{31}P NMR (121 MHz, $CDCl_3$): δ 30.8; HRMS(FAB^+) calculated for [$C_{12}H_{20}O_3P$] ($[M+H]^+$) 243.1150; Found 243.1158.

Diethyl 3-(4-cyanophenyl)propyl-1-phosphonate (**2d**): Colorless liquid; ^{1}H NMR (300 MHz, $CDCl_3$): δ 7.51-7.55 (m, 2H), 7.23-7.27 (m 2H), 3.96-4.13 (m, 4H), 2.73 (t, $J=7.5$ Hz, 2H), 1.83-1.97 (m, 2H), 1.63-1.74 (m, 2H), 1.27 (t, $J=7.1$ Hz, 6H); ^{13}C NMR (75 MHz, $CDCl_3$): δ 146.7, 132.2, 129.3, 118.9, 110.0, 61.6, 61.5, 36.5, 36.3, 25.9, 24.0, 23.8, 23.7, 16.5, 16.4; ^{31}P NMR (121 MHz, $CDCl_3$): δ 31.3; HRMS (FAB^+) calculated for [$C_{14}H_{21}O_3NP$] ($[M+H]^+$) 282.1259; Found 282.1262.

Reference

Kedrowski, S. M. A., and Dougherty, D. A. (2010). Room-temperature alternative to the Arbuzov reaction: the reductive deoxygenation of acyl phosphonates. *Org. Lett.*, **12**, 3990–3993.

C-P bond_3

Types of reaction: C-P bond formation
Reaction conditions: Dichloromethane, room temperature
Synthetic strategy: One-step transformation
Catalyst: Indim(III) chloride ($InCl_3$)
Keywords: *N*-Benzoyloxycarbonylamino sulfones, triethyl phosphite, $InCl_3$, dichloromethane, room temperature, Arbuzov reaction, *N*-Benzoyloxycarbonylamino phosphonates

General reaction scheme

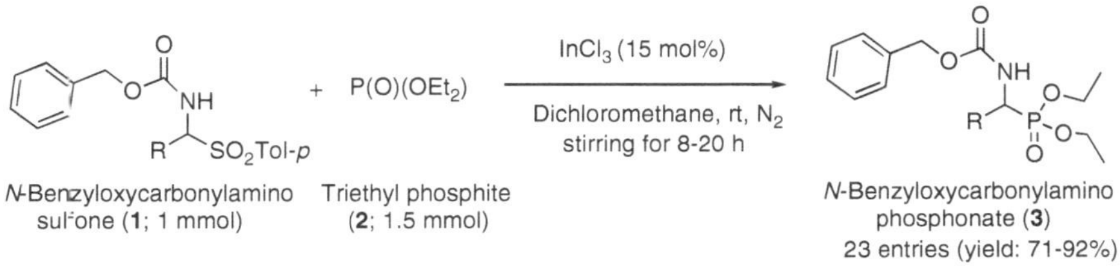

R = C_3H_5, 4-$CH_3C_6H_4$, 4-$(CH_3)_2CHC_6H_4$, 4-ClC_6H_4, 3-ClC_6H_4, 2,4-di-Cl-C_6H_3, 4-BrC_6H_4, 4-FC_6H_4, 3-Cl, 4-FC_6H_3, 2,4,6-tri-F-ClC_6H_2, 4-$OCH_3C_6H_4$, 3,4,5-tri-OCH_3-ClC_6H_2, 4-$NO_2C_6H_4$, 3-$NO_2C_6H_4$, 3,4-(-OCH_2O-)C_6H_3, vanilyl, 2-naphthyl, 2-furyl, 2-thienyl, benzyl, *n*-propyl, *n*-pentyl, isobutyl

Representative entries

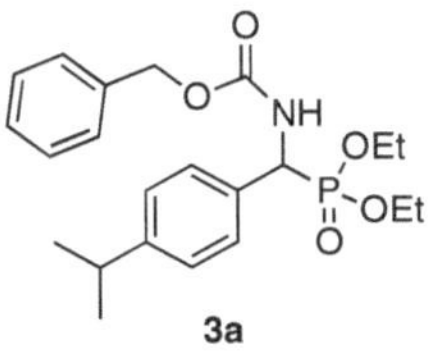

3a

Benzyl ((diethoxyphosphoryl)(4-isopropylphenyl) methyl)carbamate (**3a**): White solid, mp 92 -94 °C; reaction time: 9 h; yield: 91%

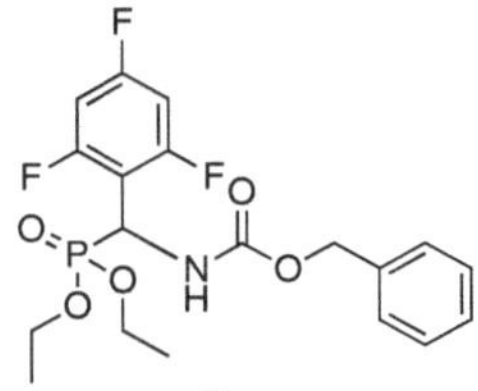

3b

Benzyl ((diethoxyphosphoryl)(2,4,6-trifluorophenyl)methyl)carbamate (**3b**): White solid, mp 72 -74 °C; reaction time: 12 h; yield: 84%

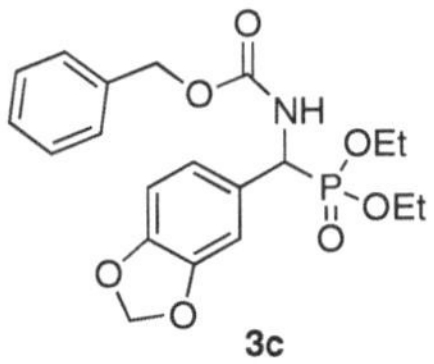

3c

Benzyl (benzo[*d*][1,3]dioxol-5-yl (diethoxyphosphoryl)methyl)carbamate (**3c**): White solid, mp 123 -125 °C; reaction time: 8 h; yield: 91%

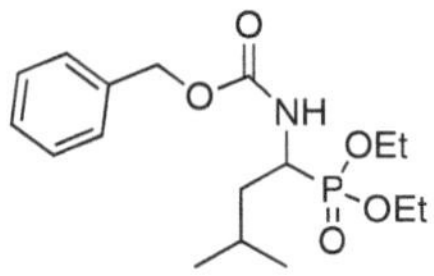

3d

Benzyl (1-(diethoxyphosphoryl)-3-methylbutyl) carbamate (**3d**): Colorless liquid, reaction time: 11 h; yield: 84%

Experimental procedure

Triethyl phosphite (**2**, 1.5 mmol) was added drop-wise to a solution of *N*-benzyloxycarbonyl amino sulfone (**1**; 1 mmol), and $InCl_3$ (15 mol%) in dichloromethane (10 mL) under nitrogen atmosphere. The mixture was then stirred (8-20 h) at room temperature and the reaction was monitored by TLC. After completion, distilled water (5 mL) was added, and the mixture was extracted with ethyl acetate (3 × 10 mL). The combined organic portions were washed with water (2 × 10 mL), dried over anhydrous sodium sulfate, and concentrated under vacuum. The residue was subjected to column chromatography on silica gel (20-35% EtOAc/hexane) to obtain pure protected α-amino phosphonate **3** with good yields (71-92%). All the products were characterized from their spectral studies including IR, ^{1}H NMR, ^{13}C NMR, and HRMS.

Characterization data of two representative compounds

Benzyl ((diethoxyphosphoryl)(4-isopropylphenyl) methyl)carbamate (**3a**): White solid, mp 92 -94 °C; IR (KBr): 3229, 1714, 1546, 1252, 1034 cm^{-1}; ^{1}H NMR (200 MHz, $CDCl_3$): δ 7.39-7.21 (m, 7H), 7.15 (d, J=8.0 Hz, 2H), 6.17 (dd, J=10.0, 4.0 Hz, 1H), 5.18-4.98 (m, 3H), 4.12-4.00 (m, 2H), 3.88 (m, 1H), 3.67 (m, 1H), 2.89 (m, 1H), 1.24 (t, J=7.0 Hz, 3H), 1.22 (d, J=7.0 Hz, 6H), 1.05 (t, J=7.0 Hz, 3H); ^{13}C NMR (50 MHz, $CDCl_3$): δ 155.9 (d, J=9.0 Hz), 149.0, 136.1, 132.7, 128.8, 128.2, 128.0, 126.9, 67.1, 63.3 (d, J=6.2 Hz), 63.1 (d, J=6.5 Hz), 52.2 (d, J=150.0 Hz), 34.0, 24.0, 16.2 (d, J=6.0 Hz), 16.0 (d, J=6.0 Hz); ESIMS: *m/z* 442 $[M+Na]^+$; HRMS (ESI): Calcd for $C_{22}H_{30}NO_5PNa$ $[M+Na]^+$: 442.1759; Found: 442.1765.

Benzyl (benzo[*d*][1,3]dioxol-5-yl(diethoxyphosphoryl)methyl)carbamate (**3c**): White solid, mp 123 -125 °C; IR (KBr): 3227, 1717, 1551, 1469, 1256, 1029 cm^{-1}; ^{1}H NMR (200 MHz, $CDCl_3$): δ 7.29 (brs, 5H), 6.99 (d, J=2.0 Hz, 1H), 6.89 (dd, J=8.0, 2.0 Hz, 1H), 6.72 (d, J=8.0 Hz, 1H), 6.18 (dd, J=10.0, 4.0 Hz, 1H), 5.92 (s, 2H), 5.12 (d, J=12.0 Hz, 1H), 5.01 (d, J=12.0 Hz, 1H), 4.99 (dd, J=24.0, 10.0 Hz, 1H), 4.17-4.01 (m, 2H), 3.93 (m, 1H), 3.70 (m, 1H), 1.25 (t, J=7.0 Hz, 3H), 1.11 (t, J=7.0 Hz, 3H); ^{13}C NMR (50 MHz, $CDCl_3$): δ 155.8 (d, J=9.0 Hz), 148.0, 147.6, 136.2, 129.2, 128.5, 128.2, 121.4, 108.3, 101.1, 67.3,

63.5 (d, $J=6.5$ Hz), 62.8 (d, $J=6.5$ Hz), 52.1 (d, $J=155.0$ Hz), 16.1 (d, $J=6.0$ Hz), 16.0 (d, $J=6.0$ Hz); ESIMS: m/z 422 $[M+H]^+$; HRMS (ESI): Calcd for $C_{20}H_{25}NO_7P$ $[M+H]^+$: 422.1368; Found: 422.1365.

Reference

Das, B., Damodar, K., and Bhunia, N. (2009). A simple and efficient access to α-amino phosphonates from *N*-benzyloxycarbonylamino sulfones using Indium(III) chloride. *J. Org. Chem.*, **74**, 5607–5609.

C-P bond_4

Types of reaction: C-P and C-C bond formation
Reaction conditions: Solvent-free, room temperature
Synthetic strategy: One-pot multicomponent reaction
Catalyst: 5-Hydroxypentylammonium acetate (5-HPAA; reusable task-specific ionic liquid)
Keywords: Alkyl/aryl/heteroaryl aldehydes, nitromethane, trialkylphosphites, 5-HPAA, solvent-free, room temperature, one-pot tandem Henry-Michael reaction, multicomponent reaction, terminal vinylphosphonates

General reaction scheme

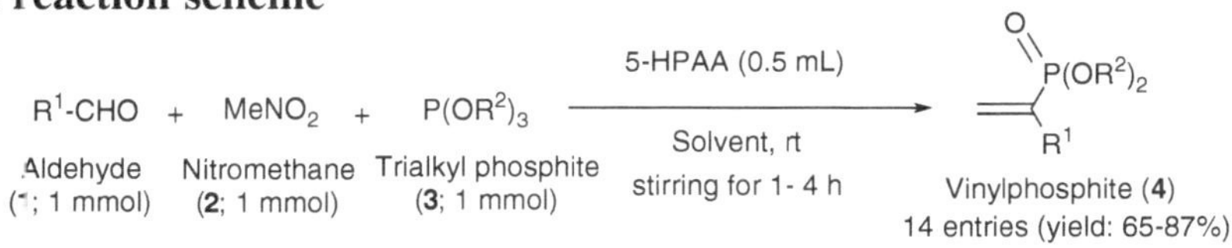

R^1 = C_6H_5, 4-$CH_3C_6H_4$, 4-ClC_6H_4, 4-BrC_6H_4, 4-$OCH_3C_6H_4$, 3-$OCH_3C_6H_4$, 2-$OCH_3C_6H_4$, 1-naphthyl, 3-indolyl, 2-thienyl, *n*-propyl, isopropyl
R^2 = Et, Me, *i*-Pr

Representative entries

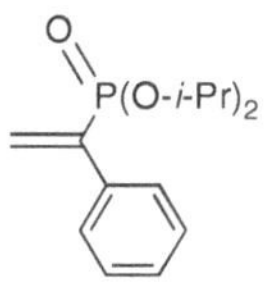

4a

Di-*iso*-propyl (1-phenylvinyl)phosphonate (**4a**): Colorless oil, reaction time: 1.2 h; yield: 85%

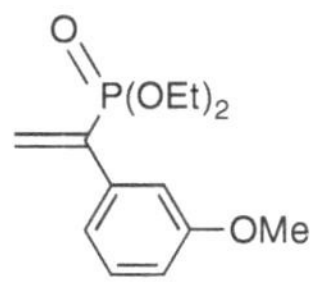

4b

Diethyl (1-(3-methoxyphenyl)vinyl) phosphonate (**4b**): Yellow oil, reaction time: 1 h; yield: 81%

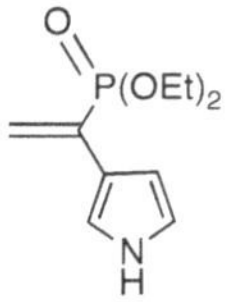

4c

Diethyl 1-(1*H*-indol-3-yl)ethenylphosphonate (**4c**): Yellow oil, reaction time: 1.5 h; yield: 83%

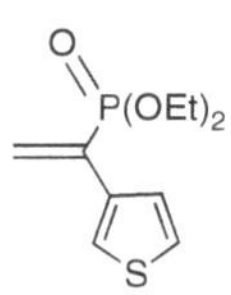

4d

Diethyl 1-(1*H*-indol-3-yl)ethenylphosphonate (**4d**): Yellow oil, reaction time: 2 h; yield: 80%

Experimental procedure

To a stirred mixture of aldehyde (**1**; 1 mmol), nitromethane (**2**; 1 mmol) and trialkylphosphite (**3**; 1 mmol) was added with 5-hydroxypentylammonium acetate (5-HPAA; 05 mL). After stirring the resulting mixture at room temperature for the appropriate time (1-4 h), H_2O (10 mL) was added and the reaction mixture was extracted with ethyl acetate (3 × 10 mL). The combined organic layers were dried over Na_2SO_4 and filtered. Evaporation of the filtrate gave an almost pure product, which was purified further by chromatography on silica (*n*-hexane–EtOAc, 1:2). The separated aqueous layer containing 5-HPAA was evaporated, dried under reduced pressure at 100 °C for 24 h and reused in subsequent reactions. The products were characterized on the basis of spectral analyses.

Characterization data of two representative compounds

Diethyl (1-(3-methoxyphenyl)vinyl) phosphonate (**4b**): Yellow oil; IR (neat): 1621, 1510, 1235, 1011, 803 cm^{-1}; ^1H NMR (400 MHz): δ 7.29 (t, *J*=4.0 Hz, 1H, Ar), 7.12 (d, *J*=8.8 Hz, 2H, Ar), 6.89 (d, *J*=8.0 Hz, 1H, Ar), 6.35 (d, *J*=21.6 Hz, 1H, $=CH_2$), 6.18 (d, *J*=45.6 Hz, 1H, $=CH_2$), 4.04-4.20 (m, 4H, OCH_2CH_3), 3.83 (s, 3H, OCH_3), 1.30 (t, *J*=7.2 Hz, 6H, OCH_2CH_3).

Diethyl 1-(1*H*-indol-3-yl)ethenylphosphonate (**4c**): Yellow oil; IR (neat): 1601, 1499, 1222, 1001, 821 cm^{-1}; ^1H NMR (400 MHz): δ 8.91 (brs, 1H, N*H*), 8.86 (s, 1H, Ar), 7.62 (d, *J*=7.6 Hz, 1H, Ar), 7.27 (d, *J*=7.6 Hz, 1H, Ar), 7.09-7.13 (m, 2H, Ar), 6.35 (d, *J*=48.0 Hz, 1H, $=CH_2$), 6.23 (d, *J*=22.8 Hz, 1H, $=CH_2$), 3.99-4.09 (m, 4H, OCH_2CH_3), 0.96 (t, *J*=7.2 Hz, 6H, OCH_2CH_3).

Reference

Sobhani, S., and Honarmand, M. (2013). A simple and efficient method for one-pot, three-component synthesis of terminal vinylphosphonates using a task-specific ionic liquid. *Synlett*, **24**, 236–240.

C-P bond_5

Types of reaction: C-P bond formation
Reaction conditions: Dichloromethane, room temperature
Synthetic strategy: Hydrophosphorylation of imines
Catalyst: *p*-Tosyl chloride (*p*-TsCl)
Keywords: Imines, diethyl phosphate, *p*-TsCl, dichloromethane, room temperature, hydrophosphorylation, α-aminophosphonates

General reaction scheme

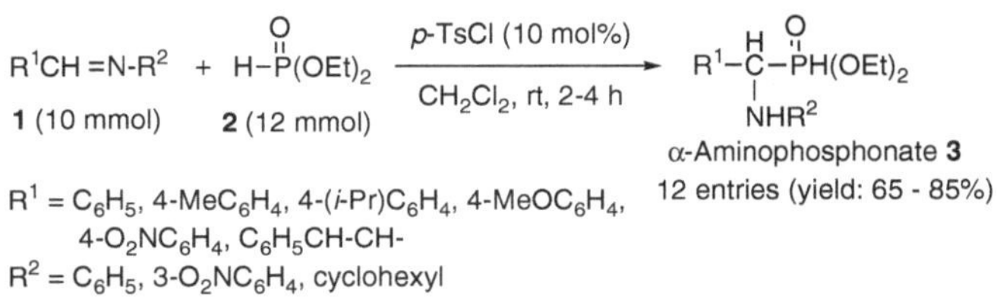

Plausible mechanism

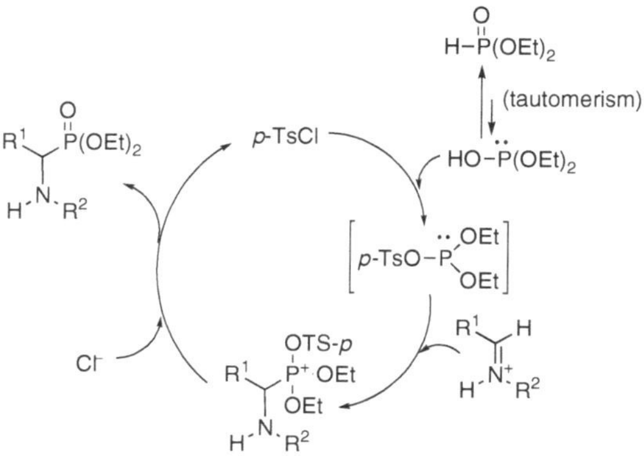

Representative entries

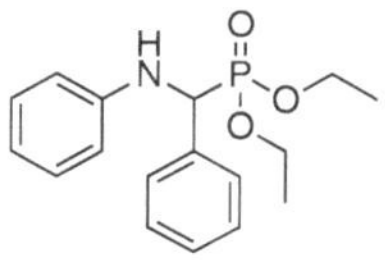

3a

Diethyl[phenyl(phenylamino)methyl]phosphonate (**3a**): Reaction time: 2 h; yield: 83%

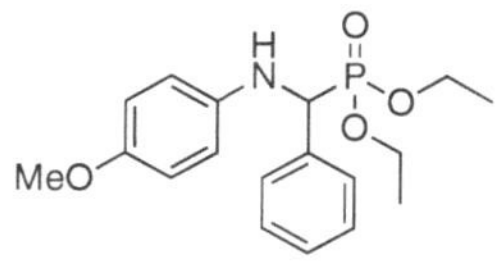

3b

Diethyl (((4-methoxyphenyl)amino) (phenyl) methyl)phosphonate (**3b**): Reaction time: 3 h; yield: 72%

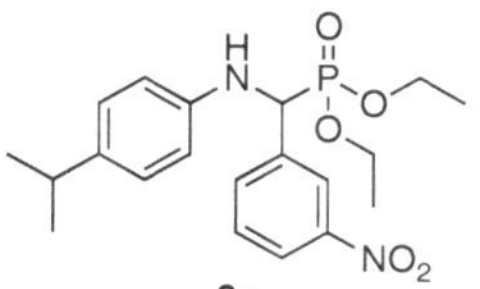

3c

Diethyl (((4-isopropylphenyl)amino)(3-nitrophenyl)methyl)phosphonate (**3c**;): Reaction time: 4 h; yield: 68%)

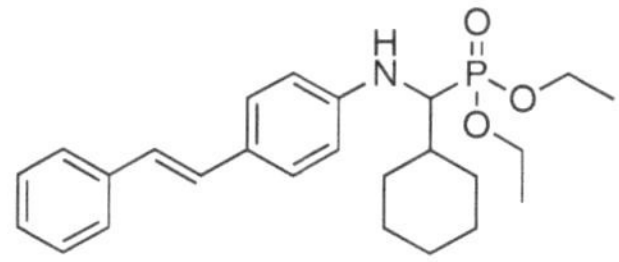

3d

(*E*)-Diethyl (cyclohexyl((4-styrylphenyl)amino) methyl)phosphonate (**3d**): Reaction time: 3 h; yield: 73%

Experimental procedure

p-TsCl (1 mmol) was added to a stirred mixture of diethyl phosphite (**2**; 12 mmol) in dichloromethane (10 mL) at room temperature. Then 10 mmol of imine (**1**) was added to the reaction mixture and the mixture was stirred for 2-4 h at room temperature. On completion of the reaction, water (50 mL) was added to the reaction mixture and the organic layer was extracted with EtOAc (4 × 50 mL), dried with $CaCl_2$, and the solvent was evaporated under reduced pressure to afford the crude product of α-aminophosphonate **3**, which was purified using column chromatography on silica gel (EtOAc-*n*-hexane mixture as eluent). All the products were characterized by spectral analyses.

Characterization data of one representative compound

Diethyl[phenyl(phenylamino)methyl]phosphonate (**3a**): ^{1}H NMR (250 MHz, $CDCl_3$): δ 1.13 (t, $J=7.0$ Hz, 3H), 1.29 (t, $J=7.0$ Hz, 3H), 3.60-3.75 (m, 1H), 3.85-4.01 (m, 1H), 4.03-4.22

(m, 2H), 4.76 (d, J_{HP}=24.2 Hz, 1H), 6.59 (d, J=8.5 Hz, 2H), 6.69 (t, J=7.2 Hz, 1H), 7.11 (t, J=8.3 Hz, 2H), 7.22-7.52 (m, 5H); ^{31}P NMR (101.2 MHz, $CDCl_3$, H_3PO_4): δ 23.31; ^{13}C NMR (62.9 MHz, $CDCl_3$): δ 16.2 (d, J_{PC}=6.3 Hz), 16.4 (d, J_{PC}=6.3 Hz), 56.1 (d, J_{PC}=150.3 Hz), 63.1, 63.2, 113.8, 118.3, 127.9 (d, J_{PC}=5.7 Hz), 128.5 (d, J_{PC}=2.5 Hz), 129.2, 136.0 (d, J_{PC}=2.5 Hz), 146.3 (d, J_{PC}=14.5 Hz).

Reference

Kabaoudin, B., and Jafari, E. (2008). Hydrophosphorylation of imines catalyzed by tosyl chloride for the synthesis of α-aminophosphonates. *Synlett*, 1837–1839.

C-P bond_6

Types of reaction: C-P and C-C bond formation
Reaction conditions: Water, room temperature
Synthetic strategy: One-pot multi-component hydrophosphorylation
Catalyst: Indium (In)/ HCl
Keywords: Nitro-compounds, carbonyl compounds, dialkyl or trialkyl phosphites, Indium (In) metal, hydrochloric acid, water, room temperature, reduction of nitro-group, imines, hydrophosphorylation, α-aminophosphonates

General reaction scheme

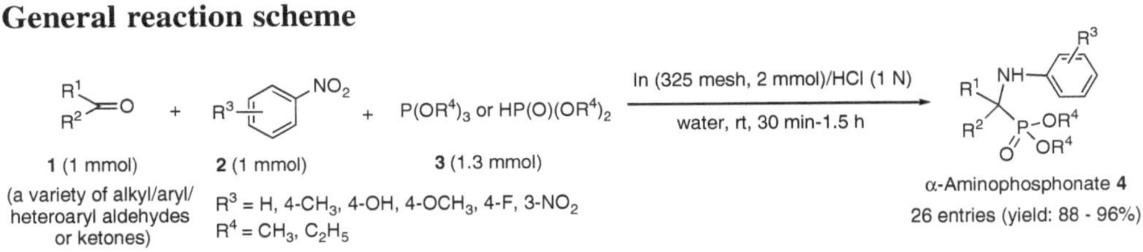

Representative entries

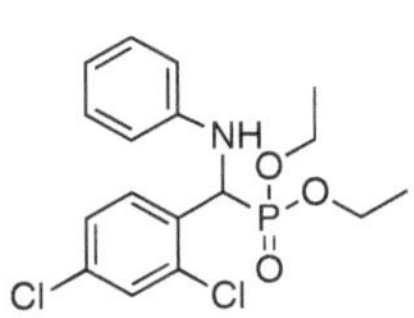

4a

Diethyl (2,4-dichlorophenyl)(phenylamino) methylphosphonate (**4a**): Reaction time: 40 min; yield: 92%

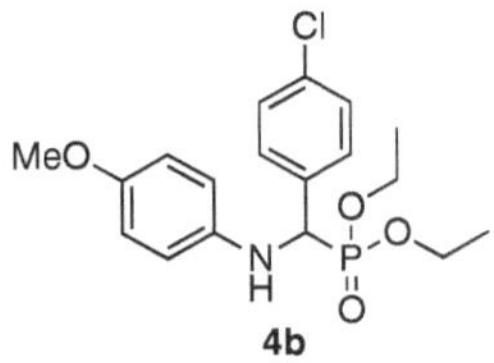

4b

Diethyl (4-chlorophenyl)(4-methoxyphenylamino)methylphosphonate (**4b**): Reaction time: 40 min; yield: 95%

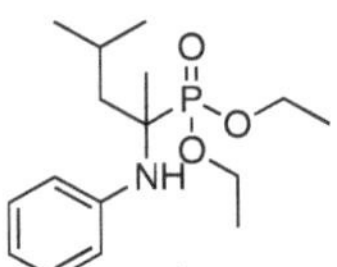

4c

Diethyl 4-methyl-2-(phenylamino)pentan-2-ylphosphonate (**4c**): Reaction time: 90 min; yield: 93%)

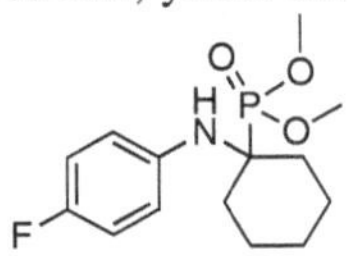

4d

Dimethyl 1-(4-fluorophenylamino) cyclohexylphosphonate (**4d**): Reaction time: 75 min; yield: 92%

Experimental procedure

To a mixture of nitro compound (**2**; 1.0 mmol) were added indium powder (325 mesh; 2 mmol), 1 N aqueous HCl (1 mL), carbonyl compound (**1**; 1.0 mmol), alkyl phosphite (**3**; 1.3 mmol), and water (5 mL). The mixture was then stirred at room temperature. On completion of the reaction (as monitored by TLC), it was washed with saturated sodium bicarbonate solution (3 × 5 mL) and water (3 × 5 mL) and extracted with ethyl acetate (3 × 5 mL). The extract was concentrated, and the residue was subjected to column chromatography (silica gel, hexane/EtOAc) to obtain pure α-amino phosphonate (**4**). All the products were characterized by spectral analyses.

Characterization data of two representative compounds

Diethyl (2,4-dichlorophenyl)(phenylamino)methylphosphonate (**4a**): IR: 3309, 1607, 1508, 1247 cm^{-1}; ^{1}H NMR (200 MHz, $CDCl_3$): δ 7.51 (d, J=8.0 Hz, 1H), 7.40 (d, J=2.0 Hz, 1H), 7.22 (dd, J=8.0, 2.0 Hz, 1H), 7.06 (t, J=8.0 Hz, 2H), 6.64 (t, J=8.0 Hz, 1H), 6.48 (d, J=8.0 Hz, 2H), 5.23 (dd, J=24.0 & 10.0 Hz, 1H), 4.82 (t, J=10.0 Hz, 1H), 4.25-4.12 (m, 2H), 3.91 (m, 1H), 3.67 (m, 1H), 1.35 (t, J=7.0 Hz, 3H), 1.11 (t, J=2.0 Hz, 3H); ^{13}C NMR (50 MHz, $CDCl_3$): δ 146.2 (d, J=14.0 Hz), 130.2, 129.8, 128.2, 119.1, 113.5, 96.2, 63.4 (d, J=6.5 Hz), 63.2 (d, J=6.5 Hz), 51.6 (d, J=155.2 Hz), 16.6 (d, J=6.5 Hz), 16.2 (d, J=6.5 Hz); ESIMS: m/z 387 $[M]^+$, 388, 390, 392 $[M+H]_+$; HRMS (ESI): Calcd. for $C_{17}H_{20}Cl_2NO_3PNa$: m/z 410.0455; Found: m/z 410.0459 $[M+Na]^+$.

Diethyl 4-methyl-2-(phenylamino)pentan-2-ylphosphonate (**4c**): IR: 3319, 1602, 1500, 1322, 1225 cm^{-1}; ^{1}H NMR (200 MHz, $CDCl_3$): δ 7.11(t, J=8.0 Hz, 2H), 6.90 (d, J=8.0 Hz, 2H), 6.75 (t, J=8.0 Hz, 1H), 4.12-3.90 (m, 4H), 3.52 (brs, 1H), 2.01-3.82 (m, 2H), 1.69 (m, 1H), 1.50 (d, J=15.0 Hz, 3H) 1.24 (t, J=7.0 Hz, 3H), 1.20 (t, J=7.0 Hz, 3H), 0.98 (d, J=7.0 Hz, 3H), 0.91 (d, J=7.0 Hz, 3H); ^{13}C NMR (50 MHz, $CDCl_3$): δ 146.2, 128.9, 119.5, 118.8, 62.5 (d, J=6.2 Hz), 62.2 (d, J=6.5 Hz), 57.9 (d, J=155.0 Hz), 43.4, 25.0, 23.4 (d, J=6.5 Hz), 21.2, 16.1 (d, J=6.0 Hz); ESIMS: m/z 314 $[M+H]^+$; HRMS (ESI): Calcd. for $C_{16}H_{28}NO_3PNa$: m/z 336.1704; Found: m/z 336.1711 $[M+Na]^+$.

Reference

Das, B., Satyalakshmi, G., Suneel, K., and Damodar, K. (2009). Organic reactions in water: A distinct novel approach for an efficient synthesis of α-amino phosphonates starting directly from nitro compounds. *J. Org. Chem.*, **74**, 8400–8402.

C-P bond_7

Types of reaction: C-P bond formation
Reaction conditions: Ionic liquid (IL), room temperature
Synthetic strategy: One-pot multi-component reaction
Catalyst: 3-(*N*,*N*-dimethyldodecylammonium)propanesulfonic acid ([DDPA][HSO_4])
Keywords: Aromatic aldehydes, anilines, triethylphosphite/diethylphosphite, [DDPA][HSO_4] (IL), room temperature, One-pot multi-component reaction, α-aminophosphonates

General reaction scheme

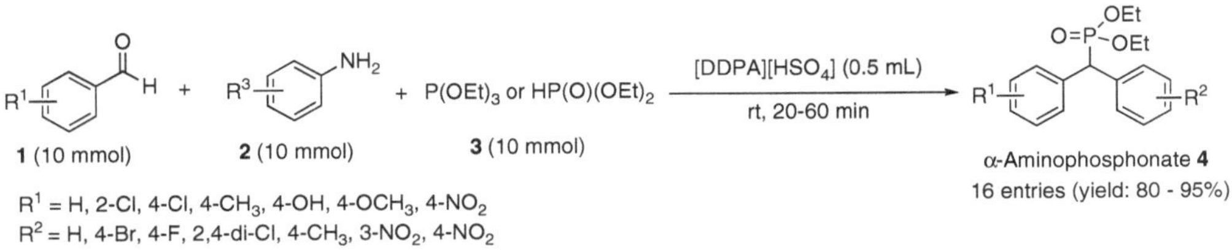

Representative entries

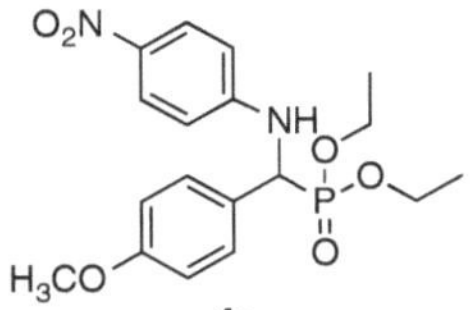

4a

Diethyl ((4-methoxyphenyl)((4-nitrophenyl) amino)methyl)phosphonate) (**4a**): Reaction time: 60 min; yield: 93%

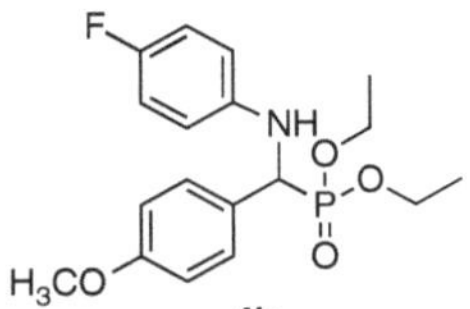

4b

Diethyl (((4-fluorophenyl)amino)(4-methoxyphenyl)methyl)phosphonate (**4b**): Reaction time: 60 min; yield: 83%

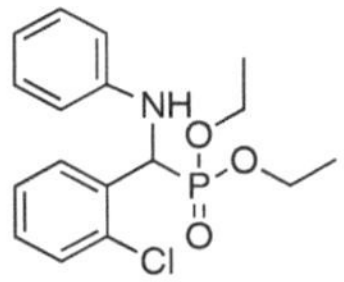

4c

Diethyl ((2-chlorophenyl) (phenylamino)methyl) phosphonate (**4c**): Reaction time: 20 min; yield: 95%

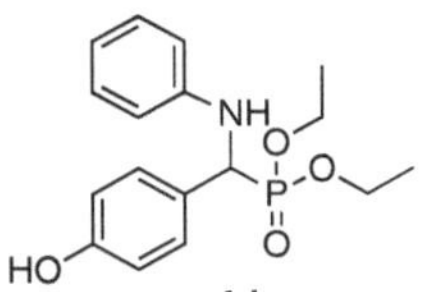

4d

Diethyl ((4-hydroxyphenyl) (phenylamino)methyl) phosphonate (**4d**): Reaction time: 30 min; yield: 91%

Experimental procedure

To a round-bottomed flask charged with aldehyde (**1**; 10 mmol) and aniline (**2**; 10 mmol) was added [DDPA][HSO_4] (0.5 mmol) under stirring. The mixture was stirred at room temperature for 5 min and then triethylphosphate/diethylphosphite (**3**; 10 mmol) was added and stirring was continued for stipulated time (20-60 min). On completion of the reaction as monitored by TLC, 1 mL water was added to extract the IL-catalyst from the mixture, and the products, α-aminophosphonates (**4**), were obtained by filtration and dried under vacuum. The products were identified by comparison of their physical and spectral data with those reported for authentic samples in literature cited in this article.

Reference

Fang, D., Cao, Y., and Yang, J. (2013). Clean procedure for the synthesis of α-aminophosphonates catalyzed by biodegradable ionic liquid. *Phosphorus Sulfur Silicon Relat. Elem.*, **188**, 826–832.

C-P bond_8

Types of reaction: C-P bond formation
Reaction conditions: Solvent-free, room temperature
Synthetic strategy: One-pot multi-component reaction
Catalyst: Nano-TiO_2
Keywords: Aromatic aldehydes, anilines, triethylphosphite, TiO_2-nanoparticles, solvent-free, room temperature, one-pot multi-component reaction, α-aminophosphonates

General reaction scheme

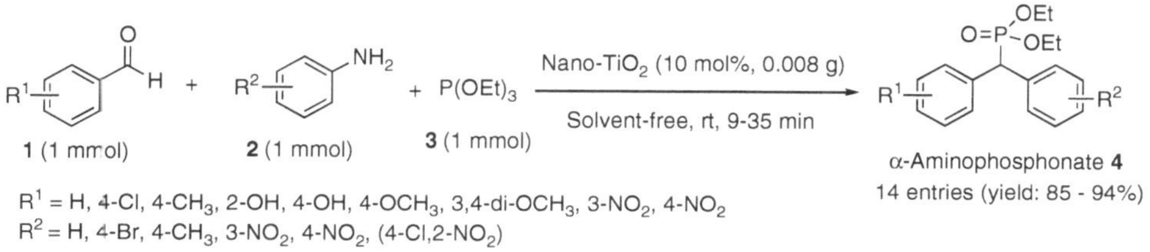

Representative entries

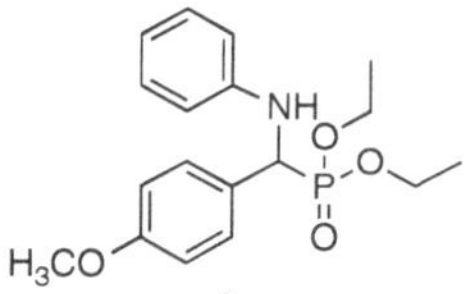

4a

Diethyl ((4-methoxyphenyl)(phenylamino) methyl) phosphonate (**4a**): Solid, mp 110-111 °C; reaction time: 18 min; yield: 90%

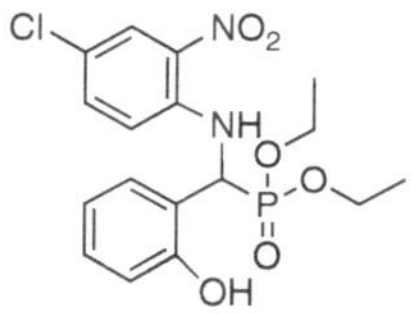

4b

Diethyl (((4-chloro-2-nitrophenyl)amino)(2-hydroxyphenyl)methyl)phosphonate (**4b**): Solid, mp 127-128 °C; reaction time: 20 min; yield: 92%

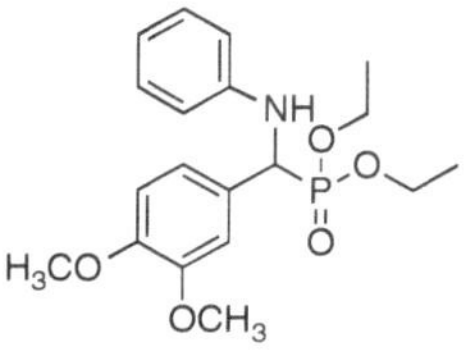

4c

Diethyl ((3,4-dimethoxyphenyl)(phenylamino) methyl)phosphonate (**4c**): Solid, mp 76-78 °C; reaction time: 18 min; yield: 94%

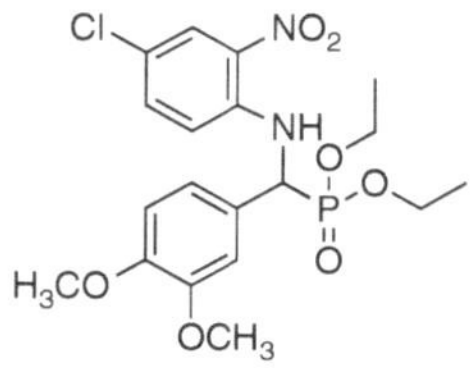

4d

Diethyl (((4-chloro-2-nitrophenyl)amino)(3,4-dimethoxyphenyl)methyl)phosphonate (**4d**): Solid, mp 108 °C; reaction time: 19 min; yield: 93%

Experimental procedure

Arylaldehyde (**1**; 1 mmol), aromatic amine (**2**; 1 mmol), triethylphosphite (**3**; 1 mmol), and nano-TiO_2 (10 mol%, 0.008 g) were mixed together under solvent-free and ambient conditions, and then was continued to stir at room temperature for stipulated time (9-35 min). On completion of the reaction (monitored by TLC), dichloromethane was added to the reaction mixture and the catalyst was removed from the reaction mixture by centrifugation, washed with dichloromethane, and dried. The recovered nano-TiO_2 catalyst was reused for four cycles without any significant loss in its activity.

The organic solvent was removed under reduced pressure, and the crude residue was subjected to column chromatographic purification on silica gel (ethyl acetate/*n*-hexane 20:80 as eluent) so as to obtain pure products of α-aminophosphonates **4**. All products were characterized by NMR and FTIR spectral data and compared with the known compounds in the literature cited in this article.

Reference

Shaterian, H. R., Farbodeh, J., and Mohammadnia, M. (2013). Nano-TiO_2: An eco-friendly and clean reusable heterogeneous catalyst for preparation of α-aminophosphonates under ambient and solvent-free conditions. *Phosphorus Sulfur Silicon Relat. Elem.*, **188**, 850–854.

C-P bond_9

Types of reaction: C-P bond formation
Reaction conditions: Dichloromethane, room temperature
Synthetic strategy: One-pot multi-component reaction
Catalyst: $TiCl_4$
Keywords: Pyridine-3-carboxyaldehyde (nicotinaldehyde), anilines, triethylphosphite, $TiCl_4$, dichloromethane, room temperature, one-pot multi-component reaction, diphenyl 1-(arylamino)(pyridine-3-yl)methylphosphonates

General reaction scheme

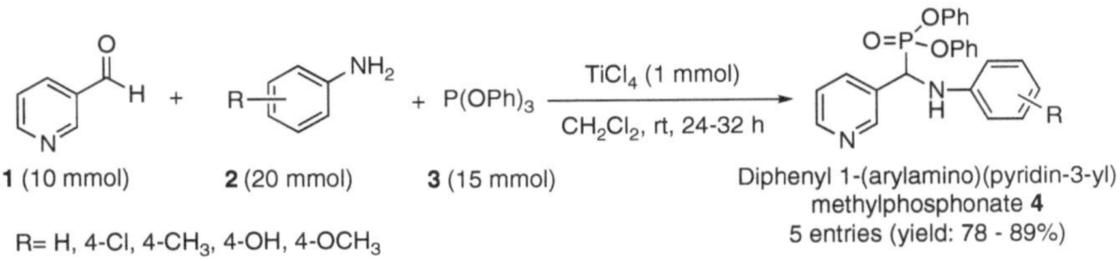

Representative entries

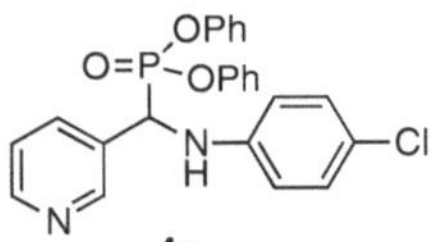

4a

Diphenyl (((4-chlorophenyl)amino)(pyridin-3-yl) methyl)phosphonate (**4a**): White solid, mp 135-137 °C; reaction time: 28 h; yield: 78%

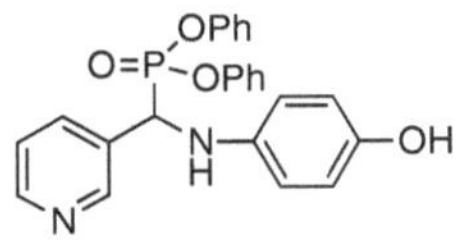

4b

Diphenyl (((4-hydroxyphenyl)amino) (pyridin-3-yl)methyl)phosphonate (**4b**): White solid, mp 205-207 °C; reaction time: 24 h; yield: 81%

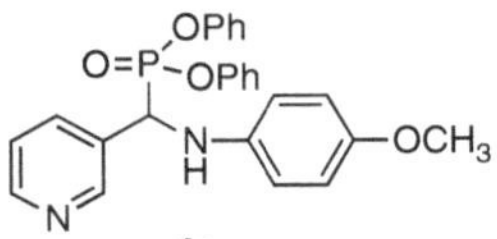

4c

Diphenyl (((4-methoxyphenyl)amino)(pyridin-3-yl)methyl)phosphonate (**4c**): White solid, mp 160-162 °C; reaction time: 30 h; yield: 88%

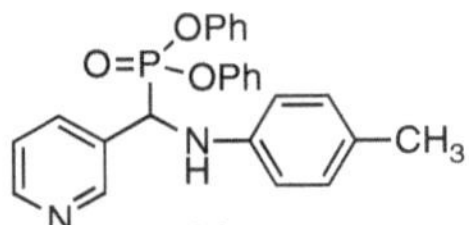

4d

Diphenyl (pyridin-3-yl(p-tolylamino)methyl) phosphonate (**4d**): White solid, mp 194-196 °C; reaction time: 32 h; yield: 89%

Experimental procedure

A mixture of nicotinaldehyde (**1**; 10.0 mmol, 1.07 g), aromatic amine (**2**; 20.0 mmol), triphenylphosphite (**3**; 15.0 mmol, 4.65 g), and titanium tetrachloride ($TiCl_4$; 1.0 mmol, 0.19 g) in dichloromethane (10 mL) was stirred at room temperature for 24-32 h; the progress of the reaction was monitored by TLC. After the completion of reaction, the resulting residue obtained on removal of the solvent under reduced pressure was treated with methanol (20 mL), and then filtered to remove the solid materials. A mixture of water (10 mL) and dichloromethane (20 mL) was added to the filtrate, and the organic phase was separated and dried over anhydrous sodium sulfate. The solvent was then removed under reduced pressure to give the crude product purified on recrystallization from ethanol to give the pure products of diphenyl 1-(arylamino)(pyridine-3-yl)methylphosphonates **4** as white solids. All products were characterized on the basis of elemental analyses and detailed spectral studies.

Characterization data of two representative entries

Diphenyl (((4-chlorophenyl)amino)(pyridin-3-yl)methyl)phosphonate (**4a**): Solid, mp 135-137 °C; IR (KBr): 3305 (NH), 1588 (C=N), 1361 (P=O), 820 (P–O–C) cm^{-1}; ^{1}H NMR (400 MHz, DMSO-d_6): δ 8.88-6.97 (m, 19H, Ar-H, and NH), 5.75 (d, $J=13$ Hz, 1H, CH); ^{31}P NMR (202.48 MHz, DMSO-d_6): δ 16.1; AP$^+$–MS: *m/z* (rel.%) 453 ([M^{37}Cl+H]$^+$, 33), 451 ([M^{35}Cl+H]$^+$, 100), 417 (12), 385 (27), 368 (30), 327 (32), and 249 (11); HRMS (AP$^+$): *m/z* [M+H]$^+$ calcd for $C_{24}H_{21}{}^{35}ClN_2O_3P$: 451.0978; Found: 451.0963. Anal calcd for $C_{24}H_{20}ClN_2O_3P$ (450.85): C, 63.94; H, 4.47; N, 6.21; P, 6.87; Found: C, 64.20; H, 4.21; N, 6.16; P, 6.85%.

Diphenyl (((4-methoxyphenyl)amino)(pyridin-3-yl)methyl)phosphonate (**4c**): Solid, mp 160-162 °C; IR (KBr): 3422 (NH), 1582 (C=N), 1325 (P=O), and 801 (P–O–C) cm^{-1}; ^{1}H NMR (400 MHz, DMSO-d_6): δ 8.93-6.71 (m, 19H, Ar–H, and NH); 5.69 (d, $J=16$ Hz, 1H, CH), 4.17 (s, 3H, OCH_3); ^{31}P NMR (202.48 MHz, DMSO-d_6): δ 15.1; AP$^+$–MS: *m/z* (rel.%) 447 ([M+H]$_+$, 3), 415 (2), 294 (5), 193 (6), 140 (8), 117 (22), 100 (100), 97 (57), and 77 (47); HRMS (AP$^+$): *m/z* [M+H]$^+$ calcd for $C_{25}H_{24}N_2O_4P$: 447.1474; Found: 447.1478. Anal calcd for $C_{25}H_{24}N_2O_4P$ (446.43): C, 67.26; H, 5.19; N, 6.27; P, 6.94; Found: C, 67.22; H, 5.02; N, 6.57; P, 6.90%.

Reference

Abdel-Megeed, M. F., El-Hiti, G. A., Badr, B. E., and Azaam, M. M. (2013). Synthesis of a series of diphenyl (arylamino)(pyridin-3-yl)methylphosphonates as potential antimicrobial agents. *Phosphorus Sulfur Silicon Relat. Elem.*, **188**, 879–885.

C-P bond_10

Types of reaction: C-P bond formation
Reaction conditions: Neat condition, room temperature, ultrasound irradiation
Synthetic strategy: One-pot multi-component reaction
Catalyst: No catalyst
Keywords: Aromatic aldehydes, anilines, triethylphosphite, neat, room temperature, ultrasound irradiation, one-pot multi-component reaction, α-aminophosphonates

General reaction scheme

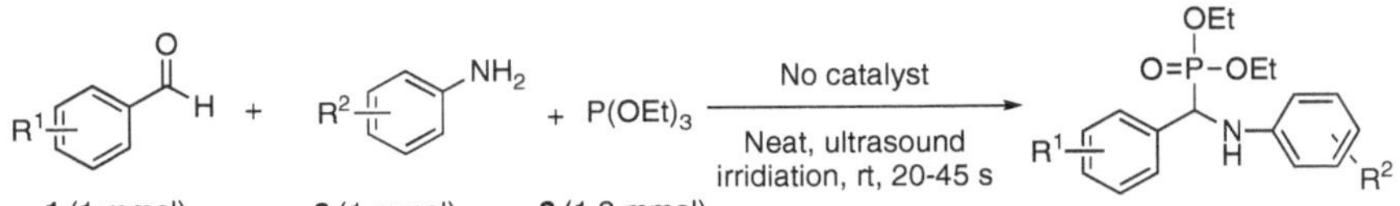

1 (1 mmol) 2 (1 mmol) 3 (1.3 mmol)

α-aminophosphonate **4**
17 entries (yield: 80-99%)

R^{1} = H, 4-Cl, 4-F, 2-Br, 4-OCH_3, 2-OH, 4-$N(CH_3)_2$, 2,6-di-Cl, 3,5-di-F, 3,4-O-CH_2-O-, 2,4,5-tri-OCH_3 (besides, 1-naphthaldehyde)

R^2 = 4-Cl, 4-F, 4-CH_3, 4-OCF_3

Representative entries

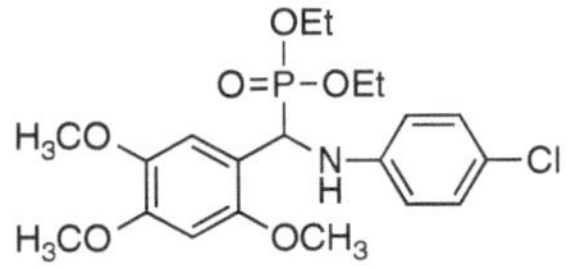

4a

Diethyl (((4-chlorophenyl)amino)(2,4,5-trimethoxyphenyl)methyl)phosphonate (**4a**): Reaction time: 45 s; yield: 86%

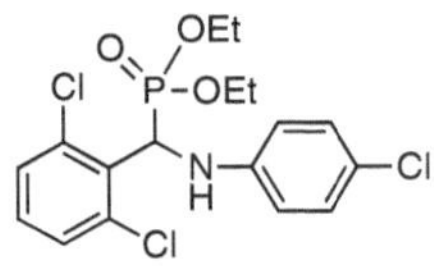

4b

Diethyl (((4-chlorophenyl)amino)(2,6-dichlorophenyl)methyl)phosphonate (**4b**): Reaction time: 20 s; yield: 92%

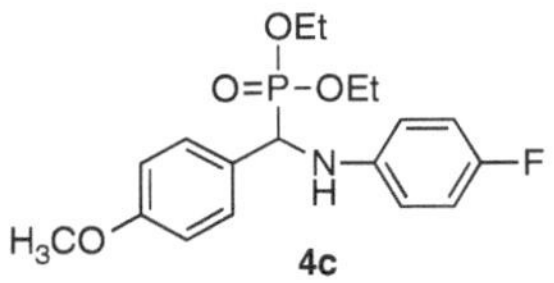

4c

Diethyl (((4-fluorophenyl)amino)(4-methoxyphenyl)methyl)phosphonate (**4c**): Reaction time: 30 s; yield: 88%

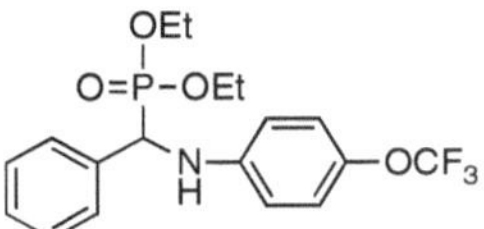

4d

Diethyl (phenyl((4-(trifluoromethoxy)phenyl) amino)methyl)phosphonate (**4d**): Reaction time: 20 s; yield: 87%

Experimental procedure

In a 5 mL glass test tube, a mixture of aldehyde (**1**; 1 mmol), aniline (**2**; 1 mmol) and triethylphosphite (**3**; 1.3 mmol) was stirred at room temperature for specified time (20-45 s) under ultrasonic irradiation. Progress of the reaction was monitored by TLC (silica gel) using ethyl acetate-hexane as eluent. After completion of the reaction (as indicated by TLC), the mixture was quenched with water and extracted with ethyl acetate, dried over magnesium sulfate, and then concentrated under reduced pressure to afford the crude product. The crude product thus obtained was purified by the process of crystallization to furnish pure α-amniophosphates **4** with excellent yield (80-99%). All products were characterized on the basis of detailed spectral studies, and from comparisons of all physical and spectral data with those of authentic samples reported in literature.

Reference

Dar, B., Singh, A., Sahu, A., Patidar, P., Chakraborty, A., Sharma, M., and Singh, B. (2012). Catalyst and solvent-free, ultrasound promoted rapid protocol for the one-pot synthesis of α-aminophosphonates at room temperature. *Tetrahedron*, **53**, 5497–5502.

C-P bond_11

Types of reaction: C-P and C-C bond formation
Reaction conditions: Neat condition, room temperature, ultrasound irradiation
Synthetic strategy: One-pot multi-component reaction
Catalyst: Clay-supported heteropolyacid (HPA)
Keywords: Aromatic aldehydes, malononitrile, triethylphosphite, clay-supported heteropolyacid (HPA), heterogeneous catalyst, neat, room temperature, ultrasound irradiation, one-pot multi-component reaction, β-phosphonomalononitriles

General reaction scheme

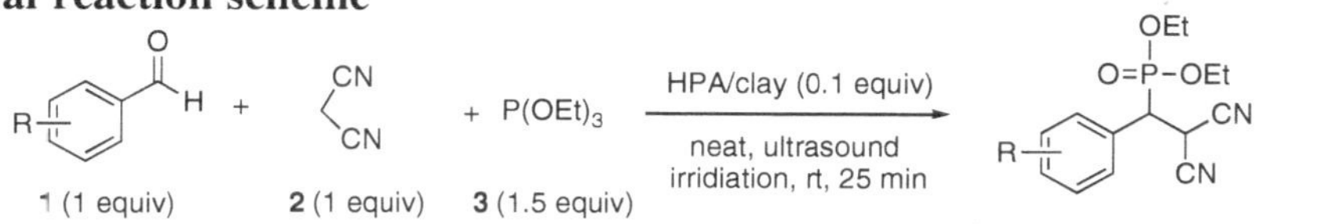

β-phosphonomalononitrile **4**
17 entries (yield: 58-97%)

R= H; 4-Br; 4-Cl; 4-F; 2-OCH_3; 3-NO_2; 3,4-di-Cl; 3,5-di-F; 3-Br,4-OCH_3; 3-Br, 4-F; 3,4-O-CH_2-O-; 3-OH,4-OCH_3; 2,3-di-OCH_3; 2,4,5-tri-OCH_3; (besides, naphthalene-1-aldehyde, naphthalene-2-aldehyde, furan-2-carbaldehyde, nicotinaldehyde)

Representative entries

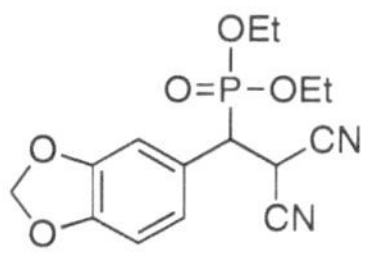

4a

Diethyl (1-(benzo[*d*][1,3]dioxol-5-yl)-2,2-dicyanoethyl)phosphonate (**4a**): Yield: 91%

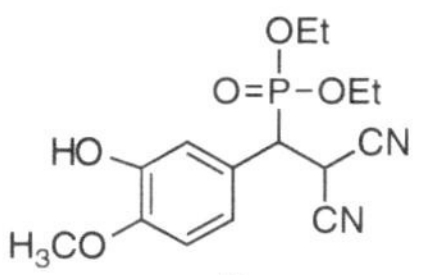

4b

Diethyl (2,2-dicyano-1-(3-hydroxy-4-methoxyphenyl)ethyl)phosphonate (**4b**): Yield: 78%

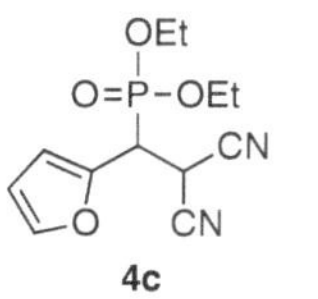

4c

Diethyl (2,2-dicyano-1-(furan-2-yl)ethyl) phosphonate (**4c**): Yield: 58%

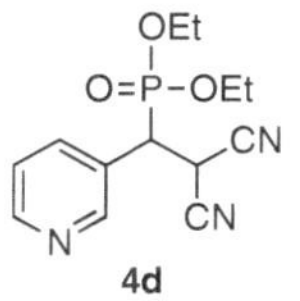

4d

Diethyl (2,2-dicyano-1-(pyridin-3-yl)ethyl) phosphonate (**4d**): Yield: 61%

Experimental procedure

In a 5 mL glass test tube, a mixture of an aldehyde (**1**; 1 equiv), malononitrile (**2**; 1 equiv), triethylphosphite (**3**; 1.5 equiv), and the catalyst (0.1 equiv) was stirred at room temperature for 25 min under ultrasonic irradiation. After completion of the reaction (monitored by TLC), the mixture was dissolved in ethyl acetate and filtered to separate the catalyst. The organic layer was dried over magnesium sulfate and concentrated under reduced pressure to obtain the crude product, followed by purification by crystallization technique to furnish pure β-phosphonomalononitrile **4** with good yield (58-97%). All products were characterized on the basis of detailed spectral studies, and from comparisons of physical and spectral data with those of authentic samples reported in literature. The catalyst as filtered out was subsequently washed with ethyl acetate, dried, and reused several times under same reaction conditions.

Reference

Dar, B. A., Pandey, N., Singh, S., Bamezai, R. K., Sharma, M., Vishwakarma, R. A., and Singh, B. (2014). Heterogeneous reusable catalyst, ultrasound energy and no solvent: a quick and green recipe for one-pot synthesis of β-phosphonomalonates at room temperature. *Tetrahedron Lett.*, **55**, 623–628.

C-P bond_12

Types of reaction: C-P bond formation
Reaction conditions: Toluene, room temperature
Synthetic strategy: Room temperature oxidative P-alkynylation
Catalyst: 1,10-Phenanthroline
Keywords: Copper acetylides, secondary phosphine-boranes, 1,10-phenanthroline, oxygen atmosphere, toluene, room temperature, oxidative P-alkynylation, enantioselective alkynylphosphine-boranes

General reaction scheme

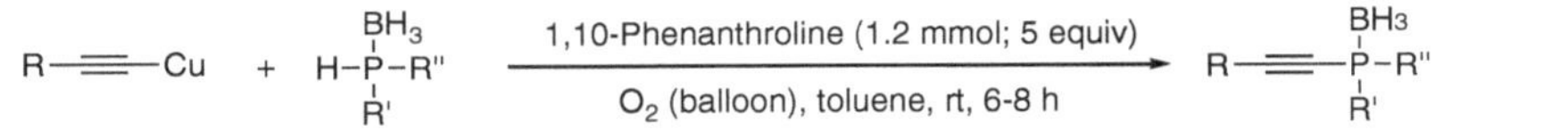

1 (0.24 mmol; 1 equiv) **2** (0.36 mmol; 1.5 equiv) Alkynylphosphine-borane **3** 27 entries (yield: 26-80%) (retention of configuration at phosphorus)

R = *n*-hexyl, *n*-dodecyl, isopentyl, *n*-propyl, TBDPSOCH$_2$, PhCH2CH$_2$, ^{t}BuOCO, propenyl, Ph, *o*-CH$_3$C$_6$H$_4$, *p*-CH$_3$C$_6$H$_4$, 2-OMeC$_6$H$_4$, 3-OMeC$_6$H$_4$, 4-OMeC$_6$H$_4$, 4-^{t}BuC$_6$H$_4$, naphthyl, 4-FC$_6$H$_4$, 2-CF$_3$C$_6$H$_4$

R', R'' = cyclohecyl, cyclohexyl; Et, Et; Me, ^{i}Pr; Me, cyclohexyl; Me, ^{t}Bu; ^{t}Bu, ^{t}Bu; Ph, Ph; Ph, Me

Representative entries

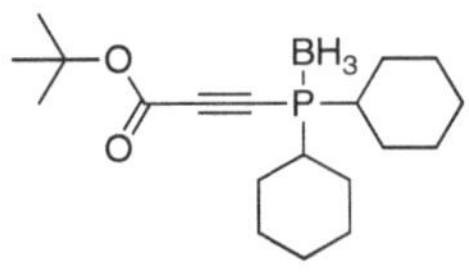

3a

(*tert*-Butoxycarbonylethynyl) dicyclohexylphosphine-borane (**3a**): White solid, mp 105 °C; yield: 50%

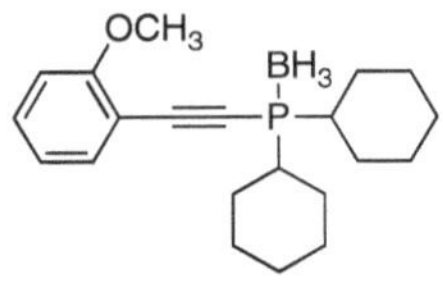

3b

Dicyclohexyl[(2-methoxyphenyl) ethynyl] phosphine-borane (**3b**): White solid, mp 121 °C; yield: 61%

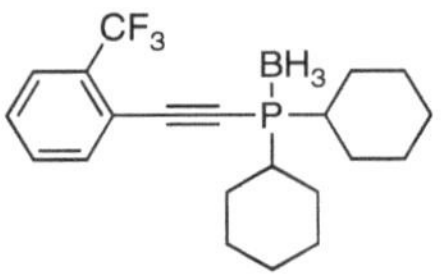

3c

Dicyclohexyl[(2-trifluoromethylphenyl)ethynyl] phosphine-borane (**3c**): White solid, mp 110 °C; yield: 57%

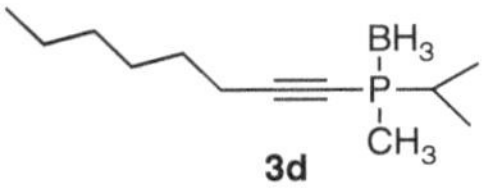

3d

Isopropylmethyl(oct-1-yn-1-yl)phosphine-borane (**3d**): Colorless oil; yield: 70%

Experimental procedure

To a solution of phosphine-borane (**2**; 0.36 mmol) in toluene (0.6 mL) taken in a 2 mL round bottom flask was added successively the alkynylcopper reagent (**1**; 0.24 mmol) and 1,10-phenanthroline (216 mg, 1.2 mmol), and the resulting brownish slurry was vigorously stirred at room temperature and under an atmosphere of oxygen (balloon). After complete disappearance of the alkynylcopper reagent (mostly within 6-8h), the crude reaction mixture was diluted with dichloromethane, filtered over a plug of silica gel and concentrated under vacuum. The residue was finally purified by flash chromatography over silica gel to afford the desired alkynylphosphine-borane **3**. All the products were characterized by means of detailed spectral studies.

Characterization data of two representative entries

(*tert*-Butoxycarbonylethynyl)dicyclohexylphosphine-borane (**3a**): White solid, mp 105 °C; IR (ATR): 2925, 2853, 2407, 2388, 2344, 1710, 1454, 1371, 1251, 1154, 1055, 1001, 968, 903, 889, 837, 821, 800, 783, 749, 707, 595, 551, 520, 459, 415 cm^{-1}; ^{1}H NMR (300 MHz, $CDCl_3$): δ 1.97-1.80 (m, 10H), 1.73 (br. s, 2H), 1.50 (obs. s, 9H), 1.52- 1.19 (m, 10H), 1.15-0.10 (br. m, 3H); ^{13}C NMR (75 MHz, $CDCl_3$): δ 151.0 (d, $J_{C\text{-}P}=3.0$ Hz), 97.3 (d, $J_{C\text{-}P}=4.2$ Hz), 85.1, 74.8 (d, $J_{C\text{-}P}=69.7$ Hz), 32.0 (d, $J_{C\text{-}P}=34.9$ Hz), 28.0, 27.1 (d, $J_{C\text{-}P}=3.8$ Hz), 26.7, 26.5 (d, $J_{C\text{-}P}=$ 2.8 Hz), 26.4, 25.7 (d, $J_{C\text{-}P}=1.0$ Hz); ^{31}P NMR (121 MHz, $CDCl_3$): δ 24.5-22.5 (m); ^{11}B NMR (96 MHz, $CDCl_3$): δ −39.0 to −46.2 (m); ESIHRMS: *m/z* calcd for $C_{19}H_{34}O_2PBNa$ $[M+Na]^+$ 359.2287, found 359.2280.

Dicyclohexyl[(2-trifluoromethylphenyl)ethynyl] phosphine-borane (**3c**): White solid, mp 110 °C; IR (ATR): νmax 2922, 2854, 2388, 2185, 1450, 1318, 1263, 1174, 1122, 1058, 1032, 862, 760, 661, 639, 589, 462 cm^{-1}; ^{1}H NMR (300 MHz, $CDCl_3$): δ 7.73-7.65 (m, 2H), 7.52 (app. quint, $J=7.6$ Hz, 2H), 2.03-1.80 (m, 10H), 1.74 (br. s, 2H), 1.66-1.19 (m, 10H), 1.12 to −0.20 (br. m, 3H); ^{13}C NMR (75 MHz, $CDCl_3$): δ 135.1 (d, $J_{C\text{-}P}=1.8$ Hz), 132.1 (q, $J_{C\text{-}F}=30.8$ Hz), 131.7, 129.8, 126.1 (q, JC-F = 5.0 Hz), 123.3 (q, $J_{C\text{-}F}=272.0$ Hz), 119.3, 101.7 (d, $J_{C\text{-}P}=8.7$ Hz), 84.7 (d, $J_{C\text{-}P}=82.2$ Hz), 32.4 (d, $J_{C\text{-}P}=36.3$ Hz), 26.9 (d, $J_{C\text{-}P}=3.7$ Hz), 26.8, 26.6 (d, $J_{C\text{-}P}=2.4$ Hz), 26.4, 25.9 (d, $J_{C\text{-}P}=0.9$ Hz); ^{31}P NMR (121 MHz, $CDCl_3$): δ 22.5–20.5 (m); ^{11}B NMR (96 MHz, CDC13): δ −39.2 to −44.6 (m); ^{19}F NMR (282 MHz, CDC13): δ −61.7 (m); ESIHRMS: *m/z* calcd for $C_{21}H_{29}F_3PBNa$ $[M+Na]^+$ 403.1950, found 403.1947.

Reference

Jouvin, K., Veillard, R., Theunissen, C., Alayrac, C., Gaumont, A.-C., and Evano, G. (2013). Unprecedented synthesis of alkynylphosphine-boranes through room-temperature oxidative alkynylation. *Org. Lett.*, **15**, 4592–4595.

C-P bond_13

Types of reaction: C-P bond formation
Reaction conditions: Ethanol, room temperature
Synthetic strategy: One-pot multicomponent
Catalyst: Indium trichloride ($InCl_3$)
Keywords: Chromene phosphonic acid diethyl ester, salicylaldehydes, malononitrile, triethyl phosphate, ethanol, room temperature, one-pot multicomponent synthesis

General reaction scheme:

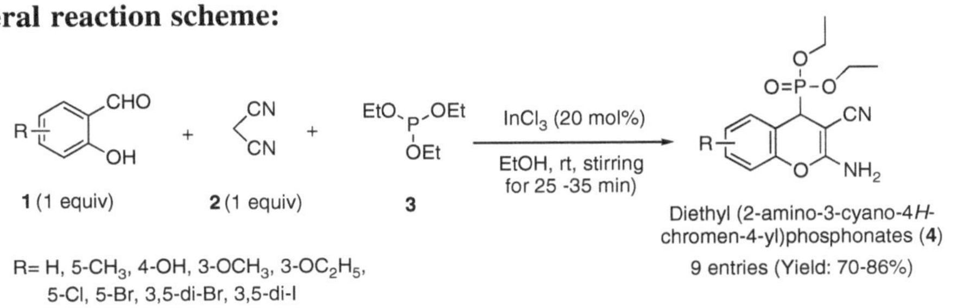

Representative entries

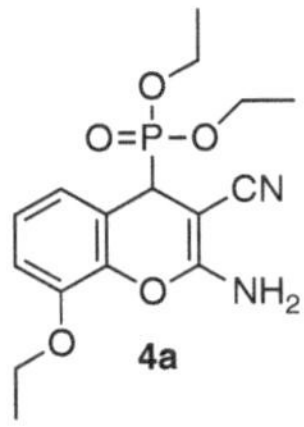

4a

Diethyl (2-amino-3-cyano-8-ethoxy-4*H*-chromen-4-yl)phosphonate (**4a**): Reaction time: 35 min; yield: 82%

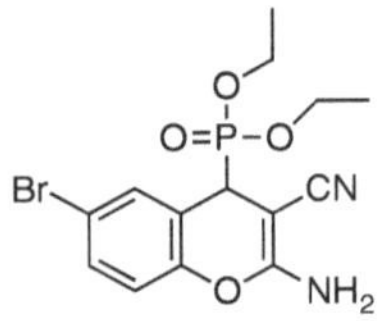

4b

Diethyl (2-amino-6-bromo-3-cyano-4*H*-chromen-4-yl)phosphonate (**4b**): Reaction time: 30 min; yield: 85%

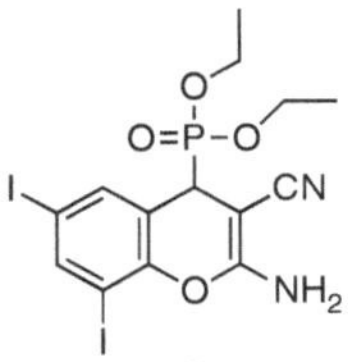

4c

Diethyl (2-amino-3-cyano-6,8-diiodo-4*H*-chromen-4-yl)phosphonate (**4c**): Reaction time: 25 min; yield: 82%

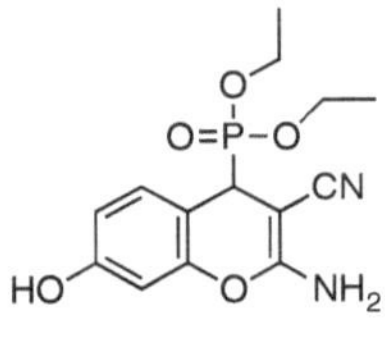

4d

Diethyl (2-amino-3-cyano-7-hydroxy-4*H*-chromen-4-yl)phosphonate (**4d**): Reaction time: 30 min; yield: 70%

Experimental procedure

To a stirred mixture of arylaldehyde (**1**; 1.0 mmol) or salicylaldehyde (1.0 mmol) and malononitrile (**2**; 1.0 mmol) in ethanol, triethyl phosphite was added followed by addition of $InCl_3$ (20 mol%). The reaction mixture was stirred at room temperature until the reaction was completed as monitored by TLC (25-35 min). After completion of the reaction, the mixture was poured into ice and extracted with EtOAc. The organic extracts were washed with brine solution, dried over Na_2SO_4, and concentrated under vacuum to yield pure diethyl (2-amino-3-cyano-4*H*-chromen-4-yl) phosphonate (**4**) in good to excellent yield (70-86%). All the products were characterized from their analytical and detailed spectral studies including IR, ^{1}H NMR, ^{13}C NMR, ^{31}P NMR and MS.

Characterization data of one representative compound

Diethyl (2-amino-6-bromo-3-cyano-4*H*-chromen-4-yl)phosphonate (**4b**): White solid, mp 180-182 °C; IR (KBr): 3353, 3156, 2190, 1655, 1417, 1241, 1040 cm^{-1}; ^{1}H NMR (300 MHz, DMSO-d_6): δ 1.13 (m, 6 H), 3.91 (m, 4 H), 4.17 (d, $^2J_{PH}=18.0$ Hz, 1H), 6.99 (d, $J=8.7$ Hz, 1H), 7.26 (br s, 2H, NH_2, D_2O exchangeable), 7.44 (m, 2 H); ^{13}C NMR (75 MHz, DMSO-d_6): δ 16.1, 16.2, 34.1 (d, $^1J_{CP}=144.0$ Hz), 47.0 (d, $^2J_{CP}=7.5$ Hz), 62.2 (d, $^2J_{CP}=7.5$ Hz), 62.4

(d, $^2J_{CP}=7.5$ Hz), 115.6, 118.1, 119.8, 120.4, 131.4, 131.8, 149.2, 162.3; ^{31}P NMR (202.5 MHz, DMSO-d_6): δ 22.64; MS: $m/z=387$ [M$^+$], 389 [M+2]. Anal. Calcd (%) for $C_{14}H_{16}BrN_2O_4P$: C, 43.43; H, 4.17; N, 7.24; Found: C, 43.38; H, 4.11; N, 7.20.

Reference

Jayashree, P., Shanthi, G., and Perumal, P. T. (2009). Indium trichloride catalyzed one-pot synthesis of new (2-amino-3-cyano-4*H*-chromen-4-yl) phosphonic acid diethyl ester. *Synlett*, 917–920.

C-P bond_14

Types of reaction: C-P bond formation
Reaction conditions: Alcohol, room temperature
Synthetic strategy: One-pot multicomponent
Catalyst: Ethylenediamine diacetate (EDDA)
Keywords: Chromene phosphonates, salicylaldehydes, malononitrile, ethyl cyanoacetate, triethyl phosphate, alcohol, ethylenediamine diacetate (EDDA); room temperature, one-pot multicomponent synthesis

General reaction schemes

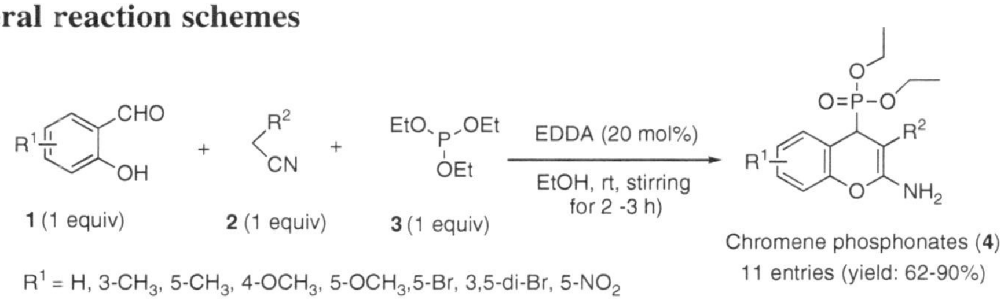

R^1 = H, 3-CH_3, 5-CH_3, 4-OCH_3, 5-OCH_3,5-Br, 3,5-di-Br, 5-NO_2
(1 more entry with -hydroxy-1-naphthaldehyde)
R^2 = CN, COOEt

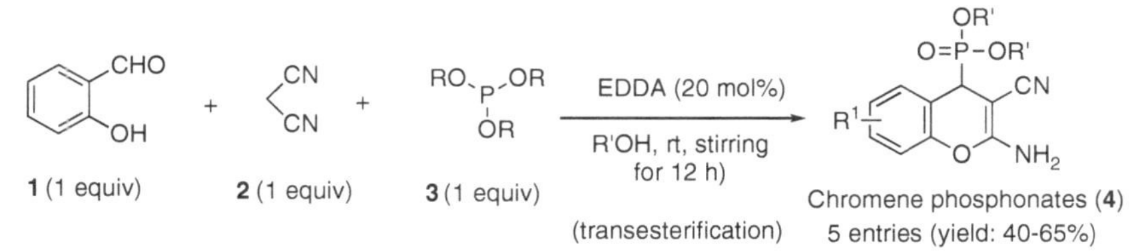

Representative entries

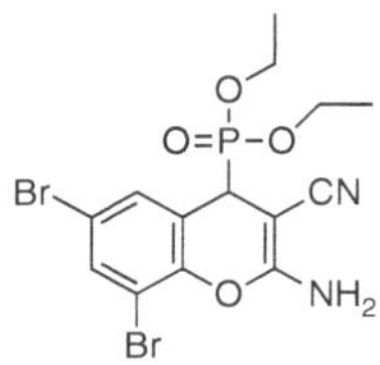

4a

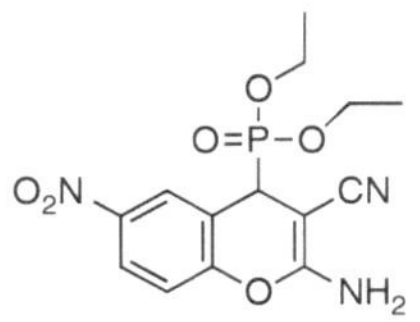

4b

Diethyl (2-amino-6,8-dibromo-3-cyano-4*H*-chromen-4-yl)phosphonate (**4a**): Solid, mp 198-200 °C, reaction time: 3 h; yield: 80%

Diethyl (2-amino-3-cyano-6-nitro-4*H*-chromen-4-yl)phosphonate (**4b**): Solid, mp 80-82 °C, reaction time: 3 h; yield: 75%

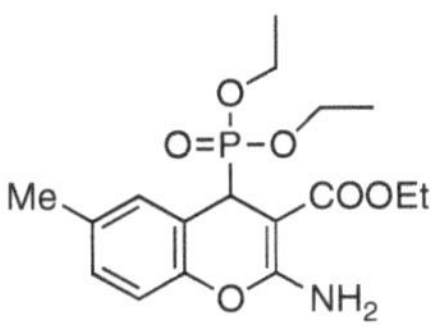

4c

Ethyl 2-amino-4-(diethoxyphosphoryl)-6-methyl-4*H*-chromene-3-carboxylate (**4c**): Oil, reaction time: 2 h; yield: 65%

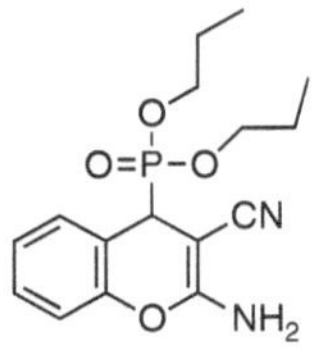

4d

Dipropyl (2-amino-3-cyano-4*H*-chromen-4-yl) phosphonate (**4d**): Reaction time: 12 h; yield: 40%

Experimental procedure

To a stirred mixture of arylaldehyde (**1**; 1.0 mmol) or salicylaldehyde (1.0 mmol), malononitrile (or ethyl cyanoacetate; **2**; 1.0 mmol) and triethyl phosphite (or phosphite derivative; **3**; 1.0 mmol) in 5 mL of ethanol (or other alcohol), ethylenediamine diacetate EDDA (0.2 mmol) was added. The resulting mixture was stirred at room temperature until completion of the reaction, as indicated by TLC. The reaction mixture was evaporated under reduced pressure and the crude product was purified by silica gel column chromatography using EtOAc/*n*-hexane as the eluent to give the desired products (**4**) in good yields. All the products were characterized from their detailed spectral studies including IR, ^{1}H NMR, ^{13}C NMR, and EI- and HR-MS.

Characterization data of two representative compounds

Diethyl (2-amino-6,8-dibromo-3-cyano-4*H*-chromen-4-yl)phosphonate (**4a**): Solid, mp 198-200 °C; IR (KBr): 3341, 3159, 2984, 2198, 1662, 1412, 1235, 1165, 1040, 861 cm^{-1}; ^{1}H NMR (300 MHz, acetone-d_6): δ 7.76 (t, J=2.1 Hz, 1H), 7.54-7.53 (m, 1H), 6.73 (br s, 2H), 4.12-3.98 (m, 5H), 1.28 (t, J=7.2 Hz, 3H), 1.20 (t, J=6.9 Hz, 3H); ^{13}C NMR (75 MHz, DMSO-d_6): δ 162.1,146.5,134.1,131.5, 122.1, 119.4, 116.0, 110.6, 62.7 (d, $^{2}J_{CP}$=7.1 Hz), 62.6 (d, $^{2}J_{CP}$=7.1 Hz), 47.5 (d, $^{2}J_{CP}$=7.7 Hz), 34.6 (d, $^{1}J_{CP}$=144.5 Hz), 16.3, 16.2; EIMS (%relative intensity): *m/z* 466 (M^{+}, 6), 331 (48), 330 (12), 329 (100), 327 (51), 293 (14), 169 (9), 149 (37), 85 (4), 71 (7).

Ethyl 2-amino-4-(diethoxyphosphoryl)-6-methyl-4*H*-chromene-3-carboxylate (**4c**): Oil; IR (neat): 3388, 3288, 2982, 2931, 1676, 1627, 1532, 1492, 1412, 1308, 1204, 1046 cm^{-1}; ^{1}H NMR (300 MHz, acetone-d_6): δ 7.14 (s, 1H), 7.06 (d, J=8.1 Hz, 1H,), 6.88 (d, J=8.1 Hz, 1H), 4.27 (d, $^{2}J_{PH}$=19.5 Hz, 1H,), 4.23-3.77 (m, 6H), 2.30 (s, 3H), 1.28 (t, J=6.9 Hz, 3H), 1.19 (t, J=6.9 Hz, 3H), 1.11 (t, J=7.2 Hz, 3H); ^{13}C NMR (75 MHz, acetone-d_6): δ 163.5, 149.9, 134.3, 130.8, 130.7, 129.6, 129.5, 121.1, 116.3, 62.8 (d, $^{2}J_{CP}$=7.6 Hz), 62.7 (d, $^{2}J_{CP}$=7.1 Hz), 59.9, 35.9 (d, $^{1}J_{CP}$=143.9 Hz), 20.7, 16.9, 16.8, 15.0; HRMS: *m/z* (M^{+}) calcd for $C_{17}H_{24}NO_6P$: 369.1341; Found: 369.1340.

Reference

Rao Kolla, S., and Lee, Y. R. (2012). Efficient one-pot synthesis of β-phosphono malonates and 2-amino-4*H*-chromen-4-ylphosphonate derivatives by ethylenediamine diacetate-catalyzed three-component reactions. *Tetrahedron*, **68**, 226–237.

Further information

Synthesis of (2-amino-3-cyano-4*H*-chromen-4-yl) phosphonic acid diethyl ester has also been reported to be catalyzed by potassium phosphate (K_3PO_4; 20 mol%) at room temperature.

Reference

Gaikwad, D. S., Undale, K. A., Shaikh, T. S., and Pore, D. M. (2011). An efficient multi-component synthesis of (2-amino-3-cyano-4*H*-chromen-4-yl) phosphonic acid diethyl ester. *Competes Rendus Chimie*, **14**, 865–868.

C-P bond_15

Types of reaction: C-P and C-O bond formation
Reaction conditions: Ethanol-water (1:1 v/v), room temperature
Synthetic strategy: One-pot multicomponent reaction
Catalyst: Nanocrystalline MgO
Keywords: Salicylaldehydes, malononitrile/ethyl cyanoacetate, triethylphosphite, nanocrystalline MgO heterogeneous catalyst, ethanol-water (1:1), room temperature, one-pot three-component synthesis, (2-amino-3-cyano-4*H*-chromen-4-yl)phosphonic acid diethyl esters

General reaction scheme

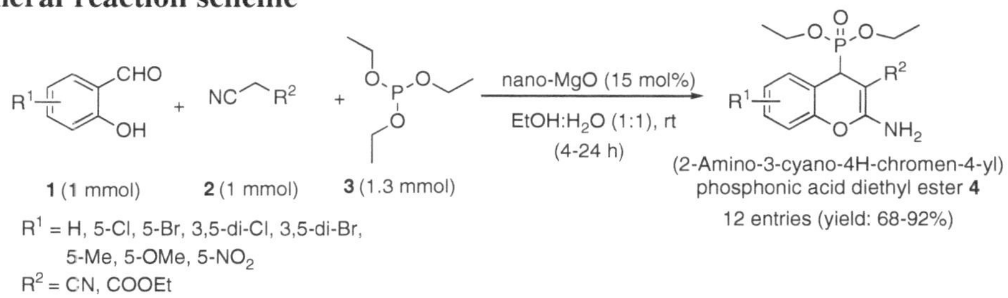

Proposed mechanism

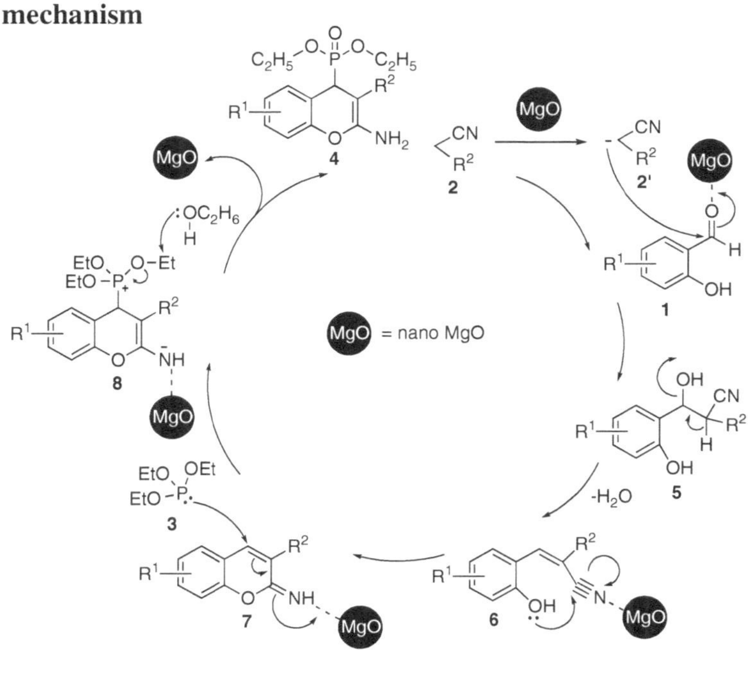

Representative entries

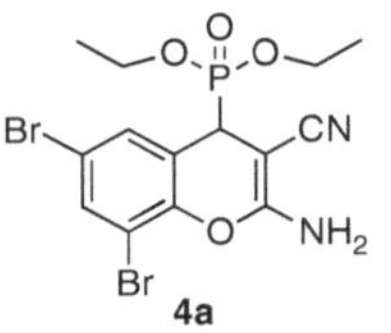

4a

Diethyl (2-amino-6,8-dibromo-3-cyano-4*H*-chromen-4-yl)phosphonate (**4a**): White solid, mp 195-197 °C; reaction time: 17 h; yield: 82%

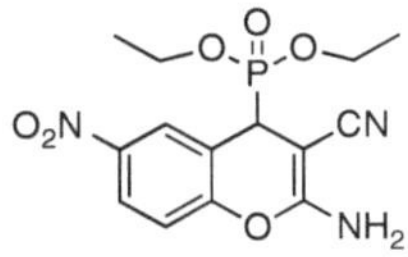

4b

Diethyl (2-amino-3-cyano-6-nitro-4*H*-chromen-4-yl)phosphonate (**4b**). Yellow solid, mp 210–212 °C.; reaction time: 24 h; yield: 68%

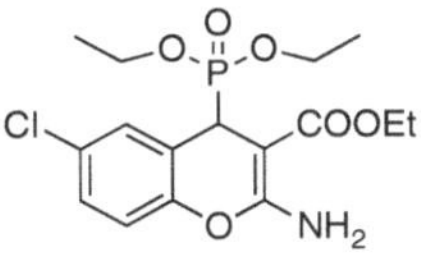

4c

Ethyl 2-amino-6-chloro-4-(diethoxyphosphoryl)-4*H*-chromene-3-carboxylate (**4c**): White solid, mp 195-197 °C; reaction time: 7 h; yield: 81%

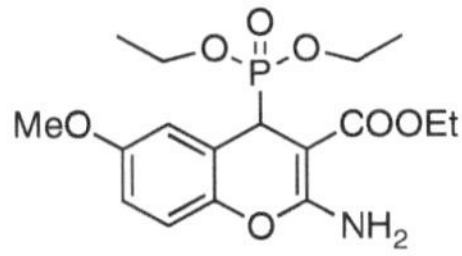

4d

Ethyl 2-amino-4-(diethoxyphosphoryl)-6-methoxy-4*H*-chromene-3-carboxylate (**4d**): Oil; reaction time: 8 h; yield: 81%

Experimental procedure

MgO nano-powder (15 mol%; 0.0065 g) was placed within 3 mL of EtOH:H_2O (1:1 v/v) in an oven-dried screw cap test tube (10 mL) and was sonicated for few minutes. To this well-dispersed catalytic medium was added with salicylaldehyde (**1**; 1 mmol), malononitrile/ethyl cyanoacetate (**2**; 1 mmol) and triethylphosphite (**3**; 1.3 mmol) in a sequential manner. The whole reaction mixture was stirred vigorously with the aid of an external bar magnet at room temperature for stipulated time (4-24 h) as indicated in table 2. The progress of the reaction was monitored by TLC. On completion of the reaction, the solvent was evaporated out and then the resulting mixture was extracted with ethyl acetate (or dichloromethane) followed by centrifugation to separate out the nano-catalyst. The organic layer was dried over anhydrous sodium sulphate, and the solvent was removed with the help of a rotary evaporator to obtain the crude product of (2-amino-3-cyano-4*H*-chromen-4-yl)phosphonic acid diethyl ester (**4**) which was further purified by means of column chromatography using petrol ether-ethyl acetate as eluent. The structure of each purified phosphonate ester functionalized 2-amino-3-cyano-4*H*-chromene scaffold was confirmed by analytical as well as spectral studies including FT-IR, ^{1}H NMR, ^{13}C NMR and TOF-MS. The recovered nano-MgO catalyst was washed with acetone followed by drying in an oven at 100 °C and reused as such for the subsequent reactions with fresh batch of reactants up to 4th run.

Characterization data of two representative compounds

Diethyl (2-amino-6,8-dibromo-3-cyano-4*H*-chromen-4-yl)phosphonate **(4a).** White solid, mp 195-197 °C; IR (KBr): 3342, 3309, 3151, 2976, 2858, 2193, 1661, 1610, 1591, 1558, 1454, 1412, 1285, 1234, 1047, 860, 797, 727, 652, 613, 509 cm^{-1}; ^{1}H-NMR (400 MHz, $CDCl_3$): δ 1.27 (t, J=6.8 Hz, 7.2 Hz, 3H, -OCH_2CH_3), 1.37 (t, J=7.2 Hz, 3H, -OCH_2CH_3), 3.86 (d, J_{PH}=18.8 Hz, 1H, C_4-H), 4.06-4.12 (m, 2H, -OCH_2CH_3), 4.13-4.19 (m, 2H, -OCH_2CH_3), 5.18 (s, 2H, -NH_2), 7.65 (t, J=2 Hz, 1H, Ar-H), 7.43 (t, J=2.4 Hz, 2.2 Hz, 1H, Ar-H). ^{13}C-NMR (100 MHz, $CDCl_3$): δ 161.32 (d, J_{CP}=3 Hz), 146.09 (d, J_{CP}=5 Hz), 135.01 (d, J_{CP}=3 Hz), 131.31 (d, J_{CP}=4 Hz), 120.22 (d, J_{CP}=6 Hz), 118.49 (CN), 117.32 (d, J_{CP}=2 Hz), 111.35 (d, J_{CP}=3 Hz), 63.58 (d, J_{CP}=7 Hz, -OCH_2CH_3), 63.39

(d, $J_{CP}=7$ Hz, -OCH$_2$CH$_3$), 51.73 (d, $J_{CP}=7$ Hz), 35.69 (d, $J_{CP}=148$ Hz, C-4), 16.36 (d, $J_{CP}=5$ Hz, -OCH$_2$$CH_3$), 16.26 (d, $J_{CP}=15$ Hz, -OCH$_2$$CH_3$); TOF-MS: 486.9036 $[M+Na]^+$. Anal. Calcd for $C_{14}H_{15}Br_2N_2O_4P$: C, 36.08; H, 3.24; N, 6.01. Found: C, 36.11; H, 3.22; N, 6.02.

Ethyl 2-amino-4-(diethoxyphosphoryl)-6-methoxy-4*H*-chromene-3-carboxylate **(4d).** Oil; IR (KBr): 3358, 3265, 3192, 2974, 2905, 1678, 1624, 1533, 1477, 1227, 1040, 957, 770 cm^{-1}; ^{1}H-NMR (400 MHz, $CDCl_3$): δ 1.12 (t, $J=6.8$ Hz, 7.2 Hz, 3H, -COOCH$_2$CH_3), 1.24 (t, $J=7.2$ Hz, 3H, -OCH$_2$CH_3), 1.30 (t, $J=7.2$ Hz, 3H, -OCH$_2$CH_3), 3.77 (s, 3H, -OCH_3), 3.79 - 3.81 (m, 1H, -COOCH_2CH$_3$), 3.87 - 3.93 (m, 1H, -COOCH_2CH$_3$), 3.99 - 4.06 (m, 2H, -OCH_2CH$_3$), 4.13 - 4.25 (m, 2H, -OCH_2CH$_3$), 4.36 (d, $J_{PH}=19.6$ Hz, 1H, C$_4$-H), 6.58 (br s, 2H, -NH_2), 6.74 (dt, $J=8.7$ Hz, 2.4 Hz & 2.8 Hz, 1H, Ar-H), 6.87 (dd, $J=9.4$ Hz, 9.2 Hz & 1.6 Hz, 2H, Ar-H); ^{13}C-NMR (100 MHz, $CDCl_3$): δ 168.58 (C=O), 162.11, 155.99 (d, $J_{CP}=2$ Hz), 144.71 (d, $J_{CP}=3$ Hz), 120.51 (d, $J_{CP}=6$ Hz), 116.55 (d, $J_{CP}=3$ Hz), 114.35 (d, $J_{CP}=4$ Hz), 113.45 (d, $J_{CP}=4$ Hz), 70.15, 62.71 (d, $J_{CP}=7$ Hz, -OCH$_2$CH$_3$), 62.30 (d, $J_{CP}=7$ Hz, -OCH$_2$CH$_3$), 59.58 (-COOCH$_2$CH$_3$), 55.61 (-OCH$_3$), 35.35 (d, $J_{CP}=145$ Hz, C-4), 16.38 (d, $J_{CP}=6$ Hz, -OCH$_2$$CH_3$), 16.32 (d, $J_{CP}=7$ Hz, -OCH$_2$$CH_3$), 14.50 (-COOCH$_2$$CH_3$); TOF-MS: 408.1191 $[M+Na]^+$. Elemental analysis: Calcd. (%) for $C_{17}H_{24}NO_7P$: C, 52.99; H, 6.28; N, 3.63; found: C, 52.92; H, 6.26; N, 3.65.

Reference

Brahmachari, G., and Laskar, S. (2014). Nano-MgO-catalyzed one-pot synthesis of phosphonate ester functionalized 2-amino-3-cyano-4*H*-chromene scaffolds at room temperature. *Phosphorus, Sulfur, and Silicon and the Related Elements*, **189**, 873–888.

Chapter 6

Carbon – Halogen Bond Forming Reactions at Room Temperature

Carbon – Fluorine Bond Forming Reactions: C-F bond_1

Type of reaction: C-F bond formation
Reaction conditions: Acetonitrile, room temperature
Synthetic strategy: Direct fluorination of $C(sp^3)$-H group
Catalyst: Vanadium(III) oxide (V_2O_3)
Keywords: Alkanes, SelectFluor, Vanadium(III) oxide, acetonitrile, room temperature, Direct fluorination of $C(sp^3)$-H group, fluorinated alkanes

General reaction scheme

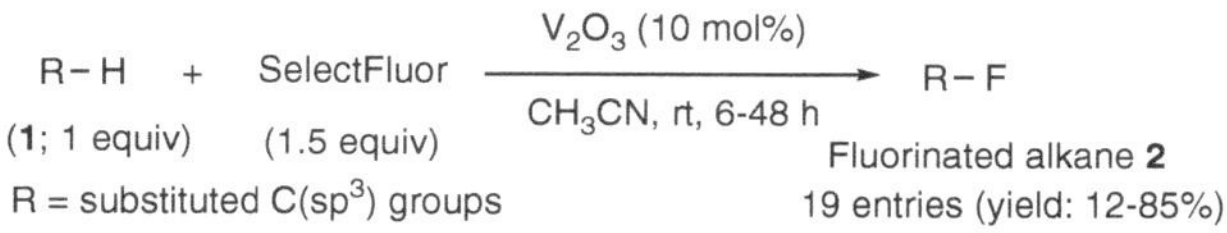

Representative entries

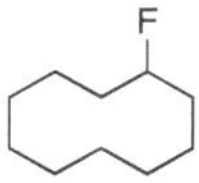

2a

Fluorocyclodecane (**2a**): Reaction time: 20 h yield: 61%

2b

1-(2-Fluoropropan-2-yl)-4-methyl-7-oxabicyclo [2.2.1]heptane (**2b**): Colorless oil; reaction time: 24 h yield: 53%

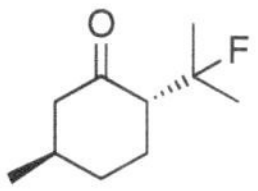

2c

(2*R*,5*R*)-2-(2-Fluoropropan-2-yl)-5-methylcyclohexanone (**2c**): Colorless oil; reaction time: 30 h yield: 75%

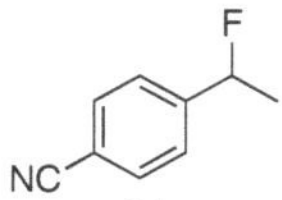

2d

4-(1-Fluoroethyl)benzonitrile (**2d**): Colorless oil; reaction time: 24 h yield: 24%

Experimental procedure

To a 4 mL clear vial charged with vanadium(III) oxide (3.0 mg, 0.02 mmol, 10 mol%), Selectfluor (106.3 mg, 0.3 mmol, 1.5 equiv) in anhydrous acetonitrile (2.0 mL), and the alkane (**1**; 0.2 mmol, 1.0 equiv) were added. The reaction mixture was degassed thrice by *Freeze-Pump-Thaw* cycles and then stirred at room temperature for 6-48 h. Upon completion, the reaction mixture was poured into diethyl ether (20 mL), filtrated, concentrated and purified by silica gel flash column chromatography using diethyl ether/pentane as the eluent to afford fluorinated alkane **2**, characterized by spectral studies.

Characterization data of representative compounds

1-(2-Fluoropropan-2-yl)-4-methyl-7-oxabicyclo[2.2.1]heptane (**2b**): Colorless oil; ^{1}H NMR (400 MHz, C_6D_6): δ 1.76-1.84 (m, 2H), 1.24-1.46 (m, 15H); ^{13}C NMR (100 MHz, C_6D_6): δ

Room Temperature Organic Synthesis. http://dx.doi.org/10.1016/B978-0-12-801025-9.00006-1

94.9 (d, $J=172.0$ Hz), 89.6 (d, $J=21.1$ Hz), 83.7, 37.2, 32.5 (d, $J=5.3$ Hz), 23.5 (d, $J=25.6$ Hz), 21.3; ^{19}F NMR (376 MHz, C_6D_6): δ –146.24 (hept, $J=21.5$ Hz, 1 F); MS(EI): Calcd for $C_{12}H_{17}FO$ $[M]^+$ 172.1, Found 172.2.

(2*R*,5*R*)-2-(2-Fluoropropan-2-yl)-5-methylcyclohexanone (**2c**): Colorless oil; ^{1}H NMR (400 MHz, C_6D_6): δ 2.27-2.33 (m, 1H), 2.09-2.19 (m, 2H), 1.55 (d, $J=23.2$ Hz, 3H), 1.44 (d, $J=22.8$ Hz, 3H), 1.31-1.45 (m, 3H), 1.18-1.29 (m, 1H), 0.78-0.88 (m, 1H), 0.62 (d, $J=6.0$ Hz, 3H); ^{13}C NMR (100 MHz, C_6D_6): δ 207.9 (d, $J=11.0$ Hz), 95.3 (d, $J=165.5$ Hz), 58.8 (d, $J=22.5$ Hz), 51.2 (d, $J=4.2$ Hz), 35.7, 33.6, 27.7 (d, $J=4.8$ Hz), 27.5 (d, $J=23.3$ Hz), 22.3 (d, $J=23.3$ Hz), 22.1; ^{19}F NMR (376 MHz, C_6D_6): δ –133.75 (heptd, $J=23.0$, 7.5 Hz, 1 F); MS(ESI): Calcd for $C_{10}H_{17}FONa$ $[M+Na]^+$ 195.1, Found 195.1.

Reference

Xia, J.-B., Ma, Y., and Chen, C. (2014). Vanadium-catalyzed C(sp^3)-H fluorination reactions. *Org. Chem. Front.*, **1**, 468–472.

Carbon – Fluorine Bond Forming Reactions: C-F bond_2

Type of reaction: C-F and C-N bond formation
Reaction conditions: Acetonitrile-water (1:2), room temperature
Synthetic strategy: Regioselective azidofluorination of unactivaied alkenes
Catalyst: Trifluoroacetic acid (TFA)
Keywords: Subtituted alkenes, trimethylsilyl azide ($TMSN_3$), SelectFluor, trifluoroacetic acid (TFA), aqueous acetonitrile, room temperature, regioselective azidofluorination, β-fluorinated azides

General reaction scheme

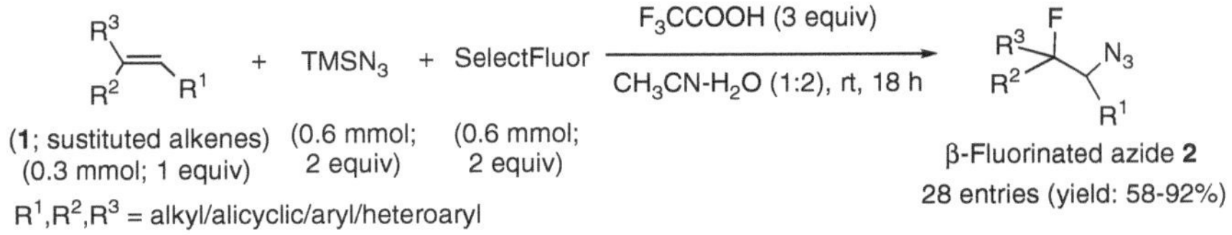

Representative entries

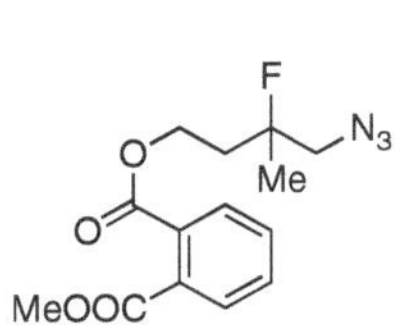

2a

4-Azido-3-fluoro-3-methylbutyl methyl phthalate (**2a**): Yellow oil; yield: 92%

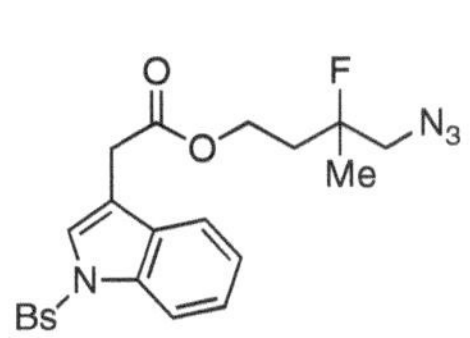

2b

4-Azido-3-fluoro-3-methylbutyl 2-(1-((4-bromophenyl)sulfonyl)-1*H*-indol-3-yl)acetate (**2b**): Yellow oil; yield: 65%

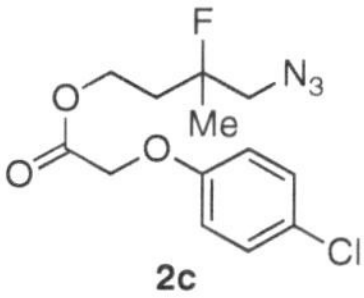

2c

4-Azido-3-fluoro-3-methylbutyl 2-(4-chlorophenoxy)acetate (**2c**): Yellow oil; yield: 91%

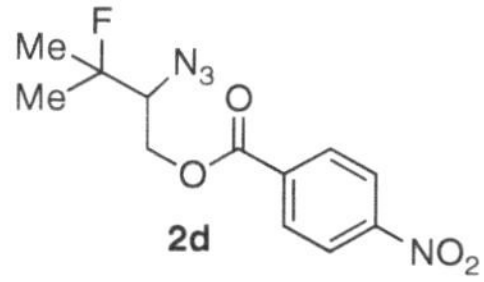

2d

2-Azido-3-fluoro-3-methylbutyl 4-nitrobenzoate (**2d**): Yellow oil; yield: 58%

Experimental procedure

Alkene (**1**; 1 equiv, 0.3 mmol) and Selectfluor (212 mg, 0.6 mmol) were placed in a Schlenk tube under a nitrogen atmosphere. Trifluoroacetic acid (69 μL, 0.9 mmol), $TMSN_3$ (78 μL, 0.6 mmol), acetonitrile (1 mL) and water (2 mL) were then added successively, and the reaction mixture was stirred at room temperature for 18 h. Upon completion, the resulting mixture was extracted with dichloromethane (3 x 15 mL), and the combined organic layer was dried over anhydrous sodium sulfate. After the removal of the solvent under reduced pressure, the crude product was purified by column chromatography on silica gel with hexane-ethyl acetate (7 : 1, v/ v) as the eluent to afford the pure product **2**, followed by characterization by detailed spectral studies.

Characterization data of representative compounds

4-Azido-3-fluoro-3-methylbutyl methyl phthalate (**2a**): Yellow oil; IR (neat): 2954, 2104, 1728, 1595, 1434, 1277, 1126, 1076, 959, 746 cm^{-1}; ^1H NMR (400 MHz, $CDCl_3$): δ 7.73-7.76 (m, 1H), 7.67-7.71 (m, 1H), 7.53-7.57 (m, 2H), 4.48 (t, $J=6.8$ Hz, 2H), 3.91 (s, 3H), 3.31-3.44 (m, 2H), 2.05-2.25 (m, 2H), 1.45 (d, $J=21.6$ Hz, 3H); ^{13}C NMR (100 MHz, $CDCl_3$): δ 167.7, 167.5, 132.1, 131.5, 131.3, 131.1, 129.0, 128.6, 95.4 (d, $J=173.1$ Hz), 60.8 (d, $J=6.8$ Hz), 58.1 (d, $J=24.2$ Hz), 52.6, 35.8 (d, $J=22.0$ Hz), 22.4 (d, $J=23.6$ Hz); ^{19}F NMR (282 MHz, $CDCl_3$): δ −150.4 (m, 1 F); ESI-MS: (*m*/*z*) 332 (M^++Na); HRMS calcd for $C_{14}H_{16}FN_3O_4Na$ (M+Na): 331.1017, Found 331.1017.

4-Azido-3-fluoro-3-methylbutyl 2-(1-((4-bromophenyl)sulfonyl)-1*H*-indol-3-yl)acetate (**2b**): Yellow oil; IR (neat): 3065, 2982, 2104, 1732, 1447, 1372, 1277, 1174, 979, 748 cm^{-1}; ^1H NMR (400 MHz, $CDCl_3$): δ 7.99 (d, $J=8.4$ Hz, 1H), 7.89 (d, $J=7.6$ Hz, 2H), 7.42-7.57 (m, 5H), 7.34 (t, $J=7.6$ Hz, 1H), 7.24-7.28 (m, 1H), 4.25 (t, $J=6.4$ Hz, 2H), 3.70 (s, 3H), 3.21 (d, $J=18.8$ Hz, 2H), 1.89-2.08 (m, 2H), 1.28 (d, $J=21.6$ Hz, 3H); ^{13}C NMR (100 MHz, $CDCl_3$): δ 170.2, 138.2, 135.0, 133.8, 130.3, 129.3, 126.8, 125.0, 124.6, 123.4, 119.5, 115.0, 113.7, 95.3 (d, $J=173.1$ Hz), 60.2 (d, $J=6.9$ Hz), 58.0 (d, $J=24.3$ Hz), 35.7 (d, $J=22.8$ Hz), 31.0, 22.3 (d, $J=23.5$ Hz); ^{19}F NMR (282 MHz, $CDCl_3$): δ −150.1 (m, 1 F); ESI-MS: (*m*/*z*) 467 (M^++Na); HRMS calcd for $C_{21}H_{21}FN_4O_4SNa$ (M+Na): 467.1160, Found 467.1154.

Reference

Li, Z., Zhang, C., Zhu, L., Liu, C., and Li, C. (2014). Transition-metal-free, room-temperature radical azidofluorination of unactivated alkenes in aqueous solution. *Org. Chem. Front.*, **1**, 100–104.

Carbon – Fluorine Bond Forming Reactions: C-F bond_3

Type of reaction: C-F and C-halo bond formation
Reaction conditions: Dichloromethane, room temperature
Synthetic strategy: One-step halofluorination of alkenes
Catalyst: Trihaloisocyanuric acid (TXCA) / HF-pyridine

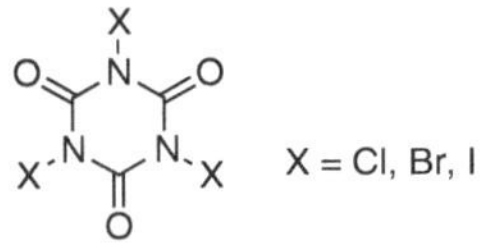

(TXCA)

Keywords: Alkenes, trihaloisocyanuric acid (TXCA), HF-pyridine, dichloromethane, room temperature, regioselective halofluorination of alkenes, vicinal halofluoroalkanes

General reaction scheme

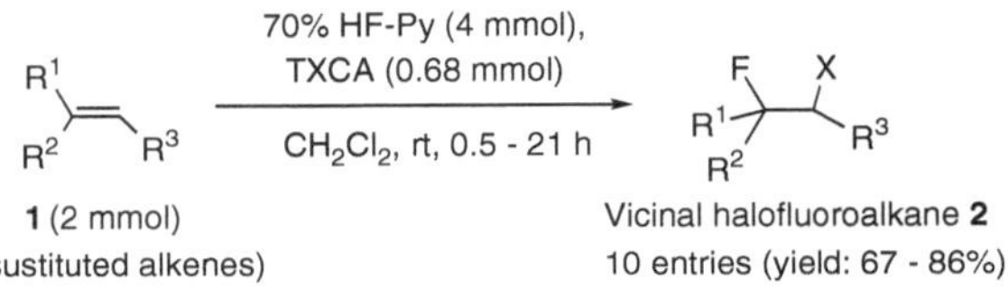

Representative entries

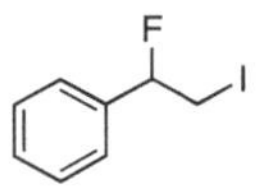

2a

(1-Fluoro-2-iodoethyl)benzene (**2a**): Yellowish liquid; reaction time: 21 h; yield: 83%

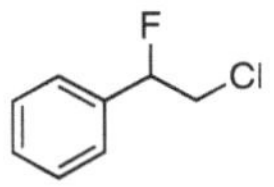

2b

(2-Chloro-1-fluoroethyl)benzene (**2b**): Yellowish liquid; reaction time: 0.5 h; yield: 80%

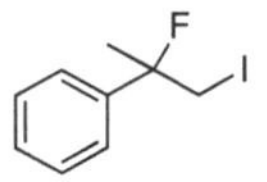

2c

(2-Fluoro-1-iodopropan-2-yl)benzene (**2c**): Yellowish liquid; reaction time: 2 h; yield: 77%

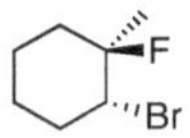

2d

(1*R*,2*R*)-2-Bromo-1-fluoro-1-methylcyclohexane (**2d**): Yellowish liquid; reaction time: 7 h; yield: 67%

Experimental procedure

A 50 mL single-necked PTFE flask equipped with a septum containing a well-stirred mixture of alkene (**1**; 2 mmol) and 70% HF-Py (4 mmol) in anhydrous dichloromethane (10 mL) under argon, was treated with trihaloisocyanuric acid (0.68 mmol) at 0 °C. After the addition, the ice bath was removed, and the stirring continued at room temperature for stipulated period (0.5-21 h). After completion, the reaction was quenched with distilled water (10 mL), and the product was extracted with CH_2Cl_2 (3 × 10 mL). The combined organic layers were subsequently washed with saturated aqueous $NaHCO_3$ (10 mL) and dried over Na_2SO_4. After filtration and evaporation of the solvent, vicinal halofluoroalkane derivatives **2** were obtained, followed by characterization based on spectral studies.

Characterization data of representative compounds

(2-Fluoro-1-iodopropan-2-yl)benzene (**2c**): Yellowish liquid; ^{1}H NMR (200 MHz, $CDCl_3$): δ 1.9 (d, $J=21.8$ Hz, 3 H), 3.6 (d, $J=20.3$ Hz, 2 H), 7.4 (m, 5 H); ^{13}C NMR (50 MHz, $CDCl_3$): δ = 15.7 (CH_3, d, $J=28.2$ Hz), 27.1 (CH_2, d, $J=24.5$ Hz), 94.6 (C, d, $J=178.5$ Hz), 124.4 (2 CH, d, $J=9.1$ Hz), 128.3 (2 CH), 128.7 (CH), 141.9 (C, d, $J=22.0$ Hz); ^{19}F NMR (282 MHz, $CDCl_3$): δ −142.6 (sext, $J=21.0$ Hz); MS: *m/z* (%)264 (M^+, 2), 244 (4), 137 (100), 123 (43), 103 (30), 77 (24), 51 (32).

(1*R*,2*R*)-2-Bromo-1-fluoro-1-methylcyclohexane (**2d**): Yellowish liquid; ^{1}H NMR (200 MHz, $CDCl_3$): δ 1.5-1.6 (d, $J=22.5$ Hz, 3 H), 1.4-2.3 (m, 8 H), 4.2 (td, $J=7.4$, 3.9 Hz, 1H); ^{13}C NMR (50 MHz, $CDCl_3$): δ 22.1 (CH_2, d, $J=6.3$ Hz), 22.9 (CH_2), 23.8 (CH_3, d, $J=24.6$ Hz), 33.2 (CH_2), 34.8 (CH_2, d, $J=17.5$ Hz), 57.2 (*C*H, d, $J=26.3$ Hz), 95.6 (*C*, d, $J=172.6$ Hz); ^{19}F NMR (282 MHz, $CDCl_3$): δ −137.3 (m); MS: *m/z* (%) 194 and 196 [M^+ and $(M+2)^+$, 1], 134 (2), 132 (2), 115 (11), 99 (5), 95 (100), 73 (49).

Reference

Crespo, L. T. C., da Ribeiro, R. S., de Mattos, M. C. S., and Esteves, P. M. (2010). Halofluorination of alkenes using trihaloisocyanuric acids and HF-pyridine. *Synthesis*, 2379–2382.

Carbon – Iodine Bond Forming Reactions: C-I bond_1

Type of reaction: C-I and C-Nu bond formation
Reaction conditions: Acetonitrile, room temperature
Synthetic strategy: Co-iodination of alkenes
Catalysts: (Diacetoxyiodo)benzene/iodine
Keywords: Alkenes, (diacetoxyiodo)benzene (DAIB), iodine, nucleophiles (Nu), acetonitrile, room temperature, co-iodination reaction, iodo-alkanes

General reaction scheme

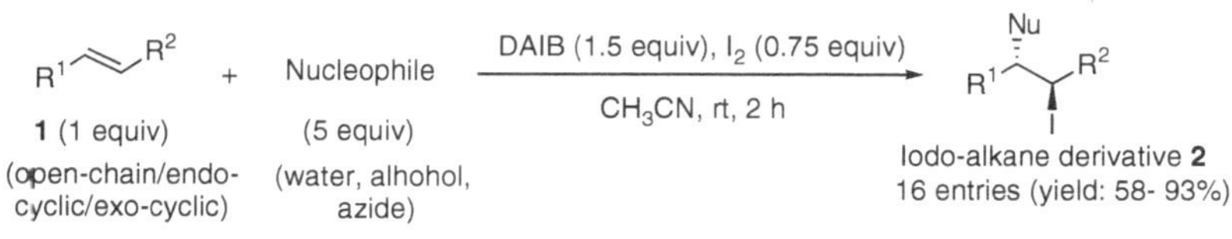

Representative entries

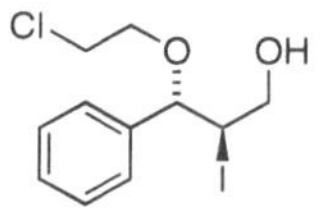

2a

(2*R*,3*S*)-3-(2-Chloroethoxy)-2-iodo-3-phenylpropan-1-ol (**2a**): Yield: 62%

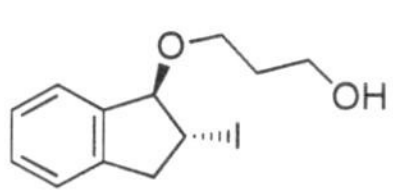

2b

3-(((1*R*,2*R*)-2-Iodo-2,3-dihydro-1*H*-inden-1-yl)oxy)propan-1-ol (**2b**): Yield: 72%

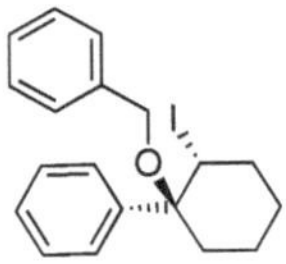

2c

((1*S*,2*R*)-1-(Benzyloxy)-2-iodocyclohexyl) benzene (**2c**): Yield: 60%

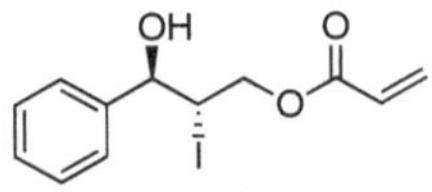

2d

(2*S*,3*R*)-3-Hydroxy-2-iodo-3-phenylpropyl acrylate (**2d**): Yield: 72%

Experimental procedure

To a solution of alkene (**1**; 2.0 mmol) and the nucleophilic additive (10 mmol) in acetonitrile (12 mL) were added (diacetoxyiodo)benzene (DAIB; 3.0 mmol), and iodine (1.5 mmol) in one portion, and the mixture was stirred at room temperature for 2 h for completion. The reaction mixture was then poured in water (40 mL) and extracted with dichloromethane (2 × 30 mL), followed by decolorizing the reaction mixture with the addition of required amount of solid $NaHSO_3$. The combined organic extract was then dried over anhydrous magnesium sulfate and evaporated to yield the crude product, which was purified by column chromatography (petroleum ether/dichloromethane) on silica to afford iodo-alkane **2**. Each of the products was characterized by spectral studies.

Characterization data of representative compounds

(2*R*,3*S*)-3-(2-Chloroethoxy)-2-iodo-3-phenylpropan-1-ol (**2a**): ^{1}H NMR (300 MHz, $CDCl_3$): δ 7.36 (m, 5H, Ar-H), 4.66 (d, $J = 9.0$ Hz, 1H, C_3-H), 4.36 (m, 1H, C_2-H), 3.91 (m, 2H, -CH_2OH), 3.59 (m, 4H, -OCH_2-CH_2Cl), 2.83 (t, $J = 6.0$ Hz, 1H, -CH_2OH); ^{13}C NMR (75 MHz, $CDCl_3$): δ 138.7, 128.9, 128.7, 127.6, 86.5, 69.6, 66.5, 43.1, 38.9; HRMS (EI): *m*/*z* calcd. for $C_{11}H_{14}O_2ICl$ 339.97273, Found 339.97193.

(2*S*,3*R*)-3-Hydroxy-2-iodo-3-phenylpropyl acrylate (**2d**): ^{1}H NMR (300 MHz, $CDCl_3$): δ 7.37-7.36 (m, 10H, Ar-H), 6.39 (dd, 1H, $J = 1.0$, 18.0 Hz, -COCH=CH_2), 6.08 (m, 1H, –COCH=CH_2), 5.87 (dd, $J = 1.0$, 12.0 Hz, (-COCH=CH_2), 4.91(m, 1H, -CHOH), 4.63-4.58 (m, 2H, -CH_2O), 2.89 (d, $J = 4.0$ Hz, 1H, -OH, D_2O exchangeable); ^{13}C NMR (75 MHz, $CDCl_3$): δ 165.8,140.2, 131.9, 128.6, 127.8, 126.6, 75.73, 65.6, 36.4; HRMS (EI): *m*/*z* calcd. for $C_{12}H_{13}O_3I$ 331.9909, Found 331.9921.

Reference

Gottam, H. and Vinod, T. K. (2011). Versatile and iodine atom-economic co-iodination of alkenes. *J. Org. Chem.*, **76**, 974–977.

Carbon – Iodine Bond Forming Reactions: C-I bond_2

Type of reaction: C-I bond formation
Reaction conditions: $CDCl_3$, room temperature
Synthetic strategy: Regioselective hydroiodination of alkynes
Catalyst: $I_2/Ph_2P(O)H$
Keywords: Alkynes, iodine/hydrophosphine binary system, $CDCl_3$, room temperature, regioselective hydroiodination, Markovnikov-type addition, vinyl iodides

General reaction scheme

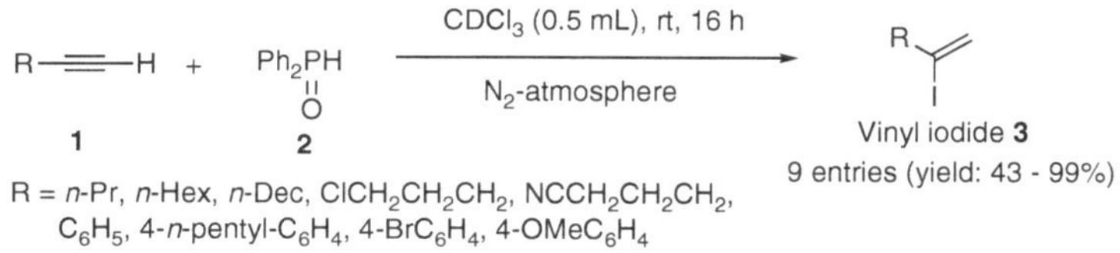

Conditions:

Aliphatic alkynes: **1** (0.2 mmol), **2** (0.3 mmol), iodine (0.15 mmol), $CDCl_3$ (0.5 mL), rt
Aromatic alkynes: **1** (0.6 mmol), **2** (0.2 mmol), iodine (0.1 mmol), $CDCl_3$ (0.5 mL), 0 °C to rt

Representative entries

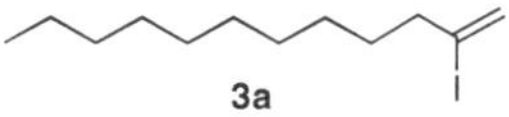

3a

2-Iodododec-1-ene (**3a**): Colorless oil; yield: 73%

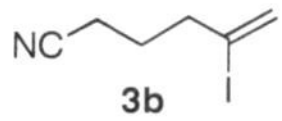

3b

5-Iodohex-5-enenitrile (**3b**): Colorless oil; yield: 43%

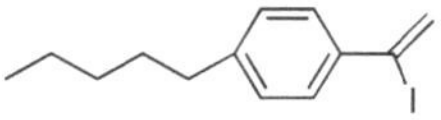

3c

1-(1-Iodovinyl)-4-pentylbenzene (**3c**): Colorless oil; yield: 77%

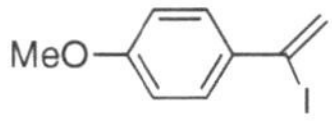

3d

1-(1-Iodovinyl)-4-methoxybenzene (**3d**): Colorless oil; yield: 81%

Experimental procedure

As per the reaction conditions as mentioned above, diphenylphosphine oxide (**2**) was placed in a Schlenk tube at room temperature, followed by $CDCl_3$, iodine, and alkyne (**1**) under a nitrogen atmosphere. The mixture was stood statically for 16 h at room temperature, and then EtOH (100 mL) was added to quench the reaction. After a while, the crude solution turned colorless from pale purple and white solid of $Ph_2P(O)OH$ was precipitated. The precipitate was filtered, and the crude solution was purified by preparative TLC (silica gel) to furnish vinyl iodide **3**. All the products were characterized on the basis of spectral studies.

Characterization data of representative compounds

5-Iodohex-5-enenitrile (**3b**): Colorless oil; ^1H NMR (400 MHz, $CDCl_3$): δ 1.89 (qin, $J=6.9$ Hz, 2H), 2.36 (t, $J=7.1$ Hz, 2H), 2.55 (t, $J=7.0$ Hz, 2H), 5.81 (s, 1H), 6.17 (s, 1H); ^{13}C NMR (100 MHz, $CDCl_3$) δ 15.4, 24.4, 43.3, 108.6, 119.0, 127.9; MS (EI), m/z (%) = 221 (M^+, 13), 94 ($M^+ - I$, 30), 67 ($M^+ - I - CN$, 100).

1-(1-Iodovinyl)-4-methoxybenzene (**3d**): Colorless oil; ^1H NMR (400 MHz, $CDCl_3$): δ 3.81 (s, 3H), 5.99 (d, $J=1.8$ Hz, 1H), 6.37 (d, $J=1.8$ Hz, 1H), 6.82 (d, $J=9.2$ Hz, 2H), 7.46 (d, $J=8.7$ Hz, 2H); ^{13}C NMR (100 MHz, $CDCl_3$) δ 55.5, 107.4, 113.5, 125.7, 129.5, 134.3, 160.1; MS (EI), m/z (%) = 260 (M^+, 3), 133 ($M^+ - I$, 100).

Reference

Kawaguchi, S., and Ogawa, A. (2010). Highly selective hydroiodation of alkynes using an iodine-hydrophosphine binary system. *Org. Lett.*, **12**, 1893–1895.

Carbon – Iodine Bond Forming Reactions: C-I bond_3

Type of reaction: C-I and C-O bond formation
Reaction conditions: Dichlorometahne, room temperature
Synthetic strategy: Au/Mo bimetallic catalytic synthesis of α-iodo-α,β-unsaturated carbonyls from propargylic alcohols
Catalysts: $Au(PPh_3)NTf_2$ / $MoO_2(acac)_2$ / Ph_3PO
Keywords: Propargylic alcohols, bimetallic catalysis, $Au(PPh_3)NTf_2$, $MoO_2(acac)_2$, Ph_3PO, *N*-iodosuccinimide, dichloromethane, room temperature, diastereoselectivity, α-iodo-α,β-unsaturated carbonyls

General reaction scheme

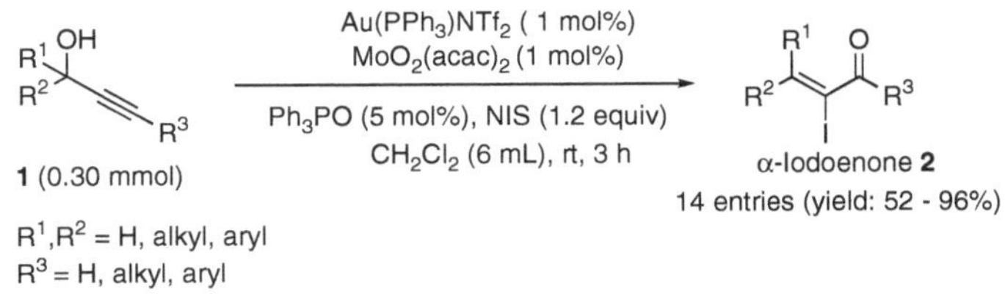

Representative entries

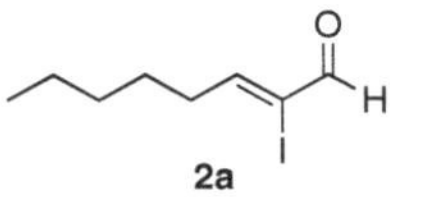

2a

(*Z*)-2-Iodooct-2-enal (**2a**): Geometrical isomers (*Z*/*E* > 50/1) in 73% yield with the exception that 2 mol % $Au(PPh_3)NTf_2$, 2 mol % of $MoO_2(acac)_2$, and 10 mol % of Ph_3PO were used.

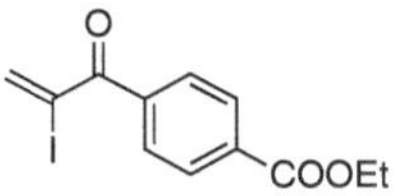

2b

Ethyl 4-(2-iodoacryloyl)benzoate (**2b**): 72% yield with the exception that 2 mol % $Au(PPh_3)NTf_2$, 2 mol % of $MoO_2(acac)_2$, and 10 mol % of Ph_3PO were used.

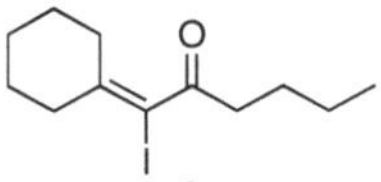

2c

1-Cyclohexylidene-1-iodohexan-2-one (**2c**): Yield: 87%

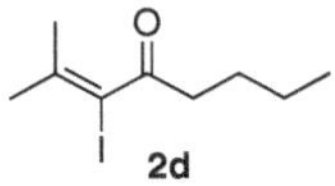

2d

3-Iodo-2-methyloct-2-en-4-one (**2d**): Yield: 96%

Experimental procedure

Ph_3PO (4.2 mg, 0.015 mmol), MoO_2(acac)2 (1.0 mg, 0.003 mmol), $Au(PPh_3)NTf_2$ (2.2 mg, 0.003 mmol), and NIS (81 mg, 0.36 mmol) were added sequentially to a solution of propargyl alcohol (**1**; 0.30 mmol) in anhydrous dichloromethane (6.0 mL, 0.05 M), and the resulting mixture was stirred at room temperature for 3 h. Upon completion of the reaction, it was quenched with saturated aqueous sodium bisulfate, and then extracted with Et_2O (3 × 8 mL). The combined organic layers were washed with water (10 mL) and brine (10 mL), dried with anhydrous magnesium sulfate, and filtered. The filtrate was concentrated, and the residue was purified through silica gel flash column chromatography (hexanes/ethyl acetate = 50/1) to yield the desired α-iodo-α,β-unsaturated carbonyl compound **2**. Each of the products was characterized by spectral studies.

Characterization data of representative compounds

Ethyl 4-(2-iodoacryloyl)benzoate (**2b**): IR (neat): 2982, 1718, 1671, 1587, 1501, 1407, 1368, 1278, 1177, 1106, 1021, 964, 870, 757, 741 cm^{-1}; ^{1}H NMR (400 MHz, $CDCl_3$): δ 8.13 (d, *J* = 8.8 Hz, 2H,), 7.83 (d, *J* = 8.8 Hz, 2H), 6.91 (d, *J* = 2.4 Hz, 1H), 6.90 (d, *J* = 2.4 Hz, 1H), 4.42 (q, *J* = 7.2 Hz, 2H), 1.42 (t, *J* = 7.2 Hz, 3H); ^{13}C NMR (100 MHz, $CDCl_3$): δ 191.1, 165.5, 139.4, 137.4, 134.2, 129.5(9), 129.5(6), 107.9, 61.5, 14.2; MS (ES^+): Calculated for $[C_{12}H_{11}NaIO_3]^+$: 353.0; found: 353.2.

1-Cyclohexylidene-1-iodohexan-2-one (**2c**): ^{1}H NMR (400 MHz, $CDCl_3$): δ 2.79 (t, *J* = 7.2 Hz, 2H), 2.41 (t, *J* = 6.0 Hz, 2H), 2.33 (t, *J* = 6.0 Hz, 2H), 1.65-1.52 (m, 8H), 1.35 (sextet, *J* = 7.2 Hz, 2H,), 0.93 (t, *J* = 7.2 Hz, 3H); ^{13}C NMR (100 MHz, $CDCl_3$): δ 202.5, 149.2, 92.4, 40.3, 39.6, 32.9, 28.0, 27.4, 26.3, 25.9, 22.3, 13.8.

Reference

Ye, L., and Zhang, L. (2009). Practical synthesis of linear α-iodo/bromo-α, β-unsaturated aldehydes/ketones from propargylic alcohols via Au/Mo bimetallic catalysis. *Org. Lett.*, **11**, 3646–3649.

Carbon – Iodine Bond Forming Reactions: C-I bond_4

Type of reaction: C-I bond formation
Reaction conditions: Acetonitrile, room temperature
Synthetic strategy: Iodination of terminal alkynes
Catalyst: CuI/Et_3N/PhI$(OAc)_2$
Keywords: Terminal alkynes, potassium iodide, copper(I) iodide, triethyl amine (Et_3N), (diacetoxyiodo)benzene, acetonitrile, room temperature, terminal iodination, 1-iodoalkenes

General reaction scheme

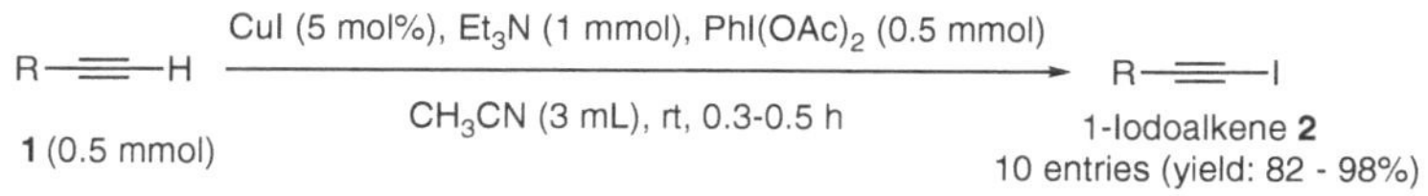

R = C_6H_5, 4-MeC_6H_4, 4-EtC_6H_4, 4-*n*-PrC_6H_4, 4-OMeC_6H_4, 4-BrC_6H_4, 4-$NO_2C_6H_4$, 2-$CF_3C_6H_4$, *n*-Bu, *n*-C_5H_{11}

Representative entries

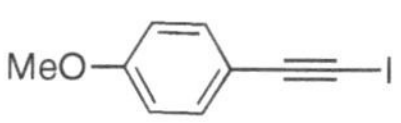

2a

1-(Iodoethynyl)-4-methoxybenzene (**2a**): Solid, mp 61-62 °C; reaction time: 0.5 h; yield: 97%

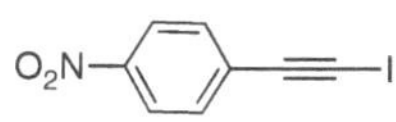

2b

1-(Iodoethynyl)-4-nitrobenzene (**2b**): Solid, mp 180-181 °C; reaction time: 0.3 h; yield: 95%

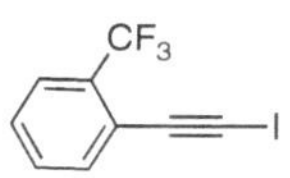

2c

1-(Iodoethynyl)-2-(trifluoromethyl)benzene (**2c**): Solid, mp 64-65 °C; reaction time: 0.3 h; yield: 87%

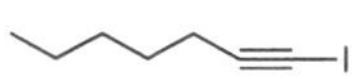

2d

1-Iodohept-1-yne (**2d**): Oil; reaction time: 0.5 h; yield: 85%

Experimental procedure

A mixture of alkyne (**1**; 0.5 mmol), (diacetoxyiodo)benzene (0.5 mmol), triethylamine (1.0 mmol), KI (1.0 mmol), and CuI (5.0 mol%) in acetonitrile (3 mL) was stirred at room temperature for 0.3-0.5 h to complete the reaction. The crude product was then separated on a silica gel plate using petrol ether as eluent to obtain 1-iodoalkyne **2** in good to excellent yield. All the products were characterized applying spectral techniques and elemental studies.

Characterization data of two representative compounds

1-(Iodoethynyl)-4-methoxybenzene (**2a**): Solid, mp 61-62 °C; IR (KBr): 2162, 824 cm^{-1}; ^{1}H NMR (400 MHz, $CDCl_3$): δ 7.37 (d, $J=8.8$ Hz, 2 H), 6.83 (d, $J=7.6$Hz, 2 H), 3.81 (s, 3 H); ^{13}C NMR (100 MHz, $CDCl_3$): δ 160.0, 133.8, 115.6, 113.8, 94.0, 55.3, 3.7; MS (EI): m/z (%) = 258 (100) [M^+].

1-(Iodoethynyl)-2-(trifluoromethyl)benzene (**2c**): Solid, mp 64-65 °C; IR (KBr): 2170, 752, 735, cm^{-1}; ^{1}H NMR (400 MHz, $CDCl_3$): δ 7.64 (d, $J=8.0$ Hz, 1 H), 7.58 (d, $J=7.6$ Hz, 1 H), 7.49 (dd, $J=7.6$, 7.2 Hz, 1 H), 7.41 (dd, $J=8.0$, 7.6 Hz, 1 H); ^{13}C NMR (100 MHz, $CDCl_3$): δ 134.9, 131.3, 128.5, 125.8, 124.6, 121.9, 121.4, 89.8, 13.2; MS (EI): m/z (%) = 296 (100) [M^+]. Anal. Calcd for $C_9H_4F_3I$: C, 36.50; H, 1.35; Found: C, 36.38; H, 1.33.

Reference

Yan, J., Li, J., and Cheng, D. (2007). Novel and efficient synthesis of 1-iodoalkynes. *Synlett*, 2442–2444.

Carbon – Iodine Bond Forming Reactions: C-I bond_5

Type of reaction: C-I bond formation
Reaction conditions: Acetonitrile, room temperature
Synthetic strategy: Aryne insertion into I-I σ-bonds
Catalysts/reagents: Cesium fluoride (CsF)
Keywords: *o*-(Trimethylsilyl)aryl triflates, iodine, cesium fluoride, acetonitrile, room temperature, aryne, insertion reaction, substituted and polycyclic *o*-diiodoarenes

General reaction scheme

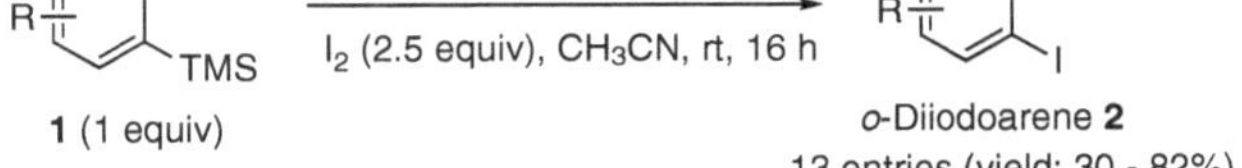

1 (1 equiv)

R = electron-donating or -withdrawing substituents / fused polycyclic systems

o-Diiodoarene **2**
13 entries (yield: 30 - 82%)

Representative entries

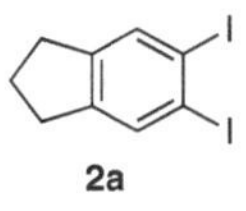

2a

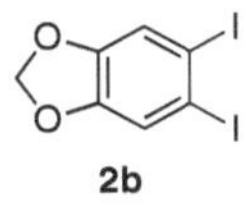

2b

5,6-Diiodo-2,3-dihydro-1*H*-indene (**2a**): Colorless oil; yield: 73%

5,6-Diiodobenzo[*d*][1,3]dioxole (**2b**): Colorless solid; yield: 78%

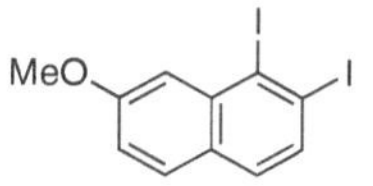
2c

1,2-Diiodo-7-methoxynaphthalene (**2c**): Colorless solid; mp 73.9-75.7 °C; yield: 70%

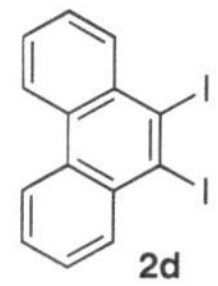
2d

9,10-Diiodophenanthrene (**2d**): Colorless solid; mp 173-174.5 °C; yield: 68%

Experimental procedure

o-(Trimethylsilyl)aryltriflate (**1**; 1 equiv) and iodine (2.5 equiv) were placed into a Schlenk flask under argon in acetonitrile (0.1 M), followed by the addition of anhydrous CsF (5 equiv). The reaction mixture was then stirred at room temperature for 16 h to complete the reaction. Upon completion, the solvent was removed in *vacuo* and the crude product was purified by using column chromatography (hexane/dichloromethane 5:1) to obtain *o*-diodoarene **2**. Each of the products was characterized by spectral studies.

Characterization data of representative compounds

5,6-Diiodobenzo[*d*][1,3]dioxole (**2b**): Colorless solid; ^{1}H NMR (400 MHz, $CDCl_3$): δ 7.26 (s, 2H), 5.95 (s, 2H); ^{13}C NMR (100 MHz, $CDCl_3$): δ 148.9 (2C), 118.8 (2CH), 102.3 (CH_2), 96.6 (2C).

1,2-Diiodo-7-methoxynaphthalene (**2c**): Colorless solid; mp 73.9-75.7 °C; ^{1}H NMR (400 MHz, $CDCl_3$): δ 7.73 (d, $J=8.5$ Hz, 1H), 7.54 (d, $J=8.9$ Hz, 1H), 7.50 (d, $J=2.4$ Hz, 1H), 7.39 (d, $J=8.5$ Hz, 1H), 7.07 (dd, $J=8.9$, 2.5 Hz, 1H), 3.93 (s, 3H); ^{13}C NMR (100 MHz, $CDCl_3$): δ 160.2 (C), 138.2 (C), 133.6 (CH), 130.3 (CH), 129.7 (CH), 127.5 (C), 119.6 (CH), 114.4 (CH), 112.2 (C), 110.0 (C), 55.6 (CH_3); HRMS (EI): $C_{11}H_8I_2O$, calculated: 409.8665; Found: 409.8664.

Reference

Rodríguez-Lojo, D., Cobas, A., Peña, D., Pérez, D., and Guitián, E. (2012). Aryne insertion into I-I σ-bonds. *Org. Lett.*, **14**, 1363–1365.

Carbon – Iodine Bond Forming Reactions: C-I bond_6

Type of reaction: C-I, C-N and C-C bond formation
Reaction conditions: THF/^{t}BuOH, room temperature
Synthetic strategy: One-pot multicomponent synthesis involving coupling-addition-cyclocondensation reactions
Catalysts/reagents: $PdCl_2(PPh_3)_2$, CuI, Et_3N, PTSA
Keywords: Acid chlorides, *N*-Boc-protected propargylamine, $PdCl_2(PPh_3)_2$, copper(I) iodide, triethylamine, *p*-toluenesulfonic acid, THF, ^{t}BuOH, one-pot multicomponent, coupling-addition-cyclocondensation reactions, *N*-Boc-iodopyrroles

General reaction scheme

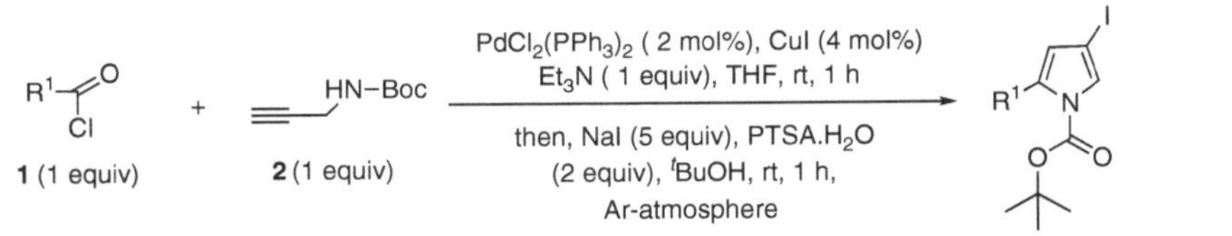

R^1 = C_6H_5, 2-MeC_6H_4, 3-MeC_6H_4, 4-MeC_6H_4, 4-OMeC_6H_4, 4-ClC_6H_4, 4-FC_6H_4, 2-thienyl, β-styryl, cyclopropyl, 1-adamantyl
(reaction time: 3 h when R^1 = cyclopropyl, and 21 h when R^1 = 1-adamantyl)

N-Boc-4-iodopyrrole **3**
11 entries (yield: 61 - 74%)

Representative entries

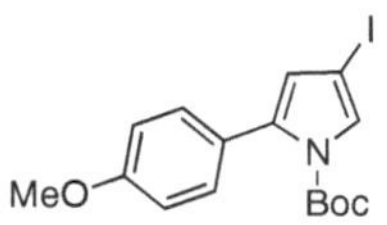

3a

tert-Butyl 4-iodo-2-(4-methoxyphenyl)-1*H*-pyrrole-1-carboxylate (**3a**): Colorless solid, mp 71-72 °C; yield: 73%

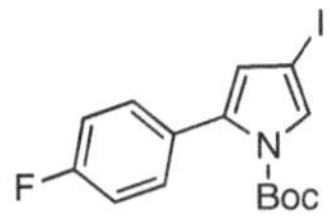

3b

tert-Butyl 2-(4-fluorophenyl)-4-iodo-1*H*-pyrrole-1-carboxylate (**3b**):): Colorless solid, mp 74°C; yield: 75%

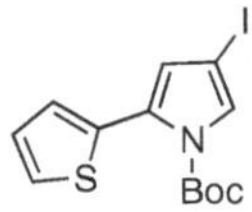

3c

tert- Butyl 4-iodo-2-(thiophen-2-yl)-1*H*-pyrrole-1-carboxylate (**3c**): Colorless solid, mp 55°C; yield: 63%

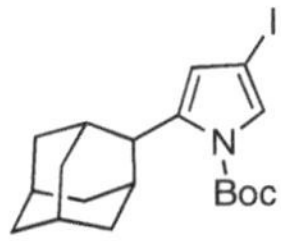

3d

tert-Butyl 2-(adamantan-1-yl)-4-iodo-1*H*-pyrrole-1-carboxylate (**3d**): Colorless solid, mp 151°C; yield: 61%

Experimental procedure

In a screw-cap vessel $PdCl_2(PPh_3)_2$ (70 mg, 0.10 mmol) and CuI (39 mg, 0.20 mmol) were placed and heated under argon atmosphere, followed by cooling to room temperature. Then, 25 mL of dry THF were added and the mixture was degassed with argon. Dry triethylamine (0.69 mL, 5.00 mmol), 5.00 mmol of the acid chloride **1** and *tert*-butyl prop-2-ynylcarbamate (**2**) (776 mg, 5.00 mmol) were successively added to the mixture which was stirred at room temperature until the conversion was complete (monitored by TLC). On completion of this first pahse reaction, sodium iodide (3.79 g, 25.0 mmol), toluene-4-sulfonic acid monohydrate (1.94 g, 10.0 mmol) and 5 mL of *tert*-butanol were successively added to the mixture which was stirred at room temperature for 1 h (monitored by TLC). The reaction mixture was diluted with 50 mL brine, the phases were separated and the aqueous phase was extracted with dichloromethane (3 x 25 mL). The combined organic layers were dried with anhydrous sodium sulfate. After removal of the solvents in *vacuo* the residue was absorbed onto celite and chromatographed on silica gel with petrol ether (40-60 °C)/ethyl acetate (PE-EE) to obtain the pyrroles **4**. All the pure products were characterized based on detailed spectral and analytical studies.

Characterization data of representative compounds

tert- Butyl 4-iodo-2-(thiophen-2-yl)-1*H*-pyrrole-1-carboxylate (**3c**): Colorless solid, mp 55°C; IR (KBr): 3152, 2979, 1750, 1655, 1638, 1560, 1543, 1509, 1475, 1420, 1371, 1345, 1305,

1295, 1255, 1219, 1158, 1141, 1081, 932, 903, 850, 818, 768, 699, 592 cm^{-1}; ^{1}H NMR (500 MHz, $CDCl_3$): δ 1.41 (s, 9 H), 6.37 (d, $J=1.9$ Hz, 1 H), 7.01 (dd, $J=5.0$ & 3.5 Hz, 1 H), 7.05 (dd, $J=3.5$ & 1.3 Hz, 1 H), 7.32 (dd, $J=5.0$ & 1.3 Hz, 1 H), 7.44 (d, $J=1.9$ Hz, 1 H); ^{13}C NMR (125 MHz, $CDCl_3$): δ 27.6 (CH_3), 64.1 (C_{quat}), 84.5 (C_{quat}), 122.6 (CH), 126.2 (CH), 126.5 (CH), 127.7 (CH), 128.3 (CH), 128.5 (C_{quat}), 133.0 (C_{quat}), 147.6 (C_{quat}); EI+MS (*m/z*; %): 375 (M^+, 2), 319 ($(M-C_4H_9+H)^+$, 5), 275 ($(M-C_5H_9O_2+H)^+$, 9), 147 ($(M-C_5H_9O_2-I)^+$, 2), 57 ($C_4H_9{}^+$, 100), 41 (20). Anal. calcd for $C_{13}H_{14}INO_2S$ (375.2): C 41.61, H 3.76, N 3.73; Found: C 41.75, H 3.82, N 3.48.

tert-Butyl 2-(adamantan-1-yl)-4-iodo-1*H*-pyrrole-1-carboxylate (**3d**): Colorless solid, mp 151°C; IR (KBr): 3166, 2980, 2908, 2849, 2677, 1748, 1543, 1489, 1452, 1395, 1368, 1316, 1297, 1255, 1234, 1163, 1137, 1105, 1089, 1011, 976, 937, 899, 849, 819, 804, 764, 679, 644, 590, 533 cm^{-1}; ^{1}H NMR (500 MHz, $CDCl_3$): δ 1.58 (s, 9 H), 1.68-1.78 (m, 6 H), 2.0 (br, 3 H), 2.05-2.08 (m, 6 H), 6.07-6.09 (m, 1 H), 7.24-7.27 (m, 1 H); ^{13}C NMR (125 MHz, $CDCl_3$): δ 28.2 (CH_3), 28.9 (CH), 35.7 (C_{quat}), 36.9 (CH_2), 40.7 (CH_2), 63.6 (C_{quat}), 83.9 (C_{quat}), 117.6 (CH), 128.0 (CH), 147.7 (C_{quat}), 148.3 (C_{quat}); EI+MS (*m/z*; %): 427 (M^+, 2), 371 ($(M-C_4H_9+H)^+$, 6), 327 ($(M-C_5H_9O_2+H)^+$, 18), 270 (4), 57 (C_4H_9 +, 100), 41 (7). Anal. calcd for $C_{19}H_{26}INO_2$ (427.3): C 53.40, H 6.13, N 3.28; Found: C 53.46, H 6.22, N 3.11.

Reference

Merkul, E., Boersch, C., Frank, W., and Müller, T. J. J. (2009). Three-component synthesis of *N*-Boc-4-iodopyrroles and sequential one-pot alkynylation. *Org. Lett.*, **11**, 2269–2272.

Carbon – Iodine Bond Forming Reactions: C-I bond_7

Type of reaction: C-I and C-C bond formation
Reaction conditions: Dichloromethane, room temperature
Synthetic strategy: Iodonium-promoted 5-*endo-dig* carbocyclization of alkynes and active methlene compounds
Catalysts/reagents: Iodine
Keywords: δ-Alkynyl-β-ketoesters, iodine, dichloromethane, room temperature, Iodonium-promoted 5-*endo-dig* carbocyclization of active methylene substrates onto alkynes; iodocyclopentenes

General reaction scheme

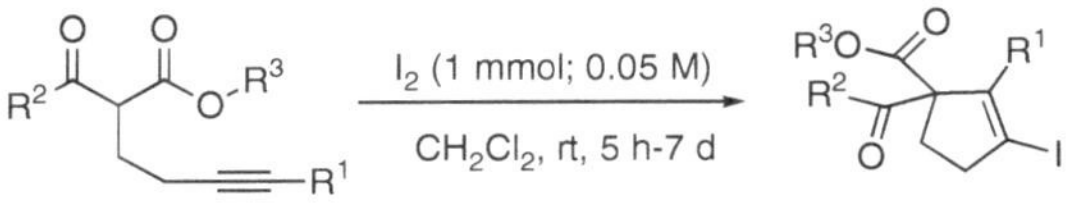

1 (1 mmol)
R^1 = H/alkyl/aryl
R^2/R^3 = alkyl/aryl/halo/alicycles

Iodocyclopentene **2**
12 entries (yield: 20 - 80%)

Representative entries

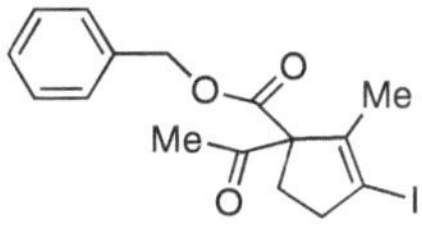

2a

Benzyl 1-acetyl-3-iodo-2-methylcyclopent-2-enecarboxylate (**2a**): Yellow oil; reaction time: 5 h; yield: 60%

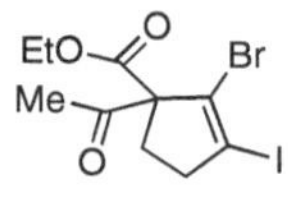

2b

Ethyl 1-acetyl-2-bromo-3-iodocyclopent-2-enecarboxylate (**2b**): Orange oil; reaction time: 5d; yield: 75%

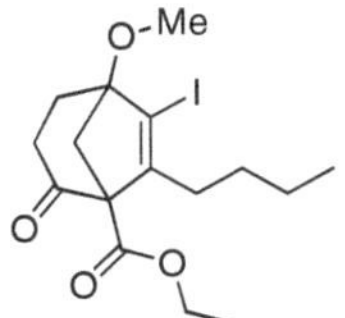

2c

Ethyl 7-butyl-6-iodo-5-methoxy-2-oxobicyclo [3.2.1]oct-6-ene-1-carboxylate (**2c**): Orange oil; reaction time: 15 h; yield: 62%

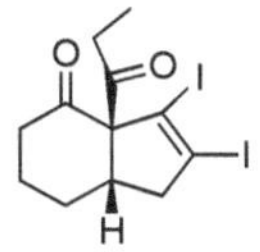

2d

(3a*S*,7a*S*)-2,3-Diiodo-3a-propionyl-5,6,7,7a-tetrahydro-1H-inden-4(3a*H*)-one (**2d**): Yellow oil; reaction time: 24 h; yield: 80%

Experimental procedure

To a solution of β-ketoester (**1**; 1 mmol, 1 equiv) in 20 mL of dichloromethane was added iodine (254 mg, 1 mmol, 1 equiv), and the resulting mixture was stirred at room temperature for stipulated time to complete the reaction. Upon completion of the reaction (monitored by TLC), the crude mixture was concentrated and purified by flash chromatography (hexanes/EtOAc) to yield iodocyclopentene **2**. Each of the products was characterized by spectral studies.

Characterization data of representative compounds

Benzyl 1-acetyl-3-iodo-2-methylcyclopent-2-enecarboxylate (**2a**): Yellow oil; IR (KBr): 1737, 1712 cm^{-1}; ^{1}H NMR (400 MHz, $CDCl_3$): δ 7.40-7.30 (m, 5H), 5.25-5.14 (d, $J=6.8$ Hz, 2H), 2.86-2.63 (m, 3H), 2.29-2.21 (m, 1H), 2.11 (s, 3H), 1.90 (t, $J=2.6$ Hz, 3H); ^{13}C NMR (100 MHz, $CDCl_3$): δ 203.5 (C), 170.5 (C), 141.0 (C), 135.6 (C), 129.1 (CH), 128.9 (CH), 128.7 (CH), 100.4 (C), 73.6 (C), 67.8 (CH_2), 42.6 (CH_2), 33.1 (CH_2), 27.0 (CH_3), 18.1 (CH_3); HRMS (EI): Calcd for $C_{16}H_{17}IO_3$: 384.0222; Found: 384.0235.

(3a*S*,7a*S*)-2,3-Diiodo-3a-propionyl-5,6,7,7a-tetrahydro-1*H*-inden-4(3a*H*)-one (**2d**): Yellow oil; IR (KBr): 1712 cm^{-1}; ^{1}H NMR (400 MHz, $CDCl_3$): δ 4.43-4.22 (m, 2H), 3.47-3.32 (m, 1H), 2.80 (dd, J =15.9 & 7.1 Hz, 1H), 2.63-2.53 (m, 1H), 2.47 (dd, J =16.1 & 7.2 Hz, 1H), 2.40-2.30 (m, 1H), 1.96-1.88 (m, 2H), 1.73-1.61 (m, 2H), 1.43 (t, $J=7.2$ Hz, 3H); ^{13}C NMR (100 MHz, $CDCl_3$): δ 205.4 (C), 169.5 (C), 113.0 (C), 105.1 (C), 74.2 (C), 62.2 (CH_2), 48.8 (CH), 47.9 (CH_2), 40.0 (CH_2), 26.9 (CH_2), 22.5 (CH_2),14.2 (CH_3); HRMS (EI): Calcd for $C_{12}H_{14}I_2O_3$: 459.9032; Found: 459.034.

Reference

Barluenga, J., Palomas, D., Rubio, E., and González, J. M. (2007). Iodocarbocyclization reaction of β-ketoesters and alkynes. *Org. Lett.*, **9**, 2823–2826.

Carbon – Iodine Bond Forming Reactions: C-I bond_8

Type of reaction: C-I and C-C bond formation
Reaction conditions: Nitromethane, room temperature
Synthetic strategy: Electrophilic iodocyclization
Catalyst/reagent: Iodine/$NaHCO_3$
Keywords: Propargylic aryl ethers, iodine, sodium bicarbonate, nitromethane, room temperature, electrophilic iodocyclization, 3,4-disubstituted 2*H*-benzopyrans

General reaction scheme

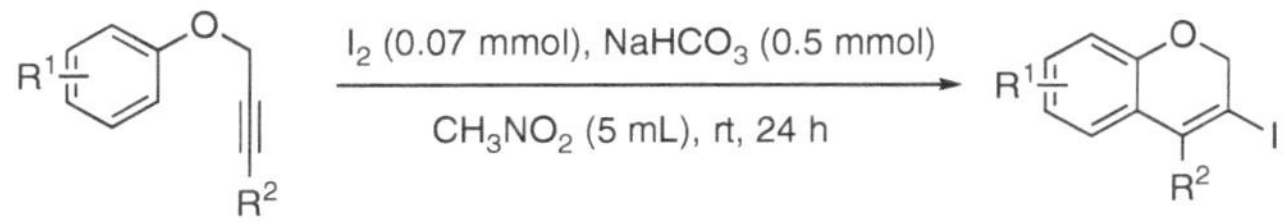

1 (0.25 mmol)
R^1 = H, 4-Cl, 4-Me, 4-CHO, 3-^{t}Bu, 4-^{t}Bu, 3,4-di-^{t}Bu
R^2 = C_6H_5, 2-OMeC_6H_4, CH_2OH, cyclopentyl

3,4-Disubstituted 2*H*-benzopyran **2**
15 entries (yield: 52 - 92%)

Representative entries

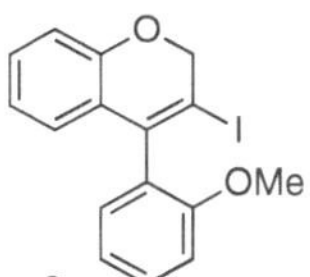

2a

3-Iodo-4-(2-methoxyphenyl)-2*H*-chromene (**2a**): Light brown solid, mp 111-113 °C; yield: 85%

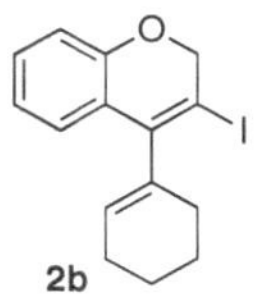

2b

4-(Cyclohex-1-en-1-yl)-3-iodo-2*H*-chromene (**2b**): Light brown oil; yield: 74%

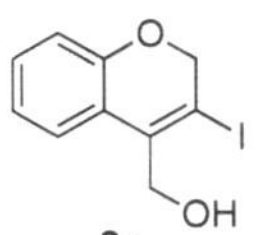

2c

(3-Iodo-2*H*-chromen-4-yl)methanol (**2c**): Yellow solid, mp 62-63 °C; yield: 79%

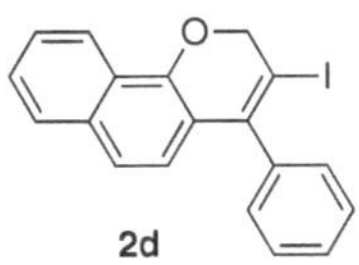

2d

3-Iodo-4-phenyl-2*H*-benzo[*h*]chromene (**2d**): Light brown solid, mp 104-105 °C; yield: 61%

Experimental procedure

To a solution of propargylic aryl (**1**; 0.25 mmol)ether in 3 mL of nitromethane, 2.0 equiv of $NaHCO_3$ and 3.0 equiv of iodine dissolved in 2 mL of nitromethane was added gradually, and the reaction mixture was allowed to stir at room temperature for 24 h. Upon completion of the reaction, excess I_2 was removed by washing with saturated aqueous sodium thiosulfate solution. The resulting mixture was then extracted by diethyl ether (3 x 10 mL), and the organic combined ether layers were dried over anhydrous sodium sulfate and concentrated under vacuum to yield the crude product, which was purified by flash chromatography on silica gel using hexanes/ethyl acetate as the eluent to afford 3,4-disubstituted 2*H*-benzopyran **2**. Each of the products was characterized by spectral studies.

Characterization data of representative compounds

3-Iodo-4-(2-methoxyphenyl)-2*H*-chromene (**2a**): Light brown solid, mp 111-113 °C; IR (neat): 2955, 2918, 1505, 1244, 1171, 1037 cm^{-1}; ^{1}H NMR (300 MHz, $CDCl_3$): δ 3.75 (s, 3H), 5.01-5.13 (m, 2H), 6.57 (dd, $J=7.8$ & 1.5 Hz, 1H), 6.74 (dt, $J=7.6$ & 1.1 Hz, 1H), 6.83 (dd, $J=7.1$ & 1.0 Hz, 1H), 6.96-7.14 (m, 4H), 7.37-7.42 (m, 1H); ^{13}C NMR (75 MHz, $CDCl_3$): δ 55.9, 74.9, 92.7, 111.6, 116.0, 121.0, 121.7, 123.8, 126.1, 129.0, 129.5, 130.0, 131.0, 139.5, 153.2, 156.9; HRMS: *m/z* 363.99644.

(3-Iodo-2*H*-chromen-4-yl)methanol (**2c**): Yellow solid, mp 62-63 °C; IR (neat): 3334, 2934, 1480, 1444, 1219, 1009 cm^{-1}; ^{1}H NMR (300 MHz, $CDCl_3$): δ 2.06 (s, 1H), 4.64 (s, 2H), 4.86 (s, 2H), 6.82 (d, $J=8.0$ Hz, 1H), 6.92-6.98 (m, 1H), 7.18 (dt, $J=8.0$ & 1.3 Hz, 1H), 7.41 (dd, $J=7.7$ & 1.1 Hz, 1H); ^{13}C NMR (75 MHz, $CDCl_3$): δ 66.0, 75.0, 93.8, 116.4, 121.8, 122.1, 124.3, 129.9, 137.3, 153.7; HRMS: *m/z* 287.96514.

Reference

Worlikar, S. A., Kesharwani, T., Yao, T., and Larock, R. C. (2007). Synthesis of 3,4-disubstituted 2*H*-benzopyrans through C-C bond formation via electrophilic cyclization. *J. Org. Chem.*, **72**, 1347–1353.

Chapter 7

Carbon – Boron Bond Forming Reactions at Room Temperature

C-B bond_1

Type of reaction: C-B and C-C bond formation
Reaction conditions: Tetrahydrofuran, room temperature
Synthetic strategy: Regioselective carboboration of alkynes
Catalyst: Cu(I)Cl, 4,5-bis(diphenylphosphino)-9,9-dimethylxanthene (xantphos; ligand), NaO^{t}Bu
Keywords: Alkynes, methyl iodide, bis(pinacolato)diboron [$B_2(pin)_2$], Cu(I)Cl, xantphos, tetrahydrofuran (THF), room temperature, regioselective carboboration of alkynes, tri- and tetra-substituted vinyl boronic esters

General reaction scheme

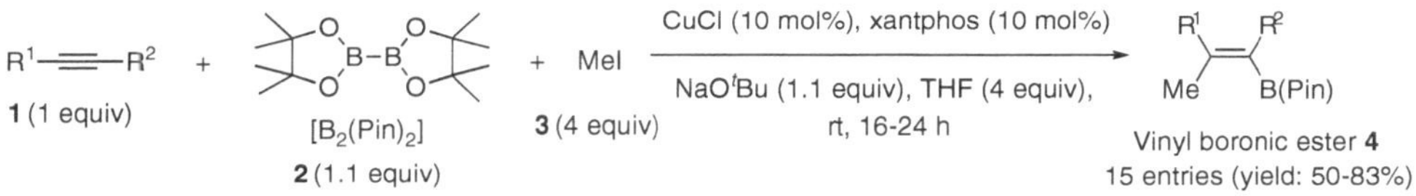

$R^1 = C_6H_5$, 3-MeC_6H_4, 4-MeC_6H_4, 4-$^tBuC_6H_4$, 2-$OMeC_6H_4$, 4-$OMeC_6H_4$, 4-BrC_6H_4, 3-FC_6H_4, 4-FC_6H_4, 4-$(CH_3CO_2CH_2)C_6H_4$, 2,4,5-$Me_3C_6H_2$, 3,5-$(CF_3)_2C_6H_3$, 2-thienyl
R^2= H, Me, C_6H_5

Representative entries

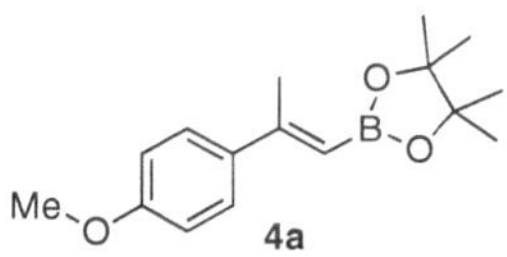
4a

(*E*)-2-(2-(4-Methoxyphenyl)prop-1-enyl)-4,4,5,5-tetramethyl-1,3,2-dioxaborolane (**4a**): Colorless oil; yield 75%

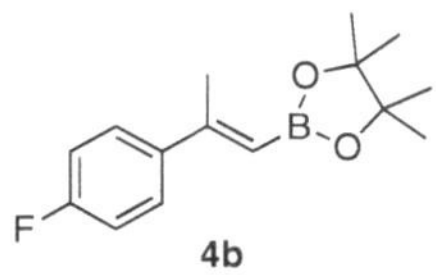
4b

(*E*)-2-(2-(4-Fluorophenyl)prop-1-enyl)-4,4,5,5-tetramethyl-1,3,2-dioxaborolane (**4b**): Colorless oil; yield 72%

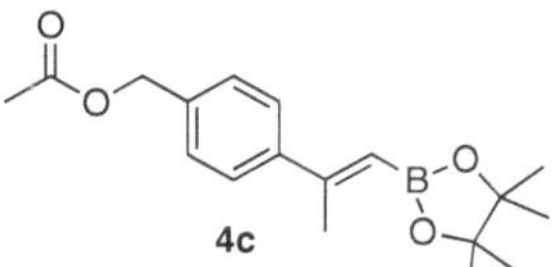
4c

(*E*)-4-(1-(4,4,5,5-Tetramethyl-1,3,2-dioxaborolan-2-yl)prop-1-en-2-yl)benzyl acetate (**4c**): Colorless oil; yield 70%

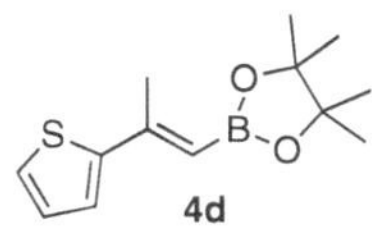
4d

(*E*)-4,4,5,5-Tetramethyl-2-(2-(thiophen-2-yl)prop-1-enyl)-1,3,2-dioxaborolane, (**4d**): Colorless oil; yield 83%

Experimental procedure

An oven-dried vial was charged with CuCl (0.04 mmol, 3.95 mg), xantphos (0.04 mmol) and bis(pinacolato)diboron (**2**: 0.44 mmol, 112 mg). After being sealed with a septum, the vial was

Room Temperature Organic Synthesis. http://dx.doi.org/10.1016/B978-0-12-801025-9.00007-3

connected to an argon-vacuum line and was evacuated and backfilled with argon. THF (0.1-0.6 M) was added and the mixture was stirred for 5 min at room temperature. Then, a 2.0 M NaO^{t}Bu solution in THF (0.22 mL) was added drop-wise. After stirring for 5 min, the alkyne or enyne (**1**; 0.4 mmol) was added (neat or dissolved in the minimal amount of THF if it is a solid) followed by methyl iodide (1.6 mmol, 0.1 mL). The septum was sealed with parafilm and the mixture was stirred at room temperature for 16-24 h. Diethyl ether and water was added, and the layers were separated. The aqueous phase was extracted with Et_2O (×2) and the combined organic layers were dried over Mg_2SO_4 and concentrated. The crude product was purified using a fluorisil column (100% pentanes) or a rapid-short SiO_2 column (2-5% EtOAc-hexanes). All the products were characterized using spectral techniques.

Characterization data of two representative compounds

(*E*)-4-(1-(4,4,5,5-Tetramethyl-1,3,2-dioxaborolan-2-yl)prop-1-en-2-yl)benzyl acetate (**4c**): Colorless oil; ^{1}H NMR (300 MHz, $CDCl_3$): δ 7.48 (d, *J*=8.4 Hz, 2H), 7.30 (d, *J*=8.5 Hz, 2H), 5.75 (q, *J*=1.0 Hz, 1H), 5.09 (s, 2H), 2.39 (d, *J*=1.0 Hz, 3H), 2.09 (s, 3H), 1.31 (s, 12H); ^{13}C NMR (75 MHz, $CDCl_3$): δ 170.8, 157.1, 143.8, 135.5, 128.0 (2C), 126.0 (2C), 82.9 (2C), 65.9, 24.8 (4C), 20.9, 20.0 [note: the carbon attached to boron was not observed due to quadrupole broadening caused by the ^{11}B nucleus]; HRMS (EI$^+$): Calcd for $C_{18}H_{25}BO_4$ $[M]^+$: 316.1846; Found: 316.1852.

(*E*)-4,4,5,5-Tetramethyl-2-(2-(thiophen-2-yl)prop-1-enyl)-1,3,2-dioxaborolane (**4d**): Colorless oil; ^{1}H NMR (300 MHz, $CDCl_3$): δ 7.21 (dd, *J*=5.1, 1.1 Hz, 1H), 7.15 (dd, *J*=3.7, 1.1 Hz, 1H), 6.98 (dd, *J*=5.1, 3.7 Hz, 1H), 5.82 (q, *J*=0.9 Hz, 1H), 2.42 (d, *J*=0.9 Hz, 1H), 1.30 (s, 12H); ^{13}C NMR (75 MHz, $CDCl_3$): δ 150.4, 148.4, 127.5, 125.7, 124.6, 82.9 (2C) 24.8 (4C), 19.7 [note: the carbon attached to boron was not observed due to quadrupole broadening caused by the ^{11}B nucleus]; HRMS (EI$^+$): Calcd for $C_{13}H_{19}BO_2S$ [M]+: 250.1199, Found: 250.1207.

Reference

Alfaro, R., Parra, A., Alemàn, J., García Ruano, J. L., and Tortosa, M. (2012). Copper(I)-catalyzed formal carboboration of alkynes: synthesis of tri- and tetrasubstituted vinylboronates. *J. Am. Chem. Soc.*, **134**, 15165–15168.

C-B bond_2

Type of reaction: C-B bond formation
Reaction conditions: Diethyl ether, room temperature
Synthetic strategy: 1,4-Hydroboration of 1,3-dienes
Catalyst: Imino-pyridine (**L1** or **L2**) – iron(II) complex (L.$FeCl_2$; **1**) in the presence of Mg(0)

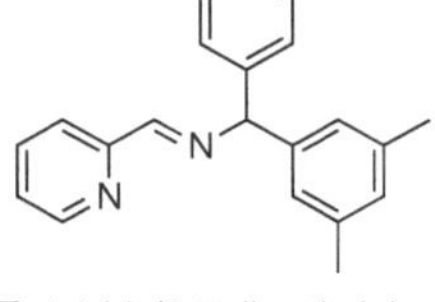

L1 [(*E*)-1,1-bis(3,5-dimethylphenyl)-*N*-(pyridin-2-ylmethylene)methanamine]

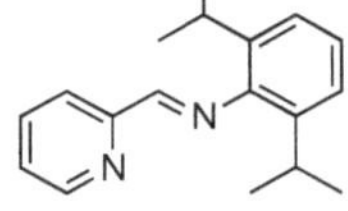

L2 [(*E*)-2,6-diisopropyl-*N*-(pyridin-2-ylmethylene)aniline]

Keywords: 1,3-Dienes, iron(II)-complex (L.$FeCl_2$,**1**), magnesium (Mg), diethyl ether, room temperature, 1,4-hydroboration, chemo-, regio- and stereoselectivity, linear (*E*)-γ-disubstituted allylboranes

General reaction scheme

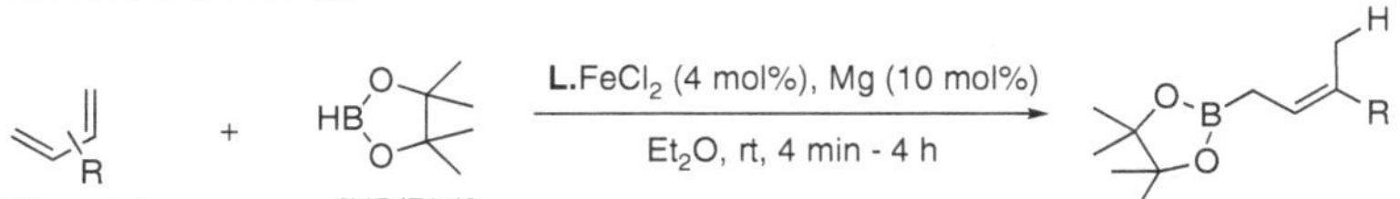

2 (1 equiv) (substituted open-chainand cyclic 1,3-dienes) [HB(Pin)] **3** (1.2 equiv)

γ-Disubstituted allyl pinacolborane **4**
11 entries (yield: 66-92%)

Representative entries

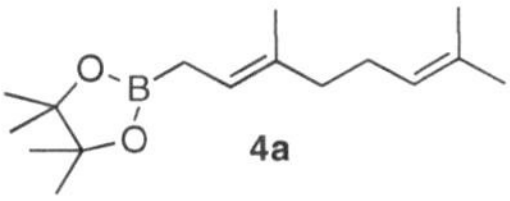

4a

(*E*)-2-(3,7-Dimethylocta-2,6-dien-1-yl)-4,4,5,5-tetramethyl-1,3,2-dioxaborolane (**4a**): Colorless oil; yield 89%

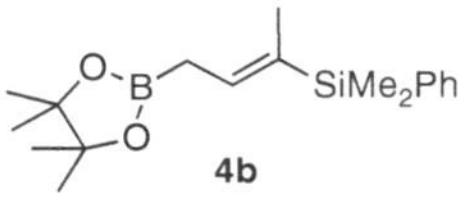

4b

(*E*)-Dimethyl(phenyl)(4-(4,4,5,5-tetramethyl-1,3,2-dioxaborolan-2-yl)but-2-en-2-yl)silane (**4b**): Colorless oil; yield 91%

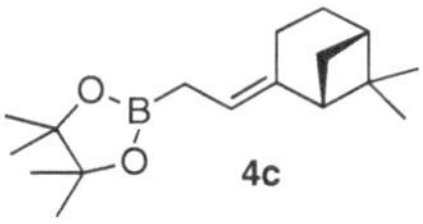

4c

2-((*E*)-2-((1*R*,5*S*)-6,6-Dimethylbicyclo[3.1.1]heptan-2-ylidene)ethyl)-4,4,5,5-tetramethyl-1,3,2-dioxaborolane (**4c**): Colorless oil; yield 91%

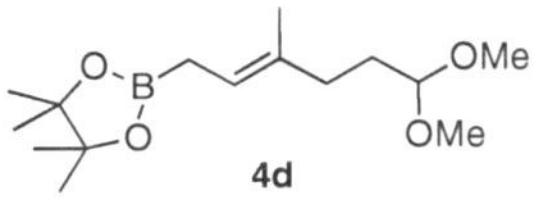

4d

(*E*)-2-(6,6-Dimethoxy-3-methylhex-2-en-1-yl)-4,4,5,5-tetramethyl-1,3,2-dioxaborolane (**4d**): Colorless oil; yield 71%

Experimental procedure

An oven-dried vial was charged with complex **L**·$FeCl_2$·CH_2Cl_2 (**1**; 11.4 mg, 0.02 mmol, 0.04 equiv), 1,3-diene (**2**; 0.5 mmol, 1 equiv) and diethyl ether (1 mL), followed by magnesium (1.2 mg, 0.050 mmol, 0.10 equiv). Pinacolborane (**3**; 77.6 mg, 88.0 μL, 0.6 mmol, 1.2 equiv) was added and the reaction mixture was stirred at room temperature for 3 h. Water (1 mL) was added to the reaction mixture and the aqueous layer was extracted with diethyl ether (3 × 1 mL). The combined organic layers were passed through a plug of silica gel, eluting with diethyl ether, and concentrated in *vacuo* to obtain the desired product **4**, characterized by means of spectral studies.

Characterization data of two representative compounds

2-((*E*)-2-((1*R*,5*S*)-6,6-Dimethylbicyclo[3.1.1]heptan-2-ylidene)ethyl)-4,4,5,5-tetramethyl-1,3,2-dioxaborolane (**3c**): Colorless oil; ^{1}H NMR (500 MHz, C_6D_6): δ 7.57-7.59 (m, 2H), 7.18-7.25 (m, 3H), 6.28 (tq, J = 7.8, 1.8 Hz, 1H), 1.98 (d, J = 7.8 Hz, 2H), 1.75 (s, 3H), 1.02 (s, 12H), 0.33 (s, 6H); ^{13}C NMR (125 MHz, $CDCl_3$): δ 139.22, 136.08, 134.00, 133.76, 128.62, 128.33, 127.54, 83.11, 24.71, 14.74, −3.32; ^{11}B NMR (96 MHz, C_6D_6): δ 33.1; HRMS-ESI (*m/z*): Calcd for [$C_{18}H_{29}BO_2Si$+Na], 339.1922. Found, 339.1920.

(*E*)-2-(6,6-Dimethoxy-3-methylhex-2-en-1-yl)-4,4,5,5-tetramethyl-1,3,2-dioxaborolane (**3d**): Colorless oil; ^{1}H NMR (500 MHz, C_6D_6): δ 5.60 (tq, J = 7.8, 1.4 Hz, 1H), 4.35 (t, J = 6.0 Hz, 1H), 3.15 (s, 6H), 2.15 (t, J = 8.7 Hz, 2H), 1.77–1.85 (m, 4H), 1.61 (d, J = 0.7 Hz, 3H), 1.03

(s, 12H); ^{13}C NMR (125 MHz, C_6D_6): δ 134.05, 119.97, 104.21, 82.90, 52.11, 35.16, 31.34, 24.85, 24.77, 15.99; ^{11}B NMR (96 MHz, C_6D_6): δ 33.4; HRMS-FIA (*m/z*): Calcd for [$C_{15}H_{29}BO_4$+Na], 307.2051; Found, 307.2050.

Reference

Wu, J. Y., Moreau, B., and Ritter, T. (2009). Iron-catalyzed 1,4-hydroboration of 1,3-dienes. *J. Am. Chem. Soc.*, **131**, 12915–12917.

C-B bond_3

Type of reaction: C-B bond formation
Reaction conditions: Tetrahydrofuran (THF), room temperature
Synthetic strategy: One-pot synthesis of pinacolboronic esters
Catalyst: Magnesium (Mg^0)
Keywords: Alkyl/aryl/heteroaryl/vinyl/allyl bromides, magnesium (Mg), tetrahydrofuran (THF), pinacolborane, room temperature, Grignard conditions/Barbier conditions, pinacolboronic esters

General reaction schemes

Grignard conditions

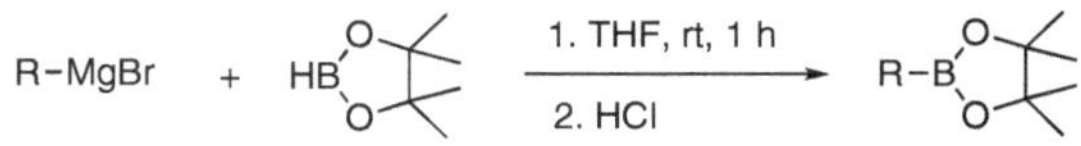

1 (4.5 mmol) **2** (4.5 mmol)
R = alkyl, aryl

Pinacolboronic ester **3**
10 entries (yield: 65-99%)

Barbier conditions

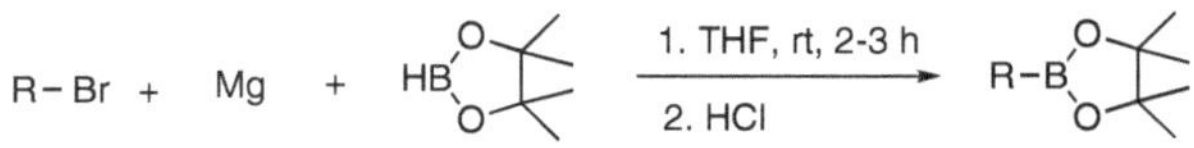

1 (2 mmol) (2.4 mmol) **2** (2 mmol)
R = alkyl/aryl/heteroaryl/vinyl/allyl
(*in case of allyl bromide, its 2 equiv (4 mmol) was required)

Pinacolboronic ester **3**
16 entries (yield: 65-95%)

Representative entries

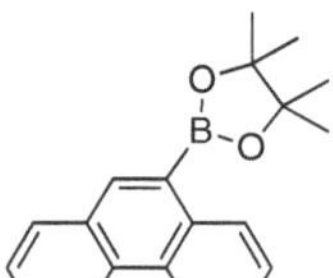

3a

4,4,5,5-Tetramethyl-2-(phenanthren-9-yl)-1,3,2-dioxaborolane (**4a**): Colorless oil; yield 85%

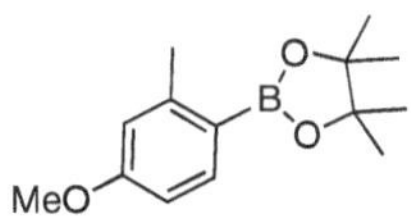

3b

2-(4-Methoxy-2-methylphenyl)-4,4,5,5-tetramethyl-1,3,2-dioxaborolane (**4b**): Clear oil; yield 97%

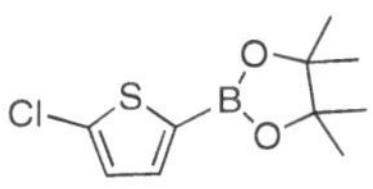

3c

2-(5-Chlorothiophen-2-yl)-4,4,5,5-tetramethyl-1,3,2-dioxaborolane (**4c**): Clear oil; yield 75%

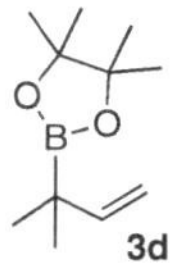

3d

4,4,5,5-Tetramethyl-2-(2-methylbut-3-en-2-yl)-1,3,2-dioxaborolane (**4d**): Clear oil; yield 95%

Experimental procedure

Under Grignard conditions

A 25 mL round-bottom flask equipped with a magnetic stir bar and fitted with a rubber septum was charged with anhydrous THF (4.0 mL) followed by pinacolborane (**2**; 0.57 g, 4.5 mmol). Alkyl/aryl bromide (**1**; 4.5 mmol, 1 M/THF) was added drop-wise over 5 min at room temperature (25 °C) with constant stirring for 1 h. Upon completion of the reaction (monitored by NMR), the reaction mixture was cooled to 0 °C (ice bath) and acidified with 3 M aqueous HCl (3 mL). After 10 min of stirring, the reaction mixture was warmed to room temperature (25 °C) and stirred for an additional 30 min. The reaction mixture was then transferred to a separatory funnel and extracted with diethyl ether (3 × 15 mL). The combined organic layers were dried over anhydrous $MgSO_4$, filtered, and dried in *vacuo* to 1,3,2-dioxaborolane derivative **3**, identified by means of spectral studies.

Under Barbier conditions

A 25 mL round-bottom flask equipped with a magnetic stir bar was charged with magnesium turnings (0.058 g, 2.4 mmol) and was activated by addition of iodine crystals and warming until iodine sublimed. The flask was cooled to 25 °C and was purged with argon. THF (3.5 mL) was added to the flask, followed by the addition of neat pinacolborane (**2**; 0.29 mL, 2.0 mmol). Alkyl/aryl/heretoaryl/vinyl/allyl bromide (**1**: 0.243 mL, 2.0 mmol) was then added drop-wise over 5 min with constant stirring at 25 °C. Upon completion of the reaction within 2-3 h (NMR monitoring), the reaction mixture was cooled to 0 ° C (ice bath) and acidified with 3 M aqueous HCl (3 mL). After 10 min of stirring the reaction mixture was warmed to room temperature (25 °C) and stirred for an additional 30 min. The reaction mixture was then transferred to a separatory funnel and extracted with diethyl ether (3 × 15 mL). The combined organic layers were dried over anhydrous $MgSO_4$, filtered, and dried in *vacuo* to pinacolborane esters **3**, characterized on the basis of spectral studies.

Characterization data of two representative compounds

2-(5-Chlorothiophen-2-yl)-4,4,5,5-tetramethyl-1,3,2-dioxaborolane (**4c**): Clear oil; IR (Nujol): 1019, 1056, 1146, 1211, 1302, 1377, 1463, 1522 cm^{-1}; ^{1}H NMR (500 MHz, $CDCl_3$): δ 1.34 (s, 12H), 6.98 (d, $J=3.5$ Hz, 1H), 7.41 (d, $J=4$ Hz, 1H); ^{13}C NMR (125.7 MHz, $CDCl_3$): δ 24.7, 84.3, 127.7, 136.9; ^{11}B NMR (160.4 MHz, $CDCl_3$): δ +28.1 (s).

4,4,5,5-Tetramethyl-2-(2-methylbut-3-en-2-yl)-1,3,2-dioxaborolane (**4d**): Clear oil; ^{1}H NMR (500 MHz, $CDCl_3$): δ 1.07 (s, 6H), 1.22 (s, 12H), 4.90 (dd, $J=0.5$, 0.5 Hz, 1H), 4.93 (dd, $J=0.3$, 0.6 Hz, 1H), 5.96 (dd, $J=5.7$, 3.3 Hz, 1H); ^{13}C NMR (125.7 MHz, $CDCl_3$): δ 23.4, 24.5, 83.2, 110.0, 146.7; ^{11}B NMR (160.4 MHz, $CDCl_3$): δ +33.9 (s).

Reference

Clary, J. W., Rettenmaier, T. J., Snelling, R., Bryks, W., Banwell, J., Wipke, W. T., and Singaram, B. (2011). Hydride as a leaving group in the reaction of pinacolborane with halides under ambient Grignard and Barbier conditions. one-pot synthesis of alkyl, aryl, heteroaryl, vinyl, and allyl pinacolboronic esters. *J. Org. Chem.*, **76**, 9602–9610.

C-B bond_4

Type of reaction: C-B bond formation
Reaction conditions: Dichloromethane/toluene, room temperature
Synthetic strategy: Hydroboration of alkynes
Catalyst: $HZrCp_2Cl$
Keywords: Monosubstituted alkynes, methylpentanediolborane (MPBH), $HZrCp_2Cl$, dichloromethane/toluene, room temperature, regio- and stereoselectivity, functionalized (*E*)-vinylboronates

General reaction scheme

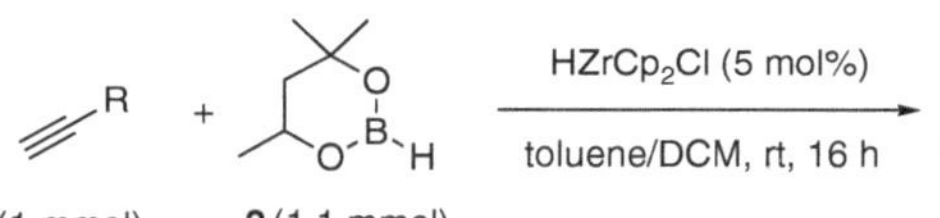

1 (1 mmol) **2** (1.1 mmol) (*E*)-Vinyl boronate **3**

11 entries (yield: 68-90%)

R = C_6H_5, CH_2Cl, CH_2Br, $(CH_2)_3Cl$, CH_2OMe, CH_2OAc, CH_2O^tBu, $CH(OEt)_2$, TMS, *n*-Bu, *c*-hex

Representative entries

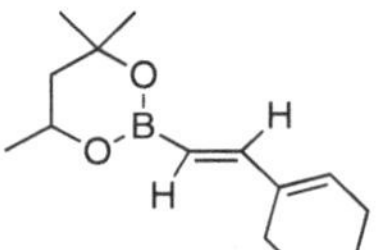

3a

(*E*)-2-(2-(Cyclohex-1-en-1-yl)vinyl)-4,4,6-trimethyl-1,3,2-dioxaborinane (**3a**): Clear oil; yield 66%

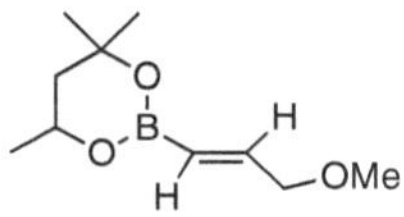

3b

(*E*)-2-(3-Methoxyprop-1-en-1-yl)-4,4,6-trimethyl-1,3,2-dioxaborinane (**3b**): Clear oil; yield 70%

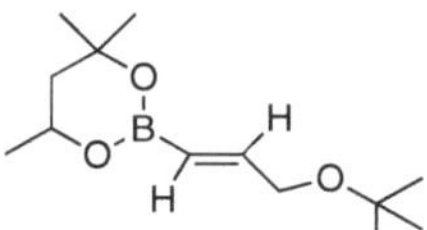

3c

(*E*)-2-(3-(*tert*-Butoxy)prop-1-en-1-yl)-4,4,6-trimethyl-1,3,2-dioxaborinane (**3c**): Clear oil; yield 82%

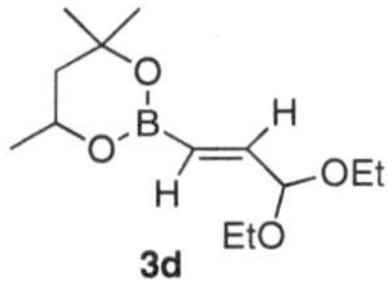

3d

(*E*)-2-(3,3-Diethoxyprop-1-en-1-yl)-4,4,6-trimethyl-1,3,2-dioxaborinane (**3d**): Clear oil; yield 80%

Experimental procedure

A 5-mL round-bottom flask equipped with a side arm and magnetic stirring was charged with $HZrCp_2Cl$ (0.05 mmol, 13 mg) and dry dichloromethane or toluene (1 mL) under a nitrogen atmosphere and cooled to 0 °C. Alkyne (**1**: 1.0 mmol) was then added drop-wise, the mixture then was stirred for 2 min, and methylpentanediolborane (MPBH; **2**; 1.1 mmol) was added drop-wise *via* a syringe. Stirring was continued for 16 h at room temperature (20 °C) for completion of the reaction (monitored by GC analysis). Removal of the solvent under reduced pressure, followed by flash column chromatography (3% ether in hexane) provided the desired product **3** as clear oil. All the products were characterized by spectral studies.

Characterization data of two representative compounds

(*E*)-2-(2-(Cyclohex-1-en-1-yl)vinyl)-4,4,6-trimethyl-1,3,2-dioxaborinane (**3a**): Clear oil; IR: 3018, 2974, 2929, 1634, 1607, 1300, 1162, 997, 795, 771 cm^{-1}; ^1H NMR (300 MHz, $CDCl_3$): δ 6.91 (d, *J*=17.6 Hz, 1H), 5.88 (br t, *J*=3.2 Hz, 1H), 5.34 (d, *J*=17.6 Hz, 1H), 4.20 (dqd, *J*=11.8, 6.2, 3.0 Hz, 1H), 2.18-2.04 (m, 2H), 1.76 (dd, *J*=13.9, 3.0 Hz, 1H), 1.68-1.53 (m, 4H), 1.47 (dd, *J*=13.9, 11.8 Hz, 1H), 1.30-1.27 (m, 8H), 1.25 (d, *J*=6.2 Hz, 3H); ^{13}C NMR (75.5 MHz, $CDCl_3$): δ 150.1, 137.1 (br), 132.4, 117.4, 70.4, 64.5, 45.9, 31.1, 28.0, 25.9, 23.9, 23.1, 22.4, 22.3; ^{11}B NMR (96 MHz, $CDCl_3$): δ 26.0; LRMS (DCI NH_3/isobutane) *m/z* (%) 235(100) $[M+H]^+$, 247(40), 248(23), 250(34), 251(4); HRMS (DCI NH_3) m/z calcd for $C_{14}H_{24}BO_2$: 235.1864; Found 235.1861.

(*E*)-2-(3,3-Diethoxyprop-1-en-1-yl)-4,4,6-trimethyl-1,3,2-dioxaborinane (**3d**): Clear oil; IR: 2971, 2933, 1648, 1421, 1054, 1000, 896, 767 cm^{-1}; ^1H NMR (300 MHz, $CDCl_3$): δ 6.34 (dd, *J*=18.0, 5.0 Hz, 1H), 5.60 (dd, *J*=18.0, 1.2 Hz, 1H), 4.78 (dd, *J*=5.0, 1.2 Hz, 1H), 4.13 (dqd, *J*=11.7, 6.2, 3.0 Hz, 1H), 3.57 (m, 2H), 3.44 (dq, *J*=9.5, 7.1 Hz, 2H), 1.71 (dd, *J*=13.8, 3.0 Hz, 1H), 1.40 (dd, *J*=13.8, 11.7 Hz, 1H), 1.21 (s, 6H), 1.18 (d, *J*=6.2 Hz, 3H), 1.13 (t, *J*=7.1 Hz, 6H); 13C NMR (75.5 MHz, $CDCl_3$): δ 145.2, 126.9 (br), 102.4, 70.7, 64.6, 61.0 (2C), 45.9, 31.1, 28.0, 23.0, 15.1(2C). ^{11}B NMR (96 MHz, $CDCl_3$): δ 25.5.

[*Note by the investigators: In the 75.5 MHz ^{13}C-NMR spectra, C atoms that are directly bonded to boron, marked (br), were approximately 250 Hz wide and can thus remain undetected*]

Reference

Ganesh, N. P., d'Hondt, S., and Chavant, P. Y. (2007). Methylpentanediolborane: easy access to new air- and chromatography-stable, highly functionalized vinylboronates. *J. Org. Chem.*, **72**, 4510–4514.

C-B bond_5

Type of reaction: C-B bond formation
Reaction conditions: Toluene, room temperature
Synthetic strategy: Catalytic enantioselective diboration of pro-chiral allenes
Catalyst: Tris(dibenzylideneacetone)dipalladium(0) [$Pd_2(dba)_3$] / (*R*,*R*)-phosphoramidite (ligand **I**)

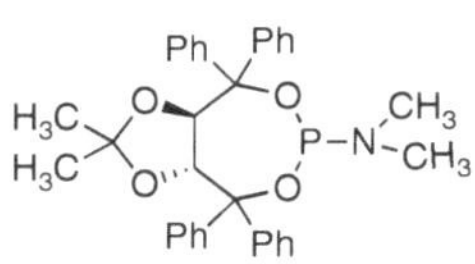

(Ligand **I**)

Keywords: Pro-chiral allenes, *bis*(pinacolato)diboron, $Pd_2(dba)_3$, (*R*,*R*)-phosphoramidite (ligand **I**), toluene, room temperature, enantioselective diboration

General reaction scheme

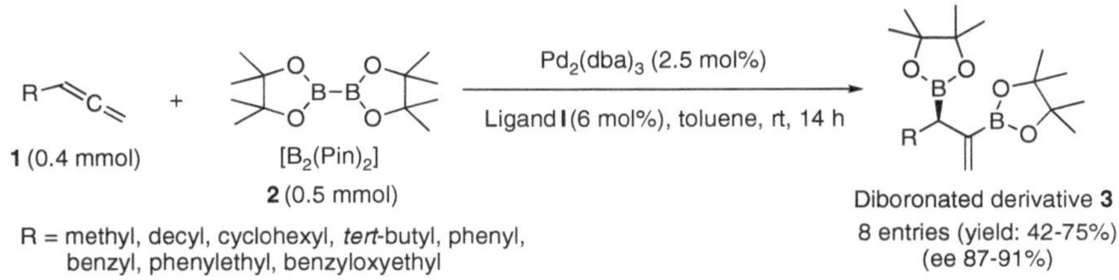

Representative entries

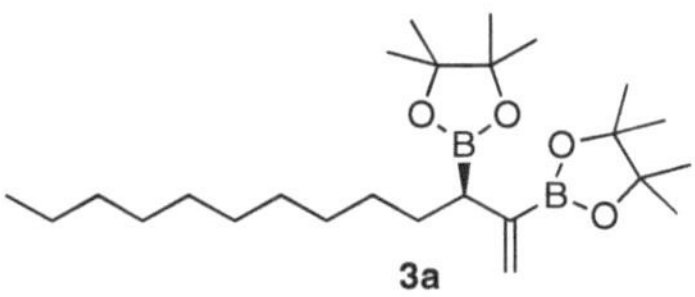

3a

(*R*)-2,2′-(Tridec-1-ene-2,3-diyl)bis(4,4,5,5-tetramethyl-1,3,2-dioxaborolane) (**3a**): Yield 61% (ee 91%)

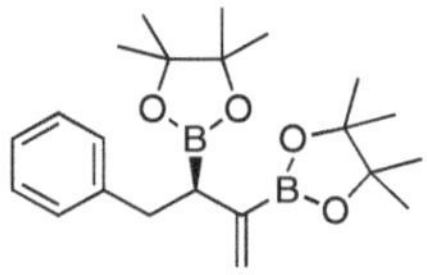

3b

(*R*)-2,2′-(4-Phenylbut-1-ene-2,3-diyl)bis(4,4,5,5-tetramethyl-1,3,2-dioxaborolane) (**3b**): Yield 65% (ee 90%)

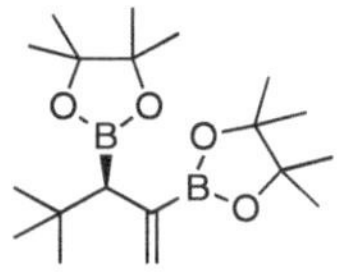

3c

(*S*)-2,2′-(4,4-Dimethylpent-1-ene-2,3-diyl)bis(4,4,5,5-tetramethyl-1,3,2-dioxaborolane) (**3c**): Yield 42% (ee 89%)

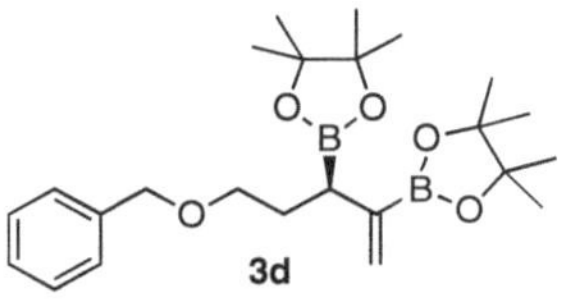

3d

(*R*)-2,2′-(5-(Benzyloxy)pent-1-ene-2,3-diyl)bis(4,4,5,5-tetramethyl-1,3,2-dioxaborolane) (**3d**): Yield 57% (ee 91%)

Experimental procedure

An oven-dried 20-mL vial was charged with tris(dibenzylideneacetone)dipalladium(0) (9.4 mg, 0.010 mmol), (*R*,*R*)-TADDOL-$PNMe_2$ (ligand **I**; 13.2 mg, 0.024 mmol), 2.5 mL toluene and a magnetic stir bar. After stirring for 30 minutes, bis(pinacolato)diboron (**2**; 0.50 mmol) was added to the mixture followed by allene (**1**; 0.4 mmol). The vial was sealed with a teflon-cap, and stirred at room temperature for 14 h. The resulting solution was concentrated under reduced pressure and then chromatographed on silica gel (2% EtOAc/hexane) to provide pure diboron derivative **3** (mostly as oil) in moderate to good yields (42-75%) with good enantioselectivity (ee 87-91%). Products were identified based on spectral studies.

Characterization data of one representative compound

(*R*)-2,2′-(Tridec-1-ene-2,3-diyl)bis(4,4,5,5-tetramethyl-1,3,2-dioxaborolane) (**3a**): IR (CH_2Cl_2): 2929 (s), 1349 (s), 1146 (m) cm^{-1}; ^1H NMR (400 MHz, $CDCl_3$): δ 5.76 (1H, d, $J=3.5$ Hz, $BCCH_2$), 5.55 (1H, d, $J=3.0$ Hz, $BCCH_2$), 1.90 (2H, dd, $J=7.8$ Hz, 7.5 Hz, CH_2CHB), 1.16-1.48 (2H, m, $BCHCH_2$), 1.24 (6H, s, $BOC(CH_3)_2$), 1.23 (6H, s, $BOC(CH_3)_2$), 1.22

(6H, s, BOC(CH_3)$_2$), 1.21 (6H, s, BOC(CH_3)$_2$), 0.86 (3H, t, $J=7$ Hz, CH_2CH_3); ^{13}C NMR (100 MHz, $CDCl_3$): δ 127.4, 83.1, 82.8, 31.8, 29.6, 29.6, 29.5, 29.5, 29.4, 29.2, 29.0, 24.7, 24.6, 24.5, 24.5, 22.5, 14.0; HRMS (ESI) Calc'd for $C_25H_{48}B_2O_4$ $(M+Na)^+$: 457.3631; Found $(M+Na)^+$: 457.3626.

Reference

Pelz, N. F., Woodward, A. R., Burks, H. E., Sieber, J. D., and Morken, J. P. (2004). Palladium-catalyzed enantioselective diboration of prochiral allenes. *J. Am. Chem. Soc.*, **126**, 16328–16329.

Chapter 8

Miscellaneous Bond Forming Reactions at Room Temperature

Carbon – Hydrogen Bond Forming Reaction: C-H bond_1

Type of reaction: C-H bond formation
Reaction conditions: Toluene, room temperature
Synthetic strategy: Catalytic semireduction of terminal alkynes
Catalyst: Copper catalyst [IPrCuO*t*-Bu]/polymethylhydrosiloxane (PMHS)
Keywords: Terminal alkynes, isobutanol, polymethylhydrosiloxane (PMHS), copper catalyst, toluene, room temperature, semireduction of terminal alkynes, alkenes

General reaction scheme

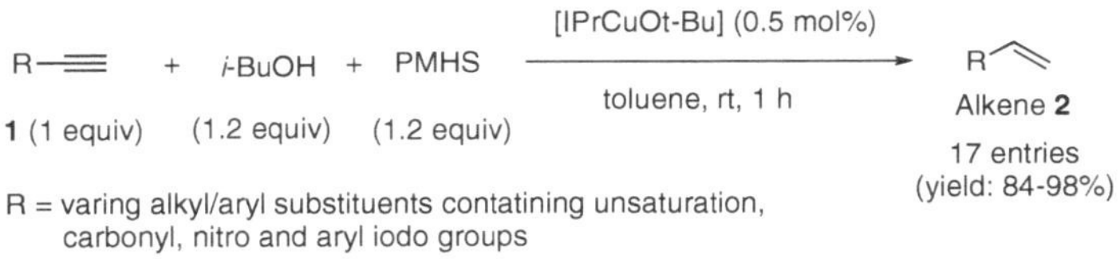

Proposed mechanism

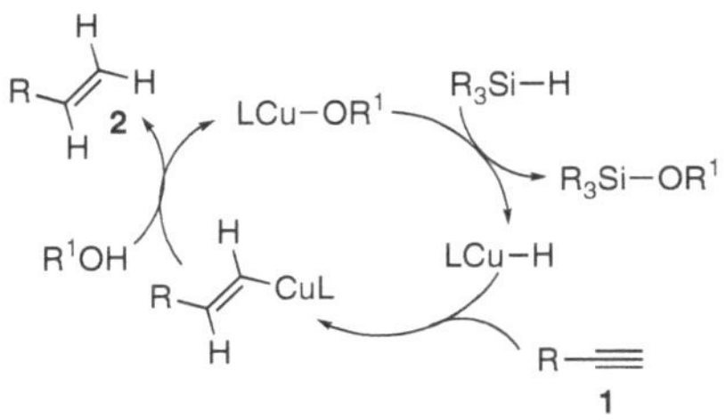

Representative entries

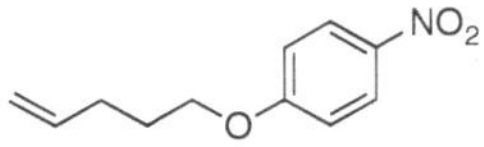

2a

1-Nitro-4-(pent-4-en-1-yloxy)benzene (**2a**): Yellow solid; yield: 94%

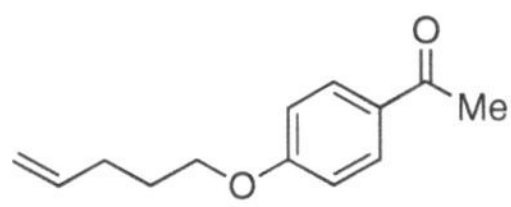

2b

1-(4-(Pent-4-en-1-yloxy)phenyl)ethanone (**2b**): Colorless oil; yield: 93%

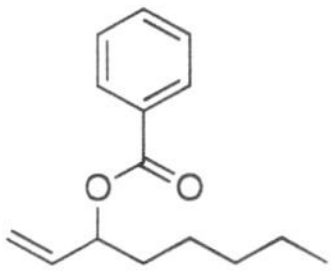

2c

Oct-1-en-3-yl benzoate (**2c**): Colorless oil; yield: 92%

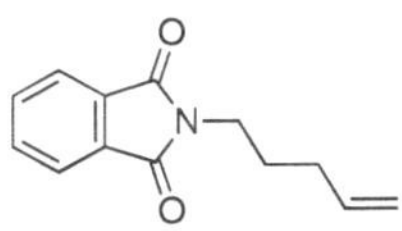

2d

2-(Pent-4-en-1-yl)isoindoline-1,3-dione (**2d**): White solid; yield: 91%

Room Temperature Organic Synthesis. http://dx.doi.org/10.1016/B978-0-12-801025-9.00008-5

Experimental procedure

In a glove box, a dram vial was charged with terminal alkyne (**1**; 1 equiv), isobutanol (1.2 equiv) polymethylhydosiloxane (PMHS, 1.2 equiv), toluene (1.0 mL), and the copper complex (0.5 mol%). The mixture was then stirred at room temperature for 1 h to obtain the crude product, followed by purification. All the products were characterized by means of spectral studies.

Characterization data of two representative entries

1-(4-(Pent-4-en-1-yloxy)phenyl)ethanone (**2b**): Colorless oil, IR (neat): 3076, 2944, 1675, 1255, 1019 cm^{-1}; ^{1}H NMR (300 MHz, $CDCl_3$): δ 7.93 (d, J=8.9 Hz, 2H), 6.92 (d, J=8.9 Hz, 2H), 5.91-5.89 (m, 1H), 5.18-4.95 (m, 2H), 4.03 (t, J=6.4 Hz, 2H), 2.56 (s, 3H), 2.43-2.17 (m, 2H), 2.09-1.79 (m, 2H); ^{13}C NMR (126 MHz, $CDCl_3$): δ 196.9, 163.1, 137.7, 130.7, 130.3, 115.6, 114.3, 67.5, 30.1, 28.4, 26.5; HRMS calculated for $[M+H]^+$ 205.1228; Found 205.1233.

2-(Pent-4-en-1-yl)isoindoline-1,3-dione (**2d**): White solid; IR (neat): 3061, 2938, 1712, 1397 cm^{-1}; ^{1}H NMR (300 MHz, $CDCl_3$): δ 7.84 (dd, J=5.3, 3.1 Hz, 2H), 7.71 (dd, J=5.5, 3.0 Hz, 2H), 5.94-5.68 (m, 1H), 5.16-4.90 (m, 2H), 3.88-3.57 (m, 2H), 2.12 (dt, J=14.1, 7.3 Hz, 2H), 1.94-1.68 (m, 2H); ^{13}C NMR (126 MHz, $CDCl_3$): δ 168.6, 137.5, 134.0, 132.3, 123.3, 115.5, 37.7, 31.1, 27.7; HRMS calculated for $[M+Na]^+$ 238.0843; Found 238.0852.

Reference

Whittaker, A. M., and Lalic, G. (2013). Monophasic catalytic system for the selective semireduction of alkynes. *Org. Lett.*, **15**, 1112–1115.

Nitrogen – Hydrogen Bond Forming Reaction: N-H bond_1

Types of reaction: N-H bond formation
Reaction conditions: Water, room temperature
Synthetic strategy: Chemoselective reduction of nitroaromatics
Catalyst: Zn dust/NH_4Cl/TPGS-750-M (surfactant)

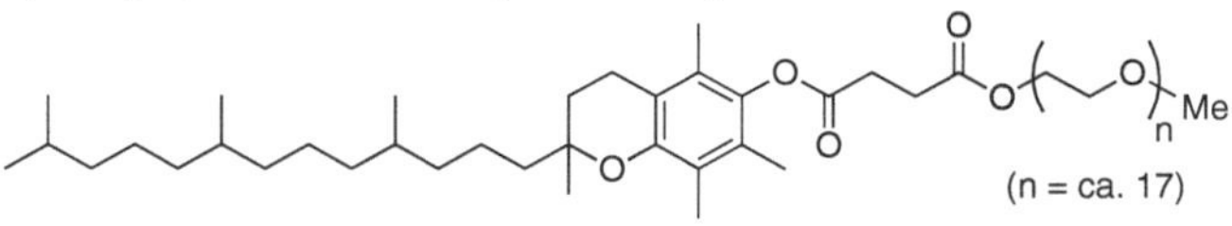

TPGS-750-M

Keywords: Aryl/heteroaryl nitro compounds, Zn dust, ammonium chloride (NH_4Cl), TPGS-750-M (surfactant), water, room temperature, chemoselective recustion of nitro group, amino derivatives

General reaction scheme

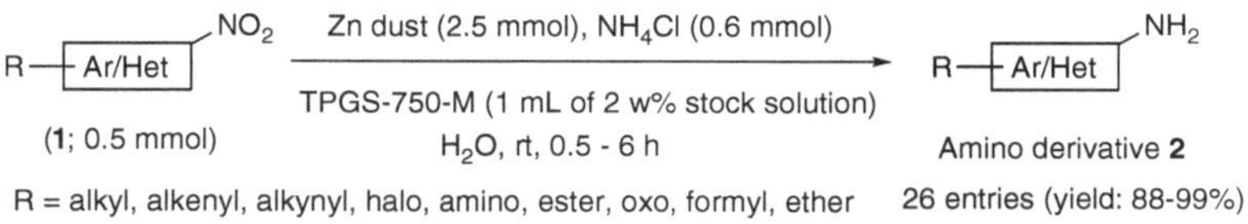

Representative entries

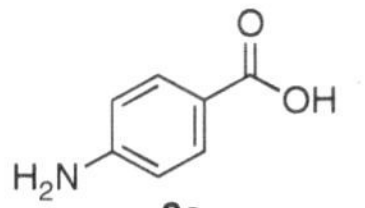

2a

4-Aminobenzoic acid (**2a**): Off-white crystals; yield: 94%

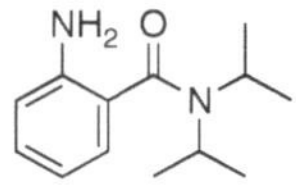

2b

2-Amino-*N*,*N*-diisopropylbenzamide (**2b**): Off-white crystals; yield: 94%

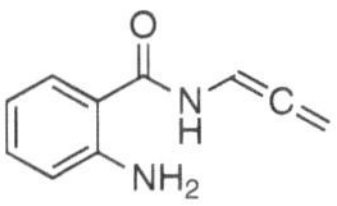

2c

2-Amino-*N*-(propa-1,2-dien-1-yl)benzamide (**2c**): Off-white solid; yield: 99%

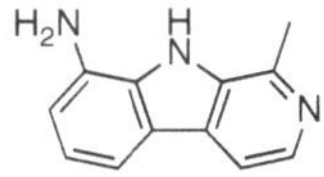

2d

8-Aminoharmane (**2d**): Yellow crystalline solid, mp 145 °C (decomp.); yield: 88%

Experimental procedure

To a 5 mL round bottom vial equipped with a magnetic stir bar was added nitroarene (**1**; 0.5 mmol) and ammonium chloride (0.6 mmol, 1.2 equiv), followed by 2 wt. % TPGS-750-M stock solution (1.0 mL). The mixture was stirred for approximately 1 min to distribute and dissolve starting nitroarene; zinc dust (2.5 mmol, 5.0 equiv) was then added in a single batch to the resulting stirred emulsion, and the reaction was allowed to continue stirring at room temperature. Upon completion of the reaction (monitored by TLC / GC-MS), it was filtered through a 1 cm silica gel plug to remove water and zinc solids, and eluted by a minimal volume of EtOAc (generally between 5 and 10 mL). Alternately, the completed reaction was extracted three times by a minimum volume (generally 250 to 500 μL) of EtOAc or diethyl ether, and dried over anhydrous sodium sulfate. The resulting organic solution was concentrated *in vacuo*, and analyzed by GC-MS and NMR spectrometry in most of the cases.

The physical and spectral properties of the synthesized compounds were found to be comparable with the reported values in literature.

Characterization data of two representative compounds

2-Amino-*N*,*N*-diisopropylbenzamide (**2b**): Off-white crystals; ^{1}H NMR (500 MHz, $CDCl_3$): δ 7.11 (td, $J=7.79$ and 1.56 Hz, 1H), 6.99 (dd, $J=7.40$ and 1.43 Hz, 1H), 6.67 - 6.74 (m, 2H), 3.55 - 4.00 (m, 1H), 1.35 (br. s., 7 H).

8-Aminoharmane (**2d**): Yellow crystalline solid, mp 145 °C (decomp.); ^{1}H NMR (500 MHz, $CDCl_3$): δ 10.06 (br.s, 1H), 8.28 (d, $J=5.45$ Hz, 1H), 7.76 (d, $J=5.45$ Hz, 1H), 7.56 (dd, $J=7.92$, 0.65 Hz, 1H), 7.12 (s, 1H), 6.89 (dd, $J=7.53$ and 0.52 Hz, 1H), 3.51 (br.s, 2H), 2.66 (s, 3H); ^{13}C NMR (125 MHz, $CDCl_3$): δ 142.43, 137.80, 134.95, 132.41, 131.10, 129.38, 122.63, 121.00, 113.52, 113.31, 112.21, 19.93.

Reference

Kelly, S. M., and Lipshutz, B. H. (2014). Chemoselective reductions of nitroaromatics in water at room temperature. *Org. Lett.*, **16**, 98–101.

Oxygen – Silicon Bond Forming Reactions: O-Si bond_1

Types of reaction: O-Si bond formation
Reaction conditions: THF, room temperature
Synthetic strategy: C_2-Symmetric copper-bound *N*-heterocyclic carbene (NHC) catalyzed enantioselective hydrosilylation of pro-chiral ketones
Catalyst: (*R*,*R*,*R*,*R*)-CuPhEt (**I**)/KO^{t}Bu

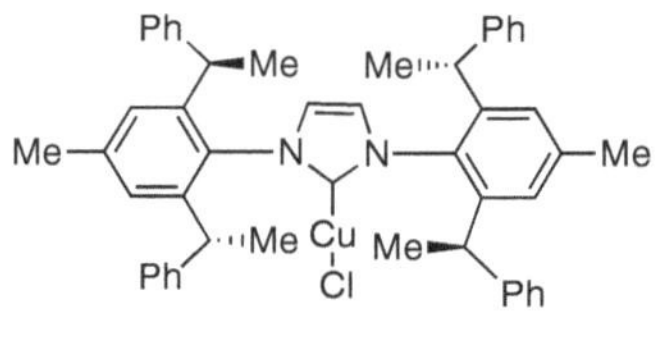

(*R*,*R*,*R*,*R*)-CuPhEt (**I**)

Keywords: Dialkyl and aryl-alkyl ketones, diethyl/diphenylsilane, CuPhEt (**I**), KO^{t}Bu, THF, room temperature, enantioselective hydroxysilylation

General reaction scheme

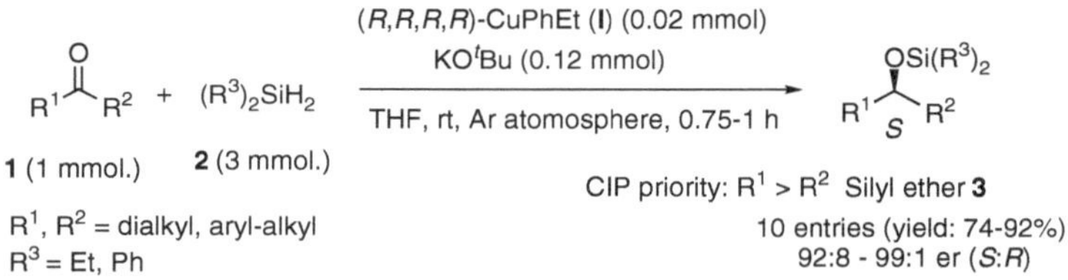

Representative entries

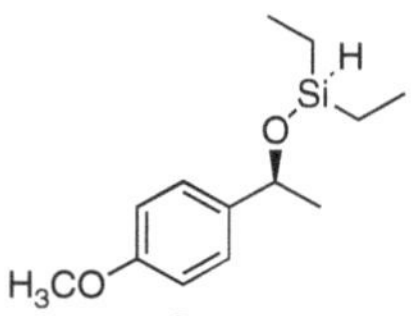

3a

(*S*)-Diethyl(1-(4-methoxyphenyl)ethoxy)silane (**3a**): Reaction time: 1 h; yield: 85% (97:3 er)

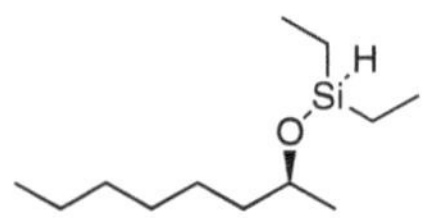

3b

(*S*)-Diethyl(octan-2-yloxy)silane (**3b**): Reaction time: 0.75 h; yield: 77% (97:3 er)

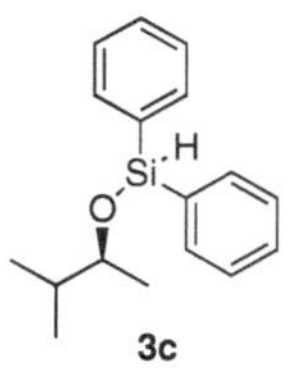

3c

(*S*)-((3-Methylbutan-2-yl)oxy)diphenylsilane (**3c**): Reaction time: 0.5 h; yield: 88% (99:1 er)

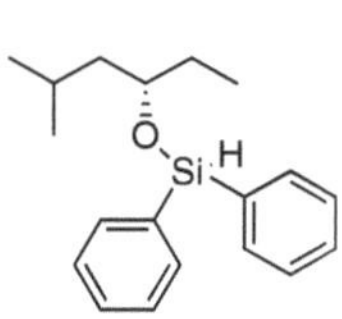

3d

(*S*)-((5-Methylhexan-3-yl)oxy)diphenylsilane (**3d**): Reaction time: 0.75 h; yield: 77% (96:4 er)

Experimental procedure

To an oven dried vial fitted with a septum screw cap and purged with argon was added copper carbenoid CuPhEt (**I**; 2 mol%) and freshly sublimed potassium *tert*-butoxide (12 mol%) in freshly distilled THF (2 mL). The mixture was stirred for 10 min at room temperature before addition cf the silane (**2**; 3.0 mmol). An immediate color change to bright yellow from cloudy white was observed and the reaction was stirred for an additional 20 min before addition of the ketone (**1**; 1.0 mmol) as a concentrated solution in THF. The stirring was continued until completion of the reaction (monitored by GC or TLC).

Upon completion, the reaction solution was treated with charcoal and filtered through a plug of celite. The celite plug was rinsed with ethyl acetate and the combined organic layers were concentrated under reduced pressure, followed by purification of the crude product thus obtained. Pure silyl ethers **3** were characterized based on analytical and spectral studies.

Characterization data of two representative compounds

(*S*)-Diethyl(1-(4-methoxyphenyl)ethoxy)silane (**3a**): ^{1}H NMR (400 MHz, $CDCl_3$): δ 0.54 (qd, J=8.02 & 2.34Hz, 4H), 0.83 (t, J=7.76 Hz, 3H), 0.90 (t, J=7.76 Hz, 3H), 1.37 (d, J=6.41 Hz, 3H), 3.72 (s, 3H), 4.33 (quintet, J=2.43 Hz, 1H), 4.75 (q, J=6.46 Hz, 1H), 6.78 (d, J=8.39 Hz, 2H), 7.18 (d, J=8.39 Hz, 2H); ^{13}C NMR (100 MHz, $CDCl_3$): δ 5.45 (2 × CH_2), 6.70 (2 × CH_3), 26.33 (CH_3), 55.21 (CH_3), 72.09 (CH), 113.50 (CH), 126.60 (CH), 138.09 (C), 158.59 (C). er (*S*:*R*)=97:3; RT_R=54.6, RT_S=55.9.

(*S*)-((5-Methylhexan-3-yl)oxy)diphenylsilane (**3d**): ^{1}H NMR (400 MHz, $CDCl_3$): δ 0.88 (m, overlapping signals, 9H), 1.28 (m, 1H), 1.58 (m, 3H), 1.78 (m, J=6.53 Hz, 1H), 3.86 (m, 1H), 5.52 (s, 1H), 7.37-7.50 (m, 6H), 7.65-7.73 (m, 4H); ^{13}C NMR (100 MHz, $CDCl_3$): δ 9.77 (CH_3), 22.11 (CH_3), 23.42 (CH_3), 24.42 (CH), 30.15 (CH2), 45.79 (CH_2), 74.09 (CH_2), 127.91, 130.15, 134.74. Anal calcd for $C_{19}H_{26}SiO$: C, 76.45; H, 8.78. Found: C, 76.24; H, 8.40. er (*S*:*R*)=96:4; RT_R=22.1, RT_S=23.3.

Reference

Albright, A., and Gawley, R. E. (2011). Application of a C_2-symmetric copper carbenoid in the enantioselective hydrosilylation of dialkyl and aryl-alkyl ketones. *J. Am. Chem. Soc.*, **133**, 19680–19683.

Oxygen – Silicon Bond Forming Reactions: O-Si bond_2

Types of reaction: O-Si bond formation
Reaction conditions: THF, room temperature
Synthetic strategy: One-step silyl enol ether formation *via* silyl transfer
Catalyst: *N*-Heterocyclic carbene (NHC) (**I**)

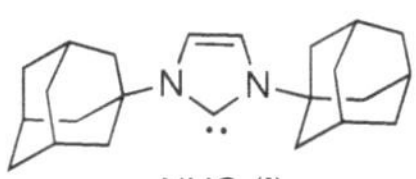

NHC (**I**)

Keywords: Ketones, trisilyl enols, THF, room temperature, silyl transfer, *N*-heterocyclic carbine (NHC), silyl enol ethers

General reaction scheme

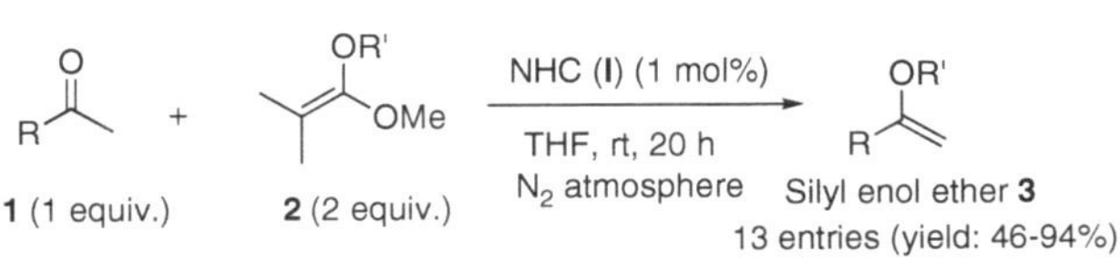

R = alkyl, alicyclic, aryl
R' = trimethylsilyl (TMS), *tert*-butyldimethylsilyl (TBS)

Representative entries

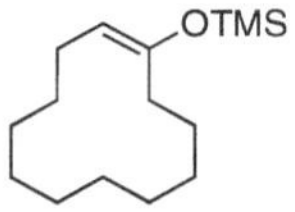

3a

(*E*)-(Cyclododec-1-en-1-yloxy)trimethylsilane (**3a**): Yield: 84%

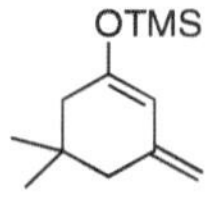

3b

((5,5-Dimethyl-3-methylenecyclohex-1-en-1-yl)oxy) trimethylsilane (**3b**): Yield: 79%

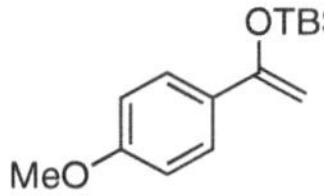

3c

tert-Butyl((1-(4-methoxyphenyl)vinyl)oxy) dimethylsilane (**3c**): Yield: 93%

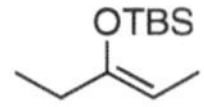

3d

(*Z*)-*tert*-Butyldimethyl(pent-2-en-3-yloxy)silane (**3d**): Yield: 76%

Experimental procedure

To the carbonyl compound (**1**; 37.2 mmol) in a dry flask under nitrogen atmosphere were added anhydrous THF (37.2 mL), methyl trimethylsilyl dimethylketene acetal/1-methoxy-1-*tert*-butyldimethylsilylethene (**2**; 55.7 mmol) and NHC (**I**; 0.37 mmol). The mixture was stirred at room temperature (23 °C) for 20 h. The volatiles were then removed via eparoation, and the residue was distilled under vacuum to afford silyl enol ether **3**. Each of the products was characterized based on spectral studies.

Characterization data of two representative compounds

(*E*)-(Cyclododec-1-en-1-yloxy)trimethylsilane (**3a**): ^{1}H NMR (400 MHz, CD_2Cl_2): δ 4.55 (t, $J=7.3$ Hz, 1H), 2.06-1.99 (m, 4H), 1.47-1.43 (m, 4H), 1.34-1.31 (m, 12H), 0.16 (s, 9H); ^{13}C NMR (100 MHz, CD_2Cl_2): δ 149.9, 111.4, 36.4, 26.9, 26.8, 26.5, 26.3, 25.3, 25.2, 25.13, 25.06, 24.0, 0.7.

tert-Butyl((1-(4-methoxyphenyl)vinyl)oxy)dimethylsilane (**3c**): ^{1}H NMR (400 MHz, CD_2Cl_2): δ 7.59-7.52 (m, 2H), 6.89-6.83 (m, 2H), 4.79 (d, $J=1.8$ Hz, 1H), 4.34 (d, $J=1.8$ Hz, 1H), 3.81 (s, 3H), 1.01 (s, 9H), 0.22 (s, 6H); ^{13}C NMR (100 MHz, CD_2Cl_2): δ 160.2, 156.1, 130.8, 126.9, 113.7, 89.4, 55.6, 26.0, 18.6, −4.5.

Reference

Song, J. J., Tan, Z., Reeves, J. T., Fandrick, D. R., Yee, N. K., and Senanayake, C. H. (2008). *N*-Heterocyclic carbene-catalyzed silyl enol ether formation. *Org. Lett.*, **10**, 877–880.

Oxygen – Silicon Bond Forming Reactions: O-Si bond_3

Types of reaction: O-Si bond formation
Reaction conditions: Toluene, room temperature
Synthetic strategy: Chemo- and stereoselective hydrosilylation
Catalyst: Indium(III) chloride ($InCl_3$)
Keywords: α,β-Enone esters, triethylsilane, Indium(III) chloride, trifluoroacetic acid (TFA), toluene, room temperature, 1,4-hydrosilylation, silyl enol ether, stereoselectivity

General reaction scheme

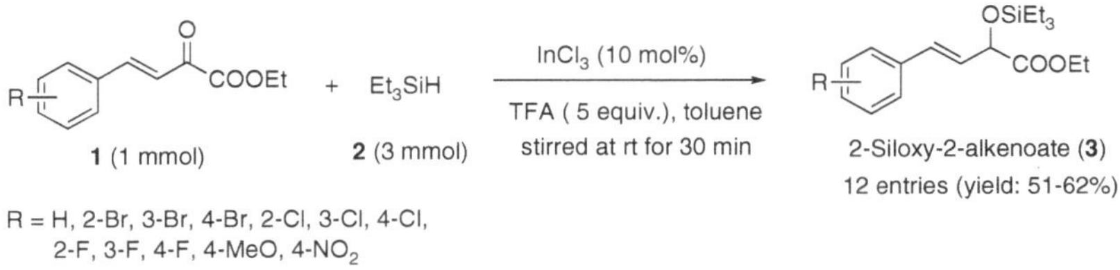

Representative entries

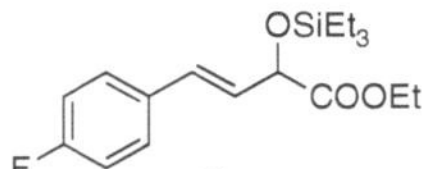

3a

(*E*)-Ethyl 4-(4-fluorophenyl)-2-((triethylsilyl)oxy) but-3-enoate (**3a**): Yellow oil; yield: 62%

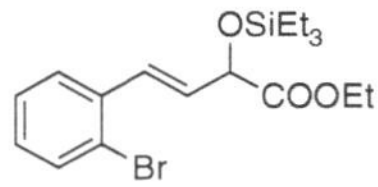

3b

(*E*)-Ethyl 4-(2-bromophenyl)-2-((triethylsilyl) oxy)but-3-enoate (**3b**): Yellow oil; yield: 51%

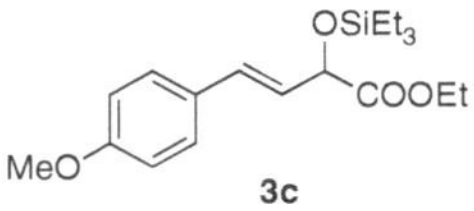

3c

(*E*)-Ethyl 4-(4-methoxyphenyl)-2-((triethylsilyl) oxy) but-3-enoate (**3c**): Yellow oil; yield: 54%

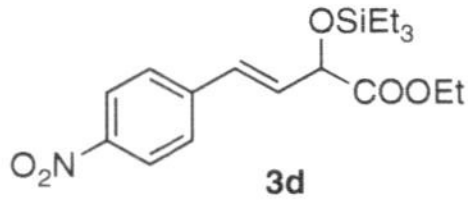

3d

(*E*)-Ethyl 4-(4-nitrophenyl)-2-((triethylsilyl)oxy) but-3-enoate (**3d**): Yellow oil; yield: 51%

Experimental procedure

To a mixture of α,β-enone ester (**1**; 1.0 mmol), $InCl_3$ (0.10 mmol), and Et_3SiH (2; 3.0 mmol) in toluene (2.0 mL) was added with TFA (5.0 mmol, 0.38 mL) at room temperature drop-wise. The solution was then stirred at the room temperature for about 30 min when the ester was exhausted completely (TLC). The resulting solution was diluted with ethyl acetate (10 mL) followed by addition of aq. $NaHCO_3$ (10 mL) and H_2O (10 mL), and then shaken well. The organic solution was separated, and the aqueous layer was extracted by EtOAc (2 × 10 mL). The combined organic layer was dried over anhydrous Na_2SO_4, concentrated to dryness, and the residue was purified by flash chromatography on silica gel to afford 2-siloxy-2-alkenoate **3**. Each of the products was characterized on the basis of detailed spectral studies.

Characterization data of two representative compounds

(*E*)-Ethyl 4-(2-bromophenyl)-2-((triethylsilyl)oxy)but-3-enoate (**3b**): Yellow oil; IR: 2957, 2911, 2877, 1723, 1644, 1250, 978, 786, 745, 704 cm^{-1}; ^1H NMR (300 MHz, $CDCl_3$): δ 7.56-7.05 (m, 4 H), 6.14 (t, $J=7.2$ Hz, 1H), 4.21 (q, $J=6.9$ Hz, 2H), 3.64 (d, $J=7.2$ Hz, 2H), 1.30 (t, $J=6.9$ Hz, 3H), 1.00 (t, $J=7.8$ Hz, 9 H), 0.75 (q, $J=7.8$ Hz, 6H); ^{13}C NMR (75 MHz, $CDCl_3$): δ 164.6, 141.7, 139.1, 132.7, 130.2, 127.9, 127.5, 124.5, 118.8, 61.1, 32.3, 14.2, 6.8, 5.5; ESI-MS: *m/z* 423.0 $[M+Na]^+$; ESI-HRMS: *m/z* calcd for $C_{18}H_{27}BrNaO_3Si$: 421.0811; Found: 421.0801.

(*E*)-Ethyl 4-(4-methoxyphenyl)-2-((triethylsilyl)oxy) but-3-enoate (**3c**): Yellow oil; IR: 2956, 2876, 2835, 1722, 1644, 1512, 1318, 1106, 978, 807 cm^{-1}; ^1H NMR (300 MHz, $CDCl_3$): δ 7.12 (d, $J=8.7$ Hz, 2H), 6.84 (d, $J=8.7$ Hz, 2H), 6.15 (t, $J=7.8$ Hz, 1H), 4.20 (q, $J=7.2$ Hz,

2H), 3.78 (s, 3 H), 3.47 (d, J = 7.5 Hz, 2 H), 1.29 (t, J = 7.2 Hz, 3 H), 1.01 (t, J = 5.2 Hz, 9 H), 0.75 (q, *J* = 5.2 Hz, 6H); ^{13}C NMR (75 MHz, $CDCl_3$): δ 164.8, 158.0, 140.9, 131.8, 129.4, 121.2, 113.9, 60.9, 55.2, 31.0, 14.2, 6.8, 5.4; ESI-MS: *m/z* 373.0 $[M+Na]^+$; ESIHRMS: *m/z* calcd for $C_{19}H_{30}NaO_4Si$: 373.1811; Found: 373.1810.

Reference

Xing, P., Zang, W., Huang, Z.-G., Zhan, Y.-X., Zhu, C.-J., and Jiang, B. (2012). A mild method for Indium (III)-catalyzed 1,4-hydrosilylation of α,β-enone esters with triethylsilane and trifluoroacetic acid. *Synlett*, **23**, 2269–2273.

Oxygen – Silicon Bond Forming Reactions: O-Si bond_4

Types of reaction: O-Si and C-C bond formation
Reaction conditions: THF, room temperature
Synthetic strategy: Installation of a 1,4-difunctionality via regioselective Ni-catalyzed reductive coupling of ynoates and aldehydes using MCR strategy
Catalyst: $Ni(COD)_2$ / $SIPr.HBF_4$ / KO^tBu / PPh_3 [COD = 1,5-cyclooctadiene]

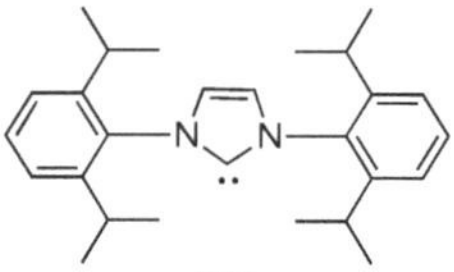

SIPr

Keywords: Ynoates, aldehydes, triethylsilane, $Ni(COD)_2$, $SIPr.HBF_4$, KO^tBu, PPh_3, THF, room temperature, regioselective reductive coupling, multicomponent reaction (MCR), silyl-protected γ-hydroxy-α,β-enoates

General reaction scheme

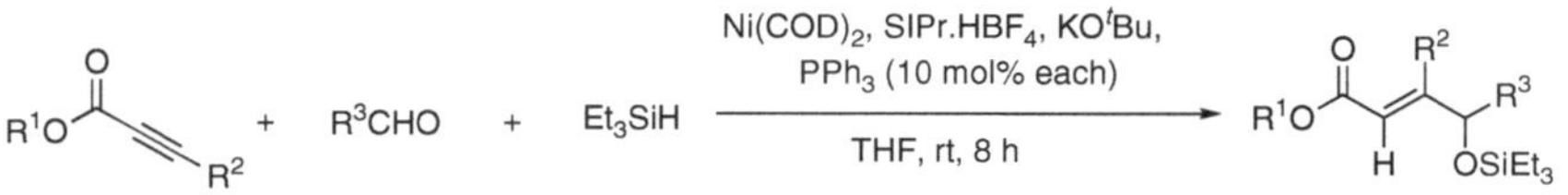

1 (0.5 mmol) **2** (0.55 mmol) **3** (0.55 mmol)

Silyl-protected γ-hydroxy- α, β-enoate **4**
18 entries (yield: 56-91%)
regioselectivity: 80:20 - 98:2

R^1 = Me, Et, ^{t}Bu, naphthyl; R^2 = H, *n*-Pr, *n*-hex, $SiMe_3$
R^3 = C_6H_5, 4-ClC_6H_4, 2-FC_6H_4, 3-FC_6H_4, 4-FC_6H_4, 2-MeC_6H_4, 4-MeC_6H_4, 4-$OMeC_6H_4$, 4-$CO_2MeC_6H_4$, naphthyl, *n*-hepthyl, $CH_2CH_2CHMe_2$, $SiMe_3$

Representative entries

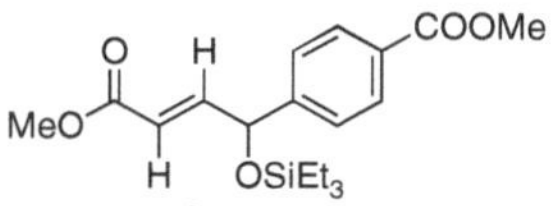

4a

(*E*)-Methyl 4-(4-methoxy-4-oxo-1-((triethylsilyl)oxy)but-2-en-1-yl)benzoate (**4a**): Colorless oil; yield: 74% (80:20)

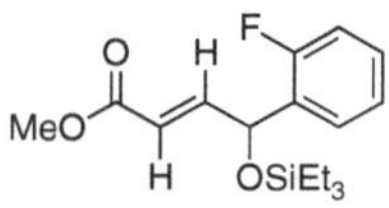

4b

(*E*)-Methyl 4-(2-fluorophenyl)-4-((triethylsilyl)oxy)but-2-enoate (**4b**): Colorless oil; yield: 85% (95:5)

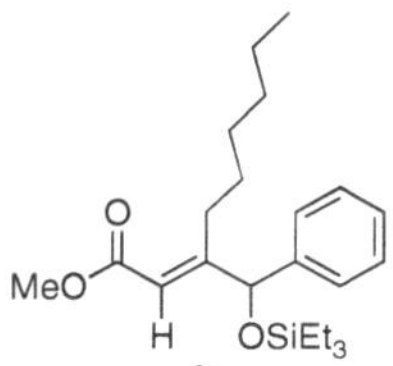

4c

(*E*)-Methyl 3-(phenyl((triethylsilyl)oxy)methyl) non-2-enoate (**3c**): Colorless oil; yield: 91% (98:2)

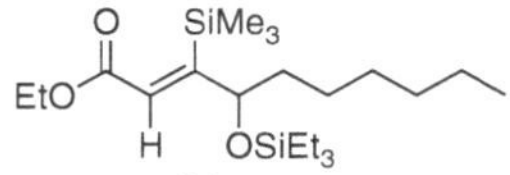

4d

(*Z*)-Ethyl 4-((triethylsilyl)oxy)-3-(trimethylsilyl) dec-2-enoate (**4d**): Colorless oil; yield: 81% (98:2)

Experimental procedure

A mixture of $Ni(COD)_2$ (13.8 mg, 0.050 mmol), SIPr·HBF_4 (23.9 mg, 0.050 mmol), PPh_3 (13.1 mg, 0.050 mmol), and KO^tBu (5.6 mg, 0.050 mmol) was added with THF (4.0 mL) in a dried Schlenk flask at 0 °C, and the resulting mixture was stirred at the same temperature for 15 min, followed by the successive addition of Et_3SiH (**3**; 88 μL, 0.55 mmol) and an aldehyde (**2**; 0.55 mmol). After the mixture was stirred at 0 °C for 5 min, a solution of an ynoate (**1**; 0.50 mmol) in 4.0 mL of THF was added at room temperature (23 °C) over a period of 8 h using a syringe pump. Upon completion of the addition, the reaction mixture was stirred at room temperature for another 1 h before being concentrated under vacuum. The ratio of the two isomers was determined from the ^1H NMR spectrum of the crude products. The desired 1,4-difunctional compound was separated from the isomeric mixture using column chromatography (with diethyl ether/hexanes as eluent). All the products were characterized on the basis of detailed spectral studies.

Characterization data of two representative compounds

(*E*)-Methyl 4-(4-methoxy-4-oxo-1-((triethylsilyl)oxy)but-2-en-1-yl)benzoate (**4a**): Colorless oil; IR (neat): 2953, 2912, 2876, 1721, 1658, 1610, 1458, 1435, 1413, 1242, 1191, 1164, 1108, 1018, 971, 888, 857, 825, 807, 769, 725, 566 cm^{-1}; ^1H NMR (400 MHz, $CDCl_3$): δ 8.01 (d, $J=7.9$ Hz, 2H), 7.40 (d, $J=7.9$ Hz, 2H), 6.96 (dd, $J=15.4$, 4.5 Hz, 1H), 6.15 (dd, $J=15.4$, 1.8 Hz, 1H), 5.36 (dd, $J=4.5$, 1.8 Hz, 1H), 3.91 (s, 3H), 3.72 (s, 3H), 0.90 (t, $J=7.9$ Hz, 9H), 0.63-0.53 (m, 6H); ^{13}C NMR (101 MHz, $CDCl_3$): δ 167.0, 166.9, 149.7, 146.9, 130.0, 129.7, 126.3, 119.3, 73.6, 52.2, 51.8, 6.8, 4.9; HRMS-ESI (*m/z*) $[M+Na]^+$ calcd for $C_{19}H_{28}O_5SiNa$ 387.1604; Found 387.1595.

(*E*)-Methyl 3-(phenyl((triethylsilyl)oxy)methyl)non-2-enoate (**3c**): Colorless oil; IR (neat): 2953, 2875, 1720, 1649, 1493, 1454, 1433, 1413, 1377, 1312, 1209, 1145, 1102, 1062, 1004, 974, 938, 915, 884, 849, 825, 725, 698 cm^{-1}; ^1H NMR (400 MHz, $CDCl_3$): δ 7.32-7.23 (m, 5H), 6.26 (s, 1H), 5.10 (s, 1H), 3.70 (s, 3H), 2.67-2.61 (m, 1H), 2.06-2.01 (m, 1H), 1.29-1.18 (m, 8H), 0.89-0.83 (m, 12H), 0.57-0.51 (m, 6H); ^{13}C NMR (101 MHz, $CDCl_3$): δ 167.4, 164.9, 142.0, 128.3, 127.9, 127.2, 113.7, 77.8, 51.1, 31.6, 29.9, 29.6, 29.5, 22.7, 14.2, 6.8, 4.9; HRMS-ESI (*m/z*) $[M+Na]^+$ calcd for $C_{23}H_{38}O_3SiNa$ 413.2488; Found 413.2492.

Reference

Rodrigo, S. K. and Guan, H. (2012). Quick installation of a 1,4-difunctionality via regioselective nickel-catalyzed reductive coupling of ynoates and aldehydes. *J. Org. Chem.*, **77**, 8303–8309.

Sulfur – Nitrogen Bond Forming Reaction: S-N bond_1

Types of reaction: S-N bond formation
Reaction conditions: Acetonitrile, room temperature
Synthetic strategy: Sulfonylation of amines
Catalyst: Indium (In) metal
Keywords: *p*-Toluene sulfonyl chloride, amines, indium metal, acetonitrile, room temperature, sulfonylation, sulfonamides

General reaction scheme

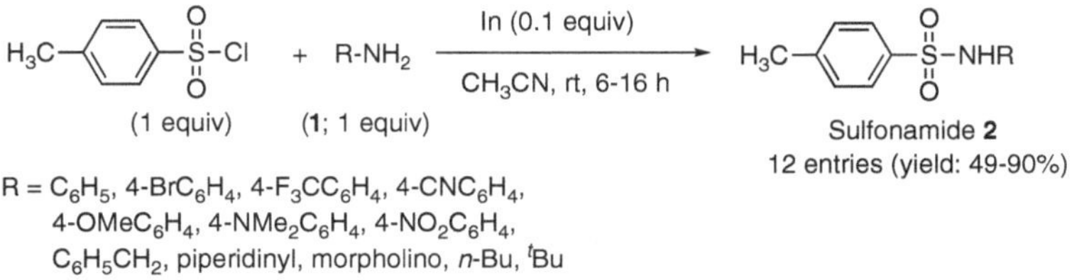

Representative entries

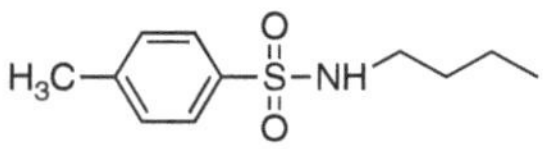

2a

N-Butyl-4-methylbenzenesulfonamide (**2a**): Reaction time: 6 h; yield: 89%

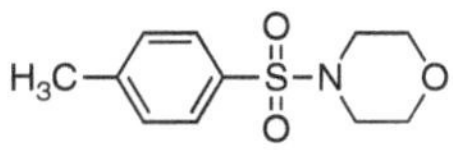

2b

4-Tosylmorpholine (**2b**): Reaction time: 8 h; yield: 80%

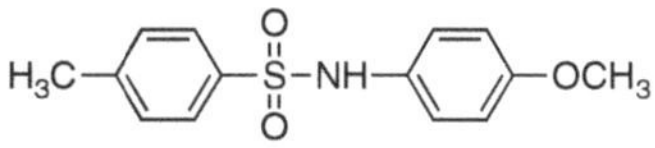

2c

N-(4-Methoxyphenyl)-4-methylbenzenesulfonamide (**2c**): Reaction time: 8 h; yield: 83%

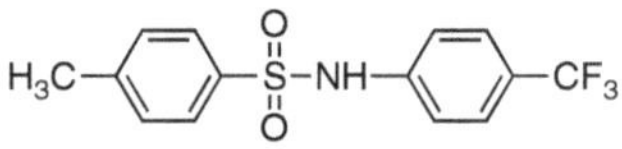

2d

4-Methyl-*N*-(4-(trifluoromethyl)phenyl) benzenesulfonamide (**2d**): Reaction time: 12 h; yield: 71%

Experimental procedure

A solution of amine (**1**; 1 mmol), *p*-toluenesulfonyl chloride (1 mmol) and indium metal (11.5 mg, 0.1 mmol) in 3 mL of acetonitrile was stirred at room temperature for stipulated time under argon atmosphere. Upon completion of the reaction, the mixture was partitioned between water and ether; the organic layer was washed with brine, dried over sodium sulfate and purified by column chromatography on silica gel eluting with hexane-EtOAc to afford sulfonamide **2**. The physical and spectral properties of the compounds were found to be comparable with the reported values in literature.

Reference

Kim, J.-G., and Jang, D. O. (2007). Mild and efficient indium metal catalyzed synthesis of sulfonamides and sulfonic esters. *Synlett*, 2501–2504.

Sulfur – Oxygen Bond Forming Reactions: S-O bond_1

Types of reaction: S-O bond formation
Reaction conditions: Acetonitrile, room temperature
Synthetic strategy: Sulfonylation of alcohols
Catalyst: Indium (In) metal
Keywords: *p*-Toluene sulfonyl chloride, alcohols, indium metal, acetonitrile, room temperature, sulfonylation, sulfonic esters

General reaction scheme

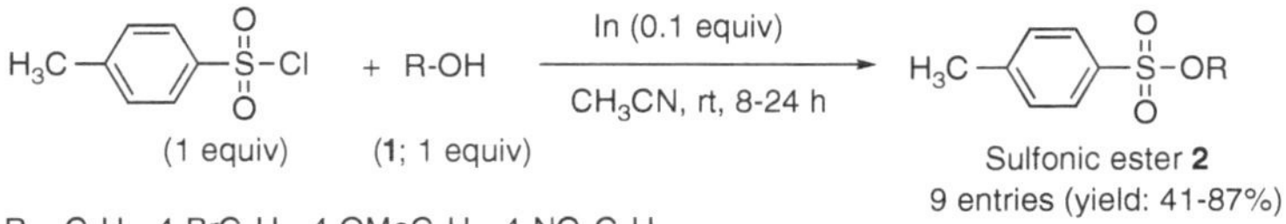

R = C_6H_5, 4-BrC_6H_4, 4-OMeC_6H_4, 4-$NO_2C_6H_4$, 2,4,6-tri-MeC_6H_2, $C_6H_5CH_2CH_2$, *n*-decyl, ^{t}Bu, (1*S*,2*R*,5*S*)-2-isopropyl-5-methylcyclohexyl

Representative entries

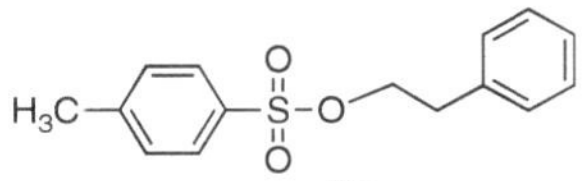

2a

Phenethyl 4-methylbenzenesulfonate (**2a**): Reaction time: 8 h; yield: 87%

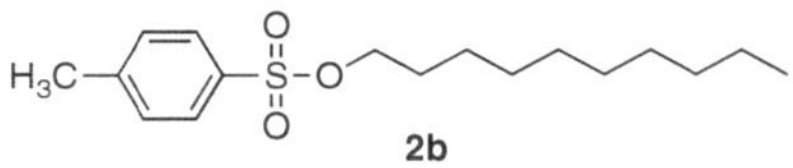

2b

Decyl 4-methylbenzenesulfonate (**2b**): Reaction time: 8 h; yield: 85%

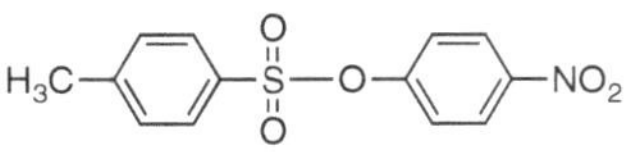

2c

4-Nitrophenyl 4-methylbenzenesulfonate (**2c**): Reaction time: 12 h; yield: 74%

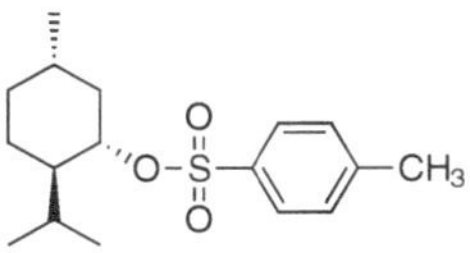

2d

(1*S*,2*R*,5*S*)-2-Isopropyl-5-methylcyclohexyl 4-methylbenzenesulfonate (**2d**): Reaction time: 12 h; yield: 81%

Experimental procedure

A solution of alcohol (**1**; 1 mmol), *p*-toluenesulfonyl chloride (1 mmol) and indium metal (11.5 mg, 0.1 mmol) in 3 mL of acetonitrile was stirred at room temperature for stipulated time under argon atmosphere. Upon completion of the reaction, the mixture was partitioned between water and ether; the organic layer was washed with brine, dried over sodium sulfate and purified by column chromatography on silica gel eluting with hexane-EtOAc to afford sulfonic ester **2**. The physical and spectral properties of the compounds were found to be comparable with the reported values in literature.

Reference

Kim, J.-G., and Jang, D. O. (2007). Mild and efficient indium metal catalyzed synthesis of sulfonamides and sulfonic esters. *Synlett*, 2501–2504.

Sulfur – Oxygen Bond Forming Reactions: S-O bond_2

Types of reaction: S-O, S-Cl and C-N bond formation
Reaction conditions: Acetonitrile, room temperature
Synthetic strategy: Direct oxidative conversion of thiols to sulfonyl chlorides, followed by sulfonamide formation
Catalyst: Hydrogen peroxide/thionyl chloride,
Keywords: Thiols, hydrogen peroxide, thionyl chloride, amines, acetonitrile, room temperature, direct oxidative sulfonylation, sulfonamides

General reaction schemes

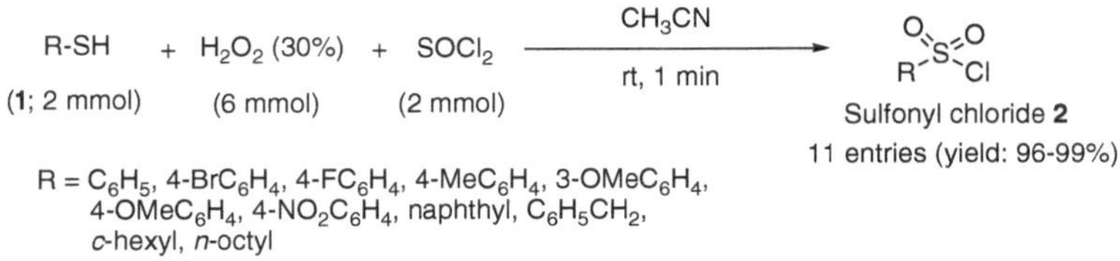

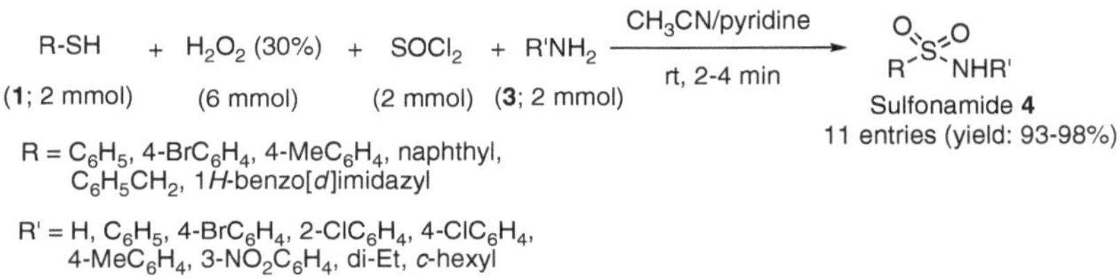

Representative entries

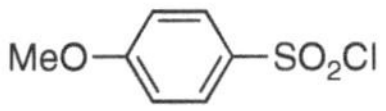

2a

4-Methoxybenzene-1-sulfonyl chloride (**2a**):
Yield: 98%

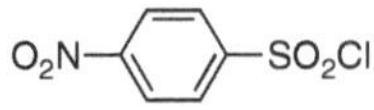

2b

4-Nitrobenzene-1-sulfonyl chloride (**2b**):
Yield: 98%

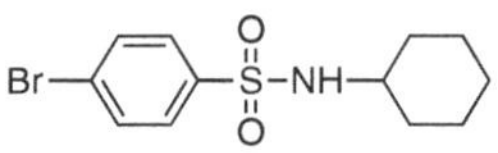

4a

4-Bromo-*N*-cyclohexylbenzenesulfonamide (**4a**): Solid, mp 100 °C; reaction time: 2 min; yield: 98%

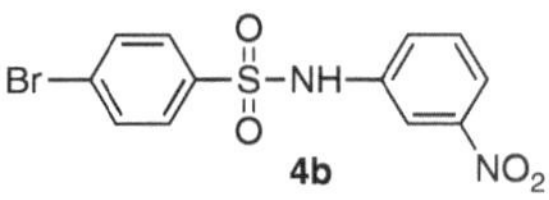

4b

4-Bromo-*N*-(3-nitrophenyl)benzenesulfonamide (**4b**): Solid, mp 118-120 °C; reaction time: 3 min; yield: 96%

Experimental procedures

Preparation of sulfonyl chloride **2**: A mixture of thiol compound (**1**; 2 mmol), 30% hydrogen peroxide (6 mmol, 0.6 mL) and thionyl chloride (2 mmol, 0.14 mL) was stirred in acetonitrile at room temperature for just 1 min. Upon completion of the reaction as indicated by TLC, the reaction mixture was quenched by adding water (10 mL), extracted with ethyl acetate (4 × 5 mL) and the extract dried with anhydrous magnesium sulfate. The filtrate was evaporated under vacuum to afford the pure product of sulfonyl chloride **2**. All of products are known compounds and were characterized by comparison of their physical and spectral (^1H & ^{13}C NMR) properties with those of authentic samples.

Preparation of sulfonamide **4**: A mixture of thiol compound (**1**; 2 mmol), 30% hydrogen peroxide (6 mmol, 0.6 mL) and thionyl chloride (2 mmol, 0.14 mL) was stirred in acetonitrile at room temperature for an appropriate time. After completion of the reaction as indicated by TLC, a solution of amine (**3**; 2 mmol) in pyridine (1 mL) was added to the reaction mixture. The resulting mixture was then continued to stir at room temperature until completion of the reaction. On completion of the reaction, the resulting mixture was acidified with 2 N HCl solution, and extracted with EtOAc. The organic layer was washed with water and brine, and dried over magnesium sulfate. The filtrate was evaporated to obtain pure sulfonamide derivative **4** crystalline solid. Recrystallization from a mixture of ethanol and water afforded analytically pure product. All the products were characterized by analytical and spectral studies.

Characterization data of one representative sulfonamide compound

4-Bromo-*N*-(3-nitrophenyl)benzenesulfonamide (**4b**): Solid, mp 118-120 °C; ^{1}H NMR (200 MHz, $CDCl_3$): δ 7.17-7.74 (m, 7H), 7.86-8.00 (m, 2H); ^{13}C NMR (50 MHz, $CDCl_3$): δ 115.4, 120.2, 126.4, 128.8, 129.1, 130.6, 132.9, 137.4, 137.6, 148.8. Anal. Calcd for $C_{12}H_9N_2SO_2Br$: C, 44.31; H, 2.76; N, 8.61; S, 9.85; Found: C, 43.65; H, 2.68; N, 8.53; S, 9.50.

Reference

Bahrami, K., Khodaei, M. M., and Soheilizad, M. (2009). Direct Conversion of thiols to sulfonyl chlorides and sulfonamides. *J. Org. Chem.*, **79**, 9287–9291.

Sulfur – Oxygen Bond Forming Reactions: S-O bond_3

Types of reaction: S-O bond formation
Reaction conditions: Neat condition, room temperature
Synthetic strategy: One-step transformation (oxidation)
Catalyst: *p*-Toluene sulfonic acid (*p*-TsOH)
Keywords: Sulfides, 30% hydrogen peroxide, *p*-TsOH, room temperature, oxidation, sulfoxides, chemoselectivity

General reaction scheme

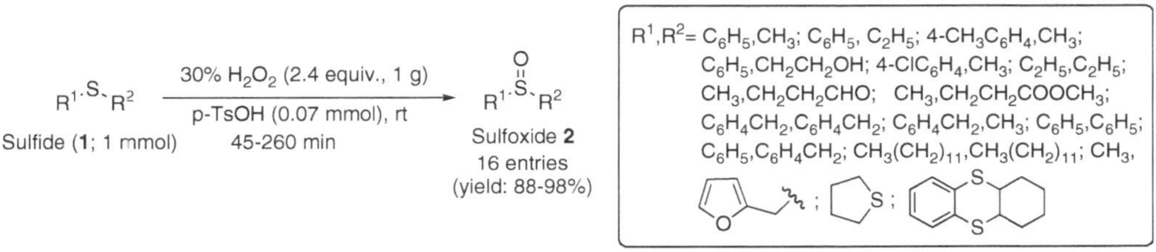

Plausible mechanisms: Two alternative mechanisms are proposed for the oxidation reaction as depicted below:

Mechansim-1

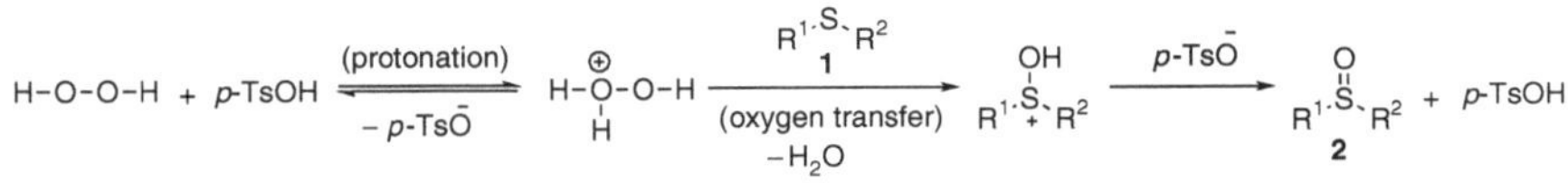

Mechansim-2

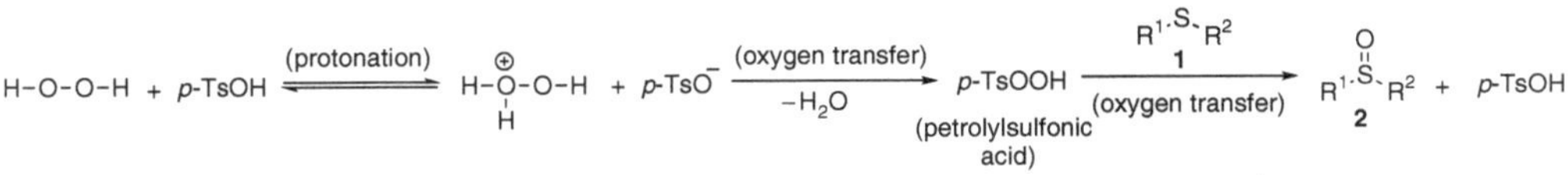

Representative entries

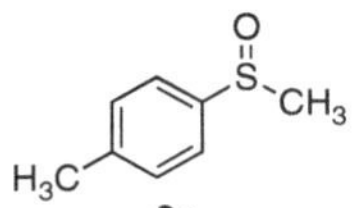

2a

1-Methyl-4-(methylsulfinyl)benzene (**2a**): Reaction time: 85 min; yield: 96%

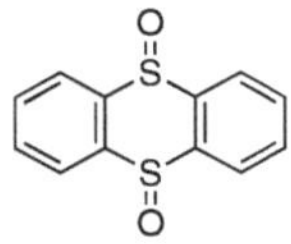

2b

Thianthrene 5,10-dioxide (**2b**): Reaction time: 270 min; yield: 98%

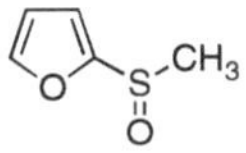

2c

2-(Methylsulfinyl)furan (**2c**): Reaction time: 130 min; yield: 95%

$CH_3(CH_2)_{11}SO(CH_2)_{11}CH_3$

2d

1-(Dodecylsulfinyl)dodecane (**4d**): Reaction time: 260 min; yield: 88%

Experimental procedure

The sulfide (**1**; 1 mmol) was added to a solution of 30% H_2O_2 (2.4 equiv, 1 g) and *p*-TsOH (0.07 mmol, 0.012 g), and the mixture was stirred at room temperature for stipulated time (45-206 min). After completion of the reaction (as monitored by TLC), the product was extracted with dichloromethane (3 × 10 mL) and the combined organic layers were washed with brine (15 mL) and dried over anhydrous sodium sulphate. The crude product, obtained on removal of the solvent under reduced pressure, was purified by short-column chromatography on silica gel with EtOAc/*n*-hexane (1/2) as eluent. All the products were characterized by IR, ^1H NMR, and ^{13}C NMR, and comparison of the spectral data as well as physical properties with those reported for authentic samples (literature cited in this article).

Reference

Rostami, A., Hassanian, F., Ghorbani-Choghamarani, A., and Saadat, S. (2013). Selective oxidation of sulfides to sulfoxides using H_2O_2 catalyzed by *p*-toluenesulfonic acid (*p*-TsOH) under solvent-free conditions. *Phosphorus Sulfur Silicon Relat. Elem.*, **188**, 833–838.

Sulfur – Sulfur Bond Forming Reactions: S-S bond_1

Types of reaction: S-S bond formation
Reaction conditions: Dimethylsulfoxide (DMSO), room temperature
Synthetic strategy: Selective oxidation of thiols to disulfides with DMSO
Catalyst: Dichlorodioxomolybdenum [$MoO_2Cl_2(DMSO)_2$]
Keywords: Thiols, dimethylsulfoxide, molybdenum(VI) complex, room temperature, oxo-transfer catalysis, oxidation, disulfides

General reaction scheme

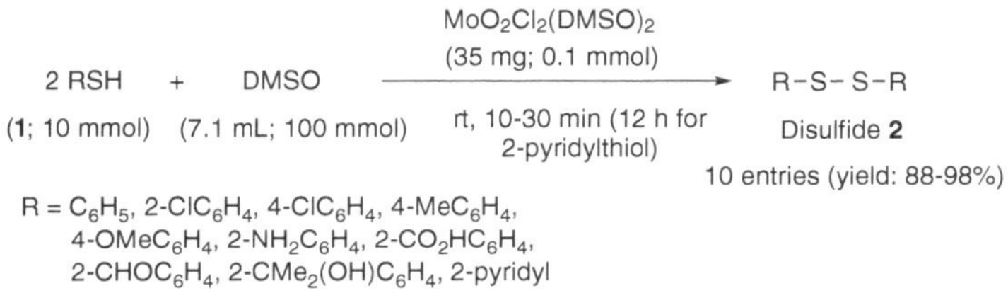

Proposed mechanism

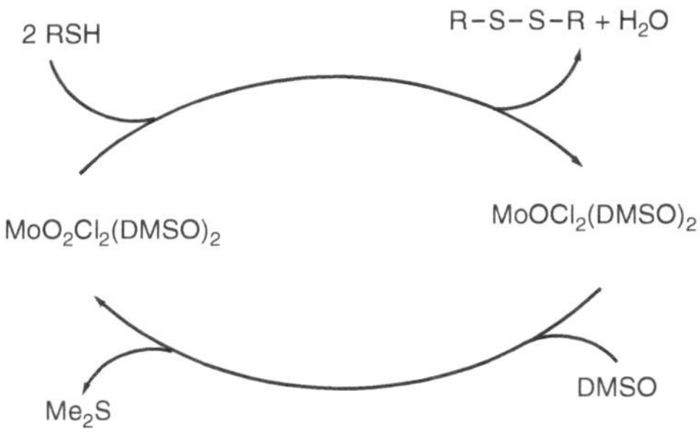

Representative entries

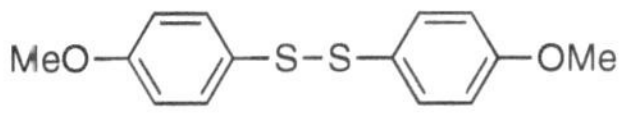

2a

1,2-Bis(4-methoxyphenyl)disulfane (**2a**): Solid, mp 42-43 °C; reaction time: 30 min; yield: 95%

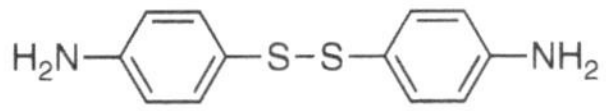

2b

4,4′-Disulfanediyldianiline (**2b**): Solid, mp 88-89 °C; reaction time: 30 min; yield: 91%

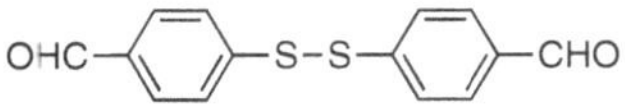

2c

4,4′-Disulfanediyldibenzaldehyde (**2c**): Solid, mp 149-150 °C; reaction time: 30 min; yield: 91%

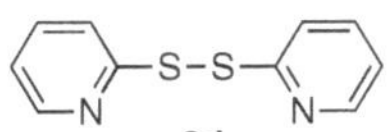

2d

1,2-Di(pyridin-2-yl)disulfane (**2d**): Solid, mp 55-57 °C; reaction time: 12 h; yield: 89%

Experimental procedure

To a 50 mL Erlenmeyer flask containing DMSO (7.1 mL, 100 mmol) and $MoO_2Cl_2(DMSO)_2$ (35 mg, 0.1 mmol) was added thiophenol (**1**; 10 mmol) and the resulting solution was stirred for stipulated time at room temperature. Upon completion of reaction (ascertained by GC/MS), the reaction mixture was treated with 2 M NaOH (25 mL). The precipitate was collected by filtration, washed with water (3×20 mL), and dried under vacuum to afford pure diaryl disulfide **2**. The physical and spectral properties of the compounds were found to be comparable with the reported values in literature.

Reference

Sanz, R., Aguado, R., Pedrosa, M. R., and Arnáiz, F. J. (2002). Simple and selective oxidation of thiols to disulfides with dimethylsulfoxide catalyzed by dichlorodioxomolybdenum(VI). *Synlett*, 856–858.

Sulfur – Sulfur Bond Forming Reactions: S-S bond_2

Types of reaction: S-S bond formation
Reaction conditions: Dichloromethane or solvent-free, room temperature
Synthetic strategy: Oxidation of thiols to disulfides
Catalyst: 1,3-Dibromo-5,5-dimethylhydantoin [DBDMH]
Keywords: Thiols, dichloromethane or solvent-free conditions, 1,3-dibromo-5,5-dimethylhydantoin (DBDMH), room temperature, oxidation, disulfides

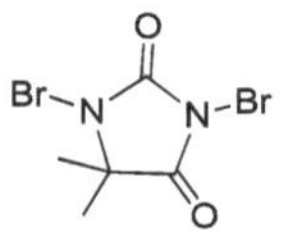

General reaction scheme

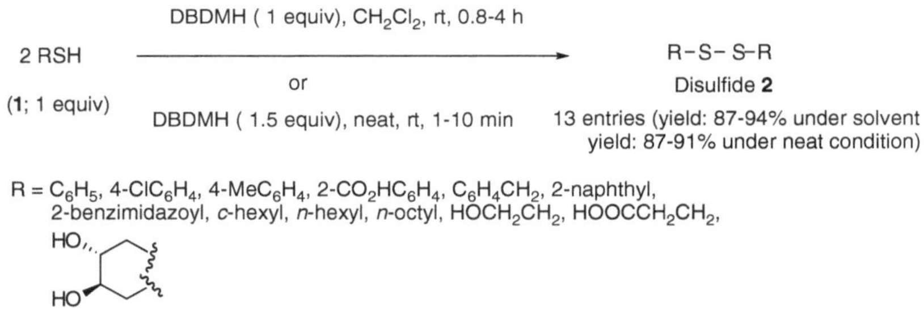

Proposed mechanism

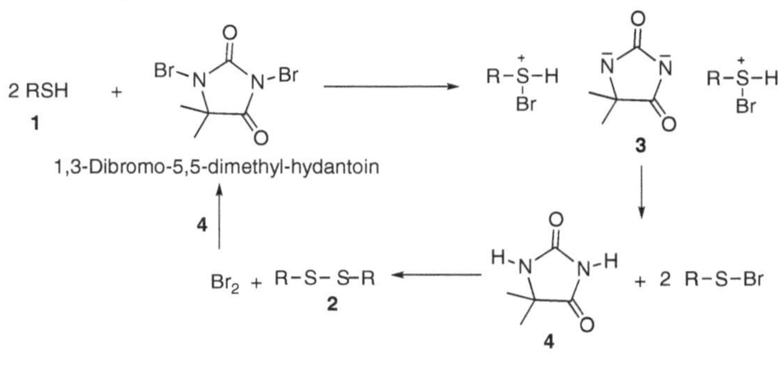

Representative entries

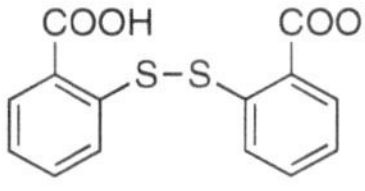

2a

2,2′-Disulfanediyldibenzoic acid (**2a**): Solid, mp 287-289 °C; reaction time: 5 h /10 min; yield: 94% /95% [respectively in DCM and neat conditions]

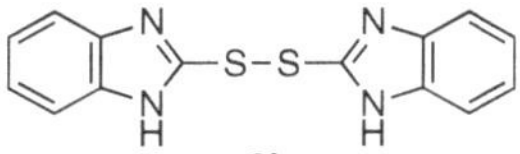

2b

1,2-Bis(1*H*-benzo[*d*]imidazol-2-yl)disulfane (**2b**): Solid, mp 198-200 °C; reaction time: 4 h / 8 min; yield: 92% / 89%

$HOOCCH_2CH_2SSCH_2CH_2COOH$

2c

3-((2-Hydroperoxyallyl)sulfinothioyl)propanoic acid (**2c**): Solid, mp 156-158 °C; reaction time: 3.5 h / 5 min; yield: 90% / 87%

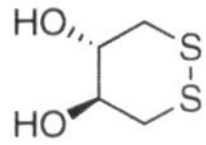

2d

(4*S*,5*S*)-1,2-Dithiane-4,5-diol (**2d**): Solid, mp 131-133 °C; reaction time: 2.8 h / 3 min; yield: 90% / 88%

Experimental procedures

Oxidation in solution: To a solution of thiol (**1**; 1 mmol) in dichloromethane (10 mL) was added DBDMH (1 mmol), and the mixture was stirred vigorously at room temperature for stipulated time (monitored by TLC). After completion of the reaction, 20 mL of dichloromethane was added and the reaction mixture was extracted with saturated sodium bi-carbonate (2 × 20 mL) and dried over anhydrous calcium chloride (1 g). Evaporation of the solvent, followed by recrystallization or chromatography on silica gel, afforded pure disulfide **2** in 87-94% yield.

Oxidation under solvent-free condition: DBDMH (1.5 mmol) was added to thiols (**1**; 1 mmol) slowly in a mortar at room temperature. The mixture turned deep red and the reaction was completed after very short time (1-10 min). To the resulting solid mixture, dichloromethane was added and the mixture was with saturated sodium bi-carbonate (2 × 20 mL) and dried over anhydrous calcium chloride (1 g). Evaporation of the solvent, followed by recrystallization or chromatography on silica gel, afforded pure disulfide **2** in 87-91% yield.

The physical and spectral properties of the synthesized compounds were found to be comparable with the reported values in literature.

Reference

Khazaei, A., Zolfigol, M. A., and Rostami, A. (2004). 1,3-Dibromo-5,5-dimethylhydantoin [dbdmh] as an efficient and selective agent for the oxidation of thiols to disulfides in solution or under solvent-free conditions. *Synlett*, 2959–2961.

Sulfur – Sulfur Bond Forming Reactions: S-S bond_3

Types of reaction: S-S bond formation
Reaction conditions: Ethyl acetate or water, room temperature
Synthetic strategy: Oxidation of thiols to disulfides (oxidative coupling)
Catalyst: Sodium iodide
Keywords: Thiols, 30% hydrogen peroxide, ethyl acetate or water medium, sodium iodide, room temperature, oxidative coupling, disulfides

General reaction scheme

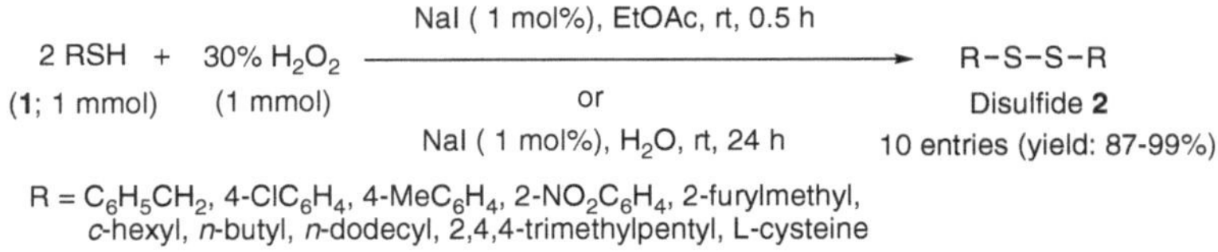

Representative entries

$CH_3(CH_2)_{11}S\text{-}S(CH_2)_{11}CH_3$

2a

1,2-Didodecyldisulfane (**2a**): Colorless crystals, mp 32 °C; yield: 97%

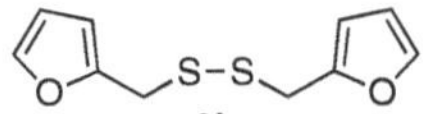

2b

1,2-*Bis*(furan-2-ylmethyl)disulfane (**2b**): Colorless oil; yield: 98%

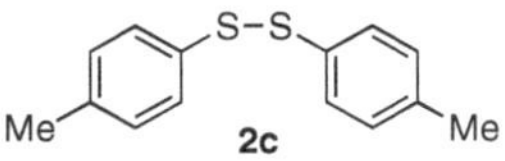

2c

1,2-Di-*p*-tolyldisulfane (**2c**): Colorless crystals, mp 44-45 °C; yield: 99%

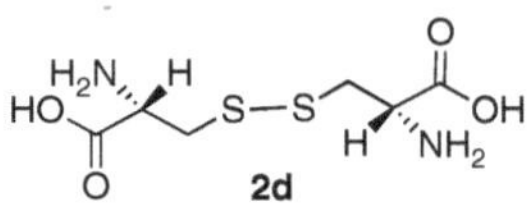

2d

(2*S*,2′*S*)-3,3′-disulfanediylbis(2-aminopropanoic acid) (**2d**): Colorless solid, mp 216-220 °C; yield: 95%

Experimental procedures

Oxidation of thiols in ethyl acetate solution: To a stirred solution of a thiol (**1**; 1 mmol) in ethyl acetate (3 mL) was added sodium iodide (1.5 mg, 1 mol%) and 30% H_2O_2 (0.11 mL, 1 mmol) and the mixture was then stirred at room temperature for 0.5 h. Saturated aqueous solution of sodium thiosulfate (15 mL) was added, and the resulting mixture was extracted with ethyl acetate (3 × 15 mL). The combined organic phases were washed with brine (15 mL) and dried over anhydrous magnesium sulfate. The solvent was evaporated, and the residue was purified by silica gel column chromatography (hexane-EtOAc) to afford the pure difulfide **2**.

Oxidation of thiols in water: To a stirred solution of a thiol (**1**; 1 mmol) in water (3 mL) was added sodium iodide (1.5 mg, 1 mol%) and 30% H_2O_2 (0.11 mL, 1 mmol) and the mixture was then stirred at room temperature for 24 h. Saturated aqueous solution of sodium thiosulfate (15 mL) was added, and the resulting mixture was extracted with ethyl acetate (3 × 15 mL). The combined organic phases were washed with brine (15 mL) and dried over anhydrous magnesium sulfate. The solvent was evaporated, and the residue was purified by silica gel column chromatography (hexane-EtOAc) to afford the pure difulfide **2**.

The physical and spectral properties of the synthesized compounds were found to be comparable with the reported values in literature.

Characterization data of two representative compounds

1,2-*Bis*(furan-2-ylmethyl)disulfane (**2b**): Colorless oil; ^{1}H NMR (400 MHz, $CDCl_3$): δ 3.69 (s, 4 H), 6.22 (d, J=3.17 Hz, 2H), 6.33 (dd, J=1.95, 2.44 Hz, 2H), 7.39 (dd, J=0.73, 1.95 Hz, 2H); ^{13}C NMR (100 MHz, $CDCl_3$): δ 35.7, 108.9, 110.7, 142.4, 150.2.

1,2-Di-*p*-tolyldisulfane (**2c**): Colorless crystals, mp 44-45 °C; ^{1}H NMR (400 MHz, $CDCl_3$): δ 2.31 (s, 6H), 7.10 (d, $J=8.05$ Hz, 4 H), 7.38 (d, $J=8.05$ Hz, 4 H); ^{13}C NMR (100 MHz, $CDCl_3$): δ 21.0, 128.6, 129.8, 133.9, 137.4.

Reference

Kirihara, M., Asai, Y., Ogawa, S., Noguchi, T., Hatano, A., and Hirai, Y. (2007). A mild and environmentally benign oxidation of thiols to disulfides. *Synlett*, 3286–3289.

Index

Note: Page numbers followed by *f* indicate figures and *t* indicate tables.

B

E

F

N

P

S

FSC
www.fsc.org
MIX
Papier | Fördert gute Waldnutzung
FSC® C083411

Kolibri 360 GmbH
Ferdinand-Jühlke-Straße 7
99095 Erfurt, Deutschland
produktsicherheit@kolibri360.de